HANDBOOK
of
DETECTION
of
ENZYMES
on
ELECTROPHORETIC
GELS

SECOND EDITION

GENNADY P. MANCHENKO, Ph.D.

Institute of Marine Biology
Russian Academy of Science
Vladivostok, Russia

CRC Press
Taylor & Francis Group
Boca Raton London New York

CRC Press is an imprint of the
Taylor & Francis Group, an **informa** business

CRC Press
Taylor & Francis Group
6000 Broken Sound Parkway NW, Suite 300
Boca Raton, FL 33487-2742

© 2003 by Taylor & Francis Group, LLC
CRC Press is an imprint of Taylor & Francis Group, an Informa business

First issued in paperback 2019

No claim to original U.S. Government works

ISBN 13: 978-0-367-45461-6 (pbk)
ISBN 13: 978-0-8493-1257-1 (hbk)

**Visit the Taylor & Francis Web site at
http://www.taylorandfrancis.com**

**and the CRC Press Web site at
http://www.crcpress.com**

Library of Congress Card Number 2002031312

Library of Congress Cataloging-in-Publication Data

Manchenko, Gennady P.
  Handbook of detection of enzymes on electrophoretic gels / by Gennady P.
Manchenko.—2nd ed.
    p. cm.
  Includes bibliographical references and index.
  ISBN 0-8493-1257-4 (alk. paper)
    1. Enzymes—Purification—Handbooks, manuals, etc. 2. Gel
electrophoresis—Handbooks, manuals, etc. I. Title.
  QP601 .M314 2003
  572'.7—dc21   2002031312
                                                                CIP

Preface

Isozymes detected by zymogram techniques presented in this book continue to be widely used as gene markers. However, principally new applications of enzyme electrophoresis and zymogram techniques have appeared which are outlined in Part I, "Introduction." These include (1) involvement of enzymatic proteins in proteome studies, (2) testing of enzyme-coding genes for their expression at protein level, and (3) identification and study of a new type of enzyme protein isoform (alternative isozymes) generated by alternative splicing. The principles of zymogram techniques have not undergone drastic development so that only several additions were made to Part II, "General Principles of Enzyme Detection on Electrophoretic Gels." Part III, "Methods of Detection of Specific Enzymes," is completely updated. Zymogram techniques are added for about 100 enzymes not previously included in the book. In total, this edition includes more than 900 different methods suitable for detection of more than 400 different enzymes. Information on the subunit structures of a majority of these enzymes is included in enzyme sheets to facilitate the interpretation of isozyme patterns developed on zymograms. Enzyme names and nomenclature numbers are given according to the last edition of the Enzyme List by the Nomenclature Committee of the International Union of Biochemistry and Molecular Biology (1992). Appendix C is added which contains information on buffer systems most commonly used for enzyme electrophoresis in starch, cellulose acetate, and polyacrylamide gels. The number of references is almost doubled throughout the book due to inclusion of a bulk of new information related to zymogram techniques, which has appeared during the last decade.

Preparation of the 2nd edition of the book was much helped by my three short sojourns (in 1995, 1999, and 2001) at Bodega Marine Laboratory (BML), University of California, Davis, where I could work with the literature at Cadet Hand Library. The library also provided me access to the Web-of-Science (Science Citation Index database, Institute for Scientific Information, Philadelphia, PA). These visits were supported by gifts from the Eugene Garfield Foundation to BML. I greatly appreciate the generous and cordial hospitality of colleagues at BML, its secretarial and technical staff. Special thanks are due to Dennis Hedgecock, who was my host there and who bore all the weight of cares about my stay and work and whose support I always felt. I am deeply indebted to the successive directors of BML, Jim Clegg and Susan Williams, for the welcome I received at the lab. Working at libraries at Davis campus I twice enjoyed the hospitality of Andrei and Irina Zalensky, my old friends since their time at the Institute of Marine Biology (IMB) in Vladivostok, Russia. Really, "old friends and old vines are the best." It is also a pleasure to thank my younger friends, former colleagues at the Genetics Lab of the IMB: Dmitri and Svetlana Zaykin (Raleigh, NC), Dmitri Churikov (Bodega Bay, CA), and Andrei and Olga Tatarenkov (Irvine, CA). It was most unselfish of them to offer me help and encouragement during my peregrination in the United States. During these visits, I also had opportunities to work at libraries of the University of Washington in Seattle. In this wonderful city, I enjoyed the hospitality of Robin Waples and his lovely family and of Fred Utter. Bright memories of my short visits with Dick Koehn (then at the University of Utah, Salt Lake City) and with Oleg Serov (then at the University of Rio de Janeiro, Brazil) greatly elevated my spirit and thus encouraged the work. I am thankful to Anton Chichvarkhin and Victoria Pankova, both of the IMB, for technical help in the final stage of getting the book ready. I appreciate the support and understanding of Vladimir Kasyanov, the director of IMB. Special thanks are due to Alexander Pudovkin of IMB, whose help and support I enjoyed during the work on this project, as well as during my entire scientific career. Preparation of this edition was partly supported by a grant from the Governor of Primorye (Russia). My last (but not the least) thanks go to the many colleagues from different countries who sent reprints of their publications, which were of great help in bringing new material into the book, that is presented now for their judgment.

Preface to the First Edition

Gel electrophoresis of enzymes is a very powerful analytical method which is at present widely used in various fields of biological and medical sciences and successfully applied in different fields of practical human activity. The tremendous expansion of the method is mainly due to its simplicity and its ability to separate isozymes and allozymes, which have proven to be very useful genetic markers.

The key step of enzyme electrophoresis is the detection of enzymes on electrophoretic gels, i.e., the procedure of obtaining enzyme electropherograms, or zymograms. Within about 35 years since the first adaptations of histochemical methods for purposes of electrophoretic zymography there were many significant advances in this field, which was extremely active in respect to development of new techniques. Information on enzyme-detection techniques, which is contained in some well-known handbooks and manuals, does not reflect these new developments and in this respect is out of date.

The purpose of this book is to bring together descriptions of numerous enzyme-detection techniques, which have been developed during the last three and a half decades, in one special volume. This book should be useful not only for those who work professionally in the field but also for those who are only starting to master techniques of electrophoretic zymography. Therefore, the book includes detailed descriptions of numerous enzyme-specific methods suitable for detection of more than 300 different enzymes, as well as descriptions of the general principles of enzyme detection on electrophoretic gels.

The Author

Gennady P. Manchenko, Ph.D., is senior researcher at the Laboratory of Genetics at the Institute of Marine Biology of the Russian Academy of Sciences at Vladivostok.

Dr. Manchenko graduated from Novosibirsk State University in 1973 with the qualification of a cytogeneticist. As a probationer, he joined the Institute of Marine Biology at Vladivostok in 1973 and was promoted to junior researcher in 1975. He obtained his Ph.D. in 1981 from the Leningrad State University. From 1983 he was employed as the senior researcher at the Laboratory of Genetics of the Institute of Marine Biology at Vladivostok.

Dr. Manchenko was a member of the American Genetic Association from 1978 to 1983 and is a member of the Vavilov Society of Geneticists and Breeders in Russia, and a member of the Scientific Council of the Institute of Marine Biology at Vladivostok.

He has been the recipient of research supports from the George Soros Foundation, Eugene Garfield Foundation, International Science Foundation, and Russian Foundation for Basic Research. Current research interests focus on theoretical and applied isozymology.

Acknowledgments

I should like to thank Dr. Alexander I. Pudovkin (Institute of Marine Biology, Vladivostok) and Dr. Robert P. Higgins (Smithsonian Institution, Washington, D.C.) for reading some parts of the manuscript and correcting my English. I should also like to thank my Russian colleagues for their encouragement and my foreign colleagues for reprints of original papers, which are referred to in the book. However, the first and most honorable position in the list of acknowledgments must be given to my first teacher in isozymology, Dr. Oleg L. Serov (Institute of Cytology and Genetics, Novosibirsk), who introduced me to this exciting field of molecular genetics two decades ago when I was still an undergraduate at the State University at Novosibirsk.

Preparation of the manuscript was partially supported by the George Soros Foundation.

To Clement L. Markert

Table of Contents

---

* More than 400 different enzymes are considered in this section, presented in numerical order according to their Enzyme Commission numbers published in 1992. For convenience, the enzyme numbers are given at the beginning of each enzyme headline. An alphabetical list of the enzymes is also given as Appendix B.

1 ———— Introduction

Detection of enzymes on electrophoretic gels means visualization of gel areas occupied by specific enzyme molecules after their electrophoretic separation.

Electrophoresis is the migration of charged particles (e.g., protein molecules) in an electrolyte under the influence of an electric field. Tiselius may be considered the father of protein electrophoresis. He developed the "moving boundary" method to separate serum proteins in solution.[1] As the result of further developments, a modified method called "zone" electrophoresis was devised in which different protein molecules were separated into distinct zones in stabilized media. In 1955 Smithies introduced starch gel as a stabilized (or supporting) medium for electrophoretically separated proteins.[2] In 1957 Kohn reported the use of cellulose acetate as a very useful supporting medium.[3] Polyacrylamide gel was introduced 2 years later by Ornstein and Davis[4] and Raymond and Weintraub.[5] At present these supporting media are the most widely used for electrophoretic separation of proteins. The pore size of starch and polyacrylamide gel matrices is of the same order-of-magnitude as the size of protein molecules. This results in the "molecular sieving" effect, which allows more effective electrophoretic separation of similarly charged protein molecules that differ in size and shape. Thus, during zone electrophoresis molecules of each protein type move at different rates along the gel according to their specific properties. Using different concentrations of polyacrylamide or starch (this influences the pore size of the gel matrix) and different electrophoretic buffer systems (this influences the net charge of the protein molecule), discrete zones on electrophoretic gels may be obtained almost for each protein type. It is beyond the scope of this book to give a full theoretical treatment of the principles underlying protein electrophoresis. This was done extremely well by Andrews.[6]

The basis for the specific enzyme detection was set in 1939 by the pioneering efforts of Gomori, who developed histochemical methods for visual identification of sites of alkaline phosphatase activity in animal tissues.[7] Further developments have led to the foundation of enzyme histochemistry as a separate field of biological investigation.[8–10]

In 1957 Hunter and Markert first applied histochemical staining procedures to starch gel after electrophoresis of crude tissue extracts for the visualization of gel areas containing esterase activity.[11] As a result, the position of the enzyme was marked by a band (or zone) of stain directly in the gel. This visual display of enzymes on electrophoretic gels has been termed a *zymogram*.[12] Since that time methods of detection of enzymes on electrophoretic gels are also known as zymographic (or zymogram) techniques.

The term *isozymes* (or *isoenzymes*) was introduced in 1959 by Markert and Möller to designate different molecular forms of the same enzyme occurring either in a single individual or in different members of the same species.[12] Numerous studies of different enzymes have shown that there are three main causes of formation of multiple molecular forms of enzymes: (1) the presence of more than one gene locus coding for the enzyme, (2) the presence of more than one allele at a single gene locus coding for the enzyme, and (3) the posttranslation modifications of the formed enzymatic polypeptides resulting in formation of nongenetic or so-called "secondary" isozymes.[13–19] According to the recommendations of the Commission on Biological Nomenclature of the International Union of Pure and Applied Chemistry and the International Union of Biochemistry (IUPAC-IUB), isozymes are defined as genetically determined, multiple molecular forms of an enzyme.[20] The term *isozymes* is usually used to denote multiple molecular forms deriving from different genetic loci, whereas the term *allozymes* is used to denote multiple molecular forms deriving from different alleles of the same genetic locus. The term *allelic isozymes* is also used by some isozymologists.

Since the advent of the zymogram method and the discovery of isozymes, enzyme electrophoresis has been used increasingly to provide useful information in a wide range of biological and biochemical fields and in different fields of practical human activities. The use of isozymes and allozymes as gene markers significantly advanced our knowledge in such areas as population and evolutionary genetics, developmental genetics, molecular evolution, and enzymology. Isoenzymes and allozymes are widely used for solving numerous problems of systematics as well as for reconstruction of phylogenetic relationships between related species. They are of considerable importance for clinical and diagnostic medicine and medical genetics, breeding control of agricultural organisms, agricultural entomology, fishery management, genetic monitoring of environmental pollution, estimation of genetic resources, forensic science, etc.

For a long period of time the major limitation of enzyme electrophoresis was a relatively small number of enzymes for which specific zymogram methods have been developed. By the middle of the 1970s, only about 50 enzymes were available for electrophoretic analysis.[21] The popular and frequently cited handbook of enzyme electrophoresis and detection techniques, written by Harris and Hopkinson at the end of the 1970s,[14] includes specific zymogram methods for only 80 of the 3200 enzymes that had been identified by 1992 by the Nomenclature Committee of the International Union of Biochemistry and Molecular Biology (NC-IUBMB).[22] New zymographic methods (predominantly histochemical and autoradiographic) were developed from time to time for different purposes. They accumulated at a slow rate and were frequently hidden in special journals, and thus were not easily available for those who work with isozymes. Since the late 1970s and early 1980s several, principally new, approaches were implemented that considerably enriched electrophoresis and enzyme detection. These were (1) bioautographic methods based on the use of a microbial reagent to locate specific enzyme activity after gel electrophoresis;[23] (2) two-dimensional spectroscopy of electrophoretic gels using special optical devices, which permitted one to analyze two-dimensional gels in a fashion analogous to one-dimensional spectroscopy of solutions;[24] and (3) immunoblotting, or the procedure for immunohistochemical visualization of an enzyme protein, which was based on the use of monoclonal antibodies specific to a certain enzyme.[25] Theoretically, these methods permit the detection of isozyme patterns of almost all the known enzymes. However, bioautographic and two-dimensional spectroscopy methods did not receive further development since their first introduction into enzyme electrophoresis. The most impressive is the expansion of immunoblotting techniques. Monoclonal and polyclonal antibodies about five tens of different enzymes are now commercially available (e.g., see Sigma Catalog, 2002/2003, Antibody Index). Nevertheless, the immunoblotting technique is still not as widely used as other zymogram techniques because of several objective reasons (see Part II, Section 6 — Immunoblotting).

This book was conceived with only the purpose to bring together in one volume all specific zymogram techniques developed since the pioneer work by Hunter and Markert in 1957.[11] Therefore, it does not cover practical aspects of gel electrophoresis, genetic interpretation of banding patterns developed on zymograms, and numerous applications of enzyme electrophoresis and zymogram techniques in biological and medical research. These areas are covered in other publications,[14,17–19,26–32] which make excellent companion references for this handbook.

The general principles of visualization of enzymes on electrophoretic gels are outlined in Part II, which includes descriptions of histochemical, autoradiographic, bioautographic, two-dimensional spectroscopic, immunoblotting, and other methods. Part III includes three sections. Section 1 describes the structure of enzyme sheets given in Section 3. General considerations, comments, and recommendations concerning the choice of support medium and zymogram technique, the recording and preservation of zymograms, resource-saving strategies, and troubleshooting and safety regulations are given in Section 2. Section 3, which comprises the main and largest part of the book, contains detailed descriptions and outlines of more than 900 zymogram techniques suitable for detection of over 400 different enzymes.

Enzyme electrophoresis remains the most simple and powerful tool for separation and identification of the second-level structural gene products. Despite the overwhelming expansion of DNA technologies and DNA markers during the last two decades, there are several important and obvious advantages of enzyme electrophoresis and isozymes as gene markers.[33] First, gel electrophoresis and zymogram methods are technically simple and not time consuming. Second, genetic polymorphisms at a large number of nuclear gene loci scattered over the genome can be studied using these techniques with relative ease, speed, and low cost. Third, genotypic interpretation of individual variation of isozyme patterns can usually be inferred directly from these patterns taking into account the codominant expression of allelic isozymes (allozymes) and the highly conserved subunit structure of enzymes.[34] Finally, the most important advantage of isozymes over DNA markers is that their study can bring molecular evolutionists much closer to the real "stuff" of adaptive evolution.[33] Thus, this handbook should prove useful for all those who traditionally use isozymes as gene markers in various fields of biological and medical research.

Besides being valuable gene markers, isozymes are structural isoforms of enzymatic proteins and thus serve as the subject of inquiry of proteomics — the hottest growth area in the postgenome biology.[35] Identification of proteins and their isoforms separated by gel electrophoresis is the main problem and the most laborious task of functional proteomics.[36] It is important to stress here that it is the combination of gel electrophoresis and enzyme-specific zymogram techniques that enables the coupling of electrophoretic separation and the detection and identification of enzyme protein isoforms in a single procedure. The goal of proteomics is fundamental analysis of a complete protein set of an organism (i.e., the proteome), including electrophoretic detection of multiple isoforms of protein molecules, determination of their function, and localization of their expression. Strikingly, almost the same purposes were recognized and formulated for isozymology by Clement Markert as early as the mid-1970s.[37] Enzymes constitute a significant part of organismal proteins.[38] Therefore, it is clear that isozymology may essentially be considered the proteomics of enzymatic proteins. It has an obvious advantage over proteomics dealing with nonenzymatic proteins. Indeed, the use of gel electrophoresis and zymogram techniques enables versatile analysis of unpurified enzyme preparations, including:

1. Detection and identification of enzyme proteins and their structural isoforms (i.e., isozymes)
2. In-gel examination of functional properties of isozymes
3. Examination of spatial and temporal expression and localization of isozymes
4. Inference of the genetic basis of isozymes from individual variation and tissue specificity of their banding patterns detected on zymograms
5. Discrimination between different enzymes with similar and overlapping substrate specificities
6. Identification of enzyme proteins with wide substrate specificities, which demonstrate several distinct catalytic functions
7. Determination of subunit structure of enzyme molecules from their allozyme and isozyme patterns
8. Estimation of molecular weights of different isozymes by combination of native and sodium dodecyl sulfate–polyacrylamide gel (SDS-PAG) electrophoresis
9. Determination of isoelectric points of isozymes by isoelectric focusing technique, etc.

These examples demonstrate the valuable but not yet fully realized application of enzyme electrophoresis and zymogram techniques in proteome investigations.

Another but related application of these techniques concerns determination of expression of DNA sequences coding for enzymes at the translational level. Testing of sequences for their expression usually involves the search of the mRNA or cDNA sequence in question for insertions, deletions, or stop codons.[39] Sometimes the testing of an enzyme-coding DNA sequence for expression of catalytically active enzyme molecules is realized using *Escherichia coli* strains deficient for the enzyme in question. The use of zymogram techniques for this purpose is more effective, cheap, and time saving. The number of applications of zymogram techniques for testing cloned enzyme-coding genes for their expression at the protein level is rapidly growing.[40–45] One commonly used strategy of gene cloning is the screening of DNA libraries with suitable DNA probes deduced from the amino acid sequences of the enzyme in question. The use of gel electrophoresis for protein separation, followed by subsequent N-terminal amino acid sequencing, is a fast method for generating the necessary amino acid sequence.[46] An absolute prerequisite for this is the specific and sensitive zymogram technique suitable for detection of the enzyme inside the gel.

The unexpectedly high diversity of protein isoforms generated by alternative splicing of pre-mRNAs transcribed from single genes represents a challenge for postgenome biology and an impetus for proteomics.[47,48] "Alternative" isozymes generated via alternative splicing represent a phenomenon of general biological significance.[49] Although electrophoretic patterns of alternative isozymes (determined by single genes) resemble those of conventional isozymes (determined by separate genes), in some special cases this type of isozyme may be identified with certainty.[49,50] Diversified electrophoretic investigation of alternative isozymes initiates a new and promising application of enzyme electrophoresis and the zymogram techniques.

REFERENCES FOR PART I

1. Tiselius, A., A new apparatus for electrophoretic analysis of colloidal mixtures, *Trans. Faraday Soc.*, 33, 524, 1937.
2. Smithies, O., Zone electrophoresis in starch gels: group variations in the serum proteins of normal human adults, *Biochem. J.*, 61, 629, 1955.
3. Kohn, J., A cellulose acetate supporting medium for zone electrophoresis, *Clin. Chim. Acta*, 2, 297, 1957.
4. Ornstein, L. and Davis, B.J., *Disc Electrophoresis*, Distillation Products Industries (Division of Eastman Kodak Co.), Kingsport, TN, 1959.
5. Raymond, S. and Weintraub, L., Acrylamide gel as a supporting medium for zone electrophoresis, *Science*, 130, 711, 1959.
6. Andrews, A.T., *Electrophoresis: Theory, Techniques, and Biochemical and Clinical Applications*, Clarendon Press, Oxford, 1988.
7. Gomori, G., Microtechnical demonstration of phosphatase in tissue sections, *Proc. Soc. Exp. Biol. Med.*, 42, 23, 1939.
8. Pearse, A.G.E., *Histochemistry: Theoretical and Applied*, Vol. I, Williams & Wilkins, Baltimore, 1968.
9. Pearse, A.G.E., *Histochemistry: Theoretical and Applied*, Vol. II, Williams & Wilkins, Baltimore, 1972.
10. Burstone, M.S., *Enzyme Histochemistry*, Academic Press, New York, 1962.
11. Hunter, R.L. and Markert, C.L., Histochemical demonstration of enzymes separated by zone electrophoresis in starch gels, *Science*, 125, 1294, 1957.
12. Markert, C.L. and Möller, F., Multiple forms of enzymes: tissue, ontogenetic, and species specific patterns, *Proc. Natl. Acad. Sci. U.S.A.*, 45, 753, 1959.
13. Kenney, W.C., Molecular nature of isozymes, *Horizons Biochem. Biophys.*, 1, 38, 1974.
14. Harris, H. and Hopkinson, D.A., *Handbook of Enzyme Electrophoresis in Human Genetics*, North-Holland, Amsterdam, 1976 (loose-leaf, with supplements in 1977 and 1978).
15. Korochkin, L.I., Serov, O.L., and Manchenko, G.P., Definition of isoenzymes, in *Genetics of Isoenzymes*, Beljaev, D.K., Ed., Nauka, Moscow, 1977, p. 5 (in Russian).
16. Rothe, G.M., A survey on the formation and localization of secondary isozymes in mammalian, *Hum. Genet.*, 56, 129, 1980.
17. Moss, D.W., *Isoenzymes*, Chapman & Hall Ltd., London, 1982.
18. Richardson, B.J., Baverstock, P.R., and Adams, M., *Allozyme Electrophoresis: A Handbook for Animal Systematics and Population Studies*, Academic Press, Sydney, 1986.
19. Buth, D.G., Genetic principles and the interpretation of electrophoretic data, in *Electrophoretic and Isoelectric Focusing Techniques in Fisheries Management*, Whitmore, D.H., Ed., CRC Press, Boca Raton, FL, 1990, p. 1.
20. IUPAC-IUB Commission on Biochemical Nomenclature, Nomenclature of multiple forms of enzymes: recommendations (1976), *J. Biol. Chem.*, 252, 5939, 1977.
21. Siciliano, M.J. and Shaw, C.R., Separation and visualization of enzymes on gels, in *Chromatographic and Electrophoretic Techniques*, Vol. 2, *Zone Electrophoresis*, Smith, I., Ed., Heinemann, London, 1976, p. 185.
22. NC-IUBMB, *Enzyme Nomenclature*, Academic Press, San Diego, 1992.
23. Naylor, S.L. and Klebe, R.L., Bioautography: a general method for the visualization of enzymes, *Biochem. Genet.*, 15, 1193, 1977.
24. Klebe, R.J., Mancuso, M.G., Brown, C.R., and Teng, L., Two-dimensional spectroscopy of electrophoretic gels, *Biochem. Genet.*, 19, 655, 1981.
25. Vora, S., Monoclonal antibodies in enzyme research: present and potential applications, *Anal. Biochem.*, 144, 307, 1985.
26. Utter, F., Aebersold, P., and Winans, G., Interpreting genetic variation detected by electrophoresis, in *Population Genetics and Fishery Management*, Reeman, N. and Utter, F., Eds., University of Washington Press, Seattle, 1987, p. 21.
27. Pasteur, N., Pasteur, G., Bonhomme, F., Catalan, J., and Britton-Davidian, J., *Practical Isozyme Genetics*, Ellis Horwood, Chichester, 1988.
28. Wendel, J.F. and Weeden, N.F., Visualization and interpretation of plant isozymes, in *Isozymes in Plant Biology*, Solties, D.E. and Solties, P.S., Eds., Dioscorides Press, Portland, OR, 1989, p. 5.
29. Morizot, D.C. and Schmidt, M.E., Starch gel electrophoresis and histochemical visualization of proteins, in *Electrophoretic and Isoelectric Focusing Techniques in Fisheries Management*, Whitmore, D.H., Ed., CRC Press, Boca Raton, FL, 1990, p. 23.
30. Murphy, R.W., Sites, J.W., Jr., Buth, D.G., and Haufler, C.H., Proteins: isozyme electrophoresis, in *Molecular Systematics*, Hillis, D.M., Moritz, C., and Mable, B.K., Eds., Sinauer, Sunderland, 1996, p. 51.
31. May, B., Starch gel electrophoresis of allozymes, in *Molecular Genetic Analysis of Populations: A Practical Approach*, 2nd ed., Hoelzel, A.R., Ed., Oxford University Press, Oxford, 1998, p. 1.
32. Acquaah, G., *Practical Protein Electrophoresis for Genetic Research*, Timber Press, Portland, OR, 1992.
33. Avise, J.C., *Molecular Markers, Natural History, and Evolution*, Plenum Press, New York, 1994.
34. Manchenko, G.P., Subunit structure of enzymes: allozymic data, *Isozyme Bull.*, 21, 144, 1988.

35. Lieber, D.C., *Proteomics*, Humana Press, Totowa, NJ, 2001.
36. Kinter, M. and Sherman, N.E., *Protein Sequencing and Identification Using Tandem Mass Spectrometry*, Wiley-Interscience, New York, 2001.
37. Markert, C.L., Biology of isozymes, in *Isozymes I: Molecular Structure*, Markert, C.L., Ed., Academic Press, New York, 1975, p. 1.
38. Venter, J.C., Adams, M.D., Myers, E.W., and 274 coauthors, The sequence of the human genome, *Science*, 291, 1304, 2001.
39. Charlesworth, D., Liu, F.-L., and Zhang L., The evolution of the alcohol dehydrogenase gene family by loss of introns in plants of the genus *Leavenworthia* (Brassicaceae), *Mol. Biol. Evol.*, 15, 552, 1998.
40. Hendriksen, P.J.M., Hoogerbrugge, J.W., Baarends, W.M., De Boer, P., Vreeburg, J.T.M., Vos, E.A., Van Der Lende, T., and Grootegoed, J.A., Testis-specific expression of a functional retroposon encoding glucose-6-phosphate dehydrogenase in the mouse, *Genomics*, 41, 350, 1997.
41. Pfeiffer-Guglielmi, B., Broer, S., Broer, A., and Hamprecht, B., Isozyme pattern of glycogen phosphorylase in the rat nervous system and rat astroglia-rich primary cultures: electrophoretic and polymerase chain reaction studies, *Neurochem. Res.*, 25, 1485, 2000.
42. May, O., Siemann, M., and Syldatk, C., A new method for the detection of hydantoinases with respect to their enantioselectivity on acrylamide gels based on enzyme activity stain, *Biotechnol. Tech.*, 12, 309, 1998.
43. Okwumabua, O., Persaud, J.S., and Reddy, P.G., Cloning and characterization of the gene encoding the glutamate dehydrogenase of *Streptococcus suis* serotype 2, *Clin. Diagn. Lab. Immun.*, 8, 251, 2001.
44. Kee, C., Sohn, S., and Hwang, J.M., Stromelysin gene transfer into cultured human trabecular cells and rat trabecular meshwork *in vivo*, *Invest. Ophthalmol. Vis. Sci.*, 42, 2856, 2001.
45. Lim, W.J., Park, S.R., Cho, S.J., Kim, M.K., Ryu, S.K., Hong, S.Y., Seo, W.T., Kim, H., and Yun, H.D., Cloning and characterization of an intracellular isoamylase gene from *Pectobacterium chrysanthemi* PY35, *Biochem. Biophys. Res. Commun.*, 287, 348, 2001.
46. Patterson, S.D., From electrophoretically separated protein to identification: strategies for sequence and mass analysis, *Anal. Biochem.*, 221, 1, 1994.
47. Black, D.L., Protein diversity from alternative splicing: a challenge for bioinformatics and post-genome biology, *Cell*, 103, 367, 2000.
48. Schmucker, D., Clemens, J.C, Shu, H., Worby, C.A, Xiao, J., Muda, M., Dixon, J.E, and Zipursky, S.L., *Drosophila* DSCAM is an axon guidance receptor exhibiting extraordinary molecular diversity, *Cell*, 101, 671, 2000.
49. Manchenko, G.P., Isozymes generated via alternative splicing of pre-mRNAs transcribed from single genes, *Gene Fam. Isozymes Bull.*, 34, 12, 2001.
50. Manchenko, G.P., Unusual isozyme patterns of glucose-6-phosphate isomerase in *Polydora brevipalpa* (Polychaeta: Spionidae), *Biochem. Genet.*, 39, 285, 2001.

II _____ General Principles of Enzyme Detection on Electrophoretic Gels

The objective of this part is to describe the general principles involved in visualization of gel areas containing specific enzymes separated during gel electrophoresis. These principles may be classified as resulting in positive or negative zymograms, as chemical or physical, as chromogenic or fluorogenic, as based on chemical coupling or enzymatic coupling, etc. It seems, however, that the mixed operational classification will be preferable for purely practical consideration. According to this classification, the main principles of enzyme visualization on electrophoretic gels are defined as:

1. Based on chromogenic reactions
2. Based on fluorogenic reactions
3. Autoradiography
4. Bioautography
5. Two-dimensional gel spectroscopy
6. Immunoblotting
7. Miscellanies

Two-dimensional spectroscopy of electrophoretic gels and immunohistochemical methods have good perspectives and are of great value for further developments in electrophoretic zymography. However, the great majority of zymographic techniques so far developed are based on the use of chromogenic and fluorogenic reactions or autoradiography.

SECTION 1
CHROMOGENIC REACTIONS

Chromogenic reactions are those that result in formation of a chromophore at sites of enzyme activities. The great majority of these reactions were adopted from well-tried histochemical or colorimetric enzyme assay methods. In a simple one-step chromogenic reaction a colorless substrate is enzymatically converted into a colored product. In a broad sense, the class of chromogenic reactions includes any reactions or set of reactions that reveal discrete zones (or bands) of enzyme activity visible at daylight. In most cases the primary product(s) of an enzyme reaction is not readily detectable, and a supplementary reagent(s) is added to the reaction mixture, which somehow reacts with the primary product(s) to form a visible secondary product(s). These secondary products can be formed as the result of spontaneous reactions or so-called chemical coupling. In some cases, however, none of the primary products can be detected by chemical coupling, and an additional enzymatic reaction(s) is needed in order to reach a detectable product(s). This procedure is called enzymatic coupling, and supplementary exogenous enzymes that are added to reaction mixtures are known as linking or auxiliary enzymes. The resultant zymogram is positively stained (colored bands on achromatic background) if the color is formed by a reaction with the product of the enzyme activity. A negatively stained zymogram (achromatic bands on colored background) is obtained if the color is formed by a reaction with the substrate.

Many different enzymes produce the same molecules (e.g., NADH, NADPH, orthophosphate, ammonia, hydrogen peroxide, etc.) or different molecules with essentially the same chemical properties (e.g., aldehydes, ketones, reducing sugars, thiols, etc.). This means that very similar or identical chromogenic reactions can be used to detect different enzymes. Thus, classification of chromogenic reactions based on properties of products that are detected seems to be the more useful and practical approach. Such classification is advantageous because it allows one to choose an adequate chromogenic reaction to visualize activity bands even of those enzymes that have not yet been detected on electrophoretic gels.

PRODUCTS REDUCING TETRAZOLIUM SALTS

The first histochemical method for the detection of enzyme activity using a tetrazolium salt was developed in 1951 by Seligman and Rutenberg.[1] Markert and Möller[2] were the first to adopt this histochemical procedure for detection of NAD(P)-dependent dehydrogenases on electrophoretic gels. These dehydrogenases produce reduced NADH or NADPH, the electron donors for reduction of tetrazolium salts, which are especially good electron acceptors. Reduction of a tetrazolium salt results in the formation of an intensely colored, water-insoluble precipitate, formazan. The reductive reaction proceeds rapidly in the presence of some electron carrier intermediaries. Initially, the diaphorase–Methylene Blue system was used for the transfer of electrons from NAD(P)H to a tetrazolium salt.[2] Further, it was found that phenazine methosulfate (PMS) is preferable as an intermediary catalyst. Molecules of PMS can accept electrons from NAD(P)H and reduce a tetrazolium salt, repeating the cycle and thus being very effective, even at low concentrations. Many NAD(P)-dependent dehydrogenases are detected on electrophoretic gels using the PMS–tetrazolium system (e.g., see 1.1.1.1 — ADH, Method 1; 1.1.1.37 — MDH, Method 1; 1.1.1.40 — ME; 1.1.1.42 — IDH, Method 1, etc.).*

---

* Referenced enzymes and methods can be found in Part III, Section 3, where they are listed in numerical order according to the EC numbers recommended in 1992 by the Nomenclature Committee of the International Union of Biochemistry and Molecular Biology.

5

The enzymes that are FMN (flavin mononucleotide)- or FAD (flavin-adenine dinucleotide)-containing flavoproteins (e.g., many oxidases) can also be detected using the PMS–tetrazolium system by a mechanism similar to that described above for NAD(P)-dependent dehydrogenases. In an oxidase detection system PMS molecules accept electrons from a reduced flavine group of an enzymatic flavoprotein and transfer them to a tetrazolium salt (e.g., see 1.1.3.15 — GOX, Method 1; 1.1.3.22 — XOX, Method 1). Being the prosthetic groups, FMN and FAD are tightly bound to the enzyme molecules and are not dissociated during electrophoresis, so they can be omitted from staining solutions used to detect oxidase activities.

Different tetrazolium salts that vary in their reduction potentials are now commercially available. Tetrazolium salts with high reduction potential are reduced with greater ease than those with low reduction potential. Among tetrazolium salts commonly used in electrophoretic zymography, methyl thiazolyl tetrazolium (MTT) is usually preferable for NAD(P)H detection because of its high reduction potential and photostability. Tetranitro blue tetrazolium (TNT) possesses a somewhat lower reduction potential than MTT, but a higher one than nitro blue tetrazolium (NBT). The lowest reduction potential is characteristic for triphenyltetrazolium chloride (TTC).

The products of enzymatic reactions containing free sulfhydryl groups (e.g., coenzyme A (CoA)–SH) can also reduce tetrazolium salts (usually NBT and MTT are preferable) in the presence of PMS or another intermediary catalyst, dichlorophenol indophenol (DCIP). An example is the detection method of citrate synthase (4.1.3.7 — CS).

When DCIP is substituted for PMS, NAD(P)H cannot effectively reduce MTT, but sulfhydryls (e.g., glutathione) can do this by the use of DCIP as an intermediary catalyst. This property of DCIP is used in detection methods developed for NAD(P)H-dependent glutathione reductase (1.6.4.2 — GSR, Method 2) and NAD(P)H diaphorases (1.6.99.1 — DIA(NADPH); 1.8.1.4 — DIA(NADH)).

Such products of enzymatic reactions as 4-imidazolone-5-propionate and β-sulfinylpyruvate can also reduce tetrazolium salts (viz., NBT) in the presence of PMS (2.6.1.1 — GOT, Method 3; 4.2.1.49 — UH). 5-Bromo-4-chloro-3-indoxyl can directly reduce NBT at room temperature without any intermediary catalyst (3.1.3.1 — ALP, Method 2).

Reducing sugars cause the reduction of TTC in an alkaline medium at room temperature (ketohexoses) or at 100°C (aldohexoses). Many enzymes producing ketosugars (e.g., D-fructose) and aldohexoses (e.g., D-mannose) are detected using this tetrazolium salt (2.4.1.7 — SP, Method 1; 4.2.1.46 — TDPGD; 3.2.1.24 — α-MAN, Method 3; 3.2.1.26 — FF, Method 4).

Some nonenzymatic proteins containing sulfhydryl groups may be nonspecifically stained on dehydrogenase zymograms obtained using the PMS–NBT or PMS–MTT system at alkaline pH. This nonspecific staining of –SH-rich proteins is sometimes erroneously attributed to so-called "nothing dehydrogenases" (1.X.X.X — NDH). Unlike real nothing dehydrogenases, however, –SH-rich proteins can also be stained in alkaline tetrazolium solutions lacking PMS.

The ability of free –SH groups to reduce NBT (or MTT) in an alkaline medium is used to detect some enzymes generating alkaline products (for details, see below, "Products that Cause a pH Change").

Many enzymes generate products that are the substrates for certain NAD(P)-dependent dehydrogenases. These enzymes can be detected using exogenous dehydrogenases as linking enzymes coupled with the PMS–MTT or PMS–NBT system. For example, the stains for hexokinase (2.7.1.1 — HK), phosphoglucomutase (5.4.2.2 — PGM, Method 1), and glucose-6-phosphate isomerase (5.3.1.9 — GPI) all use glucose-6-phosphate dehydrogenase as a linking enzyme; the stain for fumarate hydratase (4.2.1.2 — FH, Method 1) uses exogenous malate dehydrogenase; the stain for aconitase (4.2.1.3 — ACON, Method 1) uses auxiliary isocitrate dehydrogenase (1.1.1.42 — IDH, Method 1), etc. Similarly, xanthine oxidase coupled with the PMS–MTT system is used as a linking enzyme in the enzyme-linked stain for purine-nucleoside phosphorylase (2.4.2.1 — PNP).

For the detection of some enzymes, one or even more linked enzymatic reactions are used in order to reach a product detectable via dehydrogenase or oxidase. For example, mannose-6-phosphate isomerase (5.3.1.8 — MPI) is detected using two linked reactions sequentially catalyzed by auxiliary glucose-6-phosphate isomerase and glucose-6-phosphate dehydrogenase. Similarly, the stain for adenosine deaminase (3.5.4.4 — ADA, Method 1) uses exogenous purine-nucleoside phosphorylase to produce hypoxanthine, which is then detected by xanthine oxidase coupled with the PMS–NBT or PMS–MTT system. The detection system of mannokinase (2.7.1.7 — MK) includes three linked enzymatic reactions sequentially catalyzed by auxiliary mannose-6-phosphate isomerase, glucose-6-phosphate isomerase, and glucose-6-phosphate dehydrogenase. Three linking enzymes are also used in the detection system of dextransucrase (2.4.1.5 — DS, Method 3).

It should be noted that staining solutions containing PMS and a tetrazolium are light sensitive. Thus, incubation of gels in PMS–tetrazolium solutions should be carried out in the dark.

Even today, the tetrazolium dyes are of central importance in enzyme staining methods. A great majority of zymographic techniques described in Part III of this handbook are based on the use of the PMS–tetrazolium system.

PRODUCTS CAPABLE OF COUPLING WITH DIAZONIUM SALTS

Many hydrolytic enzymes (e.g., aminopeptidases, esterases, glycosidases, phosphatases) are visualized on electrophoretic gels using artificial substrates that are naphthol or naphthylamine derivatives. When naphthol or naphthylamine is liberated enzymatically, it immediately couples with a diazonium salt. An insoluble colored precipitate (an azo dye) is formed as a result of the coupling reaction in gel areas where specific hydrolase activity is located.

There are two main components of the azo coupling system. The first one is the diazonium ion. The diazonium ions are not stable and need to be in a salt form for prolonged storage.[3] A number of different stabilized diazonium salts are now commercially available. Usually these are salts or double salts of zinc chloride (e.g., Fast Blue B salt, Fast Blue BB salt, Fast Blue RR salt, Fast Violet B salt), tetrafluoroborate or sulfate (e.g., Fast Garnet GBC salt), naphthalenedisulfonate (e.g., Fast Red B salt), etc.[4] For further stabilization such inert additives as aluminum, magnesium, sodium and zinc sulfates, magnesium oxide and bicarbonate, and others can also be added to commercial diazonium salt preparations.

It should be taken into account that some enzymes may be inhibited by these additives. When the diazonium salt acts as an enzyme inhibitor, a two-step, or so-called postcoupling, staining procedure is recommended. In this procedure, the gel is initially incubated in substrate solution for an essential period of time (usually 30 min) and only then diazonium salt is added. However, the one-step azo coupling procedure is usually preferable because it considerably reduces diffusion of enzymatically liberated naphthols and naphthylamines and results in development of more sharp and distinct activity bands.

Dissolved diazonium salts give very unstable ions, so solutions should be prepared immediately before use. The stability of diazonium ions depends on the pH of the staining solution. For example, Fast Black K salt is stable in acidic pH, while Fast Blue RR salt is recommended for use at neutral and alkaline pH. High temperature also contributes to decreasing stability of diazonium salts. When the azo coupling system is used to detect the enzyme, a compromise should be achieved between the pH optimum of the enzyme activity and diazonium salt stability. If pH values extreme for diazonium salt stability must be used, it is recommended that the diazonium salt solution be replaced as often as necessary, depending on the total period of gel incubation. Because diazonium ions can themselves act as enzyme inhibitors, the use of optimal concentrations of 1 mg/ml is recommended.[3]

The second component of the azo coupling system is an enzyme-specific artificial substrate that is an amide, an ester, or a glucoside of the coupling agent (naphthol or naphthylamine). The substrates that are derivatives of substituted naphthol (e.g., 6-bromo-2-naphthol) or substituted naphthylamine (e.g., 1-methoxy-3-naphthylamine) may be preferable; this is because some substituted naphthols coupled with diazonium ions give azo dyes of greater insolubility. The use of derivatives of 1-methoxy-3-naphthylamine, which complexes much more quickly with diazonium ions than does 2-naphthylamine, reduces the problem of product diffusion during color formation.

Examples of the use of the azo coupling system in electrophoretic zymography are the detection of esterases (3.1.1 ... — EST, Method 1) and acid phosphatase (3.1.3.2 — ACP, Method 3), with α-naphthol as the coupling agent; the detection of β-glucuronidase (3.2.1.31 — β-GUS, Method 2), with naphthol-AS-BI or 6-bromo-2-naphthol as coupling agents; the detection of leucine aminopeptidase (3.4.11.1 — LAP, Method 1), with 2-naphthylamine as the coupling agent; and the detection of some proteinases (3.4.21–24 ... — PROT, General Principles of

Detection), with 1-methoxy-3-naphthylamine as the coupling agent.

Diazonium ions also couple with such enzymatic products as oxaloacetate, which serves as a coupling agent of the azo coupling system in the detection of some nonhydrolytic enzymes, e.g., glutamic–oxaloacetic transaminase (2.6.1.1 — GOT, Method 1), phosphoenolpyruvate carboxylase (4.1.1.31 — PEPC), and pyruvate carboxylase (6.4.1.1 — PC).

The azo coupling system was the first enzyme-staining system successfully adopted from histochemistry for purposes of electrophoretic zymography.[5] A great number of zymographic techniques described in Part III are based on the use of this staining system.

PRODUCTS THAT CAUSE A pH CHANGE

Local acidic–alkaline pH change takes place in areas of acid electrophoretic gels where enzymes producing ammonia ions are localized. These enzymes are detected by incubation of electrophorized gels in staining solutions containing the substrate and an appropriate pH indicator dye, e.g., Phenol Violet, which is light orange at acidic pH and becomes dark blue at alkaline pH. Examples are the detection of urease (3.5.1.5 — UR, Method 1), adenosine deaminase (3.5.4.4 — ADA, Method 3), and AMP deaminase (3.5.4.6 — AMPDA, Method 3).

The local pH increase due to enzymatic production of ammonia ions is used in catalyzing silver deposition from neutral nitrate solution in the presence of photographic developers, 4-hydroxyphenol and 4-hydroxyaniline and the reducing thiol reagent 2-mercaptoethanol.[6] 2-Mercaptoethanol is used to increase the difference of the reduction potential between ammonia-producing enzyme bands and gel background. In the absence of this reducing agent the intensity of the enzyme activity bands is sensibly decreased and the gel background increased. Optimal contrast between enzymatic bands and the background is due to 4-hydroxyaniline, which reduces the initially oxidized 4-hydroxyphenol, which reduces faster and with more contrast to the black metallic form than the silver ions to the black metallic form. The photographic developers initiate silver deposition, which further proceeds in an autocatalyzed way. Thus, the method is based on the faster deposition of metallic silver from a neutral silver solution in gel areas where the ammonia-producing enzyme is localized than in enzyme-free gel areas. The example of successful application of the silver deposition method is urease (3.5.1.5 — UR, Method 2).

Tetrazolium salts may also be used to detect enzymes that produce ammonia ions and so elevate pH in those gel areas where they are localized. In these cases the elevated pH increases the rate of reduction of tetrazolium salt by the sulfhydryl compound dithiothreitol, which is included in the staining solution along with the substrate and a tetrazolium salt. The slightly less-sensitive NBT is recommended for use in these stains in preference to MTT.[7] Examples of applications of this detection system are staining methods for arginase (3.5.3.1 — ARG, Method 1) and cytidine deaminase (3.5.4.5 — CDA, Method 1).

Local alkaline–acidic pH changes can also occur in electrophoretic gels as a result of the catalytic activity of some enzymes. These local pH changes are detected using such pH indicator dyes as Bromothymol Blue (blue at pH > 7.6; yellow at pH < 6.0) or Phenol Red (red at pH > 8.2; yellow at pH < 6.8). Examples of this are the detection of trypsin (3.4.21.4 — T; Other Methods, A), esteroprotease (3.4.21.40 — EP, Method 2), and carbonic anhydrase (4.2.1.1 — CA, Method 1). The orthophosphate, liberated by enzymatic activity, treated with calcium ions, liberates protons during the formation of calcium phosphate gel. So, phosphatase activity bands can also be detected by the color change of the pH indicator dye. The example is alkaline phosphatase (3.1.3.1 — ALP, Method 6).

The following conditions should be kept when any method of the local pH change detection is applied: (1) nonbuffered staining solutions should be used, and (2) gel buffers should be of minimal concentrations and desirable pH values to allow the pH change to occur as the result of an enzyme action.

The main disadvantage of all staining methods using pH indicator dyes is the diffused character of developed enzyme activity bands. This is not the case, however, for the NBT–dithiothreitol method, which results in formation of practically nondiffusable formazan.

ORTHOPHOSPHATE

Many enzymes liberate orthophosphate as the reaction product. Several different methods were developed for the visualization of phosphate-liberating enzymes. These include the lead sulfide method and its reduced modification, the calcium phosphate method, the acid phosphomolybdate method and its Malachite Green modification, the pH indicator method (described above), and the enzymatic method. All these methods may be applied only when electrophoresis and gel staining are carried out in phosphate-free buffers.

Gomori[8] was the first to use calcium ions to precipitate orthophosphate released by the action of phosphate-liberating enzymes. Calcium phosphate precipitate, however, is soluble in acidic solution and so cannot be used for the detection of some phosphate-liberating enzymes with acidic pH optimum, e.g., acid phosphatase. In order to avoid this drawback, a lead is used that forms insoluble lead phosphate.[9] This phosphate salt is colorless and therefore not easily recognizable; however, treating it with sulfide results in the formation of a brownish black insoluble precipitate of lead sulfide. Examples of the application of the lead sulfide method in electrophoretic zymography are the detection methods for alkaline phosphatase (3.1.3.1 — ALP, Method 7) and acylphosphatase (3.6.1.7 — AP). The main disadvantage of the method is the need for sequential treatment of electrophorized gel with several solutions in order to achieve a colored end product. When this method is used, it should be remembered that lead also precipitates borate ions. So, borate-containing buffers should not be used for electrophoresis or staining. The silver nitrate may be used to form silver phosphate, which is then reduced by sodium hydroxide treatment to black metallic silver. This method has some advantages over the lead sulfide method.

The example of application of this method is aldolase (4.1.2.13 — ALD, Other Methods).

The reduced Gomori method can be applied to transparent electrophoretic gels (e.g., polyacrylamide gel (PAG)). It is based on the formation of white calcium phosphate precipitate, which is insoluble in alkaline conditions. The white bands of calcium phosphate precipitation are clearly visible when the stained gel is viewed against a dark background.[10] For example, this method is used to detect fructose bisphosphatase (3.1.3.11 — FBP, Method 2), dehydroquinate synthase (4.6.1.3 — DQS), and chorismate synthase (4.6.1.4 — CHOS). When more opaque electrophoretic gels are used (e.g., starch or acetate cellulose), calcium phosphate may be subsequently stained with Alizarin Red S.

The acid phosphomolybdate method is based on the formation of phosphomolybdate and its subsequent reduction by aminonaphthol sulfonic acid[11] or ascorbic acid,[12] which results in a blue stain. The example is the detection method for inorganic pyrophosphatase (3.6.1.1 — PP, Method 1). The phosphomolybdate method, however, suffers from the diffusion of both orthophosphate and phosphomolybdate. Isozyme patterns visualized by this method fade quickly. To reduce diffusion of the colored end product, the use of an additional stain ingredient, Malachite Green, is recommended.[13] The Malachite Green–phosphomolybdate complex formed is almost fully insoluble, so zymograms obtained by this method may be stored for several months (for details, see 3.6.1.1 — PP, Method 1, *Notes*).

The enzymatic method[14] has proven to be the most sensitive for the detection of phosphate-liberating enzymes. The principle of the method involves employing the phosphate-requiring phosphorolytic cleavage of inosine catalyzed by auxiliary purinenucleoside phosphorylase to produce hypoxanthine, which is then detected using a second linking enzyme, xanthine oxidase, coupled with the PMS–NBT or PMS–MTT system. The colored end product is blue formazan. The enzymatic method, compared to other methods for the detection of orthophosphate, is advantageous due to its high sensitivity and generation of a nondiffusible formazan. The example of this method is the detection of alkaline phosphatase (3.1.3.1 — ALP, Method 8).

Other enzymatic methods for detection of orthophosphate exist that use linked enzymatic reactions involving either the glyceraldehyde-3-phosphate dehydrogenase[15] or the phosphorylase[16] reactions. Both these methods include readily reversible enzymatic reactions. In addition, the first one requires glyceraldehyde-3-phosphate, which is unstable, while the other involves three enzymatic steps subsequently catalyzed by auxiliary phosphorylase, phosphoglucomutase, and glucose-6-phosphate dehydrogenase, and so is very complex. Because of these disadvantages, neither enzymatic method is in wide use.

Some enzymes that produce phosphorus-containing compounds also can be detected by appropriate phosphate-detecting methods described above after the cleavage of orthophosphate by auxiliary alkaline phosphatase (e.g., see 3.1.4.17 — CNPE, Method 2; 3.1.4.40 — CMP-SH; 3.6.1.X — NDP). In these cases, however, it should be taken into account that auxiliary alkaline phosphatase must not cleave orthophosphate from phosphorous-containing substrates.

Orthophosphate can also be cleaved from glucose 1-phosphate and ribose-1-phosphate by treatment with H_2SO_4 and subsequently detected by the acid phosphomolybdate method or the Malachite Green phosphomolybdate modification of this method. As a result, the positively stained zymograms are obtained for enzymes that produce these phosphosugars (5.4.2.7 — PPM, Method 3), whereas negative zymograms are obtained when these phosphosugars are used as substrates (5.4.2.2 — PGM, Method 2). Orthophosphate can be liberated from glycerone phosphate and glyceraldehyde-3-phosphate by treatment with iodacetamide or iodacetate and then detected by a suitable phosphate-detecting method. For example, see aldolase (4.1.2.13 — ALD, Other Methods).

Since many enzymes liberate orthophosphate as the reaction product or produce phosphorus-containing products from which orthophosphate can be readily cleaved chemically or enzymatically, the phosphate-detecting methods are of great importance for electrophoretic zymography.

PYROPHOSPHATE

Pyrophosphate-liberating enzymes can be visualized by incubation of transparent electrophoretic gels with buffered solutions containing certain substrates and subsequently treating the gels with calcium ions.[10] As a result, white bands of calcium pyrophosphate precipitation are clearly visible when the stained gel is viewed against a dark background. Examples are detection methods for DNA-directed RNA polymerase (2.7.7.6 — DDRP) and UDPglucose pyrophosphorylase (2.7.7.9 — UGPP, Method 3).

Manganese ions also can be used to detect the product pyrophosphate on transparent electrophoretic gels. The bands formed as a result of manganese pyrophosphate precipitation are not apparent when stained gel is placed directly on a dark background, but they are very distinct when gel is held a few inches from a dark background and lighted indirectly. An example is the detection method for hypoxanthine phosphoribosyltransferase (2.4.2.8 — HPRT, Method 3).

A coupled reaction catalyzed by auxiliary inorganic pyrophosphatase from yeast can be used to convert the product pyrophosphate into orthophosphate, which is then visualized by an appropriate phosphate-detecting method (see the section on orthophosphate, above). Yeast inorganic pyrophosphatase is a specific catalyst for the hydrolysis of pyrophosphate in the presence of magnesium ions, and several organic pyrophosphates, such as ATP and ADP, are not attacked. This method may not be used for detection of pyrophosphate-releasing enzymes, which use substrates also hydrolyzed by auxiliary pyrophosphatase. An example of successful application of this method is sulfate adenylyltransferase (2.7.7.4 — SAT).

Pyrophosphate can also be detected using three coupled reactions sequentially catalyzed by auxiliary enzymes pyrophosphate-fructose-6-phosphate 1-phosphotransferase (PFPPT), aldolase, and NAD-dependent glyceraldehyde-3-phosphate

dehydrogenase coupled with the PMS–MTT system. The use of bacterial PFPPT is recommended because the plant enzyme requires for its activity fructose-2,6-bisphosphate, which is very expensive. Triose-phosphate isomerase may be used as the fourth auxiliary enzyme to accelerate coupling reactions and to enhance the intensity of colored bands. This enzymatic method is adopted from the detection method for PFPPT (2.7.1.90 — PFPPT). Fluorogenic modification of the method may also be found in the same place.

HYDROGEN PEROXIDE

Hydrogen peroxide is the obligatory product of almost all oxidases that use oxygen as an acceptor. It also serves as a proton-accepting substrate for peroxidase. So, a coupled peroxidase reaction is widely used to detect enzymes producing hydrogen peroxide. There are many redox dyes that can be used as proton donors in the peroxidase reaction. The more frequently used dyes are 3-amino-9-ethyl carbazole, o-dianisidine dihydrochloride, and tetramethyl benzidine (see 1.11.1.7 — PER, Method 1). When oxidized, these redox dyes change in both color and solubility. They are soluble when reduced but become insoluble upon oxidation. The former goes from a yellow or light brown dye to a red or dark brown compound, while the latter two change from colorless to orange and blue, respectively.

The chromogenic peroxidase reaction also is used as the last step in some enzyme-linked detection systems. The more widely used one is the staining method for peptidases (3.4.11 or 13 ... — PEP, Method 1), where the peroxidase reaction is used together with the first-step coupled reaction catalyzed by auxiliary L-amino acid oxidase. Another example is the enzyme-linked detection system for β-fructofuranosidase (3.2.1.26 — FF, Method 3), in which the chromogenic peroxidase reaction is used coupled with a hydrogen peroxide-generating reaction catalyzed by the other auxiliary enzyme glucose oxidase.

Hydrogen peroxide also takes part in chemical chromogenic reactions, which usually are used in detection methods for enzymes utilizing hydrogen peroxide as the substrate. In these cases, negatively stained zymograms are obtained. Two such chromogenic reactions are used to detect catalase (1.11.1.6 — CAT, Methods 1 and 2). The first one is based on the formation of a dark green compound as a result of the chemical reaction between hydrogen peroxide and potassium ferricyanide in the presence of Fe^{3+} ions. The second chromogenic reaction proceeds between potassium iodide and starch in the presence of hydrogen peroxide and thiosulfate. In areas of the starch gel (or PAG-containing starch) where CAT is localized, hydrogen peroxide is enzymatically destroyed. Upon exposure to potassium iodide, wherever hydrogen peroxide is not destroyed, the iodide is oxidized to iodine and an intense blue starch–iodine chromatophore is formed. Thiosulfate is inactivated in gel areas where hydrogen peroxide is presented and remains active in areas of CAT activity where hydrogen peroxide is destroyed. The function of thiosulfate is to reduce any iodine that escapes into solution during the staining procedure and settles on the gel areas occupied by CAT.[17]

The highly sensitive lanthanide luminescence method has been developed based on the photoassisted oxidation of phenantroline dicarboxylic acid dihydrazide (PDAdh) by hydrogen peroxide. The resulting PDA interacts with europium ions and luminescent Eu:PDA complex forms, which can be observed under ultraviolet (UV) light (for more details, see "Products that Form Luminescent Lanthanide Chelates" in Section 2).

Hydrogen peroxide can be detected using cerium and diaminobenzidine (DAB). Cerium chloride reacts with hydrogen peroxide, forming the insoluble cerium perhydroxide of a pale yellow color. This pale reaction product can be converted to a readily visible precipitate by the addition of DAB. Cerium perhydroxide oxidizes DAB to its insoluble polymeric form of a dark brown color. The cerium–DAB method was successfully used to detect sulfhydryl oxidase (see 1.8.3.2 — TO), L-2-hydroxyacid oxidase (see 1.1.3.15 — HAOX, Method 3), and amino acid oxidases (see 1.4.3.2 — LAOX, Method 2; 1.4.3.3 — DAOX, Method 4) on native electrophoretic gels and nitrocellulose blots.

CARBONATE IONS

Carbon dioxide–producing enzymes are visualized using the calcium carbonate method.[10] This method is based on the formation of a white calcium carbonate precipitate in gel areas where the enzyme producing carbon dioxide is localized. The method is applicable only for transparent gels and is applied as a one-step procedure using buffered staining solution containing the enzyme-specific substrate(s) and Ca^{2+} ions. After an appropriate period of incubation in staining solution the gel is viewed against a dark background for white bands of calcium carbonate precipitation. The examples are detection methods for isocitrate dehydrogenase (1.1.1.42 — IDH, Method 2) and pyruvate decarboxylase (4.1.1.1 — PDC, Method 1).

The calcium carbonate method is essentially the same as the calcium phosphate and calcium pyrophosphate methods (see above), but about ten times less sensitive. An enzymatic method for detection of carbon dioxide-producing enzymes also exists. This method uses a linked enzymatic reaction catalyzed by auxiliary phosphoenolpyruvate carboxylase to convert phosphoenolpyruvate and carbon dioxide into orthophosphate and oxaloacetate. Oxaloacetate is then coupled with diazonium salt to form insoluble azo dye (4.1.1.31 — PEPC).

The modification of this method is possible by the use of an additional linked reaction catalyzed by auxiliary malate dehydrogenase in place of the azo coupling reaction. In this modification oxaloacetate and NADH are converted by malate dehydrogenase into malate and NAD. Gel areas where NADH-into-NAD conversion occurs may be registered in long-wave UV light as dark (nonfluorescent) bands visible on a light (fluorescent) background. When bands are well developed, the gel may be treated with PMS–MTT to obtain a zymogram with achromatic bands on a blue background.

Further modification of the enzymatic method may consist of the application of different phosphate-detecting methods (see above) to visualize areas of orthophosphate production by linked-enzyme phosphoenolpyruvate carboxylase. The calcium phosphate method, however, may not be used because calcium carbonate precipitate will also form and block the linked reaction of phosphoenolpyruvate carboxylase.

The use of plant phosphoenolpyruvate carboxylase is preferable because it does not require acetyl-CoA for its activity.

COLORED PRODUCTS

In some cases the products of enzymatic reactions are readily visible due to the acquisition of some properties not found in the substrates. For instance, chromogenic synthetic substrates that are derivatives of *para*-nitrophenol or *para*-nitroaniline are used to detect some hydrolytic enzymes. These substrates are colorless; however, hydrolytically cleaved *para*-nitrophenol and *para*-nitroaniline are yellow (e.g., 3.1.3.1 — ALP, Method 4; 3.4.21.5 — THR). Colored bands developed by the use of these chromogenic substrates may be too faint for visual quantitative analysis and are usually registered spectrophotometrically on transparent gels using a scanning spectrophotometer. The treatment of acidic gels with alkali enhances the band coloration caused by *para*-nitrophenol (3.1.16.1 — SE, Method 1). Yellow *para*-nitroaniline may be converted into a readily visible red azo dye after diazotation with *N*-(1-naphthyl)ethylenediamine.[18] The examples are detection methods for trypsin (3.4.21.4 — T, Method 3) and chymotrypsin (3.4.21.1 — CT, Method 3).

Another example of the detection methods based on formation of a colored product is the use of phenolphthalein phosphate as a substrate for acid phosphatase (3.1.3.2 — ACP, Method 1). In this method, released phenolphthalein leads to the formation of pink zones after treatment of the gel with alkali.

Synthetic chromogenic substrates that are derivatives of the indoxyl are also used to detect different hydrolytic activities. When indoxyl derivatives (e.g., indoxyl esters) are hydrolyzed, the liberated indoxyl is oxidized by atmospheric oxygen to an intensely colored indigo dye. This oxidation, however, occurs very slowly, especially in acidic conditions. So, undesirable diffusion of soluble indoxyl may be significant. Derivatives of 5-bromo-4-chloro-3-indoxyl, ammonium 5-indoxyl, and 5-bromoindoxyl are usually used in place of simple indoxyl derivatives (e.g., 3.1.3.1 — ALP, Method 2; 3.1.4.1 — PDE-1, Method 2; 3.2.1.31 — β-GUS, Method 3). Hydrolytically cleaved 5-bromo-4-chloro-3-indoxyl forms a ketone, which dimerizes under alkaline conditions into dehydroindigo and releases hydrogen ions. If NBT is added, it is reduced by these ions to blue formazan. Thus, the formation of insoluble formazan and an indigo dye occurs simultaneously, resulting in development of sharp colored bands. An example is the detection method for alkaline phosphatase (3.1.3.1 — ALP, Method 2).

PRODUCTS BEARING REDUCED THIOL GROUPS

There are at least three different methods for the detection of products bearing reduced thiol groups. The disulfide 5,5'-dithiobis(2-nitrobenzoic acid) (DTNB) reacts with a free thiol at pH 8.0 and yields a compound of yellow color. This reaction is used in the detection method for glutathione reductase (1.6.4.2 — GSR, Method 3). DTNB attached to polyacrylamide was synthesized and shown to be useful for detection of thiol-producing enzymes after PAG electrophoresis (3.1.1.7 — ACHE, Method 2; 3.1.3.1 — ALP, Method 9).[19]

The products bearing free thiol groups reduce ferricyanide to ferrocyanide, which further reacts with mercury and yields a reddish brown precipitating complex. For instance, this reaction is used to detect acetyl-CoA hydrolase (3.1.2.1 — ACoAH) and malate synthase (4.1.3.2 — MS).

Free thiol groups can also reduce tetrazolium salts via the intermediary catalyst dichlorophenol indophenol, resulting in formation of insoluble formazan (see also "Products Reducing Tetrazolium Salts," above). Examples include the detection methods for glutathione reductase (1.6.4.2 — GSR, Method 2) and citrate synthase (4.1.3.7 — CS).

A starch–iodine chromogenic reaction also can be used to detect enzymes that use or produce compounds bearing reduced thiol groups (for details, see the paragraph below).

PRODUCTS THAT INFLUENCE THE STARCH–IODINE REACTION

The starch–iodine chromogenic reaction has been known for more than a century. However, iodine (I_2) is insoluble in water. Its solubility increases considerably in the presence of iodide (I^-), so KI is usually included in iodine solutions. Some compounds influence the starch–iodine reaction due to their ability to reduce iodine to iodide, which fails to react with starch. Such reductive properties toward iodine are displayed by some compounds bearing free thiol groups, e.g., by reduced glutathione. On the contrary, oxidized glutathione does not preclude the formation of the starch–iodine chromophore. Thus, positively stained zymograms can be obtained after starch gel electrophoresis (or electrophoresis in PAG containing soluble starch) by the use of the starch–iodine reaction for those enzymes that convert reduced glutathione into its oxidized form. The examples are detection methods for glutathione transferase (2.5.1.18 — GT, Method 1) and glyoxalase I (4.4.1.5 — GLO, Method 3). Using the starch–iodine reaction, negatively stained zymograms will be obtained for enzymes that convert oxidized glutathione into its reduced form. In principle, glutathione reductase can serve as a good candidate for application of the starch–iodine detection method; however, several more sensitive and practical methods are available for detection of this enzyme (1.6.4.2 — GSR).

Hydrogen peroxide influences the starch–iodine reaction due to its ability to oxidize iodide in acid conditions into iodine, which then takes part in formation of the starch–iodine colored complex. This property of hydrogen peroxide is used to obtain negatively stained zymograms of catalase (for details, see "Hydrogen Peroxide," above).

Enzymes that release iodine can be detected after electrophoresis in starch-containing gels using an acid peroxide reagent.[20] The only enzyme detected by this method is glutathione transferase (2.5.1.18 — GT, Method 2). In this case the detection depends on the release of iodine from synthetic substrate (which is the iodide derivative) as a result of its conjugation with reduced glutathione and subsequent oxidation of the iodide to iodine. In gel areas where iodide is enzymatically liberated, blue staining develops due to starch–iodine complex formation.

PRODUCTS OF POLYMERIZING ENZYMES

Some enzymes catalyze reactions of polymerization using specific polymeric molecules as primers. Detection of these enzymes is based on the elevation of concentration of polymeric molecules in those gel areas where polymerizing enzymes are located. The use of stains specific to polymeric molecules usually results in the appearance of intensely stained bands visible on a less intensely stained background. For instance, iodine solution is used as a specific stain for detection of starch or glycogen synthesized by phosphorylase (2.4.1.1 — PHOS, Method 1). Methylene Blue and Pyronin B are used as specific stains for poly(A)ribonucleotide molecules synthesized by polyribonucleotide nucleotidyltransferase (2.7.7.8 — PNT) and mRNA replicas produced by RNA-directed RNA polymerase (2.7.7.48 — RNAP).

In some cases, when direct chromogenic staining of polymeric products is not possible, additional chemical treatments may be needed to visualize polymerase activity bands (e.g., see 2.4.1.5 — DS, Method 2).

PRODUCTS OF DEPOLYMERIZING ENZYMES

There are many hydrolytic enzymes that display depolymerizing activity toward such polymeric molecules as polypeptides, RNA, DNA, polysaccharides, and mucopolysaccharides. The general principle of the methods used to detect depolymerizing enzymes is the incorporation of a specific polymeric substrate into a reactive agarose plate, which is held in contact with an electrophorized gel. After incubation of this "sandwich" for a sufficient period of time, a substrate-containing agarose plate is treated with a substrate-precipitating agent, washed to remove the low-molecular-weight products of depolymerization, and specifically stained for general proteins, nucleic acids, polysaccharides, or mucopolysaccharides. Depolymerase activity areas appear as achromatic bands on colored backgrounds. Examples of the "sandwich" method are detection procedures for endopeptidases (3.4.21–24 ... — PROT), endonucleases (3.1.21.1 — DNASE; 3.1.27.5 — RNASE, Method 1), and endoglucanases (3.2.1.6 — EG).

In some cases depolymerase activity bands become visible just after treatment with a substrate-precipitating agent, so that no additional staining of the substrate-containing gel is needed. The example is the detection method for polygalacturonase (3.2.1.15 — PG, Method 1).

Different modifications of the "sandwich" method have been developed. Some of them consist of methods using colored substrates and are based on decoloration of the substrate-containing agarose gels in those areas that are in contact with regions of the depolymerizing enzyme located in separating gels. Examples are detection methods for cellulase (3.2.1.4 — CEL, Method 2) and endoxylanase (3.2.1.8 — EX, Method 1). The inclusion of polymeric substrates into the separating gel matrix may be preferable for detection of depolymerizing enzymes, which may be renatured after electrophoresis in sodium dodecyl sulfate (SDS)-containing gels, where SDS is used to inactivate depolymerizing enzymes and prevent substrate hydrolysis during the run. This modification is advantageous because separated enzymes do not need to diffuse out of the separating gel, and therefore, isozymes with slight differences in electrophoretic mobility can be readily differentiated (e.g., 3.2.1.4 — CEL, Method 2, *Notes*; 3.1.21.1 — DNASE, *Notes*).

The separating starch gel itself serves as a substrate for amylase (3.2.1.1 — α-AMY, Method 1) and phosphorylase (2.4.1.1 — PHOS, Method 3).

All methods described above are negative in the sense that color is formed by a chromogenic reaction with the substrate. These methods are usually used in a two-step procedure. The exceptions are those methods where colored substrates are used.

SECTION 2
FLUOROGENIC REACTIONS

Fluorogenic reactions are those that result in formation of a fluorochrome at sites of enzymatic activities in electrophoretic gels. Fluorochromes are molecules that, being irradiated by UV light, reradiate with emission of light of a longer wavelength. Positive fluorescent methods depend on the generation of a highly fluorescent product by an enzyme action on a nonfluorescent substrate(s). On the contrary, negative fluorescent methods depend on the generation of a nonfluorescent product from a fluorescent substrate.

The more widely used in electrophoretic zymography are the natural fluorochromes NADH and NADPH, which lose their fluorescent properties upon oxidation. There is also a diverse group of artificial fluorochromes. Among these, 4-methylumbelliferon is the most widely used.

NADH and NADPH

Positive fluorescent zymograms may be obtained for almost all NAD(P)-dependent dehydrogenases that generate NAD(P)H from NAD(P). These dehydrogenases are usually detected using the tetrazolium method, which is more convenient because it results in the development of dehydrogenase activity bands visible in daylight (see "Products Reducing Tetrazolium Salts" in Section 1). However, positive fluorescent stains are less expensive than tetrazolium stains because they do not require PMS and a tetrazolium salt. These stains are valuable for detection of some dehydrogenases that are inhibited by PMS or require reduced glutathione for their activity, which causes nonspecific formation of formazan (1.1.1.205 — IMPDH; 1.2.1.1 — FDHG; 2.4.2.8 — HPRT, Method 2). For some dehydrogenases activated by manganese ions the positive fluorescent stains may be more sensitive than the tetrazolium stains because of the inhibitory effect of manganese ions on the action of PMS as an electron

carrier intermediary (1.1.1.40 — ME, *Notes*; 1.1.1.42 — IDH, General Notes).

Negative fluorescent stains are less sensitive and less convenient than positive ones. This is because the concentration of fluorescent substrates (NADH or NADPH) needs to be adjusted very carefully so as to enable detection of nonfluorescent products (NAD or NADP) against a fluorescent background. If NADH or NADPH concentrations are too low, the fluorescence of the gel background may not be sufficiently intense to detect nonfluorescent bands of NAD or NADP generated by backward dehydrogenase reactions. On the contrary, if NADH or NADPH concentrations are too high, dehydrogenase isozymes of low activity may not be able to oxidize all NADH or NADPH molecules and, thus, to be detected as nonfluorescent bands on a fluorescent background. Nevertheless, negative fluorescent stains are important and valuable for detection of the NADH- and NADPH-dependent activities of some reductases (e.g., 1.1.1.1 — ADH, Method 2; 1.1.1.2 — ADH(NADP), Method 2 and General Notes; 1.6.4.2 — GSR, Method 1). These stains are very useful for detecting some dehydrogenases using backward reactions if the substrates for the forward reactions are not available or are too expensive (e.g., see 1.1.1.17 — M-1-PDH, Method 2; 1.5.1.17 — ALPD, Method 2 and General Notes). Finally, the backward NADH- or NADPH-dependent reactions of exogenous dehydrogenases are widely used as the endpoint reactions in enzyme-linked detection methods. In these cases the negative fluorescent zymograms are also obtained. Many different enzymes are detected by enzyme-linked negative fluorescent methods. For example, lactate dehydrogenase is used as the linking enzyme to detect pyruvate-producing enzymes. The stains for those enzymes depend on detection of nonfluorescent NAD against a background of fluorescent NADH (e.g., 2.6.1.2 — GPT, Method 1; 2.7.1.40 — PK, Method 1). Exogenous lactate dehydrogenase in combination with exogenous pyruvate kinase is widely used for negative fluorescent detection of ADP-producing enzymes (e.g., 2.7.1.6 — GALK, Method 2; 2.7.1.30 — GLYCK, Method 3; 2.7.1.35 — PNK; 2.7.3.2 — CK, Method 1; 2.7.3.3 — ARGK, Method 2; 2.7.4.3 — AK, Method 2).

In general, when the backward NADH- or NADPH-dependent reaction of any exogenous dehydrogenase is used to detect the end product, negative fluorescent zymograms are obtained. Many different dehydrogenases are used as linking enzymes in negative fluorescent stains. These are glycerol-3-phosphate dehydrogenase (e.g., 2.2.1.2 — TALD, Method 1), glyceraldehyde-3-phosphate dehydrogenase (e.g., 2.7.2.3 — PGK), malate dehydrogenase (e.g., 2.6.1.1 — GOT, Method 2), octopine dehydrogenase (e.g., 2.7.3.3 — ARGK, Method 3), saccharopine dehydrogenase (e.g., 4.1.1.20 — DAPD), glutamate dehydrogenase (e.g., 4.3.2.1 — ASL, Method 1), etc.

4-Methylumbelliferone

The positive fluorescent stains that are based on the use of the artificial fluorochrome, 4-methylumbelliferone, are very sensitive and especially valuable for the detection of hydrolytic enzymes.

Such derivatives of this fluorochrome as esters and glycosides do not fluoresce. However, being freed upon hydrolytic cleavage of the ester or glycosidic linkages, 4-methylumbelliferone restores its fluorescent properties. 4-Methylumbelliferone exhibits optimum fluorescence in alkaline conditions, while the great majority of hydrolases exhibit optimal activity in acid conditions. Thus, the positive fluorescent stains that employ derivatives of 4-methylumbelliferone as substrates usually are applied in two-step procedures. The first step is the incubation of electrophorized gel with acid substrate solution, and the second step is making the surface of the gel alkaline by using ammonia vapor or an alkaline buffer. Examples of positive fluorescent stains based on the use of 4-methylumbelliferone derivatives are the detection methods for esterases (3.1.1 ... — EST, Method 2), phosphatases (3.1.3.1 — ALP, Method 3; 3.1.3.2 — ACP, Method 2), phosphodiesterase I (3.1.4.1 — PDE-I, Method 1), arylsulfatase (3.1.6.1 — ARS, Method 1), cellulase (3.2.1.4 — CEL, Method 3), sialidase (3.2.1.18 — SIA), α-glucosidase (3.2.1.20 — α-GLU, Method 1), β-galactosidase (3.2.1.23 — β-GAL, Method 1), etc.

PRODUCTS THAT FORM LUMINESCENT LANTHANIDE CHELATES

Luminescent lanthanide chelates are compounds that consist of a lanthanide ion (e.g., Tb^{3+} or Eu^{3+}) bound to an organic chelating molecule capable of absorbing ultraviolet light and then transferring the excitation energy to the bound lanthanide ion, which emits luminescence in the green-to-red region of the spectrum.[21]

Artificial substrates were synthesized for alkaline phosphatase, β-galactosidase, and xanthine oxidase, which are converted by enzymatic action into the product salicylic acid, which then forms the luminescent lanthanide chelate after treatment with Tb^{3+} ions and ethylene diamine tetraacetate (EDTA) in alkaline conditions (3.2.1.23 — β-GAL, Method 5). The lanthanide chelate luminescence is easily visible to the naked eye under a UV lamp emitting in the midultraviolet range (300 to 340 nm) and can be photographed on Polaroid instant film using a time-resolved photographic camera.[22]

The hydrogen peroxide molecules produced by numerous oxidases cannot interact directly with a lanthanide ion and form a luminescent lanthanide chelate. However, in the presence of light and hydrogen peroxide the 1,10-phenantroline-2,9-dicarboxylic acid dihydrazide (PDAdh) is converted into the product, 1,10-phenantroline-2,9-dicarboxylic acid (PDA), which forms a luminescent lanthanide chelate in the presence of Eu^{3+} ions. This conversion is due to photoassisted oxidation of PDAdh to PDA by hydrogen peroxide, followed by formation of the Eu:PDA luminescent complex. In principle, the use of this phenomenon allows detection of numerous oxidases that produce hydrogen peroxide (see "Hydrogen Peroxide" in Section 1). This approach, however, is not widely used now because PDAdh is not yet commercially available and should be synthesized under labo-

ratory conditions.[22] Glucose oxidase (1.1.3.4 — GO, Method 2) is the only enzyme that can be detected by this method at present.

The lanthanide luminescent methods have several disadvantages that restrict their wide application. The first one is that the luminescent products readily diffuse in a solution. To minimize undesirable diffusion, the staining solutions should be applied as 1% agar overlays. The problem of diffusion of a lanthanide chelate product was solved for alkaline phosphatase (3.1.3.1 — ALP, Method 10) by the use of the 5-tert-octyl- substituent of the substrate salicyl phosphate. The bulky, hydrophobic nature of this substituent minimizes diffusion of the product 5-tert-octylsalicylic acid, especially when alkaline phosphatase transferred on a nylon membrane is detected.[21] Another disadvantage of the lanthanide luminescent methods is that some of the substrates involved in luminescent detection are not yet commercially available and should be synthesized under laboratory conditions (3.2.1.23 — β-GAL, Method 5, Notes). Finally, the detection of enzyme activity by this method cannot be carried out in a one-step procedure when derivatives of salicylic acid are used as substrates. This is because highly alkaline conditions (pH 12.5) are required for optimal formation of the luminescent ternary complex salicylic acid:Tb^{3+}:EDTA. These conditions are not compatible with those required for normal action of any enzyme under analysis. Fortunately, this is not the case for enzymes producing hydrogen peroxide, where formation of the luminescent lanthanide complex occurs at considerably lower pH values so that the enzyme detection can be carried out in a one-step procedure (1.1.3.4 — GO, Method 2).

The main advantage of the lanthanide chelate luminescent methods is their high sensitivity, which in some cases is comparable to or even higher than that obtained with radioisotopic detection methods.[21] Again, these methods are of potential value because they can be used for detection of a great number of hydrolases and oxidoreductases after synthesis of convenient substrates suitable for formation of luminescent lanthanide chelates.[22]

MISCELLANY

There is also a heterogeneous group of natural and artificial fluorochromes that are used in positive fluorescent stains of separate enzymes. These are oxidized homovanillic acid and bieugenol (1.11.1.7 — PER, Method 1, Notes), 5-dimethylaminonaphthalene-1-sulfonyl; 2.8.1.1 — TST, Method 2; 2.8.1.3 — TTST; 4.2.1.1 — CA, Other Methods, B), fluorescein (3.1.1 ... — EST, Method 3; 4.2.1.1 — CA, Method 2, Notes), Bromocresol Blue (4.2.1.1 — CA, Method 1, Notes), naphthylamine (3.4.19.3 — OPP; 3.4.21–24 ... — PROT), antranilate (4.1.3.27 — ANS), uroporphyrin (4.3.1.8 — UPS), and flavanone (5.5.1.6 — CI, Notes).

Ethidium bromide is the only fluorochrome that is used in negative fluorescent stains (3.1.21.1 — DNASE, Notes; 3.1.27.5 — RNASE, Method 1, Notes).

SECTION 3
AUTORADIOGRAPHY

There is a group of positive zymographic methods that depend on the detection of radioactive products by conventional autoradiography on x-ray film. These methods are based on separation of the radioactively labeled products from their labeled precursors. There are several different approaches to separate radioactive substrates and products.

The first approach is to precipitate the radioactive product in a gel matrix and wash away the substrate. Lanthanum chloride is used to precipitate such phosphorylated products as phosphosugars (e.g., 2.4.2.3 — UP; 2.7.1.6 — GALK, Method 1), glycerone phosphate (2.7.1.30 — GLYCK, Method 1), and nucleoside monophosphates (e.g., 2.7.1.20 — ADK; 2.7.1.74 — DCK; 2.7.1.113 — DGK). Such products as phosphoproteins and aminoacyl tRNAs are precipitated using trichloroacetic acid (e.g., 2.7.1.37 — PROTK; 6.1.1.2 — TRL), and DNA is precipitated by streptomycin (2.7.7.7 — DDDP, Method 1).

Another approach is to use an ion exchange process to bind the radioactive products to special papers that incorporate an ion exchanger and wash away the radioactive substrates. The ion exchange papers used for this purpose are acetate cellulose (e.g., 2.1.1.6 — CMT), DEAE cellulose (e.g., 2.4.2.7 — APRT), and nitrocellulose (e.g., 3.4.21.73 — UK, Method 1).

Radioactively labeled streptavidin is used as a specific probe to detect enzymes that are biotinyl proteins. This method is specific for plant acetyl-CoA carboxylase (6.4.1.2 — ACC), which is the only known biotinyl protein in plants.

The ability of some enzymes to transfer the product molecules to their own molecules provides a good opportunity to detect their activity in electrophoretic gels using radioactively labeled substrates. Separation of a labeled substrate from a labeled product is achieved by precipitations of the enzyme-product complex with trichloracetic acid (e.g., 2.4.2.30 — NART).

Perhaps only one negative enzyme-detecting method based on precipitation of the radioactively labeled substrate and washing away the labeled product is now in practical use (3.1.3.5 — 5'-N, Method 1), although in the past such methods were used for detection of some depolymerizing enzymes.

As Harris and Hopkinson[7] have pointed out, the radioactive methods can be applied to the analysis of a variety of enzymes. They are of good resolution and sensitivity. However, they are disadvantageous because there is inevitably a time lag of a week or more between carrying out electrophoresis and seeing the results. At present, the radioactive methods are usually used only for those enzymes that cannot be detected by any other more practical procedure.

SECTION 4
BIOAUTOGRAPHY

The term *bioautography* was introduced for the first time to designate a detection system used to identify vitamins after paper chromatography.[23] After separation of vitamins, chromatographic paper was placed in contact with an agar layer containing mutant bacteria requiring certain vitamins for their growth and proliferation. The areas of bacterial proliferation were easily visible on a transparent agar layer.

A similar bioautographic procedure was developed for visualization of enzymes on electrophoretic gels using bacterial stocks with different auxotrophic mutations and naturally occurring stocks of fastidious yeast.[24,25] Deficient bacteria or yeast that cannot utilize the substrate of the enzyme to be detected and require the product of the enzyme reaction for growth are used. After an indicator agar containing mutant bacteria, minimal medium, and a certain substrate(s) is placed in contact with the electrophorized gel and incubated for a sufficient period of time, the areas of a certain enzyme being detected become visible as opaque bands on a translucent background of the indicator agar (e.g., see 6.3.4.5 — ARGS).

The most important step in the application of the bioautographic method for the detection of certain enzymes is the selection of an appropriate auxotrophic mutant. A strain of bacteria deficient in a particular enzyme is the logical choice to detect corresponding plant or animal enzymes. However, the substrate of the enzyme under study must not be metabolized by alternative pathways in the bacterium. Another requirement is that the end product of the enzyme reaction must be transported into the bacteria (e.g., some phosphorylated products do not penetrate the bacterial cell wall). Finally, the bacterial strain must have a low frequency of reverse mutations that lead to prototrophy.

In order to test whether or not a bioautographic procedure can be developed with a given bacterial strain, the following test is recommended.[25] About 10^7 bacteria should be inoculated into 5 ml of each of the following media: (1) medium lacking the nutrient, i.e., the end product of the enzyme reaction; (2) medium lacking the nutrient but containing the substrate for the enzyme; (3) medium containing the nutrient; and (4) medium containing both the substrate and the nutrient. The bacteria are suitable for bioautography if they grow only in the medium containing the nutrient. The compositions of media for *Escherichia coli* and *Pediococcus cerevisiae* are presented in Appendixes A-1 and A-2, respectively.

Enzymes detected by bioautography, the stocks of bacteria used, and related information are presented in the table on the next page. It should be pointed out that bioautographic methods are not in wide use. This is because bioautography is not practical and because visualized bands of enzymatic activities usually are not sufficiently sharp due to diffusion of enzymatic products. However, bioautography is of great value for visualization of enzymes that are difficult or impossible to detect using other methods.

SECTION 5
TWO-DIMENSIONAL GEL SPECTROSCOPY

An optical technique termed *two-dimensional gel spectroscopy* has been developed that permits one to analyze two-dimensional electrophoretic gels in a fashion that is essentially identical to the routine spectrophotometric or fluorometric methods of analysis of enzymes in solution.[26] The principle of two-dimensional gel spectroscopy involves the irradiation of electrophorized PAG (or translucent agarose replica) at a desired wavelength, and subsequent analysis of the gel at a desired wavelength for absorption or fluorescent bands that arise as a result of enzymatic conversion of a substrate to an optically detectable product.

A special optical apparatus may be constructed from commercially available optics.[26] The apparatus can both generate and detect monochromatic light at any wavelength from 200 nm in the ultraviolet to greater than 4500 nm in the infrared. Highly monochromatic light of a particular wavelength is selected by the proper choice of interference filters. This light is of spectral purity comparable to that obtained with a spectrophotometer that utilizes a grating monochromator. The light emanating from the gel is analyzed using a catadioptric lens, which has a spectral transmission from 200 to 4500 nm and utilizes first surface mirrors in order to image objects on the film plane. The use of mirrors permits the transmission of ultraviolet and infrared light that would be absorbed by glass lenses. A standard f/1.2, 55-mm camera lens is used at wavelengths above 400 nm.

Photographic films suitable for recording signals from approximately 200 to 1100 nm are commercially available.[27,28] For the recording of fluorescence, Kodak Tri-X (ASA 400) film is employed. A high-contrast image is obtained with Kodak Technical Pan 2415 film in situations in which light is not a limiting factor. Exposure times are determined empirically. This is because each filter combination requires a different amount of light for proper exposure.

By means of two-dimensional gel spectroscopy, enzymes can be detected in transparent gels under conditions that are identical to those employed in spectrophotometric and fluorometric methods under which enzymes are operationally defined. The choice of the proper interference filter(s) depends on the spectral characteristics of the substrates and products and the intensity of mercury lines in the desired area of the spectrum. Thus, the conditions to be employed in the detection of a certain enzyme are predictable either from the literature or from routine spectrophotometric assays.

At present, the method is not widely used because the apparatus needed for two-dimensional gel spectroscopy is not commercially available. Only three enzymes have so far been detected by two-dimensional gel spectroscopy. These are lactate dehydrogenase (1.1.1.27 — LDH, Other Methods, C), monoamine oxidase (1.4.3.4 — MAOX, Other Methods, A), and trypsin (3.4.21.4 — T, Other Methods, C). However, since many published spectrophotometric or fluorometric methods can be used virtually without change, two-dimensional gel spectroscopy is of great potential value for electrophoretic zymography of many previously undetectable enzymes.

Deficient Bacteria Employed in Bioautography of Some Enzymes

Enzyme to Be Detected	Bacterial Stock[a]	Bacterial Mutation	Bacterial Enzyme Defect	Nutritional Requirement[b]
2.4.2.7 — APRT	*E. coli* (PC 0273)	pur A⁻	Adenylosuccinate synthase	Adenine
3.5.4.4 — ADA	*E. coli* (PC 0273)	pur A⁻	Adenylosuccinate synthase	Adenine
6.3.4.5 — ARGS	*E. coli* (AB 1115)	arg G⁻	Argininosuccinate synthase	Arginine
4.3.2.1 — ASL	*E. coli* (AT 753)	arg H⁻	Argininosuccinate lyase	Arginine
	P. cerevisiae (ATCC 8081)	— [c]	Arginine biosynthesis	Arginine
3.4.21.4 — T	*E. coli* (AT 753)	arg H⁻	Argininosuccinate lyase	Arginine
1.1.1.27 — LDH	*E. coli* (CB 482)	lct⁻	L-Lactate dehydrogenase	Pyruvate
3.5.1.14 — AA	*P. cerevisiae* (ATCC 8081)	— [c]	Aminoacylase	Methionine
2.6.1.42 — BCT	*P. cerevisiae* (ATCC 8081)	— [c]	Branched-chain amino acid biosynthesis	Isoleucine (or leucine or valine)
1.5.1.3 — DHFR	*P. cerevisiae* (ATCC 8081)	— [c]	Dihydrofolate reductase	Folinic acid

[a] *E. coli* K12 stocks from the *E. coli* Genetic Stock Center at Yale University Medical School; *P. cerevisiae* ATCC 8081 (American Type Culture Collection, strain 8081).

[b] Each bacterial stock may have other requirements.

[c] Naturally occurring auxotroph.

From Naylor, S.-L. and Kelbe, R.-J., *Biochem. Genet.*, 15, 1193, 1977 and Naylor, S.-L., *Isozmes: Current Topics in Biological and Medical Research*, Vol. 4, Rattazzi, M.-C., Scandalios, J.-G., and Whitt, G.-S., Eds., Alan R. Liss, New York, 1980, p. 69. With permission.

SECTION 6
IMMUNOBLOTTING

In the present context, immunoblotting (Western blotting) refers to techniques for transferring separated zones of proteins from electrophoretic gels to immobilization matrices, and subsequent detection of certain immobilized proteins using specific antibodies and anti-antibodies that are radiolabeled or linked with an indicator enzyme (see the flowchart for immunoblotting on the next page).[29-31]

IMMOBILIZATION MATRICES

One of the most widely used immobilizing matrices is nitrocellulose (NC). A nitrocellulose membrane with a 0.45-μm pore size is most commonly employed. However, NC membranes with other porosities may also be of value. For example, material with a 0.22-μm pore size may be preferable for low-molecular-weight proteins. In addition to the NC, the nylon-based membrane Zetabind (ZB), available from AMF Cuno Division (Meriden, Connecticut), has been introduced for protein blotting. This material is also marketed as Zeta Probe by Bio-Red Lab (Richmond, California) and Hybond-N by Amersham (Arlington Heights, Illinois). The ZB membrane possesses several advantages. It is as thin and fine-grained as an NC membrane but mechanically stronger, and has about a sixfold higher protein-binding capacity. In addition, the highly charged cationic nature of the ZB matrix results in better transfer of proteins from SDS-containing gels in which proteins are present as complexes with anionic SDS molecules.

A number of other immobilizing matrices were also introduced, but they are not in as wide use for protein blotting as NC and ZB membranes.[32]

TRANSFER OF PROTEINS

There are two main ways of transferring proteins from electrophoretic gels onto the immobilizing matrix: (1) capillary blotting, and (2) electroblotting.

Capillary blotting consists in transferring of proteins by means of solvent flow.[33] In this method the transfer is achieved by placing the gel onto several thicknesses of filter paper or Whatman 3MM paper wetted with transfer buffer and supported

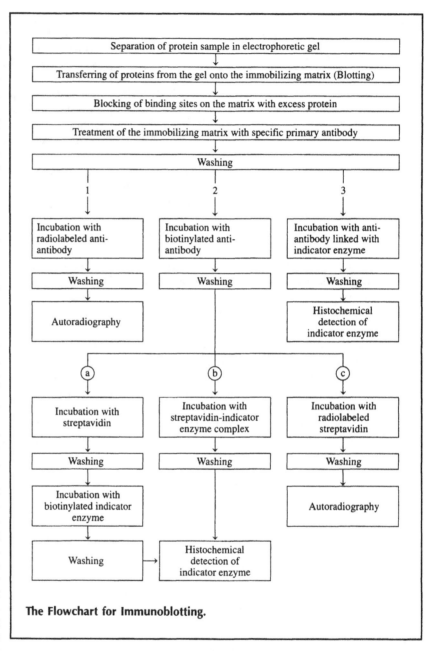

The Flowchart for Immunoblotting.

on a piece of sponge with the same buffer. A sheet of transfer membrane is placed on top of the gel, and then several pieces of dry filter paper and paper towels are placed above the transfer membrane and held together with a light weight (usually 0.5 to 1.0 kg). The buffer flows from the sponge and wet filter paper through the gel carrying the protein molecules onto the transfer membrane, through which the buffer passes and is absorbed by dry filter papers. Overnight blotting is usually enough to transfer sufficient amounts of proteins to be detected.

Electroblotting employs electrophoretic elution of the proteins.[29] The procedure consists of wetting a sheet of transfer membrane and placing it on the gel, which is supported by a porous pad (e.g., foam sponge or layers of wet filter paper) mounted on a stiff plastic grid. No air bubbles should be trapped between the gel and transfer membrane. A second porous pad and plastic grid are placed above the transfer membrane. The "sandwich" is then strengthened with rubber bands and placed in a tank of transfer buffer between two electrodes.

Electrophoretic gel buffers of low ionic strength are usually used as transfer buffers for capillary blotting and electroblotting. The voltage gradients of 5 to 10 V/cm enable transfers to be essentially complete in 1 to 3 h. The use of lower gradients requires longer times for complete transfer. For transfer of proteins from SDS-containing gels, neutral or weakly basic buffers are used. In these conditions the proteins migrate as anions and the NC membrane must be on the anode side. In general, it is important to know the net charge of the proteins to be transferred under the conditions used to choose the correct side of the gel on which the transfer membrane must be placed.

One reason for the increased effectiveness of ZB membranes in protein transfer relates to the lack of requirement for methanol in the transfer buffer. Methanol was introduced because it was found to increase the binding capacity of NC membranes for proteins. However, methanol also reduces the efficiency with which proteins are eluted from gels, and electroblotting must therefore be carried out for at least 12 h to obtain efficient transfer of high-molecular-weight proteins, especially when 10% PAG is used. When ZB membranes are used, methanol need not be present in the transfer buffer. This increases the detection sensitivity and reduces diffusion of protein molecules because the time of electroblotting can be reduced considerably.

It is possible to eliminate the blotting step and to detect the resolved proteins directly within electrophoretic PAG. The in-gel immunodetection proved preferable for large or difficult-to-transfer proteins.[34]

SPECIFIC ANTIBODIES AND LABELED ANTI-ANTIBODIES

Specific primary antibodies against certain enzymes can be obtained by routine immunization procedures using pure enzyme preparations and subsequent separation of immunoglobulins, which are essentially polyclonal antibodies specific to different immunogenic determinants of the same enzyme molecule.

Monoclonal primary antibodies specific to certain enzymes are of special value and of potential value in the future for immunohistochemical or immunoautoradiographic detection of enzymes immobilized on NC or ZB membranes.[35] Monoclonal antibodies specific to certain proteins are produced by specific hybridomas.[36] The construction of a specific hybridoma includes the following steps:

1. Immunization of the mouse with a crude protein preparation
2. Separation of lymphocytes from the spleen of the immunized mouse
3. Hybridization of lymphocytes with myeloma cells (the malignant cells of the immune system)

4. Cloning of somatic cell hybrids
5. Selection of the clone(s) that produces antibodies specific to certain enzymes or isozymes

Among all the known advantages of monoclonal antibodies, the feasibility of using unpurified protein preparations to produce monospecific antibodies is very useful for specific detection of the enzyme proteins immobilized on the supporting matrix. Those who need more information concerning theory, production, purification, testing, storage, labeling, and application of monoclonal antibodies are referred to special manuals.[37,38]

A relatively simple and rapid method of immunization that gives polyclonal antibodies of high specificity was developed.[39] This method includes the following steps: (1) separation of proteins by two-dimensional gel electrophoresis, (2) transfer of proteins onto an NC membrane by a blotting technique, and (3) immunization of rabbits by specific proteins bound to an NC matrix after the cutting out of corresponding protein spots on the NC membrane and the powdering of the membrane pieces by sonication.

Species-specific second antibodies, or anti-antibodies, can be obtained by routine immunization of sheep, goat, rabbit, or donkey with species-specific immunoglobulins. When monoclonal antibodies produced by hybridomas are used as primary antibodies, anti-antibodies specific to mouse immunoglobulins should be used as second antibodies. Different species-specific anti-antibodies are now commercially available, e.g., from Sigma or from Amersham. These are (1) radiolabeled anti-antibodies, (2) biotinylated anti-antibodies, and (3) anti-antibodies linked with an indicator enzyme (see the flowchart). Biotinylated anti-antibodies are used coupled with (a) streptavidin and a biotinylated indicator enzyme, (b) a streptavidin–indicator enzyme complex, or (c) radiolabeled streptavidin or protein A. Peroxidase, β-galactosidase, and alkaline phosphatase are usually used as indicator enzymes.

DETECTION OF TRANSFERRED PROTEINS ON BLOTS

After blotting, any given protein can be visualized on an immobilizing membrane by the methods pictured in the flowchart. An important step in immunological detection of transferred proteins is the blocking of all additional binding sites on the matrix with an immunologically neutral protein. Usually 1 to 5% bovine serum albumin, ovalbumin, hemoglobin, or fish skin gelatin dissolved in buffered saline solution is used to saturate excess protein-binding sites. This process was termed *quenching*. Buffered Tween 20 was also shown to be a very useful agent for quenching NC membranes.[40] After quenching, an immobilizing membrane is incubated with the appropriate dilution of specific primary antibody in the quenching solution and is rinsed in the quenching solution. After this treatment three different methods may be used to visualize the antigen–primary antibody complex (flowchart). The most direct ways (as designated on the flowchart) are (1) the treatment of the membrane with a radiolabeled anti-antibody, and subsequent washing and autoradiography, or (3) the treatment of the membrane with an anti-antibody linked with an indicator enzyme, washing, and subsequent histochemical detection of an indicator enzyme using appropriate histochemical procedures described in Part III for peroxidase, β-galactosidase, or alkaline phosphatase, which are commonly used as indicator enzymes in immunoblotting procedures. Another, more complex way (2) includes the treatment of the membrane with a biotinylated anti-antibody and subsequent washing of the membrane. After this, three different procedures may be employed. The first one includes incubation of the membrane with streptavidin to form a streptavidin bridge, washing of the membrane, incubation with a biotinylated indicator enzyme, a second washing, and histochemical detection of an indicator enzyme. The second method includes treatment of the membrane with a streptavidin–indicator enzyme complex, washing, and subsequent histochemical detection of an indicator enzyme. The third method includes incubation of the membrane with radiolabeled streptavidin, washing, and autoradiography.

A highly sensitive enhanced chemiluminescence (ECL) procedure is now widely used to detect proteins immobilized on nitrocellulose or polyvinylidene difluoride (PVDF) membranes (available from NEN™ Life Science Products, Boston, Massachusetts). This procedure is based on an enhanced version of a chemiluminescence reaction in which horseradish peroxidase conjugated to primary or secondary antibodies catalyzes light emission (428 nm) from the oxidation of luminol in the presence of hydrogen peroxide. The phenol derivatives, p-iodophenol and p-phenylphenol, are usually used as enhancers of chemiluminescence. The use of enhancers increases the emission approximately 1000-fold.[41] The emitted light is captured on x-ray film, and areas of antigen localization are then detected as dark bands on the developed x-ray film (see 1.11.1.7 — PER, Other Methods, D).

Immunoblotting in conjunction with the use of monoclonal antibodies is valuable for the detection of specific enzymatic and nonenzymatic proteins on electrophoretic gels. However, considerable time, energy, and expense are involved in the production of hybridoma antibodies. This is an important reason why the diversity of commercially available enzyme protein-specific monoclonal antibodies is still not sufficiently rich, and why immunoblotting is not as widely used in electrophoretic zymography as histochemical methods based on chromogenic and fluorogenic reactions. Indeed, specific antibodies (mostly monoclonal) are now commercially available for only three and a half tens of enzymes (e.g., see Sigma Catalog, 2002/2003). Again, these are antibodies against enzymes from single organisms. The phylogenetic range of application of such antibodies is restricted. As a rule, reactivity of such antibodies is restricted by a single species serving as a source of immunogen. Only when synthesized peptides with phylogenetically conserved amino acid sequences are used as immunogens can the range of reactivity of produced antibodies be extended over several related species. The narrow phylogenetic range of reactivity of antibodies is another reason why immunoblotting methods are still not as widely used as histochemical ones. These methods are cited in

enzyme sheets (Part III, Section 3) as "Other Methods" only for those enzymes for which more practical alternative methods are also available for detection. Examples are carbonic anhydrase (4.2.1.1 — CA, Other Methods, C), 6-phosphofructokinase (2.7.1.11 — PFK, Other Methods), pyruvate kinase (2.7.1.40 — PK, Other Methods), urease (3.5.1.5 — UR, Other Methods, C), etc. Additional data concerning the immunoblotting technique may be found in special guides.[42,43]

SECTION 7
MISCELLANIES

This section describes heterogeneous principles of specific visualization of enzymatic and nonenzymatic proteins not covered by preceding sections.

DETECTION BY INCORPORATING WATER-INSOLUBLE SUBSTRATES INTO SEPARATING GELS

The detection of hydrolytic enzymes degrading water-insoluble substrates that have been fractionated by PAG (or SDS-PAG) electrophoresis usually is achieved through the substrate-containing agar replica technique. This technique, however, suffers from a loss of resolution due to the inefficient diffusion of enzyme molecules from PAGs. To overcome this problem, a simple method was developed for incorporating water-insoluble polysaccharides into the separating PAG.[44] This method is based on the ability of many water-insoluble polysaccharides to dissolve in alkaline solutions. After mixing an alkaline polysaccharide solution with a polyacrylamide solution, the mixture can be neutralized with HCl just before the addition of ammonium persulfate and TEMED. A polysaccharide, although it becomes insoluble after neutralization, remains in a uniform suspension in the polymerized gel. For example, this method was used to detect xylanase activity bands (see 3.2.1.32 — XYL).

DETECTION OF LIPID-METABOLIZING ENZYMES

Some water-insoluble substrates (e.g., sphingomyelin) can be immobilized on PVDF membranes. Proteins separated by PAG electrophoresis are transferred onto a PVDF membrane and a corresponding hydrolytic activity detected negatively by autoradiography using radiolabeled substrates (e.g., see 3.1.4.12 — SPD) or positively using monoclonal antibodies specific to the reaction products (e.g., see 3.2.1.18 — SIA, Other Methods, D; 2.4.1.90 — AGS, Other Methods, B). The method based on the use of substrate-immobilized PVDF membranes for the detection of lipid-metabolizing enzymes is termed *Far Eastern blotting* and is expected to contribute to the development of a new approach to the lipid biochemistry field.[45]

DETECTION BY A TWO-DIMENSIONAL ELECTROPHORESIS PROCEDURE

This technique was developed to detect various DNA-metabolizing enzymes based on the use of a two-dimensional gel system.[46,47] In the first dimension enzyme proteins are resolved using either native or SDS–polyacrylamide gels containing defined 3' or 5' end-labeled oligonucleotides annealed to single-stranded M13 DNA. After electrophoretic separation in the first dimension, enzymes catalyze in-gel reactions that elongate, degrade, or modify their DNA substrates embedded in the gel. The ^{32}P-labeled substrates and products are then resolved according to size in the second dimension of electrophoresis through a denaturing DNA sequencing gel. Identification and characterization of enzymes is based on the position and size of the DNA products detected on processed gels by autoradiography.

The procedure of detection can be described in five steps:

1. Defined ^{32}P-labeled DNA substrates are embedded in PAG.
2. Enzyme proteins are separated by routine PAG electrophoresis (SDS-PAG electrophoresis is followed by SDS extraction and enzyme renaturation steps).
3. The sample line is sliced vertically into narrow gel strips. Each strip is placed in a test tube containing buffers, cofactors, and precursors necessary for specific enzyme reaction.
4. The processed gel strips are reoriented by a 90° rotation and cast within a denaturing DNA sequencing PAG. ^{32}P-labeled reaction products are resolved by size in a second dimension of electrophoresis.
5. The location of reaction products in DNA sequencing PAG is determined by autoradiography.

The versatility of this technique is extended by the use of defined labeled oligonucleotides annealed to M13 DNA that act as dual substrates for various DNA-metabolizing enzymes. Modification of DNA substrates within the gel either before or after in-gel enzyme reactions can create intermediate substrates. Restriction endonucleases, polynucleotide kinase, acid phosphatase, and the Lys-Trp-Lys tripeptide, which acts as an apurinic–apyrimidinic endonuclease, can penetrate PAG and modify immobilized DNA substrates. This approach may be used to detect enzymes of a complete DNA metabolic pathway or to dissect the enzymatic steps of a certain pathway, e.g., replication, repair, recombination, etc.[47]

This technique was applied to detect DNA polymerase (see 2.7.7.7 — DDDP, Other Methods, C), exonucleases (e.g., see 3.1.11.2 — EXO III), DNA ligases (e.g., see 6.5.1.1 — DL(ATP), Method 2), uracyl N-glycosilase (see 3.2.2.X — UDG), and different restriction endonucleases (see 3.1.21.4 — RESTR).

A detailed description of the principles of the activity gel, overlay gel, and activity blotting techniques used to detect various DNA-modifying (repair) enzymes may be found in the excellent review by Seki et al.[48]

DETECTION OF ENZYMATIC PROTEINS BY SPECIFIC PROBES

Labeled Aptamers and Aptamer Beacons as Specific Probes

Aptamers are oligonucleotides (typically 15 to 60 nucleotides long) that have been selected for specific binding to different molecular targets, ranging from small organics to proteins.[49–51] For example, DNA aptamers have been selected that can specifically bind to serine protease thrombin,[52] and RNA aptamers have been selected that can specifically bind to isozymes of protein kinase C.[53] Aptamers have affinities and specificities that are similar to those of antibodies. The most important advantage of aptamers over antibodies is that once an aptamer sequence and the corresponding template have been obtained, the radiolabeled or fluorescently labeled aptamer can be easily produced in the lab using routine molecular biology techniques. Thus, aptamers may represent a new type of reagent that could be used to detect proteins directly in electrophoretic gels or on NC membranes after blotting. It may be possible to substitute aptamers in many applications that currently rely on antibodies, particularly in immunoblotting. Aptamers act as a regenerable resource that can be stored indefinitely in a single tube in a freezer, ready to be applied anytime when necessary.[53]

A new class of molecules termed *aptamer beacons* was designed for detecting a wide range of ligands.[54] Aptamer beacons can adopt two or more conformations, one of which allows specific ligand binding. A fluorescence-quenching pair was used to report conformational changes induced by ligand binding. An anti-thrombin aptamer was engineered into an anti-thrombin aptamer beacon by the addition to the 5' end of nucleotides that are complementary to nucleotides at the 3' end of the anti-thrombin aptamer. In the absence of thrombin, the added nucleotides form a duplex with nucleotides of the 3' end and force the aptamer beacon into a stem-loop structure. In the presence of thrombin, the anti-thrombin aptamer beacon changes conformation and forms the ligand-binding structure. This causes a change in the distance between a fluorophore attached to the 5' end and a quencher attached to the 3' end, and results in an easily detectable change in fluorescence intensity.[54]

Although aptamers or aptamer beacons have not yet been used for direct detection of certain enzymes on electrophoretic gels, they are of potential value in this respect at least for some enzymes (e.g., thrombin and protein kinase C).

Labeled Inhibitors as Specific Probes

Biotinylated aprotinin (bovine pancreatic trypsin inhibitor) was used as a specific probe for the detection of serine proteases (neutrophil elastase and cathepsin G, pancreatic trypsin and chimotrypsin, and plasma plasmin) on nitrocellulose blots after SDS-PAG electrophoresis.[55] The biotinylated aprotinin detection method coupled with the use of the avidin–alkaline phosphatase conjugate system proved more sensitive than the routine zymography technique based on the use of substrates copolymerized in the resolving gel. The detection limit of the method is as little as 0.2 ng of trypsin. When used in conjunction with proteinase-specific irreversible inhibitors the method can provide additional information on the identity of the serine proteinases being examined. Similar results were obtained with a biotinylated potato chymotrypsin inhibitor used as a probe on Western blots for the detection and quantitation of chymotrypsin and the detection of chymotrypsin-like serine proteinase from ovine articular chondrocytes.[56] This detection system could detect as little as 0.1 ng of chymotrypsin. The principle used in both detection systems outlined above can be applied to the detection of other proteinases using appropriate inhibitory proteins.

Examples of application of this technique are given in trypsin (see 3.4.21.4 — T, Other Methods, D) and thrombin (see 3.4.21.5 — THR, Other Methods).

SPECIFIC DETECTION OF SOME NONENZYMATIC PROTEINS

The main advantage of enzymes over nonenzymatic proteins is that they are detected using highly specific methods based on the specificity of enzymatic reactions. This allows detection of certain enzymes on certain electrophoretic gels, and thus unequivocally identifies the gene products under study. Some nonenzymatic proteins are specific to certain organisms and tissues (e.g., hemoglobins, albumins, parvalbumins, haptoglobins, cristallins) and therefore can be easily identified using the "general proteins" stain. There are also some nonenzymatic proteins that can be detected by specific detection methods based on the use of specific probes. Enzymes, metal ions, and other compounds may be used as probes specific to nonenzymatic proteins. Some examples are listed below.

Detection of Proteinase and Ribonuclease Inhibitory Proteins Using Reverse Zymography

Reverse zymography is an electrophoretic technique suitable for identification of endogenous inhibitory proteins of hydrolytic enzymes (such as proteases and ribonucleases) in electrophoretic gels. The principle of this technique consists of incubation of the electrophoretic gel with a gel plate containing exogenous hydrolase and a corresponding substrate. Areas of localization of inhibitory proteins are detected as the zones of inhibited hydrolysis. Reverse zymograms obtained using purified recombinant human gelatinase have sensitivities for tissue inhibitors of metalloproteinases that are favorable in comparison to the immunoblotting technique, but have the benefit of visualizing multiple inhibitory proteins simultaneously.[57] Areas of ribonuclease inhibitor were visualized by sandwiching the electrophoretic PAG between a cellulose acetate membrane moistened with a solution of bovine pancreatic ribonuclease A and a dried agarose film sheet containing yeast RNA and ethidium bromide. During incubation of the sandwich at 37°C the ribonuclease penetrated the PAG and digested the RNA substrate in the agarose film. The enzyme was inactivated in PAG areas where

the inhibitory protein was present. Fluorescent bands corresponding to the ribonuclease inhibitor were observed in UV light on a dark background of the gel.[58]

Although inhibitory proteins are not enzymes, their detection in electrophoretic gels by reverse zymography is based on the high specificity of protein–protein intermolecular interaction, similar to the specific interaction between enzyme protein and substrate molecules. Using different proteinases and nucleases, a wide variety of different inhibitory proteins can be detected by the reverse zymography technique, and thus involved in a fruitful electrophoretic analysis.

Detection of Proteinase Inhibitory Proteins Using Biotinylated Proteinases as Specific Probes

The method of reverse zymography described above has some inherent weaknesses. For example, the proteins are not fixed within an electrophoretic gel and thus diffuse, resulting in a diminution in the resolution and detection limits for inhibitory proteins. Moreover, the optimal concentration of the exogenous enzyme and incubation conditions for the digestion step must be determined empirically for each sample. The copolymerized substrate-resolving PAG system is not appropriate for comparative analysis of samples of inhibitory proteins with widely differing specific activities. To overcome the disadvantages of reverse zymography, biotinylated trypsin was prepared and used as a specific probe to detect trypsin inhibitory proteins separated by SDS-PAG and electroblotted to nitrocellulose.[59] The method could detect 0.15 ng of the serine proteinase inhibitor, aprotinin. Densitometric examination of Western blots indicated that there was a linear relationship between the amount of active aprotinin electrophorized (1 to 50 ng) and the intensity of the blot obtained with biotinylated trypsin and the avidin–alkaline phosphatase conjugate system. This method applied to gradient SDS-PAG was successfully used to detect and study different molecular forms of the serine proteinase inhibitor from ovine chondrocytes.[60]

Biotinylated Hyaluronan as a Specific Probe for Hyaluronan-Binding Proteins

Hyaluronan-binding proteins (hyaloadherins) were visualized on Western blots after SDS-PAG electrophoresis using biotinylated hyaluronan and an avidin–alkaline phosphatase conjugate in conjunction with NBT–5-bromo-4-chloro-3-indolyl phosphate.[61] The method is specific and sensitive. For example, it can detect ≥2 ng of such hyaloadherins as the bovine nasal cartilage link-protein, aggrecan hyaluronan–binding region, and human fibroblast hyaluronan receptor CD-44.

Transferrin

Transferrin is a protein that binds any free iron in the blood. It is specifically detected using radioactive iron and autoradiography. However, the "general proteins" stain is suitable for routine typing of transferrin bands once their location has been confirmed using radioactive iron.[62]

REFERENCES FOR PART II

1. Seligman, A.M. and Rutenberg, A.M., The histochemical demonstration of succinic dehydrogenase, *Science*, 113, 317, 1951.
2. Markert, C.L. and Möller, F., Multiple forms of enzymes: tissue, ontogenetic, and species specific patterns, *Proc. Natl. Acad. Sci. U.S.A.*, 45, 753, 1959.
3. Pearse, A.G.E., *Histochemistry: Theoretical and Applied*, Vol. I, Williams & Wilkins, Baltimore, 1968.
4. Burstone, M.S., *Enzyme Histochemistry*, Academic Press, New York, 1962.
5. Hunter, R.L. and Markert, C.L., Histochemical demonstration of enzymes separated by zone electrophoresis in starch gels, *Science*, 125, 1294, 1957.
6. Martin de Llano, J.J., Garcia-Segura, J.M., and Gavilanes, J.G., Selective silver staining of urease activity in polyacrylamide gels, *Anal. Biochem.*, 177, 37, 1989.
7. Harris, H. and Hopkinson, D.A., *Handbook of Enzyme Electrophoresis in Human Genetics*, North-Holland, Amsterdam, 1976 (loose-leaf, with supplements in 1977 and 1978).
8. Gomori, G., Microtechnical demonstration of phosphatase in tissue sections, *Proc. Soc. Exp. Biol. Med.*, 42, 23, 1939.
9. Gomori, G., Distribution of acid phosphatase in the tissues under normal and under pathologic conditions, *Arch. Pathol.*, 32, 189, 1941.
10. Nimmo, H.G. and Nimmo, G.A., A general method for the localization of enzymes that produce phosphate, pyrophosphate, or CO_2 after polyacrylamide gel electrophoresis, *Anal. Biochem.*, 121, 17, 1982.
11. Fiske, C.H. and Subbarow, Y., The colorimetric determination of phosphorus, *J. Biol. Chem.*, 66, 375, 1925.
12. Fisher, R.A., Turner, B.M., Dorkin, H.L., and Harris, H., Studies on human erythrocyte inorganic pyrophosphatase, *Ann. Hum. Genet.*, 37, 341, 1974.
13. Zlotnick, G.W. and Gottlieb, M., A sensitive staining technique for the detection of phosphohydrolase activities after polyacrylamide gel electrophoresis, *Anal. Biochem.*, 153, 121, 1986.
14. Klebe, R.J., Schloss, S., Mock, L., and Link, C.R., Visualization of isozymes which generate inorganic phosphate, *Biochem. Genet.*, 19, 921, 1981.
15. Cornell, N.W., Leadbetter, M.G., and Veech, R.L., Modifications in the enzymatic assay for inorganic phosphate, *Anal. Biochem.*, 95, 524, 1979.
16. Lowry, O.H., Schulz, D.W., and Passonneau, J.V., Effects of adenylic acid on the kinetics of muscle phosphorylase a, *J. Biol. Chem.*, 239, 1947, 1964.
17. Thorup, O.A., Strole, W.B., and Leavell, B.S., A method for the localization of catalase on starch gels, *J. Lab. Clin. Med.*, 58, 122, 1961.
18. Ohlsson, B.G., Weström, B.R., and Karlsson, B.W., Enzymoblotting: a method for localizing proteinases and their zymogens using *para*-nitroanilide substrates after agarose gel electrophoresis and transfer to nitrocellulose, *Anal. Biochem.*, 152, 239, 1986.
19. Harris, R.B. and Wilson, I.B., Polyacrylamide gels which contain a novel mixed disulfide compound can be used to detect enzymes that catalyze thiol-producing reactions, *Anal. Biochem.*, 134, 126, 1983.
20. Clark, A.G., A direct method for the visualization of glutathione S-transferase activity in polyacrylamide gels, *Anal. Biochem.*, 123, 147, 1982.
21. Templeton, E.F.G., Wong, H.E., Evangelista, R.A., Granger, T., and Pollak, A., Time-resolved fluorescence detection of enzyme-amplified lanthanide luminescence for nucleic acid hybridization assays, *Clin. Chem.*, 37, 1506, 1991.
22. Evangelista, R.A., Pollak, A., and Templeton, E.F.G., Enzyme-amplified lanthanide luminescence for enzyme detection in bioanalytical assays, *Anal. Biochem.*, 197, 213, 1991.
23. Usdin, E., Shockman, G.D., and Toennies, G., Tetrazolium bioautography, *Appl. Microbiol.*, 2, 29, 1954.
24. Naylor, S.L. and Klebe, R.J., Bioautography: a general method for the visualization of enzymes, *Biochem. Genet.*, 15, 1193, 1977.
25. Naylor, S.L., Bioautographic visualization of enzymes, in *Isozymes: Current Topics in Biological and Medical Research*, Vol. 4, Rattazzi, M.C., Scandalios, J.G., and Whitt, G.S., Eds., Alan R. Liss, New York, 1980, p. 69.
26. Klebe, R.J., Mancuso, M.G., Brown, C.R., and Teng, L., Two-dimensional spectroscopy of electrophoretic gels, *Biochem. Genet.*, 19, 655, 1981.
27. Kodak, *Kodak Plates and Films for Scientific Photography*, Kodak Technical Publication P-315, Eastman Kodak Co., Rochester, NY, 1973.
28. West, W., The spectral sensitivity of emulsions spectral sensitization, desensitization and other photographic effects of dyes, in *Neblette's Handbook of Photography and Reprography*, Sturge, J.M., Ed., Van Nostrand Reinhold, New York, 1977, p. 73.
29. Towbin, H., Staehelin, T., and Gordon, J., Electrophoretic transfer of proteins from polyacrylamide gels to nitrocellulose sheets: procedure and some applications, *Proc. Natl. Acad. Sci. U.S.A.*, 76, 4350, 1979.
30. Burnette, W.N., "Western blotting": electrophoretic transfer of proteins from sodium dodecyl sulfate–polyacrylamide gels to unmodified nitrocellulose and radiographic detection with antibody and radiolabeled protein A, *Anal. Biochem.*, 112, 195, 1981.
31. Gershoni, J.M. and Palade, G.E., Protein blotting: principles and applications, *Anal. Biochem.*, 131, 1, 1983.
32. Andrews, A.T., *Electrophoresis: Theory, Techniques, and Biochemical and Clinical Applications*, Clarendon Press, Oxford, 1988.
33. Southern, E.M., Detection of specific sequences among DNA fragments separated by gel electrophoresis, *J. Mol. Biol.*, 98, 503, 1975.
34. Desai, S., Dworecki, B., and Cichon, E., Direct immunodetection of antigens within the precast polyacrylamide gel, *Anal. Biochem.*, 297, 94, 2001.
35. Vora, S., Monoclonal antibodies in enzyme research: present and potential applications, *Anal. Biochem.*, 144, 307, 1985.

36. Köhler, G. and Milstein, C., Continuous cultures of fused cells secreting antibody of predefined specificity, *Nature*, 256, 495, 1975.

37. Goding, J.W., *Monoclonal Antibodies: Principles and Practice*, 3rd ed., Academic Press, San Diego, 1996.

38. Shepherd, P.S. and Den, C., *Monoclonal Antibodies: A Practical Approach*, Oxford University Press, Oxford, 2000.

39. Diano, M., Le Bivic, A., and Hirn, M., A method for the production of highly specific polyclonal antibodies, *Anal. Biochem.*, 166, 224, 1987.

40. Blake, M.S., Johnston, K.H., Russell-Jones, G.J., and Gotschlich, E.C., A rapid, sensitive method for detection of alkaline phosphatase-conjugated anti-antibody on Western blots, *Anal. Biochem.*, 136, 175, 1984.

41. Thorpe, G.H.G., Kricka, L.J., Moseley, S.B., and Whitehead, T.P., Phenols as enhancers of the chemiluminescent horseradish peroxidase–luminol–hydrogen peroxide reaction: application in luminescence-monitored enzyme immunoassays, *Clin. Chem.*, 31, 1335, 1985.

42. Bjerrun, O.G. and Heegaard, N.H.H., Eds., *CRC Handbook of Immunoblotting of Proteins*, CRC Press, Boca Raton, FL, 1988.

43. Pound, J., Ed., *Immunochemical Protocols*, Humana Press, Totowa, NJ, 1998.

44. Chen, P. and Buller, C.S., Activity staining of xylanases in polyacrylamide gels containing xylan, *Anal. Biochem.*, 226, 186, 1995.

45. Taki, T. and Ishikawa, D., TLC blotting: application to microscale analysis of lipids and as a new approach to lipid–protein interaction, *Anal. Biochem.*, 251, 135, 1997.

46. Longley, M.J. and Mosbaugh, D.W., Characterization of DNA metabolizing enzymes *in situ* following polyacrylamide gel electrophoresis, *Biochemistry*, 30, 2655, 1991.

47. Longley, M.J. and Mosbaugh, D.W., *In situ* detection of DNA-metabolizing enzymes following polyacrylamide gel electrophoresis, *Methods Enzymol.*, 218, 587, 1993.

48. Seki, S., Akiyama, K., Watanabe, S., and Tsutsui, K., Activity gel and activity blotting methods for detecting DNA-modifying (repair) enzymes, *J. Chromatogr.*, 618, 147, 1993.

49. Gold, L., Conformational properties of oligonucleotides, *Nucleic Acids Symp. Ser.*, 33, 20, 1995.

50. Osborne, S.E., Matsumura, I., and Ellington, A.D., Aptamers as therapeutic and diagnostic reagents: problems and prospects, *Curr. Opin. Chem. Biol.*, 1, 5, 1997.

51. Famulok, M. and Mayer, G., Aptamers as tools in molecular biology and immunology, *Curr. Top. Microbiol. Immunol.*, 243, 123, 1999.

52. Bock, L.C., Griffin, L.C., Latham, J.A., Vermaas, E.H., and Toole, J.J., Selection of single-stranded DNA molecules that bind and inhibit human thrombin, *Nature*, 355, 564, 1992.

53. Conrad, R. and Ellington, A.D., Detecting immobilized protein kinase C isozymes with RNA aptamers, *Anal. Biochem.*, 242, 261, 1996.

54. Hamaguchi, N., Ellington, A., and Stanton, M., Aptamer beacons for the direct detection of proteins, *Anal. Biochem.*, 294, 126, 2001.

55. Melrose, J., Ghosh, P., and Patel, M., Biotinylated aprotinin: a versatile probe for the detection of serine proteinases on Western blots, *Int. J. Biochem. Cell Biol.*, 27, 891, 1995.

56. Rodgers, K.J., Melrose, J., and Ghosh, P., Biotin-labeled potato chymotrypsin inhibitor-1: a useful probe for the detection and quantitation of chymotrypsin-like serine proteinases on Western blots and its application in the detection of a serine proteinase synthesized by articular chondrocytes, *Anal. Biochem.*, 227, 129, 1995.

57. Oliver, G.W., Leferson, J.D., Stetler-Stevenson, W.G., and Kleiner, D.E., Quantitative reverse zymography: analysis of picogram amounts of metalloproteinase inhibitors using gelatinase A and B reverse zymograms, *Anal. Biochem.*, 244, 161, 1997.

58. Nadano, D., Yasuda, T., Takeshita, H., and Kishi, K., Activity staining of mammalian ribonuclease inhibitors after electrophoresis in sealed vertical slab polyacrylamide gels, *Anal. Biochem.*, 227, 210, 1995.

59. Melrose, J., Rodgers, K., and Ghosh, P., The preparation and use of biotinylated trypsin in Western blotting for detection of trypsin inhibitory proteins, *Anal. Biochem.*, 222, 34, 1994.

60. Rodgers, K.J., Melrose, J., and Ghosh, P., Biotinylated trypsin and its application as a sensitive, versatile probe for the detection and characterisation of an ovine chondrocyte serine proteinase inhibitor using Western blotting, *Electrophoresis*, 17, 213, 1996.

61. Melrose, J., Numata, Y., and Ghosh, P., Biotinylated hyaluronan: a versatile and highly sensitive probe capable of detecting nanogram levels of hyaluronan binding proteins (hyaloadherins) on electroblots by a novel affinity detection procedure, *Electrophoresis*, 17, 205, 1996.

62. Richardson, B.J., Baverstock, P.R., and Adams, M., *Allozyme Electrophoresis: A Handbook for Animal Systematics and Population Studies*, Academic Press, Sydney, 1986.

III ⎯⎯⎯⎯⎯⎯ Methods of Detection of Specific Enzymes

This part is central for the book. It contains detailed descriptions and outlines of more than 900 enzyme-specific zymogram techniques suitable for visualization on electrophoretic gels of more than 400 different enzymes. The information presented herein is traditionally arranged around the "enzyme sheet," which includes basic information required for the detection of a particular enzyme. At the beginning of this part, additional information of general relevance to the enzyme sheets is given. It includes a description of the structure of an enzyme sheet, general considerations, and comments and recommendations concerning application of zymogram techniques, as well as resource-saving strategies, troubleshooting, and safety measures. Abbreviations commonly used in the enzyme sheets are also given.

SECTION 1
THE STRUCTURE OF ENZYME SHEETS

The enzymes considered in this part are presented in numerical order according to the enzyme code (EC) numbers recommended by the Nomenclature Committee of the International Union of Biochemistry and Molecular Biology (NC-IUBMB) in 1992.[1] Each enzyme sheet is presented in a set format and includes information on enzyme reaction, enzyme source, subunit structure of enzyme molecules, specific zymogram methods (with four types of information given for each method: visualization scheme, staining solution, procedure, and notes), additional methods, general notes, and references.

As a rule, the recommended enzyme name is given in the enzyme sheet headline, while additional enzyme names are presented for cross-reference purposes in the section "Other Names." The enzyme symbols used are the same as the enzyme abbreviations most commonly encountered in the literature. Some symbols (while absent from the literature) are constructed as uppercase abbreviations in accordance with specific recommendations.[2,3] Thus, the enzyme sheet heading includes the EC number and the full and abbreviated names of the enzyme, for example: 1.1.1.1 — Alcohol Dehydrogenase; ADH. An alphabetical list of all the enzymes considered is given for convenience in Appendix B.

The enzyme reaction adapted from the enzyme list published by the Nomenclature Committee of the International Union of Biochemistry and Molecular Biology[1] is given in each enzyme sheet under the subheading "Reaction."

Groups of living organisms that are sources of the specific enzyme are listed in the section "Enzyme Source" to address more directly the application of specific zymogram methods to certain groups of organisms. This information was selected from Dixon and Webb's book, *Enzymes*,[4] current literature, current reference books, catalogs of chemical companies, and the Science Citation Index database of the Institute for Scientific Information (Philadelphia).

The knowledge on the subunit structure of enzymatic molecules is of fundamental importance for an adequate interpretation of banding patterns developed on zymograms by methods given in the book. Information on the subunit structure, if available, is given in the section "Subunit Structure." An enzyme is indicated as a monomer if its native molecules are represented by single polypeptides. The subunit structures of homomeric enzyme molecules (i.e., those consisting of several identical subunits) are indicated as dimer, trimer, tetramer, etc., and those of heteromeric enzyme molecules as heterodimer, heterotrimer, heterotetramer, etc. The subunit composition of heteromeric enzyme molecules, if available, is indicated in parentheses (e.g., heterodimer ($\alpha\beta$), heterotetramer ($\alpha\beta\gamma\delta$), heterotetramer ($2\alpha2\beta$), etc.) or described in more detail in the section "General Notes." Some authors believe that the subunit structure of most enzymes is highly conserved throughout the organismal evolution so that the subunit structure known for vertebrates may serve as a reliable indication for all other animal groups and even plants.[5] Although this is true for many enzymes, some enzymes demonstrate different subunit structures in different higher taxa and even within them.[6,7] This suggests that information on subunit structure should be as complete as possible to make the interpretation of banding patterns detected on zymograms adequate and reliable. Higher taxons in which the subunit structure of the enzyme is determined are listed in parentheses. If the subunit structure of an enzyme varies within a higher taxon (a situation common in bacteria), more detailed taxonomic information may be given. Conventional biochemical methods necessary to detect the subunit structure of enzyme molecules require purified enzyme preparations that are laborious and time consuming, and therefore not suitable for wide phylogenetic surveys. The combination of enzyme electrophoresis and the zymogram technique is of great advantage in this respect because it allows the determination of the subunit structure of enzyme molecules quickly and comparatively easily, using unpurified enzyme preparations (e.g., tissue homogenates).[8] This is due to a well-defined relationship between electrophoretic patterns of enzyme activity bands revealed on zymograms and the subunit structure of native enzyme molecules.[5,9–11] The subunit numbers can be inferred from allozyme patterns observed in heterozygotes as well as from isozyme patterns detected in interspecific F_1 hybrids, somatic cell hybrids, and *in vitro* hybridization experiments. Such information is now available for about one third of enzymes included in the book (author's unpublished data). For many of these enzymes the subunit structure is determined in a wide phylogenetic range of living organisms (e.g., see 1.1.1.37 — MDH; 1.1.1.42 — IDH; 1.1.1.44 — PGD; 2.6.1.1 — GOT; 5.3.1.9 — GPI; 5.4.2.2 — PGM). The "Subunit Structure" section contains two types of information. The first type includes data obtained by conventional methods (mostly by combined native PAG and SDS-PAG electrophoresis of purified enzyme preparations). These data are obtained from the Science Citation Index database (Web-of-Science) of the Institute for Scientific Information (Philadelphia). They are marked by the superscript # (i.e., monomer#, dimer#, trimer#, etc.). The second type includes data inferred from banding patterns detected on zymograms. This information is given without any marking (i.e., monomer, dimer, trimer, etc.).

More than one principally different zymogram technique is given for many enzymes. Information on each separate technique is given in the section "Method," which is subdivided into four subsections: "Visualization Scheme," "Staining Solution," "Procedure," and "Notes."

The diagram of the reaction sequence involved in the enzyme detection is schematically pictured in the subsection "Visualization Scheme." A general principle was used to construct such diagrams for the great majority of the enzyme detection methods described in this book. According to this principle, arrows indicate the direction of the reactions leading to visualization of the enzyme activity bands on the electrophoretic gels. All the participants of the enzyme detection reactions are given. Reagents included in the staining solution are given in bold type. All enzyme names are abbreviated, with the abbreviated names of the linking enzymes deciphered in the recipe of the staining solution. The abbreviated name of the enzyme being detected is enclosed in a box. The final product that enables the enzyme to be visualized is indicated by UV when registered in ultraviolet light or VIS when registered in daylight; for those compounds that change color during the reaction, VIS′ indicates the initial color and VIS indicates the color after the change. When the visualization procedure is a compound process that should be carried out sequentially step-by-step, different stages of the procedure are indicated on the reaction diagram. In some cases (e.g., when the scheme of an autoradiographic procedure is given) different stages depicted on the diagram are supplied with additional information for convenience (e.g., Stage 1: Enzyme reaction; Stage 2: Washing the gel; etc.).

The subsection "Staining Solution" ("Reaction Mixture" and "Indicator Agar" for autoradiographic and bioautographic methods, respectively) gives the recipe for the reaction mixture used to visualize areas occupied by the enzyme on the electrophoretic gel. The reference for the given recipe is also presented. Efforts have been made to obtain the recipes from the original articles in which they were first published. This goal, however, was not always achieved because many of the original recipes were modified throughout the years and the accumulated changes were not always well documented. The recipe may include several subrecipes, designated A, B, C, etc., when the compound-visualization procedure should be carried out in a step-by-step

fashion or when the preparation of the staining solution should be carried out by successive mixing of solutions prepared separately according to subrecipes. As a rule, the recipes are given in the form in which they are presented in the cited paper. As a consequence, in some recipes the involved reagents are given in absolute quantities, while in others they are given in concentrations. Each of the two modes of presentation of reagents in a recipe has its own advantages and disadvantages; therefore, no attempt has been made for overall standardization of the recipe presentation.

The subsection "Procedure" contains a detailed description of sequential procedures leading to the development of enzyme activity bands on an electrophoretic gel and thus represents a peculiar "know-how" of the method. It may be subdivided when a compound procedure of the enzyme visualization is described. Information on the mode of documentation of results obtained and some peculiarities of preservation of stained gels is also given there.

Diverse information that is of value for more successful application of the method and more adequate interpretation of obtained results is given under the subheadline "Notes." This information (if any) includes remarks concerning possible artifacts generated by the method and recommendations on control gel stainings, when they are needed. It also includes recommendations concerning possible counterstaining procedures used to make stained bands more contrasting and clearly visible, highlights some important details of the mechanism involved in the enzyme detection system, etc.

The section "Other Methods" initially included those methods for which the author had information concerning only general principles of the enzyme visualization. However, during the writing of the book this section was filled with methods that seemed supplementary or not widely used. For example, such detection methods as bioautography, two-dimensional spectroscopy, and immunoblotting were referred to as "Other Methods" when more simple and practical or commonly used zymogram methods were available for the enzyme under consideration. Methods that seemed adequate but had not been used so far to detect the enzyme under consideration were also given under "Other Methods," as recommended for application. It is not excluded for some enzymes that such methods may prove even more useful than the methods given in the "Method" section.

"General Notes" are given after the description of all the main zymogram methods developed for the enzyme under consideration. These notes give additional information that is of value for all the methods described for the enzyme. These are remarks on the enzyme substrate specificity, specific inhibitors and activators, comparative analysis of different detection methods, etc.

The list of "References" is given for each enzyme at the end of the enzyme sheet. The original papers are included in the list when possible to indicate the priority in development of the enzyme detection methods. In some cases where there are many references, those that contain reviews are preferred.

REFERENCES

1. NC-IUBMB, *Enzyme Nomenclature*, Academic Press, San Diego, 1992.

2. Shows, T.B. and 24 coauthors, Guidelines for human gene nomenclature: an international system for human gene nomenclature (ISGN, 1987), *Cytogenet. Cell Genet.*, 46, 11, 1987.

3. Shaklee, J.B., Allendorf, F.W., Morizot, D.C., and Whitt, G.S., Gene nomenclature for protein-coding loci in fish, *Trans. Am. Fish. Soc.*, 119, 2, 1990.

4. Dixon, M. and Webb, E.C., *Enzymes*, Academic Press, San Diego, 1979.

5. Richardson, B.J., Baverstock, P.R., and Adams, M., *Allozyme Electrophoresis: A Handbook for Animal Systematics and Population Studies*, Academic Press, Sydney, 1986.

6. Manchenko, G.P., Subunit structure of enzymes: allozymic data, *Isozyme Bull.*, 21, 144, 1988.

7. Ward, R.D., Skibinski, D.O.F., and Woodwark, M., Protein heterozygosity, protein structure, and taxonomic differentiation, *Evol. Biol.*, 26, 73, 1992.

8. Shaw, C.R., The use of genetic variation in the analysis of isozyme structure, *Brookhaven Symp. Biol.*, 17, 117, 1964.

9. Harris, H. and Hopkinson D.A., *Handbook of Enzyme Electrophoresis in Human Genetics*, North-Holland, Amsterdam, 1976.

10. Korochkin, L.I., Serov, O.L., and Manchenko, G.P., Isozyme conception, in *Genetics of Isoenzymes*, Beljaev, D.K., Ed., Nauka, Moscow, 1977, p. 5 (in Russian).

11. Buth, D.G., Genetic principles and the interpretation of electrophoretic data, in *Electrophoretic and Isoelectric Focusing Techniques in Fisheries Management*, Whitmore, D.H., Ed., CRC Press, Boca Raton, FL, 1990, p. 1.

SECTION 2
GENERAL CONSIDERATIONS, COMMENTS, AND RECOMMENDATIONS

THE CHOICE OF SUPPORT MEDIUM FOR ENZYME ELECTROPHORESIS AND DETECTION

The principles of detection of enzymes on electrophoretic gels are not dependent upon the type of support medium used for enzyme separation. However, in practice, the adequate choice of support medium is of great importance. This is because:

1. The quality of zymograms obtained via enzyme detection methods depends on characteristics that differ significantly in different support media.

2. A certain support medium may be preferable when a certain biological object is used as the source of the enzyme.

3. The choice of a certain support medium may depend on the question to be answered and the resources available to the researcher.

4. A number of zymographic methods are adapted or are recommended to be preferably applied to a certain support medium.

At present the most popular support media are starch, cellulose acetate, and polyacrylamide. Many of the characteristics that should be taken into account during the choice of support medium have been discussed by many authors.[1-5] The most important of them are considered below.

Cellulose Acetate Gel

Cellulose acetate gel requires only 0.5 to 2 μl per sample per enzyme run, and thus is the medium of preference when many enzymes must be detected from single small samples. The use of cellulose acetate may be highly desirable when small samples are more easily obtained than large ones (e.g., such a situation is common for isozyme studies of cell cultures or cultures of microorganisms). It is also preferable when enzyme stains containing very expensive ingredients should be used. This is because the 30 × 15 cm cellulose acetate sheets usually require not more than 3 ml of the staining solution. Cellulose acetate gels are commercially available and are ready for loading with samples after 10 min of soaking in an appropriate electrophoretic buffer. Only 1 h is usually needed to run cellulose acetate gels using relatively low run voltages. This allows, if necessary, the electrophoretic separation of enzymes to be carried out at room temperature without special cooling devices. Enzyme activity bands develop on cellulose acetate gels quickly because of the high porosity of the gel matrix and since the ingredients of the staining solution easily diffuse into the gel. This advantage is especially valuable when the procedure for the enzyme detection involves an exogenous linking enzyme(s) of high molecular weight(s). Electrophorized cellulose acetate gels can be frozen before or after the completion of the staining and thus can be available for further reference. Finally, cellulose acetate gels are durable and flexible, allowing the stained gel to be handled without unnecessary caution.

On the other hand, cellulose acetate gels cannot be sliced into a number of thin gels in the fashion of starch gel blocks. They are more expensive than equivalent starch or polyacrylamide gels, not translucent, and so cannot be quantified by densitometric methods without additional treatment. Cellulose acetate gels display reduced stain intensity for isozymes with low activity because only small volumes of samples can be applied to the gel. This support medium does not have the ability to cause a molecular sieving effect because of its coarse pore structure. So it separates proteins primarily on charge, with little or no separation on size. Finally, electroendosmosis does occur to an appreciable extent with just this support medium. In practice, however, at least some of these disadvantages are not critical. Indeed, electroendosmosis usually does not affect the relative mobility of allozymes and isozymes. Again, it may be reduced by the use of relatively more concentrated electrophoretic buffers. The absence of the molecular sieving effect does not play any important role in separation of allozymes, which usually do not differ markedly in molecular weight or shape. Gel slicing may offer no advantage in the peculiar situations that commonly occur during biochemical systematic studies. For example, during interspecific comparisons, each compared locus may require specimens from different species to be applied to a gel in a locus-specific order, placing side by side just those allozymes that need to be tested for identity or difference. As to the relatively high cost of the cellulose acetate gels, it should be pointed out that equivalent starch or polyacrylamide gels are actually more costly when labor costs are included. If densitometric scanning is desired, stained cellulose acetate gels can be cleared by washing them in 5% acetic acid, treating with 95% ethanol for 1 min, and immersing in 10% acetic acid and 90% ethanol mixture for 5 min. Before scanning, the gel should be dried and heated on a glass plate at 60 to 70°C for 20 min.

It should also be kept in mind that the use of cellulose acetate strips (i.e., the nongel form) does not give as good resolution of allozymes and isozymes as cellulose acetate gels do. Cellulose acetate gels are produced by a number of manufacturers.

The excellent handbook by Richardson et al.[4] contains comprehensive information on the application of cellulose acetate gels for isozyme and allozyme analysis and is recommended for more detailed consultation.

Polyacrylamide Gel

This support medium is formed by the vinyl polymerization of acrylamide monomers and cross-linking of the formed long polyacrylamide chains by the bifunctional co-monomer N,N'-methylene-bis-acrylamide in anaerobic conditions in the presence of such catalysts as ammonium persulfate and N,N,N',N'-tetramethylethylenediamine, or 3-dimethylaminopropionitrile. The main advantages of polyacrylamide gels are as follows:

1. The composition of the gel can be modified in a controlled way to achieve the best separation of the isozymes under question due to the optimal molecular sieving effect.
2. The gel matrix is highly homogeneous.
3. The gel is clear and can be directly subjected to quantification of enzyme activity bands by densitometry.
4. The results obtained with polyacrylamide gels are highly reproducible.
5. The use of this support medium allows detection of enzymes after electrophoresis of very dilute samples (e.g., some biological fluids, extracts of algal tissues, etc.).
6. The denaturing SDS–polyacrylamide gel system is the only electrophoretic medium suitable for analysis of many nonwater-soluble monomeric enzymes that can be renatured after appropriate treatments.
7. The sharpest protein bands are obtained with this support medium.
8. The polyacrylamide gels are rigid and thus are convenient for handling.

It is commonly acknowledged at present that the polyacrylamide gel is the best medium for separation of different classes of nonenzymatic proteins. This is, however, not always true for enzymatic proteins because of at least several pronounced disadvantages of this gel toward electrophoretic analysis of enzymes.

The first and the most critical disadvantage is the formation of nongenetic secondary isozymes as a result of the action of residual nonreacted persulfate, an oxidizing agent, which can cause structural modifications of enzyme molecules or the loss of their catalytic activity. Polyacrylamide gels always contain a small residual fraction of unpolymerized acrylamide monomers, which also can react with enzymatic molecules. This drawback may be eliminated by preelectrophoresis treatment, but only if continuous electrophoretic buffer systems are used. When the use of discontinuous buffer systems is needed, the polymerized gel should be soaked for a period of days in several changes of an appropriate gel buffer. This diminishes the value of the polyacrylamide gel in large-scale electrophoretic analysis of enzymes.

The next disadvantage of the polyacrylamide gel is the almost total impossibility to slice gel blocks into thin gel plates. This considerably limits the multilocus analysis from one electrophoretic run.

When vertical polyacrylamide gels are used, only isozymes moving toward one electrode (usually anode) are detected. Hence, two separate runs are needed to detect both anodally and cathodally moving isozymes. This problem can be avoided by the use of horizontal gel slabs, which allows the placing of samples anywhere in the slab. Positioning samples in the middle of the gel causes isozymes migrating toward either or both electrodes to be observed on the same gel. The use of horizontal gel blocks, however, does not allow exploitation of the most important advantage of vertical polyacrylamide gels, i.e., its value for analysis of very dilute samples with low enzyme concentration.

A problem with detection of some enzymes on polyacrylamide gels can occur when visualization mechanisms involve linking enzymes of large molecular weight. This is because of the slow diffusion of such enzymes into the polyacrylamide gel matrix. Indeed, most separations of enzymes are carried out with 5 to 12% polyacrylamide gels, which optimally separate proteins with a molecular weight range of 20,000 to 150,000.[5] So, diffusion of such routine linking enzymes as, for example, xanthine oxidase from milk (mol wt 290,000) or glutamate dehydrogenase from liver (mol wt 1,000,000) into commonly used 7.5% gels will be very slow, if at all.

Finally, both acrylamide and bis-acrylamide are highly toxic, and even very dilute solutions of these monomers can cause skin irritation and disturbance of the central nervous system. Polymerized gel, however, is relatively nontoxic and can be handled safely.

Starch Gel

At present starch gel is the most popular support medium for enzyme electrophoresis in population genetics and biochemical systematics studies. This is due to some critical advantages of starch gel over other support media.

The main advantage of this support medium is that starch gel blocks can be easily sliced into several gel slices. As many as eight to ten slices can be obtained from a gel block 1 cm thick. This affords a researcher the ability to reveal the genotype of an individual for a much larger number of genic loci in one run than is possible with any other support medium. This characteristic of starch gel is especially valuable for large-scale isozyme and allozyme screenings. It allows rapid assessment of the genetic composition of a population, and the multilocus identity of individuals.

Being a natural biological product, starch does not contain any undesirable admixtures capable of inactivating enzymes or causing *in vitro* generation of nongenetic secondary isozymes, as is characteristic of polyacrylamide gel. This simplifies isozymal patterns displayed on zymograms and makes them more easily interpretable in genetic terms, especially for those enzymes that are represented by multiple isozymal forms. The degree of resolution attainable by electrophoresis on a starch gel is exceeded only by that attainable with a polyacrylamide gel. A starch gel works well with samples the size of a fruit fly. Some zymographic techniques (e.g., those based on the use of the starch–iodine chromogenic reaction) were developed specifically for the starch gel.

At the same time, the starch gel has some disadvantages; however, they are not critical and do not considerably diminish its value for large-scale isozyme and allozyme surveys. When compared with polyacrylamide and cellulose acetate, the following disadvantages of the starch gel are usually listed:

1. Different lots of commercially available hydrolyzed starch supplied by different (and even the same) manufacturers usually differ in composition and may contain differing proportions of amylose and amylopectin, which can affect gelling ability and resolution. Thus, it is essential that each new lot of starch be thoroughly tested in order to calibrate it before use.

2. Starch gel is not translucent and so not suitable for direct quantitative measurement by densitometry. However, it can be relatively easily rendered transparent by soaking it in hot (70°C) glycerol for a few minutes, or by soaking it first in water:glycerol:acetic acid (5:5:1 by volume) and then overnight in pure glycerol. The lack of uniformity of the starch gel matrix also makes it less suitable for densitometric quantitation than the polyacrylamide gel. Again, this disadvantage is not critical for population genetics surveys because the great majority of electrophoretically detectable intra- and interspecific allozymic and isozymal differences are qualitative rather than quantitative.

3. Conventional starch gel stains usually involve volumes of 25 to 50 ml, which is about one order more than is required for staining of cellulose acetate gels of the same size. The amount of staining solution, however, can be reduced considerably by the use of an appropriate procedure for stain application. For example, only 6 ml of the stain is just enough to cover the cut surface of the gel (30 × 20 cm) if applied dropwise with a Pasteur pipette.

4. Starch gel is more friable than polyacrylamide or cellulose acetate gels and is not easy to handle; however, there are usually no serious problems with handling starch gel for those who have spent a day practicing.

The choice of the most appropriate support medium is a very important step in any survey that uses enzyme electrophoresis. Usually no single support medium is *a priori* superior to any other. Each of the three media discussed above has its own particular characteristics that should be taken into account, depending on the problem to be solved, the biological object to be used as the enzyme source, the battery of enzymes to be analyzed, the resources available to the researcher, and other factors and circumstances. The correct choice of support medium will allow one to save time and money and to obtain more adequate data for solving the problem under investigation.

STRATEGIES OF GEL STAINING

Several important questions that should be optimally resolved usually arise in connection with staining electrophoretic gels. The main ones are:

1. The choice of a more appropriate zymogram method, when more than one detection method is available for the enzyme under question

2. The choice of procedure for the faster and more correct preparation of functional staining solution

3. The choice of the mode of application of staining solution to the electrophoretic gel

4. The choice of methods allowing an increase in the staining intensity of the enzyme activity bands

5. The choice of the adequate staining of control gels for the assurance that the chosen zymogram method is specific for the enzyme under analysis

Unfortunately, each choice depends on many factors that may not be predicted with certainty beforehand. The most important characteristics of these factors are given below, with the purpose of allowing one to choose an optimal strategy of gel staining depending on the circumstances.

The Choice of Zymogram Method

This book comprises more than 900 different zymogram methods developed for more than 400 different enzymes. Thus, for many enzymes two or more principally different zymogram methods are available. Different methods developed for the same enzyme may differ in:

- Their applicability to a certain support medium
- Sets of reagents and special equipment needed
- Sensitivity
- Stain compatibility in double staining of the same gel
- Time and labor demands
- Methods of quantification of the resulting zymograms
- Sharpness and stability of stained bands
- Modes of band registration and zymogram preservation
- The cost of information obtained, etc.

These differences should be taken into account when making the choice of the more optimal zymogram method, depending on species and tissues that are thought to be used as the enzyme source, the support medium that is planned to be used or has already been chosen, the problem under study, the resources available, the kind and quality of information that must be obtained, etc. It is obvious that in each situation the choice of an optimal zymogram method will be the result of a complex compromise.

Preparation of Staining Solution

Each staining solution is buffered with the staining buffer at a specific pH. The pH value of the staining buffer used for the enzyme stain is a compromise between (1) the pH optimum of the enzyme activity, (2) the pH optimum of the staining reaction(s) used in the detection method, (3) the pH value of the buffer used for preparation of the gel, and (4) the pH optimum of the linking enzymes, if those are involved in staining reaction(s).

Many substrates (e.g., those used by dehydrogenases) are acids. In such cases stronger buffering capacity is required. Therefore, the use of high-concentration staining buffers is recommended. In some cases the substrates are such strong acids that their solutions must be prepared and brought to neutral pH before being added to staining solutions. Some commercially available substrates are unstable or very expensive. Solutions of such substrates can be prepared enzymatically in the laboratory. An example is glyceraldehyde-3-phosphate (see 1.2.1.12 — GA-3-PD, Method 1). It is a good practice to add substrates and other ingredients to staining buffers, but not vice versa. This prevents the possible sharp decrease of the pH value of the staining mixture to a level at which the enzyme under analysis will not function or other ingredients of the stain will be inactivated. For example, the reduced forms of NAD and NAD(P) cofactors denature in acid conditions very quickly. Dry chemicals should be taken out of the refrigerator or freezer about half

an hour before preparation of staining solutions. This lets them warm up and prevents condensation. It is a general rule to add ingredients to the staining buffer in the sequence given in the recipe of the stain. The intermediate and final pH values should be checked for staining solutions that include acids or that are prepared for the first time.

For large-scale population surveys of isozymes, the preparation of stock solutions of reagents that are used regularly is recommended. The use of stock solutions allows one to prepare staining solutions quickly and accurately, to use only minimal quantities of expensive reagents, and to reduce the number of times the refrigerated chemicals are opened. This mode of stain preparation is especially valuable when large numbers of samples are being run for only a few enzymes. On the contrary, when as large as possible a number of enzymes is needed to be stained only once, the use of dry reagents is preferable. Many workers with much experience in stain preparation use an "analytical spatula" and simple eye control to weigh most dry reagents, unless they are very expensive or proportions of the stain ingredients are critical. For example, the amounts of NADH or NADPH added to the negative fluorescent stains are critical. If too little amounts are added, the background will not fluoresce sufficiently, but when too much is added, the detected enzyme (or NAD(P)H-dependent linking enzyme) will not be able to convert sufficient quantities of NAD(P)H into NAD(P) for nonfluorescent bands to become visible. Substrates at a high concentration can sometimes inhibit enzymatic activity. An example is the brain isozyme of octopine dehydrogenase in cuttlefish. When a negative fluorescent method is used to detect this enzyme, the use of a pyruvate concentration higher than 2 mM is not recommended (see 1.5.1.11 — ONDH, Method 2). Again, special care should be taken with some couplers and dyes that can display inhibitory effects on catalytic activity of some enzymes when they are above certain critical concentrations or even at relatively low levels. Examples are some diazo dyes, which have inhibitory effects on acid phosphatase (see 3.1.3.2 — ACP, Method 3, *Notes*), and the PMS–MTT system, which inhibits some dehydrogenases (e.g., see 1.1.1.22 — UGDH; 1.1.1.138 — MD(NADP)). Many of the staining mixtures tolerate, to some extent, variations in the amounts of their ingredients. It should be remembered, however, that use of substrate concentrations that are only somewhat lower than saturation levels can cause weak staining of enzymatic bands. On the other hand, substrate concentrations that are sufficiently higher than saturation levels usually do not reduce the staining intensity of enzymatic bands. Therefore, when unique and precious material is analyzed or when the optimal substrate concentration for the enzyme under analysis is not known, the use of higher substrate concentrations is recommended. The same is true for linking enzymes. The concentration of linking enzymes should also be increased when their activity is decreased as a result of storage. However, when a particular staining solution is used regularly, the possibility of reduction of the quantity of expensive ingredients usually exists. The recipes of staining solutions presented in this part often involve the addition of excess amounts of reagents. Thus, it is usually possible to reduce the amounts of the more expensive reagents, and thus the total costs of stains.

As a rule, the use of allozymes and isozymes as genic markers does not require quantitative measurements. Thus, the exact quantitative standardization is usually not necessary when preparing the same staining solution to detect allele frequencies or to compare isozyme patterns in different populations or in samples taken from the same population at different times.

It is a general rule that staining solutions should be prepared as quickly as possible and just before use. The speed of the stain preparation is often more important than the precision. This is especially true for large-scale surveys in which many enzymes are to be stained. The speed of stain preparation may in some cases be the most important limiting condition, essential for the production of functional staining solutions.

The main limitation of the use of reagents as stock solutions is their instability. Stock solutions usually involve only those reagents that are stable in solutions for at least several weeks. When two or more reagents have been mixed in a solution, the stain should be further prepared as quickly as possible. On the other hand, the mixtures of dry reagents essential for visualization of some enzymes can be stored in a refrigerator for a long time while being desiccated. The use of dry reagent mixtures is advantageous for electrophoretic enzyme assays that are to be carried out under field conditions. At present some producers are beginning to supply enzyme detection kits that are ready for use. For example, Innovative Chemistry, Inc. (P.O. Box 90, Marshfield, MA 02050) supplies kits suitable for detection of 18 different human enzymes. The only thing that must be done with the kit is to reconstitute one vial of enzyme reagent with a certain volume of deionized water and to pour the resulting staining solution over the surface of the electrophoretic gel.

It should always be remembered that the use of reagents (especially substrates and linking enzyme preparations) of high purity will allow one to avoid many problems caused by the use of low-quality reagents containing concomitants, which can interfere with other compounds and make the staining mixture nonfunctional or result in development of artifactual or nonspecifically stained bands. In many cases the cost of high-quality reagents is of secondary concern, because the most expensive "ingredients" in enzyme electrophoretic surveys are labor, time, and the precious collected material.

Amounts of staining solutions sufficient for gel staining vary considerably, depending on the size of the gel to be stained and the mode of application of the staining solution to the gel.

Modes of Application of Staining Solutions

Staining solutions can be applied directly to the surface of electrophoretic gels or used as filter paper or agar overlays. A standard staining solution method is the earliest and the most widely used. It consists of placing an electrophoretic gel (or gel slice) in a special staining tray, adding the staining solution until the gel is completely covered by fluid, and incubating the gel at room temperature or 37°C (usually in the dark) until enzyme activity bands are visible. Specifically designed staining trays

made of glass or Plexiglas are usually used for this purpose. It is better to use a staining tray that exactly fits the size of the gel to be stained. If the trays for gel staining are larger than the gel, the amount of staining solution will have to be increased. This is not desirable because stains are usually expensive. For the staining tray that exactly fits the size of the 300 cm^2 gel, only about 50 ml of staining solution is needed to cover the top surface of the gel with fluid. However, this method is not always practical since many stains include reagents that are too expensive to be maintained at effective concentrations in large volumes. The problem can be overcome by preparing the concentrated staining solution in a small total volume that is just enough to flood the gel surface if applied dropwise with a pipette. Automatic adjustable-volume pipettes with disposable plastic tips or simple Pasteur pipettes are usually used for this purpose. The stain diffuses directly into the gel where the visualization reaction takes place. When the dropwise application method is used, only about 3 ml of staining solution is needed to stain a gel of 300 cm^2 in size. Another method of application may also be recommended. It comprises placing a glass rod on the edge of the gel, pouring the staining solution on the gel above the glass rod, and spreading the solution evenly over the gel surface with the glass rod. When dealing with cellulose acetate gels, small amounts of staining solution can be applied by spreading it evenly on the glass plate to the width of the gel, and subsequent spreading of the stain over the gel surface by placing the gel, with the porous side down, onto the solution, avoiding the formation of air bubbles between the gel and glass plate. Methods that use small amounts of concentrated staining solutions are applicable to cellulose acetate gels and cut surfaces of starch and (less frequently) polyacrylamide gels. It is important to point out that application and spreading of small amounts of staining solutions should be done as quickly as possible to prevent uneven entrapping of the fluid by the gel matrix and subsequent uneven staining intensity of the enzyme activity bands in different parts of the gel. After the stain is absorbed (usually after 1 to 5 min) the gel should be covered with a sheet of plastic wrap sufficiently large to fully enclose the gel and protect it from drying during incubation.

Many stains, predominantly those that are based on positive or negative fluorescence, are applied to the gel using the filter paper overlay method. This method requires considerably less stain than the standard staining solution method based on the use of staining trays and requiring the immersion of the whole gel into the staining solution. When the time period of gel incubation is expected to be longer than 30 min, the paper–gel sandwich should be covered with a sheet of plastic wrap of appropriate size. A filter and chromatographic paper (Whatman No. 1 and 3MM) are usually used in this method. Cellulose acetate strips are also used sometimes and work even better than filter paper, but they are more expensive. The filter paper overlay method is useful for negative staining of enzymatic bands using negative fluorescent stains containing NAD(P)H coupled with subsequent counterstaining of the gel with a PMS–MTT mixture. In this case the areas of the gel occupied by the enzyme are indicated by achromatic bands on a blue background of the gel and paper. Moreover, the precipitation of blue formazan in the paper is preferential. Thus, the stained paper may be easily washed, fixed, dried, and stored for a permanent record.

Staining solutions can also be applied as a 1:1 mixture with 2% molten agar (or agarose) cooled to 50 to 60°C. After preparation the mixture is quickly poured uniformly over the gel and allowed to cool and solidify. It should be remembered when using this method that if the agar solution is too hot, it can inactivate some reagents, e.g., linking enzymes, but if it is too cold, it can solidify just in the process of mixing with the staining solution. The agar overlay method has the advantage of bringing the reagents into intimate contact with the enzyme molecules to be stained for, and it also prevents diffusion of the stained bands, especially when the final product of the enzyme-visualizing reaction is soluble or when coupled enzymatic reactions generating soluble intermediates are involved in the enzyme-visualizing mechanism. Usually, the bands of enzymatic activity are well visible on the transparent agar layer. It can easily be taken off from the electrophoretic gel and the developed bands quantified by a scanning densitometer. Placed on a sheet of filter paper and dried, the developed agar layer may be stored for a long time and used as a permanent record of the zymogram. The agar overlay method is the only method suitable for detection of enzymes on nontransparent gels by two-dimensional gel spectroscopy. At present this method is one of the most popular methods of stain application in starch gel electrophoresis of enzymes.

The reactive agarose plate method is a modification of the agar overlay method. In this method 1% agarose solution containing all reagents needed for visualization of the enzyme activity is poured into a tray that fits the size of the electrophoretic gel to be stained. After agarose solidification the electrophoretic gel is laid over the agarose in order to expose the reactive agarose to the gel-entrapped enzyme. This method is used for visualization of enzymes by bioautography. Thin, substrate-containing agarose plates are also widely used for detection of many hydrolases with depolymerizing activity.

Each of the three main methods of application of staining solutions described above has its own advantages and disadvantages that depend on the type of support medium used for enzyme electrophoresis, the zymogram method chosen for visualization of the enzyme activity bands, the character and quality of information desired, and financial resources available. In general, however, the agar overlay technique gives better results than any other method and is always preferable when isozymes with high and low activity are presented on the same gel.

Modes of Enhancement of Staining Intensity of Enzyme Activity Bands

These modes can be divided into two main groups. The first group is represented by methods that are applicable before the procedure of detection of enzyme activity bands. It includes methods of protection and activation of enzymes during preparation of enzyme-containing samples and their electrophoretic run. Some modifications of the sample application procedure can increase the sample amount applied onto the gel, and thus increase the enzyme amount in the gel and the staining intensity of enzymatic bands on zymograms. These methods, however, have no direct relation to the procedure and mechanisms of the enzyme visualization and so are beyond the scope of this book. If interested, the reader may find more detailed information on the matter in other manuals on enzyme electrophoresis.[1,2,4,6–9]

The second group comprises methods that are applied during or after the procedure of visualization of enzyme activity bands. One of these methods is the so-called postcoupling technique. It is recommended for use when the staining solution contains a particular reagent(s) that inhibits catalytic activity of the enzyme to be stained. For example, acid phosphatase is inhibited by some Fast diazo dyes (see 3.1.3.2 — ACP, Method 3, *Notes*), while some dehydrogenases (see 1.1.1.22 — UGDH; 1.1.1.138 — MD(NADP)) are supposed to be inhibited by PMS or MTT, or both. In such cases all reagents except the potential inhibitor should be applied to the gel at the first step in the usual way. After incubation for an appropriate period of time (depending on the expected activity of the detected enzyme), the remaining reagent should be added to the stain. The main disadvantage of the postcoupling technique is that it is not always possible to monitor the intensity of the visualizing reaction over time. Indeed, the product of the phosphatase reaction, naphthol, becomes visible only after coupling with diazo dye. Fortunately, the reaction catalyzed by NAD(P)-dependent dehydrogenases can be monitored under a UV lamp because the reaction products NADH or NADPH fluoresce in long-wave UV light. Thus, when the bands of NAD(P)H fluorescence are well developed, a PMS–MTT mixture should be added to the stain to make dehydrogenase activity bands visible in daylight. Another disadvantage of the postcoupling technique is the diffuse character of the developed enzymatic bands. This is because of relatively small sizes of enzyme product molecules, which readily diffuse through the gel matrix before insoluble colored precipitates are formed.

Some NAD(P)-dependent dehydrogenases are activated by manganese ions. When the PMS–MTT system is used to detect such enzymes, weak staining intensity of the enzymatic bands can be observed because of the inhibitory effect of manganese ions toward PMS as an electron acceptor (e.g., see 1.1.1.42 — IDH, Method 1, *Notes*). To overcome this problem, diaphorase should be used in place of PMS (see 2.4.1.90 — AGS, Method 1, *Notes*), or the amount of PMS included in the stain should be increased several times. Another way is to observe the enzyme activity bands in long-wave UV light after incubation of the gel in staining solution lacking PMS and MTT. When applying PMS–tetrazolium stains it should also be kept in mind that PMS is not stable at high-alkaline pH values.[10] This can cause weak staining intensity of activity bands of dehydrogenases with high pH optima or when staining solutions with erroneously high pH values are used. The positive fluorescent methods may be preferable over PMS–tetrazolium methods when the endogenous enzyme superoxide dismutase (also known as tetrazolium oxidase) comigrates with NAD(P)-dependent dehydrogenases and considerably reduces or even fully prevents development of their bands via the PMS–tetrazolium stain.[11]

It may sometimes be desirable to increase the concentration of certain reagents (usually substrates, cofactors, and linking enzymes) in order to enhance the intensity of staining of the enzyme activity bands. The use of the agar overlay method for detection of weak-activity enzymes on cellulose acetate gels is preferable because it allows not only an increase in the quantities of reagents applied, but also a prolonging of the gel incubation time.

In some cases weak staining of enzyme activity bands may be the consequence of inhibitory action of the enzyme reaction product toward the detected enzyme. To increase the speed of the enzyme reaction the product of enzymatic reaction should be trapped. An auxiliary enzyme that utilizes the product of the detected enzyme as its own substrate can be used for this purpose (see 2.6.1.1 — GOT, Method 4, *Notes*). Some original zymographic methods are based on the same mechanism (e.g., see 5.4.2.4 — BPGM). In some cases the excess products may be trapped and the intensity of band staining enhanced by the use of specific reagents. For example, such carbonyl-trapping reagents as hydrazine hydrate and aminooxiacetic acid are used to trap 2-oxoglutarate with the purpose of displacing the glutamate dehydrogenase reaction toward the production of NAD(P)H, which serves as a fluorescent indicator of the enzyme reaction.[12] As a result, the accelerated production of NAD(P)H intensifies the staining of glutamate dehydrogenase bands by the tetrazolium method (see 1.4.1.2–4 — GDH, Method 1, *Notes*).

The apparent activity of some enzymes can be doubled by linking it to two sequential reactions or to dehydrogenase 1/isomerase/dehydrogenase 2 sequential reactions, which produce two molecules of a reduced or oxidized dehydrogenase cofactor. For example, two NADP-dependent sequentially acting dehydrogenases — glucose-6-phosphate dehydrogenase and 6-phosphogluconate dehydrogenase — are used to double the production of fluorescent NADPH in the positive fluorescent method developed for detection of galactokinase (see 2.7.1.6 — GALK, Method 3). Another example is the glycerol-3-phosphate dehydrogenase/triose-phosphate isomerase/glyceraldehyde-3-phosphate dehydrogenase system of sequentially acting auxiliary enzymes, which can work in both directions and may be used to double the production of NADH (the forward direction) or NAD (the backward direction). Examples of the use of this linking enzyme system are detection methods for glycerol kinase (2.7.1.30 — GLYCK, Method 2), phosphoglycerate mutase (5.4.2.1 — PGLM, Method 1), and phosphoglycerate kinase (2.7.2.3 — PGK).

Enzymatic cycling is a method widely used for amplifying the signal of enzymatic reaction. This method uses two different enzymes coupled in the opposite direction, so that the substrate of one of them is the product of the other, and vice versa. Under such conditions, the reaction turns out with no consumption of recycling substrates, while other products of enzymatic reactions are accumulated with each turn of the cycle. The use of enzymatic cycling is of special value for detecting enzymes present in low concentrations and producing low levels of detectable signals. The method of enzymatic cycling was proposed for determining L-glutamate, which involves auxiliary enzymes L-glutamate oxidase (EC 1.4.3.11), alanine transaminase (EC 2.6.1.2), and horseradish peroxidase (EC 1.11.1.7).[13] This method may be applied to detect enzymes producing L-glutamate (e.g., see 3.5.1.2 — GLUT, Other Methods, C; 4.1.3.27 — AS, Other Methods, B). ATP-ADP cycling reactions catalyzed by auxiliary enzymes adenylate kinase (2.7.4.3 — AK), pyruvate kinase (2.7.1.40 — PK), and hexokinase (2.7.1.1 — HK) are used coupled with auxiliary glucose-6-phosphate dehydrogenase (1.1.1.49 — G-6-PD) to detect 5'-AMP produced by 3',5'-cyclic-nucleotide phosphodiesterase (e.g., see 3.1.4.17 — CNPE, Method 3; 4.6.1.1 — AC).[14] These reactions may be used to detect activity bands of other enzymes producing 5'-AMP, particularly amino acid–tRNA ligases (6.1.1.2 — TRL; 6.1.1.4 — LRL; 6.1.1.9 — VRL; 6.1.1.16 — CRL), using the second step of an aminoacyl–tRNA ligase reaction.

Fluorescence of 4-methylumbelliferone, which is the product of hydrolytic cleavage of 4-methylumbelliferone derivatives by numerous hydrolases, may be enhanced considerably by treating the developed acidic gels with an alkaline buffer (pH 9 to 10) or ammonia vapor.

The miniaturization of developed polyacrylamide gels by treatment with hot (70°C) 50% (w/v) polyethyleneglycol (PEG 2000) increases the sensitivity of enzyme detection methods five to ten times.[15] Methanol-containing fixatives may be used to miniaturize (although to a lesser extent) developed starch gels (for details, see "Recording and Preservation of Zymograms," below).

Specificity of Zymogram Methods and Some Related Problems

Most of the zymogram methods presented in this part are enzyme specific. However, some methods can detect more than one enzyme. This results from the ability of some enzymes to utilize one or more of the applied reagents, or to use certain buffer constituents coupled with the applied staining solution reagents. Another reason for the development of unexpected enzymatic bands is the combined effect of two comigrating endogenous enzymes, one of which acts as a linking enzyme for the other in the presence of certain necessary reagents contained in the applied staining solution. Some enzymes can form complexes with cofactors or substrates that are stable during electrophoresis, or can use residual amounts of their substrates that are contained in some reagents as concomitants. Therefore, when other reagents needed for visualization reactions of such enzymes are available from applied staining solutions, the nonspecifically stained additional bands will appear. Finally, several cases are known where nonenzymatic protein bands or even nonprotein bands are developed by the use of certain stains.

Thus, when a new enzyme detection method is developed or when an approved method is used to detect a certain enzyme from a new enzyme source, it should be tested for its specificity for the enzyme under analysis. This is especially necessary for complex methods involving one or more reactions catalyzed by exogenous linking enzymes. A standard procedure for testing specificity of the method to be used includes a series of experimental stainings in which each individual constituent of the stain is omitted one at a time in order to determine if any stained bands appear after treating control gels with incomplete stains. The specificity of the enzyme detection method can also be tested by the use of alternative methods, if they are available for the enzyme under investigation, or if they may be devised by using a backward reaction, or by using a principally different mechanism of visualization, which involves different substrates, cofactors, sets of linking enzymes, etc.

Nonspecific staining of additional unexpected bands on zymograms of some enzymes is illustrated by the following examples.

Bands of adenylate kinase (see 2.7.4.3 — AK, Method 1) can develop on zymograms of pyruvate kinase (2.7.1.40 — PK, Method 2), creatine kinase (2.7.3.2 — CK, Method 2), and arginine kinase (2.7.3.3 — ARGK, Method 1). This is because all the reagents needed for AK detection (i.e., ADP, glucose, NADP, MTT, PMS, hexokinase, and glucose-6-phosphate dehydrogenase) are contained in staining solutions used to detect the enzymes listed above.

It is well established that adenylate kinase molecules have a monomeric subunit structure that is highly conserved during organismal evolution.[16] In this connection, the description of unusual allozymic polymorphisms of dimeric adenylate kinase revealed in some invertebrate and vertebrate animals[17–19] is the consequence of misleading interpretation of allozymic variation of nonspecifically stained bands of dimeric glucose dehydrogenase (see 1.1.1.47 — GD, Method 1), which is widely distributed among different animal groups. To be developed, this dehydrogenase requires just those reagents (i.e., glucose, NADP, MTT, and PMS) that are present in the adenylate kinase stain.

When negative fluorescent methods are used to detect opine dehydrogenases (e.g., 1.5.1.11 — ONDH, Method 2; 1.5.1.17 — ALPDH, Method 2), the bands caused by lactate dehydrogenase activity can also develop in some invertebrate species containing all these dehydrogenases.

Some phosphatases of invertebrates and vertebrates can use phosphoenolpyruvate as a substrate and thereby produce pyruvate. Thus, the bands of phosphoenolpyruvate phosphatase activity can develop on pyruvate kinase zymograms obtained by a negative fluorescent method (2.7.1.40 — PK, Method 1, *Notes*) and on zymograms of many other enzymes obtained using detection methods involving pyruvate kinase and lactate dehydrogenase as linking enzymes (e.g., see 2.7.3.2 — CK, Method 1; 2.7.3.3 — ARGK, Method 2; 2.7.4.3 — AK, Method 2; 2.7.4.4 — NPK; 2.7.4.8 — GUK, Method 1; 2.7.6.1 — RPPPK, Method 1).

It has been found that human aconitase isozymes can sometimes appear on zymograms obtained using staining solution lacking both the enzyme substrate, *cis*-aconitate, and the linking enzyme, isocitrate dehydrogenase.[2] This unexpected phenomenon was shown to occur only when electrophoresis was carried out using a citrate-containing buffer. It is known that aconitase is capable of catalyzing two reactions:

1. citrate = *cis*-aconitate + H_2O
2. *cis*-aconitate + H_2O = isocitrate (see 4.2.1.3 — ACON)

Thus, aconitase can use citrate from a citrate-containing gel buffer as a substrate and thereby produce isocitrate. When aconitase comigrates with isocitrate dehydrogenase, staining occurs at those gel sites where two endogenous enzymes occur together. When one or both of these enzymes are polymorphic, the individual differences in development of unexpected additional bands may be observed, because under these conditions both enzymes migrate to overlapping positions only in some individuals.

When a standard staining solution method is used to detect hexokinase (see 2.7.1.1 — HK), bands of 6-phosphogluconate dehydrogenase (1.1.1.44 — PGD) may also develop, especially when the gel is subjected to prolonged incubation. This is because the linking enzyme glucose-6-phosphate dehydrogenase utilizes the glucose-6-phosphate generated by endogenous hexokinase and produces glucono-1,5-lactone 6-phosphate, which spontaneously turns into 6-phosphogluconate, the substrate of 6-phosphogluconate dehydrogenase. The molecules of this substrate freely diffuse in the liquid stain and become available for endogenous 6-phosphogluconate dehydrogenase, for which bands may further develop in the presence of NADP, PMS, and MTT involved in the hexokinase stain. Of course, this is the case only when an NADP-dependent glucose-6-phosphate dehydrogenase preparation is used in the hexokinase stain.

Additional unexpected bands are frequently observed on zymograms of NAD(P)-dependent dehydrogenases. These bands develop even when staining solutions lacking any dehydrogenase substrates are used. This effect has been referred to as "nothing dehydrogenase" (see I.X.X.X — NDH). Two main reasons are known to cause the NDH phenomenon: (1) binding of endogenous substrates by some dehydrogenase molecules, and (2) contamination of some commercial preparations with substances that serve as substrates for some dehydrogenases. Two enzymes — alcohol dehydrogenase and lactate dehydrogenase — usually are the most probable candidates for NDH. The bands of NDH activity may be easily identified by their repetitive occurrence on gels stained for different dehydrogenases or other enzymes detected using linked dehydrogenase reactions.

Some nonenzymatic proteins can also cause the development of false bands on gels specifically stained for some enzymatic activities. At least four examples are known. The first one concerns SH-rich proteins, which are capable of reducing a tetrazolium salt even in the absence of NAD(P) and PMS.[20,21] The second example is the interference of albumins with the starch–iodine color reaction, resulting in the development of false amylase bands on zymograms obtained by the negative starch–iodine method.[22] Albumins are also known to be able to cause the appearance of additional artifactual bands on acid phosphatase[23] and alkaline phosphatase[24] zymograms obtained with the azo coupling methods. In the last case it is shown that the alkaline phosphatase–like activity in the albumin zone is an artifact due to a bilirubin–albumin complex that nonspecifically couples with a diazo compound.

When the samples used for electrophoresis contain high levels of endogenous reduced glutathione or other sulfhydryl compounds (e.g., 2-mercaptoethanol or dithiothreitol) added to the samples prior to electrophoresis to stabilize or activate the enzyme under analysis, the appearance of additional stained bands should be expected on zymograms obtained by the use of staining solutions containing MTT–PMS. The bands caused by electrically neutral 2-mercaptoethanol and dithiothreitol molecules are usually detected in the cathodal part of starch gels because of the effect of electroendosmosis. Negatively charged molecules of reduced glutathione usually migrate to the very anodal part of the gel. The bands caused by all these thiol reagents are monomorphic, diffuse, and stained fairly weakly, and are observed on zymograms of different enzymes detected for the same samples by tetrazolium methods.

Two different enzymes with overlapping substrate specificity are able to catalyze the same reaction and thus to display their activity bands on the same zymogram. For example, it was shown that some isozymes of alcohol dehydrogenase from *Drosophila melanogaster* also oxidize L(+)-lactate or D(–)-lactate with NAD as a cofactor, and intraspecific electrophoretic variation observed on lactate dehydrogenase zymograms could be attributed to the presence of alcohol dehydrogenase.[25] This phenomenon was called pseudopolymorphism. Another example is the existence of some isozymes in the snail *Cepaea nemoralis* that can utilize either malate or lactate as a substrate, converting both into pyruvate.[26] A special study has shown[27] that there are many enzymes that are listed in the enzyme list[28] under different code numbers, although they are coded by one and the same gene in a wide range of organisms that represent phylogenetically distinct groups. As to dehydrogenases, identical allozymic patterns were revealed on zymograms of octanol dehydrogenase (1.1.1.73 — ODH), formaldehyde dehydrogenase (1.2.1.1 — FDHG), and D-lactaldehyde dehydrogenase (1.1.1.78 — DLADH) from the sipunculid *Phascolosoma japonicum*. Identical allozymic variations were observed on zymograms of ODH and FDHG in each of four bivalve species of the genus *Macoma* that were examined. In the phoronid *Phoronopsis harmeri* the allozymic pattern of the ODH-1 isozyme was found to be identical to those of DLADH, while the allozymic pattern of the ODH-2 isozyme was identical to that observed on the FDHG zymogram. The same allozymic patterns were revealed on ODH, FDHG (FDHG-2 isozyme), and DLADH zymograms in the mushroom *Boletus edulis*. All these examples are evidence that certain dehydrogenases that are believed to be different proteins are really a single enzyme protein with broad substrate specificity. The same bands can also be observed on zymograms of acid and alkaline phosphatases, and on zymograms of different peptidases. It was shown, for example, that numerous phosphatases from the sipunculid intestine display identical allozymic patterns in the same sample of conspecific individuals.[29] Similar results were obtained with hexokinase (2.7.1.1 — HK) and fructokinase (2.7.1.4 — FK) from a mushroom, nemertine, domestic fly, starfish, and sea urchin.[27] Thus, a number of cases are known where enzymes that are believed to be different are encoded by a single genic locus, and the possibility exists to erroneously score the same locus twice or even more.

RECORDING AND PRESERVATION OF ZYMOGRAMS

There are at least three different methods of recording the enzyme patterns developed on electrophoretic gels: (1) schematic recording of zymograms, (2) photography of zymograms, and (3) the tracing of the band position on paper overlays or cellulose acetate gels. After recording, zymograms can be stored in special fixative solutions or dried.

Schematic recording of zymograms involves a two-dimensional representation of the banding patterns observed on developed gels. Although such representation suggests some degree of simplification and subjectivity, it can be excused during routine surveys of well-known enzyme polymorphisms.

Another technique that allows the keeping of a permanent record of the results of each electrophoretic experiment is photography. The main advantage of photographs is their objectivity. They are relatively rapidly produced and easy to store. Stained cellulose acetate and starch gels are photographed with uniform lighting from above, while stained translucent agar overlays and polyacrylamide gels are photographed on a light box with lighting transmitted from below. Sometimes special conditions are required for photographing stained gels. For example, polyacrylamide gels stained using the method of calcium phosphate precipitation should be photographed by reflected light against a dark background. In order to produce permanent records of indicator agar plates developed by the method of bioautography, an indirect lighting system is used that is a large version of the lighting system employed for photography of immunodiffusion plates. For photographing fluorescent bands (or nonfluorescent bands on fluorescent background), reflected UV light is used in combination with a yellow filter (e.g., Wratte Gelatin Filter No. 2E, Eastman Kodak Company). The exposure for fluorescent gels is usually 20 to 30 sec. The use of an automatic-exposure camera considerably facilitates the photographing of such gels. The use of a yellow filter also enhances the contrast of blue bands and is routinely employed in photographing all gels stained with tetrazolium stains. Black-and-white films such as Kodak Panatomic-X, Plus-X, or Technical Pan are recommended for photographing bands visible in daylight. For photographing fluorescent bands (or nonfluorescent bands on fluorescent background), high-speed films such as Polaroid film type 55 or 57 or regular film ASA 400 are recommended. Color slides of good quality can be obtained with Kodachrome ASA 25.[30,31]

Another, more rapid way of photorecording is direct photocopying of the stained gels using a photocopier. This method does not possess the lag time between taking a photograph and obtaining usable information. It is less expensive than the photography techniques. The main disadvantage of this method is that it gives photocopies of slightly inferior quality, compared to photographs.[4]

The position of stained bands can be traced directly on cellulose acetate gels and filter paper overlays or on transparent protective coverings (e.g., polyethylene sheets) by careful tracing over the bands with a ballpoint pen or a marker. This technique is especially useful for recording electrophoretic patterns obtained using fluorescent detection methods or for recording bands that fade after a short period of time.

Stained gels can be fixed and stored for a permanent record or dried after appropriate treatments. Cellulose acetate gels are usually fixed in 5% formalin or 10% acetic acid for 10 to 15 min and then stored in 10% glycerol or placed in airtight plastic bags for convenient handling. Starch gels are usually fixed in 7% acetic acid or 25% ethanol and then stored in special trays filled with a 5:5:1 (v/v) mixture of methanol:water:acetic acid. This fixative toughens and shrinks the gels and makes them opaque. The alcohol gel wash recommended for starch gels consists of a 5:4:2:1 (v/v) mixture of ethanol:water:acetic acid:glycerol.[32] It not only stops the staining but also toughens the gel and helps bleach out some of the background. The use of 50 to 100% glycerol does not harden the starch gel as the alcohol-containing fixatives do; however, it helps to maintain gel integrity and makes the gel translucent and suitable for densitometric measurements. A 5% glycerol–7% acetic acid mixture is also frequently used for fixing and storing stained starch gels. Fixed gels can also be stored in airtight plastic bags. Stained polyacrylamide gels are usually fixed in 7% acetic acid or glycerol–acetic acid mixtures. The fixed gels can then be stored in a refrigerator for several months.

The method that gives a more permanent record of isozyme patterns is gel drying. For example, a simple method was developed for the preparation of fully transparent, flexible, dry sheets of stained starch gels.[33] In this method the stained gel is kept in 7% acetic acid, washed twice with water to remove excess reagents, and kept in 5% glycerol for 15 to 30 min. The gel is then placed on a glass plate of appropriate size, previously covered with a second cellophane sheet. Both sheets must be sufficiently large to allow at least a 2-cm margin on all sides of the glass plate. Both cellophane sheets must be soaked in 5% glycerol solution for 15 to 30 min before use. The formation of air bubbles between the gel and the sheets should be avoided. The four edges of the double cellophane are then folded over the back of the glass plate, the excess amount of glycerol solution removed by filter paper, and the processed plate dried at room temperature or at 60 to 80°C. After the gel and cellophane have been dried, the cellophane sheets are cut with a razor blade, about 1 cm from the border of the gel. The dried transparent and flexible gel covered with cellophane can then easily be stripped off the glass plate. This method allows storage of stained starch gels for up to 9 years with full preservation of isozyme patterns. A similar procedure may be used to obtain dried polyacrylamide slab gels, except 65% methanol containing 0.5% glycerol should be substituted for the 5% glycerol solution. Dry polyacrylamide slab gels suitable for autoradiographic detection of [^{14}C]- and [^{3}H]-labeled zones can also be obtained by placing the thin slab onto a glass plate large enough to allow the formation of at least a 2-cm margin all around the gel, subsequent covering of the gel with 2% agarose solution, and drying of the "sandwich" slowly and evenly, at first using an infrared lamp and then at room temperature. The stained polyacrylamide slab gels pretreated for 1 h with a 7.5% acetic acid–1.5% glycerol solution also can be dried after being covered with a 5% aqueous solution of gelatin or sprinkled with aquacide I, II, or III.[5,34] The instruction for a treatment that converts cellulose acetate gel into a film similar to cellophane is supplied with Cellogel by the manufacturer.

The above-described techniques and facilities developed to record and preserve zymograms are now being intensively substituted with digital photo cameras, scanners, and PC softwares to produce, process, and store images of zymograms.

RESOURCE-SAVING STRATEGIES

The cost of information obtained by enzyme detection methods can be a limiting factor in some electrophoretic surveys of enzymes, e.g., in large-scale population studies. Several different ways exist to overcome this problem.

As outlined above, the choice of adequate and less-expensive zymogram methods, the reduction of quantities of the most expensive reagents in the staining solutions used, and the use of the most economical modes of stain application can help one to decrease expenses considerably. For example, when detecting glutamate–oxaloacetate transaminase (see 2.6.1.1 — GOT) in a population survey there is no need to use quantitative enzymatic Method 4 because the routine diazo coupling Method 1 gives quite satisfactory results and is about two orders less expensive than the former one. For many histochemical detection methods the expenditure of staining solutions can be decreased more than ten times when applied on a filter paper overlay or dropwise. Some other specific tactics also can be used to save resources during electrophoretic studies of enzymes. The most important of them are (1) simultaneous detection of two or even more enzymes on the same gel, (2) successive detection of different enzymes on the same gel, (3) reusing of staining solutions, (4) the use of several origins on the same gel, (5) multiple replication of running gels by electroblotting, and (6) the use of semipreparative gels as the sources of linking enzymes.

Simultaneous Detection of Several Enzymes on the Same Gel

Theoretically, any two or more different enzymes can be simultaneously stained on the same gel when their detection methods are based on the same visualizing mechanism.[4] In practice, however, the combined stains are applicable only to those enzymes displaying nonoverlapping isozymal spectra under the electrophoretic conditions used.

Good examples of stain compatibility based on identity of the detection mechanisms involved are many NAD(P)-dependent dehydrogenase stains involving the PMS–tetrazolium system and some other stains using any NAD(P)-dependent dehydrogenase as a linking enzyme to catalyze the last step in the visualizing reaction, i.e., the production of colored formazan.

Different dehydrogenase stains are also compatible when positive or negative fluorescent detection of their bands is based on an NAD(P)-into-NAD(P)H or NAD(P)H-into-NAD(P) conversion mechanism, respectively, or when detection of some nondehydrogenase enzymes is based on this same mechanism implemented through a linked reaction catalyzed by an appropriate NAD(P)-dependent dehydrogenase.

A combination of stains for different hydrolases that use 4-methylumbelliferone derivatives as substrates is possible and can be successful, especially when hydrolases with similar pH optima are combined.

The main disadvantage of the stain combinations listed above and of some other possible combinations is that different sets of bands displayed by different enzymes are of the same color and hence may be confused. Thus, before being combined, different enzymes should be stained on separate gels electrophorized under the same conditions or stained on different gel slices obtained from the same gel block to identify relative positions of bands displayed by enzymes, the stainings of which are supposed to be combined. However, when compatible stains of polymorphic enzymes with different subunit structures of their molecules are combined, there is no need to stain enzymes separately because allozymic patterns displayed by different enzymes can be easily identified through the number of allozymes displayed in heterozygous individuals.[2,35]

The combination of compatible enzyme stains is particularly suited to large-scale population surveys where the range of expected genotypes is known. This tactic also allows staining of more enzymes when only extremely small sample volumes are available that are not sufficient for loading several different gels. It not only saves expensive reagents involved in the process of enzyme staining but also decreases the gel and time expense.

Simultaneous staining of different enzymes on the same gel is possible even when their detection methods are incompatible. This can be achieved through application of incompatible staining solutions in different strips of agar or acetate cellulose overlays applied to different parts of the gel. The tactic of by-strip application is justifiable only when gel areas occupied by different enzymes to be detected were identified as a result of previous stainings of enzymes on separate gels. The use of the by-strip application method prevents mixing of incompatible stains, and interaction or interference of ingredients from different stains can occur only in a narrow contact zone where ingredients from different stains mix due to restricted diffusion.

Successive Detection of Different Enzymes on the Same Gel

This approach can be implemented not only when the stain compatibility of different enzymes is based on identity of mechanisms of their detection (i.e., in situations described above), but also when compatibility is based on tolerance of different detection mechanisms.[4] The latter situation assumes that none of the reagents from the first stain interact or interfere with any reagent from the next one. In this case the activity bands caused by different enzymes usually are of different colors and so can be readily distinguished. A particular sequence of applied stains may not be reversible. For example, successful double stains are a purine-nucleoside phosphorylase tetrazolium stain (see 2.4.2.1

— PNP) followed by a glutamate–oxaloacetate transaminase stain based on the diazo coupling mechanism (see 2.6.1.1 — GOT, Method 1), and an esterase fluorescent stain (see 3.1.1 ... — EST, Method 2) followed by a peptidase stain involving L-amino acid oxidase and peroxidase as linking enzymes (see 3.4.11 or 13 ... — PEP, Method 1). In both cases the reverse sequence of stain application will not be successful simply because the products of the first-step staining reactions will mask the products of the second-step staining reactions.

In some cases different stains are incompatible because an ingredient of one stain is able to react with an ingredient in another. Examples are a glycerol-3-phosphate dehydrogenase tetrazolium stain (see 1.1.1.8 — G-3-PD) followed by a glycerol kinase tetrazolium stain (see 2.7.1.30 — GLYCK, Method 2), or a glucose–phosphate isomerase stain (see 5.3.1.9 — GPI) followed by a mannose–phosphate isomerase stain (see 5.3.1.8 — MPI), where the substrates of the first stains will react with linking enzymes of the following ones. The same is true for negative fluorescent stains involving NAD(P)H followed by a tetrazolium stain where NAD(P)H of the first stain will interact with the PMS–tetrazolium system of the following one, and thus will result in formation of colored formazan across the whole gel surface. The problem of stain incompatibility can be overcome by careful washing of the gel after the first staining, with the purpose of removing soluble ingredients capable of interaction with ingredients in the following stain.[4] This tactic, however, is applicable only when undesirable compounds of the first stain are soluble. Most suitable for implementation of this tactic are coarse-pored cellulose acetate gels, thin slices of starch gel blocks, and thin blocks of low-percentage polyacrylamide gels. The use of just those gels allows small molecules of undesirable reagents to diffuse readily into the washing solution, while large enzyme molecules remain in their final run position, being incorporated into the gel matrix. In some cases this tactic may be the only possible one for detecting different enzymes on the same gel when incompatible detection mechanisms are involved in different stains. An example is a situation when banding patterns produced by different incompatible stains overlap, and thus the by-strip application method cannot be used.

The successive detection of different enzymes using compatible stains may also sometimes be preferable because it does not require preliminary identification staining of different enzymes on separate gels. An example is a situation when the final detected products of successive stains are the same color and the banding patterns displayed by the different enzymes cannot be identified on the basis of differences of their subunit structures (e.g., when all enzymes are monomorphic or when their molecules are of the same subunit structure).

When compatibility of successive stains is based on the identity of the detection mechanisms involved, both reagent and gel saving may be achieved because only the addition of a specific substrate is needed for detection of each following enzyme. Only gel saving is achieved through double staining of the same gel if different detection mechanisms are involved in different successive stains. Again, successive double stainings are of great value when extremely small volumes of samples that are not sufficient for loading more than one gel are available, and gel areas occupied by different enzymes overlap.

The Reuse of Staining Solutions

When a large-scale population survey is carried out using the standard staining solution method of enzyme detection, some staining solutions may be used repeatedly immediately after the first use, after 1 day of storage in a refrigerator, or after several days of storage in a freezer. The staining solutions that easily endure this tactic are those used for tetrazolium detection of numerous NAD(P)-dependent dehydrogenases and many other enzymes that are detected through the use of dehydrogenases as linking enzymes to produce a final visible product, formazan. The repeated addition of MTT or NBT and a linking enzyme (if involved) to the previously used staining solution is sometimes desirable. The tactic of reusing tetrazolium-containing stains is of great value, especially when expensive substrates and cofactors are involved in visualizing reactions. The potential reagent savings make this tactic a desirable addition to the resource-saving strategies of a laboratory.

The Use of Several Origins on One Gel

Two or more different sets of samples may be loaded in different origin positions on the same gel.[4] This is a very useful tactic, especially when large-scale population screening of single-locus enzyme polymorphisms is carried out. The use of multiple origins is most effective for polymorphic enzymes with a small number of allelic variants of low electrophoretic mobility. Preliminary test runs may be needed to find optimal relative positions of origins to be sure that allozyme sets from different origins do not overlap. It should be taken into consideration that separation of the same allozymes started from different origins may be different. This disadvantage is especially pronounced when discontinuous buffer systems are used for electrophoresis. However, there are usually no problems with genetic interpretation of enzyme polymorphisms with known genotypes. The multiple-origin tactic is the most effective one in a search for rare electrophoretic variants. The advantages of this tactic are the obvious cost savings in gel media, fine biochemicals, and time.

Multiple Replication of Electrophoretic Gels by Electroblotting

Electrophoretic transfer of enzymes from a single running gel onto immobilizing matrices (e.g., Zeta Pore or Hybond-N membranes and Whatman DE 81 paper) allows one to obtain multiple replicas, which can then be developed separately using different enzyme stains.[36] It is important to remember when using this tactic that the faster-migrating enzymes should be detected on the first replica, while the slower-migrating ones should be on the last replica. Thus, use of the replication tactic requires preliminary knowledge of relative positions of isozymal sets displayed by different enzymes under analysis. Multiple replication and multiple successive staining tactics may be combined, resulting in a considerable increase in the number of different enzymes analyzed through one gel electrophoretic run. The electrophoretic transfer of enzymes requires additional expensive materials, special designs, and time. Thus, it is not as practical as some other resource-saving tactics described above. At the same time, it may be potentially very valuable when only extremely small volumes of each sample are available and obtaining additional material is very labor- and time-consuming, or impossible.

The Use of Semipreparative Gels for Production of Preparations of Linking Enzymes

A great number of enzyme-visualizing methods listed below in this part involve exogenous linking enzymes, some of which are not yet commercially available. Again, in some cases the purchase of even one vial of very expensive enzyme preparation may not be reasonable, e.g., when only one or a few gel stainings are supposed to be carried out. In such situations the semipreparative gel electrophoresis presents a good opportunity to obtain the enzyme needed in small but sufficient amounts and purity. In principle, all the enzymes that can be detected through any visualizing methods and thus located on electrophoretic gels may be used as linking enzymes. One critical condition must be fulfilled thereat: the preparation of the enzyme to be used as the linking one must be free of the enzyme to be detected. The use of the linking enzyme preparation contaminated with the enzyme under detection will result in background staining without any distinctly visible bands. The tactic of using semipreparative gels for the production of linking enzyme preparations was tested by the author and proved to be a good addition to his zymographic "cookery."

The starch gel blocks routinely used for analytic electrophoresis may be easily adapted for semipreparative enzyme electrophoresis. After completion of electrophoretic separation the position of a desired enzyme on the gel is located by an appropriate detection method, and the gel area occupied by the enzyme is cut out. The use of starch gel as the supporting medium is preferable because a simple "freezing–squeezing" procedure may be used to remove the enzyme from the gel matrix. This procedure, for example, was successfully used by the author to obtain preparation of isocitrate dehydrogenase involved as a linking enzyme in an aconitase stain (see 4.2.1.3 — ACON, Method 1). A more effective method, electrophoretic elution, may also be used to remove enzymes from a gel matrix, but it is more time-consuming and requires an additional device.[5]

An electrophoretic procedure very similar to that used for electrophoretic transfer of proteins may sometimes be even more effective than that described above. In this case proteins move in the transverse direction after being applied on the upper surface of a starch gel block by the agar overlay method. After an appropriate period of time electrophoresis is stopped. The position of the desired linking enzyme within the gel block is located by a specific detection method. For this purpose the gel slice obtained by transverse slicing through the gel block is used. A longitudinal slice is then made through the zone occupied by the enzyme, and two reactive gel plates with enzyme-containing cut surfaces are obtained. The staining solution containing all necessary ingredients except the linking enzyme is applied dropwise to the surface of the gel to be stained. The reactive gel plates are then placed above, the cut surfaces down, avoiding formation of air bubbles. The linking enzyme molecules embedded in the gel matrix of the reactive plates usually work sufficiently well due to the diffusion of small molecules of reagents involved in the stain and those produced by the enzyme under detection. This variant of the method, for example, was successfully used by the author to obtain the reactive starch gel plates containing glucose–phosphate isomerase involved in the mannose–phosphate isomerase stain (see 5.3.1.8 — MPI) as the linking enzyme.

Of course, the use of semipreparative gel electrophoresis for production of auxiliary enzymes is not appropriate for large-scale isozyme survey; however, it provides a good opportunity for new developments in electrophoretic zymography. This tactic should not be considered now as an effective resource-saving one. Rather, it can be used to help develop and test new enzyme detection methods and thus is of great potential value.

TROUBLESHOOTING

During practical application of enzyme detection methods some problems concerning the enzyme staining originate from time to time. One may hope that a good knowledge of the detection mechanisms given in Part II will help to solve these problems. Nevertheless, some useful advice concerning stain troubleshooting is given below. These pieces of advice are organized into a list including the most common causes of stain failures and recommendations for their avoidance:

- **Some reagents are erroneously omitted from the staining solution:** Check the recipe. Add missing reagents or prepare a new solution.
- **Unstable reagents deteriorated during storage under inappropriate conditions:** First of all, check the reactivity of the linking enzymes, if involved, by a rough qualitative enzyme assay using appropriate staining solutions. Replace linking enzyme preparations if they have deteriorated. If the stain still does not work, check which reagents of the stain other than the linking enzymes are not functional. This may be done through simple analysis of functioning of other stains that include reagents common to the stain under question, or through one-by-one replacement of suspected reagents with fresh ones. If reserve vials of suspect reagents are not available, try to use suitable enzyme preparations and substrates to produce suspect reagents enzymatically. If available, use any alternative detection method to be sure that the enzyme under detection is active in your samples. Use control samples that show activity of the enzyme under detection with certainty. Follow storage instructions depicted on containers. Store most chemicals in a refrigerator or freezer in a special airtight box with a desiccant. Do not leave refrigerated

chemicals open at room temperature. Hydrolysis following water condensation is particularly damaging to many compounds. To prevent condensation, take the chemicals out of the refrigerator about half an hour before use or warm the containers in your hands before opening. On the contrary, keep linking enzyme preparations at room temperature for as short a time as possible. Remember that some reagents are light sensitive and that high temperature also contributes to the breakdown of many reagents.

- **Intermediate or final pH value of staining solution deviates from that recommended in the recipe:** Check intermediate pH values during preparation of the staining solution and final pH before application. Prepare a new staining buffer if it has deteriorated. Use sodium or other salts rather than free strong acids or bases if they are involved in the stain. Always add reagents in the sequence given in the staining solution recipe.
- **Enzyme inhibitors are present in used water:** If the stain works well in laboratory conditions and does not work under field conditions, the bad quality of the water used is the first candidate for the reason of the stain failure. Use filtered rain or thawed snow water in the field and bidistilled or deionized water in the laboratory.

It also should be kept in mind that many errors that have been made during former stages of enzyme electrophoretic analysis (e.g., storage, preparation, and electrophoretic run of samples) manifest themselves only during the stage of enzyme detection. For example, the enzyme to be detected may be inactivated during tissue storage under inappropriate conditions (e.g., after freezing–thawing events) or may run off the anodal or cathodal end of the gel (e.g., when the enzyme moves quickly toward the anode or cathode or the running time is too long).

Some other hints for troubleshooting, especially for stages other than enzyme staining, may be found in the excellent handbook for allozyme electrophoresis by Richardson et al.[4]

SAFETY REGULATIONS

A large array of compounds is used in enzyme-visualizing procedures. A number of these compounds are known to be carcinogens, skin irritants, or poisons. Many of these compounds have not even been tested yet for their long-term effects on human health and heredity. Thus, it is good practice for those dealing with such a diverse group of chemicals to follow the general safety codex given below:

- Treat all chemicals as potentially hazardous to your health.
- Wear gloves and a dust mask when handling chemicals.
- Do not breathe dust and vapors; mix stains in a vented hood.

- Avoid contaminating other things with the chemicals being used.
- Read all labels on containers and follow the prescribed safety measures.
- Be careful when handling stained gels and disposing of used staining solutions.
- Thoroughly clean up any spills on your chemical table.
- Wash your hands after each handling.
- Do not keep food in a refrigerator together with chemicals.

REFERENCES

1. Brewer, G.J., *An Introduction to Isozyme Techniques*, Academic Press, New York, 1970.
2. Harris, H. and Hopkinson, D.A., *Handbook of Enzyme Electrophoresis in Human Genetics*, North-Holland, Amsterdam, 1976 (loose-leaf, with supplements in 1977 and 1978).
3. Gordon, A.H., *Electrophoresis of Proteins in Polyacrylamide and Starch Gels*, North-Holland, Amsterdam, 1980.
4. Richardson, B.J., Baverstock, P.R., and Adams, M., *Allozyme Electrophoresis: A Handbook for Animal Systematics and Population Studies*, Academic Press, Sydney, 1986.
5. Andrews, A.T., *Electrophoresis: Theory, Techniques, and Biochemical and Clinical Applications*, Clarendon Press, Oxford, 1988.
6. Conkle, M.T., Hodgskiss, P.D., Nunnally, L.B., and Hunter, S.C., *Starch Gel Electrophoresis of Conifer Seeds: A Laboratory Manual*, General Technical Report PSW-64, Pacific Southwest Forest and Range Experiment Station, Berkeley, CA, 1982.
7. Cheliak, W.M. and Pitel, J.A., *Techniques for Starch Gel Electrophoresis of Enzymes from Forest Tree Species*, Information Report PI-X-42, Petawawa National Forestry Institute, Canadian Forestry Service, Agriculture Canada, Chalk River, Ontario, 1984.
8. Morizot, D.C. and Schmidt, M.E., Starch gel electrophoresis and histochemical visualization of proteins, in *Electrophoretic and Isoelectric Focusing Techniques in Fisheries Management*, Whitmore, D.H., Ed., CRC Press, Boca Raton, FL, 1990, p. 23.
9. Murphy, R.W., Sites, J.W., Jr., Buth, D.G., and Haufler, C.H., Proteins: isozyme electrophoresis, in *Molecular Systematics*, Hillis, D.M., Moritz, C., and Mable, B.K., Eds., Sinauer, Sunderland, MA, 1996, p. 51.
10. Ghosh, R. and Quayle, J.R., Phenazine ethosulfate as a preferred electron acceptor to phenazine methosulfate in dye linked enzyme assays, *Anal. Biochem.*, 99, 112, 1980.
11. Andersen, E. and Christensen, K., Superoxide dismutase (E.C. 1.15.1.1) polymorphism in various breeds of dogs, *Anim. Blood Groups Biochem. Genet.*, 8 (Suppl. 1), 27, 1977.
12. Pérez-de la Mora, M., Méndez-Franco, J., Salceda, R., and Riesgo-Escovar, J.R., A glutamate dehydrogenase-based method for the assay of L-glutamic acid: formation of pyridine nucleotide fluorescent derivatives, *Anal. Biochem.*, 180, 248, 1989.
13. Valero, E. and Garcia-Carmona, F., A continuous spectrophotometric method based on enzymatic cycling for determining L-glutamate, *Anal. Biochem.*, 259, 265, 1998.
14. Sugiyama, A. and Lurie, K.G., An enzymatic fluorometric assay for adenosine 3',5'-monophosphate, *Anal. Biochem.*, 218, 20, 1994.
15. Mohamed, M.A., Lerro, K.A., and Prestwich, G.D., Polyacrylamide gel miniaturization improves protein visualization and autoradiographic detection, *Anal. Biochem.*, 177, 287, 1989.
16. Manchenko, G.P., Subunit structure of enzymes: allozymic data, *Isozyme Bull.*, 21, 144, 1988.
17. Ahmad, M., Skibinsky, D.O.F., and Beardmore, J.A., An estimate of genetic variation in the common mussel *Mytilus edulis*, *Biochem. Genet.*, 15, 833, 1977.
18. Fujio, Y. and Kato, Y., Genetic variation in fish populations, *Bull. Jpn. Soc. Sci. Fish.*, 45, 1169, 1979.
19. Fujio, Y., Yamanaka, R., and Smith, P.J., Genetic variation in marine molluscs, *Bull. Jpn. Soc. Sci. Fish.*, 49, 1809, 1983.
20. Somer, H., "Nothing dehydrogenase" reaction as an artefact in serum isoenzyme analyses, *Clin. Chim. Acta*, 60, 223, 1975.
21. Sri Venugopal, K.S. and Adiga, P.R., Artifactual staining of proteins on polyacrylamide gels by nitro blue tetrazolium chloride and phenazine methosulfate, *Anal. Biochem.*, 101, 215, 1980.
22. Zimniak-Przybylska, Z. and Przybylska, J., Interference of *Pisum* seed albumins with detecting amylase activity on electropherograms: an apparent relationship between protein patterns and amylase zymograms, *Genet. Pol.*, 17, 133, 1976.
23. Zimniak-Przybylska, Z. and Przybylska, J., Relationships between electrophoretic seed protein patterns and zymograms of amylases and acid phosphatases in *Pisum*, *Genet. Pol.*, 15, 435, 1974.
24. Hardin, E., Passey, R.B., and Fuller, J.B., Artifactual alkaline phosphatase isoenzyme band, caused by bilirubin, on cellulose acetate electropherograms, *Clin. Chem.*, 24, 178, 1978.
25. Onoufriou, A. and Alahiothis, S.N., Enzyme specificity: "pseudopolymorphism" of lactate dehydrogenase in *Drosophila melanogaster*, *Biochem. Genet.*, 19, 277, 1981.
26. Gill, P.D., Nongenetic variation in isoenzymes of lactate dehydrogenase of *Cepaea nemoralis*, *Comp. Biochem. Physiol.*, 59B, 271, 1978.
27. Manchenko, G.P., The use of allozymic patterns in genetic identification of enzymes, *Isozyme Bull.*, 23, 102, 1990.
28. NC-IUBMB, *Enzyme Nomenclature*, Academic Press, San Diego, 1992.
29. Manchenko, G.P., Allozymic variation and substrate specificity of phosphatase from sipunculid *Phascolosoma japonicum*, *Isozyme Bull.*, 21, 168, 1988.
30. Vallejos, C.E., Enzyme activity staining, in *Isozymes in Plant Genetics and Breeding*, Part A, Tanskley, S.D. and Orton, T.J., Eds., Elsevier, Amsterdam, 1983, p. 469.

31. Shaklee, J.B. and Keenan, C.P., *A Practical Laboratory Guide to the Techniques and Methodology of Electrophoresis and Its Application to Fish Fillet Identification*, Report 177, CSIRO Marine Laboratories, Hobart, Australia, 1986.

32. Siciliano, M.J. and Shaw, C.R., Separation and visualization of enzymes on gels, in *Chromatographic and Electrophoretic Techniques*, Vol. 2, *Zone Electrophoresis*, Smith, I., Ed., Heinemann, London, 1976, p. 185.

33. Numachi, K.-I., A simple method for preservation and scanning of starch gels, *Biochem. Genet.*, 19, 233, 1981.

34. Remtulla, M.A. and Boyde, T.R.C., Method for the preservation of polyacrylamide slabs after electrophoresis, *Lab. Pract.*, 23, 484, 1974.

35. Buth, D.G., Genetic principles and the interpretation of electrophoretic data, in *Electrophoretic and Isoelectric Focusing Techniques in Fisheries Management*, Whitmore, D.H., Ed., CRC Press, Boca Raton, FL, 1990, p. 1.

36. McLellan, T. and Ramshaw, J.A.M., Serial electrophoretic transfers: a technique for the identification of numerous enzymes from single polyacrylamide gels, *Biochem. Genet.*, 19, 647, 1981.

SECTION 3
ENZYME SHEETS

ABBREVIATIONS COMMONLY USED IN THE ENZYME SHEETS

ADP	Adenosine 5'-diphosphate
AMP	Adenosine 5'-monophosphate
ATP	Adenosine 5'-triphosphate
CDP	Cytidine 5'-diphosphate
CMP	Cytidine 5'-monophosphate
CoA	Coenzyme A
CTP	Cytidine 5'-triphosphate
DNA	Deoxyribonucleic acid
EDTA	Ethylenediaminetetraacetic acid
FAD	Flavin–adenine dinucleotide
FMN	Flavin mononucleotide
GDP	Guanosine 5'-diphosphate
GMP	Guanosine 5'-monophosphate
GTP	Guanosine 5'-triphosphate
IMP	Inosine 5'-monophosphate
INT	p-Iodonitrotetrazolium Violet
ITP	Inosine 5'-triphosphate
MTT	Methyl thiazolyl tetrazolium (thiazolyl blue)
NAD	Nicotinamide–adenine dinucleotide (oxidized)
NADH	Nicotinamide–adenine dinucleotide (reduced)
NADP	Nicotinamide–adenine dinucleotide phosphate (oxidized)
NADPH	Nicotinamide–adenine dinucleotide phosphate (reduced)
NBT	Nitro blue tetrazolium
PAG	Polyacrylamide gel
PMS	Phenazine methosulfate
RNA	Ribonucleic acid
SDS	Sodium dodecyl sulfate
Tris	Tris(hydroxymethyl) aminomethane
UDP	Uridine 5'-diphosphate
UMP	Uridine 5'-monophosphate
UTP	Uridine 5'-triphosphate
UV	Ultraviolet

1.1.1.1 — Alcohol Dehydrogenase; ADH

OTHER NAMES	Aldehyde reductase
REACTION	Alcohol + NAD = aldehyde or ketone + NADH
ENZYME SOURCE	Bacteria, fungi, plants, protozoa, invertebrates, vertebrates
SUBUNIT STRUCTURE	Dimer (fungi, plants, invertebrates, vertebrates)

METHOD 1

Visualization Scheme

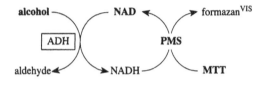

Staining Solution[1] (modified)

0.05 M Tris–HCl buffer, pH 8.5	50 ml
Ethanol	2 ml
NAD	40 mg
MTT	10 mg
PMS	1 mg

Procedure

Incubate the gel in staining solution in the dark at 37°C until dark blue bands appear. Rinse stained gel in water and fix in 25% ethanol.

Notes: A large number of primary and secondary straight- and branched-chain aliphatic and aromatic alcohols and hemiacetals may be used as substrates to examine additional ADH isozymes. The animal enzyme acts on cyclic secondary alcohols while the yeast enzyme does not.

Some ADH isozymes from various invertebrate and vertebrate species display high substrate specificity to octanol. These ADH forms are included in the enzyme list under the separate name of octanol dehydrogenase (see 1.1.1.73 — ODH).[2]

ADH bands often appear on other NAD-dependent dehydrogenase zymograms (see 1.X.X.X — NDH). Pyrazole is usually used as a specific inhibitor of mammalian ADH. The addition of pyrazole (final concentration of 1 to 2 mg/ml) does not interfere with any of the stains so far tested.[3]

Some ADH allozymes of *Drosophila melanogaster* display specificity toward sarcosine, choline, and dihydroorotate.[4]

METHOD 2

Visualization Scheme

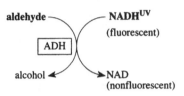

Staining Solution[5]

A.	0.05 M Tris–phosphate buffer, pH 7.0	25 ml
	Acetaldehyde	0.1 ml
	NADH	10 mg
B.	2% Agar solution (60°C)	25 ml

Procedure

Mix A and B components of the staining solution and pour the mixture over the surface of the gel. Incubate the gel at 37°C and view under long-wave UV light after 10 min to about 1 h. When dark bands are clearly visible on a light (fluorescent) background, photograph the gel using a yellow filter.

To make these bands visible in daylight (white bands on blue background), cover the processed gel with a second 1% agar overlay containing MTT (or NBT) and PMS.

Notes: If too much NADH is added to the staining solution, the enzyme should not be able to convert enough NADH into NAD for nonfluorescent bands to become visible.

Many different aldehydes may be used as substrates to examine NADH-dependent aldehyde reductase activity of ADH.

GENERAL NOTES

Human liver cytosol ADH ($\gamma\gamma$ isozyme) was identified as the only enzyme protein demonstrating NAD-dependent iso-bile acid 3 β-hydroxysteroid dehydrogenase activity in the human liver.[6]

Structural phylogenetic analysis suggests that glutathione-dependent formaldehyde dehydrogenase (see 1.2.1.1 — FDHG) is a progenitor of plant and animal ADH.[7]

REFERENCES

1. Brewer, G.J., *An Introduction to Isoenzyme Techniques*, Academic Press, New York, 1970, p. 117.
2. NC-IUBMB, *Enzyme Nomenclature*, Academic Press, San Diego, 1992, p. 33 (EC 1.1.1.73).
3. Richardson, B.J., Baverstock, P.R., and Adams, M., *Allozyme Electrophoresis: A Handbook for Animal Systematics and Population Studies*, Academic Press, Sydney, 1986, p. 221.
4. Eisses, K.Th., Schoonen, G.E.J., Scharloo, W., and Thörig, G.E.W., Evidence for a multiple function of the alcohol dehydrogenase allozyme ADH[71k] of *Drosophila melanogaster*, *Comp. Biochem. Physiol.*, 82B, 863, 1985.
5. Smith, M., Hopkinson, D.A., and Harris, H., Studies on the properties of the human alcohol dehydrogenase isozymes determined by the different loci ADH_1, ADH_2, ADH_3, *Ann. Hum. Genet.*, 37, 49, 1973.
6. Marschall, H.U., Oppermann, U.C.T., Svensson, S., Nordling, E., Persson, B., Hoog, J.J.O., and Jornvall, H., Human liver class I alcohol dehydrogenase γγ isozyme: the sole cytosolic 3 β-hydroxysteroid dehydrogenase of iso-bile acids, *Hepatology*, 31, 990, 2000.
7. Fliegmann, J. and Sandermann, H., Maize glutathione-dependent formaldehyde dehydrogenase cDNA: a novel plant gene of detoxification, *Plant Mol. Biol.*, 34, 843, 1997.

1.1.1.2 — Alcohol Dehydrogenase (NADP); ADH(NADP)

OTHER NAMES	Aldehyde reductase (NADPH)
REACTION	Alcohol + NADP = aldehyde + NADPH
ENZYME SOURCE	Bacteria, invertebrates, vertebrates
SUBUNIT STRUCTURE	Monomer (plants, invertebrates)

METHOD 1

Visualization Scheme

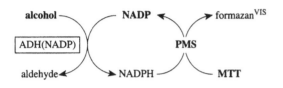

Staining Solution[1]

0.1 M Glycine–NaOH buffer, pH 9.5	50 ml
NADP	20 mg
NBT	20 mg
PMS	2 mg
1,2-Propanediol	5 ml

Procedure

Incubate the gel in staining solution in the dark at 37°C until dark blue bands appear. Rinse the stained gel in water and fix in 25% ethanol.

Notes: Many different alcohols may be used as substrates for ADH(NADP). Bacterial enzymes oxidize only primary alcohols, while others act also on secondary alcohols. Some ADH(NADP) isozymes of invertebrate and vertebrate species display relatively high specificity toward certain alcohols and aldehydes and are included in the enzyme list under separate names (see 1.1.1.19 — GLR; 1.1.1.72 — GLYD; 1.1.1.80 — IPDH).

METHOD 2

Visualization Scheme

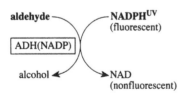

Staining Solution[2]

A.	0.2 M Sodium–phosphate buffer, pH 7.0	8 ml
	5 mg/ml NADPH	5 ml
	10 mM p-Nitrobenzaldehyde	5 ml
B.	1.4% Agar solution (60°C)	20 ml

Procedure

Mix A and B components of the staining solution and pour the mixture over the surface of the gel. Incubate the gel at 30°C and view under long-wave UV light. When dark bands are clearly visible on a light (fluorescent) background, photograph the gel using a yellow filter. To make these bands visible in daylight (white bands on blue background), cover the processed gel with a second 1% agar overlay containing MTT (or NBT) and PMS. Wash the negatively stained gel in water and fix in 25% ethanol.

Notes: Many different aldehydes may be used as substrates to examine the NADPH-dependent aldehyde reductase activity of ADH(NADP). Some NADPH-dependent aldehyde reductases catalyze the reaction only in one (reductase) direction. For example, it is the case for the mouse liver enzyme highly specific to p-nitrobenzaldehyde. Like mammalian ADH(NADP), the enzyme from *Drosophila* displays highest activity toward acet-aldehyde and p-nitrobenzaldehyde. It also demonstrates high activity with D-glyceraldehyde, o-nitrobenzaldehyde, and pyridine-3-carboxaldehyde as substrates.[3] The amount of NADPH added to the staining solution is critical for the development of nonfluorescent bands of ADH(NADP). If too much NADPH is added to the staining solution, the enzyme should not be able to convert enough NADPH into NADP for nonfluorescent bands to become visible.

GENERAL NOTES

Aldehyde reductase is a generic term used to describe the activities of a number of enzymes catalyzing the reduction of various aliphatic, aromatic, and biogenic aldehydes in the presence of NADPH. Much of the problem in identifying aldehyde reductase forms appears to lie in conflicts in terminology, arising from overlapping substrate specificity and an absence of agreement about the physiological roles of these enzymes. Clear classification of aldehyde reductases requires further biochemical and genetic studies.[2]

REFERENCES

1. Turner, A.J. and Tipton, K.F., The characterization of two reduced nicotinamide–adenine dinucleotide phosphate-linked aldehyde reductases from pig brain, *Biochem. J.*, 130, 765, 1972.
2. Duley, J.A. and Holmes, R.S., Biochemical genetics of aldehyde reductase in the mouse: *Ahr-1* — a new locus linked to the alcohol dehydrogenase gene complex on chromosome 3, *Biochem. Genet.*, 20, 1067, 1982.
3. Atrian, S. and Gonzàlez-Duarte, R., An aldo-keto reductase activity in *Drosophila melanogaster* and *Drosophila hydei*: a possible function in alcohol metabolism, *Comp. Biochem. Physiol.*, 81B, 949, 1985.

1.1.1.3 — Homoserine Dehydrogenase; HSDH

REACTION   L-Homoserine + NAD(P) = L-aspartate 4-semialdehyde + NAD(P)H

ENZYME SOURCE   Bacteria, fungi, plants, invertebrates (insects)

SUBUNIT STRUCTURE   Dimer# (fungi, plants); also see *Notes*

METHOD

Visualization Scheme

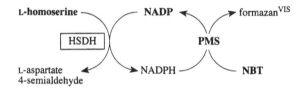

Staining Solution[1]

100 mM Tris–HCl buffer, pH 8.9
50 mM L-Homoserine
1 mM NADP
0.25 mg/ml NBT
0.025 mg/ml PMS
400 mM KCl

Procedure

Incubate the gel in staining solution in the dark at 37°C until dark blue bands appear. Wash the developed gel in water and fix in 25% ethanol.

Notes: The enzyme from some sources (e.g., yeast) acts most rapidly with NAD, while from others (e.g., *Neurospora*) with NADP. The enzyme from *E. coli* catalyzes the phosphorylation of L-aspartate (the reaction of EC 2.7.2.4) in addition to its homoserine dehydrogenase activity.[2] The enzyme from spinach leaves is a tetramer that also exhibits aspartate kinase (EC 2.7.2.4) activity.[3]

REFERENCES

1. Ogilvie, J.M., Sightler, J.H., and Clark, R.B., Homoserine dehydrogenase of *Escherichia coli* K 12 λ: I. Feedback inhibition by L-threonine and activation by potassium ions, *Biochemistry*, 8, 3557, 1969.
2. NC-IUBMB, *Enzyme Nomenclature*, Academic Press, San Diego, 1992, p. 24 (EC 1.1.1.3, Comments).
3. Pavagi, S., Kochhar, S., Kochhar, V.K., and Sane, P.V., Purification and characterization of homoserine dehydrogenase from spinach leaves, *Biochem. Mol. Biol. Int.*, 36, 649, 1995.

1.1.1.6 — Glycerol Dehydrogenase; GLYCD

REACTION   Glycerol + NAD = glycerone + NADH

ENZYME SOURCE   Bacteria, fungi

SUBUNIT STRUCTURE   Tetramer# (bacteria), dimer# (fungi)

METHOD

Visualization Scheme

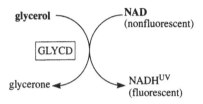

Staining Solution[1]

A. 0.2 M Tris–HCl, pH 8.0
    1 mM NAD
B. 0.2 M Tris–HCl, pH 8.0
    0.1 M Glycerol

Procedure

Following electrophoresis, soak the gel in solution A for ca. 3 min. Then rinse the gel quickly in water to remove any cofactor from the gel surface, and after a few minutes, view under long-wave UV light to detect any nonsubstrate-dependent fluorescent bands. Overlay the gel with a filter paper strip soaked in solution B. Observe fluorescent bands of GLYCD in UV light. Record the zymogram or photograph using a yellow filter.

Notes: When a zymogram that is visible in daylight is required, counterstain the processed gel with MTT–PMS solution to obtain blue bands in the gel areas where GLYCD activity is present. Fix the stained gel in 25% ethanol.

REFERENCES

1. Seymour, J.L. and Lazarus, R.A., Native gel activity stain and preparative electrophoretic method for the detection and purification of pyridine nucleotide-linked dehydrogenases, *Anal. Biochem.*, 178, 243, 1989.

1.1.1.8 — Glycerol-3-Phosphate Dehydrogenase (NAD); G-3-PD

OTHER NAMES	α-Glycerophosphate dehydrogenase
REACTION	*sn*-Glycerol-3-phosphate + NAD = glycerone phosphate + NADH
ENZYME SOURCE	Bacteria, fungi, protozoa, invertebrates, vertebrates
SUBUNIT STRUCTURE	Dimer (fungi, invertebrates, vertebrates)

METHOD

Visualization Scheme

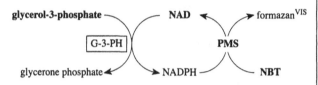

Staining Solution[1]

0.1 *M* Tris–HCl buffer, pH 8.0	85 ml
NAD	50 mg
NBT	30 mg
PMS	2 mg
1 *M* Glycerol-3-phosphate, pH 7.0	10 ml
0.1 *M* NaCN	5 ml

Procedure

Incubate the gel in staining solution in the dark at 37°C until dark blue bands appear. Wash the developed gel in water and fix in 25% ethanol.

Notes: NaCN is included in the stain to prevent development of superoxide dismutase (see 1.15.1.1 — SOD) bands, which can interfere with G-3-PD bands. Pyruvate and pyrazole may be added in a staining solution to a final concentration of 1 to 2 mg/ml of each to inhibit nonspecific NAD-into-NADH conversion by LDH and ADH, respectively. The enzyme activity is highly dependent on substrate concentration. When a low concentration of glycerol-3-phosphate is used, the product glycerone phosphate, which inhibits the forward G-3-PD reaction, should be taken away using two exogenous linking enzymes: triose-phosphate isomerase (5.3.1.1 — TPI) and glyceraldehyde-3-phosphate dehydrogenase (1.2.1.12 — GA-3-PD). The latter enzyme doubles NADH production and formation of the formazan. Only those preparations of the linking enzymes that are substantially free of G-3-PD impurity should be used for this purpose.

REFERENCE

1. Shaw, C.R. and Prasad, R., Starch gel electrophoresis of enzymes: a compilation of recipes, *Biochem. Genet.*, 4, 297, 1970.

1.1.1.10 — L-Xylulose Reductase; XR

REACTION	Xylitol + NADP = L-xylulose + NADPH
ENZYME SOURCE	Vertebrates
SUBUNIT STRUCTURE	Tetramer (vertebrates)

METHOD

Visualization Scheme

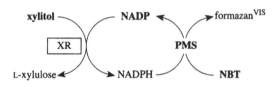

Staining Solution[1]

0.1 *M* Tris–HCl buffer, pH 8.0	50 ml
Xylitol	150 mg
NADP	10 mg
NBT	10 mg
PMS	1.5 mg

Procedure

Incubate the gel in staining solution in the dark at 37°C until dark blue bands appear. Wash the stained gel in water and fix in 25% ethanol.

REFERENCES

1. Bell, L.J., Moyer, J.T., and Numachi, K.-I., Morphological and genetic variation in Japanese populations of the anemonefish *Amphiprion clarkii*, *Mar. Biol.*, 72, 99, 1982.

OTHER NAMES   L-Iditol 2-dehydrogenase (recommended name), polyol dehydrogenase

REACTIONS   1. L-Iditol + NAD = L-sorbose + NADH

                     2. D-Sorbitol + NAD = D-fructose + NADH

ENZYME SOURCE   Bacteria, plants, invertebrates, vertebrates

SUBUNIT STRUCTURE   Monomer (invertebrates), dimer (invertebrates), tetramer (invertebrates, vertebrates)

METHOD

Visualization Scheme

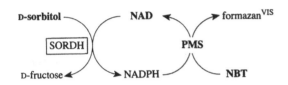

Staining Solution[1]

0.05 M Tris–HCl buffer, pH 8.0	100 ml
D-Sorbitol	500 mg
NAD	10 mg
NBT	15 mg
PMS	2 mg

Procedure

Incubate the gel in staining solution in the dark at 37°C until dark blue bands appear. Wash the stained gel in water and fix in 25% ethanol.

Notes: Pyruvate and pyrazole may be included in a staining solution to a final concentration of 1 to 2 mg/ml of each to inhibit LDH and ADH, respectively, and to prevent appearance of nonspecifically stained bands. The enzyme from some sources (e.g., *Candida utilis*) is more active with xylitol. Many sugars may be used as substrates, but NAD cannot be replaced by NADP. NADP-dependent SORDH activity has been reported only in *Drosophila melanogaster*.[2]

REFERENCES

1. Lin, C.C., Schipmann, G., Kittrell, W.A., and Ohno, S., The predominance of heterozygotes found in wild goldfish of Lake Erie at the gene locus for sorbitol dehydrogenase, *Biochem. Genet.*, 3, 603, 1969.
2. Bischoff, W.L., Ontogeny of sorbitol dehydrogenases in *Drosophila melanogaster*, *Biochem. Genet.*, 16, 485, 1978.

REACTION   D-Mannitol-1-phosphate + NAD = D-fructose-6-phosphate + NADH

ENZYME SOURCE   Bacteria, fungi

SUBUNIT STRUCTURE   Tetramer[#] (bacteria)

METHOD 1

Visualization Scheme

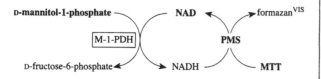

Staining Solution[1]

0.2 M Tris–HCl buffer, pH 8.0	50 ml
Mannitol-1-phosphate	5 mg
NAD	20 mg
MTT	12.5 mg
PMS	5 mg

Procedure

Incubate the gel in staining solution in the dark at 37°C until dark blue bands appear. Wash the stained gel in water and fix in 25% ethanol.

METHOD 2

Visualization Scheme

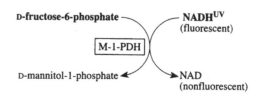

Staining Solution*

A.	0.02 M Tris–HCl buffer, pH 8.0	25 ml
	D-Fructose-6-phosphate (Ba salt)	20 mg
	NADH	10 mg
B.	2% Agarose solution (60°C)	25 ml

Procedure

Mix A and B components of the staining solution and pour over the surface of the gel. Incubate the gel at 37°C and view under long-wave UV light after 10 to 60 min. When dark bands are clearly visible on a light (fluorescent) background, photograph the gel using a yellow filter. To make these bands visible in daylight (white bands on blue background), cover the processed gel with a second 1% agarose overlay containing MTT (or NBT) and PMS. Wash the negatively stained gel in water and fix in 25% ethanol.

Notes: The amount of NADH added to the staining solution is critical for development of M-1-PDH bands. If too much NADH is added, the enzyme should not be able to convert enough NADH into NAD for non-fluorescent bands to become visible.

The staining solution recommended in Method 2 is about 100 times less expensive than that in Method 1.

GENERAL NOTES

The enzyme is highly specific to both the substrate and the cofactor.

REFERENCES

1. Selander, R.K., Caugant, D.A., Ochman, H., Musser, J.J.M., Gilmour, M.N., and Whittam, T.S., Methods of multilocus enzyme electrophoresis for bacterial population genetics and systematics, *Appl. Environ. Microbiol.*, 51, 873, 1986.

* New; recommended for use.

1.1.1.19 — Glucuronate Reductase; GLR

OTHER NAMES	Hexonate dehydrogenase
REACTION	L-Gulonate + NADP = D-glucuronate + NADPH
ENZYME SOURCE	Vertebrates
SUBUNIT STRUCTURE	Unknown or no data available

METHOD

Visualization Scheme

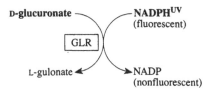

Staining Solution[1]

0.03 M Potassium phosphate buffer, pH 7.0	5 ml
D-Glucuronate (Na salt)	5 mg
NADPH	1.5 mg

Procedure

Apply the staining solution to the gel on a filter paper overlay, and after incubation at 37°C (30 to 60 min), view the gel under long-wave UV light. Dark (nonfluorescent) bands of GLR are visible on the light (fluorescent) background of the gel. Photograph the developed gel using a yellow filter.

These bands may be made visible in daylight (white bands on blue background) after treating the processed gel with MTT–PMS solution applied as a filter paper or 1% agarose overlay.

Notes: The amount of NADPH added to the staining solution is critical for development of nonfluorescent GLR bands. If too much NADPH is added, the enzyme should not be able to convert enough NADPH into NADP for nonfluorescent bands to become visible.

The enzyme also reduces D-galacturonate. A mammalian enzyme may be identical with NADP-dependent alcohol dehydrogenase (see 1.1.1.2 — ADH(NADP), Method 1, *Notes*, and General Notes).

REFERENCES

1. Baker, C.M.A. and Manwell, C., Heterozygosity of the sheep: polymorphism of "malic enzyme," isocitrate dehydrogenase (NADP+), catalase and esterase, *Aust. J. Biol. Sci.*, 30, 127, 1977.

1.1.1.22 — UDPglucose 6-Dehydrogenase; UGDH

REACTION	UDPglucose + 2 NAD + H_2O = UDPglucuronate + 2 NADH
ENZYME SOURCE	Invertebrates, vertebrates
SUBUNIT STRUCTURE	Unknown or no data available

METHOD

Visualization Scheme

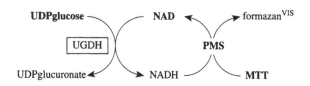

Staining Solution[1]

0.03 M Potassium phosphate buffer, pH 7.0	10 ml
UDPglucose	2 mg
NAD	4 mg
PMS	1 mg
MTT	10 mg

Procedure

Apply the staining solution to the gel on a filter paper overlay and incubate at 37°C in a dark, moistened chamber until dark blue bands appear. Wash the developed gel in water and fix in 25% ethanol.

Notes: Staining solution lacking PMS and MTT may also be used to detect UGDH activity bands (fluorescent bands visible in long-wave UV light). The fluorescent detection system is more sensitive than the PMS–MTT system because of the presumed inhibitory effect of PMS and MTT on UGDH activity.[1]

The reaction catalyzed by UGDH is essentially irreversible.

REFERENCES

1. Baker, C.M.A. and Manwell, C., Heterozygosity of the sheep: polymorphism of "malic enzyme," isocitrate dehydrogenase (NADP+), catalase and esterase, *Aust. J. Biol. Sci.*, 30, 127, 1977.

1.1.1.23 — Histidinol Dehydrogenase; HISDH

REACTION	L-Histidinol + 2 NAD = L-histidine + 2 NADH (see *Notes*)
ENZYME SOURCE	Bacteria, fungi, plants
SUBUNIT STRUCTURE	Dimer* (bacteria, plants)

METHOD

Visualization Scheme

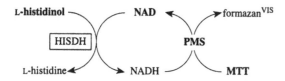

Staining Solution[1]

0.1 M Tris–HCl buffer, pH 8.0	50 ml
L-Histidinol	40 mg
NAD	20 mg
MTT	10 mg
PMS	1 mg

Procedure

Incubate the gel in staining solution in the dark at 37°C until dark blue bands appear. Wash the developed gel in water and fix in 25% ethanol.

Notes: L-Histidinol is oxidized in two steps:

1. L-Histidinol + NAD = L-histidinal + NADH
2. L-Histidinal + NAD = L-histidine + NADH

GENERAL NOTES

The *Neurospora* enzyme also catalyzes the reactions of EC 3.5.4.19 and EC 3.6.1.31.[2]

REFERENCES

1. Creaser, E.H., Bennett, D.J., and Drysdale, R.B., The purification and properties of histidinol dehydrogenase from *Neurospora crassa*, *Biochem. J.*, 103, 36, 1967.
2. NC-IUBMB, *Enzyme Nomenclature*, Academic Press, San Diego, 1992, p. 27 (EC 1.1.1.23, Comments).

1.1.1.25 — Shikimate 5-Dehydrogenase; SHDH

REACTION	Shikimate + NADP = 5-dehydro-shikimate + NADPH
ENZYME SOURCE	Bacteria, fungi, plants
SUBUNIT STRUCTURE	Monomer (plants), dimer (fungi)

METHOD

Visualization Scheme

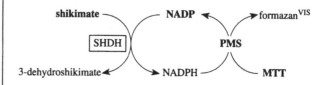

Staining Solution[1]

0.1 M Tris–HCl buffer, pH 7.1	30 ml
Shikimic acid	25 mg
NADP	5 mg
MTT	6 mg
PMS (see Procedure)	1 mg

Procedure

Incubate the gel 30 min at 37°C in the dark in a staining solution lacking PMS; then add PMS and incubate until dark blue bands appear. Wash the developed gel in water and fix in 4% formaldehyde or 25% ethanol.

REFERENCES

1. Van Dijk, H. and Van Delden, W., Genetic variability in *Plantago* species in relation to their ecology: Part 1: genetic analysis of the allozyme variation in *P. major* subspecies, *Theor. Appl. Genet.*, 60, 285, 1981.

1.1.1.27 — L-Lactate Dehydrogenase; LDH

OTHER NAMES   Lactic acid dehydrogenase

REACTION   L-Lactate + NAD = pyruvate + NADH

ENZYME SOURCE   Bacteria, fungi, plants, protozoa, invertebrates, vertebrates

SUBUNIT STRUCTURE   Monomer (invertebrates), dimer (protozoa, invertebrates), tetramer (plants, invertebrates, vertebrates)

METHOD

Visualization Scheme

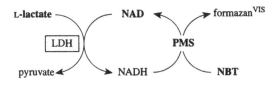

Staining Solution[1]

0.2 M Tris–HCl buffer, pH 8.0	60 ml
0.5 M L-Lactate, pH 8.0	12 ml
10 mg/ml NAD	2.7 ml
1 mg/ml NBT	6.7 ml
1 mg/ml PMS	6.7 ml

Procedure

Incubate the gel in staining solution in the dark at 37°C until dark blue bands appear. Wash the developed gel in water and fix in 25% ethanol.

Notes: The LDH activity bands can also be detected, although weakly, even in the absence of lactate from the staining solution.[2] This is the so-called "nothing dehydrogenase" (see 1.X.X.X — NDH) phenomenon, which is probably due to lactate bound to the enzyme molecules.

A mammalian LDH-X isozyme specific to a postpubertal testis and sperm is relatively more active with α-hydroxybutyrate and α-hydroxy-valeriate than with lactate.[3]

Use D-lactate in the staining solution in place of L-lactate to detect D-lactate dehydrogenase (EC 1.1.1.28).

OTHER METHODS

A. The reverse reaction can also be used to detect LDH activity bands.[3] The staining solution includes 20 ml of 0.05 M Tris–HCl buffer (pH 8.0), 100 mg of sodium pyruvate, and 10 mg of NADH. Dark (nonfluorescent) bands of LDH activity are observed under long-wave UV light.

B. A bioautographic procedure for visualization of LDH activity bands was developed using *E. coli* strain CB482 deficient in LDH as the microbial reagent.[4] For growth and proliferation of *E. coli* (CB482), pyruvate is required as the source of carbon. An indicator agar for LDH bioautography contains 10^8 microorganisms/ml, 2.5 mg/ml L-lactate, and 2.5 mg/ml NAD in minimal medium (see Appendix A-1) lacking a carbon source. When an electrophoretic gel containing zones of LDH activity is placed in contact with the indicator agar, bacteria proliferate at the regions where L-lactate is converted to pyruvate by exogenous LDH. Bands of bacterial proliferation are observed by transmitted light. Bioautographic detection of LDH is significantly more complex and time consuming than the NBT–PMS method and therefore less appropriate for routine laboratory use.

C. Two-dimensional spectroscopy of electrophoretic gels also permits detection of LDH activity bands.[5] This procedure, however, requires a special optical device that is not commercially available.

GENERAL NOTES

The enzyme also oxidizes other L-2-hydroxymonocarboxylic acids. The bacterial enzyme is specific to NAD. The animal enzyme can also use NADP, but more slowly.[6]

REFERENCES

1. Whitt, G.S., Developmental genetics of the lactate dehydrogenase isozymes of fish, *J. Exp. Zool.*, 175, 1, 1970.
2. Falkenberg, F., Lehmann, F.-G., and Pfleiderer, G., LDH (lactate dehydrogenase) isozymes as cause of nonspecific tetrazolium salt staining in gel enzymograms ("nothing dehydrogenase"), *Clin. Chim. Acta*, 23, 265, 1969.
3. Harris, H. and Hopkinson, D.A., *Handbook of Enzyme Electrophoresis in Human Genetics*, North-Holland, Amsterdam, 1976 (loose-leaf, with supplements in 1977 and 1978).
4. Naylor, S.L. and Klebe, R.J., Bioautography: a general method for the visualization of isozymes, *Biochem. Genet.*, 15, 1193, 1977.
5. Klebe, R.J., Mancuso, M.G., Brown, C.R., and Teng, L., Two-dimensional spectroscopy of electrophoretic gels, *Biochem. Genet.*, 19, 655, 1981.
6. NC-IUBMB, *Enzyme Nomenclature*, Academic Press, San Diego, 1992, p. 27 (EC 1.1.1.27, Comments).

1.1.1.29 — Glycerate Dehydrogenase; G-2-DH

OTHER NAMES   Glycerate-2-dehydrogenase

REACTION   D-Glycerate + NAD = hydroxypyruvate + NADH

ENZYME SOURCE   Fungi, plants, invertebrates, vertebrates

SUBUNIT STRUCTURE   Dimer (plants)

METHOD

Visualization Scheme

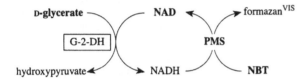

Staining Solution[1]

0.04 M Tris–HCl buffer, pH 8.0	50 ml
D,L-Glyceric acid (hemicalcium salt)	100 mg
NAD	25 mg
NBT	15 mg
PMS	1 mg

Procedure

Incubate the gel in staining solution in the dark at 37°C until dark blue bands appear. Wash the developed gel in water and fix in 25% ethanol.

Notes: Because LDH can also react with D-glycerate, an additional gel should be stained for LDH as a control.

G-2-DH activity in plants is an associated activity of glyoxylate reductase (EC 1.1.1.26) and is not a different protein moiety.

The rat liver enzyme is equally effective with reduced NAD or NADP in the reverse reaction, but in the forward reaction NAD is more effective than NADP.

REFERENCES

1. Siciliano, M.J. and Shaw, C.R., Separation and visualization of enzymes on gels, in *Chromatographic and Electrophoretic Techniques*, Vol. 2, *Zone Electrophoresis*, Smith, I., Ed., Heinemann, London, 1976, p. 185.

1.1.1.30 — 3-Hydroxybutyrate Dehydrogenase; HBDH

OTHER NAMES   β-Hydroxybutyrate dehydrogenase

REACTION   D-β-Hydroxybutanoate + NAD = acetoacetate + NADH

ENZYME SOURCE   Bacteria, fungi, protozoa, invertebrates, vertebrates

SUBUNIT STRUCTURE   Dimer (invertebrates, vertebrates)

METHOD

Visualization Scheme

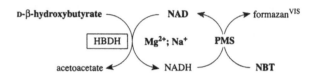

Staining Solution[1] (modified)

0.1 M Phosphate buffer, pH 7.4	90 ml
1 M D,L-β-Hydroxybutyrate	10 mg
NAD	30 mg
NBT	25 mg
PMS	2.5 mg
MgCl$_2$ (Anhydrous)	10 mg
NaCl	575 mg

Procedure

Incubate the gel in staining solution in the dark at 37°C until dark blue bands appear. Wash the developed gel in water and fix in 25% ethanol.

Notes: The addition of NAD to the electrophoretic gel is often beneficial. The addition of 1 to 2 mg/ml pyruvate or 1 to 2 mg/ml pyrazole to the staining solution may be desirable if the tissue chosen as the source of HBDH has high levels of LDH or ADH, respectively.

The *Drosophila* enzyme may be identical to 1.1.1.45 — GUDH and 1.1.1.69 — GNDH (see below).

The beef heart mitochondrial enzyme has a specific and absolute requirement for lecithin.

REFERENCES

1. Shaw, C.R. and Prasad, R., Starch gel electrophoresis of enzymes: a compilation of recipes, *Biochem. Genet.*, 4, 297, 1970.

1.1.1.35 — 3-Hydroxyacyl-CoA Dehydrogenase; HADH

OTHER NAMES	β-Hydroxyacyl dehydrogenase, β-keto-reductase
REACTION	L-3-Hydroxyacyl-CoA + NAD = 3-oxoacyl-CoA + NADH
ENZYME SOURCE	Vertebrates
SUBUNIT STRUCTURE	Dimer (vertebrates)

METHOD

Visualization Scheme

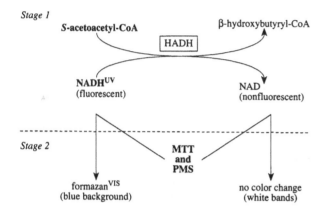

Staining Solution[1]

A. 0.1 *M* Citrate–phosphate buffer, pH 5.3
    1.7 m*M* S-Acetoacetyl-CoA
    2 m*M* NADH

B. 0.1 *M* Citrate–phosphate buffer, pH 5.3
    2.5 mg/ml MTT
    0.02 mg/ml PMS

Procedure

Apply solution A to the gel surface dropwise and incubate the gel at 37°C in a moist chamber. View the gel under long-wave UV light. Dark bands of HADH activity are visible on the light (fluorescent) background of the gel. When the bands are clearly visible, apply solution B to reveal HADH activity as white bands on a blue background. Rinse negatively stained gel under hot tap water for about 30 sec to increase the intensity of the background color. Fix the developed gel in 25% ethanol.

Notes: The enzyme also oxidizes L-3-hydroxyacyl-*N*-acylthioethanolamine and L-3-hydroxyacylhydrolipoate; HADH from some sources acts more slowly with NADP.

REFERENCES

1. Craig, I., Tolley, E., and Bobrow, M., A preliminary analysis of the segregation of human hydroxyacyl coenzyme A dehydrogenase in human–mouse somatic cell hybrids, *Cytogenet. Cell Genet.*, 16, 114, 1976.

1.1.1.37 — Malate Dehydrogenase; MDH

OTHER NAMES	Malic dehydrogenase
REACTION	L-Malate + NAD = oxaloacetate + NADH
ENZYME SOURCE	Bacteria, fungi, algae, plants, protozoa, invertebrates, vertebrates
SUBUNIT STRUCTURE	Dimer (fungi, algae, plants, protozoa, invertebrates, vertebrates)

Method 1

Visualization Scheme

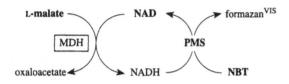

Staining Solution[1]

0.1 M Tris–HCl buffer, pH 8.0	100 ml
L-Malic acid (disodium salt)	250 mg
NAD	30 mg
NBT	25 mg
PMS	2 mg

Procedure

Incubate the gel in staining solution in the dark at 37°C until dark blue bands appear. Wash the developed gel in water and fix in 25% ethanol.

Notes: The adding of 1 to 2 mg/ml pyruvate or 1 to 2 mg/ml pyrazole to the staining solution may be desirable if the tissue chosen as the source of MDH has high levels of LDH or ADH, respectively.

Method 2

Visualization Scheme

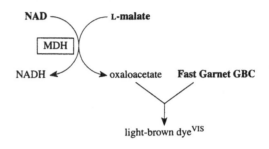

Staining Solution[2]

0.1 M Tris–HCl buffer, pH 8.0	50 ml
Fast Garnet GBC salt	150 mg
L-Malic acid (sodium salt)	125 mg
NAD	60 mg

Procedure

Incubate the gel in staining solution in the dark at 37°C until light brown bands appear. Wash the developed gel in water and fix in 50% glycerol.

Notes: Greater sensitivity of the method may be obtained using a post-coupling technique. All ingredients of the staining solution except Fast Garnet GBC (the potential inhibitor of MDH) are applied to the gel. The gel is incubated for a period of time, and then Fast Garnet GBC (dissolved in a minimal volume of stain buffer) is added.

General Notes

Method 1 is more commonly used and more sensitive than Method 2. It does, however, allow the nonspecific staining of NAD-dependent "nothing dehydrogenase" bands (see 1.X.X.X — NDH), whereas Method 2 avoids this problem.

References

1. Shaw, C.R. and Prasad, R., Starch gel electrophoresis of enzymes: a compilation of recipes, *Biochem. Genet.*, 4, 297, 1970.
2. Richardson, B.J., Baverstock, P.R., and Adams, M., *Allozyme Electrophoresis: A Handbook for Animal Systematics and Population Studies*, Academic Press, Sydney, 1986, p. 201.

1.1.1.39 — Malate Dehydrogenase (Decarboxylating); MDHD

OTHER NAMES	NAD-dependent "malic" enzyme, pyruvic–malic decarboxylase
REACTION	L-Malate + NAD = pyruvate + CO_2 + NADH
ENZYME SOURCE	Invertebrates (*Ascaris*), vertebrates
SUBUNIT STRUCTURE	Tetramer (vertebrates)

METHOD

Visualization Scheme

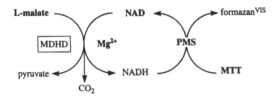

Staining Solution[1]

0.1 M Tris–HCl, pH 7.4	50 ml
1 M L-Malic acid (disodium salt, pH 8.0)	10 ml
10 mg/ml NAD	10 ml
1 M MgCl$_2$	10 ml
10 mg/ml MTT	10 ml
5 mg/ml PMS	10 ml

Procedure

Incubate the gel in staining solution in the dark at 37°C until dark blue bands appear (however, see *Notes*). Wash the stained gel in water and fix in 25% ethanol.

Notes: Activity bands of MDHD from muscle tissues of salmonid fish species are not detected with MDH (see 1.1.1.37 — MDH) and ME (see 1.1.1.40 — ME) stains at pH 8.0. However, its bands develop in a modified MDH stain (pH 7.0 to 7.4; added MgCl$_2$) and a modified ME stain (pH 7.0 to 7.4; added NAD). The modified MDH stain detects MDH and MDHD bands, while the modified ME stain detects ME and MDHD bands. Importantly, MDH and ME bands develop less intensely at the lower pH values than at pH 8.0. These differences may be used for differentiation between MDHD bands and MDH and ME bands. Conditions for specific detection of only MDHD bands are not yet identified (however, see Other Methods, below).

The enzyme bands are not developed in modified stains when oxaloacetate is used as a substrate in place of malate. This provides strong evidence that the enzyme is indeed EC 1.1.1.39, but not EC 1.1.1.38, NAD-dependent "malic" enzyme.

The activity of MDHD from the muscle tissue of salmonid fish is associated with the mitochondrial fraction. Like ME (EC 1.1.1.40), MDHD is a tetramer. However, no hybrids between MDHD and the mitochondrial form of ME are formed, providing evidence that these are structurally very different enzyme proteins.

OTHER METHODS

A bioautographic procedure for specific visualization of MDHD activity bands is recommended. This is a modification of the bioautographic procedure used for visualization of LDH (see 1.1.1.27 — LDH, Other Methods, B). An indicator agar for MDHD bioautography should include 10^8 cells/ml *E. coli* (CB482), 2.5 mg/ml L-malate, and 2.5 mg/ml NAD in minimal medium (see Appendix A-1) lacking a carbon source. Because pyruvate is required for growth and proliferation of *E. coli* (strain CB482), bacteria in the indicator agar applied to the electrophoretic gel will proliferate only at the gel regions where L-malate is converted to pyruvate by exogenous MDHD. After incubation of the electrophoretic gel with the indicator agar, bands of bacterial proliferation are observed by transmitted light. Although minimal medium contains Mg^{2+} ions, it may prove desirable to add some more for sufficient MDHD activity.

REFERENCES

1. Verspoor, E. and Jordan, W.C., Detection of an NAD+-dependent malic enzyme locus in the Atlantic salmon *Salmo salar* and other salmonid fish, *Biochem. Genet.*, 32, 105, 1994.

1.1.1.40 — Malate Dehydrogenase (NADP); ME

OTHER NAMES — Malate dehydrogenase (oxaloacetate decarboxylating) (NADP) (recommended name), "malic" enzyme, pyruvic–malic carboxylase

REACTION — L-Malate + NADP = pyruvate + CO_2 + NADPH

ENZYME SOURCE — Bacteria, fungi, plants, protozoa, invertebrates, vertebrates

SUBUNIT STRUCTURE — Monomer (plants, invertebrates), dimer (invertebrates), tetramer (plants, invertebrates, vertebrates)

Method

Visualization Scheme

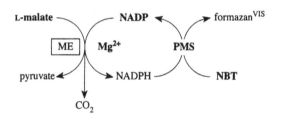

Staining Solution[1]

0.05 M Tris–HCl buffer, pH 8.0	50 ml
L-Malic acid (sodium salt)	700 mg
NADP	15 mg
NBT	15 mg
PMS	1 mg
$MgCl_2$ (Anhydrous)	50 mg

Procedure

Incubate the gel in staining solution in the dark at 37°C until dark blue bands appear. Rinse the stained gel in water and fix in 25% ethanol.

Notes: In some cases increased activity of ME can be obtained using $MnCl_2$ in place of $MgCl_2$ in the staining solution. It should be remembered, however, that in the presence of Mn^{2+} ions PMS acts less effectively as an electron-carrier intermediary. When Mn^{2+} ions are used, a positive fluorescent stain may be preferable. This stain uses a staining solution similar to that described above, but lacking PMS and NBT and containing $MnCl_2$ in place of $MgCl_2$. After an appropriate period of time of incubating the gel with staining solution, fluorescent bands of ME activity can be observed under long-wave UV light.

References

1. Siciliano, M.J. and Shaw, C.R., Separation and visualization of enzymes on gels, in *Chromatographic and Electrophoretic Techniques*, Vol. 2, *Zone Electrophoresis*, Smith, I., Ed., Heinemann, London, 1976, p. 185.

1.1.1.41 — Isocitrate Dehydrogenase (NAD); IDH(NAD)

OTHER NAMES	β-Ketoglutaric–isocitric carboxylase, isocitric dehydrogenase
REACTION	Isocitrate + NAD = 2-oxoglutarate + CO_2 + NADH
ENZYME SOURCE	Fungi, plants, invertebrates, vertebrates
SUBUNIT STRUCTURE	See General Notes

METHOD

Visualization Scheme

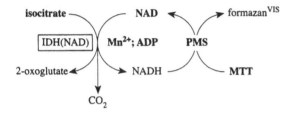

Staining Solution[1]

0.1 M Tris–HCl buffer, pH 8.4	100 ml
Isocitric acid	125 mg
ADP	7.5 mg
$MnCl_2$	65 mg
NAD	6 mg
MTT	8 mg
PMS	2 mg

Procedure

Incubate the gel in staining solution in the dark at 37°C until dark blue bands appear. Rinse the developed gel in water and fix in 25% ethanol.

Notes: It is recommended that $MgCl_2$ be used in place of $MnCl_2$ because Mn^{2+} ions significantly lower the PMS action as an electron-carrier intermediary. The enzyme is an allosteric one specifically activated by ADP. In the absence of ADP, the activity of IDH(NAD) is very low.

The isomer of isocitrate involved is (1R,2S)-1-hydroxypropane-1,2,3-tricarboxylate, formerly termed *threo*-D$_s$-isocitrate. The enzyme does not decarboxylate oxalosuccinate.[2]

GENERAL NOTES

The enzyme from yeast mitochondria is an octamer composed of four each of two nonidentical but related subunits (catalytic and regulatory).[3]

The plant enzyme is represented by one type of catalytic subunit and two types of regulatory subunits. The catalytic subunit associates with each type of regulatory subunit to form different types of heterodimers.[4]

Different types of IDH(NAD) subunits are also known in mammals.[5]

REFERENCES

1. Menken, S.B.J., Allozyme Polymorphism and the Speciation Process in Small Ermine Moths (Lepidoptera, Yponomeutidae), Ph.D. thesis, University of Leiden, The Netherlands, 1980.
2. NC-IUBMB, *Enzyme Nomenclature*, Academic Press, San Diego, 1992, p. 29 (EC 1.1.1.41, Comments).
3. Panisko, E.A. and McAlister-Henn, L., Subunit interactions of yeast NAD(+)-specific isocitrate dehydrogenase, *J. Biol. Chem.*, 276, 1204, 2001.
4. Lancien, M., Gadal, P., and Hodges, M., Molecular characterization of higher plant NAD-dependent isocitrate dehydrogenase: evidence for a heteromeric structure by the complementation of yeast mutants, *Plant J.*, 16, 325, 1998.
5. Nichols, B.J., Perry, A.C.F., Hall, L., and Denton, R.M., Molecular cloning and deduced amino-acid sequences of the α- and β-subunits of mammalian NAD$^{(+)}$ isocitrate dehydrogenase, *Biochem. J.*, 310, 917, 1995.

1.1.1.42 — Isocitrate Dehydrogenase (NADP); IDH

OTHER NAMES	Oxalosuccinate decarboxylase
REACTION	Isocitrate + NADP = 2-oxoglutarate + CO_2 + NADPH
ENZYME SOURCE	Bacteria, fungi, algae, plants, protozoa, invertebrates, vertebrates
SUBUNIT STRUCTURE	Dimer (fungi, algae, plants, protozoa, invertebrates, vertebrates)

METHOD 1

Visualization Scheme

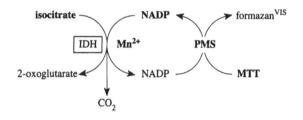

Staining Solution[1]

0.1 M Tris–HCl buffer, pH 8.0	100 ml
0.1 M Isocitric acid (trisodium salt)	3 ml
NADP	20 mg
NBT	10 mg
PMS	3 mg
0.25 M $MnCl_2$	0.4 ml

Procedure

Incubate the gel in staining solution in the dark at 37°C until dark blue bands appear. Wash the stained gel in water and fix in 25% ethanol.

Notes: Addition of NADP to the gel or acetate cellulose soaking buffer usually proves beneficial. It is recommended that $MgCl_2$ be used instead of $MnCl_2$ because Mn^{2+} ions significantly lower the PMS action as an electron-carrier intermediary.

METHOD 2

Visualization Scheme

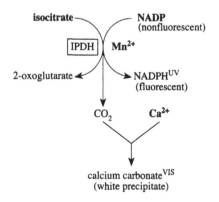

Staining Solution[2]

100 mM Tris–HCl buffer, pH 7.5
1 mM Isocitrate
1 mM NADP
100 mM Mn^{2+}
100 mM Ca^{2+}

Procedure

Place the electrophorized PAG in the stain buffer at 37°C. After 20 to 30 min transfer the gel to the staining solution and incubate at 37°C. View the gel against a dark background, and when an activity stain of sufficient intensity is obtained, remove the gel from the staining solution and store in 50 mM glycine–KOH buffer (pH 10.0), containing 5 mM Ca^{2+}, either at 5°C or at room temperature in the presence of an antibacterial agent. Under these conditions, the stained gel can be stored for several months with little deterioration.

Notes: This method was developed for PAG. It is, however, also applicable to acetate cellulose and starch gels where IDH activity bands can be observed in long-wave UV light as fluorescent zones on a dark (nonfluorescent) background.

GENERAL NOTES

Method 1 is more commonly used than Method 2. It does, however, allow the staining of NADP-dependent "nothing dehydrogenase" (see 1.X.X.X — NDH), whereas the use of Method 2 avoids this problem.

Manganese ions are the best activators of IDH; however, their use in a tetrazolium stain is not effective because of the inhibitory effect toward PMS. So, a positive fluorescent stain using a staining solution similar to those presented in Methods 1 and 2, but lacking PMS–NBT and Ca^{2+}, respectively, may be preferable when preparations with low IDH activity are studied.

The isomer of isocitrate involved is (1R,2S)-1-hydroxypropane-1,2,3-tricarboxylate, formerly termed *threo*-D_s-isocitrate. The enzyme also decarboxylates oxalosuccinate.[3]

REFERENCES

1. Henderson, N.S., Isozymes of isocitrate dehydrogenase: subunit structure and intracellular location, *J. Exp. Zool.*, 158, 263, 1965.
2. Nimmo, H.G. and Nimmo, G.A., A general method for the localization of enzymes that produce phosphate, pyrophosphate, or CO_2 after polyacrylamide gel electrophoresis, *Anal. Biochem.*, 121, 17, 1982.
3. NC-IUBMB, *Enzyme Nomenclature*, Academic Press, San Diego, 1992, p. 29 (EC 1.1.1.42, Comments).

1.1.1.44 — Phosphogluconate Dehydrogenase (Decarboxylating); PGD

OTHER NAMES	Phosphogluconic acid dehydrogenase, 6-phosphogluconate dehydrogenase, 6-phosphogluconic dehydrogenase, 6-phosphogluconic carboxylase
REACTION	6-Phospho-D-gluconate + NADP = D-ribulose 5-phosphate + CO_2 + NADPH
ENZYME SOURCE	Bacteria, green algae, fungi, plants, protozoa, invertebrates, vertebrates
SUBUNIT STRUCTURE	Dimer (fungi, plants, invertebrates, vertebrates)

METHOD

Visualization Scheme

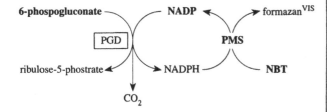

Staining Solution[1]

0.05 M Tris–HCl buffer, pH 8.0	100 ml
6-Phosphogluconic acid (trisodium salt)	200 mg
NADP	20 mg
NBT	25 mg
PMS	2 mg

Procedure

Incubate the gel in staining solution in the dark at 37°C until dark blue bands appear. Wash the developed gel in water and fix in 25% ethanol.

Notes: Addition of MgCl$_2$ to the staining solution (25 mM final concentration) accelerates development of PGD activity bands. The amounts of substrate and cofactor in the staining solution may be decreased five and two times, respectively.

The enzyme from some sources can use NAD as well as NADP.[2]

REFERENCES

1. Shaw, C.R. and Prasad, R., Starch gel electrophoresis of enzymes: a compilation of recipes, *Biochem. Genet.*, 4, 297, 1970.
2. NC-IUBMB, *Enzyme Nomenclature*, Academic Press, San Diego, 1992, p. 29 (EC 1.1.1.44, Comments).

1.1.1.45 — L-Gulonate 3-Dehydrogenase; GUDH

OTHER NAMES	L-3-Aldonate dehydrogenase, L-gluconate dehydrogenase, β-L-hydroxyacid dehydrogenase, 3-oxoacid dehydrogenase
REACTION	L-Gulonate + NAD = 3-dehydro-L-gulonate + NADH; also oxidizes other L-3-hydroxyacids
ENZYME SOURCE	Invertebrates
SUBUNIT STRUCTURE	Dimer (invertebrates)

METHOD

Visualization Scheme

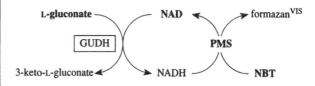

Staining Solution[1]

0.05 M Tris–phosphate buffer, pH 8.2, containing 40 mM pyrazole	100 ml
L-Gluconate	200 mg
NAD	20 mg
NBT	10 mg
PMS	0.4 mg

Procedure

Incubate the gel in staining solution in the dark at 37°C until dark blue bands appear. Wash the stained gel in water and fix in 25% ethanol.

Notes: β-Hydroxybutyrate also may be used as a substrate for *Drosophila* GUDH. D-Gluconate is also oxidized by *Drosophila* GUDH. *Drosophila* GUDH may be identical to 1.1.1.30 — HBDH (see above) or 1.1.1.69 — GNDH (see below).

REFERENCES

1. Tobler, J.E. and Grell, E.H., Genetics and physiological expression of β-hydroxy acid dehydrogenase in *Drosophila*, *Biochem. Genet.*, 16, 333, 1978.

OTHER NAMES	Galactose-6-phosphate dehydrogenase, hexose-6-phosphate dehydrogenase
REACTIONS	1. β-D-Glucose + NAD(P) = D-glucono-1,5-lactone + NAD(P)H
	2. D-Galactose-6-phosphate + NADP = D-galactono-1,4-lactone-6-phosphate + NADPH (presumed reaction)
ENZYME SOURCE	Bacteria, fungi, invertebrates, vertebrates
SUBUNIT STRUCTURE	Dimer (vertebrates), tetramer (fungi)

Method 1

Visualization Scheme

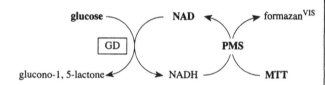

Staining Solution[1] (modified)

0.05 M Phosphate buffer, pH 7.3	50 ml
D-Glucose	5 g
NAD	30 mg
MTT	8 mg
PMS	3 mg

Procedure

Incubate the gel in staining solution in the dark at 37°C until dark blue bands appear. Wash the developed gel in water and fix in 25% ethanol.

Notes: In some situations (e.g., when *Drosophila* male whole-body homogenates are used as the source of enzyme) glucose oxidase (see 1.1.3.4 — GO) bands are also visualized by Method 1.[2]

Method 2

Visualization Scheme

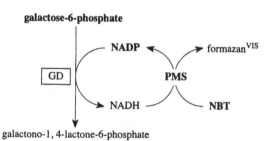

Staining Solution[3] (modified)

0.05 M Tris–HCl buffer, pH 7.0	45 ml
1 M Galactose-6-phosphate	2.5 ml
NADP	20 mg
NBT	25 mg
PMS	2 mg

Procedure

Incubate the gel in staining solution in the dark at 37°C until dark blue bands appear. Wash the developed gel in water and fix in 25% ethanol.

Notes: The enzyme is also capable of using glucose-6-phosphate as a substrate in some taxa. It may appear as a stain artifact on hexokinase (2.7.1.1 — HK), adenylate kinase (2.7.4.3 — AK, Method 1), and glucose-6-phosphate dehydrogenase (1.1.1.49 — G-6-PD) zymograms.

General Notes

The addition of NAD(P) to electrophoretic gels and acetate cellulose soaking buffers often proves beneficial.

D-Xylose also may be used as substrate for GD.[4]

References

1. Harris, H. and Hopkinson, D.A., *Handbook of Enzyme Electrophoresis in Human Genetics*, North-Holland, Amsterdam, 1976 (loose-leaf, with supplements in 1977 and 1978).
2. Cavener, D.R., Genetics of male-specific glucose oxidase and the identification of other unusual hexose enzymes in *Drosophila melanogaster*, *Biochem. Genet.*, 18, 929, 1980.
3. Shaw, C.R. and Koen, A.L., Glucose-6-phosphate dehydrogenase and hexose-6-phosphate dehydrogenase of mammalian tissues, *Ann. N.Y. Acad. Sci.*, 151, 149, 1968.
4. NC-IUBMB, *Enzyme Nomenclature*, Academic Press, San Diego, 1992, p. 30 (EC 1.1.1.47, Comments).

1.1.1.48 — Galactose 1-Dehydrogenase; GALDH

REACTION   D-Galactose + NAD = D-galactono-1,4-lactone + NADH

ENZYME SOURCE   Bacteria, vertebrates

SUBUNIT STRUCTURE   Unknown or no data available

Method

Visualization Scheme

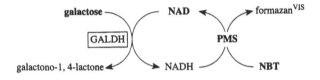

Staining Solution[1]

0.1 M Tris–HCl buffer, pH 8.4	100 ml
D(+)Galactose	900 mg
NAD	30 mg
NBT	20 mg
PMS	4 mg

Procedure

Incubate the gel in staining solution in the dark at 37°C until dark blue bands appear. Wash the developed gel in water and fix in 25% ethanol.

References

1. Cuatrecasas, P. and Segal, S., Electrophoretic heterogeneity of mammalian galactose dehydrogenase, *Science*, 154, 533, 1966.

OTHER NAMES	Glucose-6-phosphate dehydrogenase
REACTION	D-Glucose-6-phosphate + NADP = D-glucono-1,5-lactone-6-phosphate + NADPH
ENZYME SOURCE	Bacteria, green algae, fungi, plants, protozoa, invertebrates, vertebrates
SUBUNIT STRUCTURE	Monomer (plants, invertebrates), dimer (plants, invertebrates, vertebrates), tetramer (invertebrates, vertebrates)

Method

Visualization Scheme

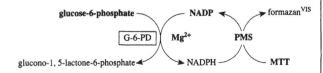

Staining Solution[1] (modified)

0.1 M Tris–HCl buffer, pH 8.0	45 ml
Glucose-6-phosphate (disodium salt)	20 mg
NADP	10 mg
MTT	10 mg
PMS	1 mg
0.2 M MgCl$_2$	5 ml

Procedure

Incubate the gel in staining solution in the dark at 37°C until dark blue bands appear. Wash the developed gel in water and fix in 25% ethanol.

Notes: The addition of NADP to electrophoretic gels and acetate cellulose soaking buffers often proves beneficial. The bands of glucose dehydrogenase (see 1.1.1.47 — GD, Method 2) also can develop on G-6-PD zymograms.[2] Thus, an additional slice of the same starch gel block, or additional acetate cellulose strip, should, in some situations, be stained for GD as controls.

Other Methods

The immunoblotting procedure (for details, see Part II) based on the utility of monoclonal antibodies specific to rat[3] and human[4] G-6-PD can also be used to detect the enzyme protein on electrophoretic gels. This procedure is not as practical as that described above; however, it may be of great value in special (biochemical, immunological, phylogenetic, and genetic) analyses of G-6-PD. Antibodies specific to Bakers yeast G-6-PD are now commercially available from Sigma.

References

1. Shaw, C.R. and Prasad, R., Starch gel electrophoresis of enzymes: a compilation of recipes, *Biochem. Genet.*, 4, 297, 1970.
2. Shaw, C.R. and Koen, A.L., Glucose-6-phosphate dehydrogenase and hexose-6-phosphate dehydrogenase of mammalian tissues, *Ann. N.Y. Acad. Sci.*, 151, 149, 1968.
3. Dao, M.L., Johnson, B.C., and Hartman, P.E., Preparation of monoclonal antibody to rat liver glucose-6-phosphate dehydrogenase and the study of its immunoreactivity with native and inactivated enzyme, *Proc. Natl. Acad. Sci. U.S.A.*, 79, 2860, 1980.
4. Damiani, G., Frascio, M., Benatti, U., Morelli, A., Zocchi, E., Fabbi, M., Bargellesi, A., Pontremoli, S., and De Flora, A., Monoclonal antibodies to human erythrocyte glucose-6-phosphate dehydrogenase, *FEBS Lett.*, 119, 169, 1980.

1.1.1.51 — 3(or 17)β-Hydroxysteroid Dehydrogenase; 3(17)β-HSD

REACTION	Testosterone + NAD(P) = androst-4-ene-3,17-dione + NAD(P)H
ENZYME SOURCE	Bacteria, vertebrates
SUBUNIT STRUCTURE	Heterotetramer# (bacteria; see *Notes*)

METHOD

Visualization Scheme

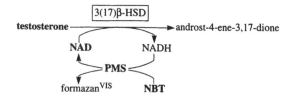

Staining Solution[1]

0.1 μM Potassium phosphate buffer, pH 7.3
5 μM Testosterone
4% Dioxane
15 μM NAD
3.1 μM NBT
0.8 μM PMS

Procedure

Incubate the gel in staining solution in the dark at 37°C until dark blue bands appear. Wash the stained gel in water and fix in 25% ethanol.

Notes: The enzyme from *Pseudomonas testosteroni* is a heterotetramer consisting of two different subunits (A and B) of similar molecular weight. The most abundant 3(17)β-HSD isoform of *P. testosteroni* is represented by AB₃ molecules. Using staining solution similar to that given above, but containing pregnenolone as a substrate, O'Conner et al. failed to detect substrate-specific bands of 3(17)β-HSD after PAG electrophoresis of human placental preparations. These authors stressed that the histochemical method detected identical banding patterns in the presence and in the absence of the substrate. The formation of formazan in the apparent absence of the substrate was attributed to the "nothing dehydrogenase" phenomenon.[2]

REFERENCES

1. Schultz, R.M., Groman, E.V., and Engel, L.L., 3(17)β-Hydroxysteroid dehydrogenase of *Pseudomonas testosteroni*, *J. Biol. Chem.*, 252, 3775, 1977.
2. O'Conner, J.L., Edwards, D.P., and Brandsome, E.D., Jr., Localizing steroid dehydrogenase activity on acrylamide gels: the perils of "histochemistry," *Anal. Biochem.*, 78, 205, 1977.

1.1.1.53 — (R)-20-Hydroxysteroid Dehydrogenase; HSDH

OTHER NAMES	3α (or 20β)-Hydroxysteroid dehydrogenase (recommended name), cortisone reductase
REACTIONS	1. Androstan-3α,17β-diol + NAD = 17β-hydroxyandrostan-30ne + NADH
	2. (20R)-17α,20,21-Trihydroxypregn-4-ene-3,11-dione + NAD = cortisone + NADH (see also *Notes*)
ENZYME SOURCE	Bacteria
SUBUNIT STRUCTURE	Unknown or no data available

METHOD

Visualization Scheme

(20R)-17α, 20, 21-trihydroxypregn-4-ene-3, 11-dione

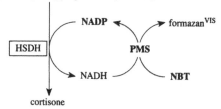

cortisone

Staining Solution[1]

0.1 M Tris–HCl buffer, pH 7.1
0.5 mM NAD
0.02 mg/ml 5α-Pregnan-20β-ol-3-on (dissolved in minimal volume of isopropanol)
0.3 mg/ml NBT
0.02 mg/ml PMS

Procedure

Incubate the gel in staining solution in the dark at 37°C until dark blue bands appear. Fix the stained gel in 25% ethanol.

Notes: Alcohol dehydrogenase bands can also become apparent on HSDH zymograms obtained by this method. Thus, an additional gel should be stained for alcohol dehydrogenase (see 1.1.1.1 — ADH, Method 1) as a control.

The enzyme also acts on other 20-keto-steroids containing different substituents of the steroid system. The 3α-hydroxyl group or 20β-hydroxyl group of pregnane and androstane steroids can also act as donor.[2]

REFERENCES

1. Blomquist, C.H., The molecular weight and substrate specificity of 20β-hydroxysteroid dehydrogenase from *Streptomyces hydrogenans*, *Arch. Biochem. Biophys.*, 159, 590, 1973.
2. NC-IUBMB, *Enzyme Nomenclature*, Academic Press, San Diego, 1992, p. 31 (EC 1.1.1.53, Comments).

REACTION	Estradiol-17β + NAD(P) = estrone + NAD(P)H
ENZYME SOURCE	Bacteria, vertebrates
SUBUNIT STRUCTURE	Unknown or no data available

METHOD

Visualization Scheme

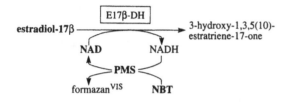

Staining Solution[1,2]

0.1 μM Potassium phosphate buffer, pH 7.3
5 μM Estradiol-17β
4% Dioxane
15 μM NAD
3.1 μM NBT
0.8 μM PMS

Procedure

Incubate the gel in staining solution in the dark at 37°C until dark blue bands appear. Wash the stained gel in water and fix in 25% ethanol.

Notes: Multiple activity bands of E17β-DH were detected after PAG electrophoresis of crude extracts of *Pseudomonas testosteroni* induced with testosterone. However, only one of these bands was specific to estradiol-17β. Other bands were also developed with testosterone, providing evidence that they are actually bands of 3(or 17)β-hydroxysteroid dehydrogenase (see 1.1.1.51 — 3(17)β-HSD). Thus, to identify activity bands of E17β-DH, an additional gel should be stained for 3(17)β-HSD using testosterone as the substrate.

GENERAL NOTES

It is established that the two hydrogen atoms at carbon 4 of the dihydropyridine ring of NAD(P) are not equivalent in that the hydrogen is transferred stereospecifically. The two faces of the dihydropyridine ring are denoted A and B. The two groups of enzymes that display stereospecificity with respect to different faces of the ring are referred to as A specific and B specific, respectively. The E17β-DH is B specific with respect to NAD(P).[3]

REFERENCES

1. Schultz, R.M., Groman, E.V., and Engel, L.L., 3(17)β-Hydroxysteroid dehydrogenase of *Pseudomonas testosteroni*, *J. Biol. Chem.*, 252, 3775, 1977.
2. Groman, E.V. and Engel, L.L., Hydroxysteroid dehydrogenases of *Pseudomonas testosteroni*: separation of a 17β-hydroxysteroid dehydrogenase from the 3(17)β-hydroxysteroid dehydrogenase and comparison of the two enzymes, *Biochim. Biophys. Acta*, 485, 249, 1977.
3. NC-IUBMB, *Enzyme Nomenclature*, Academic Press, San Diego, 1992, p. 9 (Class 1, Oxidoreductases), p. 31 (EC 1.1.1.53, Comments).

1.1.1.67 — Mannitol 2-Dehydrogenase; MD(NAD)

REACTION   D-Mannitol + NAD = D-fructose + NADH

ENZYME SOURCE   Bacteria, fungi

SUBUNIT STRUCTURE   Monomer[#] (bacteria – *Pseudomonas*), tetramer[#] (fungi – *Cladosporium fulvum*)

METHOD

Visualization Scheme

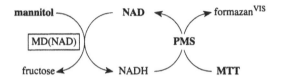

Staining Solution*

A.	0.1 *M* Tris–HCl buffer, pH 8.5	25 ml
	NAD	20 mg
	D-Mannitol	200 mg
	MTT	10 mg
	PMS	1 mg
B.	2% Agarose solution (60°C)	25 ml

Procedure

Mix A and B components and pour the mixture over the surface of the gel. Incubate the gel at 37°C in the dark until dark blue bands appear. Fix the stained agarose plate in 25% ethanol.

Notes: This method was used by the author to detect MD(NAD) from mushroom *Boletus edulis* (unpublished data).

---

* New; recommended for use.

1.1.1.69 — Gluconate 5-Dehydrogenase; GNDH

OTHER NAMES	5-Keto-D-gluconate reductase, D-gluconate dehydrogenase, 5-keto-D-gluconate 5-reductase
REACTION	D-Gluconate + NAD(P) = 5-dehydro-D-gluconate + NAD(P)H
ENZYME SOURCE	Bacteria, invertebrates, vertebrates
SUBUNIT STRUCTURE	Unknown or no data available

METHOD 1

Visualization Scheme

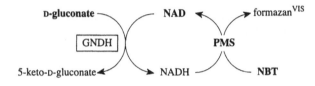

Staining Solution[1]

0.2 M Tris–HCl buffer, pH 8.0	25 ml
80 mg/ml D-Gluconate, pH 8.0	25 ml
10 mg/ml NAD	1 ml
10 mg/ml NBT	0.5 ml
10 mg/ml PMS	0.5 ml

Procedure

Incubate the gel in staining solution in the dark at 37°C until dark blue bands appear. Wash the stained gel in water and fix in 25% ethanol.

Notes: To prepare D-gluconate solution, dissolve 2.0 g of D-gluconic acid lactone in 25 ml of H₂O and adjust the pH to 12.5 with sodium hydroxide pellets. After 30 min of incubation at room temperature, readjust the pH to 8.0 by adding HCl.

D-2-Hydroxyacid dehydrogenase (EC 1.1.99.6) activity bands can also appear on GNDH zymograms as a result of the interchange between the oxidized and reduced states of D-2-hydroxyacid dehydrogenase flavoprotein. The reduced forms of this enzyme from some sources can chemically reduce NBT via PMS without the need for NAD.

METHOD 2

Visualization Scheme

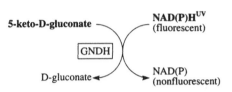

Staining Solution[2]

A. 0.2 M Tris–HCl, pH 7.0
   1 mM NADH (or NADPH)
B. 0.2 M Tris–HCl, pH 7.0
   0.1 M 5-Keto-D-gluconate

Procedure

Following electrophoresis, soak the gel in solution A for ca. 3 min. Then rinse the gel quickly in water to remove any cofactor from the gel surface, and after a few minutes, view under longwave UV light to detect any nonsubstrate-dependent dark (nonfluorescent) reductase bands on the fluorescent background of the gel. Overlay the gel with a filter paper strip soaked in solution B. Observe the dark nonfluorescent bands of GNDH on the fluorescent background of the gel. Record the zymogram or photograph using a yellow filter.

Notes: When a zymogram that is visible in daylight is required, counterstain the processed gel with MTT–PMS solution to obtain achromatic bands of GNDH on a blue background of the gel. Fix the stained gel in 25% ethanol.

GENERAL NOTES

The *Drosophila* enzyme may be identical to 1.1.1.45 — GUDH (see above).

REFERENCES

1. Buth, D.G., Staining procedures for D-2-hydroxyacid dehydrogenase as applied to studies of lower vertebrates, *Isozyme Bull.*, 13, 115, 1980.
2. Seymour, J.L. and Lazarus, R.A., Native gel activity stain and preparative electrophoretic method for the detection and purification of pyridine nucleotide-linked dehydrogenases, *Anal. Biochem.*, 178, 243, 1989.

1.1.1.72 — Glycerol Dehydrogenase (NADP); GLYD

OTHER NAMES   Glyceraldehyde reductase
REACTION   Glycerole + NADP = D-glyceraldehyde + NADPH
ENZYME SOURCE   Fungi, invertebrates, vertebrates
SUBUNIT STRUCTURE   Unknown or no data available

Method 1

Visualization Scheme

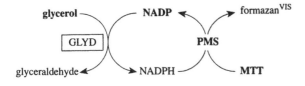

Staining Solution[1]

0.1 M Tris–HCl buffer, pH 8.4	95 ml
Glycerol	5 ml
NADP	8 mg
MTT	10 mg
PMS	2 mg

Procedure

Incubate the gel in staining solution in the dark at 37°C until dark blue bands appear. Wash the stained gel in water and fix in 25% ethanol.

Notes: The specificity of GLYD staining should be verified by exclusion of glycerol from the staining solution.

Method 2

Visualization Scheme

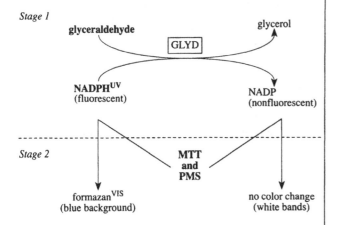

Staining Solution[2]

A. 80 mM Sodium phosphate buffer, pH 7.0
    6 mM NADPH
    110 mM D-Glyceraldehyde
B. 260 mM Tris–HCl buffer, pH 8.0
    0.6 mM PMS
    1.8 mM MTT
C. 2% Agar solution (60°C)

Procedure

Mix equal volumes of A and C solutions and pour the mixture over the gel surface. Incubate the gel for 30 min at 37°C and view under long-wave UV light. Dark bands of GLYD activity are visible on a light (fluorescent) background. When the bands are clearly visible, lift away the agar overlay. Mix equal volumes of B and C solutions and pour over the gel surface. White bands clearly visible on a blue background appear almost immediately. Rinse the stained gel under hot tap water for about 30 sec to increase the intensity of the background color. Fix the developed gel in 25% ethanol.

Notes: It should be kept in mind that the amount of NADPH added to the staining solution is critical for development of nonfluorescent GLYD bands. When too much NADPH is added, the enzyme should not be able to convert enough NADPH into NADP for nonfluorescent bands to become visible.

After electrophoresis of mammalian tissue preparations some isozymes of NADP-dependent alcohol dehydrogenase (see 1.1.1.2 — ADH(NADP)), as well as glucuronate reductase (see 1.1.1.19 — GLR) bands, may also appear on GLYD zymograms obtained by this method.

References

1. Menken, S.B.J., Allozyme Polymorphism and the Speciation Process in Small Ermine Moths (Lepidoptera, Yponomeutidae), Ph.D. thesis, University of Leiden, The Netherlands, 1980.
2. Mather, P.B. and Holmes, R.S., Aldehyde reductase isozymes in the mouse: evidence for two new loci and localization of *Ahr-3* on chromosome 7, *Biochem. Genet.*, 23, 483, 1985.

1.1.1.73 — Octanol Dehydrogenase; ODH

REACTION   Octanol + NAD = octanal + NADH
ENZYME SOURCE   Fungi, invertebrates, vertebrates
SUBUNIT STRUCTURE Dimer (fungi, invertebrates, vertebrates)

METHOD

Visualization Scheme

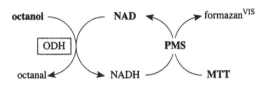

Staining Solution[1]

0.1 *M* Tris–HCl buffer, pH 8.4	100 ml
Octanol	0.25 ml
NAD	8 mg
MTT	10 mg
PMS	2 mg

Procedure

Incubate the gel in staining solution in the dark at 37°C until dark blue bands appear. Wash the stained gel in water and fix in 25% ethanol.

Notes: Usually octanol is dissolved in 1 to 2 ml of ethanol before being included in a staining solution. However, this requires control staining of an additional gel for alcohol dehydrogenase (see 1.1.1.1 — ADH, Method 1).

REFERENCES

1. Menken, S.B.J., Allozyme Polymorphism and the Speciation Process in Small Ermine Moths (Lepidoptera, Yponomeutidae), Ph.D. thesis, University of Leiden, The Netherlands, 1980.

1.1.1.78 — D-Lactaldehyde Dehydrogenase; DLAD

OTHER NAMES   Methylglyoxal reductase
REACTION   (D)-Lactaldehyde + NAD = methylglyoxal + NADH
ENZYME SOURCE   Fungi, invertebrates
SUBUNIT STRUCTURE Dimer (fungi, invertebrates; see *Notes*)

METHOD

Visualization Scheme

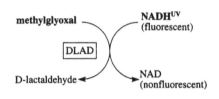

Staining Solution*

0.2 *M* Tris–HCl, pH 8.5	15 ml
NADH	7 mg
Methylglyoxal	50 µl

Procedure

Apply the staining solution to gel on a filter paper overlay and incubate at 37°C until dark (nonfluorescent) bands visible in long-wave UV light appear on a light (fluorescent) background. Record the zymogram or photograph using a yellow filter.

These bands may be made visible in daylight (white bands on a blue background) after treating the processed gel with MTT–PMS solution applied as a filter paper or 1% agar overlay.

Notes: In some living organisms DLAD may be identical to octanol dehydrogenase (see 1.1.1.73 — ODH) or formaldehyde dehydrogenase (see 1.2.1.1 — FDHG). For example, identical allozyme patterns were revealed on zymograms of ODH, FDHG, and DLAD in the mushroom *Boletus edulis*, the phoronid *Phoronopsis harmeri*, the sipunculid *Phascolosoma japonicum*, and in the four bivalve species of the genus *Macoma* (*M. irus*, *M. incongrua*, *M. baltica*, and *M. contabulata*).[1]

REFERENCES

1. Manchenko, G.P., The use of allozymic patterns in genetic identification of enzymes, *Isozyme Bull.*, 23, 102, 1990.

* New; recommended for use.

1.1.1.80 — Isopropanol Dehydrogenase (NADP); IPDH

REACTION	Propan-2-ol + NADP = acetone + NADPH
ENZYME SOURCE	Invertebrates
SUBUNIT STRUCTURE	Unknown or no data available

METHOD

Visualization Scheme

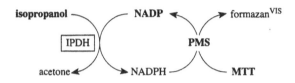

Staining Solution[1]

0.1 M Tris–HCl buffer, pH 8.4	100 ml
Isopropanol	4 ml
NADP	8 mg
MTT	10 mg
PMS	2 mg

Procedure

Incubate the gel in staining solution in the dark at 37°C until dark blue bands appear. Rinse the stained gel in water and fix in 25% ethanol.

Notes: The enzyme also acts on other short-chain secondary alcohols, and, slowly, on primary alcohols.[2]

REFERENCES

1. Menken, S.B.J., Allozyme Polymorphism and the Speciation Process in Small Ermine Moths (Lepidoptera, Yponomeutidae), Ph.D. thesis, University of Leiden, The Netherlands, 1980.
2. NC-IUBMB, *Enzyme Nomenclature*, Academic Press, San Diego, 1992, p. 34 (EC 1.1.1.80, Comments).

1.1.1.90 — Aryl-Alcohol Dehydrogenase; AAD

OTHER NAMES	*p*-Hydroxybenzyl alcohol dehydrogenase
REACTION	Aromatic alcohol + NAD = aromatic aldehyde + NADH (see General Notes)
ENZYME SOURCE	Fungi, plants
SUBUNIT STRUCTURE	Dimer (plants)

METHOD

Visualization Scheme

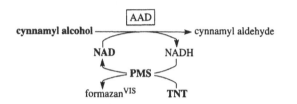

Staining Solution[1]

0.1 M Tris–HCl buffer, pH 8.8, saturated with cynnamyl alcohol (1.6 µl/ml)	25 ml
5 mg/ml NAD	2 ml
2 mg/ml Tetranitro blue tetrazolium (TNT)	2 ml
2.5 mg/ml PMS	0.2 ml

Procedure

Incubate the gel in staining solution in the dark at room temperature until colored bands appear. Wash the stained gel in water and fix in 25% ethanol.

Notes: Activity of AAD was found to be severely inhibited when extracts for electrophoresis were made using phosphate, citrate, or maleate buffers at pH values 5.4 to 5.8. Complete isozyme patterns of the enzyme in rye and triticale species were obtained with Tris buffers at pH values 8.0 to 8.8.[2]

GENERAL NOTES

AADs are a group of enzymes with broad specificity toward primary alcohols with an aromatic or cyclohex-1-ene ring, but with low or no activity toward short-chain aliphatic alcohols.[3]

REFERENCES

1. Jaaska, V., NAD-dependent aromatic alcohol dehydrogenase in wheats (*Triticum* L.) and goatgrasses (*Aegilops* L.): evolutionary genetics, *Theor. Appl. Genet.*, 67, 535, 1984.
2. Jaaska, V. and Jaaska, V., Isoenzymes of aromatic alcohol dehydrogenase in rye and triticale, *Biochem. Physiol. Pflanzen*, 179, 21, 1984.
3. NC-IUBMB, *Enzyme Nomenclature*, Academic Press, San Diego, 1992, p. 35 (EC 1.1.1.90, Comments).

REACTION	Aromatic alcohol + NADP = aromatic aldehyde + NADPH
ENZYME SOURCE	Fungi, plants
SUBUNIT STRUCTURE	Oligomer# (fungi), monomer (plants)

METHOD

Visualization Scheme

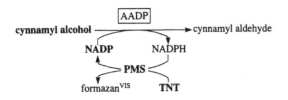

Staining Solution[1]

0.1 M Tris–HCl buffer, pH 8.8, saturated with cynnamyl alcohol (1.6 µl/ml)	25 ml
2.5 mg/ml NADP	2 ml
2 mg/ml Tetranitro blue tetrazolium (TNT)	2 ml
2.5 mg/ml PMS	0.2 ml

Procedure

Incubate the gel in staining solution in the dark at room temperature until colored bands appear. Wash the stained gel in water and fix in 25% ethanol.

Notes: The AADP isozyme patterns were found to be dependent on the composition of the homogenization buffer used to prepare enzyme extracts for electrophoresis. Thus, extracts made in phosphate, citrate, or maleate buffers at pH values 5.4 to 5.8 displayed no AADP activity bands. The enzyme activity could not be restored by the alkalizing of acid enzyme extracts with pH values below 5.0, providing evidence that the enzyme is irreversibly denatured at pH values below 5.0. Complete isozyme patterns of the enzyme in rye and triticale species were obtained with Tris buffers at pH values 8.0 to 8.8.[2]

GENERAL NOTES

The enzyme also acts on some aliphatic aldehydes, but cinnamaldehyde is the best substrate found.[3]

REFERENCES

1. Jaaska, V., NADP-dependent aromatic alcohol dehydrogenase in polyploid wheats and their diploid relatives: on the origin and phylogeny of polyploid wheats, *Theor. Appl. Genet.*, 53, 209, 1978.
2. Jaaska, V. and Jaaska, V., Isoenzymes of aromatic alcohol dehydrogenase in rye and triticale, *Biochem. Physiol. Pflanzen*, 179, 21, 1984.
3. NC-IUBMB, *Enzyme Nomenclature*, Academic Press, San Diego, 1992, p. 35 (EC 1.1.1.91, Comments).

1.1.1.95 — Phosphoglycerate Dehydrogenase; PGLYD

REACTION	3-Phosphoglycerate + NAD = 3-phosphohydroxypyruvate + NADH
ENZYME SOURCE	Bacteria, algae, invertebrates, vertebrates
SUBUNIT STRUCTURE	Tetramer[#] (bacteria)

METHOD

Visualization Scheme

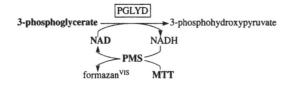

Staining Solution (adapted from 1.1.1.1 — ADH, Method 1)

50 mM Tris–HCl buffer, pH 8.5	50 ml
3-Phosphoglyceric acid	150 mg
NAD	30 mg
MTT	10 mg
PMS	1 mg

Procedure

Incubate the gel in staining solution in the dark at 37°C until dark blue bands appear. Wash the stained gel in water and fix in 25% ethanol.

Notes: The enzyme from chicken liver catalyzes the reverse reaction seven times faster than the forward reaction. However, the commercially available substrate of the reverse PGLYD reaction, hydroxypyruvic acid dimethylketal phosphate (Sigma), is expensive and its preparation should be completed at laboratory conditions.

An allozymic variation of PGLYD was revealed in two species of swallowtail butterflies.[1] Unfortunately, allozyme patterns in homo- and heterozygous individuals were not described by the authors, so the subunit structure of PGLYD molecules can not be inferred from these data.

REFERENCES

1. Hagen, R.H. and Scriber, J.M., Systematics of the *Papilio glaucus* and *P. troilus* species groups (Lepidoptera: Papilionidae): inferences from allozymes, *Ann. Entomol. Soc. Am.*, 84, 380, 1991.

1.1.1.96 — Aromatic α-Keto Acid Reductase; AKAR

OTHER NAMES	Diiodophenylpyruvate reductase (recommended name), α-keto acid reductase
REACTION	3-(3,5-Diiodo-4-hydroxyphenyl) lactate + NAD = 3-(3,5-diiodo-4-hydroxyphenyl) pyruvate + NADH
ENZYME SOURCE	Vertebrates
SUBUNIT STRUCTURE	Unknown or no data available

METHOD

Visualization Scheme

p-hydroxyphenylpyruvic acid

AKAR

NADH^{UV}
(fluorescent)

NAD
(nonfluorescent)

p-hydroxyphenyllactic acid

Staining Solution[1]

A. 0.2 *M* Tris–HCl buffer, pH 7.6   25 ml
    NADH   10 mg
    p-Hydroxyphenylpyruvic acid   25 mg
    L-Lactic acid   25 mg
B. 2% Agar solution (60°C)   25 ml

Procedure

Mix A and B components of the staining solution and pour the mixture over the gel surface. Incubate the gel at 37°C for 1 to 3 h. Dark (nonfluorescent) bands visible in long-wave UV light on a light (fluorescent) background indicate areas of AKAR localization. Record the zymogram or photograph using a yellow filter.

Notes: If too much NADH is added, the enzyme should not be able to convert enough NADH into NAD for nonfluorescent bands to become visible.

A negative zymogram (white bands on a blue background) visible in daylight may be obtained after treatment of the processed gel with the PMS–MTT solution.

Lactic acid is added to the staining solution to inhibit lactate dehydrogenase, which also displays low activity toward p-hydroxyphenylpyruvic acid.

The addition of L-malic acid (final concentration of 11 mg/ml; neutralized before addition) is also recommended to prevent development of cytoplasmic malate dehydrogenase bands.[2]

The reverse reaction, using p-hydroxyphenyllactic acid as a substrate and NAD as a cofactor, results in bands similar to those seen with the forward reaction. However, lactate dehy-drogenase is very prominent, even in the presence of pyruvic acid (inhibitor of the forward LDH reaction), and the AKAR bands are poorly defined after prolonged incubation of the gel.

GENERAL NOTES

AKAR substrates contain an aromatic ring with a pyruvate side chain. Halogenated derivatives are the most active AKAR substrates.[3]

REFERENCES

1. Donald, L.J., A description of human aromatic α-keto acid reductase, *Ann. Hum. Genet.*, 46, 299, 1982.
2. Friedrich, C.A., Morizot, D.C., Siciliano, M.J., and Ferrell, R.E., The reduction of aromatic alpha-keto acids by cytoplasmic malate dehydrogenase and lactate dehydrogenase, *Biochem. Genet.*, 25, 657, 1987.
3. NC-IUBMB, *Enzyme Nomenclature*, Academic Press, San Diego, 1992, p. 35 (EC 1.1.1.96, Comments).

1.1.1.103 — L-Threonine 3-Dehydrogenase; TRDH

REACTION	L-Threonine + NAD = L-2-amino-3-oxobutanoate + NADH
ENZYME SOURCE	Bacteria, vertebrates
SUBUNIT STRUCTURE	Tetramer# (bacteria), dimer# (vertebrates)

METHOD

Visualization Scheme

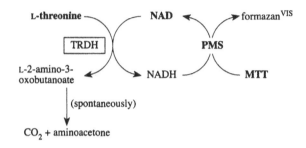

(spontaneously)

CO_2 + aminoacetone

Staining Solution[1]

0.015 M Sodium phosphate buffer, pH 7.0	50 ml
L-Threonine	50 mg
1.25% MTT	1 ml
1% PMS	0.5 ml
1% NAD	2 ml

Procedure

Incubate the gel in staining solution in the dark at 37°C until dark blue bands appear. Rinse the developed gel in water and fix in 25% ethanol.

GENERAL NOTES

The product spontaneously decarboxylates to aminoacetone.[2]

REFERENCES

1. Selander, R.K., Caugant, D.A., Ochman, H., Musser, J.J.M., Gilmour, M.N., and Whittam, T.S., Methods of multilocus enzyme electrophoresis for bacterial population genetics and systematics, *Appl. Environ. Microbiol.*, 51, 873, 1986.
2. NC-IUBMB, *Enzyme Nomenclature*, Academic Press, San Diego, 1992, p. 36 (EC 1.1.1.103, Comments).

1.1.1.105 — Retinol Dehydrogenase; RDH

REACTION	Retinol + NAD = retinal + NADH
ENZYME SOURCE	Vertebrates
SUBUNIT STRUCTURE	Unknown or no data available

METHOD

Visualization Scheme

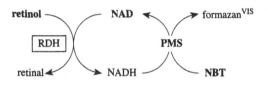

Staining Solution[1]

0.01 M Phosphate buffer, pH 7.0	100 ml
NAD	66 mg
NBT	35 mg
PMS	2 mg
Retinol (dissolved in minimal volume of acetone)	100 mg

Procedure

Incubate the gel in staining solution in the dark at 37°C until dark blue bands appear. Rinse the developed gel in water and fix in 25% ethanol.

REFERENCES

1. Koen, A.L. and Shaw, C.R., Retinol and alcohol dehydrogenases in retina and liver, *Biochim. Biophys. Acta*, 128, 48, 1966.

1.1.1.122 — L-Fucose Dehydrogenase; FUCDH

OTHER NAMES	D-*threo*-Aldose dehydrogenase (recommended name), (2*S*,3*R*)-aldose dehydrogenase
REACTION	D-*threo*-Aldose + NAD = D-*threo*-aldono-1,5-lactone + NADH
ENZYME SOURCE	Bacteria, invertebrates, vertebrates
SUBUNIT STRUCTURE	Monomer (invertebrates)

METHOD

Visualization Scheme

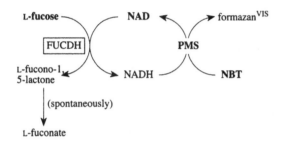

Staining Solution[1]

10 mM Tris–HCl buffer, pH 8.0
30 mM L-Fucose
6 mM NAD
0.08% NBT
0.014% PMS

Procedure

Incubate the gel in staining solution in the dark at 37°C until dark blue bands appear. Wash the stained gel in water and fix in 25% ethanol.

Notes: The enzyme acts on L-fucose, D-arabinose, L-xylose, and D-lyxose. The animal enzyme also acts on L-galactose, but shows the fastest rate with L-fucose. The enzyme from *Pseudomonas* acts on L-glucose.[1,2]

REFERENCES

1. Schachter, H., Sarney, J., McGuire, E.J., and Roseman, S., Isolation of diphosphopyridine nucleotide-dependent L-fucose dehydrogenase from pork liver, *J. Biol. Chem.*, 244, 4785, 1969.
2. NC-IUBMB, *Enzyme Nomenclature*, Academic Press, San Diego, 1992, p. 38 (EC 1.1.1.122, Comments).

1.1.1.138 — Mannitol 2-Dehydrogenase (NADP); MD(NADP)

REACTION	D-Mannitol + NADP = D-fructose + NADPH
ENZYME SOURCE	Fungi
SUBUNIT STRUCTURE	Unknown or no data available

METHOD

Visualization Scheme

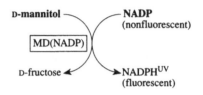

Staining Solution[1]

0.05 M Tris–HCl buffer, pH 8.5	10 ml
NADP	10 mg
D-Mannitol	100 mg

Procedure

Lay a piece of filter paper saturated with the staining solution on top of the gel and incubate at 37°C for 30 to 60 min. Remove the filter paper and view the gel under long-wave UV light. Fluorescent bands of MD(NADP) are visible on the dark (nonfluorescent) background of the gel. Record the zymogram or photograph using a yellow filter.

Notes: When a final zymogram easily visible in daylight is required, counterstain the processed gel with MTT–PMS solution. This will result in the appearance of dark blue bands of MD(NADP). The postcoupling technique is recommended because of the presumed inhibitory effect of PMS or MTT (or both) on MD(NADP) activity.

REFERENCES

1. Royse, D.J. and May, B., Use of isozyme variation to identify genotypic classes of *Agaricus brunnescens*, *Mycologia*, 74, 93, 1982.

1.1.1.141 — 15-Hydroxyprostaglandin Dehydrogenase (NAD); HPD(NAD)

REACTION   (5Z,13E)-(15S)-11α,15-Dihydroxy-9-oxoprost-13-enoate + NAD = (5Z,13E)-11α-hydroxy-9,15-dioxoprost-13-enoate + NADH

ENZYME SOURCE   Vertebrates

SUBUNIT STRUCTURE Unknown or no data available

Method 1

Visualization Scheme

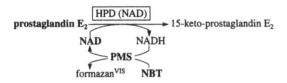

Staining Solution[1]

A. 100 mM Tris–HCl buffer, pH 9.5
    20 µg/ml Prostaglandin E_2
    3 mM NAD

B. 100 mM Tris–HCl buffer, pH 9.5
    0.5 mg/ml NBT
    0.05 mg/ml PMS

Procedure

Incubate the gel in solution A for 30 min at 37°C and then in solution B for 30 min in the dark until dark blue bands appear. Wash the stained gel in water and fix in 25% ethanol.

Notes: This method was used to detect activity bands of purified HPD(NAD) from human placenta after electrophoresis in PAG containing 20% (v/v) glycerol. The convenience of this method for detecting the enzyme after electrophoresis of crude enzyme preparations is doubtful.

Method 2

Visualization Scheme

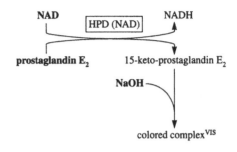

Staining Solution[1]

A. 100 mM Tris–HCl buffer, pH 9.5
    20 µg/ml Prostaglandin E_2
    3 mM NAD

B. 0.5 N NaOH

Procedure

Incubate the gel in solution A for 30 min at 37°C and then in solution B at room temperature until colored bands appear.

Notes: The method is based on the development of the alkali-derived chromophore of 15-keto-prostaglandin E_2.[2] This method was used to detect activity bands of purified HPD(NAD) from human placenta after electrophoresis in PAG containing 20% (v/v) glycerol. The convenience of this method for detecting the enzyme after electrophoresis of crude enzyme preparations is doubtful.

General Notes

Method 2 is as sensitive as Method 1, but more specific because it directly detects the product of enzyme reaction. Both methods can be adapted to detect activity bands of NADP-dependent 15-hydroxyprostaglandin dehydrogenase (EC 1.1.1.197).[1]

The enzyme acts on prostaglandin E_2, $E_{2\alpha}$, and B_1, but not on prostaglandin D_2.[3]

References

1. Tanaka, T. and Mori, N., Specific activity staining for prostaglandin metabolizing enzymes on polyacrylamide gel, *Prostaglandins Leukotriens Med.*, 23, 267, 1986.
2. Anggard, E., Larsson, C., and Samuelsson, B., The distribution of 15-hydroxyprostaglandin dehydrogenase and prostaglandin-13delta-reductase in tissues of the swine, *Acta Physiol. Scand.*, 81, 396, 1971.
3. NC-IUBMB, *Enzyme Nomenclature*, Academic Press, San Diego, 1992, p. 41 (EC 1.1.1.141, Comments).

1.1.1.145 — 3β-Hydroxy-Δ⁵-Steroid Dehydrogenase; 3β-HSD

OTHER NAMES	Progesterone reductase
REACTION	3β-Hydroxy-Δ⁵-steroid + NAD = 3-oxo-Δ⁵-steroid + NADH
ENZYME SOURCE	Vertebrates
SUBUNIT STRUCTURE	Unknown or no data available

METHOD

Visualization Scheme

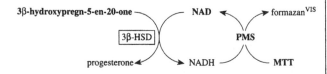

Staining Solution[1]

0.1 M Tris–HCl buffer, pH 8.0	50 ml
3β-Hydroxypregn-5-en-20-one (dissolved in minimal volume of ethanol or isopropanol)	50 mg
NAD	20 mg
MTT	10 mg
PMS	1 mg

Procedure

Incubate the gel in staining solution in the dark at 37°C until dark blue bands appear. Wash the stained gel in water and fix in 25% ethanol.

Notes: An additional gel should be stained for alcohol dehydrogenase (see 1.1.1.1 — ADH) as a control. The enzyme acts on 3β-hydroxyandrost-5-en-17-one to form androst-4-ene-3,17-dione and on 3β-hydroxypregn-5-en-20-one to form progesterone.[2]

REFERENCES

1. Engel, W., Frowein, J., Krone, W., and Wolf, U., Induction of testis alcohol dehydrogenase in prepubertal rats: I. The effects of human chorion gonadotropine (HCG), theophylline, and dibutyryl cyclic AMP, *Clin. Gen.*, 3, 34, 1971.
2. NC-IUBMB, *Enzyme Nomenclature*, Academic Press, San Diego, 1992, p. 41 (EC 1.1.1.145, Comments).

1.1.1.179 — D-Xylose 1-Dehydrogenase (NADP); XD(NADP)

REACTION	D-Xylose + NADP = D-xylono-1,5-lactone + NADPH
ENZYME SOURCE	Vertebrates
SUBUNIT STRUCTURE	Dimer# (vertebrates)

METHOD

Visualization Scheme

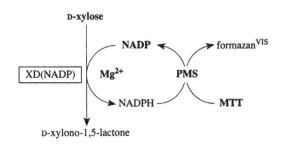

Staining Solution[1]

A. 1 M Tris–HCl buffer, pH 8.0
   12 mM MgCl₂
   0.48 mM NADP
   2 mM MTT
   0.6 mM PMS
   0.2 M D-Xylose
B. 1.8% Agar solution (60°C)

Procedure

Mix A and B components of the staining solution and pour the mixture over the gel surface. Incubate the gel at 37°C in the dark until dark blue bands appear. Fix the stained gel in 25% ethanol.

Notes: The enzyme also acts, more slowly, on L-arabinose, D-ribose, D-glucose, D-galactose, L-fucose, D-fucose, 6-deoxy-D-glucose, and 2-deoxy-D-glucose. No activity was detected with ethanol, D-erythrose, D-fructose, glycerol, D-mannose, ribitol, sorbitol, sucrose, xylitol, D-glucose 1-phosphate, D-glucose-6-phosphate, and D-ribose 5-phosphate.

GENERAL NOTES

The mammalian enzyme is shown to be identical to EC 1.3.1.20.[2]

REFERENCES

1. Newton, M.F., Nash, H.R., Peters, J., and Andrews, S.J., Xylose dehydrogenase-1, a new gene on mouse chromosome 7, *Biochem. Genet.*, 20, 733, 1982.
2. Aoki, S., Ishikura, S., Asada, Y., Usami, N., and Hara, A., Identity of dimeric dihydrodiol dehydrogenase as NADP(+)-dependent D-xylose dehydrogenase in pig liver, *Chem. Biol. Interact.*, 130, 775, 2001.

REACTIONS	1. Hypoxanthine + NAD + H_2O = xanthine + NADH
	2. Xanthine + NAD + H_2O = urate + NADH
ENZYME SOURCE	Green algae, fungi, plants, invertebrates, vertebrates
SUBUNIT STRUCTURE	Dimer (invertebrates, vertebrates)

Method

Visualization Scheme

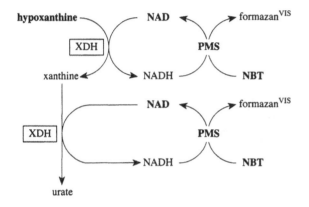

Staining Solution[1]

0.1 M Tris–HCl buffer, pH 8.0	97 ml
M Hypoxanthine (dissolved in water by hitting or in minimal volume of 1 M KOH)	3 ml
NAD	60 mg
NBT	30 mg
PMS	2 mg

Procedure

Incubate the gel in staining solution in the dark at 37°C until dark blue bands appear. Wash the stained gel in water and fix in 25% ethanol.

Notes: Xanthine oxidase (see 1.1.3.22 — XOX) bands can also develop in XDH staining solution. Therefore, an additional gel should be stained for XOX as a control. In invertebrates much of the activity of isozymes developed on gels stained for XDH can be attributed to XOX. Moreover, there is some doubt that XDH exists in vertebrates.[2]

General Notes

The enzyme acts on a variety of purines and aldehydes. Animal XDH can be interconverted to oxidase form (see 1.1.3.22 — XOX). In animal liver XDH exists *in vivo* mainly in the dehydrogenase form, but can be converted into the oxidase form by storage at –20°C, by treatment with organic solvents or proteolytic agents, or by thiol reagents (e.g., Cu^{2+}, *N*-methylmaleimide, or 4-hydroxymercuribenzoate). The effect of thiol reagents can be reversed by thiols such as 1,4-dithioerythritol. XDH can also be converted into XOX by an enzyme–thiol transhydrogenase (EC 1.8.4.7) catalyzing reaction: XDH + oxidized glutathione = XOX + reduced glutathione. In other animal tissues the enzyme exists almost entirely in the oxidase form, but can be converted into the dehydrogenase form by 1,4-dithioerythritol.[3]

References

1. Shaw, C.R. and Prasad, R., Starch gel electrophoresis of enzymes: a compilation of recipes, *Biochem. Genet.*, 4, 297, 1970.
2. Adams, M., Baverstock, P.R., Watts, C.H.S., and Gutman, G.A., Enzyme markers in inbred rat strains: genetics of new markers and strain profiles, *Biochem. Genet.*, 22, 611, 1984.
3. NC-IUBMB, *Enzyme Nomenclature*, Academic Press, San Diego, 1992, p. 49 (EC 1.1.1.204, Comments).

1.1.1.205 — IMP Dehydrogenase; IMPDH

REACTION	Inosine-5'-phosphate + NAD + H_2O = xanthosine 5'-phosphate + NADH
ENZYME SOURCE	Bacteria, plants, protozoa, invertebrates, vertebrates
SUBUNIT STRUCTURE	Tetramer# (bacteria, protozoa, vertebrates)

METHOD

Visualization Scheme

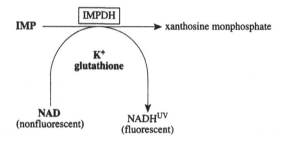

Staining Solution[1]

165 mM Tris–HCl buffer, pH 8.1
300 mM KCl
6 mM Glutathione (reduced)
6 mM NAD
3 mM Inosine-5'-phosphate

Procedure

Lay a piece of filter paper saturated with the staining solution on top of the gel and incubate at 37°C in a moist chamber for 30 to 60 min. Remove the filter paper and view the gel under long-wave UV light. Light (fluorescent) bands of IMPDH are visible on the dark (nonfluorescent) background of the gel. Record the zymogram or photograph using a yellow filter.

Notes: When a zymogram visible in daylight is required, counterstain the processed gel with MTT–PMS solution to obtain dark blue bands on a light blue background. The postcoupling technique is recommended because of nonspecific reduction of MTT via reduced glutathione. Reduced glutathione is required for activity of bacterial IMPDH. When the enzyme from plant or animal sources is studied, reduced glutathione may be omitted from the staining solution and PMS and MTT added directly to the staining solution to develop IMPDH activity bands visible in daylight by a routine one-step procedure.

REFERENCES

1. Van Diggelen, O.P. and Shin, S., A rapid fluorescence technique for electrophoretic identification of hypoxanthine phosphoribosyltransferase allozymes, *Biochem. Genet.*, 12, 375, 1974.

1.1.1.215 — Gluconate 2-Dehydrogenase; 2-KGR

OTHER NAMES	2-Keto-D-gluconate reductase, 2-ketoaldonate reductase
REACTIONS	1. D-Gluconate + NAD(P) = 2-keto-D-gluconate + NAD(P)H
	2. 5-Keto-D-gluconate + NAD(P) = 2,5-diketo-D-gluconate + NAD(P)H
ENZYME SOURCE	Bacteria
SUBUNIT STRUCTURE	Unknown or no data available

METHOD 1

Visualization Scheme

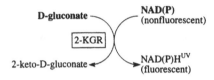

Staining Solution[1]

A. 0.2 *M* Tris–HCl, pH 8.0
1 m*M* NAD (or NADP)
B. 0.2 *M* Tris–HCl, pH 8.0
0.1 *M* D-Gluconate

Procedure

Following electrophoresis, soak the gel in solution A for ca. 3 min. Then rinse the gel quickly in water to remove any cofactor from the gel surface, and after a few minutes, view under long-wave UV light to detect any nonsubstrate-dependent fluorescent bands. Overlay the gel with a filter paper strip soaked in solution B. Observe fluorescent bands of 2-KGR in UV light. Record the zymogram or photograph using a yellow filter.

Notes: When a zymogram that is visible in daylight is required, counterstain the processed gel with MTT–PMS solution to obtain blue bands in the gel areas where 2-KGR activity is present. Fix the stained gel in 25% ethanol.
5-Keto-D-gluconate can be used as a substrate in place of D-gluconate.

METHOD 2

Visualization Scheme

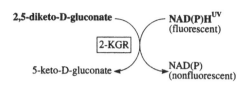

Staining Solution[1]

A. 0.2 *M* Tris–HCl, pH 7.0
1 m*M* NADH (or NADPH)
B. 0.2 *M* Tris–HCl, pH 7.0
0.1 *M* 2,5-Diketo-D-gluconate

Procedure

Following electrophoresis, soak the gel in solution A for ca. 3 min. Then rinse the gel quickly in water to remove any cofactor from the gel surface, and after a few minutes, view under long-wave UV light to detect any nonsubstrate-dependent dark (nonfluorescent) reductase bands on the fluorescent background of the gel. Overlay the gel with a filter paper strip soaked in solution B. Observe dark nonfluorescent bands of 2-KGR on the fluorescent background of the gel. Record the zymogram or photograph using a yellow filter.

Notes: When a zymogram that is visible in daylight is required, counterstain the processed gel with MTT–PMS solution to obtain achromatic bands of 2-KGR on a blue background of the gel. Fix the stained gel in 25% ethanol.
2-Keto-D-gluconate can be used as a substrate in place of 2,5-diketo-D-gluconate.

GENERAL NOTES

The enzyme also acts on L-idonate, D-galactonate, and D-xylonate.[2]

REFERENCES

1. Seymour, J.L. and Lazarus, R.A., Native gel activity stain and preparative electrophoretic method for the detection and purification of pyridine nucleotide-linked dehydrogenases, *Anal. Biochem.*, 178, 243, 1989.
2. NC-IUBMB, *Enzyme Nomenclature*, Academic Press, San Diego, 1992, p. 50 (EC 1.1.1.215, Comments).

REACTION (supposed)   Choline + NAD = betaine aldehyde + NADH

ENZYME SOURCE   Plants

SUBUNIT STRUCTURE   Unknown or no data available

METHOD

Visualization Scheme

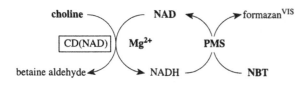

Staining Solution[1]

0.05 M Tris–HCl buffer, pH 7.5	25 ml
Choline chloride	175 mg
$MgSO_4$	25 mg
10 mg/ml NAD	1.5 ml
10 mg/ml NBT	0.5 ml
5 mg/ml PMS	0.5 ml

Procedure

Incubate the gel in staining solution in the dark at 37°C until dark blue bands appear. Wash the developed gel in water and fix in 25% ethanol.

Notes: This enzyme is not yet included in the enzyme list.[2] It may be that the activity bands developed in the staining solution presented above are caused by choline oxidase (EC 1.1.3.17), which uses oxygen as an acceptor. This enzyme contains a tightly bound FAD group that can interchange between the oxidized and reduced states. The reduced flavoprotein can chemically reduce NBT via PMS without the need for NAD. Unlike CD(NAD), choline oxidase produces hydrogen peroxide, which can be detected using the linked peroxidase reaction (see 1.11.1.7 — PER). Another enzyme, choline dehydrogenase (EC 1.1.99.1), can use PMS as an acceptor, and thus can also be developed in a staining solution for CD(NAD). The bands caused by this enzyme can be identified by the use of the staining solution identical to that presented above but lacking NAD.

In *Drosophila melanogaster*, activity bands of CD(NAD) are caused by alcohol dehydrogenase (see 1.1.1.1 — ADH).[3]

REFERENCES

1. Cheliak, W.M. and Pitel, J.A., Techniques for Starch Gel Electrophoresis of Enzymes from Forest Tree Species, Information Report PI-X-42, Petawawa National Forestry Institute, Canadian Forestry Service, Agriculture Canada, Chalk River, Ontario, 1984.

2. NC-IUBMB, *Enzyme Nomenclature*, Academic Press, San Diego, 1992.

3. Eisses, K.Th., Schoonen, G.E.J., Scharloo, W., and Thörig, G.E.W., Evidence for a multiple function of the alcohol dehydrogenase allozyme ADH[71k] of *Drosophila melanogaster*, *Comp. Biochem. Physiol.*, 82B, 863, 1985.

1.1.1.X' — 2,5-Diketo-ᴅ-Gluconate Reductase; DGR

REACTION   2-Keto-ᴅ-gluconate + NAD(P) = 2,5-diketo-ᴅ-gluconate + NAD(P)H

ENZYME SOURCE   Bacteria

SUBUNIT STRUCTURE   Monomer[#] (bacteria)

Method

Visualization Scheme

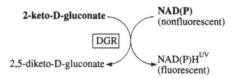

Staining Solution[1]

A.   0.2 *M* Tris–HCl, pH 8.0
     1 m*M* NAD (or NADP)

B.   0.2 *M* Tris–HCl, pH 8.0
     0.1 *M* 2-Keto-ᴅ-gluconate

Procedure

Following electrophoresis, soak the gel in solution A for ca. 3 min. Then rinse the gel quickly in water to remove any cofactor from the gel surface, and after a few minutes, view under long-wave UV light to detect any nonsubstrate-dependent fluorescent bands. Overlay the gel with a filter paper strip soaked in solution B. Observe fluorescent bands of DGR in UV light. Record the zymogram or photograph using a yellow filter.

Notes: When a zymogram that is visible in daylight is required, counterstain the processed gel with MTT–PMS solution to obtain blue bands in the gel areas where DGR activity is present. Fix the stained gel in 25% ethanol.

References

1. Seymour, J.L. and Lazarus, R.A., Native gel activity stain and preparative electrophoretic method for the detection and purification of pyridine nucleotide-linked dehydrogenases, *Anal. Biochem.*, 178, 243, 1989.

1.1.3.4 — Glucose Oxidase; GO

OTHER NAMES	Glucose oxyhydrase
REACTION	β-D-Glucose + O_2 = D-glucono-1,5-lactone + H_2O_2
ENZYME SOURCE	Fungi, plants, invertebrates, vertebrates
SUBUNIT STRUCTURE	Monomer (invertebrates)

METHOD 1

Visualization Scheme

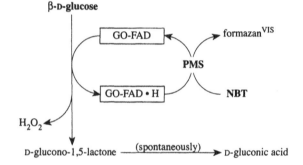

Staining Solution[1]

0.05 *M* Tris–HCl buffer, pH 8.0
37 m*M* β-D-Glucose
0.24 m*M* NBT
0.16 m*M* PMS

Procedure

Incubate the gel in staining solution in the dark at 37°C until dark blue bands appear. Wash the stained gel in water and fix in 25% ethanol.

Notes: An additional gel should be stained in a staining solution lacking glucose as a control. GO activity bands also may develop on glucose dehydrogenase (1.1.1.47 — GD) and hexokinase (2.7.1.1 — HK) zymograms, as well as on the zymograms of other enzymes where exogenous hexokinase and glucose-6-phosphate dehydrogenase are included in the staining solution to catalyze coupled reactions in the presence of MTT (or NBT) and PMS.

METHOD 2

Visualization Scheme

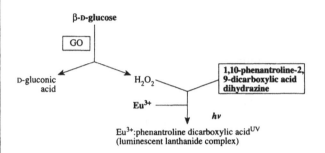

Staining Solution[2] (adapted)

0.1 *M* Acetate buffer, pH 6.0
10 m*M* β-D-Glucose
0.01 m*M* 1,10-Phenantroline-2,9-dicarboxylic acid dihydrazide
0.02 m*M* $EuCl_3$

Procedure

Apply the staining solution to the gel surface using a filter paper or 1% agarose overlay. Incubate the gel at 37°C and monitor under a UV lamp emitting in the midultraviolet region, 300 to 340 nm, for fluorescent (luminescent) bands. Record the zymogram or photograph the developed gel on Polaroid instant film using a time-resolved photographic camera (e.g., TRP 100 camera produced by Kronem Systems, Inc., Mississauga, Ontario, Canada) and filters that pass 320- to 340-nm excitation and above 515-nm emission wavelengths.

Notes: The method allows detection of less than 10^{-3} U/ml GO.

The main disadvantage of the method is that 1,10-phenantroline-2,9-dicarboxylic acid dihydrazide is not yet commercially available and should be synthesized under laboratory conditions.

Applications of the staining solution with a 1% agarose overlay are preferable because it prevents rapid diffusion of a soluble lanthanide chelate.

OTHER METHODS

A. The product hydrogen peroxide can be detected using exogenous peroxidase as a linking enzyme (see 1.11.1.7 — PER).

B. Immunoblotting procedure (for details, see Part II) may also be used to detect GO. Antibodies specific to the enzyme from *Aspergillus niger* are now available from Sigma.

1.1.3.4 — Glucose Oxidase; GO (continued)

GENERAL NOTES

The enzyme is a flavoprotein (FAD).[3]

REFERENCES

1. Cavener, D.R., Genetics of male-specific glucose oxidase and the identification of other unusual hexose enzymes in *Drosophila melanogaster*, *Biochem. Genet.*, 18, 929, 1980.
2. Evangelista, R.A., Pollak, A., and Templeton, E.F.G., Enzyme-amplified lanthanide luminescence for enzyme detection in bioanalytical assays, *Anal. Biochem.*, 197, 213, 1991.
3. NC-IUBMB, *Enzyme Nomenclature*, Academic Press, San Diego, 1992, p. 56 (EC 1.1.3.4, Comments).

1.1.3.13 — Alcohol Oxidase; ALOX

REACTION — Primary alcohol + O_2 + H_2O = aldehyde + H_2O_2

ENZYME SOURCE — Fungi

SUBUNIT STRUCTURE — Octamer[#] (fungi)

METHOD

Visualization Scheme

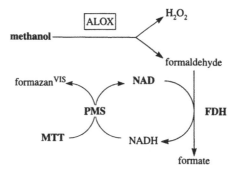

Staining Solution[1] (adapted)

0.05 *M* Potassium phosphate buffer, pH 7.5	50 ml
Methanol	1 ml
Formaldehyde dehydrogenase (FDH; EC 1.2.1.46; Sigma)	5 U
NAD	30 mg
MTT	10 mg
PMS	1 mg

Procedure

Incubate the gel in staining solution in the dark at room temperature until dark blue bands appear. Fix the stained gel in 25% ethanol.

Notes: This method is an adaptation of a histochemical method developed to detect trimethylamine-oxide aldolase (see 4.1.2.32 — TMAOA) using auxiliary formaldehyde dehydrogenase.[1]

The method can detect methanol oxidase (EC 1.1.3.31), which may be identical with ALOX.

GENERAL NOTES

The enzyme is a flavoprotein (FAD). It acts on lower primary and unsaturated alcohols, but secondary and branched-chain alcohols are not attacked.[2]

REFERENCES

1. Havemeister, W., Rehbein, H., Steinhart, H., Gonzales-Sotelo, C., Krogsgaard-Nielsen, M., and Jørgensen, B., Visualization of the enzyme trimethylamine oxide demethylase in isoelectric focusing gels by an enzyme-specific staining method, *Electrophoresis*, 20, 1934, 1999.
2. NC-IUBMB, *Enzyme Nomenclature*, Academic Press, San Diego, 1992, p. 57 (EC 1.1.3.13, Comments).

1.1.3.15 — Glycolate Oxidase; GOX

OTHER NAMES   L-2-Hydroxy acid oxidase (recommended name), hydroxyacid oxidase A, hydroxyacid oxidase B

REACTION   L-2-Hydroxy acid + O_2 = 2-oxo acid + H_2O_2

ENZYME SOURCE   Bacteria, plants, vertebrates

SUBUNIT STRUCTURE   Tetramer (vertebrates)

Method 1

Visualization Scheme

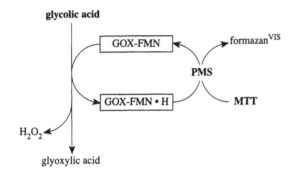

Staining Solution[1]

A.   0.5 M Tris–HCl buffer, pH 7.5   25 ml
     Glycolic acid   50 mg
     5 mg/ml MTT   1 ml
     5 mg/ml PMS   1 ml
B.   2% Agar solution (60°C)   25 ml

Procedure

Mix A and B components of the staining solution and pour the mixture over the surface of the gel. Incubate the gel in the dark at 37°C until dark blue bands appear. Fix the stained gel in 25% ethanol.

Notes: Human GOX also works well with α-hydroxyisocaproic acid as a substrate. The rat enzyme also oxidizes phenyllactic and D-lactic (2-hydroxypropionic) acids.[2]

To decrease nonspecific staining of the gel background, the use of acid stain buffer (pH 6.5 to 6.8) is recommended.

Method 2

Visualization Scheme

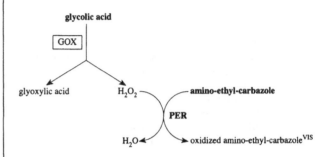

Staining Solution[3] (modified)

A.   0.5 M Tris–HCl buffer, pH 7.4
     8 U/ml Peroxidase (PER)
     2 mg/ml Glycolic (or α-hydroxyisocaproic) acid
     1 mg/ml 3-Amino-9-ethyl-carbazole (dissolved in minimal volume of acetone)
B.   2% Agarose solution (60°C)

Procedure

Mix equal volumes of A and B components of the staining solution and pour the mixture over the surface of the gel. Incubate the gel at 37°C until reddish brown bands appear. Fix the stained gel in 50% glycerol.

Notes: The resolution and sensitivity of this method are not as good as those of the MTT–PMS stain used in Method 1.

o-Dianisidine dihydrochloride may be used instead of amino-ethyl-carbazole. It should be remembered, however, that this dye is a possible carcinogen and should be handled with great care.

METHOD 3

Visualization Scheme

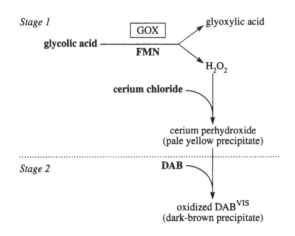

Stage 1

glycolic acid —— GOX / FMN —→ glyoxylic acid

↘ H_2O_2

cerium chloride ——

cerium perhydroxide
(pale yellow precipitate)

Stage 2 DAB ——

oxidized DAB[VIS]
(dark-brown precipitate)

Staining Solution[4]

A. 200 mM Tris–maleate buffer, pH 8.3
10 mM Glycolic acid
1 µM FMN
3 mM Cerium chloride ($CeCl_3$)

B. 100 mM Acetate buffer, pH 5.5
1 mM 3,3'-Diaminobenzidine (DAB)
20 µg/ml $CoCl_2 \times 6H_2O$
20 µg/ml $(NH_4)_2Ni(SO_4)_2 \times 6H_2O$

Procedure

Preequilibrate the electrophorized gel in 200 mM Tris–maleate buffer (pH 8.3) for 5 to 10 min, and then incubate in solution A for 30 min. Rinse the gel twice in 100 mM acetate buffer (pH 5.5) for 5 min each, and incubate in solution B for 10 to 20 min. Carry out all the procedures at 37°C. After staining is completed, photograph or dry the zymogram instantly.

Notes: The method is based on the ability of cerium chloride to react with hydrogen peroxide, forming the insoluble cerium perhydroxide of a pale yellow color. Cerium perhydroxide is then converted to a dark brown precipitate by the addition of DAB, which is oxidized by cerium perhydroxide to its polymeric form.

The method can be applied to native electrophoretic gels and nitrocellulose blots.

Glycolic acid is used as a substrate for the detection of GOX from the liver. To detect activity of the GOX isozyme specific for the kidney, α-hydroxybutiric acid should be used instead of glycolic acid (see also General Notes).

Less than 0.1 U of GOX per band can be detected by the cerium–DAB method.

GENERAL NOTES

The enzyme is a flavoprotein (FMN). It exists as two major isoenzymes: the A form preferentially oxidizes short-chain aliphatic hydroxy acids and was previously listed as glycolate oxidase (EC 1.1.3.1, deleted entry); the B form preferentially oxidizes long-chain and aromatic hydroxy acids. The rat isoenzyme B also acts as L-amino acid oxidase (EC 1.4.3.2).[5]

With respect to the enzyme from the rat liver and kidney, the cerium–DAB method (Method 3) was found to be more sensitive than that based on the use of linking peroxidase reaction (Method 2).[4]

REFERENCES

1. Harris, H. and Hopkinson, D.A., *Handbook of Enzyme Electrophoresis in Human Genetics*, North-Holland, Amsterdam, 1976 (loose-leaf, with supplements in 1977 and 1978).
2. Feinstein, R.N. and Lindahl, R., Detection of oxidases on polyacrylamide gels, *Anal. Biochem.*, 56, 353, 1973.
3. Duley, J. and Holmes, R.S., α-Hydroxyacid oxidase in the mouse: evidence for two genetic loci and a tetrameric subunit structure for the liver isozyme, *Genetics*, 76, 93, 1974.
4. Seitz, J., Keppler, C., Fahimi, H.D., and Völkl, A., A new staining method for the detection of activities of H_2O_2-producing oxidases on gels and blots using cerium and 3,3'-diaminobenzidine, *Electrophoresis*, 12, 1051, 1991.
5. NC-IUBMB, *Enzyme Nomenclature*, Academic Press, San Diego, 1992, p. 57 (EC 1.1.3.15, Comments).

1.1.3.22 — Xanthine Oxidase; XOX

OTHER NAMES   Hypoxanthine oxidase

REACTIONS

1. Hypoxanthine $+$ H_2O $+$ O_2 $=$ xanthine $+$ H_2O_2
2. Xanthine $+$ H_2O $+$ O_2 $=$ urate $+$ H_2O_2 (see also General Notes)

ENZYME SOURCE   Bacteria, vertebrates

SUBUNIT STRUCTURE   Heterotrimer[#] (bacteria – *Pseudomonas putida*), dimer[#] (mammals)

METHOD 1

Visualization Scheme

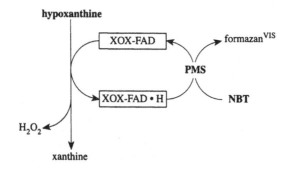

Staining Solution[1]

0.1 M Phosphate buffer, pH 6.8	60 ml
0.01 M Hypoxanthine (dissolved by hitting or in minimal volume of 1 M KOH)	10 ml
1 mg/ml NBT	5 ml
0.5 mg/ml PMS	5 ml

Procedure

Incubate the gel in staining solution in the dark at 37°C until dark blue bands appear. Wash the developed gel in water and fix in 25% ethanol.

Notes: An additional (control) gel should be stained for aldehyde oxidase (see 1.2.3.1 — AOX), for which activity bands also can be developed by this method.

A simple and sensitive method for the detection of XOX activity was developed using N,N,N',N'-tetramethylethylenediamine (TEMED) in place of PMS. Detection of 0.02 μU of XOX on electrophoretic PAG was possible. However, the mechanism of NBT reduction in this method is uncertain. Again, because the method was used to detect partially purified XOX, its specificity also remains uncertain. The authors stressed that although the addition of TEMED is not essential for activity staining, it increases the rate and the intensity of staining of XOX activity bands.[2]

METHOD 2

Visualization Scheme

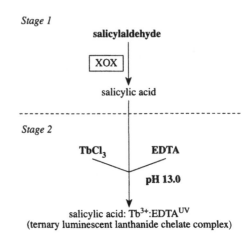

Staining Solution[3] (adapted)

A. 0.05 M HEPES, pH 7.4
   1 mM Salicylaldehyde (dissolved in ethylene glycol dimethyl ether)
B. 0.01 M HCl
   5 mM TbCl$_3$
   5 mM EDTA (tetrasodium salt)
C. 2.5 M Tris, pH 13.0

Procedure

Apply substrate solution A to the gel surface with a filter paper or 1% agarose overlay. Incubate the gel with application at 37°C for 30 min. Remove the first application and apply the next one containing developing solution (one part of solution B, one part of solution C, and three parts of deionized water). Observe luminescent XOX bands under 300- to 400-nm UV light. Photograph the developed gel on Polaroid instant film with a TRP 100 time-resolved photographic camera (Kronem Systems, Inc., Mississauga, Ontario, Canada) using a filter combination that provides excitation in the range of 320 to 400 nm and measures emission above 515 nm with a measurement time delay and gate of 440 μsec and 4.1 msec, respectively.

OTHER METHODS

An immunoblotting procedure (for details, see Part II) using monoclonal antibodies specific to bovine XOX[4] can also be used to localize the enzyme protein on electrophoretic gels. This procedure has some disadvantages but may be of great value in special analyses of XOX.

GENERAL NOTES

The enzyme also oxidizes some other purines and pterins. It has a broad substrate specificity, being able to oxidize a variety of compounds containing aldehyde moieties. XOX from bacteria and XOX from mammals are not related proteins.

The enzyme is an iron-molybdenum flavoprotein (FAD), but that from *Micrococcus* can use ferredoxin as an acceptor. Besides xanthine and hypoxanthine, it oxidizes some other purines and pterins, and aldehydes (i.e., it possesses activity of aldehyde oxidase (EC 1.2.3.1)); it probably acts on the hydrated derivatives of these substrates. Under some conditions the product is mainly superoxide ($O_2^{\cdot-}$) rather than peroxide (H_2O_2). XOX from animal tissues can be interconverted to the dehydrogenase form (see 1.1.1.204 — XDH). In animal liver the enzyme exists *in vivo* mainly in the dehydrogenase form, but can be converted into the oxidase form by storage at $-20^{\circ}C$, by treatment with organic solvents or proteolytic agents, or by thiol reagents (e.g., Cu^{2+}, *N*-methylmaleimide, or 4-hydroxymercuribenzoate). The effect of thiol reagents can be reversed by thiols such as 1,4-dithioerythritol. XDH can also be converted into XOX by enzyme–thiol transhydrogenase (EC 1.8.4.7).[5]

A considerable fraction of the enzyme can interact and move together with the dye Bromophenol Blue. This XOX retains its activity and therefore can be detected on electrophoretic gels at the dye-front. The amount of XOX activity at the dye-front is dependent on the contact time of XOX with the dye at room temperature.[2] A similar situation was described for glutathione *S*-transferase (see 2.5.1.18 — GT, General Notes).

REFERENCES

1. Feinstein, R.N. and Lindahl, R., Detection of oxidases on polyacrylamide gels, *Anal. Biochem.*, 56, 353, 1973.
2. Özer, N., Muftüoglu, M., and Ögus, I.H., A simple and sensitive method for the activity staining of xanthine oxidase, *J. Biochem. Biophys. Methods*, 36, 95, 1998.
3. Evangelista, R.A., Pollak, A., and Templeton, E.F.G., Enzyme-amplified lanthanide luminescence for enzyme detection in bioanalytical assays, *Anal. Biochem.*, 197, 213, 1991.
4. Mather, I.H., Nace, C.S., Johnson, V.G., and Goldsby, R.A., Preparation of monoclonal antibodies to xanthine oxidase and other proteins of bovine milk-fat globule membrane, *Biochem. J.*, 188, 925, 1980.
5. NC-IUBMB, *Enzyme Nomenclature*, Academic Press, San Diego, 1992, p. 58 (EC 1.1.3.22, Comments).

1.1.3.23 — Thiamin Dehydrogenase; TDH

OTHER NAMES	Thiamin oxidase (recommended name)
REACTION	Thiamin + 2 O_2 = thiaminacetic acid + 2 H_2O_2
ENZYME SOURCE	Bacteria
SUBUNIT STRUCTURE	Unknown or no data available

METHOD

Visualization Scheme

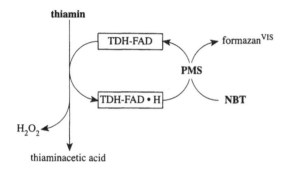

Staining Solution[1]

50 mM Sodium phosphate buffer, pH 7.0
0.015% Thiamin
0.04% NBT
0.004% PMS

Procedure

Incubate the gel in staining solution in the dark at 37°C until dark blue bands appear. Wash the stained gel in water and fix in 25% ethanol.

GENERAL NOTES

The enzyme is a flavoprotein (FAD). The two-step oxidation proceeds without the release of the intermediate aldehyde from the enzyme. The product differs from thiamin in replacement of –CH_2·CH_2·OH by –CH_2·COOH.[2]

REFERENCES

1. Neal, R.A., Bacterial metabolism of thiamine: III. Metabolism of thiamine to 3-(2'-methyl-4'-amino-5'-pyrimidylmethyl)-4-methyl-thiazole-5-acetic acid (thiamine acetic acid) by a flavoprotein isolated from a soil microorganism, *J. Biol. Chem.*, 245, 2599, 1970.
2. NC-IUBMB, *Enzyme Nomenclature*, Academic Press, San Diego, 1992, p. 58 (EC 1.1.3.23, Comments).

1.1.99.5 — Glycerol-3-Phosphate Dehydrogenase (FAD); G-3-PD(FAD)

OTHER NAMES	Glycerol-3-phosphate dehydrogenase (recommended name)
REACTION	*sn*-Glycerol-3-phosphate + acceptor = glycerone phosphate + reduced acceptor
ENZYME SOURCE	Invertebrates, vertebrates
SUBUNIT STRUCTURE	Dimer (vertebrates)

METHOD

Visualization Scheme

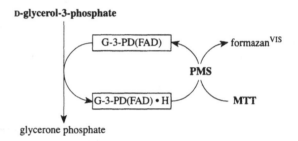

Staining Solution[1]

100 mM Sodium phosphate buffer, pH 7.5
370 mM DL-Glycerol-3-phosphate
0.4 mM PMS
0.6 mM MTT

Procedure

Incubate the gel in staining solution in the dark at 37°C until dark blue bands appear. Wash the developed gel in water and fix in 25% ethanol.

Notes: The enzyme is a flavoprotein tightly bound to mitochondrial membranes. Preparation of enzyme samples for electrophoresis includes purification of mitochondria, their disruption by sonication, and subsequent dissolving of mitochondrial membranes in 0.125% Triton X-100. Starch gels used for G-3-PD(FAD) electrophoresis contain 0.5% Triton X-100 and 0.3% egg yolk L-α-phosphatidylcholine. Before being added to the heated gel just prior to degassation, phosphatidylcholine is suspended in 10 ml of gel buffer by sonication. Triton X-100 is added after degassation of starch gel.

REFERENCES

1. Shaw, M.-A., Edwards, Y.H., and Hopkinson, D.A., Human mitochondrial glycerol phosphate dehydrogenase (GPDm) isozymes, *Ann. Hum. Genet.*, 46, 11, 1982.

1.1.99.8 — Alcohol Dehydrogenase (Acceptor); ADHA

REACTION   Primary alcohol + acceptor = alde-
           hyde + reduced acceptor
ENZYME SOURCE Bacteria
SUBUNIT STRUCTURE Unknown or no data available

METHOD

Visualization Scheme

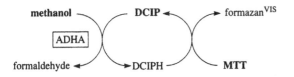

Staining Solution[1] (adapted)

20 mM Tris–HCl buffer, pH 9.0	60 ml
Methanol	0.5 ml
2,6-Dichlorophenol indophenol (DCIP)	0.3 mg
MTT	15 mg

Procedure

Incubate the gel in staining solution in the dark at 37°C until purple bands appear on a light blue background. When ADHA bands are well developed, treat the gel with 1 M HCl solution to remove a blue background. Wash the stained gel in water and fix in 25% ethanol.

GENERAL NOTES

The enzyme (a quinoprotein) acts on a wide range of primary alcohols, including methanol.[2] 2,6-Dichlorophenol indophenol can act as an acceptor.

REFERENCES

1. Wojciechowski, C.L. and Fall, R., A continuous fluorometric assay for pectin methylesterase, *Anal. Biochem.*, 237, 103, 1996.
2. NC-IUBMB, *Enzyme Nomenclature*, Academic Press, San Diego, 1992, p. 62 (EC 1.1.99.8, Comments).

1.2.1.1 — Formaldehyde Dehydrogenase (Glutathione); FDHG

REACTION	Formaldehyde + glutathione + NAD = S-formylglutathione + NADH
ENZYME SOURCE	Bacteria, fungi, plants, invertebrates, vertebrates
SUBUNIT STRUCTURE	Dimer (fungi, invertebrates, vertebrates)

METHOD

Visualization Scheme

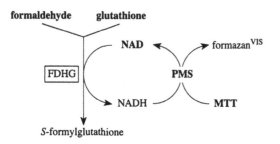

Staining Solution[1]

0.05 M Tris–HCl buffer, pH 8.0	100 ml
40% Formaldehyde	0.08 ml
Glutathione (reduced)	300 mg
NAD	50 mg
MTT	6 mg
PMS	4 mg

Procedure

Incubate the gel in staining solution in the dark at 37°C until dark blue bands appear. Wash the stained gel in water and fix in 25% ethanol.

Notes: The presence of reduced glutathione in the staining solution causes nonspecific reduction of MTT and staining of the gel background. To avoid this problem PMS and MTT may be omitted from the stain and fluorescent bands of FDHG activity observed in long-wave UV light.[2]

Unlike FDHG, the enzyme from *Pseudomonas* (see 1.2.1.46 — FDH) does not need reduced glutathione.

GENERAL NOTES

Phylogenetic analysis suggests that this enzyme is a progenitor of ethanol-consuming alcohol dehydrogenases (see 1.1.1.1 — ADH) in plants and animals. The high structural conservation of FDHG in bacteria, plants, and animals is consistent with a universal importance of this enzyme as a detoxifier.[3]

REFERENCES

1. Lush, I.E., Genetic variation of some aldehyde-oxidizing enzymes in the mouse, *Anim. Blood Groups Biochem. Genet.*, 9, 85, 1978.
2. Balakirev, E.S. and Zaykin, D.V., Allozyme variability of formaldehyde dehydrogenase in marine invertebrates, *Genetika (USSR)*, 24, 1504, 1988 (in Russian).
3. Fliegmann, J. and Sandermann, H., Maize glutathione-dependent formaldehyde dehydrogenase cDNA: a novel plant gene of detoxification, *Plant Mol. Biol.*, 34, 843, 1997.

1.2.1.2 — Formate Dehydrogenase; FD

REACTION   Formate + NAD = CO_2 + NADH
ENZYME SOURCE   Bacteria, fungi, plants
SUBUNIT STRUCTURE Dimer (plants)

METHOD

Visualization Scheme

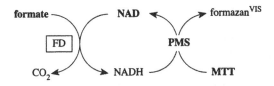

Staining Solution[1]

0.05 M Tris–HCl buffer, pH 8.0	50 ml
Formic acid (sodium salt)	200 mg
NAD	30 mg
MTT	10 mg
PMS	2 mg

Procedure

Incubate the gel in staining solution in the dark at 37°C until dark blue bands appear. Wash the developed gel in water and fix in 25% ethanol.

REFERENCES

1. Royse, D.J. and May, B., Use of isozyme variation to identify genotypic classes of *Agaricus brunnescens*, *Mycologia*, 74, 93, 1982.

1.2.1.3 — Aldehyde Dehydrogenase (NAD); ALDH

REACTION	Aldehyde + NAD + H_2O = acid + NADH
ENZYME SOURCE	Bacteria, fungi, plants, invertebrates, vertebrates
SUBUNIT STRUCTURE	Dimer (vertebrates), see also General Notes

METHOD

Visualization Scheme

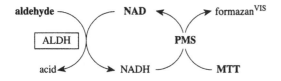

Staining Solution[1]

0.05 M Tris–HCl buffer, pH 8.5	100 ml
Salicylaldehyde	0.15 ml
NAD	50 mg
MTT	12 mg
PMS	4 mg

Procedure

Incubate the gel in staining solution in the dark at 37°C until dark blue bands appear. Wash the stained gel in water and fix in 25% ethanol.

Notes: Possible alternative substrates are acetaldehyde, benzaldehyde, and propionaldehyde. The human ALDH-1 isoenzyme also acts on pyruvaldehyde and furfuraldehyde and is activated by Mg^{2+} and Ca^{2+} ions.[2] The related plant enzyme (see 1.2.1.X — IADH) catalyzes conversion of indol-3-acetaldehyde into indol-3-acetic acid (heteroauxin), which plays an important role in plant organogenesis.[3] Animal ALDH is part of a very complex system of aldehyde-metabolizing enzymes.

A gel stained for ALDH can also exhibit aldehyde oxidase (see 1.2.3.1 — AOX), xanthine oxidase (see 1.1.3.22 — XOX), and alcohol dehydrogenase (see 1.1.1.1 — ADH) activities. The AOX bands can be identified by omitting NAD from the staining solution. The addition of 1 to 2 mg/ml pyrazole to the stain can be used to prevent development of ADH bands.

The addition of NAD to electrophoretic gels or to acetate cellulose soaking buffers often proves beneficial for the detection of ALDH activity bands.

GENERAL NOTES

ALDH is a polymorphic enzyme, responsible for oxidation of various aldehydes, including acetaldehyde and aldehydic neurotransmitter catabolites, to the corresponding carboxylic acids, utilizing NAD (EC 1.2.1.3) or NADP (EC 1.2.1.4) or both (EC 1.2.1.5).[4]

In mammals, two tetrameric liver isozymes are known as ALDH-1 (cytosolic) and ALDH-2 (mitochondrial). ALDH-1 displays broad substrate specificity, while ALDH-2 is primarily responsible for oxidation of short-chain aliphatic aldehydes, including acetaldehyde. The dimeric ALDH-3 isozyme is mainly expressed in the stomach and lungs. This isozyme primarily oxidizes long- and medium-chain aliphatic and aromatic aldehydes and can use either NAD or NADP. In humans, liver ALDH displays very low activity with NADPH, while the enzyme from the stomach utilizes NAD and NADP with comparable rates.[5]

REFERENCES

1. Lush, I.E., Genetic variation of some aldehyde-oxidizing enzymes in the mouse, *Anim. Blood Groups Biochem. Genet.*, 9, 85, 1978.
2. Teng, Y.-S., Human liver aldehyde dehydrogenase in Chinese and Asiatic Indians: gene deletion and its possible implications in alcohol metabolism, *Biochem. Genet.*, 19, 107, 1981.
3. Ballal, S.K. and Harris, J. W., Differential expression of isozymes in relation to organogenesis in two closely related species, *Experientia*, 44, 255, 1988.
4. NC-IUBMB, *Enzyme Nomenclature*, Academic Press, San Diego, 1992, p. 65 (EC 1.2.1.3, EC 1.2.1.4, EC 1.2.1.5).
5. Wierzchowski, J., Wroczynski, P., Laszuk, K., and Interewicz, E., Fluoremetric detection of aldehyde dehydrogenase activity in human blood, saliva, and organ biopsies and kinetic differentiation between class I and class III isozymes, *Anal. Biochem.*, 245, 69, 1997.

1.2.1.8 — Betaine-Aldehyde Dehydrogenase; BADH

REACTION   Betaine aldehyde + NAD + H_2O = betaine + NADH
ENZYME SOURCE   Bacteria, plants, vertebrates
SUBUNIT STRUCTURE   Tetramer[#] (bacteria), heterodimer[#] (plants)

METHOD

Visualization Scheme

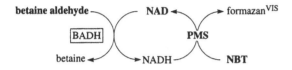

Staining Solution[1]

0.125 M Potassium phosphate buffer, pH 8.0
0.065 mM PMS
0.37 mM NBT
0.5 mM NAD
0.5 mM Betaine aldehyde

Procedure

Incubate the gel in staining solution in the dark at 30°C until dark blue bands appear. Fix the stained gel in 7% acetic acid or 25% ethanol.

Notes: When transgenically expressed, plant BADH has been shown to oxidize ω-aminoaldehydes to some extent.[2,3] NADP can also be used as a cofactor for bacterial BADH.[4]

REFERENCES

1. Arakawa, K., Mizuno, K., Kishitani, S., and Takabe, T., Immunological studies of betaine aldehyde dehydrogenase in barley, *Plant Cell Physiol.*, 33, 833, 1992.
2. Incaroensakdi, A., Matsuda, N., Hibino, T., Meng, Y.L., Ishikawa, H., Hara, A., Funaguma, T., and Takabe, T., Overproduction of spinach betaine aldehyde dehydrogenase in *Escherichia coli*, *Eur. J. Biochem.*, 267, 7015, 2000.
3. Trossart, C., Rathinasabapathi, B., and Hanson, A.D., Transgenically expressed betaine aldehyde dehydrogenase efficiently catalyzes oxidation of dimethylsulfoniopropionaldehyde and ω-aminoaldehydes, *Plant Physiol.*, 113, 1457, 1997.
4. Mori, N., Fuchigami, S., and Kitamoto, Y., Purification and properties of betaine aldehyde dehydrogenase with high affinity for NADP from *Arthrobacter globiformis*, *J. Biosci. Bioeng.*, 93, 130, 2002.

1.2.1.12 — Glyceraldehyde-3-Phosphate Dehydrogenase; GA-3-PD

OTHER NAMES	Triosephosphate dehydrogenase
REACTION	D-Glyceraldehyde-3-phosphate + orthophosphate (or arsenate) + NAD = 3-phospho-D-glyceroyl phosphate (or arsenate) + NADH
ENZYME SOURCE	Bacteria, algae, fungi, plants, invertebrates, vertebrates
SUBUNIT STRUCTURE	Monomer (algae), dimer (plants, invertebrates), tetramer (invertebrates, vertebrates)

METHOD 1

Visualization Scheme

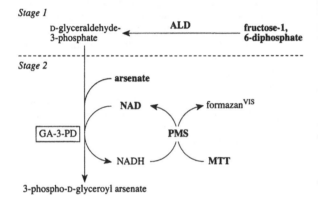

Staining Solution[1] (modified)

A. 0.1 M Tris–HCl buffer, pH 7.5   2 ml
Fructose-1,6-diphosphate (sodium salt)   50 mg
Aldolase (ALD)   50 U
B. 0.1 M Tris–HCl buffer, pH 7.5 18 ml
NAD   20 mg
MTT   6 mg
PMS   2 mg
Sodium arsenate   50 mg
C. 2% Agarose solution (60°C)   20 ml

Procedure

Incubate mixture A at 37°C for 1 h before use. Mix A and B components of the staining solution and then add agar solution (C). Pour the mixture over the gel surface. Incubate the gel in the dark at 37°C until dark blue bands appear. Fix the stained gel together with the stained agarose overlay in 25% ethanol.

Notes: The addition of NAD to the electrophoretic gel or to acetate cellulose soaking buffer often proves beneficial for GA-3-PD detection. The use of 1 to 2 mg/ml pyruvate and pyrazole in the stain is sometimes desirable to prevent or minimize nonspecific development of LDH and ADH activity bands, respectively.

The substrate D-glyceraldehyde-3-phosphate can be prepared directly from the diethylacetal barium salt using Dowex 50x4-200R, following the instructions supplied by the manufacturer (Sigma Chemical Company), or from di(cyclohexylammonium) salt using 2 M H$_2$SO$_4$.

METHOD 2

Visualization Scheme

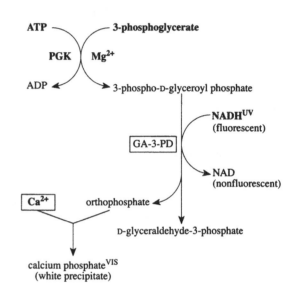

Staining Solution[2]

100 mM Tris–HCl buffer, pH 8.8
150 mM NADH
1 mM ATP
5 mM 3-Phospho-D-glycerate
1 mM Dithiothreitol
10 μg/ml Phosphoglycerate kinase (PGK)
20 mM Mg^{2+}
20 mM Ca^{2+}

Procedure

Before staining, presoak the electrophorized gel in 100 mM Tris–HCl buffer (pH 8.8) for 20 to 30 min. Incubate the gel in staining solution at 37°C until bands of white precipitate appear. The stained gel can be stored for several months in 50 mM glycine–KOH buffer (pH 10.0), containing 5 mM Ca^{2+}, either at 5°C or at room temperature in the presence of an antibacterial agent.

Notes: This method is developed for PAG. However, it is also applicable to acetate cellulose and starch gels, where GA-3-PD bands can be observed under long-wave UV light as dark (nonfluorescent) bands on a light (fluorescent) background. In this case Ca^{2+} ions should be omitted from the staining solution. It should also be taken into account that the amount of NADH added to the staining solution is critical for development of nonfluorescent bands. If too much NADH is added, the enzyme should not be able to convert enough NADH into NAD for nonfluorescent bands to become visible.

When unclean gels (e.g., starch or acetate cellulose) are used for electrophoresis, the zones of calcium phosphate precipitation can be counterstained with Alizarin Red S.

REFERENCES

1. Siciliano, M.J. and Shaw, C.R., Separation and visualization of enzymes on gels, in *Chromatographic and Electrophoretic Techniques*, Vol. 2, *Zone Electrophoresis*, Smith, I., Ed., Heinemann, London, 1976, p. 185.

2. Nimmo, H.G. and Nimmo, G.A., A general method for the localization of enzymes that produce phosphate, pyrophosphate, or CO_2 after polyacrylamide gel electrophoresis, *Anal. Biochem.*, 121, 17, 1982.

OTHER NAMES	Triosephosphate dehydrogenase (NADP)
REACTION	D-Glyceraldehyde-3-phosphate + orthophosphate (or arsenate) + NADP = 3-phospho-D-glyceroyl phosphate (or arsenate) + NADPH
ENZYME SOURCE	Bacteria, plants, invertebrates (insects)
SUBUNIT STRUCTURE	Tetramer[#] (bacteria, plant chloroplasts)

METHOD

Visualization Scheme

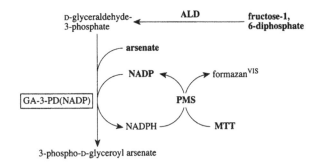

Staining Solution[1]

0.2 M Tris–HCl buffer, pH 8.0	40 ml
Fructose-1,6-diphosphate (cyclohexylammonium salt)	100 mg
Aldolase (ALD)	10 U
Sodium arsenate (7 H_2O)	50 mg
1.25% MTT	1 ml
1% PMS	0.5 ml
1% NADP	2 ml

Procedure

Incubate the gel in staining solution in the dark at 37°C until dark blue bands appear. Wash the stained gel in water and fix in 25% ethanol.

Notes: The enzyme from some sources can utilize both NADP and NAD as cofactors. However, since the enzyme requires about 100 times more NAD than NADP for equal activity, it is supposed that NADP is the physiological cofactor of GA-3-PD(NADP).

Another form of NADP-dependent D-glyceraldehyde-3-phosphate dehydrogenase (EC 1.2.1.9)[2] is also characteristic of some plants (e.g., sugar beet). This enzyme, however, does not require orthophosphate (or arsenate) and thus can be identified on a separate gel stained using the staining solution lacking arsenate.

OTHER METHODS

The reverse reaction can be used to detect GA-3-PD(NADP) activity bands using a staining solution like that given in Method 2 for NAD-dependent GA-3-PD, but containing NADPH in place of NADH (see 1.2.1.12 — GA-3-PD, Method 2).

REFERENCES

1. Selander, R.K., Caugant, D.A., Ochman, H., Musser, J.J.M., Gilmour, M.N., and Whittam, T.S., Methods of multilocus enzyme electrophoresis for bacterial population genetics and systematics, *Appl. Environ. Microbiol.*, 51, 873, 1986.
2. NC-IUBMB, *Enzyme Nomenclature*, Academic Press, San Diego, 1992, p. 66 (EC 1.2.1.9).

1.2.1.16 — Succinate-Semialdehyde Dehydrogenase; SSDH

OTHER NAMES	Succinate-semialdehyde dehydrogenase (NAD(P)) (recommended name)
REACTION	Succinate-semialdehyde + NAD(P) + H_2O = succinate + NAD(P)H
ENZYME SOURCE	Bacteria
SUBUNIT STRUCTURE	Tetramer[#] (bacteria – *Brevibacterium helvolum*)

METHOD

Visualization Scheme

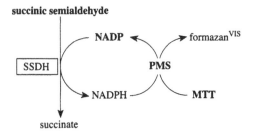

Staining Solution[1] (adapted)

0.05 M Tris–HCl buffer, pH 8.0	80 ml
NADP	15 mg
Succinic-semialdehyde (Sigma)	0.1 ml
MTT	10 mg
PMS	1 mg

Procedure

Incubate the gel in staining solution in the dark at 37°C until dark blue bands appear. Wash the developed gel in water and fix in 25% ethanol.

Notes: The reaction catalyzed by SSDH is essentially irreversible.

The enzyme from *Pseudomonas fluorescens* is about eight times more active with NADP than NAD.

REFERENCES

1. Akers, E. and Aronson, J.N., Detection on polyacrylamide gels of L-glutamic acid decarboxylase activities from *Bacillus thuringiensis*, *Anal. Biochem.*, 39, 535, 1971.

1.2.1.23 — 2-Oxoaldehyde Dehydrogenase (NAD); ODNAD

OTHER NAMES	α-Ketoaldehyde dehydrogenase, methylglyoxal dehydrogenase
REACTION	2-Oxoaldehyde + NAD + H_2O = 2-oxoacid + NADH
ENZYME SOURCE	Vertebrates
SUBUNIT STRUCTURE	Tetramer (vertebrates)

METHOD

Visualization Scheme

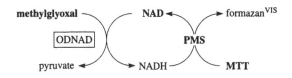

Staining Solution[1]

0.2 M Phosphate buffer, pH 6.8	15 ml
NAD	15 mg
Methylglyoxal	500 μl
MTT	10 mg
PMS	1 mg

Procedure

Apply the staining solution to the gel surface on a filter paper overlay. Incubate the gel with application at 37°C until dark blue bands appear.

Notes: Staining of the gel directly in the staining solution may prove beneficial.

Two ODNAD isozymes are detected in laboratory inbred strains of the rat *Rattus norvegicus*. One isozyme is polymorphic and represented by five-banded allozyme phenotypes in heterozygotes, providing evidence that enzyme molecules are tetramers. However, no interisozyme hybrids are observed, suggesting that different ODNAD isozymes are localized in different intracellular compartments (e.g., cytosol and mitochondria) or that they have different subunit structures.

Activity of the rat enzyme is present only in the liver and kidney.

GENERAL NOTES

A similar reaction is catalyzed by the NADP-dependent enzyme (see 1.2.1.49 — ODNADP) that is not identical with ODNAD.[2]

REFERENCES

1. Bender, K., Seibert, R.T., Wienker, T.F., Kren, V., Pravenec, M., and Bissbort, S. Biochemical genetics of methylglyoxal dehydrogenases in the laboratory rat (*Rattus norvegicus*), *Biochem. Genet.*, 32, 147, 1994.
2. NC-IUBMB, *Enzyme Nomenclature*, Academic Press, San Diego, 1992, p. 67 (EC 1.2.1.23, Comments).

1.2.1.26 — 2,5-Dioxovalerate Dehydrogenase; DOVDH

OTHER NAMES   α-Ketoglutaric semialdehyde dehydrogenase

REACTION   2,5-Dioxopentanoate + NADP + H_2O = 2-oxoglutarate + NADPH

ENZYME SOURCE   Bacteria

SUBUNIT STRUCTURE   Unknown or no data available

METHOD

Visualization Scheme

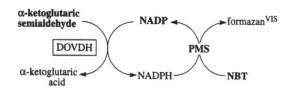

Staining Solution[1]

A. 20 mM Pyrophosphate buffer, pH 8.5
   0.5 mM α-Ketoglutaric semialdehyde
   (2,5-dioxovalerate)
   1 mM NADP

B. 0.2% PMS (several drops per stain)
   0.4% NBT (several drops per stain)

Procedure

Place the gel in solution A for several minutes, and then add solution B and incubate in the dark at 37°C until colored bands appear. Fix the stained gel in 25% ethanol.

Notes: The inclusion of NADP in the gel at a final concentration of 0.1 to 0.2 mM greatly increases the stability of the enzyme during electrophoresis.

REFERENCES

1. Adams, E. and Rosso, G., α-Ketoglutaric semialdehyde dehydrogenase of *Pseudomonas*, *J. Biol. Chem.*, 242, 1802, 1967.

1.2.1.46 — Formaldehyde Dehydrogenase; FDH

REACTION   Formaldehyde + NAD + H_2O = formate + NADH

ENZYME SOURCE   Bacteria

SUBUNIT STRUCTURE   Tetramer[a] (bacteria)

METHOD

Visualization Scheme

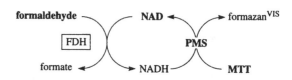

Staining Solution[1] (adapted)

50 mM Tris–HCl buffer, pH 7.8
25 mM Formaldehyde
10 mM NAD
0.02 mg/ml PMS
0.5 mg/ml MTT

Procedure

Incubate the gel in staining solution in the dark at 37°C until colored bands appear. Fix the stained gel in 25% ethanol.

Notes: Unlike the glutathione-dependent enzyme (see 1.2.1.1 — FDHG), the enzyme from bacteria does not need reduced glutathione.[2]

NADP can also be used as a cofactor by FDH from some bacterial sources.

REFERENCES

1. Havemeister, W., Rehbein, H., Steinhart, H., Gonzales-Sotelo, C., Krogsgaard-Nielsen, M., and Jørgensen, B., Visualization of the enzyme trimethylamine oxide demethylase in isoelectric focusing gels by an enzyme-specific staining method, *Electrophoresis*, 20, 1934, 1999.
2. NC-IUBMB, *Enzyme Nomenclature*, Academic Press, San Diego, 1992, p. 70 (EC 1.2.1.46, Comments).

1.2.1.49 — 2-Oxoaldehyde Dehydrogenase (NADP); ODNADP

OTHER NAMES	α-Ketoaldehyde dehydrogenase, methylglyoxal dehydrogenase
REACTION	2-Oxoaldehyde + NADP + H_2O = 2-oxoacid + NADPH
ENZYME SOURCE	Vertebrates
SUBUNIT STRUCTURE	Unknown or no data available

METHOD

Visualization Scheme

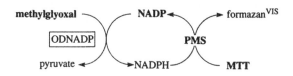

Staining Solution[1]

0.2 M Phosphate buffer, pH 6.8	15 ml
NADP	15 mg
Methylglyoxal	500 µl
MTT	10 mg
PMS	1 mg

Procedure

Apply the staining solution to the gel surface on a filter paper overlay. Incubate the gel with application at 37°C until dark blue bands appear.

Notes: Staining of the gel directly in the staining solution may prove beneficial.

Activity of the rat enzyme is present only in the heart muscle.

GENERAL NOTES

A similar reaction is catalyzed by the NAD-dependent enzyme (see 1.2.1.23 — ODNAD) that is not identical with ODNADP.[2]

REFERENCES

1. Bender, K., Seibert, R.T., Wienker, T.F., Kren, V., Pravenec, M., and Bissbort, S., Biochemical genetics of methylglyoxal dehydrogenases in the laboratory rat (*Rattus norvegicus*), *Biochem. Genet.*, 32, 147, 1994.
2. NC-IUBMB, *Enzyme Nomenclature*, Academic Press, San Diego, 1992, p. 70 (EC 1.2.1.49, Comments).

1.2.1.X — Indoleacetaldehyde Dehydrogenase; IADH

REACTION	Indole-3-acetaldehyde + NAD + H_2O = indole-3-acetic acid + NADH
ENZYME SOURCE	Plants
SUBUNIT STRUCTURE	Unknown or no data available

METHOD

Visualization Scheme

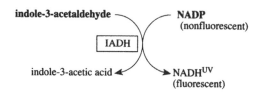

Staining Solution[1]

0.1 M Tris–HCl buffer, pH 8.8
60 mM Indole-3-acetaldehyde
5 mM NAD
20% Sucrose

Procedure

Cover the gel with staining solution and incubate in a humid chamber at room temperature. View the gel under long-wave UV light for fluorescent bands. Record the zymogram or photograph using a yellow filter.

Notes: This is the original method developed for acetate cellulose gel. It can also be applied to PAG and starch gel.

The addition of PMS and MTT (or NBT) to the staining solution will allow development of IADH activity bands visible in daylight. In some cases the addition of PMS and a tetrazolium salt should be made only when fluorescent IADH bands visible under UV light are well developed. This postcoupling technique allows avoidance of the possible inhibitory effect of PMS or a tetrazolium salt on IADH activity. When a tetrazolium method is used, sucrose may be omitted from the staining solution.

The enzyme may be identical to aldehyde dehydrogenase (NAD) (see 1.2.1.3 — ALDH).

REFERENCES

1. Ballal, S.K. and Harris, J.W., Differential expression of isozymes in relation to organogenesis in two closely related species, *Experientia*, 44, 255, 1988.

1.2.1.X′ — Aminoaldehyde Dehydrogenase; AMADH

REACTION   3-Aminopropionaldehyde + NAD +
                        H₂O = β-alanine + NADH (see also
                        General Notes)
ENZYME SOURCE   Plants
SUBUNIT STRUCTURE   Tetramer# (plants)

Method

Visualization Scheme

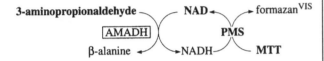

Staining Solution[1]

0.15 *M* Tris–HCl, pH 8.5
0.15 m*M* PMS
0.1 m*M* MTT
1 m*M* NAD
1 m*M* 3-Aminopropionaldehyde

Procedure

Incubate the gel in staining solution in the dark at 30°C until dark blue bands appear. Fix the stained gel in 7% acetic acid or 25% ethanol.

Notes: Nonspecific aldehyde dehydrogenases can also participate in the oxidation of aminoaldehydes.[2] A control stain should be used to differentiate between AMADH and nonspecific aldehyde dehydrogenases using propionaldehyde or butyraldehyde as a substrate.

NBT can be used in the staining solution in place of MTT. However, activity staining performed with MTT is more sensitive and much faster with respect to the color development.[1]

General Notes

The best substrate of pea AMADH is 3-aminopropionaldehyde. However, AMADH can also metabolize 4-aminobutyraldehyde and 4-guanidinobutyraldehyde to the corresponding amino acids. The enzymes oxidizing these aminoaldehydes have been classified as aminobutyraldehyde dehydrogenase (EC 1.2.1.19) and γ-guanidinobutyraldehyde dehydrogenase (EC 1.2.1.54), respectively.[3]

In plants, however, both these activities belong to AMADH. Pea AMADH is inhibited by SH reagents, several elementary aldehydes, and metal-binding agents. This enzyme does not oxidize betaine aldehyde; however, its *N*-terminal amino acid sequence shows a high degree of homology with that of plant betaine aldehyde dehydrogenase (see 1.2.1.8 — BADH) from spinach, sugar beet, and amaranth.[4]

References

1. Šebela, M., Luhová, L., Brauner, F., Galuszka, P., and Peč, P., Light microscopic localization of aminoaldehyde dehydrogenase activity in plant tissues using nitro blue tetrazolium-based staining method, *Plant Physiol. Biochem.*, 39, 831, 2001.
2. Trossart, C., Rathinasabapathi, B., and Hanson, A.D., Transgenically expressed betaine aldehyde dehydrogenase efficiently catalyzes oxidation of dimethylsulfoniopropionaldehyde and ω-aminoaldehydes, *Plant Physiol.*, 113, 1457, 1997.
3. NC-IUBMB, *Enzyme Nomenclature*, Academic Press, San Diego, 1992, p. 67 (EC 1.2.1.19), p. 71 (EC 1.2.1.54).
4. Šebela, M., Brauner, F., Radová, A., Jacobsen, S., Havliš, J., Galuszka, P., and Peč, P., Characterization of a plant aminoaldehyde dehydrogenase, *Biochim. Biophys. Acta*, 1480, 329, 2000.

1.2.3.1 — Aldehyde Oxidase; AOX

REACTION | Aldehyde + H_2O + O_2 = acid + $O_2^{\cdot-}$
Enzyme Source

Invertebrates, vertebrates

SUBUNIT STRUCTURE | Monomer (invertebrates), dimer (invertebrates)

METHOD

Visualization Scheme

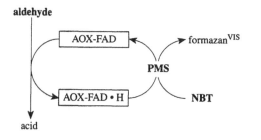

Staining Solution[1]

0.1 M Phosphate buffer, pH 6.8	60 ml
0.01 M 2-Furaldehyde (or benzaldehyde)	10 ml
1 mg/ml NBT	5 ml
0.5 mg/ml PMS	5 ml

Procedure

Incubate the gel in staining solution in the dark at 37°C until dark blue bands appear. Wash the stained gel in water and fix in 25% ethanol.

Notes: Other substrates can be used such as acetaldehyde, *p*-anisaldehyde, butyraldehyde, 2-ethylbutyraldehyde, heptaldehyde, 2-methylbutyraldehyde, propionaldehyde, and salicylaldehyde.

Gels stained for AOX may also exhibit bands of xanthine oxidase (see 1.1.3.22 — XOX) activity.

In *Drosophila*, the artificial substrate *p*-(dimethyl-amino)benzaldehyde can be oxidized by both AOX and pyridoxal oxidase (EC 1.2.3.8).[2] To differentiate AOX, XOX, and pyridoxal oxidase bands on *Drosophila* AOX zymograms, the use of different mutants deficient in expression of correspondent structural genes is recommended.[3] In *Drosophila*, the pyridoxal oxidase bands can also be identified using specific antibodies.[4]

REFERENCES

1. Feinstein, R.N. and Lindahl, R., Detection of oxidases on polyacrylamide gels, *Anal. Biochem.*, 56, 353, 1973.
2. Cypher, J.J., Tedesco, J.J.L., Courtright, J.J.B., and Barman, A.K., Tissue-specific and substrate-specific detection of aldehyde and pyridoxal oxidase in larval and imaginal tissues of *Drosophila melanogaster*, *Biochem. Genet.*, 20, 315, 1982.
3. Dickinson, W.J. and Gaughan, S., Aldehyde oxidases of *Drosophila*: contributions of several enzymes to observed activity patterns, *Biochem. Genet.*, 19, 567, 1981.
4. Warner, C.K., Watts, D.T., and Finnerty, V., Molybdenum hydroxylases in *Drosophila*: 1. Preliminary studies of pyridoxal oxidase, *Mol. Gen. Genet.*, 180, 449, 1980.

1.2.3.4 — Oxalate Oxidase; OO

OTHER NAMES	Germin (in cereal embryos)
REACTION	Oxalate + O_2 = 2 CO_2 + H_2O_2 (see General Notes)
ENZYME SOURCE	Plants
SUBUNIT STRUCTURE	Hexamer[#] (plants)

METHOD 1

Visualization Scheme

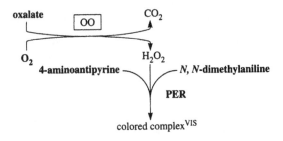

Staining Solution[1]

A. 2 mM Oxalate in 100 mM succinate buffer,
pH 3.5   5 ml
B. 100 mM Sodium phosphate buffer, pH 5.5   3 ml
4-Aminoantipyrine   0.25 mg
N,N-Dimethylaniline   0.6 µl
Peroxidase (PER)   40 U

Procedure

After electrophoresis in SDS-PAG, transfer the separated proteins to a nitrocellulose filter using the electroblotting procedure. Immerse a section of the blotted filter in a mixture of A and B components of the staining solution and incubate at room temperature. After color development (about 20 min), rinse the filter with water.

METHOD 2

Visualization Scheme

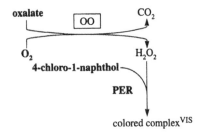

Staining Solution[2]

2 mM Oxalic acid–NaOH, pH 4.0
50 mM Succinic acid–NaOH, pH 4.0
30 mg/100 ml 4-Chloro-1-naphthol
8 mg/100 ml Peroxidase (PER) (Sigma P-8125; 120 U/mg)

Procedure

After electrophoresis in SDS-PAG, transfer the separated proteins to a nitrocellulose filter using the blotting procedure. Immerse the filter in the staining solution and incubate at room temperature for 12 h or until black bands appear. Rinse the stained filter with water.

Notes: Using a very similar method, activity bands of OO can be detected directly in SDS-PAG, however, only in the presence of 60% ethanol.[3] The ethanol-dependent method is about ten times more sensitive than that described above. The mechanism behind the ethanol effect is unclear, but it may be due to the immobilization of the enzyme in protein precipitates as a result of the treatment with ethanol. The immobilization may mimic the conditions in the plant tissue, where OO is located in the cell wall.

A tenfold increase of OO activity is observed in barley leaves from 48 h after inoculation with the powdery mildew fungus, *Blumeria* (syn. *Erysiphe*) *graminis* f.sp. *hordei*.

GENERAL NOTES

The enzyme is widely distributed in plants. It is a water-soluble, SDS- and ethanol-tolerant, pepsin-resistant flavoglycoprotein.

Plant germin is a manganese-containing protein with OO and superoxide dismutase (see 1.15.1.1 — SOD) activities.[4]

REFERENCES

1. Lane, B.G., Dunwell, J.J.M., Ray J.J.A., Schmitt, M.R., and Cumming, A.C., Germin, a protein marker of early plant development, is an oxalate oxidase, *J. Biol. Chem.*, 268, 12239, 1993.
2. Zhang, Z., Collinge, D.B., and Thordal-Christensen, H., Germin-like oxalate oxidase, a H_2O_2-producing enzyme, accumulates in barley attacked by the powdery mildew fungus, *Plant J.*, 81, 139, 1995.
3. Zhang, Z., Yang, J., Collinge, D.B., and Thordal-Christensen, H., Ethanol increases sensitivity of oxalate oxidase assay and facilitates direct activity staining in SDS gels, *Plant Mol. Biol. Reporter*, 14, 266, 1996.
4. Woo, E.J., Dunwell, J.J.M., Goodenough, P.W., Marvier, A.C., and Pickersgill, R.W., Germin is a manganese-containing homohexamer with oxalate oxidase and superoxide dismutase activities, *Nat. Struct. Biol.*, 7, 1036, 2000.

1.2.4.2 — Oxoglutarate Dehydrogenase; OGDH

OTHER NAMES	Oxoglutarate dehydrogenase (lipoamide) (recommended name), oxoglutarate decarboxylase, α-ketoglutaric dehydrogenase
REACTION	2-Oxoglutarate + NAD + CoA = succinyl-CoA + NADH + CO_2
ENZYME SOURCE	Bacteria, fungi, plants, invertebrates, vertebrates
SUBUNIT STRUCTURE	One subunit in the multienzyme 2-oxoglutarate dehydrogenase complex (bacteria, plant mitochondria)

METHOD

Visualization Scheme

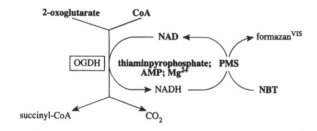

Staining Solution[1]

0.1 M Tris–HCl buffer, pH 8.0
10 mM MgCl$_2$
1 mM EDTA
0.1 mM Thiaminpyrophosphate
5 mM 2-Oxoglutarate
0.1 mM CoA
0.5 mM NAD
0.5 mM AMP
0.1 mg/ml PMS
0.3 mg/ml NBT

Procedure

Incubate the gel in staining solution in the dark at 37°C until dark blue bands appear. Fix the stained gel in 25% ethanol.

Notes: OGDH is a component of the multienzyme 2-oxoglutarate dehydrogenase complex; it requires thiamin diphosphate.[2]

REFERENCES

1. Parker, M.G. and Weitzman, D.J., The purification and regulatory properties of α-oxoglutarate dehydrogenase from *Acinetobacter lwoffi*, *Biochem. J.*, 135, 215, 1973.
2. NC-IUBMB, *Enzyme Nomenclature*, Academic Press, San Diego, 1992, p. 73 (EC 1.2.4.2, Comments).

1.3.1.2 — Dihydrouracil Dehydrogenase (NADP); DHUD

OTHER NAMES	Dihydropyrimidine dehydrogenase (NADP) (recommended name), dihydrothymine dehydrogenase
REACTION	5,6-Dihydrouracil + NADP = uracil + NADPH
ENZYME SOURCE	Invertebrates, vertebrates
SUBUNIT STRUCTURE	Unknown or no data available

METHOD

Visualization Scheme

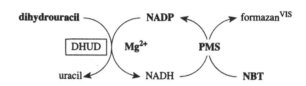

Staining Solution[1]

A. 0.25 M Potassium phosphate buffer, pH 7.5
15 mM Dihydrouracil
1.5 mM NADP
15 mM MgCl$_2$

B. 1 mg/ml NBT	25 ml
1 mg/ml PMS	2.5 ml
0.1 M NaCl	10 ml

Procedure

Mix 15 ml of A with B. Incubate the gel in the mixture in the dark at 37°C until dark blue bands appear. Wash the stained gel in water and fix in 25% ethanol.

Notes: The enzyme is inactivated during electrophoresis in Tris–barbitone buffer.
Dihydrothymine can also be used as substrate.[2]

REFERENCES

1. Hallock, R.O. and Yamada, E.W., Visualization of dihydrouracil dehydrogenase (NADP+) activity after disc gel electrophoresis, *Anal. Biochem.*, 56, 84, 1973.
2. NC-IUBMB, *Enzyme Nomenclature*, Academic Press, San Diego, 1992, p. 76 (EC 1.3.1.2, Comments).

1.3.1.14 — Dihydroorotate Dehydrogenase; DHOD

OTHER NAMES	Orotate reductase (NADH) (recommended name)
REACTION	L-Dihydroorotate + NAD = orotate + NADH
ENZYME SOURCE	Bacteria, fungi, protozoa, vertebrates
SUBUNIT STRUCTURE	Heterodimer# (α and β subunits; bacteria), heterotetramer# (2α and 2β subunits; bacteria)

METHOD

Visualization Scheme

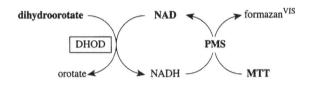

Staining Solution[1] (adapted)

0.1 M Tris–HCl buffer, pH 8.0	50 ml
L-Dihydroorotic acid	50 mg
NAD	30 mg
MTT	10 mg
PMS	1 mg

Procedure

Incubate the gel in staining solution in the dark at 37°C until dark blue bands appear. Wash the stained gel in water and fix in 25% ethanol.

Notes: In *Drosophila*, activity bands of DHOD were shown to be determined by the ADH[71k] allozyme.[2]

REFERENCES

1. Gaal, Ö., Medgyesi, G.A., and Vereczkey, L., *Electrophoresis in the Separation of Biological Macromolecules*, John Wiley & Sons, Chichester, 1980.
2. Eisses, K.Th., Schoonen, G.E.J., Scharloo, W., and Thörig, G.E.W., Evidence for a multiple function of the alcohol dehydrogenase allozyme ADH[71k] of *Drosophila melanogaster*, *Comp. Biochem. Physiol.*, 82B, 863, 1985.

1.3.1.24 — Biliverdin Reductase; BLVR

REACTION	Bilirubin + NAD(P) = biliverdin + NAD(P)H
ENZYME SOURCE	Vertebrates
SUBUNIT STRUCTURE	Monomer (vertebrates)

METHOD

Visualization Scheme

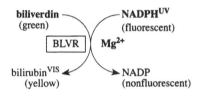

Staining Solution[1]

A. 0.36 M Tris–HCl buffer, pH 8.0	1 ml
Biliverdin	1 mg
B. 0.36 M Tris–HCl buffer, pH 8.0	0.5 ml
NADPH	2 mg
1 M MgCl$_2$	0.2 ml

Procedure

Mix A and B components of the staining solution and apply the mixture dropwise to the gel surface. Incubate the gel at 37°C in a moist chamber until yellow bands against a light green background appear.

Notes: The original method is developed for cellulose acetate gel. However, there are no obvious reasons that can restrict its application to starch gel or PAG.

The enzyme bands can also be observed under UV light (nonfluorescent bands on the fluorescent background of the gel).

REFERENCES

1. Meera Khan, P., Rijken, H., Wijnen, J.J.Th., Wijnen, L.M.M., and De Boer, L.E.M., Red cell enzyme variation in the orang utan: electrophoretic characterization of 45 enzyme systems in cellogel, in *The Orang Utan: Its Biology and Conservation*, De Boer, L.E.M., Ed., Dr. W. Junk Publishers, The Hague, 1982, p. 61.

1.3.3.6 — Palmitoyl-CoA Oxidase; PCO

OTHER NAMES   Acyl-CoA oxidase (recommended name)

REACTION   Palmitoyl-CoA + O_2 = hexadecenoyl-CoA + H_2O_2

ENZYME SOURCE   Vertebrates

SUBUNIT STRUCTURE   Heterodimer[#] (vertebrates)

METHOD

Visualization Scheme

Stage 1

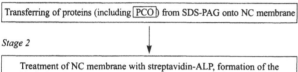

Transferring of proteins (including PCO) from SDS-PAG onto NC membrane

Stage 2

Treatment of NC membrane with streptavidin-ALP, formation of the biotinylated protein/streptavidin-ALP complexes, and elimination of nonspecific biotin signals by histochemical staining for ALP resulted in formation of a blue dye[VIS]

Stage 3

Treatment of NC membrane with biotinylated PTS1-receptor protein and formation of PTS1-containing protein (including PCO)/biotinylated PTS1-receptor protein complexes.

Stage 4

Incubation of NC membrane with streptavidin-ALP and formation of PTS1-containing protein (including PCO)/biotinylated PTS1-receptor protein/ streptavidin-ALP complexes. Histochemical staining for ALP resulted in formation of a red dye[VIS']

Visualization Procedure[1]

Stage 1. Transfer proteins from electrophoretic SDS-PAG onto nitrocellulose (NC) membrane using the routine capillary blotting procedure (see Part II, Section 6).

Stage 2. Block the NC membrane containing the blotted proteins for 1 h with 5% (w/v) dried milk powder (Nutricia) in Tris-buffered saline containing 0.05% (v/v) Tween 20 (TBST). Incubate the membrane for 1 h with streptavidin–alkaline phosphatase (ALP) diluted 1500 times in 1% (w/v) dried milk powder in TBST. Wash the membrane five times for 5 min with TBST and stain with naphthol-AS-GR-phosphate (0.2 mg/ml)/Fast Blue BN (0.35 mg/ml) in alkaline phosphatase buffer (10 mM Tris–HCL, pH 9.2; 100 mM NaCl; 5 mM MgCl$_2$).

This stage is required to eliminate any biotin signals on the membrane. In this way, endogenous biotinylated proteins are shielded by the precipitation of the water-insoluble blue dye formed as a result of coupling of the diazonium salt (Fast Blue BN) with the naphthol-AS-GR produced by alkaline phosphatase.

Stage 3. Block the NC membrane for 30 min with 5% (w/v) dried milk powder in TBST and incubate for 1 h in 1% (w/v) dried milk powder in TBST containing approximately 2 μg/ml bacterially expressed PTS1-receptor (purified or present in a crude bacterial lysate).

Stage 4. Wash the NC membrane three times with TBST and incubate for 1 h with streptavidin–ALP diluted 5000 times in 1% (w/v) dried milk powder in TBST. Wash the membrane five times for 5 min with TBST and stain with naphthol-AS-TR-phosphate (0.05 mg/ml)/Fast Blue BN (0.2 mg/ml) in alkaline phosphatase buffer.

Red bands on the developed NC membrane correspond to localization of PST1-containing proteins (including PCO) on the gel. For details concerning the identification of PCO bands, see General Notes.

Notes: The obtained Western blot simultaneously displays easily distinguishable blue and red bands. The blue bands indicate biotinylated peroxisomal proteins, and the red bands indicate PTS1-containing peroxisomal proteins (including PCO).

OTHER METHODS

Methods for specific detection of PCO are still absent. Such methods, however, can be easily developed. There are several methods suitable for specific detection of the product hydrogen peroxide (see "Hydrogen Peroxide" in Part II, Section 1) that can be easily adapted for PCO. Two such methods are outlined below:

A. The product hydrogen peroxide can be detected using exogenous peroxidase and amino-ethyl-carbazole (e.g., see 1.4.3.3 — DAOX, Method 1) or other chromogenic peroxidase substrates (see 1.11.1.7, Method 1).

B. Cerium chloride can also be used for specific detection of hydrogen peroxide (see 1.4.3.2 — LAOX, Method 2). This method is based on the ability of cerium chloride to react with hydrogen peroxide, forming the insoluble cerium perhydroxide of a pale yellow color. Cerium perhydroxide is then converted to a dark brown precipitate by the addition of diaminobenzidine, which is oxidized by cerium perhydroxide to its polymeric form. The method can be applied to electrophoretic gels and nitrocellulose blots.

GENERAL NOTES

The enzyme is a flavoprotein (FAD) acting on CoA derivatives of fatty acids with chain lengths from C_8 to C_{18}.[2]

The method detailed above is not specific for PCO. It is developed to visualize proteins containing a peroxisomal targeting sequence (PTS1) consisting of the tripeptide serine–lysine–leucine. This sequence is specifically recognized by the PTS1-receptor protein responsible for the import of proteins containing PTS1 into peroxisomes. A bacterially expressed biotinylated C-terminal part of the PTS1-receptor protein is used to specifically recognize peroxisomal proteins. Streptavidin–ALP is then used to detect histochemically the peroxisomal protein–PTS1-receptor protein complexes. The method also detects some other abundant peroxisomal enzymes (urate

1.3.3.6 — Palmitoyl-CoA Oxidase; PCO (continued)

oxidase, catalase, and alanine–glyoxylate transaminase (see 2.6.1.44 — AGTA)). When 100 µg of proteins, present in purified rat liver peroxisomes, is subjected to electrophoresis in SDS-PAG, the restricted number of red bands are developed that correspond (in order of decreasing electrophoretic mobility) to the 23-kDa subunit of PCO, urate oxidase, AGTA, catalase, and the 70-kDa subunit of PCO. Thus, the method is suitable for detection and reliable identification of the enzyme from the rat liver only. However, if molecular weights of the enzyme polypeptides listed above are not changed dramatically during the divergent evolution of vertebrates, one can expect the same arrangement of these polypeptides on SDS-PAGs after electrophoresis of protein preparations obtained from liver peroxisomes of other vertebrates, or at least mammals.

REFERENCES

1. Fransen, M., Brees, C., Van Veldhoven, P.P., and Mannaerts, G.P., The after visualization of peroxisomal proteins containing a C-terminal targeting sequence on Western blot by using the biotinylated PTS1-receptor, *Anal. Biochem.*, 242, 26, 1996.
2. NC-IUBMB, *Enzyme Nomenclature*, Academic Press, San Diego, 1992, p. 83 (EC 1.3.3.6, Comments).

1.3.99.1 — Succinate Dehydrogenase; SUDH

OTHER NAMES	Fumarate reductase, fumaric hydrogenase
REACTION	Succinate + acceptor = fumarate + reduced acceptor
ENZYME SOURCE	Bacteria, fungi, plants, protozoa, invertebrates, vertebrates
SUBUNIT STRUCTURE	Heterotetramer($\alpha,\beta, \gamma, \delta$)^{e} (bacteria, protozoa)

METHOD

Visualization Scheme

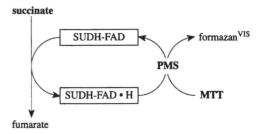

Staining Solution[1]

A. 200 mM Potassium phosphate buffer, pH 7.5   5 ml
    200 mM Succinic acid (neutralized with NaOH)   2 ml
    10 mM KCN   2 ml
    32.5 mg/ml PMS   0.5 ml
    5 mg/ml MTT   0.5 ml
B. 2% Agar solution (60°C)   9.5 ml

Procedure

Mix A and B components of the staining solution and pour the mixture over the surface of the gel. Incubate the gel in the dark at 37°C until dark blue bands appear. Fix the stained gel in 25% ethanol.

Notes: KCN is added in the staining solution to prevent development of superoxide dismutase (1.15.1.1 — SOD) bands. PMS is used as an acceptor.

The enzyme is a flavoprotein bound to mitochondrial membranes. Preparation of enzyme samples for electrophoresis includes purification of mitochondria, their disruption by sonication, and subsequent dissolving of mitochondrial membranes in 0.125% Triton X-100. Starch gels used for SUDH electrophoresis contain 0.5% Triton X-100 and 0.144% phosphatidyl-ethanolamine. Before being added to the heated gel (just prior to degassation), phosphatidylethanolamine is suspended in 10 ml of gel buffer by sonication. Triton X-100 is added after degassation of the starch gel.

1.3.99.1 — Succinate Dehydrogenase; SUDH (continued)

GENERAL NOTES

The enzyme is a flavoprotein (FAD) containing iron–sulfur centers. It is a component of the enzyme complex (EC 1.3.5.1) present in mitochondria. This complex uses ubiquinone as an acceptor and can be degraded to form SUDH, which no longer reacts with ubiquinone.[2]

REFERENCES

1. Shaw, M.-A., Edwards, Y., and Hopkinson, D.A., Human succinate dehydrogenase: biochemical and genetic characterization, *Biochem. Genet.*, 19, 741, 1981.
2. NC-IUBMB, *Enzyme Nomenclature*, Academic Press, San Diego, 1992, p. 84 (EC 1.3.99.1, Comments).

1.3.99.2 — Butyryl-CoA Dehydrogenase; BCD

OTHER NAMES	Butyryl dehydrogenase, unsaturated acyl-CoA reductase
REACTION	Butanoyl-CoA + acceptor = 2-butenoyl-CoA + reduced acceptor
ENZYME SOURCE	Bacteria, vertebrates
SUBUNIT STRUCTURE	Tetramer[*] (vertebrates)

METHOD

Visualization Scheme

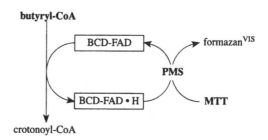

Staining Solution[1]

A. 0.2 *M* Tris–HCl buffer, pH 8.0
   0.3 m*M* Butyryl-CoA
   1.8 m*M* MTT
   0.6 m*M* PMS
   0.5 m*M* KCN
B. 2% Agar solution (60°C)

Procedure

Mix equal volumes of A and B components of the staining solution. Pour the mixture over the surface of the gel. Incubate the gel in the dark at 37°C until dark blue bands appear. Fix the stained gel in 25% ethanol.

Notes: KCN is added in the staining solution to prevent development of superoxide dismutase (1.15.1.1 — SOD) bands. PMS is used as an acceptor.
   A control stain should be used to reveal nonspecific staining of additional bands in the absence of a substrate. Acyl-CoA dehydrogenase (see 1.3.99.3 — ACD) bands also develop on BCD zymograms when butyryl-CoA is used as a substrate. These bands also develop when octanoyl-CoA is used as substrate, whereas BCD bands do not.

GENERAL NOTES

The enzyme is a flavoprotein that is a part of the mitochondrial electron-transferring system reducing ubiquinone and other acceptors.[2]

REFERENCES

1. Seeley, T.-L. and Holmes, R.S., Genetics and ontogeny of butyryl CoA dehydrogenase in the mouse and linkage of *Bcd*-1 with *Dao*-1, *Biochem. Genet.*, 19, 333, 1981.
2. NC-IUBMB, *Enzyme Nomenclature*, Academic Press, San Diego, 1992, p. 85 (EC 1.3.99.2, Comments).

1.3.99.3 — Acyl-CoA Dehydrogenase; ACD

OTHER NAMES — Acyl dehydrogenase

REACTION — Acyl-CoA + acceptor = 2,3-dehydroacyl-CoA + reduced acceptor

ENZYME SOURCE — Vertebrates

SUBUNIT STRUCTURE — Dimer# (vertebrates); see General Notes

METHOD

Visualization Scheme

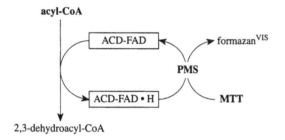

Staining Solution[1]

A. 0.2 M Tris–HCl buffer, pH 8.0
    0.3 mM Octanoyl-CoA
    1.8 mM MTT
    0.6 mM PMS
    0.5 mM KCN
B. 2% Agar solution (60°C)

Procedure

Mix equal volumes of A and B components of the staining solution. Pour the mixture over the surface of the gel. Incubate the gel in the dark at 37°C until dark blue bands appear. Fix the stained gel in 25% ethanol.

Notes: KCN is used to inhibit superoxide dismutase (1.15.1.1 — SOD) activity. PMS is used as an acceptor.

A control stain should be used to reveal nonspecific staining of additional bands in the absence of a substrate.

GENERAL NOTES

Three categories of acyl-CoA dehydrogenases are recognized, which exhibit overlapping substrate specificities: butyryl-CoA dehydrogenase (see 1.3.99.2 — BCD) acts on C_4–C_6 acyl-CoA; general or acyl-CoA dehydrogenase (ACD) acts on C_4–C_{16} acyl-CoA, with peak activity toward C_{10} acyl-CoA; and long-chain acyl-CoA dehydrogenase (EC 1.3.99.13) acts on C_6–C_{18} acyl-CoA, with peak activity toward C_{14} acyl-CoA.

The enzyme is a flavoprotein that is a part of the mitochondrial electron-transferring system reducing ubiquinone and other acceptors.[2]

Very-long-chain acyl-CoA dehydrogenase is a homodimer associated with the inner mitochondrial membrane.[3] Short- or branched-chain acyl-CoA dehydrogenase (use 2-methylbutyryl-CoA as substrate) is a homotetrameric mitochondrial enzyme.[4] This enzyme may be identical with butyryl-CoA dehydrogenase (see 1.3.99.2 — BCD).

REFERENCES

1. Seeley, T.-L. and Holmes, R.S., Genetics and ontogeny of butyryl CoA dehydrogenase in the mouse and linkage of *Bcd-1* with *Dao-1*, *Biochem. Genet.*, 19, 333, 1981.
2. NC-IUBMB, *Enzyme Nomenclature*, Academic Press, San Diego, 1992, p. 85 (EC 1.3.99.3, Comments).
3. Souri, M., Aoyama, T., Hoganson, G., and Hashimoto, T., Very-long-chain acyl-CoA dehydrogenase subunit assembles to the dimer form on mitochondrial inner membrane, *FEBS Lett.*, 426, 187, 1998.
4. Andersen, B.S., Christensen, E., Corydon, T.J., Bross, P., Pilgaard, B., Wanders, R.J.J.A., Ruiter, J.J.P.N., Simonsen, H., Winter, V., Knudsen, I., Schroeder, L.D., Gregersen, N., and Skovby, F., Isolated 2-methylbutyrylglycinuria caused by short/branched-chain acyl-CoA dehydrogenase deficiency: identification of a new enzyme defect, resolution of its molecular basis, and evidence for distinct acyl-CoA dehydrogenases in isoleucine and valine metabolism, *Am. J. Hum. Genet.*, 67, 1095, 2000.

1.4.1.1 — Alanine Dehydrogenase; ALADH

REACTION   L-Alanine + H_2O + NAD = pyruvate
                        + NH_3 + NADH
ENZYME SOURCE   Bacteria
SUBUNIT STRUCTURE   Hexamer[#] (bacteria)

METHOD

Visualization Scheme

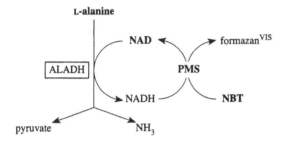

Staining Solution[1]

0.1 M Phosphate buffer, pH 7.0	100 ml
D,L-Alanine	100 mg
NAD	50 mg
NBT	30 mg
PMS	2 mg

Procedure

Incubate the gel in staining solution in the dark at 37°C until dark blue bands appear. Wash the stained gel in water and fix in 25% ethanol.

Notes: Some isozymes of mammal NAD(P)-dependent glutamate dehydrogenase (see 1.4.1.2–4 — GDH) also display ALADH activity.[2]

REFERENCES

1. Shaw, C.R. and Prasad, R., Starch gel electrophoresis of enzymes: a compilation of recipes, *Biochem. Genet.*, 4, 297, 1970.
2. Lan, N., Frieden, E.H., and Rawitch, A.B., Activity staining for glutamate and alanine dehydrogenases in polyacrylamide gels of varying porosity, *Enzyme*, 24, 416, 1979.

OTHER NAMES   Glutamic dehydrogenase

REACTION   L-Glutamate + H_2O + NAD(P) = 2-oxoglutarate + NH_3 + NAD(P)H

ENZYME SOURCE   Bacteria, algae, fungi, plants, protozoa, invertebrates, vertebrates

SUBUNIT STRUCTURE   Dimer (invertebrates, vertebrates), hexamer (plants)

METHOD 1

Visualization Scheme

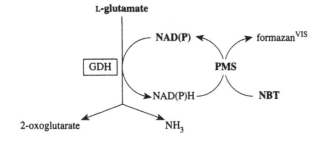

Staining Solution[1]

0.125 M Phosphate buffer, pH 7.0	25 ml
1 M L-Glutamate	5 ml
NAD	60 mg
NBT	30 mg
PMS	2 mg

Procedure

Incubate the gel in staining solution in the dark at 37°C until dark blue bands appear. Wash the stained gel in water and fix in 25% ethanol.

Notes: In higher organisms, GDH is capable of using either NAD or NADP as a cofactor and is activated by ADP. NADP is the cofactor of choice for detection of the enzyme in vertebrates, since its use avoids the problem of artifacts caused by NAD-dependent "nothing dehydrogenase" activities.[2] If NAD is chosen, the inclusion of 1 to 2 mg/ml pyruvate or pyrazole in the stain may be desirable to prevent nonspecific development of LDH or ADH bands, respectively. The addition of ADP (2 mM final concentration) in the stain is usually beneficial.

The GDH reaction is usually displaced toward the production of L-glutamate. To displace the reaction toward production of NAD(P)H and 2-oxoglutarate, the latter should be trapped using such carbonyl-trapping reagents as hydrazyne hydrate or aminooxyacetic acid. When reduced positive fluorescent modification of the method is used, the latter reagent is usually preferable because the use of hydrazyne hydrate can influence the NAD(P)H fluorescence. Therefore, the use of staining buffer (pH 8.6) containing 0.5 M glycine and 0.05 M aminooxyacetic acid is recommended to enhance the sensitivity of the positive fluorescent method.[3]

METHOD 2

Visualization Scheme

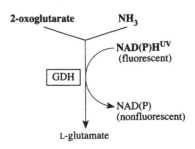

Staining Solution[4]

0.5 M Tris–HCl buffer, pH 7.6	10 ml
2-Oxoglutarate	25 mg
NAD(P)H	10 mg
NH_4Cl	10 mg

Procedure

Apply the staining solution to the gel surface on a filter paper overlay and incubate the gel in a moist chamber. After 20 to 30 min of incubation, monitor the gel under long-wave UV light for dark (nonfluorescent) bands visible on a light (fluorescent) background. Record the zymogram or photograph using a yellow filter.

Notes: If too much NAD(P)H is added to the staining solution or when GDH activity is too low, the enzyme should not be able to convert enough NAD(P)H into NAD(P) for nonfluorescent bands to become visible.

GENERAL NOTES

Three types of GDH are recognized, which exhibit different specificities to NAD and NADP as cofactors:

1. GDH (EC 1.4.1.2) is specific to NAD and predominates in plants and invertebrates. Plant GDH-NAD is activated by Ca^{2+} ions.
2. GDH (EC 1.4.1.4) is specific to NADP. This enzyme was found in some bacteria, fungi, some plants, and in coelenterates. GDH-NADP is not activated by ADP and L-leucine and is not inhibited by GTP.
3. GDH (EC 1.4.1.3) capable of using either NAD or NADP is described in some invertebrates and is predominant in vertebrates. It is activated by ADP and L-leucine and inhibited by GTP.

Some isozymes of bovine liver and placental NAD-dependent GDH also display alanine dehydrogenase (see 1.4.1.1 — ALADH) activity.[5] This activity of GDH is stimulated by GTP and inhibited by ADP. The enzyme from bovine liver exhibits catalytic activity toward L-methionine (see 2.6.1.64 — GTK, Method, *Notes*).

The NADP-dependent bacterial enzyme was found to be highly sensitive to pH variations. This enzyme was stable only in the alkaline pH range, but no activity was detected in the neutral and acidic pH range, despite appropriate buffering of gels for the specific staining procedure.[6]

Unusual NAD(P)H-dependent GDH was revealed in *Streptococcus suis* serotype 2. The enzyme is unusual in that it utilizes L-glutamate rather than 2-oxoglutarate as the substrate in the backward reaction. Activity staining of native PAG for GDH activity was used to distinguish highly virulent strains of *S. suis* type 2 from moderately virulent and avirulent strains.[7]

REFERENCES

1. Shaw, C.R. and Prasad, R., Starch gel electrophoresis of enzymes: a compilation of recipes, *Biochem. Genet.*, 4, 297, 1970.
2. Richardson, B.J., Baverstock, P.R., and Adams, M., *Allozyme Electrophoresis: A Handbook for Animal Systematics and Population Studies*, Academic Press, Sydney, 1986, p. 182.
3. Pérez-de la Mora, M., Mendez-Franco, J., Salceda, R., and Riesgo-Escovar, J.J.R., A glutamate dehydrogenase-based method for the assay of L-glutamic acid: formation of pyridine nucleotide fluorescent derivatives, *Anal. Biochem.*, 180, 248, 1989.
4. Harris, H. and Hopkinson, D.A., *Handbook of Enzyme Electrophoresis in Human Genetics*, North-Holland, Amsterdam, 1976 (loose-leaf, with supplements in 1977 and 1978).
5. Lan, N., Frieden, E.H., and Rawitch, A.B., Activity staining for glutamate and alanine dehydrogenases in polyacrylamide gels of varying porosity, *Enzyme*, 24, 416, 1979.
6. Picard, B., Krishnamoorthy, R., and Goullet, P., Enzyme polymorphism in bacteria: study of molecular relatedness by combined isoelectric focusing–electrophoresis, *Electrophoresis*, 8, 149, 1989.
7. Okwumabua, O., Persaud, J.S., and Reddy, P.G., Cloning and characterization of the gene encoding the glutamate dehydrogenase of *Streptococcus suis* serotype 2, *Clin. Diagn. Lab. Immun.*, 8, 251, 2001.

1.4.1.9 — Leucine Dehydrogenase; LEUDH

REACTION	L-Leucine + H_2O + NAD = 4-methyl-2-oxopentanoate + NH_3 + NADH
ENZYME SOURCE	Bacteria
SUBUNIT STRUCTURE	Octamer[#] (bacteria)

METHOD

Visualization Scheme

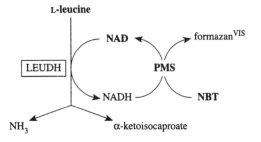

Staining Solution[1]

0.1 M Phosphate buffer, pH 7.0	100 ml
L-Leucine	50 mg
NAD	50 mg
NBT	30 mg
PMS	2 mg

Procedure

Incubate the gel in staining solution in the dark at 37°C until dark blue bands appear. Wash the stained gel in water and fix in 25% ethanol.

Notes: The enzyme also acts on isoleucine, valine, norvaline, and norleucine.[2]

REFERENCES

1. Shaw, C.R. and Prasad, R., Starch gel electrophoresis of enzymes: a compilation of recipes, *Biochem. Genet.*, 4, 297, 1970.
2. NC-IUBMB, *Enzyme Nomenclature*, Academic Press, San Diego, 1992, p. 88 (EC 1.4.1.9, Comments).

1.4.1.15 — Lysine Dehydrogenase; LYSDH

REACTION	L-Lysine + NAD = 1,2-didehydropiperidine-2-carboxylate + NH_3 + NADH
ENZYME SOURCE	Bacteria
SUBUNIT STRUCTURE	Unknown or no data available

METHOD

Visualization Scheme

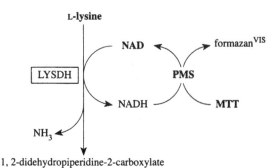

1, 2-didehydropiperidine-2-carboxylate

Staining Solution[1]

0.015 M Sodium phosphate buffer, pH 7.0	50 ml
L-Lysine	5 mg
1.25% MTT	1 ml
1% PMS	0.5 ml
1% NAD	2 ml

Procedure

Incubate the gel in staining solution in the dark at 37°C until dark blue bands appear. Wash the stained gel in water and fix in 25% ethanol.

REFERENCES

1. Selander, R.K., Caugant, D.A., Ochman, H., Musser, J.J.M., Gilmour, M.N., and Whittam, T.S., Methods of multilocus enzyme electrophoresis for bacterial population genetics and systematics, *Appl. Environ. Microbiol.*, 51, 873, 1986.

1.4.1.X — Aspartate Dehydrogenase; ASPDH

REACTION (presumed)   L-Aspartate + H_2O + NAD = oxalo-
                                   acetate + NH_3 + NADH

ENZYME SOURCE   Bacteria, fungi

SUBUNIT STRUCTURE   Dimer[#] (bacteria, fungi)

Method

Visualization Scheme

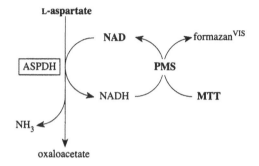

Staining Solution[1]

0.015 M Sodium phosphate buffer, pH 7.0	50 ml
L-Aspartic acid (sodium salt)	50 mg
1.25% MTT	1 ml
1% PMS	0.5 ml
1% NAD	2 ml

Procedure

Incubate the gel in staining solution in the dark at 37°C until dark blue bands appear. Wash the stained gel in water and fix in 25% ethanol.

Notes: This enzyme is not yet included in the enzyme list.[2] The exact reaction catalyzed by ASPDH is not known. Thus, the proposed product, oxaloacetate, may be misleading. It is not excluded that ASPDH is identical to the L-amino acid dehy-drogenase (EC 1.4.1.5).

References

1. Selander, R.K., Caugant, D.A., Ochman, H., Musser, J.J.M., Gilmour, M.N., and Whittam, T.S., Methods of multilocus enzyme electrophoresis for bacterial population genetics and systematics, *Appl. Environ. Microbiol.*, 51, 873, 1986.
2. NC-IUBMB, *Enzyme Nomenclature*, Academic Press, San Diego, 1992.

1.4.3.1 — D-Aspartate Oxidase; DASOX

OTHER NAMES	Aspartic oxidase
REACTION	D-Aspartate + H_2O + O_2 = oxalo-acetate + NH_3 + H_2O_2
ENZYME SOURCE	Vertebrates
SUBUNIT STRUCTURE	Monomer (vertebrates)

METHOD 1

Visualization Scheme

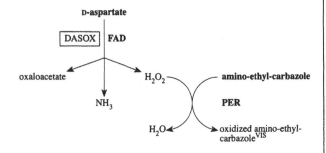

Staining Solution[1]

0.5 M Tris–HCl buffer, pH 8.0	50 ml
D-Aspartic acid (monosodium salt)	200 mg
FAD	8 mg
Peroxidase (PER; 100 U/mg)	5 mg
3-Amino-9-ethyl-carbazole (dissolved in a few drops of acetone)	25 mg

Procedure

Incubate the gel in staining solution at 37°C until red-brown bands appear. Wash the stained gel in water and fix in 50% glycerol.

Notes: With prolonged staining, weak additional bands due to superoxide dismutase (see 1.15.1.1 — SOD) activity can also develop in tissues such as the mammalian liver and kidney. Development of these bands depends on the presence of amino-ethyl-carbazole and FAD in the staining solution, but not on the presence of D-aspartate. Presumably, under these conditions of staining, sufficient amounts of the superoxide radical are generated to allow the SOD isozymes to catalyze the oxidation of amino-ethyl-carbazole, and hence be detected as red-brown bands.

METHOD 2

Visualization Scheme

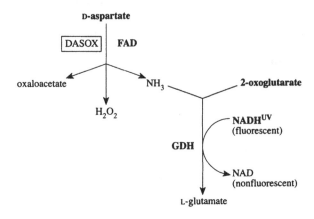

Staining Solution[2]

0.1 M Tris–HCl buffer, pH 7.6	5 ml
D-Aspartic acid	100 mg
FAD	0.1 mg
2-Oxoglutarate	25 mg
NADH	10 mg
Glutamate dehydrogenase (GDH; 500 U/ml)	0.05 ml

Procedure

Apply the staining solution to the gel surface on a filter paper overlay and incubate at 37°C in a moist chamber. After 20 to 30 min of incubation, monitor the gel under long-wave UV light. Dark (nonfluorescent) bands of DASOX are visible on a light (fluorescent) background. Record the zymogram or photograph using a yellow filter.

Notes: When a zymogram that is visible in daylight is required, the processed gel should be counterstained with MTT–PMS solution. This treatment results in the appearance of white bands of DASOX on a blue background of the gel.

METHOD 3

Visualization Scheme

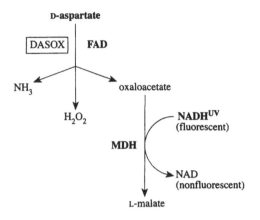

D-aspartate

DASOX   FAD

NH₃   oxaloacetate

H₂O₂   NADH^UV (fluorescent)

MDH

NAD (nonfluorescent)

L-malate

Staining Solution[1]

0.5 *M* Tris–HCl buffer, pH 8.0	5 ml
D-Aspartic acid (monosodium salt)	20 mg
FAD	1 mg
NADH	10 mg
Malate dehydrogenase (MDH)	100 U

Procedure

Apply the staining solution to the gel surface on a filter paper overlay and incubate the gel with application at 37°C in a moist chamber. After 20 to 30 min of incubation, monitor the gel under long-wave UV light. Dark (nonfluorescent) bands of DASOX are visible on a light (fluorescent) background. Record the zymogram or photograph using a yellow filter.

Notes: When a zymogram that is visible in daylight is required, the processed gel should be counterstained with MTT–PMS solution. This treatment results in the appearance of white bands of DASOX on a blue background of the gel.

GENERAL NOTES

The enzyme is a flavoprotein with FAD as a prosthetic group.[3] It was shown, however, that development of human DASOX isozymes was not dependent significantly on the presence of FAD in the stain.[1] The enzyme–FAD complex perhaps does not dissociate during the course of electrophoresis; therefore, FAD may be omitted from the staining solutions.

The enzyme from the human liver and rabbit kidney is specific to D-aspartic acid and does not act on other D-amino acids.

REFERENCES

1. Barker, R.F. and Hopkinson, D.A., The genetic and biochemical properties of the D-amino acid oxidase in human tissues, *Ann. Hum. Genet.*, 41, 27, 1977.
2. Nelson, R.L., Povey, S., Hopkinson, D.A., and Harris, H., The detection after electrophoresis of enzymes involved in ammonia metabolism using L-glutamate dehydrogenase as a linking enzyme, *Biochem. Genet.*, 15, 1023, 1977.
3. NC-IUBMB, *Enzyme Nomenclature*, Academic Press, San Diego, 1992, p. 89 (EC 1.4.3.1, Comments).

1.4.3.2 — L-Amino Acid Oxidase; LAOX

OTHER NAMES   Ophio-amino acid oxidase
REACTION   L-Amino acid + H_2O + O_2 = 2-oxo
                   acid + NH_3 + H_2O_2
ENZYME SOURCE   Bacteria, fungi, vertebrates
SUBUNIT STRUCTURE Monomer (vertebrates)

METHOD 1

Visualization Scheme

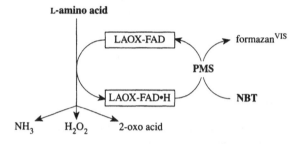

Staining Solution[1]

65 mM Phosphate buffer, pH 6.8	60 ml
10 mM L-Amino acid (see General Notes)	10 ml
1 mg/ml NBT	5 ml
0.5 mg/ml PMS	5 ml

Procedure

Incubate the gel in staining solution in the dark at 37°C until dark blue bands appear. Wash the developed gel in water and fix in 25% ethanol.

METHOD 2

Visualization Scheme

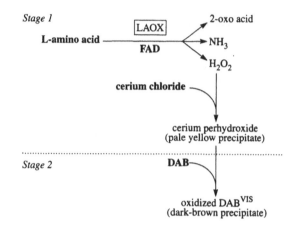

Staining Solution[2]

A. 200 mM Tris–maleate buffer, pH 8.3
   10 mM L-Amino acid (e.g., L-phenylalanine or L-methionine; see also General Notes)
   1 µM FAD
   3 mM Cerium chloride ($CeCl_3$)
B. 100 mM Acetate buffer, pH 5.5
   1 mM 3,3'-Diaminobenzidine (DAB)
   20 µg/ml $CoCl_2 \times 6H_2O$
   20 µg/ml $(NH_4)_2Ni(SO_4)_2 \times 6H_2O$

Procedure

Preequilibrate the electrophorized gel in 200 mM Tris–maleate buffer (pH 8.3) for 5 to 10 min and then incubate in solution A for 30 min. Rinse the gel twice in 100 mM acetate buffer (pH 5.5) for 5 min each and incubate in solution B for 10 to 20 min. Carry out all the procedures at 37°C. After staining is completed, photograph or dry the zymogram instantly.

Notes: The method is based on the ability of cerium chloride to react with hydrogen peroxide, forming the insoluble cerium perhydroxide of a pale yellow color. Cerium perhydroxide is then converted to a dark brown precipitate by the addition of DAB, which is oxidized by cerium perhydroxide to its polymeric form.

The method can be applied to native electrophoretic gels and nitrocellulose blots.

OTHER METHODS

A. The product hydrogen peroxide can be detected using exogenous peroxidase and amino-ethyl-carbazole (e.g., see 1.4.3.3 — DAOX, Method 1).

B. The product ammonia can be detected using exogenous glutamate dehydrogenase in a negative fluorescent stain (e.g., see 1.4.3.3 — DAOX, Method 2).

C. A procedure has been developed for spectrophotometric detection of LAOX on PAG using gel scanning.[3] The method is based on the strong absorbance at 296 nm of aminoethyl cysteine ketimine, which originates as a result of oxidation of *S*-aminoethyl-L-cysteine by LAOX. Photocopies of gels stained by this method can be obtained using a special optical device for two-dimensional spectroscopy of electrophoretic gels.[4]

GENERAL NOTES

The enzyme from *Neurospora*, the rat kidney, and snake venom displays the highest rate of reaction when L-amino acids with five or six C atoms are used as substrates. L-Leucine is the substrate more frequently used for LAOX from vertebrates. Good results were also obtained with L-phenylalanine, L-arginine, and L-glutamine as substrates for snake venom LAOX. This enzyme displays trace activity with L-aspartate and L-proline and no histochemically detectable activity with L-glutamate,

L-cysteine, and L-serine. The enzyme from the liver and kidney also acts on some 2-oxo acids, while that from snake venom does not. The LAOX from venom of *Bothrops atrox* (9 mU) was readily detected by the cerium–DAB method (Method 2), while the use of the linking peroxidase reaction (e.g., see 1.4.3.1 — DASOX, Method 1) to detect hydrogen peroxide produced by LAOX proved unsuccessful.[2]

The enzyme is a flavoprotein (FAD).[5] However, development of LAOX activity bands usually is not dependent significantly on the presence of FAD in the stain.

REFERENCES

1. Feinstein, R.N. and Lindahl, R., Detection of oxidases on polyacrylamide gels, *Anal. Biochem.*, 56, 353, 1973.
2. Seitz, J., Keppler, C., Fahimi, H.D., and Völkl, A., A new staining method for the detection of activities of H$_2$O$_2$-producing oxidases on gels and blots using cerium and 3,3'-diaminobenzidine, *Electrophoresis*, 12, 1051, 1991.
3. Ricci, G., Caccuri, A.M., Lo Bello, M., Solinas, S.P., and Nardini, M., Ketimine rings: useful detectors of enzymatic activities in solution and on polyacrylamide gel, *Anal. Biochem.*, 165, 356, 1987.
4. Klebe, R.J., Mancuso, M.G., Brown, C.R., and Teng, L., Two-dimensional spectroscopy of electrophoretic gels, *Biochem. Genet.*, 19, 655, 1981.
5. NC-IUBMB, *Enzyme Nomenclature*, Academic Press, San Diego, 1992, p. 89 (EC 1.4.3.2, Comments).

1.4.3.3 — D-Amino Acid Oxidase; DAOX

REACTION   D-Amino acid + H_2O + O_2 = 2-oxo acid + NH_3 + H_2O_2

ENZYME SOURCE   Fungi, vertebrates

SUBUNIT STRUCTURE   Monomer (vertebrates)

METHOD 1

Visualization Scheme

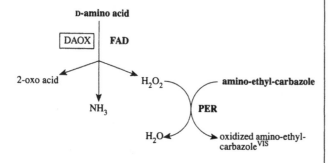

Staining Solution[1] (modified)

A. 0.018 M Tris–HCl buffer, pH 7.4
    0.0125 M D-Phenylalanine
    0.001 M FAD
B. 3-Amino-9-ethyl-carbazole (dissolved in
    a few drops of acetone)   3 mg
    Peroxidase (PER)   100 U
C. 2% Agar solution (60°C)

Procedure

Add B components to 12.5 ml of solution A and mix with 12.5 ml of solution C. Pour the mixture over the surface of the gel and incubate at 37°C until brown bands appear. Fix the stained gel in 50% glycerol.

Notes: With prolonged staining, weak additional bands due to superoxide dismutase (see 1.15.1.1 — SOD) activity can also be developed in tissues such as the liver and kidney. Development of these bands depends on the presence of amino-ethyl-carbazole and FAD in the stain, but not on the presence of D-amino acid. Presumably, under these conditions of staining, sufficient amounts of the superoxide radical are generated to allow the SOD isozymes to catalyze the oxidation of amino-ethyl-carbazole, and hence be detected as brown bands.

METHOD 2

Visualization Scheme

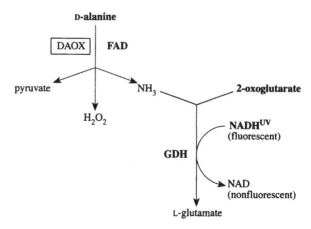

Staining Solution[2]

0.1 M Tris–HCl buffer, pH 7.6	5 ml
D-Alanine	100 mg
FAD	0.1 mg
2-Oxoglutarate	25 mg
NADH	10 mg
Glutamate dehydrogenase (GDH; 500 U/ml)	0.05 ml

Procedure

Apply the staining solution to the gel surface on a filter paper overlay and incubate the gel at 37°C in a moist chamber. After 20 to 30 min of incubation, monitor the gel under long-wave UV light. Dark (nonfluorescent) bands of DAOX are visible on a light (fluorescent) background. Record the zymogram or photograph using a yellow filter.

Notes: When a zymogram that is visible in daylight is required, the processed gel should be counterstained with MTT–PMS solution. This treatment results in the appearance of white bands of DAOX on a blue background of the gel.

It should be taken into account when using this method that the amount of NADH added to the staining solution is critical for development of nonfluorescent DAOX bands. If too much NADH is added, the enzyme should not be able to convert enough NADH into NAD for nonfluorescent bands to become visible.

METHOD 3

Visualization Scheme

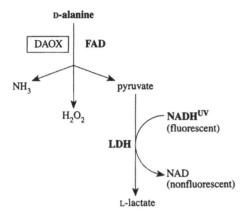

Staining Solution[3]

A. 0.5 *M* Tris–HCl buffer, pH 8.0
   30 m*M* D-Alanine
B. FAD   1 mg
   NADH   10 mg
   Lactate dehydrogenase (LDH)   50 U

Procedure

Add B components to 5 ml of solution A and apply the mixture to the gel surface on a filter paper overlay. Incubate the gel with application at 37°C in a moist chamber for 20 to 30 min and view under long-wave UV light. Dark (nonfluorescent) bands of DAOX are visible on a light (fluorescent) background. Record the zymogram or photograph using a yellow filter.

Notes: When a zymogram that is visible in daylight is required, the processed gel should be counterstained with MTT–PMS solution. This treatment results in the appearance of white bands of DAOX on a blue background of the gel.

METHOD 4

Visualization Scheme

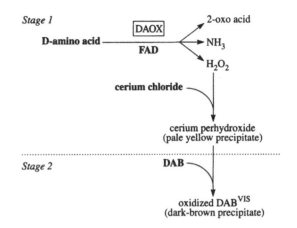

Staining Solution[4]

A. 200 m*M* Tris–maleate buffer, pH 8.3
   10 m*M* D-Proline (see also General Notes)
   1 μ*M* FAD
   3 m*M* Cerium chloride (CeCl$_3$)
B. 100 m*M* Acetate buffer, pH 5.5
   1 m*M* 3,3′-Diaminobenzidine (DAB)
   20 μg/ml CoCl$_2$ × 6H$_2$O
   20 μg/ml (NH$_4$)$_2$Ni(SO$_4$)$_2$ × 6H$_2$O

Procedure

Preequilibrate the electrophorized gel in 200 m*M* Tris–maleate buffer (pH 8.3) for 5 to 10 min and then incubate in solution A for 30 min. Rinse the gel twice in 100 m*M* acetate buffer (pH 5.5) for 5 min each and incubate in solution B for 10 to 20 min. Carry out all the procedures at 37°C. After staining is completed, photograph or dry the zymogram instantly.

Notes: The method is based on the ability of cerium chloride to react with hydrogen peroxide, forming the insoluble cerium perhydroxide of a pale yellow color. Cerium perhydroxide is then converted to a dark brown precipitate by the addition of DAB, which is oxidized by cerium perhydroxide to its polymeric form.

The method can be applied to native electrophoretic gels and nitrocellulose blots.

GENERAL NOTES

The resolution and sensitivity of the positive stain (Method 1) is better than that obtained with any of the negative fluorescent stains (Methods 2 and 3).

Human DAOX acts on many D-amino acids, except D-aspartic acid, D-cysteine, D-cystine, and D-glutamic acid. The highest activity was revealed with D-phenylalanine, D-leucine, D-alanine,

1.4.3.3 — D-Amino Acid Oxidase; DAOX (continued)

D-isoleucine, D-norleucine, D-proline, and D-tyrosine. The animal enzyme also acts on glycine.

The enzyme is a flavoprotein (FAD).[5] Better staining of DAOX bands can be obtained with higher concentrations of FAD in the staining solution. The staining intensity of DAOX can also be enhanced if FAD (final concentration of 0.2 mM) is added to the gel and bridge buffers prior to electrophoresis. Under these conditions, however, the enzyme from the human liver exhibits a single band of higher activity and greater anodal mobility than either of the two bands seen on the zymogram after electrophoresis in the absence of FAD. This effect of enhanced activity and altered electrophoretic mobility is not observed when homogenates are treated with FAD (final concentration of 0.5 mM) prior to electrophoresis.[3]

The DAOX from the porcine kidney (0.2 U) was readily detected by the cerium–DAB method (Method 4), while the use of Method 1 based on the coupled peroxidase reaction proved unsuccessful.[4]

REFERENCES

1. Duley, J. and Holmes, R.S., α-Hydroxyacid oxidase in the mouse: evidence for two genetic loci and a tetrameric subunit structure for the liver isozyme, *Genetics*, 76, 93, 1974.
2. Nelson, R.L., Povey, S., Hopkinson, D.A., and Harris, H., The detection after electrophoresis of enzymes involved in ammonia metabolism using L-glutamate dehydrogenase as a linking enzyme, *Biochem. Genet.*, 15, 1023, 1977.
3. Barker, R.F. and Hopkinson, D.A., The genetic and biochemical properties of the D-amino acid oxidase in human tissues, *Ann. Hum. Genet.*, 41, 27, 1977.
4. Seitz, J., Keppler, C., Fahimi, H.D., and Völkl, A., A new staining method for the detection of activities of H$_2$O$_2$-producing oxidases on gels and blots using cerium and 3,3'-diaminobenzidine, *Electrophoresis*, 12, 1051, 1991.
5. NC-IUBMB, *Enzyme Nomenclature*, Academic Press, San Diego, 1992, p. 90 (EC 1.4.3.3, Comments).

1.4.3.4 — Monoamine Oxidase; MAOX

OTHER NAMES	Amine oxidase (flavin containing) (recommended name), tyramine oxidase, tyraminase, amine oxidase, adrenaline oxidase
REACTION	$RCH_2NH_2 + H_2O + O_2 = RCHO + NH_3 + H_2O_2$ (acts on primary amines, and usually also on secondary and tertiary amines with small substituents)
ENZYME SOURCE	Bacteria, fungi, plants, invertebrates, vertebrates
SUBUNIT STRUCTURE	Dimer[#] (bacteria), hexamer[#] (vertebrates); see General Notes

METHOD 1

Visualization Scheme

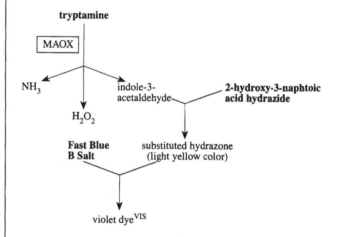

Staining Solution[1]

A.	KH$_2$PO$_4$	3.71 g
	0.5 M NaOH	46.8 ml
	H$_2$O	953 ml
B.	40% H$_2$SO$_4$	50 ml
	1 M NaOH	1 ml
	0.2 M K$_2$HPO$_4$	20 ml
C.	2-Hydroxy-3-naphtoic acid hydrazide	25 mg
D.	0.1 M Tryptamine hydrochloride	10 ml
E.	Fast Blue B salt	200 mg
	Solution A, pH 8.0	100 ml

Procedure

Dissolve the C component in heated solution B (80 to 90°C). Filtrate cooled mixture and add solution D. Incubate the gel in the resulting mixture at 37°C until light yellow bands appear. Place the gel in solution E for 10 min and then transfer it in 7% acetic acid. Violet bands clearly visible on a blue background appear after about 30 min.

METHOD 2

Visualization Scheme

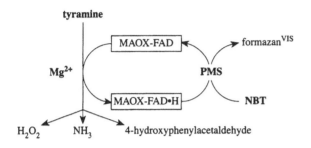

Staining Solution[2]

0.5 M Phosphate buffer, pH 7.4	5 ml
0.2 M Tyramine (or benzylamine, or isoamylamine, or tryptamine)	2 ml
5 mM MgCl$_2$	2 ml
1 mg/ml NBT	5 ml
· 1 mg/ml PMS	0.5 ml
10 mM KCN	2 ml

Procedure

Incubate the gel in staining solution in the dark at 37°C until dark blue bands appear. Wash the stained gel in water and fix in 25% ethanol.

Notes: KCN is used to prevent development of superoxide dismutase (see 1.15.1.1 — SOD) bands. It should be remembered, however, that MAOX is also inhibited by cyanide ions.

OTHER METHODS

A. Two-dimensional spectroscopy of the electrophoretic gel (see Part II for details) can be used to visualize MAOX activity bands by purely optical means.[3] In this procedure MAOX is detected due to its formation of H$_2$O$_2$ as a reaction product. With the use of peroxidase as a coupling enzyme and homovanillic acid as a peroxidase substrate, a highly fluorescent product is formed, which is detected fluorimetrically in a substrate-containing agarose overlay at an excitation of 313 nm and an emission of 421 nm. A special optical device permits photocopies of the developed agarose overlay to be obtained. This method is not widely used because several simpler procedures for visualization of MAOX activity bands on electrophoretic gels are available.

B. A linked peroxidase reaction also can be used to develop MAOX bands visible in daylight (see, for example, 1.4.3.3 — DAOX, Method 1).

C. A linked glutamate dehydrogenase reaction can be used in a negative fluorescent stain to detect ammonia produced by MAOX (see, for example, 1.4.3.3 — DAOX, Method 2).

D. The immunoblotting procedure (for details, see Part II) using monoclonal antibodies specific to human MAOX[4] can also be used for localization of the enzyme protein on electrophoretic gels. This procedure, however, is time-consuming and very expensive, and thus unsuitable for routine laboratory use in MAOX electrophoretic studies.

GENERAL NOTES

The enzyme is a flavoprotein with FAD as the prosthetic group.[5] The flavoprotein complex does not dissociate during electrophoresis because FAD is tightly bound to the enzyme molecule. Therefore, the addition of FAD to MAOX staining solutions is not needed.

Bovine liver MAOX molecules are hexameric complexes that contain threefold rotational symmetry. The individual complexes have globular morphology and the hexamers appear to be composed of a trimer of MAOX homodimers.[6]

REFERENCES

1. Shaw, C.R. and Prasad, R., Starch gel electrophoresis of enzymes: a compilation of recipes, *Biochem. Genet.*, 4, 297, 1970.
2. Allen, J.M. and Beard, M.E., α-Hydroxy acid oxidase: localization in renal microbodies, *Science*, 149, 1507, 1965.
3. Klebe, R.J., Mancuso, M.G., Brown, C.R., and Teng, L., Two-dimensional spectroscopy of electrophoretic gels, *Biochem. Genet.*, 19, 655, 1981.
4. Denney, R.M., Fritz, R.R., Patel, N.T., and Abell, C.W., Human liver MAO-A and MAO-B separated by immunoaffinity chromatography with MAO-B specific monoclonal antibody, *Science*, 215, 1400, 1982.
5. NC-IUBMB, *Enzyme Nomenclature*, Academic Press, San Diego, 1992, p. 90 (EC 1.4.3.4, Comments).
6. Shiloff, B.A., Behrens, P.Q., Kwan, S.W., Lee, J.J.H., and Abell, C.W., Monoamine oxidase B isolated from bovine liver exists as large oligomeric complexes *in vitro*, *Eur. J. Biochem.*, 242, 41, 1996.

1.4.3.6 — Amine Oxidase (Copper Containing); AMOX

OTHER NAMES	Diamine oxidase, diamino oxyhydrase, histaminase, amine oxidase
REACTION	$RCH_2NH_2 + H_2O + O_2 = RCHO + NH_3 + H_2O_2$ (see General Notes)
ENZYME SOURCE	Bacteria, fungi, plants, invertebrates, vertebrates
SUBUNIT STRUCTURE	Dimer# (bacteria, fungi, plants)

METHOD 1

Visualization Scheme

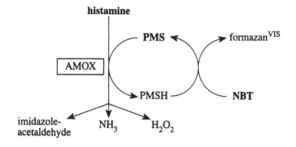

Staining Solution[1]

65 mM Phosphate buffer, pH 6.8	60 ml
10 mM Histamine	10 ml
1 mg/ml NBT	5 ml
0.5 mg/ml PMS	5 ml

Procedure

Incubate the gel in staining solution in the dark at 37°C until dark blue bands appear. Wash the stained gel in water and fix in 25% ethanol.

METHOD 2

Visualization Scheme

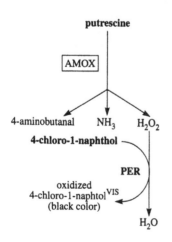

Staining Solution[2]

100 mM Phosphate buffer, pH 7.2
10 mM Putrescine
0.1 mM 4-Chloro-1-naphthol
10 mg/ml Horseradish peroxidase (PER; Sigma, type IV)

Procedure

Incubate the gel in staining solution (gel:staining solution volume of 1:20) overnight (18 h) with gentle agitation. Bands of black color indicate localization of AMOX activity in the gel.

Notes: Diaminobenzidine inhibits AMOX activity and therefore can not be used as a chromogenic substrate for auxiliary peroxidase. The use of phosphate buffer is preferable because inhibition of the staining reaction of 4-chloro-1-naphtol in a Tris-containing buffer was observed.

METHOD 3

Visualization Scheme

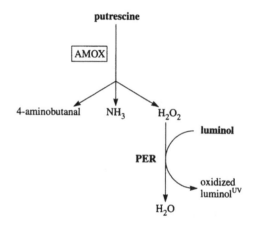

Staining Solution[3]

5 mM Phosphate buffer, pH 7.2	2 ml
Putrescine	64.4 mg
Luminol solution (enhanced chemiluminescence (ECL)–Western blotting detection reagent 2; Amersham)	2 ml
Horseradish peroxidase (PER)	40 μg

Procedure

Blot electrophorized agarose gel to a 0.45-μm pore size nitrocellulose (NC) membrane (10 × 20 cm). Perform this by placing the NC membrane, 5 pieces of Whatman No. 1 paper, and 20 pieces of cell paper on the agarose gel, and then apply gravitational pressure for 60 min. Equilibrate the membrane in 5 mM phosphate buffer (pH 7.2) for 5 min, and then press between two filter papers for 5 min. Apply the staining solution to a 10 × 20 cm filter paper placed in a plastic folder and layer the NC membrane on top of the filter paper. Place the plastic folder in a film cassette with an ECL hyperfilm (Amersham) on top and incubate overnight at room temperature. Develop the exposed ECL hyperfilm using a developer (LX 24, x-ray) and fixer (AL 4, x-ray) from Kodak. The zones of AMOX activity develop as black bands on a dark background of the film.

Notes: The method is developed for agarose gels; however, it can be easily adapted to polyacrylamide and starch gels.

Luminol is also known as 5-amino-2,3-dihydro-1,4-phthalazinedione or 3-aminophthalhydrazide.

The method allows detection of less than 0.1 U of the enzyme in electrophoresis gels and is suitable for electrophoretic analysis of AMOX in unpurified enzyme preparations, e.g., in sera. It should be kept in mind, however, that an unspecifically stained black band corresponding to serum albumin of different vertebrate species always develops on AMOX zymograms obtained using this method. This band disappears when PER or luminol is omitted from the staining solution, indicating that the serum albumin is interacting directly with a component in the staining solution. The nature of this interaction remains unclear.

Dot blots made by applying 3-μl samples to the NC membrane (after blotting the membrane with the electrophoretic gel) indicate that the AMOX activity in some sera samples is decreased (relative to the dot blot) during electrophoresis, providing evidence that AMOX is separated from a component needed to give maximal enzyme activity. This component remains unknown.

Histamine, spermine, spermidine, and 1,8-diaminooctane can also be used as substrates suitable to detect AMOX by the enhanced chemiluminescence method described above. With 1,8-diaminooctane as the substrate, more bands are detected and the strength of the bands is higher than with detection using other substrates, including putrescine. A drawback in the use of 1,8-diaminooctane is a relatively high background, as well as the possibility that activity bands of monoamine oxidase (see 1.4.3.4 — MAOX) may be also developed. To eliminate a high background, the solution containing 1,8-diaminooctane should be filtered immediately before use. To identify possible bands of MAOX, 1,8-diaminooctane should be used only in comparison with another AMOX substrate, e.g., putrescine.

OTHER METHODS

A. A coupled peroxidase reaction can be used to detect hydrogen peroxide produced by AMOX (see, for example, 1.4.3.3 — DAOX, Method 1).

B. A coupled glutamate dehydrogenase reaction can be used in a negative fluorescent stain to detect ammonia produced by AMOX (see, for example, 1.4.3.3 — DAOX, Method 2).

C. 2-Hydroxy-3-naphtoic acid hydrazide can be used to detect imidazoleacetaldehyde produced by AMOX. This method is based on the capture of aldehyde by a hydrazide and subsequent coupling of the product with Fast Blue B salt, resulting in violet dye (for details, see 1.4.3.4 — MAOX, Method 1).

GENERAL NOTES

The enzyme isolated from a number of sources exhibits low substrate specificity. Primary monoamines, diamines, and histamine are all oxidized by the enzyme.[4]

When AMOX from the porcine kidney was subjected to electrophoresis in nondenaturing PAG, it was detected as a high-molecular-weight complex that hardly entered the gel. Addition of Triton X-100, NP-40, and Chaps to AMOX samples before electrophoresis did not dissociate the complex, but Tween 20 did dissociate it.[2]

REFERENCES

1. Feinstein, R.N. and Lindahl, R., Detection of oxidases on polyacrylamide gels, *Anal. Biochem.*, 56, 353, 1973.

2. Houen, G. and Leonardsen, L., A specific peroxidase-coupled activity stain for diamine oxidases, *Anal. Biochem.*, 204, 296, 1992.

3. Bruun, L. and Houen, G., *In situ* detection of diamine oxidase activity using enhanced chemiluminescence, *Anal. Biochem.*, 233, 130, 1996.

4. NC-IUBMB, *Enzyme Nomenclature*, Academic Press, San Diego, 1992, p. 90 (EC 1.4.3.6, Comments).

1.4.3.11 — L-Glutamate Oxidase; LGO

REACTION   L-Glutamate + O_2 + H_2O = 2-oxo-
                   glutarate + NH_3 + H_2O_2

ENZYME SOURCE   Bacteria

SUBUNIT STRUCTURE Heterotrimer[#] (bacteria); see General
                   Notes

METHOD

Visualization Scheme

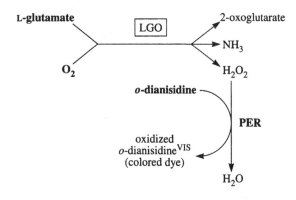

Staining Solution[1] (adapted)

50 mM Tris–HCl buffer, pH 7.5
20 mM L-Glutamate
1 U/ml Peroxidase (PER)
0.3 mg/ml o-Dianisidine dihydrochloride (dissolved in a few
  drops of ethyl alcohol)

Procedure

Incubate the gel in staining solution until colored bands appear.
Record or photograph the zymogram.

Notes: This method is an adaptation of a continuous cycling method
developed for spectrophotometric determination of L-glutamate.

GENERAL NOTES

The mature extracellular enzyme from *Streptomyces platensis* is
a heterotrimer composed of three different subunits encoded by
a single gene.[2]

REFERENCES

1. Valero, E. and Garcia-Carmona, F., A continuous spectrophotometric method based on enzymatic cycling for determining
   L-glutamate, *Anal. Biochem.*, 259, 265, 1998.
2. Chen, C.Y., Wu, W.T., Huang, C.J., Lin, M.H., Chang, C.K.,
   Huang, H.J., Liao, J.J.M., Chen, L.Y., and Liu, Y. T., A common precursor for the three subunits of L-glutamate oxidase
   encoded by *gox* gene from *Streptomyces platensis* NTU3304,
   Can. J. Microbiol., 47, 269, 2001.

1.4.3.X — Cytokinin Oxidase; CKO

REACTION   Isopentenyl adenine + H_2O + O_2 = adenine + 3-methyl-2-butenal + H_2O_2

ENZYME SOURCE   Plants

SUBUNIT STRUCTURE   Unknown or no data available

METHOD

Visualization Scheme

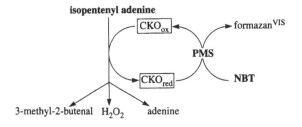

Staining Solution[1]

0.1 M Imidazole buffer, pH 6.5
0.2 mM PMS
0.1 mg/ml NBT
1 mM Isopentenyl adenine

Procedure

Incubate the gel in staining solution at 37°C in the dark until dark blue bands appear. Fix the zymogram in 25% ethanol.

Notes: Activity staining of CKO is based on amine oxidase activity (see 1.4.3.6 — AMOX) of the enzyme. There is, however, evidence that CKO is an FAD-dependent oxidoreductase (see General Notes).

OTHER METHODS

A. The product hydrogen peroxide can be detected using auxiliary peroxidase (e.g., see 1.4.3.3 — DAOX, Method 1).
B. A colorimetric detection method based on the formation of a Schiff's base between 3-methyl-2-butenal and *p*-aminophenol[2] may be adapted to detect CKO activity on electrophoretic gels.

GENERAL NOTES

The enzyme from *Phaseolus vulgaris* was classified as a copper-containing amine oxidase (see 1.4.3.6 — AMOX).[3] However, recent reports provide evidence that CKO from maize kernels is an FAD-dependent oxidase.[4,5] This has lead to a reevaluation of the classification of CKO as a copper-containing amine oxidase. It was suggested that the enzyme belongs to the family of FAD-dependent amine oxidases.[6,7] It is not excluded, however, that a copper amine oxidase type of CKO and an FAD-dependent type of CKO are present in different tissues or intracellular compartments of the same species.[1] Both CKO types are glycoproteins.[1,4]

REFERENCES

1. Kulkarni, M.J., Theerthaprasad, D., Sudharshana, L., Prasad, T.G., and Sashidhar, V.R., A novel colorimetric assay and activity staining for cytokinin oxidase based on the copper amine oxidase property of the enzyme, *Phytochem. Anal.*, 12, 180, 2001.
2. Libreros-Minotta, C.A. and Tipton, P.A., A colorimetric assay for cytokinin oxidase, *Anal. Biochem.*, 231, 339, 1995.
3. Chatfield, J.M. and Armstrong, D.J., Cytokinin oxidase from *Phaseolus vulgaris* callus tissue: enhanced *in vitro* activity of the enzyme in the presence of copper and imidazole complexes, *Plant Physiol.*, 84, 726, 1987.
4. Morris, R.O., Bilyeu, K.D., Laskey, J.J.G., and Cheikh, N.N., Isolation of a gene encoding a glycosylatyed cytokinin oxidase from maize, *Biochem. Biophys. Res. Commun.*, 225, 328, 1999.
5. Houba-Herin, N., Pethe, C., d'Alayer, J., and Laloue, M., Cytokinin oxidase from *Zea mays*: purification, cDNA cloning and expression in moss protoplasts, *Plant J.*, 17, 615, 1999.
6. Rinaldi, A.C. and Comandini, O., Cytokinin oxidase: new insight into enzyme properties, *Trends Plant Sci.*, 4, 127, 1999.
7. Rinaldi, A.C. and Comandini, O., Cytokinin oxidase strikes again, *Trends Plant Sci.*, 4, 300, 1999.

1.5.1.3 — Dihydrofolate Reductase; DHFR

OTHER NAMES | Tetrahydrofolate dehydrogenase, dihydrofolate dehydrogenase

REACTIONS | 1. 5,6,7,8-Tetrahydrofolate + NADP = 7,8-dihydrofolate + NADPH

2. 7,8-Dihydrofolate + NADP = folate + NADPH

ENZYME SOURCE | Bacteria, vertebrates

SUBUNIT STRUCTURE | Tetramer[#] (bacteria), dimer[#] (hyperthermophilic bacterium *Thermotoga maritima*), monomer[#] (chicken)

METHOD 1

Visualization Scheme

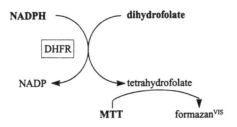

Staining Solution[1]

50 mM Phosphate buffer, pH 6.9
1 mM NADPH
1 mM Dihydrofolate
0.4 mg/ml MTT

Procedure

Incubate the gel in staining solution in the dark at 37°C until dark blue bands appear. Wash the stained gel in water and fix in 25% ethanol.

Notes: The described method was used for in-gel detection of DHFR activity bands after electrophoresis of purified enzyme preparations. Spectrophotometric experiments showed that the formation of a blue insoluble formazan is dependent on the formation of tetrahydrofolate, provided that 2-mercaptoethanol is not added to the staining solution. The mechanism of this reaction is uncertain.

METHOD 2

Visualization Scheme

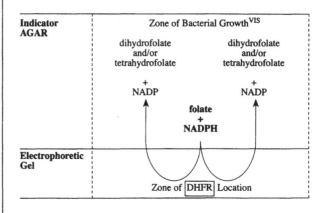

Indicator Agar[2]

1.5% indicator citrate agar containing 100 ng/ml folate, 400 ng/ml NADPH, and 10^8 bacteria/ml of *P. cerevisiae* (ATCC 8081) in citrate medium (see Appendix A-2) lacking folinic and folic acids.

Procedure

Prepare indicator agar and pour it into a sterile plate. After the bacteria-seeded indicator agar solidifies, an electrophoretic starch gel slice (cut surface down), cellulose acetate strip, or PAG is laid over the agar, avoiding the formation of air bubbles between the indicator agar and electrophoretic gels. The bands of bacterial growth become visible in transmitted light after 6 to 12 h of incubation at 37°C.

Notes: As a control for bacterial growth, tetrahydrofolate should be spotted at one corner of the indicator agar. The origin and slot locations should be marked on the indicator agar before it is removed.

Bioautographic visualization of DHFR is the only method that is able to detect the enzyme in unconcentrated crude cell extracts.

GENERAL NOTES

The enzyme from animal sources mediates the reduction of dihydrofolate and folate. Bacterial DHFR was electrophoretically separated into two structural components. The major component is a highly active "specific dihydrofolate reductase," while the minor component catalyzes the reduction of folate also.[3]

1.5.1.3 — Dihydrofolate Reductase; DHFR (continued)

References

1. Nixon, P.F. and Blakley, R.L., Dihydrofolate reductase of *Streptococcus faecium*: II. Purification and some properties of two dihydrofolate reductases from the amethopterin-resistant mutant, *Streptococcus faecium* var. *durans* strain A, *J. Biol. Chem.*, 243, 4722, 1968.
2. Naylor, S.L., Townsend, J.K., and Klebe, R.J., Bioautographic visualization of dihydrofolate reductase in enzyme over-producing BHK mutants, *Biochem. Genet.*, 18, 199, 1980.
3. Albrecht, A.M., Pearce, F.K., Suling, W.J., and Hutchison, D.J., Folate reductase and specific dihydrofolate reductase of the amethopterin-sensitive *Streptococcus faecium* var. *durans*, *Biochemistry*, 8, 960, 1969.

1.5.1.7 — Saccharopine Dehydrogenase (NAD, L-Lysine Forming); SD

OTHER NAMES	Lysine–2-oxoglutarate reductase
REACTION	N^6-(L-1,3-Dicarboxypropyl)-L-lysine + NAD + H_2O = L-lysine + 2-oxo-glutarate + NADH
ENZYME SOURCE	Fungi
SUBUNIT STRUCTURE	Unknown or no data available

Method

Visualization Scheme

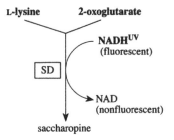

Staining Solution[1]

0.05 *M* Potassium phosphate buffer, pH 6.8
1 m*M* NADH
10 m*M* 2-Oxoglutarate
10 m*M* L-Lysine

Procedure

Apply the staining solution to the gel surface with a filter paper overlay. Incubate the gel at 37°C in a moist chamber. After 20 to 30 min of incubation, monitor the gel under long-wave UV light. Dark (nonfluorescent) bands of SD are visible on a light (fluorescent) background. Record the zymogram or photograph using a yellow filter.

Notes: If a zymogram that is visible in daylight is required, counterstain the processed gel with MTT–PMS solution. This procedure results in the appearance of white bands of SD on the blue background of the gel.

It should be remembered that if too much NADH is added to the staining solution, the enzyme should not be able to convert enough NADH into NAD for nonfluorescent bands to become visible.

References

1. Nakatani, Y., Motoji, F., and Kazuya, H., Enzymic determination of L-lysine in biological materials, *Anal. Biochem.*, 49, 225, 1972.

1.5.1.11 — D-Octopine Dehydrogenase; ONDH

OTHER NAMES   D-Octopine synthase
REACTION N^2-(D-1-Carboxyethyl)-L-arginine + NAD + H$_2$O = L-arginine + pyruvate + NADH
ENZYME SOURCE   Invertebrates (coelenterates, nemertines, priapulids, mollusks)
SUBUNIT STRUCTURE   Monomer (invertebrates)

METHOD 1

Visualization Scheme

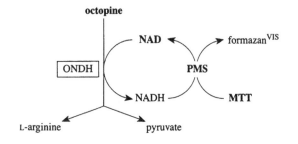

Staining Solution[1]

80 mM Tris–HCl buffer, pH 9.1
1 mM D(+)-Octopine
0.3 mM NAD
0.49 mM MTT
0.16 mM PMS

Procedure

Incubate the gel in staining solution in the dark at 37°C until dark blue bands appear. Wash the stained gel in water and fix in 25% ethanol.

METHOD 2

Visualization Scheme

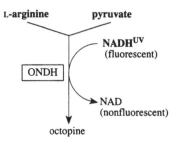

Staining Solution[1]

100 mM Triethanolamine–HCl buffer, pH 7.0
0.1 mM NADH
2 mM Pyruvate
20 mM L-Arginine

Procedure

Lay a piece of filter paper saturated with the staining solution on top of the gel and incubate at 37°C for 10 to 30 min. Remove the filter paper and view the gel under long-wave UV light. Dark (nonfluorescent) bands of ONDH are visible on a light (fluorescent) background of the gel. Record the zymogram or photograph using a yellow filter.

Notes: The staining solution may also be applied as a 1% agar overlay. When a zymogram that is visible in daylight is required, the processed gel should be counterstained with MTT–PMS solution. This results in the almost immediate appearance of white bands on a blue background.

If the concentration of NADH in the stain is too much, the enzyme should not be able to convert enough NADH into NAD for nonfluorescent ONDH bands to become visible.

In the reverse direction ONDH also acts (but more slowly) on L-ornithine, L-lysine, and L-histidine.

In some invertebrate species LDH bands can also develop on ONDH zymograms obtained by this method. Therefore, an additional gel should be stained for LDH as a control.

REFERENCES

1. Dando, P.R., Storey, K.B., Hochachka, P.W., and Storey, J.J.M., Multiple dehydrogenases in marine molluscs: electrophoretic analysis of alanopine dehydrogenase, strombine dehydrogenase, octopine dehydrogenase and lactate dehydrogenase, *Mar. Biol. Lett.*, 2, 249, 1981.

1.5.1.12 — 1-Pyrroline-5-Carboxylate Dehydrogenase; PCD

OTHER NAMES   Pyrroline dehydrogenase

REACTION   1-Pyrroline-5-carboxylate + NAD + H₂O = L-glutamate + NADH; see *Notes*

ENZYME SOURCE   Bacteria, plants, invertebrates

SUBUNIT STRUCTURE   Tetramer# (plants – *Solanum tuberosum*)

METHOD

Visualization Scheme

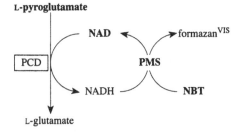

Staining Solution[1]

0.1 *M* Tris–HCl buffer, pH 8.0	100 ml
L-Pyroglutamate	50 mg
NAD	50 mg
NBT	30 mg
PMS	2 mg

Procedure

Incubate the gel in staining solution in the dark at 37°C until dark blue bands appear. Wash the stained gel in water and fix in 25% ethanol.

Notes: Although NAD is the preferred electron acceptor of plant PCD, NADP also may serve as a substrate.[2]

REFERENCES

1. Mulley, J.C. and Latter, B.D.H., Genetic variation and evolutionary relationships within a group of thirteen species of penaeid prawns, *Evolution*, 34, 904, 1980.
2. Forlani, G., Scainelli, D., and Nielsen, E., Delta(1)-pyrroline-5-carboxylate dehydrogenase from cultured cells of potato: purification and properties, *Plant Physiol.*, 113, 1413, 1997.

OTHER NAMES	D(+)-Lysopine synthase, D(+)-octopine synthase
REACTION	D(+)-Lysopine + NADP + H$_2$O = L-lysine + pyruvate + NADPH
ENZYME SOURCE	Crown Gall tumor tissues of dicotyledonous plants induced by *Agrobacterium tumefaciens* (see General Notes)
SUBUNIT STRUCTURE	Monomer# (Crown Gall tumor tissues of dicotyledonous plants)

METHOD 1

Visualization Scheme

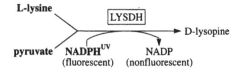

Staining Solution[1]

100 mM Phosphate buffer, pH 6.8
0.1 mM NADPH
12 mM Pyruvate
4 mM L-Lysine

Procedure

Overlay the gel with filter paper soaked in staining solution. Observe dark nonfluorescent bands of LYSDH on the fluorescent background of the gel. Record the zymogram or photograph using a yellow filter.

Notes: When a zymogram that is visible in daylight is required, counterstain the processed gel with MTT–PMS solution to obtain achromatic bands of LYSDH on a blue background of the gel. Fix the counterstained gel in 25% ethanol.

L-Arginine, L-histidine, L-glutamine, L-methionine, and L-ornithine can also be used as amino acid substrates.

METHOD 2

Visualization Scheme

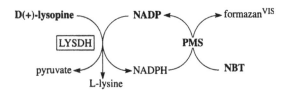

Staining Solution[2]

0.15 M Glycine-NaOH buffer, pH 9.0
0.4 mM NADP
5 mM D(+)-Lysopine
0.32 mg/ml NBT
0.04 mg/ml PMS

Procedure

Incubate the gel in staining solution in the dark at 37°C until dark blue bands appear. Wash the stained gel in water and fix in 25% ethanol.

Notes: D(+)-octopine can also be used as a substrate.

GENERAL NOTES

Crown Gall is a neoplastic disease of dicotyledonous plants induced by many different strains of *Agrobacterium tumefaciens*. This bacterium induces hypertrophies following invasion of a wound site. Neoplastic hypertrophies result following covalent integration of the tumor-inducing plasmid (Ti-plasmid) of the inducing bacterium into plant nuclear DNA. In contrast to tissues of normal plants, transformed tissues synthesize one or more opines (amino acid derivatives), which are not detected in untransformed tissues.[3] Different strains of the inducing bacterium initiate synthesis of N^2-(D-1-carboxyethyl)-amino acid or N^2-(L-1,3-dicarboxypropyl)-amino acid derivatives. Genetic information on the Ti-plasmid specifies which group of derivatives will be synthesized by the tumor.

Electrophoretic data provided evidence that there is a single enzyme protein catalyzing the reductive condensation of pyruvate with different L-amino acids, including L-arginine, L-lysine, L-histidine, L-methionine, L-glutamine, and L-ornithine.[1,2]

See also 1.5.1.19 — NOPDH, General Notes.

REFERENCES

1. Otten, L.A.B.M., Vreugdenhil, D., and Schilperoort, R.A., Properties of D(+)-lysopine dehydrogenase from Crown Gall tumor tissue, *Biochim. Biophys. Acta*, 485, 268, 1977.
2. Hack, E. and Kemp, J.D., Purification and characterization of the Crown Gall-specific enzyme, octopine synthase, *Plant Physiol.*, 65, 949, 1980.
3. Nester, E.W., Gordon, M.P., Amasino, R.M., and Yanofsky, M.F., Crown Gall: a molecular and physiological analysis, *Ann. Rev. Plant Physiol.*, 35, 387, 1984.

1.5.1.17 — Alanopine Dehydrogenase; ALPD

REACTION   2,2′-Iminodipropanoate + NAD + H$_2$O = L-alanine + pyruvate + NADH

ENZYME SOURCE   Invertebrates (coelenterates, nemertines, polychaetes, sipunculids, mollusks, brachiopods)

SUBUNIT STRUCTURE   Monomer (invertebrates)

Method 1

Visualization Scheme

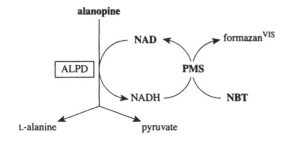

Staining Solution[1]

50 m*M* Tris–HCl buffer, pH 8.4	24 ml
50 m*M* Alanopine	6 ml
10 mg/ml NAD	6 ml
10 mg/ml NBT	2 ml
5 mg/ml PMS	2 ml

Procedure

Incubate the gel in staining solution in the dark at 37°C until dark blue bands appear. Wash the stained gel in water and fix in 25% ethanol.

Notes: The postcoupling of PMS may prove beneficial.

Method 2

Visualization Scheme

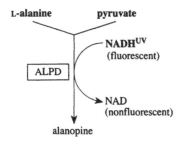

Staining Solution[2]

100 m*M* Triethanolamine–HCl buffer, pH 7.0
0.1 m*M* NADH
2 m*M* Pyruvate
20 m*M* L-Alanine

Procedure

Lay a piece of filter paper saturated with the staining solution on top of the gel and incubate at 37°C for 10 to 30 min. Remove the filter paper and view the gel under long-wave UV light. Dark (nonfluorescent) bands of ALPD are visible on a light (fluorescent) background of the gel. Record the zymogram or photograph using a yellow filter.

Notes: When a zymogram that is visible in daylight is required, counterstain the processed gel with MTT–PMS solution. This results in the appearance of white bands on a blue background.[3]

    In some invertebrate species LDH bands can also develop on ALPD zymograms obtained by this method. Therefore, an additional gel should be stained for LDH as a control.

    ALPD from some sources also catalyzes (but more slowly) the reductive imination of glycine and pyruvate to strombine (see, however, 1.5.1.22 — STRD).

General Notes

Method 2 is more sensitive than Method 1. The ratio of the maximal ALPD activities in the reverse vs. forward reactions is about 5:1.[1]

References

1. Plaxton, W.C. and Storey, K.B., Tissue specific isozymes of alanopine dehydrogenase in the channeled whelk *Busycotypus canaliculatum, Can. J. Zool.*, 60, 1568, 1982.
2. Dando, P.R., Storey, K.B., Hochachka, P.W., and Storey, J.J.M., Multiple dehydrogenases in marine molluscs: electrophoretic analysis of alanopine dehydrogenase, strombine dehydrogenase, octopine dehydrogenase and lactate dehydrogenase, *Mar. Biol. Lett.*, 2, 249, 1981.
3. Manchenko, G.P., Nonfluorescent negative stain for alanopine dehydrogenase activity on starch gels, *Anal. Biochem.*, 145, 308, 1985.

1.5.1.19 — D-Nopaline Dehydrogenase; NOPDH

OTHER NAMES	D-Nopaline synthase
REACTION	N^2-(D-1,3-Dicarboxypropyl)-L-arginine + NADP + H_2O = L-arginine + 2-oxoglutarate + NADPH
ENZYME SOURCE	Crown Gall tumor tissues of dicotyledonous plants induced by *Agrobacterium tumefaciens* (see General Notes)
SUBUNIT STRUCTURE	Tetramer# (Crown Gall tumor tissues of dicotyledonous plants)

METHOD

Visualization Scheme

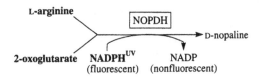

Staining Solution[1] (adapted)

100 mM Phosphate buffer, pH 6.5
14 mM 2-Mercaptoethanol
0.17 mM NADPH
12 mM Sodium α-ketoglutarate
6.7 mM L-Arginine

Procedure

Overlay the gel with filter paper soaked in staining solution or apply the staining solution as a 1% agarose overlay. Observe dark nonfluorescent bands of NOPDH on the fluorescent background of the gel. Record the zymogram or photograph using a yellow filter.

Notes: When a zymogram that is visible in daylight is required, counterstain the processed gel with MTT–PMS solution to obtain achromatic bands of NOPDH on a blue background of the gel. Fix the counterstained gel in 25% ethanol.

L-Ornithine can also be used as a substrate instead of L-arginine.

GENERAL NOTES

Crown Gall is a neoplastic disease of dicotyledonous plants induced by many different strains of *Agrobacterium tumefaciens*. This bacterium induces hypertrophies following invasion of a wound site. Neoplastic hypertrophies result following the covalent integration of a part of DNA of a large plasmid (the tumor-inducing or Ti-plasmid) into a plant nuclear genome. Transformed plant tissues synthesize one or more of a group of compounds termed *opines*, which are absent from normal plant tissues. Opines produced by various tumor types are subdivided into six classes:

1. The octopine class includes four opines (octopine, octopinic acid, lysopine, and histopine), presumably produced by the same enzyme.
2. The nopaline class includes three opines (nopaline, pyronopaline, and nopalinic acid), presumably produced by the same enzyme.
3. The agropine class includes four opines (agropine, agropinic acid, mannopine, and mannopinic acid), presumably produced by two different enzymes.
4. The leucinopine class is represented by a single opine, leucinopine.
5. The agrocinopine A and B class is represented by two corresponding opines.
6. The agrocinopine C and D class is represented by two corresponding opines.[2]

REFERENCES

1. Kemp, D.J., Sutton, D.W., and Hack, E., Purification and characterization of the Crown Gall specific enzyme nopaline dehydrogenase, *Biochemistry*, 18, 3755, 1979.
2. Nester, E.W., Gordon, M.P., Amasino, R.M., and Yanofsky, M.F., Crown Gall: a molecular and physiological analysis, *Ann. Rev. Plant Physiol.*, 35, 387, 1984.

1.5.1.22 — Strombine Dehydrogenase; STRD

REACTION N-(Carboxymethyl)-L-alanine + NAD + H_2O = glycine + pyruvate + NADH

ENZYME SOURCE   Invertebrates (sponges, coelenterates, polychaetes, mollusks, brachiopods)

SUBUNIT STRUCTURE   Monomer (invertebrates)

METHOD

Visualization Scheme

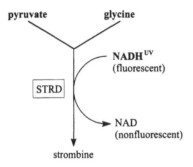

Staining Solution[1]

100 mM Triethanolamine–HCl buffer, pH 7.0
0.1 mM NADH
2 mM Pyruvate
20 mM Glycine

Procedure

Apply the staining solution to the gel surface with a filter paper overlay. Incubate the gel at 37°C in a moist chamber. After 30 min, monitor the gel under long-wave UV light. Dark (nonfluorescent) bands of STRD are visible on a light (fluorescent) background. Record the zymogram or photograph using a yellow filter.

Notes: When a zymogram that is visible in daylight is required, the processed gel should be counterstained with MTT–PMS solution. This results in the appearance of white bands on a blue background.

In some invertebrate species LDH bands can also develop on STRD zymograms obtained by this method. Therefore, an additional gel should be stained for LDH as a control.

Alanopine dehydrogenase (see 1.5.1.17 — ALPD) also catalyzes the reductive imination of glycine and pyruvate to strombine. However, it displays high specificity for L-alanine as its substrate and shows very low activity with glycine, whereas STRD shows higher glycine activity than L-alanine activity.[2] To differentiate STRD and ALPD bands, an additional control gel should be stained for ALPD.

Two polymorphic gene loci coding for enzyme proteins responsible for opine dehydrogenase activities are expressed in the adductor muscle of the oyster.[3] One locus (*Alpdh-2*) codes for true alanopine dehydrogenase specific toward α-L-alanine, while the second locus (*Alpdh-1*) codes for opine dehydrogenase that demonstrates highest catalytic activity with glycine, but is also capable of using α-L-alanine, L-serine, and β-alanine as substrates of the backward reaction.

REFERENCES

1. Dando, P.R., Storey, K.B., Hochachka, P.W., and Storey, J.J.M., Multiple dehydrogenases in marine molluscs: electrophoretic analysis of alanopine dehydrogenase, strombine dehydrogenase, octopine dehydrogenase and lactate dehydrogenase, *Mar. Biol. Lett.*, 2, 249, 1981.
2. Plaxton, W.C. and Storey, K.B., Tissue specific isozymes of alanopine dehydrogenase in the channeled whelk *Busycotypus canaliculatum*, *Can. J. Zool.*, 60, 1568, 1982.
3. Manchenko, G.P., McGoldrick, D.J., and Hedgecock, D., Genetic basis of opine dehydrogenase activities in the Pacific oyster, *Crassostrea gigas*, *Comp. Biochem. Physiol.*, 121B, 251, 1998.

1.5.1.23 — Tauropine Dehydrogenase; TAUDH

REACTION   Tauropine + NAD + H_2O = taurine + pyruvate + NADH

ENZYME SOURCE   Red algae, invertebrates (sponges, sea anemones, polychaetes, mollusks, brachiopods, echinoderms)

SUBUNIT STRUCTURE Monomer[#] (red algae), monomer (echinoderms)

METHOD

Visualization Scheme

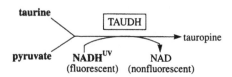

Staining Solution[1]

100 mM Phosphate buffer, pH 7.0
0.1 mM NADH
2 mM Pyruvate
20 mM Taurine

Procedure

Overlay the gel with filter paper soaked in staining solution. Observe dark nonfluorescent bands of TAUDH on the fluorescent background of the gel. Record the zymogram or photograph using a yellow filter.

Notes: When a zymogram that is visible in daylight is required, counterstain the processed gel with MTT–PMS solution to obtain achromatic bands of TAUDH on a blue background of the gel. Fix the counterstained gel in 25% ethanol.

Activity bands of hypotaurine dehydrogenase (EC 1.8.1.3) can also be developed by this method. Stain the control gel in the staining solution lacking pyruvate to detect hypotaurine dehydrogenase.

Low concentration of pyruvate (0.25 to 0.35 mM) in the staining solution may prove optimal for the enzyme from sponges.[2]

REFERENCES

1. Manchenko, G.P., Allozymic variation of tauropine dehydrogenase in the sea star, *Asterina pectinifera*, *GFI Bull.*, 31, 37, 1998.
2. Kan-No, N., Sato, M., Nagahisha, E., and Sato, Y., Purification and characterization of tauropine dehydrogenase from the marine sponge *Halichondria japonica* Kadota (Demospongia), *Fish. Sci.*, 63, 414, 1997.

1.5.1.24 — N^5-(Carboxyethyl)ornithine Synthase; CEOS

REACTION N^5-(L-1-Carboxyethyl)-L-ornithine + NADP + H_2O = L-ornithine + pyruvate + NADPH

ENZYME SOURCE   Bacteria

SUBUNIT STRUCTURE Dimer[#] (bacteria), tetramer[#] (bacteria)

METHOD

Visualization Scheme

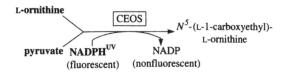

Staining Solution[1]

100 mM Tris–HCl buffer, pH 8.0
1 mM NADPH
2.5 mM Pyruvate
10 mM L-Ornithine

Procedure

Overlay the gel with filter paper soaked in staining solution. Observe dark nonfluorescent bands of CEOS on the fluorescent background of the gel. Record the zymogram or photograph using a yellow filter.

Notes: When a zymogram that is visible in daylight is required, counterstain the processed gel with NBT–PMS solution to obtain achromatic bands of CEOS on a purple background of the gel. Fix the developed gel in 25% ethanol.

The enzyme also mediates the reductive condensation between pyruvate and L-lysine to form N^6-(L-1-carboxyethyl)-L-lysine.

REFERENCES

1. Thompson, J., N^5-(L-1-Carboxyethyl)-L-ornithine:NADP+ oxidoreductase from *Streptococcus lactis*, *J. Biol. Chem.*, 264, 9592, 1989.

REACTION β-Alanopine NAD + H₂O = β-alanine + pyruvate + NADH

ENZYME SOURCE Invertebrates (coelenterates, polychaetes, mollusks, brachiopods,
sipunculids)

SUBUNIT STRUCTURE Monomer (invertebrates)

METHOD

Visualization Scheme

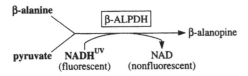

Staining Solution[1] (modified)

100 mM Phosphate buffer, pH 7.0
0.1 mM NADH
2 mM Pyruvate
20 mM β-alanine

Procedure

Overlay the gel with filter paper soaked in staining solution or apply the staining solution as a 1% agarose overlay. Monitor the gel under UV light for dark (nonfluorescent) bands of β-ALPDH visible on a light (fluorescent) background. Record the zymogram or photograph using a yellow filter.

Notes: When a zymogram that is visible in daylight is required, counterstain the processed gel with MTT–PMS solution to obtain achromatic bands of β-ALPDH on a blue background. Fix the counterstained gel in 25% ethanol.

Using allozyme patterns as gene-specific characters, it was shown that one and the same enzyme protein from the adductor muscle is responsible for multiple opine dehydrogenase activities in the Pacific oyster.[1] The same allozyme patterns (in order of decreasing intensity of staining) were detected in this species using glycine, α-L-alanine, L-serine, and β-alanine as substrates.

Similar opine dehydrogenase from foot and adductor muscles of the bivalve, *Scapharca broughtonii*, is also able to use different amino acids as substrates in its backward reaction, but demonstrates the highest activity with β-alanine (unpublished data of the author).

REFERENCES

1. Manchenko, G.P., McGoldrick, D.J., Hedgecock D., Genetic basis of opine dehydrogenase activity in the Pacific oyster, *Crassostrea gigas*, *Comp. Biochem. Physiol.*, 121B, 251, 1998.

OTHER NAMES	Phenylalanine–pyruvate reductase, methiopine dehydrogenase
REACTION	N-[1-D-(Carboxyl)ethyl]-L-phenylalanine + NAD + H_2O = L-phenylalanine + pyruvate + NADH
ENZYME SOURCE	Bacteria
SUBUNIT STRUCTURE	Dimer[#] (bacteria)

METHOD 1

Visualization Scheme

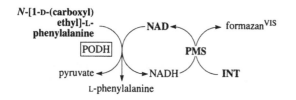

Staining Solution[1] (modified)

0.1 M Tris–HCl buffer, pH 9.5
0.42 mM NAD
10 mM N-[1-D-(Carboxyl)ethyl]-L-phenylalanine
0.60 mM 2-(p-Iodophenyl)-3-(p-nitrophenyl)-5-phenyltetrazolium chloride (INT)
0.33 mM PMS

Procedure

Incubate the gel in staining solution in the dark at room temperature until stained bands appear. Wash the stained gel in water and fix in 25% ethanol.

Notes: N-[1-(Carboxyl)ethyl]methionine (methiopine) can also be used as a substrate instead of N-[1-D-(carboxyl)ethyl]-L-phenylalanine.
MTT or NBT can be used instead of INT.

METHOD 2

Visualization Scheme

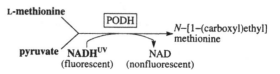

Staining Solution[1] (adapted)

100 mM 0.1 M Tris–HCl buffer, pH 8.0
0.1 mM NADH
10 mM Pyruvate
10 mM L-Methionine

Procedure

Overlay the gel with filter paper soaked in staining solution or apply the staining solution as a 1% agarose overlay. Monitor the gel under UV light for dark (nonfluorescent) bands of PODH visible on a light (fluorescent) background. Record the zymogram or photograph using a yellow filter.

Notes: When a zymogram that is visible in daylight is required, counterstain the processed gel with MTT–PMS solution to obtain achromatic bands of PODH on a blue background. Fix the counterstained gel in 25% ethanol.
The enzyme can also use L-isoleucine, L-valine, L-phenylalanine, L-leucine, and some other L-amino acids as substrates in the reductive reaction.

REFERENCES

1. Asano, Y., Yamaguchi, K., and Kondo, K., A new NAD+-dependent opine dehydrogenase from *Arthrobacter* sp. strain 1C, *J. Bacteriol.*, 171, 4466, 1989.

1.5.3.1 — Sarcosine Oxidase; SOX

REACTION	Sarcosine + H_2O + O_2 = glycine + formaldehyde + H_2O_2
ENZYME SOURCE	Bacteria, vertebrates
SUBUNIT STRUCTURE	Heterotetramer(α, β, γ, δ)[#] (bacteria – *Corynebacterium* sp.)

METHOD

Visualization Scheme

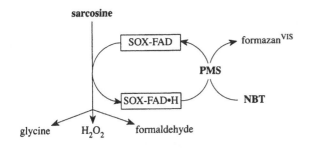

Staining Solution[1]

65 m*M* Phosphate buffer, pH 6.8	70 ml
10 m*M* Sarcosine	10 ml
NBT	5 mg
PMS	2.5 mg

Procedure

Incubate the gel in staining solution in the dark at 37°C until dark blue bands appear. Wash the stained gel in water and fix in 25% ethanol.

GENERAL NOTES

The enzyme is a flavoprotein (FAD). The flavin is either covalently or noncovalently bound in a molar ratio of 1:1.[2]

The enzyme from *Corynebacterium* sp. P-l contains noncovalently bound FAD and NAD and covalently bound FMN, attached to (His173) of the β-subunit.[3]

REFERENCES

1. Feinstein, R.N. and Lindahl, R., Detection of oxidases on polyacrylamide gels, *Anal. Biochem.*, 56, 353, 1973.
2. NC-IUBMB, *Enzyme Nomenclature*, Academic Press, San Diego, 1992, p. 96 (EC 1.5.3.1, Comments).
3. Eschenbrenner, M., Chlumsky, L.J., Khanna, P., Strasser, F., and Jorns, M.S., Organization of the multiple coenzymes and subunits and role of the covalent flavin link in the complex heterotetrameric sarcosine oxidase, *Biochemistry*, 40, 5352, 2001.

1.6.4.2 — Glutathione Reductase; GSR

OTHER NAMES	Glutathione reductase (NADPH) (recommended name)
REACTION	NADPH + GSSG (oxidized glutathione) = NADP + 2 GSH (reduced glutathione)
ENZYME SOURCE	Bacteria, fungi, plants, invertebrates, vertebrates
SUBUNIT STRUCTURE	Dimer (fungi, invertebrates, vertebrates)

METHOD 1

Visualization Scheme

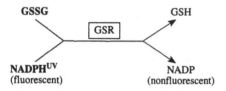

Staining Solution[1]

A. 0.3 M Tris–HCl buffer, pH 8.0   10 ml
    NADPH   7 mg
    Oxidized glutathione (GSSG)   30 mg
B. 2% Agar solution (60°C)   10 ml

Procedure

Mix A and B components of the staining solution and pour the mixture over the surface of the gel. Incubate the gel at 37°C and view under long-wave UV light. Dark (nonfluorescent) bands of GSR are visible on the light (fluorescent) background of the gel. Record the zymogram or photograph using a yellow filter.

METHOD 2

Visualization Scheme

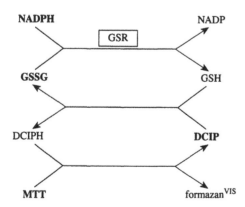

Staining Solution[2]

0.25 M Tris–HCl buffer, pH 8.4	20 ml
Oxidized glutathione (GSSG)	40 mg
NADPH	10 mg
2,6-Dichlorophenol indophenol (DCIP)	0.2 mg
MTT	10 mg

Procedure

Incubate the gel in staining solution in the dark at 37°C until purple bands appear on a blue background. When the bands have reached a satisfactory intensity, counterstain the gel with 1 M HCl solution to remove a blue background and to allow GSR bands to become more obvious.[3] Wash the stained gel in water and fix in 25% ethanol.

Notes: The stain also detects NADPH diaphorase (see 1.6.99.1 — DIA(NADPH)). Its bands can be identified by removing GSSG from the staining solution.

This method was modified by using alkaline PAG (pH 9.4 instead of pH 8.8) to resolve and detect plant GSR.[4] The modified method detects only GSR activity bands. Activity bands of diaphorase are not detected because diaphorases from plant sources displays maximum activity between pH values 8.5 and 8.8 and possibly is inactivated at pH 9.4. Importantly, the plant GSR activity is the same at either pH 8.8 or 9.4.

METHOD 3

Visualization Scheme

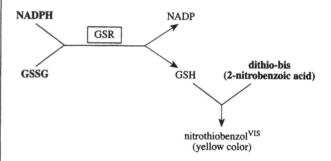

Staining Solution[5]

133 mM Tris–HCl buffer, pH 8.0
5.5 mM Oxidized glutathione (GSSG)
0.32 mM NADPH
0.88 mM 5,5′-Dithio-bis(2-nitrobenzoic acid)
33 mM EDTA
1% Agar

Procedure

Add EDTA and 2-nitrobenzoic acid to half of the Tris–HCl buffer and heat to bring the nitrobenzoic acid into solution. When the nitrobenzoic acid is completely dissolved, allow the solution to cool. After the mixture reaches a temperature of 45°C, add the NADPH and GSSG. Then add the agar to the remaining half of the Tris–HCl buffer and heat to boiling. After the agar solution is cooled to 45°C, mix the two solutions and pour the mixture over the surface of the gel. Incubate the gel at 37°C until yellow bands of GSR appear.

Notes: Additional nonspecifically stained bands can also develop on GSR zymograms obtained by this method. Thus, it is important to do control staining of an additional gel with a staining solution from which GSSG is absent.

METHOD 4

Visualization Scheme

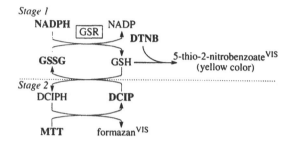

Staining Solution[6] (modified)

A. 0.25 *M* Tris–HCl buffer, pH 8.4
    1.5 m*M* NADPH
    4.0 m*M* Oxidized glutathione (GSSG)
    2 m*M* 5,5′-Dithio-bis(2-nitrobenzoic acid) (DTNB)
B. 2 m*M* NBT (filtered through Whatman No. 1 filter paper)
    0.08 m*M* Dichlorophenol indophenol (DCIP)

Procedure

Soak the electrophorized gel in solution A until yellow bands of GSR appear (usually after 5 min). Add an equal volume of solution B and incubate the gel in the dark at room temperature. Sharp dark blue bands of GSR first appear after 2 h. Maximum intensity is usually obtained after an overnight incubation of the gel. Fix the zymogram in 25% ethanol.

Notes: The blue staining of the gel background is absent when this method is used (see also General Notes).

Thioredoxin reductase (TRR; EC 1.6.4.5) can also be stained by this method. Activity bands of GSR and TRR can be distinguished by following the kinetics of color development during the first stage of the method. GSR reacts 25 times faster with GSSG than with DTNB in the presence of NADPH, while TRR mostly reacts with DTNB. Thus, when NADPH–DTNB solution lacking GSSG is used in the first stage, the yellow bands are mostly due to the activity of TRR.

GENERAL NOTES

Chromogenic Methods 2 and 3 have specific disadvantages. Method 2 generates permanently stained activity bands due to formation of insoluble formazan, but it generates a number of extraneous, nonglutathione reductase activity bands. Mammalian thioredoxin reductase and NADPH-dependent diaphorase activity bands are also developed by this method. Method 3, although much more specific for GSR, results in ephemeral and quickly diffusing faint yellow bands. Method 4 uses advantages of Methods 2 and 3 and overcomes main disadvantages of these methods. The use of DTNB in the initial staining step inhibits enzymes other than GSR that can be stained with the following DCIP–MTT counterstain. DTNB directly reacts with enzyme protein thiols and converts enzyme thiols to the disulfides. This is thought to be the reason for the lack of staining of the gel background with Method 4. DTNB is also a substrate for GSR, although considerably less efficient than GSSG. This explains why GSR activity bands can be developed by the sequential staining of the gel with DTNB and DCIP–MTT even in the absence of GSSG. Method 4 was shown to be the most sensitive chromogenic method for the detection of GSR activity bands on electrophoretic gels.[6]

REFERENCES

1. Nichols, E.A. and Ruddle, F.H., Polymorphism and linkage of glutathione reductase in *Mus musculus*, *Biochem. Genet.*, 13, 323, 1975.
2. Kaplan, J.C. and Beutler, E., Electrophoretic study of glutathione reductase in human erythrocytes and leukocytes, *Nature*, 217, 256, 1968.
3. Richardson, B.J., Baverstock, P.R., and Adams, M., *Allozyme Electrophoresis: A Handbook for Animal Systematics and Population Studies*, Academic Press, Sydney, 1986, p. 192.
4. Okpodu, C.M. and Waite, K.L., Method for detecting glutathione reductase activity on native activity gels which eliminates the background diaphorase activity, *Anal. Biochem.*, 244, 410, 1997.
5. Brewer, G.J., *An Introduction to Isozyme Techniques*, Academic Press, New York, 1970, p. 113.
6. Ye, B., Gitler, C., and Gressel, J., A high-sensitivity, single-gel, polyacrylamide gel electrophoresis method for the quantitative determination of glutathione reductase, *Electrophoresis*, 246, 159, 1997.

1.6.6.2 — Nitrate Reductase (NAD(P)H); NAR(NAD(P)H)

OTHER NAMES	Assimilatory nitrate reductase
REACTION	NAD(P)H + nitrate = NAD(P) + nitrite + H_2O
ENZYME SOURCE	Bacteria, fungi, plants
SUBUNIT STRUCTURE	Heterodimer[#] (bacteria)

METHOD

Visualization Scheme

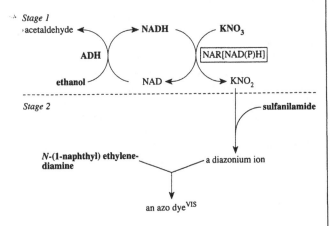

Staining Solution[1]

A. 0.1 M Potassium phosphate buffer, pH 7.5   100 ml
    KNO_3 150 mg
    95% Ethanol   2.5 ml
    NADH   30 mg
    Alcohol dehydrogenase (ADH)   100 U
B. 1% Sulfanilamide in 1 N HCl   50 ml
    0.01% N-(1-Naphthyl)ethylenediamine
      dihydrochloride in 0.1 M potassium phosphate
      buffer, pH 7.5   50 ml

Procedure

Incubate the gel in solution A at 30°C for 30 min. Discard solution A and rinse the gel with water. Place the gel in solution B. In a few minutes pink bands develop, indicating zones of NAR(NAD(P)H) activity.

Notes: Alcohol dehydrogenase and ethanol are used to regenerate NADH. When NADPH is used instead of NADH, NADP-dependent glucose-6-phosphate dehydrogenase (EC 1.1.1.49) and glucose-6-phosphate are recommended for use instead of ADH and ethanol to regenerate NADPH.

GENERAL NOTES

The activity of NAR(NAD(P)H) from spinach is due to two different proteins: an NAD(P) reductase and a nitrate reductase. The nitrate reductase is inactive with NAD(P)H but serves methyl or benzyl viologen (but not PMS) as electron carriers. The last enzyme may be identical to nitrate reductase (see 1.7.99.4 — NAR).

    The enzyme is a flavoprotein (FAD or FMN).[2]

REFERENCES

1. Upcroft, J.A. and Done, J., Starch gel electrophoresis of plant NADH–nitrate reductase and nitrite reductase, *J. Exp. Bot.*, 25, 503, 1974.
2. NC-IUBMB, *Enzyme Nomenclature*, Academic Press, San Diego, 1992, p. 103 (EC 1.6.6.2, Comments).

1.6.6.9 — Trimethylamine-*N*-Oxide Reductase; TMAOR

REACTION   NADH + trimethylamine-*N*-oxide =
NAD + trimethylamine + H_2O

ENZYME SOURCE   Bacteria

SUBUNIT STRUCTURE   Dimer# (bacteria – *Roseobacter denitrificans*)

Method

Visualization Scheme

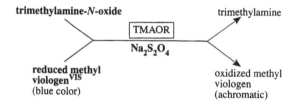

Staining Solution[1]

A. 100 m*M* Potassium phosphate buffer, pH 6.5   100 ml
    200 m*M* Methyl viologen   1 ml
    50 m*M* Sodium dithionite ($Na_2S_2O_4$)   50 μl

B. 100 m*M* Potassium phosphate buffer, pH 6.5
    2 *M* Trimethylamine-*N*-oxide

Procedure

Place the electrophorized gel in a nitrogen atmosphere in solution A. Then incubate the gel, uniformly colored blue by reduced methyl viologen, in a nitrogen atmosphere in solution B. The reduction of the substrate is coupled with the oxidation of reduced methyl viologen, and the resulting achromatic bands on a blue background indicate the areas of TMAOR activity on the gel. Record the zymogram or photograph using a yellow filter.

Notes: Methyl viologen is used instead of NADH. Sodium dithionite is used as a powerful reducing agent to reduce methyl viologen, which is of blue color.

Dimethylsulfoxide, tetrahydrothiophene-1-oxide, and pyridine-*N*-oxide can also be used as substrates.

When a permanent negative zymogram is required, the gel should be additionally stained in 2.5% solution of triphenyl tetrazolium chloride (TTC). This counterstaining results in the appearance of white TMAOR bands on a red background. The gel counterstained with TTC can be stored in 25% ethanol for a long time. The procedure of staining the gel in a nitrogen atmosphere and counterstaining it with TTC is described in detail in the procedure of staining gels for nitrate reductase (see 1.7.99.4 — NAR).

References

1. Silvestro, A., Pommier, J., Pascal, M.-C., and Giordano, G., The inducible trimethylamine *N*-oxide reductase of *Escherichia coli* K12: its localization and inducers, *Biochim. Biophys. Acta*, 999, 208, 1989.

OTHER NAMES	NADPH dehydrogenase (recommended name)
REACTION	NADPH + acceptor = NADP + reduced acceptor
ENZYME SOURCE	Bacteria, plants, invertebrates, vertebrates
SUBUNIT STRUCTURE	Monomer (invertebrates)

METHOD

Visualization Scheme

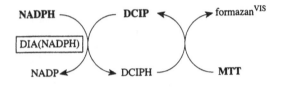

Staining Solution[1]

0.15 M Tris–HCl buffer, pH 8.2	50 ml
NADPH	10 mg
2,6-Dichlorophenol indophenol (DCIP)	2 mg
MTT	7 mg

Procedure

Incubate the gel in staining solution in the dark at 37°C until purple bands appear on a blue background. When the bands have reached a satisfactory intensity, counterstain the gel with 1 M HCl solution to remove a blue background and to allow DIA(NADPH) bands to become more obvious.[2] Wash the stained gel in water and fix in 25% ethanol.

OTHER METHODS

Alternative methods of DIA(NADPH) visualization based on defluorescence of NADPH[3] and decoloration of DCIP[4] are also described, but the positive tetrazolium method described above is usually preferred.

GENERAL NOTES

The term *diaphorase* refers to any enzyme that can catalyze the oxidation of NAD(P)H in the presence of an electron acceptor such as DCIP. A number of enzymes that have specific functions *in vivo* also have general diaphorase activity *in vitro*. This causes well-known difficulties in adequate enzyme classification of diaphorase activity bands on histochemically stained gels. One should beware of homology problems when diaphorase isozymes are used in phylogenetic comparisons (see also 1.6.99.2 — MR; 1.8.1.4 — DIA(NADH)).

The NADPH-dependent enzyme from *Clostridium thermoaceticum* displays activity toward many acceptors such as viologens, quinones (e.g., 1,4-benzoquinone or anthraquinone-2,6-disulfonate), DCIP, and clostridial rubredoxin. However, ferredoxin or lipoamide can not be used as substrates, indicating that the clostridial enzyme is not a diaphorase in the usual sense. This artificial mediator accepting pyridine nucleotide oxidoreductase (AMAROR) can be detected after PAG electrophoresis of crude extracts of *C. thermoaceticum* under strict anaerobic conditions using staining solution containing 2 mM NADPH, 2 mM 1,1′-carbamoylmethyl-viologen, and 5 mM neotetrazolium bromide in 0.1 M potassium phosphate buffer (pH 7.6). As many as ten electrophoretically distinct molecular forms of AMAPOR were detected using this staining solution. Two AMAPOR isozymes were isolated and found to be oligomers (200 and 400 kDa) consisting of α and β subunits ($\alpha_2\beta_2$ and $\alpha_4\beta_4$) and containing an Fe–S cluster and FAD. The partial amino acid sequence from the N-terminal end of the α subunit provided evidence that this region (presumably NADP binding) has 57% identity with pyrroline-5-carboxilate reductase (1.5.1.2) from *Bacillus subtilus*.[5]

REFERENCES

1. Frischer, H., Nelson, R., Noyes, C., Carson, P.E., Bowman, J.J.E., Rieckmann, K.H., and Ajmar, F., NAD(P) glycohydrolase deficiency in human erythrocytes and alteration of cytosol NADH–methemoglobin diaphorase by membrane NAD–glycohydrolase activity, *Proc. Natl. Acad. Sci. U.S.A.*, 70, 2406, 1973.
2. Richardson, B.J., Baverstock, P.R., and Adams, M., *Allozyme Electrophoresis: A Handbook for Animal Systematics and Population Studies*, Academic Press, Sydney, 1986, p. 173.
3. Čepica, S. and Stratil, A., Further studies on sheep polymorphic erythrocyte diaphorase, *Anim. Blood Groups Biochem. Genet.*, 9, 239, 1978.
4. Brewer, G.J., Eaton, J.J.W., Knutsen, C.S., and Beck, C.C., A starch gel electrophoretic method for the study of diaphorase isozymes and preliminary results with sheep and human erythrocytes, *Biochem. Biophys. Res. Commun.*, 29, 198, 1967.
5. Gunther, H., Walter, K., Kohler, P., and Simon, H., On a new artificial mediator accepting NADP(H) oxidoreductase from *Clostridium thermoaceticum*, *J. Biotechnol.*, 83, 253, 2000.

OTHER NAMES	Menadione reductase, phylloquinone reductase, quinone reductase
REACTION	NAD(P)H + acceptor = NAD(P) + reduced acceptor
ENZYME SOURCE	Bacteria, fungi, plants, invertebrates, vertebrates
SUBUNIT STRUCTURE	Dimer (plants), tetramer (plants)

METHOD

Visualization Scheme

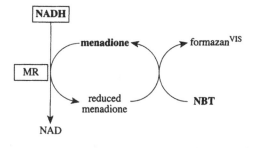

Staining Solution[1]

0.2 M Tris–HCl buffer, pH 7.0	75 ml
Menadione	25 mg
NADH	25 mg
NBT	10 mg

Procedure

Incubate the gel in staining solution in the dark at 37°C until dark blue bands appear. Wash the stained gel in water and fix in 25% ethanol.

Notes: The enzyme from some sources (e.g., bovine liver) is also active with dichlorophenol indophenol as an acceptor. Thus, its bands of activity can also develop on NAD(P)H diaphorase zymograms (see 1.6.99.1 — DIA(NADPH); 1.8.1.4 — DIA(NADH)).

The enzyme is a flavoprotein.[2]

REFERENCES

1. Cheliak, W.M. and Pitel, J.A., Techniques for Starch Gel Electrophoresis of Enzymes from Forest Tree Species, Information Report PI-X-42, Petawawa National Forestry Institute, Canadian Forestry Service, Agriculture Canada, Chalk River, Ontario, 1984.
2. NC-IUBMB, *Enzyme Nomenclature*, Academic Press, San Diego, 1992, p. 104 (EC 1.6.99.2, Comments).

REACTION	NAD(P)H + 6,7-dihydropteridine = NAD(P) + 5,6,7,8-tetrahydropteridine
ENZYME SOURCE	Bacteria, invertebrates, vertebrates
SUBUNIT STRUCTURE	Dimer (vertebrates)

METHOD

Visualization Scheme

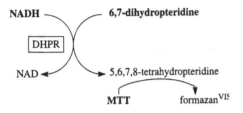

Staining Solution[1]

A. 20 mM 2-Amino-6,7,-dimethyl-4-hydroxy-5,6,7,8-tetrahydropteridine (2.3 mg/ml, make fresh daily just before use)

B. 0.1 M KOH

C. 27 mg/ml 2,6-Dichloroindophenol (make fresh weekly and store at 4°C)

D. 50 1 mM Tris–HCl buffer, pH 7.2
    1 mM NADH
    1 mM MTT

Procedure

Adjust 1 ml of solution A to pH 6.0 to 6.5 with solution B. Add solution C until just before it is in excess (i.e., before the blue color persists). 2,6-Dichloroindophenol is used to convert the tetrahydropteridine to the quinonoid dihydro form, which is a substrate for DHPR. Extract the formed reduced 2,6-dichloroindophenol with three 5-ml portions of diethyl ether. Add 1 ml of the aqueous layer of the substrate solution to 10 ml of solution D preincubated at 37°C for 10 min and apply to the gel surface. Incubate the gel at room temperature with continuous shaking. Purple bands on a blue background indicate localization of DHPR activity. Full color is developed after 30 to 60 min.

Notes: The staining solution should be prepared as rapidly as possible and at room temperature and used as soon as possible. The diethylether is extremely volatile and flammable and therefore should be used with extreme caution.

The reduction of MTT and the formation of purple formazan is influenced by the tetrahydropteridine produced by DHPR. The mechanism of this reaction is uncertain.

GENERAL NOTES

The substrate is the quinoid form of dihydropteridine.[2]

REFERENCES

1. Cotton, R.G.H. and Jennings, I., A naphthoquinone adsorbent for affinity chromatography of human dihydropteridine reductase, *Eur. J. Biochem.*, 83, 319, 1978.

2. NC-IUBMB, *Enzyme Nomenclature*, Academic Press, San Diego, 1992, p. 105 (EC 1.6.99.7, Comments).

1.7.3.3 — Urate Oxidase; UOX

OTHER NAMES	Uricase
REACTION	Uric acid + 2 H_2O + O_2 = allantoin + CO_2 + H_2O_2
ENZYME SOURCE	Bacteria, fungi, green algae, plants, invertebrates, vertebrates (present in majority of mammals, but absent in human and hominid primates)
SUBUNIT STRUCTURE	Tetramer[#] (green algae – *Chlamydomonas reinhardtii*)

METHOD

Visualization Scheme

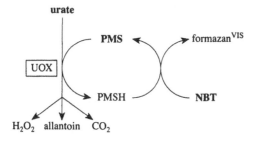

Staining Solution[1]

0.05 M Phosphate buffer, pH 6.8	70 ml
10 mM Urate	10 ml
NBT	5 mg
PMS	2.5 mg

Procedure

Incubate the gel in staining solution in the dark at 37°C until dark blue bands appear. Wash the stained gel in water and fix in 25% ethanol.

Notes: The enzyme from many sources is more active at alkaline conditions (pH 8.0 to 8.5).

OTHER METHODS

A. A linked peroxidase reaction can be used to detect the product hydrogen peroxide (see 1.11.1.7 — PER).
B. Calcium ions may be used to detect the product CO_2 after electrophoresis of UOX in PAG.[2] Calcium carbonate precipitate forms under alkaline conditions. This results in the appearance of opaque bands visible on transparent PAG (for example, see 4.1.1.1 — PDC). The calcium carbonate method, however, is not as sensitive as the tetrazolium method described above.

GENERAL NOTES

The initial products of the UOX reaction are not yet finally identified. It is well established, however, that they decompose to form allantoin.[3]

REFERENCES

1. Feinstein, R.N. and Lindahl, R., Detection of oxidases on polyacrylamide gels, *Anal. Biochem.*, 56, 353, 1973.
2. Nimmo, H.G. and Nimmo, G.A., A general method for the localization of enzymes that produce phosphate, pyrophosphate, or CO_2 after polyacrylamide gel electrophoresis, *Anal. Biochem.*, 121, 17, 1982.
3. NC-IUBMB, *Enzyme Nomenclature*, Academic Press, San Diego, 1992, p. 107 (EC 1.7.3.3, Comments).

REACTION   2 Nitric oxide + 2 H_2O + acceptor =
2 nitrite + reduced acceptor

ENZYME SOURCE   Bacteria, fungi, plants

SUBUNIT STRUCTURE   Monomer (plants)

METHOD

Visualization Scheme

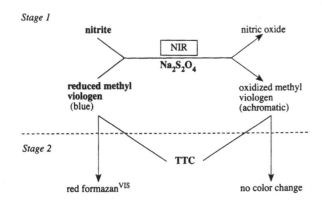

Stage 1

nitrite → NIR / $Na_2S_2O_4$ → nitric oxide

reduced methyl viologen (blue) → oxidized methyl viologen (achromatic)

Stage 2

TTC

red formazan[VIS]   no color change

Staining Solution[1]

A. 50 mM Sodium phosphate buffer, pH 8.0   100 ml
     $NaNO_2$ 100 mg

B. 0.1 M $NaHCO_3$ 3 ml
     Sodium hydrosulfite ($Na_2S_2O_4$)   120 mg

C. 50 mM Sodium phosphate buffer, pH 8.0   3 ml
     Methyl viologen   96 mg

D. H_2O 10 ml
     Triphenyl tetrazolium chloride (TTC)   250 mg

Procedure

Prepare solution A and bubble it with nitrogen (or argon) in the staining tray. Immerse the gel in solution A, cover the tray so it is airtight, and treat the gel with the gas for 5 min, using the side ports on the tray. Prepare solutions B and C. Pour C into B and mix thoroughly (the mixture should be blue). Using a syringe, inject the mixture into the tray without letting in any air. Incubate the gel until achromatic bands appear on a blue background. Then inject solution D and shake the tray gently to ensure good mixing. Incubate the gel in the dark at 5 to 10°C until the gel background becomes red. Wash the negatively stained gel in water and fix in 25% ethanol.

Notes: Sodium hydrosulfite is used to reduce methyl viologen, which turns blue in the presence of this reducing agent.

The use of NBT or MTT instead of TTC is not recommended because both give blue formazans upon reduction, thus making it difficult to detect the end point of the last reduction.

Benzyl viologen can be used instead of methyl viologen.

GENERAL NOTES

A variety of bacteria, fungi, and higher plants reduce nitrate to the level of ammonia and contain ferredoxin–nitrite reductase (EC 1.7.7.1), which produces ammonia directly from nitrite. This enzyme can also be detected by the method described above.[2]

Nitrite reductase from *Clostridium perfringens* is also detected by this method, although the product of nitrite reduction is hydroxylamine.[3]

REFERENCES

1. Vallejos, C.E., Enzyme activity staining, in *Isozymes in Plant Genetics and Breeding*, Part A, Tanskley, S.D. and Orton, T.J., Eds., Elsevier, Amsterdam, 1983, p. 469.
2. Vega, J.M. and Kamin, H., Spinach nitrite reductase: purification and properties of a siroheme-containing iron–sulfur enzyme, *J. Biol. Chem.*, 252, 896, 1977.
3. Sekiguchi, S., Seki, S., and Ishimoto, M., Purification and some properties of nitrite reductase from *Clostridium perfringens*, *J. Biochem.*, 94, 1053, 1983.

1.7.99.4 — Nitrate Reductase; NAR

OTHER NAMES	Respiratory nitrate reductase
REACTION	Nitrite + acceptor = nitrate + reduced acceptor
ENZYME SOURCE	Bacteria, fungi, plants
SUBUNIT STRUCTURE	Heterodimer# (bacteria – *Rhodobacter sphaeroides*), heterotrimer# (bacteria – *Pyrobaculum aerophilum*)

METHOD

Visualization Scheme

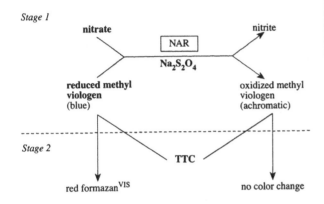

Stage 1

nitrate → NAR / $Na_2S_2O_4$ → nitrite

reduced methyl viologen (blue)

oxidized methyl viologen (achromatic)

Stage 2

TTC

red formazan^{VIS}

no color change

Staining Solution[1]

A.	50 m*M* Sodium phosphate buffer, pH 8.0	100 ml
	KNO$_3$	1 g
B.	0.1 *M* NaHCO$_3$	3 ml
	Sodium hydrosulfite (Na$_2$S$_2$O$_4$)	120 mg
C.	50 m*M* Sodium phosphate buffer, pH 8.0	3 ml
	Methyl viologen	96 mg
D.	H$_2$O	10 ml
	Triphenyl tetrazolium chloride (TTC)	250 mg

Procedure

Prepare solution A and bubble it with nitrogen (or argon) in the staining tray. Immerse the gel in solution A, cover the tray so it is airtight, and treat the gel with the gas for 5 min, using side ports on the tray. Prepare solutions B and C. Pour C into B and mix thoroughly (the mixture should be blue). Using a syringe, inject the mixture into the tray without letting in any air. Incubate the gel until achromatic bands appear on a blue background. Then inject solution D and shake the tray gently to ensure good mixing. Incubate the gel in the dark at 5 to 10°C until the gel background becomes red. Wash the negatively stained gel in water and fix in 25% ethanol.

Notes: Sodium hydrosulfite is used to reduce methyl viologen, which turns blue in the presence of this reducing agent.

The use of NBT or MTT instead of TTC is not recommended because both give blue formazans upon reduction, thus making it difficult to detect the end point of the last reduction.

Benzyl viologen can be used instead of methyl viologen.

Nitrate reductase (cytochrome) (EC 1.9.6.1) can also utilize methyl (or benzyl) viologen as an electron acceptor and thus be visualized on NAR zymograms.

The activity of nitrate reductase (NAD(P)H) (EC 1.6.6.2) from spinach (and perhaps other plants) is due to two different proteins: an NAD(P) reductase and a nitrate reductase. The nitrate reductase component is inactive with NAD(P)H, but serves methyl or benzyl viologen as an electron acceptor and thus can also develop on NAR zymograms (see 1.6.6.2 — NAR(NAD(P)H)).

REFERENCES

1. Vallejos, C.E., Enzyme activity staining, in *Isozymes in Plant Genetics and Breeding*, Part A, Tanskley, S.D. and Orton, T.J., Eds., Elsevier, Amsterdam, 1983, p. 469.

1.8.1.4 — NADH Diaphorase; DIA(NADH)

OTHER NAMES	Dihydrolipoamide dehydrogenase (recommended name), lipoamide reductase (NADH), lipoyl dehydrogenase, lipoamide dehydrogenase
REACTION	Dihydrolipoamide + NAD = lipoamide + NADH
ENZYME SOURCE	Bacteria, fungi, plants, invertebrates, vertebrates
SUBUNIT STRUCTURE	Monomer (fungi, plants, invertebrates, vertebrates)

METHOD

Visualization Scheme

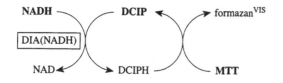

Staining Solution[1]

0.15 M Tris–HCl buffer, pH 8.2	50 ml
NADH	10 mg
2,6-Dichlorophenol indophenol (DCIP)	2 mg
MTT	7 mg

Procedure

Incubate the gel in staining solution in the dark at 37°C until purple bands appear on a blue background. When the bands have reached a satisfactory intensity, treat the gel with 1 M HCl to remove a blue background and to allow DIA(NADH) bands to become more obvious. Wash the stained gel in water and fix in 25% ethanol.

OTHER METHODS

Alternative methods of DIA(NADH) visualization based on defluorescence of NADH[2] and decoloration of DCIP[3] are also available, but the positive tetrazolium method described above is preferable.

GENERAL NOTES

This enzyme was first shown to catalyze the oxidation of NADH by Methylene Blue. This activity was called diaphorase.[4] The name diaphorase has been loosely applied to several enzymes that catalyze the oxidation of NAD(P)H in the presence of an electron acceptor such as DCIP. A number of enzymes that have specific functions *in vivo* also have general diaphorase activity *in vitro*. This causes some difficulties in adequate classification of enzymes that display diaphorase activity. For this reason, when DIA(NADH) or DIA(NADPH) isozymes are used as genic markers in phylogenetic studies, the problems of isoenzyme homology can arise (see also 1.6.99.1 — DIA(NADPH); 1.6.99.2 — MR).

The enzyme is a flavoprotein (FAD). It is a component of the multienzyme pyruvate dehydrogenase and 2-oxoglutarate dehydrogenase complexes.[4]

REFERENCES

1. Frischer, H., Nelson, R., Noyes, C., Carson, P.E., Bowman, J.J.E., Rieckmann, K.H., and Ajmar, F., NADP glycohydrolase deficiency in human erythrocytes and alteration of cytosol NADH–methemoglobin diaphorase by membrane NAD–glycohydrolase activity, *Proc. Natl. Acad. Sci. U.S.A.*, 70, 2406, 1973.
2. West, C.A., Gomperts, B.D., Huehns, E.R., Kessel, I., and Ashby, J.J.R., Demonstration of an enzyme variant in a case of congenital methaemoglobinaemia, *Br. Med. J.*, 2, 212, 1967.
3. Brewer, G.J., Eaton, J.J.W., Knutsen, C.S., and Beck, C.C., A starch gel electrophoretic method for the study of diaphorase isozymes and preliminary results with sheep and human erythrocytes, *Biochem. Biophys. Res. Commun.*, 29, 198, 1967.
4. NC-IUBMB, *Enzyme Nomenclature*, Academic Press, San Diego, 1992, p. 109 (EC 1.8.1.4, Comments).

1.8.3.1 — Sulfite Oxidase; SUOX

REACTION	Sulfite + H_2O + O_2 = sulfate + H_2O_2
ENZYME SOURCE	Bacteria, plants, invertebrates, vertebrates
SUBUNIT STRUCTURE	Dimer[#] (vertebrates)

METHOD

Visualization Scheme

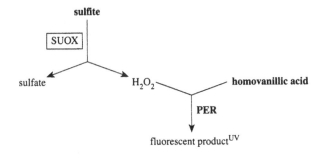

Staining Solution[1] (modified)

A. 0.1 M Tris–HCl buffer, pH 8.0
   10 mM Na$_2$SO$_3$
   2 U/ml Peroxidase (PER)
   1 mM Homovanillic acid
B. 2% Agarose solution (60°C)

Procedure

Mix equal volumes of solutions A and B and pour the mixture over the surface of the gel. Incubate the gel with an agarose application at 37°C and observe fluorescent SUOX bands under long-wave UV light. Record the zymogram or photograph using a yellow filter.

Notes: Preparation of the enzyme samples for electrophoresis includes partial purification and disruption of mitochondria.

Many chromogenic peroxidase substrates can be used instead of homovanillic acid to develop SUOX activity bands visible in daylight (see 1.11.1.7 — PER).

REFERENCES

1. Bogaart, A.M. and Bernini, L.F., The molybdoenzyme system of *Drosophila melanogaster*: I. Sulfite oxidase: identification and properties: expression of the enzyme in maroon-like (mal), low-xanthine dehydrogenase (lxd), and cinnamon (cin) flies, *Biochem. Genet.*, 19, 929, 1981.

1.8.3.2 — Thiol Oxidase; TO

OTHER NAMES	Sulfhydryl oxidase
REACTION	4 R'C(R)SH + O$_2$ = 2 R'C(R)S-S(R)CR' + 2 H$_2$O$_2$ (see General Notes)
ENZYME SOURCE	Invertebrates, vertebrates
SUBUNIT STRUCTURE	Unknown or no data available

METHOD

Visualization Scheme

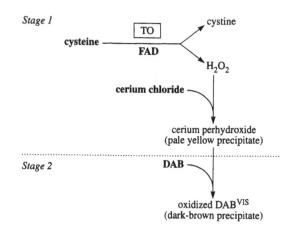

Staining Solution[1]

A. 200 mM Tris–maleate buffer, pH 8.3
10 mM L-Cysteine
1 μM FAD
3 mM Cerium chloride (CeCl$_3$)

B. 100 mM Acetate buffer, pH 5.5
1 mM 3,3'-Diaminobenzidine (DAB)
20 μg/ml CoCl$_2$ × 6H$_2$O
20 μg/ml (NH$_4$)$_2$Ni(SO$_4$)$_2$ × 6H$_2$O

Procedure

Preequilibrate the electrophorized gel in 200 mM Tris–maleate buffer (pH 8.3) for 5 to 10 min, and then incubate in solution A for 30 min. Rinse the gel twice in 100 mM acetate buffer (pH 5.5) for 5 min each and incubate in solution B for 10 to 20 min. Carry out all the procedures at 37°C. After staining is completed, photograph or dry the zymogram instantly.

Notes: The method is based on the ability of cerium chloride to react with hydrogen peroxide, forming the insoluble cerium perhydroxide of a pale yellow color. Cerium perhydroxide can be converted to a dark brown precipitate by the addition of DAB, which is oxidized by cerium perhydroxide to its polymeric form.

The method works equally well with native electrophoretic gels and corresponding nitrocellulose blots.

GENERAL NOTES

In the thiol oxidase reaction R may be = S or = O or a variety of other groups. The enzyme is not specific for R'.[2] Sulfhydryl oxidase from rat seminal vesicles is a flavoprotein (FAD) oxidizing L-cysteine and dithioerythritol equally well and generating hydrogen peroxide.[1]

REFERENCES

1. Seitz, J., Keppler, C., Fahimi, H.D., and Völkl, A., A new staining method for the detection of activities of H$_2$O$_2$-producing oxidases on gels and blots using cerium and 3,3'-diaminobenzidine, *Electrophoresis*, 12, 1051, 1991.
2. NC-IUBMB, *Enzyme Nomenclature*, Academic Press, San Diego, 1992, p. 110 (EC 1.8.3.2, Comments).

1.9.3.1 — Cytochrome *c* Oxidase; CO

OTHER NAMES   Cytochrome oxidase, cytochrome a_3, cytochrome aa_3, indophenolase (misleading name), indophenol oxidase (misleading name)

REACTION   4 Ferrocytochrome *c* + O_2 = 4 ferricytochrome *c* + 2 H_2O; also catalyzes the Nadi reaction (oxidizing reaction between α-naphthol and dimethyl-*p*-phenylenediamine)

ENZYME SOURCE   Bacteria, plants, vertebrates

SUBUNIT STRUCTURE   Heterotrimer[a] (bacteria; see General Notes), dimer[a] (bovine heart)

METHOD

Visualization Scheme

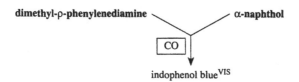

Staining Solution[1]

0.1 *M* Phosphate buffer, pH 7.0	100 ml
N,N-Dimethyl-*p*-phenylenediamine	100 mg
α-Naphthol (predissolved in 0.5 ml acetone)	150 mg

Procedure

Incubate the gel in staining solution in the dark at 37°C until colored bands appear. Wash the stained gel in water and fix in 3% acetic acid.

GENERAL NOTES

The enzyme from *Thiobacillus ferrooxidans* is a heterotrimer consisting of three polypeptides with molecular weights of 53,000, 22,000, and 17,000.[2]

REFERENCES

1. Brown, A.H.D., Nevo, E., Zohary, D., and Dagan, O., Genetic variation in natural populations of wild barley (*Hordeum spontaneum*), *Genetica*, 49, 97, 1978.
2. Kai, M., Yano, T., Tamegai, H., Fukumori, Y., and Yamanaka, T., *Thiobacillus ferrooxidans* cytochrome-*c* oxidase: purification, and molecular and enzymatic features, *J. Biochem.*, 112, 816, 1992.

1.10.3.1 — Catechol Oxidase; COX

OTHER NAMES	Diphenol oxidase, polyphenol oxidase, phenol oxidase, o-diphenolase, phenolase, tyrosinase
REACTION	2 Catechol + O_2 = 2 1,2-benzoquinone + 2 H_2O
ENZYME SOURCE	Bacteria, fungi, plants, invertebrates, vertebrates
SUBUNIT STRUCTURE	Monomer (plants, invertebrates), dimer (invertebrates)

METHOD 1

Visualization Scheme

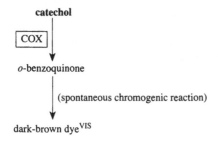

Staining Solution[1]

0.02 *M* Sodium acetate buffer, pH 4.2
0.01 *M* Catechol

Procedure

Incubate the gel in staining solution at room temperature for about 12 h. The enzyme activity is revealed by the appearance of bands of dark brown catechol melanin, presumably formed as a result of secondary nonenzymatic reactions of the *o*-benzoquinone product of catechol oxidation. Wash the stained gel in water and store in 50% glycerol.

METHOD 2

Visualization Scheme

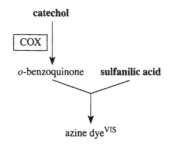

Staining Solution[2]

0.1 *M* Sodium phosphate buffer, pH 6.8	100 ml
Catechol	15 mg
Sulfanilic acid	50 mg

Procedure

Incubate the gel in staining solution at 30°C until positively stained bands appear. Wash the stained gel in water and store in 50% glycerol.

Notes: The azine dye is formed by the condensation of the *o*-quinone and the aromatic amine. Many other coupling amines and diamines can also be used instead of sulfanilic acid.

OTHER METHODS

Polyphenol oxidases of the catechol oxidase type from coffee leaves and fruit endosperm were detected after PAG electrophoresis (0.1% SDS was included in the tank buffer only) using staining solution containing 50 m*M* sodium phosphate buffer (pH 6.0), 2 m*M* 5-caffeoylquinic acid, and 0.5 m*M* *p*-phenylenediamine.[3]

GENERAL NOTES

Monophenol (tyrosine), diphenols (L-β-3,4-dihydroxyphenylalanine, dopamine, *N*-acetyldopamine, protocatechuic acid, chlorogenic acid, catechin, *m*-cresol), and polyphenol (pyrogallol) also can be used as substrates suitable for visualization of COX activity on electrophoretic gels.

The rate of staining of COX bands on gels is increased by addition to the staining solution of 1 m*M* Cu^{2+}.

The phenol oxidases are usually present as inactive proenzymes, which should be activated by incubation of the electrophorized gel in solutions containing natural activators extracted from organisms[4] or synthetic activators such as detergents, heavy metals, or alcohols,[5] or proteolytic enzymes such as chymotrypsin.[6] For example, *Drosophila* phenol oxidase is activated by soaking the gel in a 1:1 mixture of propan-2-ol and 0.1 *M* potassium phosphate buffer (pH 6.3) for 2 h at 4°C.[5]

Plant polyphenol oxidases continue to be latent even after extraction from the tissue, and activation is observed only after treatment with ammonium sulfate, proteases, detergents, etc. For example, polyphenol oxidases from coffee leaves and fruit endosperm are activated 10 to 15% by 0.035 to 1.75 m*M* SDS.[3] One explanation for this activation is that polyphenol oxidases have proenzyme forms, which are bound to membranes, and need activators.[7]

The relative activity of plant polyphenol oxidase with different substrates varies considerably from plant to plant.

This enzyme represents a group of copper proteins that also act on a variety of substituted catechols, many of which also catalyze the reaction listed under EC 1.14.18.1 (see 1.14.18.1 — MMO). This is especially true for the classical tyrosinase.[8]

REFERENCES

1. Pryor, T. and Schwartz, D., The genetic control and biochemical modification of catechol oxidase in maize, *Genetics*, 75, 75, 1973.
2. Sato, M. and Hasegawa, M., The latency of spinach chloroplast phenolase, *Phytochemistry*, 15, 61, 1976.
3. Mazzafera, P. and Robinson, S.P., Characterization of polyphenol oxidase in coffee, *Phytochemistry*, 55, 285, 2000.
4. Warner, C.K., Grell, E.H., and Jacobson, K.B., Phenol oxidase activity and the lozenge locus of *Drosophila melanogaster*, *Biochem. Genet.*, 11, 359, 1974.
5. Batterham, P. and MacKechnie, S.W., A phenol oxidase polymorphism in *Drosophila melanogaster*, *Genetica*, 54, 121, 1980.
6. Waite, J.H. and Wilbur, K.M., Phenol oxidase in the periostracum of the marine bivalve *Modiolus demissus* Dillwyn, *J. Exp. Zool.*, 195, 359, 1975.
7. Whitaker, J.R., Polyphenol oxidase, in *Food Enzymes: Structure and Mechanism*, Wong, D.W.S., Ed., Chapman & Hall, New York, 1995, p. 271.
8. NC-IUBMB, *Enzyme Nomenclature*, Academic Press, San Diego, 1992, p. 114 (EC 1.10.3.1, Comments).

OTHER NAMES	Urishiol oxidase, indophenol oxidase, *p*-diphenol oxidase, polyphenol oxidase, coniferyl alcohol oxidase
REACTION	4 Benzenediol + O_2 = 4 benzosemiquinone + 2 H_2O (see General Notes)
ENZYME SOURCE	Bacteria, fungi, plants
SUBUNIT STRUCTURE	Monomer[d] (fungi)

METHOD 1

Visualization Scheme

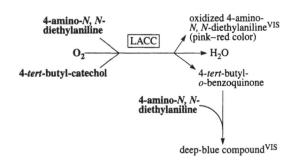

Staining Solution[1]

A. 2 m*M* Salicylhydroxamic acid
B. 10 m*M* HCl
  25 m*M* 4-Amino-*N*,*N*-diethylaniline sulfate
C. 0.1 *M* Potassium phosphate buffer, pH 6.0
D. 10 m*M* Acetic acid
  25 m*M* 4-*tert*-Butyl-catechol

Procedure

After electrophoresis, soak PAG for 5 min in solution C and incubate in solution A for 30 min at room temperature. Place the gel in solution B. Pink-red bands corresponding to LACC activity begin to develop just after this point due to the enzymatic oxidation of 4-amino-*N*,*N*-diethylaniline and the appearance of a colored semiquinonoid cation. Add an equal volume of solution D and incubate the gel for about 1 h. Pink-red bands of LACC activity turn a deep blue color.

Notes: Salicylhydroxamic acid is used to inhibit monophenol monooxigenase; activity bands can also be developed using this method (see 1.14.18.1 — MMO).

4-Amino-*N*,*N*-diethylaniline is an excellent substrate for laccase, while 4-*tert*-butyl-catechol is only slowly oxidized by LACC. However, the product 4-*tert*-butyl-*o*-benzoquinone takes part in the coupling reaction with unoxidized 4-amino-*N*,*N*-diethylaniline and in the formation of a covalent adduct of deep blue color.[2]

METHOD 2

Visualization Scheme

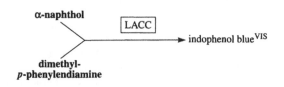

Staining Solution[3] (modified)

A. 0.1 *M* Phosphate buffer, pH 6.5
  0.01 *M* Dimethyl-*p*-phenylenediamine
B. 0.1 *M* Phosphate buffer, pH 6.5
  0.01 *M* α-Naphthol

Procedure

Mix equal volumes of A and B solutions and incubate the gel in the resulting mixture at room temperature until blue bands appear. Wash the stained gel in water and fix in 25% ethanol.

METHOD 3

Visualization Scheme

ABTS (or DAF) $\xrightarrow[\text{CAT}]{\text{LACC}}$ oxidized ABTS (or DAF)[VIS] (colored dye)

Staining Solution[4]

50 m*M* Sodium acetate buffer, pH 6.5
0.05 m*M* Diaminofluorene (DAF) or 2,2'-azinobis(3-ethylbenzothiazoline-6-sulfonate) (ABTS)
125 U/ml Catalase

Procedure

Following electrophoresis, wash the SDS-PAG extensively in 50 m*M* sodium acetate buffer (pH 6.5) to remove SDS, and then incubate in the staining solution until colored bands of LACC appear.

Notes: SDS-PAG prepared according to Laemmli[5] should be allowed to polymerize overnight to reduce the potential that free radicals remaining from the polymerization process could generate H_2O_2.

Catalase is included in the staining solution to remove any H_2O_2 species from the gel and to prevent the development of H_2O_2-dependent phenoloxidase and peroxidase activity bands.

METHOD 4

Visualization Scheme

coniferyl alcohol $\xrightarrow{\boxed{\text{LACC}}}$ oxidized coniferyl alcohol[VIS]
(opaque bands)

Staining Solution[4]

10 mM Potassium phosphate buffer, pH 6.9
11 mM Coniferyl alcohol

Procedure

Incubate PAG in the staining solution until opaque bands appear. Counterstain the processed gel with a phloroglucinol–HCl solution to obtain red-colored bands of LACC activity.

Notes: The positive reaction of opaque bands with phloroglucinol–HCl provides evidence that the product localized in the opaque gel areas is a lignin-like dehydrogenation polymer product. These bands are not developed when activity staining is carried out in the presence of cetyltrimethylammonium bromide, a known inhibitor of laccase. The obtained results suggest that LACC takes part in the ultimate step of lignin synthesis in plants. These results underline the importance of using natural substrates for in-gel enzyme detection.[4] It should be stressed, however, that coniferyl alcohol is too expensive for routine use. Coniferyl alcohol oxidase from conifer species may be detected using p-anisidine dihydrochloride (20 mM), p-phenylenediamine dihydrochloride (10 mM), and 2,7-diaminofluorene (0.68 mM).[6] See also Method 3 above.

METHOD 5

Visualization Scheme

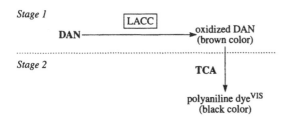

Stage 1

DAN $\xrightarrow{\boxed{\text{LACC}}}$ oxidized DAN
(brown color)

Stage 2

TCA

polyaniline dye[VIS]
(black color)

Staining Solution[7]

A. 50 mM Sodium acetate buffer, pH 5.0
B. 1 M 1,8-Diaminonaphthalene (DAN) in 100% dimethyl sulfoxide
C. 20% Trichloroacetic acid (TCA)

Procedure

Following electrophoresis, immerse the gel in a mixture of 50 ml of A and 0.1 ml of B prepared immediately prior to immersion of the gel. Incubate the gel at 40°C until brown bands of LACC appear (usually ca. 30 min). Place the processed gel in solution C to darken the DAN oxidation product.

Notes: If desirable, activity bands can be further enhanced by counterstaining the processed gel with Coomassie Brilliant Blue using protocols commonly used to detect proteins. The sensitivity of the Coomassie-enhanced DAN staining technique is more than 400 times higher than that of Coomassie staining alone.[7]

The DAN method was developed to detect fungal laccase taking into account that deficiency of the enzyme resulted in accumulation of 1,8-dihydroxynaphthalene. This observation suggested that 1,8-dihydroxynaphthalene may be a substrate of the fungal laccase. Because 1,8-dihydroxynaphthalene is not commercially available, a related compound, DAN, was tested and proved to be a very useful chromogenic substrate of the fungal LACC.[7]

GENERAL NOTES

Laccase is represented by a group of multicopper enzyme proteins of low specificity acting on both o- and p-quinols, and often acting also on aminophenols and phenylenediamine. The product semiquinone may react further either enzymatically or nonenzymatically.[8]

Phenoloxidases (see 1.10.3.1 — COX; 1.14.18.1 — MMO), including LACC, and peroxidases (see 1.11.1.7 — PER) can often make use of the same chromogenic substrates. Taking into account that some peroxidases can slowly oxidize these substrates in the absence of hydrogen peroxide, distinguishing between peroxidases and phenoloxidases can be almost impossible if done solely on the basis of enzymatic analysis of compound protein mixtures using a catalog of substrates and inhibitors. The activity staining of electrophoretic gels (zymogram technique) is an important method for assessing the complexity of phenol-oxidizing enzymes in complex protein mixtures such as tissue or whole organism homogenates.[7]

REFERENCES

1. Rescigno, A., Sanjust, E., Montanari, L., Sollai, F., Soddu, G., Rinaldi, A.C., Oliva, S., and Rinaldi, A., Detection of laccase, peroxidase, and polyphenol oxidase on a single polyacrylamide gel electrophoresis, *Anal. Lett.*, 30, 2211, 1997.
2. Rescigno, A., Sanjust, E., Pedulli, G.F., and Valgimigli, L., Spectrophotometric method for the determination of polyphenol oxidase activity by coupling of 4-*tert*-butyl-*o*-benzoquinone and 4-amino-*N*,*N*-diethylaniline, *Anal. Lett.*, 32, 2007, 1999.
3. Davenport, H.A., *Histological and Histochemical Techniques*, W.B. Saunders, Philadelphia, 1964.

1.10.3.2 — Laccase; LACC (continued)

4. Sterjiades, R., Ranocha, P., Boudet, A.M., and Goffner, D., Identification of specific laccase isoforms capable of polymerizing monolignols by an "in-gel" procedure, *Anal. Biochem.*, 242, 158, 1996.

5. Laemmli, U.K., Cleavage of structural proteins during the assembly of the head of bacteriophage T4, *Nature*, 227, 680, 1970.

6. Udagama-Randeniya, P. and Savidge, R., Electrophoretic analysis of coniferyl alcohol oxidase and related laccases, *Electrophoresis*, 15, 1072, 1994.

7. Hoopes, J.T. and Dean, J.F.D., Staining electrophoretic gels for laccase and peroxidase activity using 1,8-diaminonaphthalene, *Anal. Biochem.*, 293, 96, 2001.

8. NC-IUBMB, *Enzyme Nomenclature*, Academic Press, San Diego, 1992, p. 114 (EC 1.10.3.2, Comments).

1.10.3.3 — L-Ascorbate Oxidase; ASOX

OTHER NAMES	Ascorbase
REACTION	2 L-Ascorbate + O$_2$ = 2 dehydroascorbate + 2 H$_2$O
ENZYME SOURCE	Bacteria, plants
SUBUNIT STRUCTURE	Unknown or no data available

Method 1

Visualization Scheme

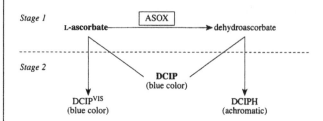

Staining Solution[1]

A.	0.1 *M* Tris–HCl buffer, pH 8.0	100 ml
	Ascorbic acid	20 mg
B.	H$_2$O	10 ml
	2,6-Dichlorophenol indophenol (DCIP)	2.5 mg

Procedure

Incubate the gel in solution A at 30°C for 15 min. Discard solution A and blot the surface of the gel to remove excess liquid. Place a piece of filter paper saturated with solution B on top of the gel. Achromatic bands of ASOX activity develop on a blue background of the gel after 5 to 10 min. Bands are ephemeral, so the zymogram should be recorded or photographed as soon as possible.

Notes: The application of solution B as a 1% agar overlay may prove preferable.

METHOD 2

Visualization Scheme

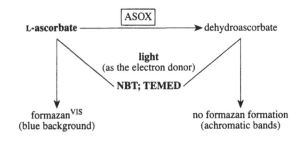

L-ascorbate ⟶ dehydroascorbate

light
(as the electron donor)
NBT; TEMED

formazan[VIS]
(blue background)

no formazan formation
(achromatic bands)

Staining Solution[2]

A. 0.1 M H_2O_2
B. 36 mM Potassium phosphate buffer, pH 7.8
C. 2.45 mM NBT
D. 36 mM Potassium phosphate buffer, pH 7.8
28 mM N,N,N',N'-Tetramethylethylenediamine (TEMED)
25 μM Ascorbate

Procedure

After electrophoresis, place PAG (1.5-mm thick) in solution A for 20 min to inhibit superoxide dismutase activity, and then wash five times (5 min each) with solution B. Afterwards soak the gel in solution C for 30 min and in solution D for a further 30 min in the dark. Finally, place the gel at a 15-cm distance from a 250-W HQ1-TS Osram lamp (at a luminous flux of 7.5 ± 0.8 mW cm^{-2} min^{-1} in the interval of 400 to 700 nm) and keep the gel temperature at 10°C. After development of the blue color, stop the reaction by washing the gel with deionized water. Achro-matic ASOX bands are visible on a blue background. The stained gel may be stored in 25% ethanol.

Notes: The method is based on the ability of a mixture of TEMED (or EDTA) and ascorbate (or riboflavin) to give a photoinduced reduction of NBT. The species directly responsible for NBT reduction is thought to be O_2. This species is absent in gel areas occupied by active ASOX.

REFERENCES

1. Vallejos, C.E., Enzyme activity staining, in *Isozymes in Plant Genetics and Breeding*, Part A, Tanskley, S.D. and Orton, T.J., Eds., Elsevier, Amsterdam, 1983, p. 469.
2. Maccarrone, M., Rossi, A., D'Andrea, G., Amicosante, G., and Avigliano, L., Electrophoretic detection of ascorbate oxidase activity by photoreduction of nitroblue tetrazolium, *Anal. Biochem.*, 188, 101, 1990.

1.11.1.6 — Catalase; CAT

REACTION	$H_2O_2 + H_2O_2 = O_2 + 2\,H_2O$
ENZYME SOURCE	Bacteria, fungi, plants, protozoa, invertebrates, vertebrates
SUBUNIT STRUCTURE	Tetramer (plants, invertebrates, vertebrates)

METHOD 1

Visualization Scheme

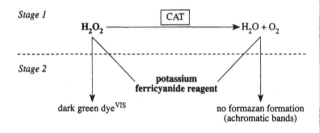

Staining Solution[1]

A. 3% H_2O_2
B. 2% Potassium ferricyanide
C. 2% Ferric chloride

Procedure

Incubate the gel in solution A for about 15 min. Pour off solution A, rinse the gel with water, and then immerse it in a 1:1 mixture of solutions B and C. Gently agitate the tray containing the gel for a few minutes. Yellow bands of CAT activity appear on a blue-green background. Wash the stained gel in water. Record or photograph the zymogram.

Notes: Prepare solutions A, B, and C on the same day as used.

METHOD 2

Visualization Scheme

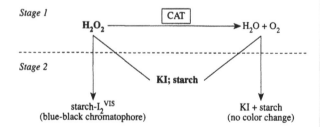

Staining Solution[2]

A.	60 mM Sodium thiosulfate ($Na_2S_2O_3$)	30 ml
B.	3% H_2O_2 (freshly prepared)	70 ml
C.	90 mM Potassium iodide (KI)	100 ml
	Acetic acid (glacial)	0.5 ml

Procedure

Mix solutions A and B quickly just before use (this is essential). Immerse the electrophorized starch gel in the mixture for 30 to 60 sec, and then place it into solution C. Agitate the staining tray gently. The gel turns a bluish black color, except at the sites of the CAT activity, which remain achromatic. Record or photograph the zymogram immediately since the bands are ephemeral.

Notes: This method was developed for starch gel, but it can also be used with PAG prepared with 0.5% soluble starch.

The method is based on the starch–iodine reaction. Thiosulfate in the staining solution is inactivated by hydrogen peroxide, except at the sites of CAT activity, where hydrogen peroxide is destroyed enzymatically. The iodide is oxidized by hydrogen peroxide to iodine, which forms a chromatophore with the starch, and sites of CAT localization remain achromatic. If any iodine diffuses into the achromatic areas, it will be reduced to iodide by thiosulfate, which remains active in these areas.

METHOD 3

Visualization Scheme

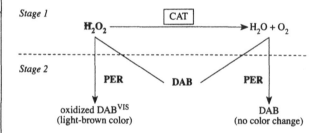

Staining Solution[3]

A. 50 mM Potassium phosphate buffer, pH 7.0
4 mg/ml 3,3′-Diaminobenzidine (DAB)
10 U/ml Peroxidase (PER)
B. 50 mM Potassium phosphate buffer, pH 7.0
20 mM H_2O_2

Procedure

Immerse the gel in solution A and incubate in the dark at room temperature for 45 min to allow peroxidase and DAB to diffuse into the gel matrix. Rinse the gel with water and place in solution B. Hydrogen peroxide is destroyed by CAT prior to the DAB oxidation by peroxide. As a result, achromatic CAT bands appear on a brown background. Record or photograph the zymogram.

Notes: The developed gel may be counterstained with $Cu(NO_3)_2$ solution to obtain a zymogram with achromatic CAT bands on a gray-black background.

DAB is a potential carcinogen. The use of an alternative chromogen (3-amino-9-ethyl-carbazole) is preferable (see 1.11.1.7 — PER, Method 1).

OTHER METHODS

A. The method of simultaneous negative staining of CAT bands and positive staining of peroxidase bands based on a modified starch–iodine reaction is also available.[4] This method is applicable to starch gel or PAG prepared with 0.5% soluble starch.

B. The immunoblotting procedure (for details, see Part II) based on the utility of monoclonal antibodies specific to the yeast enzyme[5] can also be used for immunohistochemical visualization of the enzyme protein on electrophoretic gels. This procedure is expensive and unsuitable for routine laboratory use. Monoclonal antibodies, however, may be of great value in special (biochemical, immunochemical, phylogenetic, and genetic) analyses of CAT. Monoclonal antibodies specific to the human erythrocyte CAT are now available from Sigma.

GENERAL NOTES

Mycobacteria produce two types of CAT, the heat-labile T-catalase, which also has a peroxidase-like function, and the heat-stabile M-catalase, which does not act as a peroxidase. When peroxidase stain using diaminobenzidine (DAB) as a second substrate (see 1.11.1.7 — PER, Method 1, *Notes*) is applied to PAG containing separated isozymes of mycobacterial catalases, the T-catalase isozymes appeared as brown bands. When the DAB-stained gel is then treated with the ferricyanide negative stain for catalase (e.g., see Method 1 above), the M-catalase isozymes appear as clear bands, and isozymes of T-catalase appear as blue-green bands within their respective clear zones.[6] Using enzyme activity staining of electrophoretic gels, bifunctial catalase–peroxidase and monofunctional catalase isozymes were detected in some photosynthetic bacteria.[7]

REFERENCES

1. Harris, H. and Hopkinson, D.A., *Handbook of Enzyme Electrophoresis in Human Genetics*, North-Holland, Amsterdam, 1976 (loose-leaf, with supplements in 1977 and 1978).
2. Thorup, O.A., Strole, W.B., and Leavell, B.S., A method for the localization of catalase on starch gels, *J. Lab. Clin. Med.*, 58, 122, 1961.
3. Gregory, E.M. and Fridovich, I., Visualization of catalase on acrylamide gels, *Anal. Biochem.*, 58, 57, 1974.
4. Siciliano, M.J. and Shaw, C.R., Separation and visualization of enzymes on gels, in *Chromatographic and Electrophoretic Techniques*, Vol. 2, *Zone Electrophoresis*, Smith, I., Ed., Heinemann, London, 1976, p. 185.
5. Adolf, G.R., Hartter, E., Ruis, H., and Swetly, P., Monoclonal antibodies to yeast catalase-T, *Biochem. Biophys. Res. Commun.*, 95, 350, 1980.
6. Wayne, L.G. and Diaz, G.A., A double staining method for differentiating between two classes of mycobacterial catalase in polyacrylamide electrophoresis gels, *Anal. Biochem.*, 157, 89, 1986.
7. Lim, H.K., Kim, Y.M., Lee, D.H., Kahng, H.Y., and Oh, D.C., Analysis of catalases from photosynthetic bacterium *Rhodospirillum rubrum* S1, *J. Microbiol.*, 39, 168, 2001.

REACTION	Donor + H_2O_2 = oxidized donor + $2 H_2O$
ENZYME SOURCE	Bacteria, fungi, plants, invertebrates, vertebrates
SUBUNIT STRUCTURE	Monomer (plants, invertebrates)

METHOD 1

Visualization Scheme

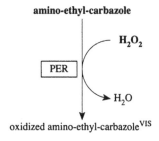

Staining Solution[1]

50 mM Sodium acetate buffer, pH 5.0	100 ml
3-Amino-9-ethyl-carbazole (dissolved in a few drops of acetone)	50 mg
3% H_2O_2 (freshly prepared)	0.75 ml

Procedure

Incubate the gel in staining solution in the dark at room temperature or at 4°C until red-brown bands appear. Wash the gel in water and fix in 50% glycerol.

Notes: The use of a higher pH of the stain buffer may give faint bands only.[2] For example, human salivary peroxidase is poorly stained in PAG prepared in the usual manner with Tris–borate buffer (pH 8.6), while good results are obtained when acidic PAG (polymerized using ferrous sulfate, ascorbic acid, and hydrogen peroxide) is used.[3]

Many other chromogenic substrates can be used instead of amino-ethyl-carbazole: o-dianisidine dihydrochloride, benzidine dihydrochloride, diaminobenzidine, tetramethylbenzidine, 3,5-dichloro-2-hydroxy-benzene sulfonate, o-tolidine, p-phenylenediamine (coupled with catechol), p-hydroxybenzene sulfonate (coupled with 4-aminoantipyrine), N-ethyl-N-(3-sulfopropyl)-m-anisidine (coupled with 4-aminoantipyrine), guaiacol, pyrogallol, 4-chloro-1-naphthol, and some others. All of them except guaiacol are carcinogens or possible carcinogens. Extreme caution should be used when handling these reagents. Amino-ethyl-carbazole is probably less hazardous, but it also should be treated with caution.

Eugenol is used as a fluorogenic peroxidase substrate. The product of the peroxidase reaction, bieugenol, fluoresces in short-wave UV light. Using this substrate, the bands of PER activity can also be observed in daylight as areas of bieugenol precipitation.[4] Homovanillic acid is also a very useful and sensitive fluorogenic substrate for PER (see 1.8.3.1 — SUOX).

METHOD 2

Visualization Scheme

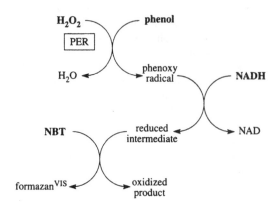

Staining Solution[5]

50 mM Phosphate buffer, pH 7.0
0.3 mg/ml NBT
2.0 mg/ml NADH
0.4 mg/ml Phenol
0.02% H_2O_2

Procedure

Prepare staining solution by dissolving 20 mg of NADH in 5 ml of 0.1 M phosphate buffer (pH 7.0) and adding to it 4 ml of 1.0 mg/ml phenol, 1 ml of 3.0 mg/ml NBT, and finally 20 μl of 10% (w/v) H_2O_2 immediately before use. Incubate the gel in staining solution at room temperature in the dark until dark blue bands appear. Fix the stained gel in 25% ethanol.

Notes: A high concentration of phenol can suppress the development of color. Control staining of additional gel should be performed to exclude misleading interpretations of nonspecifically stained bands due to diaphorase, hemoglobin, or other heme proteins.

The method was developed to detect activity of the horseradish peroxidase-labeled anti-antibodies. Peroxidase conjugated to specific monoclonal antibodies is widely used as a coupled enzyme for immunohistochemical visualization of proteins on electrophoretic gels via the immunoblotting technique (for details, see Part II).

OTHER METHODS

A. 1,8-Diaminonaphthalene (DAN) may be used as a chromogenic substrate suitable for detection of plant PER. Activity bands of PER detected by the DAN method are of brown color and can be further converted into bands of black color by treating the processed gel with trichloracetic acid. The DAN staining procedure detects plant PER activity with a sensitivity similar to that obtained with diaminobenzidine, while the health risk associated with its use is considerably reduced.[6]

B. The chromogenic guaiacol method is one of the more commonly used ones. However, this method has a disadvantage in that the orange bands produced are stable for a short period of time and gradually fade. It was found that guaiacol-stained PER bands are stabilized by Coomassie Brilliant Blue. Briefly, PAG is stained for PER activity using 150 ml of staining mixture consisting of 50 ml of 10 mM H$_2$O$_2$, 75 ml of 20 mM guaiacol, and 75 ml of a buffer (pH 4.0). When PER bands are well developed, the processed gel is placed in Coomassie-fixing solution (0.018 mg/ml Coomassie Brilliant Blue R-250 in 1% acetic acid, 24% methanol) for at least 2 days. The procedure results in good peroxidase band definition, against a pale blue background, and good staining reproducibility from gel to gel. The stained gel can be stored permanently in the Coomassie-fixing solution. Nonspecific protein bands are not developed by this method.[7]

C. PER activity bands detected using 3,3'-diaminobenzidine (DAB) can be intensified by treating the DAB-stained gel with (1) gold trichloride (acid), (2) sodium sulfide, and (3) a developer containing silver nitrate. This method gives 16- to 64-fold amplification of the conventional peroxidase–DAB staining applied to PER-conjugated anti-antibodies used in the immunoblotting procedure.[8] The principle of intensification may be based on the high affinity of DAB for metal salts and the catalyzing activity of metal sulfides for silver reduction and deposition.

D. The enhanced chemiluminescence (ECL) procedure was devised as a highly sensitive and rapid assay for the immunodetection of proteins immobilized on nitrocellulose or polyvinylidene difluoride (PVDF) membranes using horseradish PER-conjugated secondary antibodies. The horseradish PER catalyzes the oxidation of the substrate luminol, which in the presence of a chemical enhancer (e.g., p-iodophenol) produces a substantial light emission that is detected on a photographic film. Intensity of the light emission may be more than 1000-fold that of the unenhanced reaction.[9] The ECL procedure can be used to detect bacterial and mitochondrial c-type cytochromes transferred to the nitrocellulose membrane basing on their heme peroxidase activity. This assay has several advantages over the colorimetric assay: (1) it is about three times quicker than the colorimetric one; (2) the signals generated by ECL protocol are detected on a film and can be easily quantitated with a densitometer; (3) the ECL assay uses nonhazardous components; and (4) filters used in the assay can be reused (e.g., for immunoassays) after removal of the reagents.[10]

E. An immunoblotting procedure (for details, see Part II) may be used to detect the enzyme. Antibodies specific to horseradish PER are available from Sigma.

GENERAL NOTES

Chromogenic peroxidase substrates are derived from 3-amino-9-ethyl carbazole, offering access to potentially novel chromogens for the detection of peroxidase activity.[11]

Some bacterial peroxidases with a pH optimum of 8.5 are stable up to 90°C and are stable up to pH 11 at alkaline conditions.[12]

The addition of cytochrome c to plant tissue samples increased the quaiacol peroxidase activity on zymograms and proved effective in eliminating the variation in the isoperoxidase retention factor dependent on the quantity of sample applied onto the gel in a cathodal PAG electrophoresis system.[13]

A procedure was developed for double activity staining of PER and laminarinase (see 3.2.1.39 — LAM) in the same PAG using amino-ethylcarbazole for peroxidase staining and laminarin and triphenyltetrazolium for laminarinase. This procedure saves time and sample material.[14]

REFERENCES

1. Graham, R.C., Lundholm, U., and Karnovsky, M.J., Cytochemical demonstration of peroxidase activity with 3-amino-9-ethylcarbazole, *J. Histochem. Cytochem.*, 13, 150, 1964.

2. Vallejos, C.E., Enzyme activity staining, in *Isozymes in Plant Genetics and Breeding*, Part A, Tanskley, S.D. and Orton, T.J., Eds., Elsevier, Amsterdam, 1983, p. 469.

3. Azen, E.A., Salivary peroxidase (SAPX): genetic modification and relationship to the proline-rich (P$_r$) and acidic (P$_a$) proteins, *Biochem. Genet.*, 15, 9, 1977.

4. Liu, E.H., Substrate specificities of plant peroxidase isozymes, in *Isozymes*, Vol. 2, *Physiological Function*, Markert, C.L., Ed., Academic Press, New York, 1975, p. 837.

5. Takeda, K., A tetrazolium method for peroxidase staining: application to the antibody-affinity blotting of α-fetoprotein separated by lectin affinity electrophoresis, *Electrophoresis*, 8, 409, 1987.

6. Hoopes, J.T. and Dean, J.F.D., Staining electrophoretic gels for laccase and peroxidase activity using 1,8-diaminonaphthalene, *Anal. Biochem.*, 293, 96, 2001.

7. Fieldes, M.A., Using Coomassie Blue to stabilize H$_2$O$_2$-guaiacol stained peroxidases on polyacrylamide gels, *Electrophoresis*, 13, 454, 1992.

8. Iida, R., Yasuda, T., Nadano, D., and Kishi, K., Intensification of peroxidase–diaminobenzidine staining using gold–sulfide–silver: a rapid and highly sensitive method for visualization in immunoblotting, *Electrophoresis*, 11, 852, 1990.

9. Thorpe, G.H.G., Kricka, L.J., Moseley, S.B., and Whitehead, T.P., Phenols as enhancers of the chemiluminescent horseradish peroxidase–luminol–hydrogen peroxide reaction: application in luminescence-monitored enzyme immunoassays, *Clin. Chem.*, 31, 1335, 1985.

10. Vargas, C., McEwan, A.G., and Downie, J. A., Detection of c-type cytochromes using enhanced chemiluminescence, *Anal. Biochem.*, 209, 323, 1993.

11. Kreig, R., Halbhuber, K.-J., and Oehring, H., Novel chromogenic substrates with metal chelating properties for the histochemical detection of peroxidasic activity, derived from 3-amino-9-ethylcarbazole (AEC) and 3,6-diamino-9-ethylcarbazole, *Cell. Mol. Biol.*, 46, 1191, 2000.

12. Apitz, A. and van Pee, K.H., Isolation and characterization of a thermostable intracellular enzyme with peroxidase activity from *Bacillus sphaericus*, *Arch. Microbiol.*, 175, 405, 2001.

13. Jackson, P. and Ricardo, C.P.P., Cytochrome *c* aided resolution of *Lupinus albus* isoperoxidases in a cathodal polyacrylamide gel electrophoresis system, *Anal. Biochem.*, 200, 36, 1992.

14. Shimoni, M., A method for activity staining of peroxidase and β-1,3-glucanase isozymes in polyacrylamide electrophoresis gels, *Anal. Biochem.*, 220, 36, 1994.

REACTION	2 Glutathione + H_2O_2 = oxidized glutathione + 2 H_2O
ENZYME SOURCE	Vertebrates
SUBUNIT STRUCTURE	Dimer[#] (sea bass), tetramer[#] (bovine eye)

METHOD

Visualization Scheme

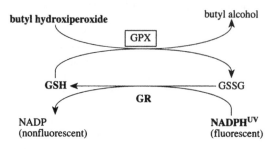

Staining Solution[1]

0.1 *M* Potassium phosphate buffer, pH 7.0	10 ml
Reduced glutathione (GSH)	30 mg
Glutathione reductase (GR)	12 U
5.4 m*M* EDTA, pH 7.0	2 ml
NADPH	15 mg
t-Butyl hydroperoxide (add just before use)	50 μl

Procedure

Apply the staining solution to the gel surface on a filter paper overlay. Incubate the gel at 37°C for 10 to 30 min. Remove the filter paper and view the gel under long-wave UV light. Dark (nonfluorescent) bands of GPX are visible on the light (fluorescent) background of the gel. Record the zymogram or photograph using a yellow filter.

Notes: Mammalian erythrocyte GPX displays activity at the pH optimum of 8.8, about ten times that at pH 7.0.

REFERENCES

1. Wijnen, L.M.M., Monteba-van Heuvel, M., Pearson, P.L., and Meera Khan, P., Assignment of a gene for glutathione peroxidase (GPX 1) to human chromosome 3, *Cytogenet. Cell Genet.*, 22, 223, 1978.

1.11.1.11 — L-Ascorbate Peroxidase; APER

REACTION   2 L-ascorbate + H_2O_2 = 2 dehydro-
                         ascorbate + 2 H_2O

ENZYME SOURCE   Cyanobacteria, algae, plants

SUBUNIT STRUCTURE   Unknown or no data available

Method

Visualization Scheme

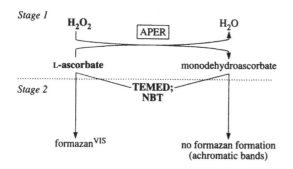

Staining Solution[1]

A. 50 mM Sodium phosphate buffer, pH 7.0
   4 mM L-Ascorbate
   2 mM H_2O_2 (added immediately prior to incubation with
   the gel)

B. 50 mM Sodium phosphate buffer, pH 7.0

C. 50 mM Sodium phosphate buffer, pH 7.8
   28 mM N,N,N',N'-Tetramethylethylenediamine
   (TEMED)
   2.45 mM NBT

Procedure

Subsequent to the electrophoretic separation, equilibrate PAG with 50 mM sodium phosphate buffer (pH 7.0) containing 2 mM ascorbate for a total of 30 min; change the equilibration buffer every 10 min. Then incubate the gel in solution A at room temperature for 20 min. Wash the gel in solution B for 1 min and immerse in solution C with gentle agitation. Achromatic bands on a purple-blue background appear within 3 to 5 min. Allow the reaction to continue for an additional 10 min; wash the gel in deionized water to stop the reaction. The zymogram may be stored in 10% acetic acid at 4°C for up to several months.

Notes: This method is based on the ability of ascorbate to reduce NBT, in the presence of TEMED, to formazan. APER, in the presence of hydrogen peroxide, is capable of preventing the accumulation of formazan due to the rapid hydrogen peroxide-dependent oxidation of ascorbate. Thus, APER activity is developed as achromatic bands on a dark blue background of the stained gel.

Since APER (especially the chloroplastic form in plants) is labile in the absence of ascorbate, include ascorbate (2 to 5 mM) in the extraction and electrophoretic buffers. Prerun the gel for 30 min to allow the ascorbate, present in the electrophoretic buffer, to enter the gel prior to the application of samples. The addition of 10% glycerol to the gel can stabilize some APER isozymes.

The method is sufficiently specific and sensitive. It can detect APER activity in a sample of 10 μg of crude protein extract of a plant leaf (less than 0.01 U). However, activity bands of ascorbate oxidase (see 1.10.3.3 — ASOX) and peroxidase (see 1.11.1.7 — PER) can also be developed by this method, but at levels greater than 0.5 and 2 to 3 U, respectively.

Some APER isozymes display activity with PER substrates (e.g., pyrogallol and 4-chloro-1-naphthol) and therefore can develop on PER zymograms obtained using these substrates.

Other Methods

A modified zymogram method for catalase was used to detect APER activity bands.[2] However, this method detects, in addition to APER activity, any catalase activity present in plant crude extracts. It is about 40 times less sensitive than the method described above.

General Notes

Chilling stress enhances the activity of APER in plant leaves.[3]

References

1. Mittler, R. and Zilinskas, B.A., Detection of ascorbate peroxidase activity in native gels by inhibition of the ascorbate-dependent reduction of nitro blue tetrazolium, *Anal. Biochem.*, 212, 540, 1993.
2. Chen, G.X. and Asada, K., Ascorbate peroxidase in tea leaves: occurrence of two isozymes and the differences in their enzymatic and molecular properties, *Plant Cell Physiol.*, 30, 987, 1989.
3. Lee, D.H. and Lee, C.B., Chilling stress-induced changes of antioxidant enzymes in the leaves of cucumber: in gel enzyme activity assays, *Plant Sci.*, 159, 75, 2000.

OTHER NAMES	Hydrogenase
REACTION	2 H$_2$ + ferricytochrome c_3 = 4 H$^+$ + ferrocytochrome c_3 (see also General Notes)
ENZYME SOURCE	Bacteria
SUBUNIT STRUCTURE	Monomer$^\#$ (bacteria)

Method

Visualization Scheme

Stage 1

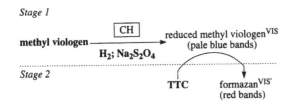

Stage 2

Staining Solution[1,2]

A. 100 mM Phosphate buffer, pH 6.8
    0.25% (w/v) Methyl viologen
B. 100 mM Phosphate buffer, pH 6.8 (air-free)
    5% (w/v) Dithionite (Na$_2$S$_2$O$_4$)
C. 2.5% (w/v) Triphenyl tetrazolium chloride (TTC)

Procedure

Place the electrophorized PAG in a hydrogen atmosphere in solution A. If immediate production of blue bands at sites of hydrogenase activity does not occur, add solution B until a faint blue color persists. This ensures that traces of oxygen trapped within the gel are removed and that the enzyme is in its reduced and most active state. Blue bands of reduced methyl viologen typically appear within 20 min after introduction of hydrogen or trace amounts of dithionite. When blue bands of reduced methyl viologen are well developed, add an equal volume of solution C. Continue incubation of the processed gel under hydrogen atmosphere until blue bands of reduced methyl viologen are totally reacted with the tetrazolium salt, giving a bright red precipitate.

General Notes

Bacterial hydrogenases may be classified by their physiological electron carriers, which are ferredoxin (EC 1.18.99.1), NAD (EC 1.12.1.2), or cytochrome c_3 (EC 1.12.2.1). Differentiation between these hydrogenases should be based on their specificity toward physiological electron carriers.

References

1. Ackrell, B.A.C., Asato, R.N., and Mower, H.F., Multiple forms of bacterial hydrogenases, *J. Bacteriol.*, 92, 828, 1966.
2. Adams, M.W.W. and Hall, D.O., Properties of the solubilized membrane-bound hydrogenase from the photosynthetic bacterium *Rhodospirillum rubrum*, *Arch. Biochem. Biophys.*, 195, 288, 1979.

1.13.11.12 — Lipoxygenase; LPX

OTHER NAMES   Lipoxidase, carotene oxidase
REACTION   Linoleate + O_2 = (9Z,11E)-(13S)-13-hydroperoxyoctadeca-9,11-dienoate
ENZYME SOURCE   Plants
SUBUNIT STRUCTURE   Monomer[#] (plants)

METHOD 1

Visualization Scheme

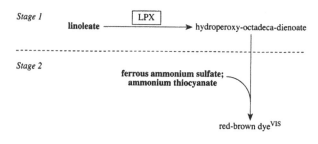

Staining Solution[1]

  A.   0.2 M Phosphate buffer, pH 6.5
       0.25% Tween 20
       7.5 mM Linoleic acid
  B.   5% Ferrous ammonium sulfate
       3% HCl
  C.   20% Ammonium thiocyanate

Procedure

Prepare suspension A by sonication for 5 min at 0°C. Immediately before using, oxygenate the substrate suspension for 20 to 30 min at room temperature. Incubate the gel in suspension A for 20 to 50 min. Rinse the gel with distilled water, immerse in solution B for 30 sec, and then immerse in solution C for 30 sec. Areas of LPX activity develop as red-brown bands. Rinse the stained gel with water. Record the zymogram or photograph because color development continues upon storage.

Notes: Substantial background staining makes detection of LPX in low-activity samples difficult.

METHOD 2

Visualization Scheme

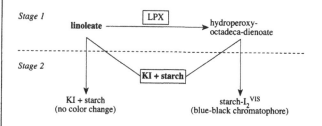

Staining Solution[2]

  A.   Linoleic acid   500 mg
       H_2O 50 ml
       Tween 80   1 drop
  B.   0.05 M Tris–HCl buffer, pH 8.3
  C.   15% Acetic acid
  D.   Saturated solution of KI in water

Procedure

Prepare stock substrate solution A by dispersing 0.5 g of linoleic acid in 25 ml of freshly deionized H_2O with the aid of a drop of Tween 80 and sonication under a stream of nitrogen. Make a total volume to 50 ml with water. Store this stock solution at 2 to 5°C in the dark under nitrogen (prepare fresh stock solution weekly). Incubate the gel (see *Notes*) for 30 min at room temperature in a freshly prepared 20-fold dilution of the stock substrate solution A in buffer B with agitation of the gel to ensure aeration. During the period of incubation of the gel prepare a C + D mixture: 100 ml of solution C purged with nitrogen and then, also under nitrogen, mixed with 5 ml of solution D. After 30 min of incubation in substrate solution, rinse the gel and gel tray with water, fill the tray completely with the C + D mixture, and cover it so that it is airtight, to prevent autoxidation of iodine into iodide and to avoid excessive background staining of the gel. Zones of LPX activity become visible as dark brown to blue bands in about 5 min and reach maximum intensity in 15 to 20 min. Record the zymogram or photograph because color development continues upon storage.

Notes: The method is applicable to starch gel or PAG containing 0.5% soluble starch. Substantial background staining makes detection of low-activity samples difficult. The high percentage of failure in the stain development with this method is due possibly to a high threshold requirement for hydroperoxide to release iodine in the gel.

Method 3

Visualization Scheme

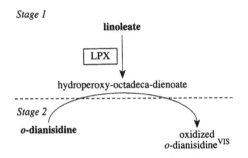

Stage 1

linoleate

LPX

hydroperoxy-octadeca-dienoate

Stage 2

o-dianisidine

oxidized o-dianisidine[VIS]

Staining Solution[3]

A. Linoleic acid   0.62 ml
    H_2O 50 ml
    Triton X-100   2 drops
B. 0.1 *M* Tris–HCl buffer, pH 8.6
C. 0.1 *M* Citrate–phosphate buffer, pH 5.8
D. 0.1% *o*-Dianisidine dihydrochloride

Procedure

Prepare stock substrate solution A by dispersing linoleic acid in water containing Triton X-100 with a sonifier for 5 min at 0°C. Dilute this solution with water to give a stock solution of 20 m*M* linoleic acid. Then prepare the working substrate solution by dilution of the stock solution to 2 m*M* either with B or C buffers. Immediately before use, oxygenate the working substrate solution for 20 to 30 min at room temperature. Rinse the gel with distilled water and place in the working substrate solution. Incubate the gel at room temperature for 30 min. Rinse the gel with distilled water and immerse in solution D. Stain the gel overnight at room temperature. Red-brown bands appear on the gel with practically no background staining. Wash the dye off and store the stained gel in water.

Notes: The stained gel may be stored in water at 4°C for as long as 18 months without color deterioration.

Heme proteins also oxidize linoleic acid and produce the hydroperoxide. Lipoxygenase is a far more powerful catalyst than heme proteins. Nevertheless, 1 m*M* sodium cyanide is usually included in the working substrate solution to inhibit any contribution of heme proteins to the formation of hydroperoxide.

When PAG prepared in the usual manner is stained by this method, dark bands usually also appear in the anodal end of the gel due to oxidation of *o*-dianisidine by ammonium persulfate.

References

1. Grossman, S., Pinsky, A., and Goldwetz, Z., A convenient method for lipoxygenase isoenzyme determination, *Anal. Biochem.*, 44, 642, 1971.
2. Hart, G.E. and Langston, P.J., Chromosomal location and evolution of isozyme structural genes in hexaploid wheat, *Heredity*, 39, 263, 1977.
3. De Lumen, B.O. and Kazeniac, S.J., Staining for lipoxygenase activity in electrophoretic gels, *Anal. Biochem.*, 72, 428, 1976.

1.14.18.1 — Monophenol Monooxygenase; MMO

OTHER NAMES	Tyrosinase, phenolase, monophenol oxidase, cresolase, polyphenol oxidase
REACTION	L-Tyrosine + L-dopa + O_2 = L-dopa + dopaquinone + H_2O
ENZYME SOURCE	Fungi, plants, invertebrates, vertebrates
SUBUNIT STRUCTURE	Unknown or no data available

METHOD

Visualization Scheme

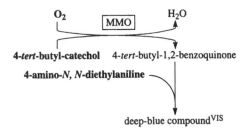

Staining Solution[1]

A. 0.1 M Potassium phosphate buffer, pH 6.0
B. 10 mM Acetic acid
   25 mM 4-tert-Butyl-catechol
C. 10 mM HCl
   25 mM 4-Amino-N,N-diethylaniline sulfate

Procedure

After electrophoresis, soak the PAG for 5 min in solution A and place in a mixture of equal volumes of solutions B and C. Incubate the gel for about 1 h at room temperature. Bands of deep blue color correspond to localization of MMO activity in the gel. Activity bands of laccase (see 1.10.3.2 — LACC) are also developed by this method. To identify activity bands of MMO, an additional (control) gel should be presoaked in 2 mM salicylhydroxamic acid (inhibitor of MMO) for 30 min and stained as described above. This gel will display only LACC activity bands.

Notes: The product 4-tert-butyl-1,2-benzoquinone takes part in the coupling reaction with 4-amino-N,N-diethylaniline. This reaction results in formation of a blue covalent adduct.[2]

GENERAL NOTES

Monophenol monooxygenases are copper-containing proteins catalyzing the *o*-hydroxylation of phenols (cresolase activity) and the oxidation of *o*-diphenols (catechol oxidase activity; see 1.10.3.1 — COX) to the corresponding *o*-quinones. These enzymes are widely distributed among living organisms, demonstrate different but frequently overlapping substrate specificities, and are commonly called tyrosinases, phenolases, phenol oxidases, or polyphenol oxidases.

REFERENCES

1. Rescigno, A., Sanjust, E., Montanari, L., Sollai, F., Soddu, G., Rinaldi, A.C., Oliva, S., and Rinaldi, A., Detection of laccase, peroxidase, and polyphenol oxidase on a single polyacrylamide gel electrophoresis, *Anal. Lett.*, 30, 2211, 1997.
2. Rescigno, A., Sanjust, E., Pedulli, G.F., and Valgimigli, L., Spectrophotometric method for the determination of polyphenol oxidase activity by coupling of 4-tert-butyl-*o*-benzoquinone and 4-amino-N,N-diethylaniline, *Anal. Lett.*, 32, 2007, 1999.

OTHER NAMES	Tetrazolium oxidase, indophenol oxidase (misleading name)
REACTION	$O_2^{.-} + O_2^{.-} + 2 H^+ = O_2 + H_2O_2$
ENZYME SOURCE	Bacteria, fungi, algae, plants, protozoa, invertebrates, vertebrates
SUBUNIT STRUCTURE	Dimer (cytoplasmic isozyme: fungi, algae, plants, invertebrates, vertebrates), tetramer (mitochondrial isozyme: plants, invertebrates, vertebrates)

METHOD

Visualization Scheme

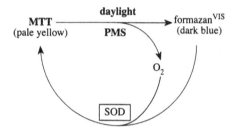

Staining Solution[1]

50 mM Tris–HCl buffer, pH 8.5	80 ml
MTT	10 mg
PMS	6 mg
MgCl$_2$·6H$_2$O	15 mg

Procedure

Incubate the gel in staining solution in daylight. The enzyme bands are seen as pale zones on a dark blue background.

Notes: When SOD activity is low, the amount of PMS should be diminished three times.

NBT may be used instead of MTT.

It is often not necessary to stain specifically for SOD, as its bands may appear on any gels stained by means of MTT(NBT)–PMS linked reactions.

OTHER METHODS

A. Francke and Taggart[2] obtained negative SOD zymograms using a modification of the method developed by Weisiger and Fridovich.[3] The gel was incubated in two changes of 50 mM phosphate buffer (pH 7.8) for a total of 10 min, then in 1 mg/ml NBT solution for 10 min, and finally in 50 mM phosphate buffer (pH 7.8) containing 0.01 mg/ml riboflavin and 3.25 mg/ml N,N,N',N'-tetramethylethylenediamine for 10 min at 24°C with gentle agitation. Areas of SOD activity remained clear after exposure of the gel to fluorescent light (see also 1.10.3.3 — ASOX, Method 2).

B. Positively stained bands of SOD can be developed using the method proposed by Misra and Fridovich.[4] According to this method, the gel is incubated in a staining solution containing 10 mM potassium phosphate buffer (pH 7.2), 2 mM o-dianisidine, and 0.1 mM riboflavin at room temperature in the dark for 1 h. Then the gel is rinsed in water and exposed to the light. Brown bands of SOD activity appear after 1 min. Intensity of the bands increases during the next 10 to 15 min of lighting the gel. Presumably, under these conditions of staining, sufficient amounts of the superoxide radical are generated to allow the SOD to catalyze the oxidation of o-dianisidine. The use of amino-ethyl-carbasole instead of o-dianisidine is recommended because o-dianisidine is known to be a carcinogen. The bands of peroxidase can also develop on SOD zymograms obtained by this method, but they only begin to develop when the SOD bands are well stained. Molecular mechanisms of this photochemical method were recently described in more detail.[5]

C. The chromogenic reaction between N,N'-diethyldithiocarbamate and copper in the copper–zinc SOD was used to visualize areas of enzyme localization in PAG as stable yellow-brown bands that are quantifiable at 448 nm.[6] This method was successfully applied to study the enzyme in red cell lysates from which hemoglobin was removed by chloroform–ethanol precipitation. Any color in gels due to the presence of trace hemoglobin is bleached overnight during staining.

GENERAL NOTES

Two phylogenetically independent isozymes of SOD are known for eukaryotes. The cytoplasmic isozyme contains both copper and zinc, while the mitochondrial isozyme contains manganese. The prokaryotic enzyme contains manganese or iron. Cytoplasmic and mitochondrial isozymes of eukaryotic SOD can be distinguished by using the inhibitory effect of 2% SDS on the mitochondrial isozyme[7] or by using differences of isozymes in their subunit structure (cytoplasmic SOD is a dimer while mitochondrial SOD is a tetramer).

SOD isozymes containing Fe, Mn, and Cu–Zn prosthetic groups can be distinguished by their sensitivity to cyanide and hydrogen peroxide. The Cu–Zn-containing SOD is inhibited by cyanide and hydrogen peroxide. The Mn-containing SOD is insensitive both to cyanide and hydrogen peroxide, and the Fe-containing SOD is resistant to cyanide and inhibited by hydrogen peroxide. Electrophoretic separation, followed by an enzymatic characterization involving use of specific inhibitors, allowed the revealing of all the three different forms of SOD in plants.[8]

Three classes of SODs with distinct molecular weights were identified in bacteria using SDS-PAG electrophoresis and a two-step staining procedure to present the achromatic activity bands of the enzymes. Activity of Cu–Zn-containing SOD was significantly enhanced by the use of this procedure.[9]

In plants, achromatic bands on dehydrogenase zymograms obtained using the tetrazolium method can be caused by peroxidase activity. The PMS–tetrazolium (MTT or NBT) system used in dehydrogenase stains tends to be unstable, and considerable background staining due to generalized reduction of a tetrazolium salt occurs under certain conditions, e.g., bright light, alkaline pH, or elevated temperature. It is suggested that some peroxidases involved in lignification apparently catalyze the peroxidative (NADH-dependent) oxidation of reduced PMS, and thus suppress the generalized reduction of NBT (or MTT). This can result in the development of achromatic bands on a blue background of plant dehydrogenase zymograms.[10] Plant germin is a manganese-containing protein demonstrating oxalate oxidase (see 1.2.3.4 — OO, General Notes) and superoxide dismutase activities.[11]

REFERENCES

1. Brewer, G.J., Achromatic regions of tetrazolium stained starch gels: inherited electrophoretic variation, *Am. J. Hum. Genet.*, 19, 674, 1967.
2. Francke, U. and Taggart, R.T., Assignment of the gene for cytoplasmic superoxide dismutase (*Sod-1*) to a region of chromosome 16 and of *Hprt* to a region of the X-chromosome in the mouse, *Proc. Natl. Acad. Sci. U.S.A.*, 76, 5230, 1979.
3. Weisiger, R.A. and Fridovich, I., Superoxide dismutase: organelle specificity, *J. Biol. Chem.*, 248, 3582, 1973.
4. Misra, H.P. and Fridovich, I., Superoxide dismutase and peroxidase: a positive activity stain applicable of polyacrylamide gel electropherograms, *Arch. Biochem. Biophys.*, 183, 511, 1977.
5. Madon, P.S., An improved photochemical method for the rapid spectrophotometric detection of superoxide dismutase, *Redox. Rep.*, 6, 123, 2001.
6. Jewett, S.L. and Rocklin, A.M., *N,N'*-Diethyldithiocarbamate as a stain for copper–zinc superoxide dismutase in polyacrylamide gels of red cell extracts, *Anal. Biochem.*, 217, 236, 1994.
7. Geller, B.L. and Winge, D.R., A method for distinguishing Cu-, Zn-, and Mn-containing superoxide dismutases, *Anal. Biochem.*, 128, 86, 1983.
8. Droillard, M.-J., Bureau, D., Paulin, A., and Daussant, J., Identification of different classes of superoxide dismutase in carnation petals, *Electrophoresis*, 10, 46, 1989.
9. Chen, J.R., Liao, C.W., Mao, S.J.J.T., Chen, T.H., and Weng, C.N., A simple technique for the simultaneous determination of molecular weight and activity of superoxide dismutase using SDS-PAG, *J. Biochem. Biophys. Methods*, 47, 233, 2001.
10. Fieldes, M.A., An explanation of the achromatic bands produced by peroxidase isozymes in polyacrylamide electrophoresis gels stained for malate dehydrogenase, *Electrophoresis*, 13, 82, 1992.
11. Woo, E.J., Dunwell, J.J.M., Goodenough, P.W., Marvier, A.C., and Pickersgill, R.W., Germin is a manganese-containing homohexamer with oxalate oxidase and superoxide dismutase activities, *Nat. Struct. Biol.*, 7, 1036, 2000.

OTHER NAMES   Ferroxidase (recommended name)

REACTION $4\ Fe(II) + 4\ H^+ + O_2 = 4\ Fe(III) + 2\ H_2O$

ENZYME SOURCE   Bacteria, invertebrates, vertebrates

SUBUNIT STRUCTURE   Monomer (invertebrates, vertebrates)

METHOD

Visualization Scheme

o-dianisidine

CP

oxidized
o-dianisidine^{VIS}

Staining Solution[1]

40 mM Acetate buffer, pH 5.5	100 ml
o-Dianisidine (dissolved in minimal volume of ethanol)	100 mg

Procedure

Incubate the gel in staining solution at 37°C until brown bands appear. Wash the gel in water and fix in 50% glycerol.

Notes: o-Dianisidine is used as an electron acceptor in place of oxygen, and Fe^{2+} ions are tightly bound by the enzyme protein molecules.

The method is specific for mammalian serum. When other mammalian tissues or other organisms are used for sample preparation, nonspecific staining of some oxidases can also occur.

The use of o-dianisidine dihydrochloride instead of o-dianisidine freebase is preferable in all instances.

The use of p-phenylenediamine instead of o-dianisidine is preferable when electrophoresis is carried out in a PAG.

It should be remembered that both o-dianisidine and p-phenylenediamine are strong carcinogens.

OTHER METHODS

The immunoblotting procedure (for details, see Part II) may also be used to detect CP. Polyclonal antibodies specific to the human enzyme are available from Sigma.

GENERAL NOTES

The enzyme is a multicopper protein called ceruloplasmin in animals and rusticyanin in bacteria.[2]

REFERENCES

1. Brewer, G.J., *An Introduction to Isozyme Techniques*, Academic Press, New York, 1970, p. 133.
2. NC-IUBMB, *Enzyme Nomenclature*, Academic Press, San Diego, 1992, p. 150 (EC 1.16.3.1, Comments).

1.X.X.X — "Nothing Dehydrogenase"; NDH

Some NAD(P)-dependent dehydrogenases have the interesting property of showing stained bands in the absence of a substrate in a staining solution, leading to the term "nothing dehydrogenase" (NDH).

The NDH phenomenon is widespread among living organisms, including bacteria, fungi, plants, protozoa, invertebrate and vertebrate animals, and humans. Several different causes of this phenomenon are known.

The NDH effect may originate as a result of the binding of endogenous substrates by NAD(P)-dependent dehydrogenase molecules.[1-4]

Another cause of the NDH effect is the contamination of some commercial preparations (e.g., acrylamide, NAD, phosphosugars) with ethanol.[4,5]

In some cases, the NDH effect may be caused by the reducing action of –SH groups present in specific proteins.[6,7] In a strict sense, this cannot be attributed to the NDH phenomenon because free –SH groups are able to reduce a tetrazolium salt in alkaline medium even in the absence of NAD and PMS. Artifact bands also can develop in MTT(NBT)–PMS-containing solutions as the result of direct chemical reaction of PMS in zones of high local pH of electrophoretic gels.[5]

Finally, the detection mechanism used may not be specific for the intended enzyme and can allow the simultaneous development of bands of other enzymes. For example, a gel stained for aldehyde dehydrogenase (see 1.2.1.3 — ALDH) can also exhibit aldehyde oxidase (1.2.3.1 — AOX), xanthine oxidase (1.1.3.22 — XOX), and alcohol dehydrogenase (1.1.1.1 — ADH) activities as well as other stained bands.[8] Lactate dehydrogenase (1.1.1.27 — LDH) and glycerol-3-phosphate dehydrogenase (1.1.1.8 — G-3-PD) bands can develop on triosephosphate isomerase zymograms (see 5.3.1.1 — TPI, Method 2).

Overlapping substrate specificity of some enzymes can also cause the appearance of additional bands on the zymograms of certain NAD-dependent dehydrogenases. It is shown, for example, that *Drosophila* alcohol dehydrogenase is capable of catalyzing the reaction of lactate oxidation[9] and chicken lactate dehydrogenase is capable of reducing oxaloacetate in the presence of NADH, i.e., to catalyze the reverse reaction of malate dehydrogenase.[10]

Fortunately, many of the stain artifacts listed above usually occur repetitively on gels stained for different enzymes. Thus, it is very important to compare zymograms obtained using similar detection mechanisms to ensure that the same bands are not being scored for supposedly different enzymes.

The observation of data presented in numerous experimental works evidences that nonspecific staining of lactate dehydrogenase and alcohol dehydrogenase in animals (especially in vertebrates) and alcohol dehydrogenase and glutamate dehydrogenase in plants is the most frequent reason for the "nothing dehydrogenase" phenomenon.

REFERENCES

1. Ressler, N. and Stitzer, K., Starch-gel investigations of the relationships between dehydrogenase proteins, *Biochim. Biophys. Acta*, 146, 1, 1967.
2. Falkenberg, F., Lehmann, F.-G., and Pfleiderer, G., LDH (lactate dehydrogenase) isozymes as cause of nonspecific tetrazolium salt staining in gel enzymograms ("nothing dehydrogenase"), *Clin. Chim. Acta*, 23, 265, 1969.
3. White, H.A., Coulson, C.J., Kemp, C.M., and Rabin, B.R., The presence of lactic acid in purified lactate dehydrogenase, *FEBS Lett.*, 34, 155, 1973.
4. Marshall, J.H., Bridge, P.D., and May, J.W., Source and avoidance of the "nothing dehydrogenase" effect, a spurious band produced during polyacrylamide gel electrophoresis of dehydrogenase enzymes from yeasts, *Anal. Biochem.*, 139, 359, 1984.
5. Wood, T. and Muzariri, C.C., The electrophoresis and detection of transketolase, *Anal. Biochem.*, 118, 221, 1981.
6. Somer, H., "Nothing dehydrogenase" reaction as an artefact in serum isoenzyme analyses, *Clin. Chim. Acta*, 60, 223, 1975.
7. Sri Venugopal, K.S. and Adiga, P.R., Artifactual staining of proteins on polyacrylamide gels by nitro blue tetrazolium chloride and phenazine methosulfate, *Anal. Biochem.*, 101, 215, 1980.
8. Holmes, R.S., Electrophoretic analyses of alcohol dehydrogenase, aldehyde dehydrogenase, aldehyde oxidase, sorbitol dehydrogenase, and xanthine oxidase from mouse tissues, *Comp. Biochem. Physiol.*, 61B, 339, 1978.
9. Onoufriou, A. and Alahiotis, S.N., Enzyme specificity: "pseudopolymorphism" of lactate dehydrogenase in *Drosophila melanogaster*, *Biochem. Genet.*, 19, 277, 1981.
10. Busquets, M., Baro, J., Cortes, A., and Bosal, J., Separation and properties of the two forms of chicken liver (*Gallus domesticus*) cytoplasmic malate dehydrogenase, *Int. J. Biochem.*, 10, 823, 1979.

OTHER NAMES   Catechol *o*-methyltransferase
REACTION   *S*-Adenosyl-L-methionine + catechol
                     = *S*-adenosyl-L-homocysteine +
                     guaiacol
ENZYME SOURCE   Vertebrates
SUBUNIT STRUCTURE   Unknown or no data available

METHOD

Visualization Scheme

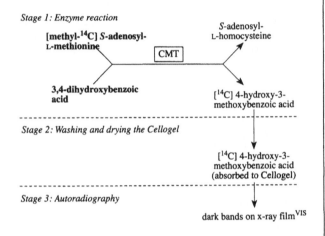

Stage 1: Enzyme reaction

[methyl-^{14}C] *S*-adenosyl-L-methionine

CMT

S-adenosyl-L-homocysteine

3,4-dihydroxybenzoic acid

[^{14}C] 4-hydroxy-3-methoxybenzoic acid

Stage 2: Washing and drying the Cellogel

[^{14}C] 4-hydroxy-3-methoxybenzoic acid (absorbed to Cellogel)

Stage 3: Autoradiography

dark bands on x-ray film^{VIS}

Reaction Mixture[1]

0.1 *M* Tris–HCl buffer, pH 8.0
1 m*M* MgCl$_2$
0.5 m*M* Dithiothreitol
1 m*M* 3,4-Dihydroxybenzoic acid
0.028 m*M* [Methyl-^{14}C] *S*-adenosyl-L-methionine
  (60 mCi/mmol; Amersham)

Procedure

Stage 1. Soak the electrophorized Cellogel strip in the reaction mixture and incubate in a moist chamber at 37°C for 2 h.

Stage 2. Wash the strip in 2 l of 0.02 *N* HCl for 4 min and dry in an oven at 60°C for about 2 h.

Stage 3. Put the dry strip directly in tight contact with medical x-ray film Kodak-X-OMAT G and expose for 8 days at room temperature or 4°C. Develop the exposed x-ray film by Kodak X-OMAT processor ME-1.

Zones of enzyme activity appear as dark bands on the x-ray film.

Notes: The method is developed for cellulose acetate gel. The soluble molecules of the unreacted [^{14}C]*S*-adenosyl-L-methio-nine can be easily removed from the gel matrix by soaking the gel in water, while the relatively insoluble radioactive product remains adsorbed to the gel at the sites of enzyme location.

The addition of adenosine deaminase to the reaction mixture does not improve the final result, though this enzyme is known to decrease the inhibitory effect of the product *S*-adenosyl-L-homocysteine on CMT.

The mammalian enzyme acts more rapidly on catecholamines (such as adrenaline or noradrenaline) than on catechols.

REFERENCES

1. Brahe, C., Crosti, N., Meera Khan, P., and Serra, A., Catechol-*o*-methyltransferase: a method for autoradiographic visualization of isozymes in Cellogel, *Biochem. Genet.*, 22, 125, 1984.

REACTION	S-Adenosyl-L-methionine + glycine = S-adenosyl-L-homocysteine + sarcosine
ENZYME SOURCE	Vertebrates
SUBUNIT STRUCTURE	Tetramer (vertebrates), see General Notes

METHOD

Visualization Scheme

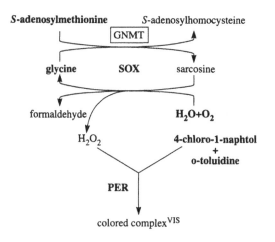

Staining Solution[1]

0.5 M Tris–HCl buffer, pH 8.5	4 ml
Glycine	28 mg
S-Adenosyl-L-methionine	20 mg
Sarcosine oxidase (SOX; Boehringer Mannheim)	4 mg
Peroxidase (PER; Sigma)	4 mg
2% (w/v) 4-Chloro-1-naphthol (in diethylene glycol)	30 µl
o-Toluidine (saturated solution in 7% acetic acid)	100 µl

Procedure

Overlay the gel with cellulose acetate foil soaked in the staining solution and incubate at room temperature until colored bands appear.

Notes: The method was designed to study GNMT isozymes in mammals.[1]

OTHER METHODS

A. The product S-adenosylhomocysteine can be detected using auxiliary enzyme adenosylhomocysteinase and detecting the outcome of adenosine with the final production of formazan (see 3.3.1.1 — AHC). This method was tested and rejected because many transmethylases existing in the liver also use S-adenosylmethionine as the methyl donor and produce S-adenosylhomocysteine. This results in nonspecific staining.[1]

B. The product sarcosine can also be detected using auxiliary sarcosine oxidase and visualizing the formaldehyde output via auxiliary NAD-dependent formaldehyde dehydrogenase (e.g., see 1.1.3.13 — ALOX).

GENERAL NOTES

GNMT is an abundant protein in mammalian livers, making up to 3% of liver cytosol proteins. The enzyme is thought to play an important role in the maintenance of the S-adenosylmethionine/S-adenosylhomocysteine ratio, which has great importance in a variety of reactions involving the methylation of biomolecules.[1]

Different subunit configurations are involved in the enzymatic roles of GNMT as a methyltransferase (tetramer) and as a polycyclic aromatic hydrocarbon (PAH)-binding protein (dimer). It is shown that posttranslational modification is involved in determining the dimeric form of GNMT, which acts as a PAH-binding receptor.[2]

REFERENCES

1. Santos, F., Amorim, A., and Kömpf, J., Specific staining of glycine N-methyltransferase, *Electrophoresis*, 16, 1898, 1995.
2. Bhat, R., Wagner, C., and Bresnick, E., The homodimeric form of glycine N-methyltransferase acts as a polycyclic aromatic hydrocarbon-binding receptor, *Biochemistry*, 36, 9906, 1997.

2.1.1.37 — DNA (Cytosine-5)-Methyltransferase; DNAMT

OTHER NAMES	DNA methyltransferase
REACTION	*S*-Adenosyl-L-methionine + DNA = *S*-adenosyl-L-homocysteine + DNA containing 5-methylcytosine
ENZYME SOURCE	Bacteria, vertebrates
SUBUNIT STRUCTURE	Unknown or no data available

Method

Visualization Scheme

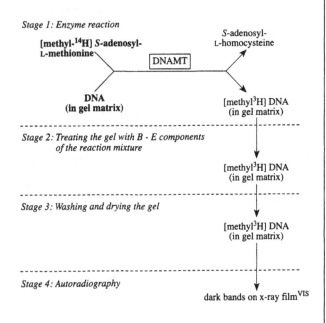

Stage 1: Enzyme reaction

[methyl-¹⁴H] *S*-adenosyl-L-methionine

DNAMT

S-adenosyl-L-homocysteine

DNA (in gel matrix)

[methyl³H] DNA (in gel matrix)

Stage 2: Treating the gel with B - E components of the reaction mixture

[methyl³H] DNA (in gel matrix)

Stage 3: Washing and drying the gel

[methyl³H] DNA (in gel matrix)

Stage 4: Autoradiography

dark bands on x-ray film[VIS]

Reaction Mixture[1]

A. 25 m*M* Tris–HCl buffer, pH 7.5
    10 μl/ml [Methyl-³H] *S*-adenosyl-L-methionine (1 mCi/ml)
    20 m*M* EDTA
    1 μ*M S*-Adenosyl-L-methionine
B. 1 mg/ml Proteinase K (Boehringer)
    1% Sodium dodecyl sulfate (SDS)
C. 5% Trichloroacetic acid
    1% Sodium pyrophosphate
D. 100% Acetic acid
E. 20% 2,5-Diphenyloxazole (dissolved in 100% acetic acid)

Procedure

Rinse the electrophorized PAG containing 0.1% SDS and 66 mg/ml DNA (mol wt >5 · 10⁶) with water, and then wash twice for 15 min in renaturing solution: 50 m*M* Tris–HCl (pH 7.5), 10 m*M* EDTA, and 10 m*M* 2-mercaptoethanol. Place the gel in renaturing solution at 0°C and wash with gentle shaking for 12 h. During this period change the renaturing solution three times.

Stage 1. After renaturing, incubate the gel in solution A at 37°C for 16 h with gentle shaking.

Stage 2. Remove the gel from solution A and place in solution B for 2 h at 37°C (this treatment of the gel is carried out for proteolytic inactivation of nucleases). Then place the gel in solution C and wash for 12 h. During the washing change solution C three times. Finally, put the gel in solution D for 30 min and then in solution E for 30 min.

Stage 3. Wash the gel 1 h in warm water, place it on a sheet of Whatman 3MM chromatographic paper, and dry.

Stage 4. Put the dry gel in tight contact with medical x-ray film and expose for an appropriate period of time. Develop the exposed x-ray film.

The enzyme activity zones appear as dark bands on the developed film.

Notes: The activity blotting technique may be adapted to detect activity bands of DNAMT on native DNA-fixed nitrocellulose membrane using [methyl-³H] *S*-adenosyl-L-methionine.[2]

References

1. Hübscher, U., Pedrali-Noy, G., Knust-Kron, B., Doerfler, W., and Spadari, S., DNA methyltransferases: acting minigel analysis and determination with DNA covalently bound to a solid matrix, *Anal. Biochem.*, 150, 442, 1985.
2. Seki, S., Akiyama, K., Watanabe, S., and Tsutsui, K., Activity gel and activity blotting methods for detecting DNA-modifying (repair) enzymes, *J. Chromatogr.*, 618, 147, 1993.

2.1.1.63 — Methylated-DNA–Protein-Cysteine S-Methyltransferase; MDPCMT

OTHER NAMES	O^6-Methylguanine-DNA methyltransferase, O-6-methylguanine-DNA transmethylase
REACTION	DNA (containing 6-O-methylguanine) + protein L-cysteine = DNA (without 6-O-methylguanine) + protein S-methyl-L-cysteine (see General Notes)
ENZYME SOURCE	Bacteria, vertebrates
SUBUNIT STRUCTURE	Unknown or no data available

METHOD

Visualization Scheme

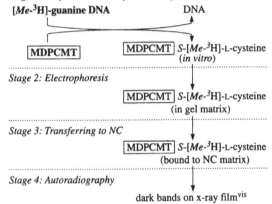

Stage 1: Enzyme reaction before electrophoresis

Stage 2: Electrophoresis

Stage 3: Transferring to NC

Stage 4: Autoradiography

Visualization Procedure[1]

Stage 1. Incubate 200 µl of cell or tissue extract prepared using 50 mM Tris–HCl (pH 8.3), 0.5 mM EDTA, and 1 mM dithiothreitol buffer with 10 µl of [Me-^{3}H]-guanine DNA substrate (200 to 500 dpm/µl) at 37°C for 90 min (although the majority of the repair reaction occurs within the first 5 min, the longer period of incubation is used to ensure completeness of reaction).

Stage 2. Subject 30 µl of the enzyme reaction mixture containing the radiolabeled enzyme, MDPCMT S-[Me-^{3}H]-L cysteine, to SDS-PAG electrophoresis according to the method of Laemmli.[2]

Stage 3. Electroblot the electrophorized gel onto a nitrocellulose membrane (NC; 0.1 µm; from Schleicher and Schuell) by the method of Towbin et al.[3] by using a 20-h electrotransfer at 4°C, but half the suggested strength of transfer buffer to lessen heating.

Stage 4. Apply the dry NC membrane to Fuji RX film and expose for 2 weeks at –75°C.

Notes: The substrate, DNA containing O^6-methylguanine with a radiolabel in the methyl group, should be prepared at laboratory conditions as described.[1]

To increase the sensitivity of ^{3}H detection, preexpose the x-ray film to a filtered flashlight to an absorbance of 0.4 at 540 nm before exposure to the NC blot.

Radiolabeled MDPCMT bands can be detected directly in the electrophorized PAG; however, the time of exposure of the gel to x-ray film should be increased several times.

GENERAL NOTES

The enzyme is involved in the repair of alkilated DNA in *E. coli*. It acts only on the alkilated DNA as an adaptive response to alkilating agents.[4]

MDPCMT transfers the methyl group from guanine to an internal cysteine residue and restores native guanine in DNA. The formation of S-methylcysteine causes the inactivation of the enzyme, which thus acts only once, and is not regenerated.[5]

REFERENCES

1. Major, G.N., Gardner, E.J., and Lawley, P.D., Direct assay for O^6-methylguanine-DNA methyltransferase and comparison of detection methods for the methylated enzyme in polyacrylamide gels and electroblots, *Biochem. J.*, 277, 89, 1991.
2. Laemmli, U.K., Cleavage of structural proteins during the assembly of the head of bacteriophage T4, *Nature*, 227, 680, 1970.
3. Towbin, H., Staehelin, T., and Gordon, J., Electrophoretic transfer of proteins from polyacrylamide gels to nitrocellulose sheets: procedure and some applications, *Proc. Natl. Acad. Sci. U.S.A.*, 76, 4350, 1979.
4. NC-IUBMB, *Enzyme Nomenclature*, Academic Press, San Diego, 1992, p. 164 (EC 2.1.1.63, Comments).
5. Pegg, A.E., Foote, R.S., Mitra, S., Perry, W., and Wiest, L., Purification and properties of O-6-methylguanine-DNA transmethylase from rat liver, *J. Biol. Chem.*, 258, 2327, 1983.

2.1.3.2 — Aspartate Carbamoyltransferase; ACT

OTHER NAMES
Carbamylaspartotranskinase, aspartate transcarbamylase

REACTION
Carbamoyl phosphate + L-aspartate = orthophosphate + N-carbamoyl-L-aspartate

ENZYME SOURCE
Bacteria, fungi, plants, vertebrates

SUBUNIT STRUCTURE
Unknown or no data available

METHOD

Visualization Scheme

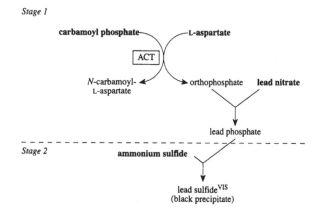

Staining Solution[1]

A. 0.5 M Tris–HCl buffer, pH 7.0   95 ml
    Carbamoyl phosphate (Li$_2$ salt)   50 mg
    L-Aspartic acid (Na salt)   100 mg
    2% Lead nitrate (aqueous solution)   5 ml
B. 1% Ammonium sulfide

Procedure

Incubate the electrophorized gel in freshly prepared filtered solution A for 10 to 20 min at room temperature. Discard solution A, rinse the gel in water, immerse it in solution B, and wash it again with water. The black zones indicate ACT activity.

OTHER METHODS

Three other methods for detection of the product orthophosphate may also be used:

A. A method that involves the formation of white calcium phosphate precipitate (e.g., see 3.1.3.2 — ACP, Method 5). This method is applicable only for clean gels (e.g., PAG).

B. A method that depends on the generation of blue phosphomolybdic acid (e.g., see 3.6.1.1 — PP, Method 1).

C. A method that involves the formation of blue formazan as a result of two coupled reactions catalyzed by the auxiliary enzymes purine-nucleoside phosphorylase and xanthine oxidase (e.g., see 3.1.3.1 — ALP, Method 8).

GENERAL NOTES

The enzymatic method (C) is preferable due to its sensitivity and generation of a nondiffusible formazan as the result of a one-step procedure.

REFERENCES

1. Shanley, M.S., Foltermann, K.F., O'Donovan, G.A., and Wild, J.J.R., Properties of hybrid aspartate transcarbamoylase formed with native subunits from divergent bacteria, *J. Biol. Chem.*, 259, 12672, 1984.

2.1.3.3 — Ornithine Carbamoyltransferase; OCT

OTHER NAMES	Citrulline phosphorylase, ornithine transcarbamylase
REACTION	Carbamoyl phosphate + L-ornithine = orthophosphate + L-citrulline
ENZYME SOURCE	Bacteria, fungi, plants, vertebrates
SUBUNIT STRUCTURE	Trimer# (bacteria, fungi, plants, vertebrates)

METHOD 1

Visualization Scheme

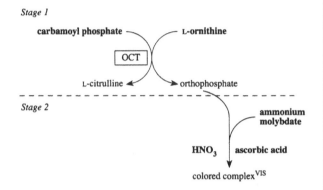

Staining Solution[1]

  A. 0.27 *M* Triethanolamine buffer, pH 7.5
      5 m*M* L-Ornithine
      15 m*M* Carbamoyl phosphate
  B. 20 m*M* Ammonium molybdate
      0.5% Nitric acid
  C. 10% Ascorbic acid

Procedure

Incubate the electrophorized gel in solution A at 37°C for 10 min. Rinse the gel in distilled water and place in solution B for 5 min. Rinse the gel again in water and immerse in solution C. Blue bands of OCT activity appear after a few minutes. Record the zymogram or photograph because the stained bands diffuse rapidly.

Notes: Malachite Green (oxalate salt or hydrochloride) may be additionally included in an ammonium molybdate solution to result in formation of a Malachite Green–phosphomolybdate colored complex.[2] This complex is more stable than the routine phosphomolybdate complex. It should be kept in mind that the use of citrate buffers for electrophoresis or staining prevents formation of the Malachite Green–phosphomolybdate colored complex.

METHOD 2

Visualization Scheme

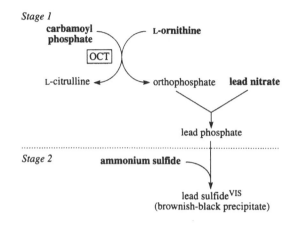

Staining Solution[3] (modified)

  A. 0.27 *M* Triethanolamine buffer, pH 7.5
      15 m*M* Carbamoyl phosphate
      5 m*M* L-Ornithine
      12 m*M* $Pb(NO_3)_2$
  B. 5% $(NH_4)_2S$

Procedure

Incubate the gel at room temperature in freshly prepared filtered solution A for 15 to 30 min or until bands of white precipitation (visible in translucent gels, e.g., PAG) appear. Discard solution A, rinse the gel in tap water for 30 min, place it in solution B for 2 min, and wash it again with water. Black bands correspond to localization of OCT activity in the gel. Wash the stained gel in 10 m*M* sodium acetate (pH 4.0) and store in 50% glycerol.

OTHER METHODS

Two other methods for the detection of the product orthophosphate may also be used:

  A. A method that involves the formation of white calcium phosphate precipitate (e.g., see 3.1.3.2 — ACP, Method 5) visible in clean gels
  B. A method that involves the formation of blue formazan as a result of two coupled reactions catalyzed by auxiliary enzymes purine-nucleoside phosphorylase and xanthine oxidase (e.g., see 3.1.3.1 — ALP, Method 8)

GENERAL NOTES

The enzymatic method (Other Methods, B) is preferable due to its sensitivity and generation of a nondiffusible formazan as the result of a one-step procedure.

2.1.3.3 — Ornithine Carbamoyltransferase; OCT (continued)

REFERENCES

1. Farkas, D.H., Skombra, C.J., Anderson., G.R., and Hughes, R.G., Jr., *In situ* staining procedure for the detection of ornithine transcarbamylase activity in polyacrylamide gels, *Anal. Biochem.*, 160, 421, 1987.
2. Zlotnick, G.W. and Gottlieb, M., A sensitive staining technique for the detection of phosphohydrolase activities after polyacrylamide gel electrophoresis, *Anal. Biochem.*, 153, 121, 1986.
3. Tsuji, S., Chicken ornithine transcarbamylase: purification and some properties, *J. Biochem.*, 94, 1307, 1983.

2.2.1.1 — Transketolase; TKET

OTHER NAMES	Glycolaldehydetransferase
REACTION	Sedoheptulose 7-phosphate + D-glyceraldehyde-3-phosphate = D-ribose 5-phosphate + D-xylulose 5-phosphate
ENZYME SOURCE	Bacteria, fungi, plants, vertebrates
SUBUNIT STRUCTURE	Dimer# (fungi — yeasts, mammals)

METHOD 1

Visualization Scheme

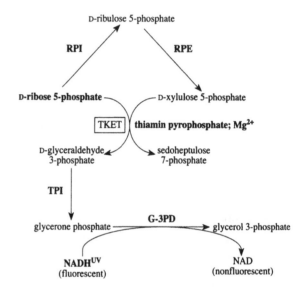

Staining Solution[1]

0.1 *M* Glycyl–glycine buffer, pH 7.4	10 ml
Ribose 5-phosphate (Na$_2$ salt)	5 mg
Thiamin pyrophosphate	1 mg
1 *M* MgCl$_2$	0.2 ml
NADH	4 mg
Ribose 5-phosphate isomerase (RPI)	50 U
Ribulose-phosphate 3-epimerase (RPE)	5 U
Glycerol-3-phosphate dehydrogenase (G-3PD) and triose-phosphate isomerase (TPI) mixture (Sigma)	40 μl

Procedure

Apply the staining solution to the gel on a filter paper overlay. Incubate the gel at 37°C for 1 to 2 h and view under long-wave UV light. Dark (nonfluorescent) bands of TKET activity are visible on a light (fluorescent) background. Record the zymogram or photograph using a yellow filter.

2.2.1.1 — Transketolase; TKET (continued)

Notes: When a zymogram that is visible in daylight is required, counterstain the processed gel with MTT–PMS solution. White bands of TKET appear on a blue background almost immediately. Wash the stained gel in water and fix in 25% ethanol.

When D-xylulose 5-phosphate is available, it may be included in the staining solution instead of ribose 5-phosphate isomerase and ribulose-phosphate 3-epimerase preparations.

METHOD 2

Visualization Scheme

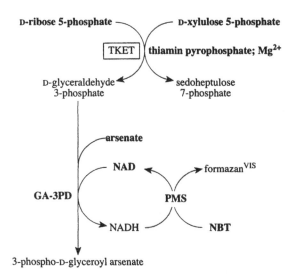

Staining Solution[2]

A. 100 mM Glycyl–glycine buffer, pH 7.4
0.4 mM D-Xylulose 5-phosphate
4 mM D-Ribose 5-phosphate
0.5 mM NAD
0.16 mM Sodium arsenate
0.76 mM EDTA
5 mM MgCl$_2$
0.6 mM Thiamin pyrophosphate
0.4 mM NBT
0.1 mM PMS
24 U/ml Glyceraldehyde-3-phosphate dehydrogenase (GA-3-PD)

B. 1% Agarose solution (60°C)

Procedure

Mix equal volumes of A and B components of the stain to prepare 0.5% indicator agarose solution. Pour the mixture over the surface of the gel. After agarose solidification, incubate the gel with the indicator agarose plate at 37°C in the dark until blue bands appear. Fix the stained indicator agarose plate in 25% ethanol.

Notes: This method was developed for detection of TKET after electrophoresis in 7.5% PAG. However, there are no obvious restrictions for application of the method for starch or acetate cellulose gels. When 7.5% PAG is used for TKET electrophoresis, the use of 9% indicator starch gel in place of 0.5% indicator agarose gel results in faster development of TKET activity bands. In both cases, colored bands appear in indicator gels because the auxiliary glyceraldehyde-3-phosphate dehydrogenase is unable to penetrate the 7.5% PAG, and the staining of an indicator gel is due to TKET diffusion out of 7.5% PAG. The use of 5.5% PAG for TKET electrophoresis does not aid in the detection of TKET activity but markedly impairs the resolution of TKET bands.

To prevent possible nonspecific development of alcohol dehydrogenase activity bands, the phosphosugars, NAD, NBT, and PMS should be prepared alcohol-free by lyophilization.

GENERAL NOTES

Transketolase–D-glyceraldehyde-3-phosphate dehydrogenase complexes are known to exist in yeasts and mammalian erythrocytes. This information should be taken into account during the interpretation of multiple TKET bands detected after electrophoresis of the enzyme preparations obtained from these organisms.

The yeast and human enzymes are dimers with the active sites located at the interface between the two identical subunits.[3,4]

REFERENCES

1. Anderson, J.E., Teng, Y.-S. and Giblett, E.R., Stains for six enzymes potentially applicable to chromosomal assignment by cell hybridization, *Cytogenet. Cell Genet.*, 14, 465, 1975.
2. Wood, T. and Muzarriri, C.C., The electrophoresis and detection of transketolase, *Anal. Biochem.*, 118, 221, 1981.
3. Meshalkina, L., Nilsson, U., Wikner, C., Kostikowa, T., and Schneider, G., Examination of the thiamin diphosphate binding site in yeast transketolase by site-directed mutagenesis, *Eur. J. Biochem.*, 244, 646, 1997.
4. Wang, J.J.L., Martin, P.R., and Singleton, C.K., Aspartate 155 of human transketolase is essential for thiamine diphosphate magnesium binding, and cofactor binding is required for dimer formation, *Biochim. Biophys. Acta*, 1341, 165, 1997.

OTHER NAMES	Dihydroxyacetonetransferase, glyceronetransferase
REACTION	Sedoheptulose 7-phosphate + D-glyceraldehyde-3-phosphate = D-erythrose 4-phosphate + D-fructose-6-phosphate
ENZYME SOURCE	Bacteria, fungi, plants, vertebrates
SUBUNIT STRUCTURE	Dimer[#] (bacteria – *E. coli*)

Method 1

Visualization Scheme

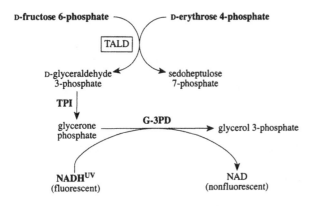

Staining Solution[1]

40 mM Tris–citrate buffer, pH 7.6	5 ml
D-Fructose-6-phosphate	6.6 mg
D-Erythrose 4-phosphate	3.3 mg
NADH	4 mg
Glycerol-3-phosphate dehydrogenase (G-3PD) and triose-phosphate isomerase (TPI) mixture (Sigma)	40 μl

Procedure

Apply the staining solution to the gel on a filter paper overlay. Incubate the gel with application in a moist chamber at 37°C for 1 to 2 h and view under long-wave UV light. Dark (nonfluorescent) bands of TALD are visible on a light (fluorescent) background. Record the zymogram or photograph using a yellow filter.

Notes: When a zymogram that is visible in daylight is required, counterstain the processed gel with MTT–PMS solution. White bands of TALD appear on a blue background almost immediately. Wash the negatively stained gel in water and fix in 25% ethanol.

Method 2

Visualization Scheme

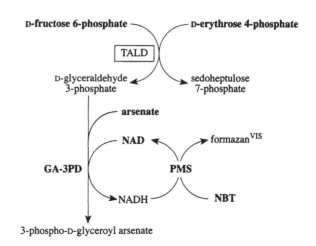

Staining Solution[2]

A. 100 mM Triethanolamine buffer, pH 7.4
1 mM EDTA
0.16 mM Sodium arsenate
3 mM D-Fructose-6-phosphate
0.4 mM D-Erythrose 4-phosphate
0.5 mM NAD
0.4 mM NBT
0.1 mM PMS
16 U/ml Glyceraldehyde-3-phosphate (GA-3PD) dehydrogenase
B. 1% Agarose solution (60°C)

Procedure

Mix equal volumes of A and B components of the staining solution and pour the mixture over the surface of the gel. Incubate the gel with agarose application at 37°C in the dark until blue bands appear. Fix the stained agarose application in 25% ethanol.

Notes: When 7.5% PAG is used for electrophoresis, colored bands of TALD activity appear only in the agarose plate. This is because the auxiliary glyceraldehyde-3-phosphate dehydrogenase is unable to penetrate the 7.5% PAG, and the staining of the agarose plate is due to TALD diffusion out of the 7.5% PAG.

To prevent nonspecific development of alcohol dehydrogenase activity bands, the phosphosugars, NAD, NBT, and PMS should be prepared alcohol-free by lyophilization.

References

1. Anderson, J.E., Teng, Y.-S., and Giblett, E.R., Stains for six enzymes potentially applicable to chromosomal assignment by cell hybridization, *Cytogenet. Cell Genet.*, 14, 465, 1975.
2. Wood, T. and Muzarriri, C.C., The electrophoresis and detection of transketolase, *Anal. Biochem.*, 118, 221, 1981.

2.3.1.85 — Fatty Acid Synthase; FAS

REACTION	Acetyl-CoA + n malonyl-CoA + 2n NADPH = long-chain fatty acid + (n + 1) CoA + n CO_2 + 2n NADP
ENZYME SOURCE	Bacteria, fungi, invertebrates, vertebrates
SUBUNIT STRUCTURE	Dimer[#] (vertebrates), see General Notes

METHOD

Visualization Scheme

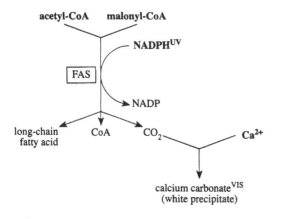

Staining Solution[1]

100 mM Tris–HCl buffer, pH 7.5
0.1 mM Acetyl-CoA
0.1 mM Malonyl-CoA
0.1 mM NADPH
100 mM Ca^{2+}

Procedure

After completion of electrophoresis in a 3.5% PAG, soak the gel in 0.1 M Tris–HCl buffer (pH 7.5) at 37°C for 20 to 30 min and transfer to the staining solution. Incubate the gel at room temperature or 37°C. View the PAG against a dark background and, when white bands of calcium carbonate precipitation of sufficient intensity are obtained, remove the gel from the staining solution and store in 50 mM glycine–KOH (pH 10), containing 5 mM Ca^{2+}, either at 5°C or at room temperature in the presence of an antibacterial agent.

Notes: Because of its very high molecular weight, FAS is not able to penetrate into a 7% PAG. Thus, the use of a 3.5% PAG is recommended for FAS electrophoresis.

OTHER METHODS

A. The product CoA-SH can reduce ferricyanide to ferrocyanide, which further reacts with mercury and yields a reddish brown precipitate (e.g., see 4.1.3.2 — MS).

B. The product CoA-SH can reduce tetrazolium salt via intermediary catalyst dichlorophenol indophenol, resulting in the formation of insoluble formazan (e.g., see 4.1.3.7 — CS).

GENERAL NOTES

The animal enzyme is a multifunctional protein that also catalyzes the reactions specified by EC 1.1.1.100, EC 1.3.1.10, EC 2.3.1.38, EC 2.3.1.39, EC 2.3.1.41, EC 3.1.2.14, and EC 4.2.1.61.[2]

The yeast enzyme is a macromolecular complex that consists of two different subunits.[3]

REFERENCES

1. Nimmo, H.G. and Nimmo, G.A., A general method for the localization of enzymes that produce phosphate, pyrophosphate, or CO_2 after polyacrylamide gel electrophoresis, *Anal. Biochem.*, 121, 17, 1982.

2. NC-IUBMB, *Enzyme Nomenclature*, Academic Press, San Diego, 1992, p. 189 (EC 2.3.1.85, Comments).

3. Kolodziej, S.J., Penczek, P.A., Schroeter, J.J.P., and Stoops, J.J.K., Structure-function relationships of the *Saccharomyces cerevisiae* fatty acid synthase: three-dimensional structure, *J. Biol. Chem.*, 271, 28422, 1996.

2.3.2.2 — Glutamyl Transpeptidase; GTP

OTHER NAMES	γ-Glutamyltransferase (recommended name), L-glutamyl transpeptidase, γ-glutamyl transpeptidase
REACTION	(5-L-Glutamyl)-peptide + amino acid = peptide + 5-L-glutamyl amino acid
ENZYME SOURCE	Bacteria, invertebrates, vertebrates
SUBUNIT STRUCTURE	Heterodimer[#] (bacteria, mammals)

METHOD 1

Visualization Scheme

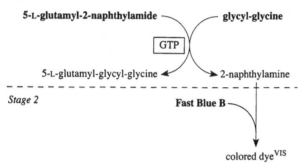

Stage 1

5-L-glutamyl-2-naphthylamide — glycyl-glycine

GTP

5-L-glutamyl-glycyl-glycine ← → 2-naphthylamine

Stage 2

Fast Blue B

colored dye[VIS]

Staining Solution[1]

A. 0.01 *M* Tris–0.05 *M* glycyl-glycine–NaOH buffer, pH 10.0 — 100 ml
B. 5-L-Glutamyl-2-naphthylamide — 100 mg
C. 0.05% Fast Blue B salt in 8% acetic acid

Procedure

Add the B component to solution A and dissolve by boiling for 15 min. The final pH of this mixture should be 9.5 at room temperature. Incubate the gel in this mixture at 37°C for 30 min and then place in solution C. Positively stained GTP bands appear after a few minutes.

Notes: The two-step staining procedure is recommended because of the presumed inhibitory effect of Fast Blue B on GTP activity.

METHOD 2

Visualization Scheme

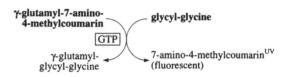

γ-glutamyl-7-amino-4-methylcoumarin — glycyl-glycine

GTP

γ-glutamyl-glycyl-glycine ← → 7-amino-4-methylcoumarin[UV] (fluorescent)

Staining Solution[2]

A. 24.2 mg/ml Tris–13.2 mg/ml glycyl-glycine buffer, pH 8.3
B. γ-Glutamyl-7-amino-4-methylcoumarin — 3 mg
   Methoxyethanol — 1 ml

Procedure

Prepare solution B by sonication. Mix 400 µl of solution B with 10 ml of solution A. Apply 1.5 ml of the resultant mixture to a cellulose acetate plate (second strip) previously kept for 30 min in Tris–barbital buffer (pH 8.6), and then store for about 10 min in the dark before use. After completion of electrophoresis, place the resolving cellulose acetate plate (first strip) on the second strip to form a "sandwich," exerting slight pressure to avoid the formation of air bubbles and to eliminate excess substrate. Incubate the sandwich in the dark at 37°C for 30 min. Separate the second strip and heat at 65°C for 3 to 4 min to stop the reaction. Observe fluorescent GTP bands in UV light (340 nm). Photograph the developed cellulose acetate plate using a yellow filter.

Notes: The fluorescence of 7-amino-4-methylcoumarin produced by the enzyme action is stable, in the dark, for at least 24 h.

GENERAL NOTES

Method 2 is sensitive, does not require a two-step staining procedure, and does not involve the use of hazardous reagents.

Neuraminidase treatment of GTP-containing preparations before electrophoresis produces a reduced anodal mobility of some human GTP isozymes, providing evidence that the association of sialic acid with GTP molecules is involved in the formation of secondary isozymes of the enzyme.[2]

REFERENCES

1. Patel, S. and O'Gorman, P., Demonstration of serum gamma-glutamyl transpeptidase isoenzymes, using Cellogel electrophoresis, *Clin. Chim. Acta*, 49, 11, 1973.
2. Sacchetti, L., Castaldo, G., and Salvatore, F., Electrophoretic behavior and partial characterization of disease-associated serum forms of gamma-glutamyltransferase, *Electrophoresis*, 10, 619, 1989.

REACTION	(5-L-Glutamyl)-L-amino acid = 5-oxoproline + L-amino acid (also see General Notes)
ENZYME SOURCE	Plants, vertebrates
SUBUNIT STRUCTURE	Monomer (vertebrates)

METHOD

Visualization Scheme

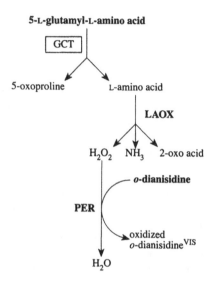

Staining Solution[1]

A.	0.05 *M* Tris–HCl buffer, pH 8.0	30 ml
	5-L-Glutamyl-L-methionine	
	(or 5-L-glutamyl-L-glutamine)	17 mg
	L-Amino acid oxidase (LAOX)	6 mg
	o-Dianisidine dihydrochloride	6 mg
	Peroxidase (PER)	12 mg
B.	2% Agar solution (60°C)	30 ml

Procedure

Mix A and B components of the staining solution and pour the mixture over the surface of the gel. Incubate the gel at room temperature or 37°C until brown bands appear. Fix the zymogram in 2 *M* acetic acid and store in 50% glycerol.

Notes: The use of 3-amino-9-ethyl-carbazole instead of *o*-dianisidine is preferable because the latter is a possible carcinogen.

GENERAL NOTES

The enzyme acts on derivatives of L-glutamate, L-2-aminobutanoate, L-alanine, and glycine.[2]

REFERENCES

1. Orlowski, M. and Meister, A., γ-Glutamyl cyclotransferase: distribution, isozymic forms, and specificity, *J. Biol. Chem.*, 248, 2836, 1973.
2. NC-IUBMB, *Enzyme Nomenclature*, Academic Press, San Diego, 1992, p. 199 (EC 2.3.2.4, Comments).

OTHER NAMES — Transamidase, transglutaminase, fibrin stabilizing factor, coagulation factor XIIIa, factor XIIIa, fibrinoligase

REACTION — Protein glutamine + alkylamine = protein N^5-alkylglutamine + NH_3 (see also General Notes)

ENZYME SOURCE — Bacteria, fungi, plants, invertebrates, vertebrates

SUBUNIT STRUCTURE — Dimer (vertebrates), see General Notes

METHOD

Visualization Scheme

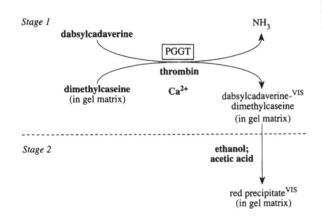

Staining Solution[1]

A. 0.05 M Tris–HCl buffer, pH 7.5
   0.2 mM Dabsylcadaverine
   20 mM $CaCl_2$
   2 mM Dithiothreitol
   14 U/ml Thrombin
B. 45% Ethanol in 10% acetic acid

Procedure

Incubate a 1% agarose electrophoretic gel containing 0.3% N,N'-dimethylcaseine in solution A at 37°C for 60 min, and then place the gel in solution B for 30 min. The zones of PGGT activity appear as red bands. To intensify positively stained bands, a few drops of diluted HCl can be added to solution B.

OTHER METHODS

A. The fluorescent activity staining method for transamidating enzymes is also available. The method is based on the transamidase-catalyzed incorporation of the fluorescent (at 254 nm) monodansylthiacadaverine or dansylcadaverine into casein.[2,3]

B. Histidine-tagged green fluorescent protein (His$_6$-Xpress-GFP; obtained by recombinant DNA technology) coupled with another protein substrate (e.g., casein) can also be used to detect PGGT activity bands on electrophoretic gels, exploiting the ability of transglutaminase to cross-link proteins through the Gln and Lys residues present in the short N-terminal extension (His$_6$-Xpress) of the His$_6$-Xpress-GFP.[4]

GENERAL NOTES

Transamidases comprise a group of Ca^{2+}-dependent thiol enzymes with the specialized function of catalyzing the formation of intermolecular γ-glutamyl-ε-lysine bridges between some native proteins. These enzymes also catalyze the incorporation of low-molecular-weight primary amines into proteins.[2]

It should be emphasized that for each new member of the transamidase family of enzymes, one should be prepared to modify conditions of the visualization procedure regarding time of enzyme reaction, pH value and concentration of staining buffers, use of activators, Ca^{2+} ion concentration, and selection of substrates other than caseine or N,N-dimethylcaseine, so as to obtain the best results.[3]

Good results can also be obtained when dimethylcaseine is added to the staining solution, but incorporation of this substrate into the gel matrix is preferable.

Thrombin is used to activate PGGT. Some other transglutaminases are not thrombin dependent.

Fibrin stabilizing factor is the precursor of the enzyme fibrinoligase. It is found in the plasma, platelets, prostate gland, placenta, uterus, and liver of mammals.[5] The fibrin stabilizing factor from plasma is comprised of two A subunits joined as a dimer and two B subunits. The B subunits do not have transglutaminase activity and possibly serve as a carrier molecule in plasma. The platelet fibrin stabilizing factor does not have B subunits and is the A_2 dimer.[6] The activation of fibrin stabilizing factor to the fibrinoligase form proceeds in two steps. The first step is cleavage of the small peptide (mol wt 4,000) from the A subunit by thrombin. The second step is activation by Ca^{2+} ions.[7] Active fibrinoligase generates intra- and intermolecular N^6-(5-glutamyl)-lysine cross-links. Transglutaminases also can form ester bonds between the Gln side chains of proteins and hydroxylipids. They stabilize protein or protein–lipid assemblies, and therefore are also known as cross-linking enzymes. In mammals, multiple transglutaminases dependent on Ca^{2+} are identified, which are encoded by separate genes, and are heterogenous in size and subunit structure. These transglutaminases undergo a number of posttranslational modifications and are involved in the performance of many important biological functions, including fibrin clot stabilization, cornified envelope formation, cataract formation, seminal vesicle coagulation, bone formation, cell growth and differentiation, apoptosis, celiac disease, maintenance of adherent junctions, lipid envelop formation, formation of the senile plagues in Alzheimer's disease, formation of the Lewy bodies in Parkinson's disease, and formation of a chemotactic factor.[4] It was found that some transglutaminases possess GTPase activity and take part in α-adrenergic receptor signaling.[8]

REFERENCES

1. Lorand, L., Parameswaran, K.N., Velasco, P.T., Hsu, L.K.-H., and Siefring, G.E., New colored and fluorescent amine substrates for activated fibrin stabilizing factor (factor XIIIa) and for transglutaminase, *Anal. Biochem.*, 131, 419, 1983.
2. Stenberg, P. and Stenflo J., A rapid and specific fluorescent activity staining procedure for transamidating enzymes, *Anal. Biochem.*, 93, 445, 1979.
3. Lorand, L., Siefring, G.E., Tong, Y.S., Bruner-Lorand, J., and Gray, A.J., Dansylcadaverine specific staining for transamidating enzymes, *Anal. Biochem.*, 93, 453, 1979.
4. Furutani, Y., Kato, A., Notoya, M., Ghoneim, M.A., and Hirose, S., A simple assay and hystochemical localization of transglutaminase activity using a derivative of green fluorescent protein as substrate, *J. Histochem. Cytochem.*, 49, 247, 2001.
5. Chung, S.I., Multiple molecular forms of transglutaminases in human and guinea pig, in *Isozymes I: Molecular Structure*, Markert, C.L., Ed., Academic Press, New York, 1975, p. 259.
6. Schwartz, M.L., Pizzo, S.V., Hill, R.L., and McKee, P.A., Human factor XIII from plasma and platelets, *J. Biol. Chem.*, 248, 1395, 1973.
7. Lorand, L., Gray, A.J., Brown, K., Credo, R.B., Cortis G.G., Domanik, R.A., and Stenberg, P., Dissociation of the subunit structure of fibrin stabilizing factor during activation of the zymogen, *Biochem. Biophys. Res. Commun.*, 56, 914, 1974.
8. Im, M.J., Russell, M.A. and Feng, J.F., Transglutaminase II: a new class of GTP-binding protein with new biological functions, *Cell. Signal.*, 9, 477, 1997.

OTHER NAMES	Muscle phosphorylase *a* and *b*, amylophosphorylase, polyphosphorylase (the recommended name is qualified in each instance by adding the name of the natural substrate, e.g., maltodextrin phosphorylase, starch phosphorylase, glycogen phosphorylase.)
REACTION	(1,4-α-D-Glucosyl)$_n$ + orthophosphate = (1,4-α-D-glucosyl)$_{n-1}$ + α-D-glucose 1-phosphate
ENZYME SOURCE	Bacteria, green algae, plants, invertebrates, vertebrates
SUBUNIT STRUCTURE	Dimer (plants), see General Notes

METHOD 1

Visualization Scheme

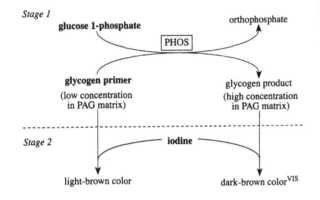

Staining Solution[1]

A. 0.1 *M* Sodium citrate buffer, pH 5.0
   25 m*M* Glucose 1-phosphate
B. 10 m*M* I$_2$
   14 m*M* KI

Procedure

Incubate the electrophorized PAG (polymerized with 0.1% glycogen primer or 0.1% starch primer) in solution A at room temperature for 5 h and then place in solution B. The gel background is stained light brown (with glycogen primer) or light blue (with starch primer), while PHOS activity bands are stained dark brown or dark blue, respectively, due to an increased concentration of glycogen or starch.

Notes: In some cases (e.g., when plant leaf preparations are used as the enzyme source), additional positively stained bands can also develop. These bands also develop when gels are incubated in solution A lacking glucose 1-phosphate and perhaps are caused by the action of the debranching enzyme (see 3.2.1.41 — DEG), which hydrolyzes the α-1,6-branch linkages of the glycogen and the starch (amylopectin form),

resulting in a greater capacity of the unit chains to adopt the helical configuration for iodine complex formation.

The colorless bands of amylase activity can also develop on gels stained using this method (see 3.2.1.1 — α-AMY, Method 1).

In contrast to the unphosphorylated (*b*) forms of the muscle (M) and brain (B) isozymes of PHOS, the *b* form of the liver (L) PHOS isozyme is not activated by AMP but is stimulated by sodium fluoride. The substrate concentration and buffer pH are also involved in activation of the *b* form of the L isozyme. Two different substrate systems were used to discriminate between phosphorylated and unphosphorylated forms of the L PHOS isozyme expressed in the rat nervous system. System 1, composed of 0.1 *M* sodium acetate buffer (pH 5.9), 40 m*M* glucose 1-phosphate, and 3 m*M* AMP, was unable to detect the *b* form of the L isozyme. System 2, composed of 50 m*M* morpholinoethane sulfonic acid buffer (pH 6.8), 100 m*M* glucose 1-phosphate, 5 m*M* AMP, 50 m*M* NaF, 149 m*M* EDTA, and 0.1 m*M* dithiothreitol, was able to detect the *b* form of the L isozyme in addition to the M and B isozymes, irrespective of their phosphorylation state.[2]

Zones of activity of potato PHOS become visible as white bands after incubation of the electrophorized PAG (containing 1% starch) in 50 m*M* Tris–citrate buffer (pH 6.2), containing 2.5 m*M* dithiothreitol, 10 m*M* glucose 1-phosphate, and 5 m*M* EDTA at 37°C for 1 h.[3]

METHOD 2

Visualization Scheme

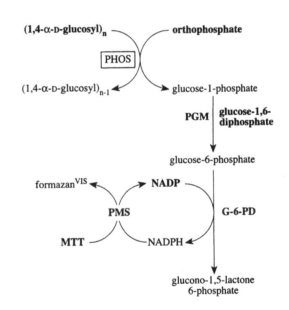

Staining Solution[4]

0.1 M Sodium phosphate buffer, pH 7.0	1 ml
Glycogen	10 mg
Glucose-6-phosphate dehydrogenase (G-6-PD)	5 U
Phosphoglucomutase (PGM)	5 U
Glucose-1,6-diphosphate	trace
NADP	2 mg
PMS	0.1 mg
MTT	0.4 mg

Procedure

Apply the staining solution dropwise to the surface of the gel and incubate in the dark at 37°C until dark blue bands appear. Fix the stained gel in 25% ethanol.

Notes: The stain is less expensive when NAD-dependent G-6-PD (Sigma G 5760 or G 5885) is used.[5]

METHOD 3

Visualization Scheme

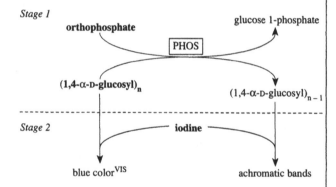

Stage 1

orthophosphate → glucose 1-phosphate

PHOS

(1,4-α-D-glucosyl)$_n$

(1,4-α-D-glucosyl)$_{n-1}$

Stage 2 —— iodine ——

blue color^{VIS}

achromatic bands

Staining Solution[6]

A. 0.1 M Sodium phosphate buffer, pH 5.1   100 ml
B. 10 mM I$_2$
    14 mM KI

Procedure

Incubate the electrophorized starch gel (or PAG containing 0.1 to 0.5% soluble starch or glycogen) in solution A at 37°C for 3 to 5 h. Discard solution A and stain the gel with solution B. The presence of achromatic to light brown bands on a blue background of starch gel marks areas of PHOS localization. When starch- or glycogen-containing PAGs are used, zones of PHOS activity are indicated by the presence of achromatic or light brown bands on a dark brown background. Record or photograph the zymogram.

Notes: Achromatic bands of amylase activity can also be developed by this method, so the control staining for amylase could be carried out in parallel (see 3.2.1.1 — α-AMY, Method 1).

GENERAL NOTES

It is known that the muscle enzyme exists in two forms: phosphorylase *a* (tetramer) and phosphorylase *b* (dimer). AMP activates both forms of the enzyme; in its absence, phosphorylase *a* possesses 60 to 70% of the maximum activity and phosphorylase *b* is inactive. Thus, the addition of AMP (about 0.25 mg/ml) to the staining solution is beneficial.

Results are obtained that provide evidence that ADPglucose can be directly used by starch phosphorylase as a substrate or metabolized to glucose 1-phosphate prior to incorporation into starch (see also 2.4.1.21 — STS, *Notes*). The mechanism of this unusual PHOS reaction remains uncertain.[3]

REFERENCES

1. Gerbrandy, S.J. and Verleur, J.D., Phosphorylase isoenzymes: localization and occurrence in different plant organs in relation to starch metabolism, *Phytochemistry*, 10, 261, 1971.
2. Pfeiffer-Guglielmi, B., Broer, S., Broer, A., and Hamprecht, B., Isozyme pattern of glycogen phosphorylase in the rat nervous system and rat astroglia-rich primary cultures: electrophoretic and polymerase chain reaction studies, *Neurochem. Res.*, 25, 1485, 2000.
3. Rammesmayer, G. and Praznik, W., Fast and sensitive simultaneous staining method of Q-enzyme, α-amylase, R-enzyme, phosphorylase and soluble starch synthase separated by starch: polyacrylamide gel electrophoresis, *J. Chromatogr.*, 623, 399, 1992.
4. Colgan, D.J., Evidence for the evolutionary significance of developmental variation in an abundant protein of orthopteran muscle, *Genetica*, 67, 81, 1985.
5. Black, W.C., IV, and Krafsur, E.S., Electrophoretic analysis of genetic variability in the house fly (*Musca domestica* L.), *Biochem. Genet.*, 23, 193, 1985.
6. Siepmann, R. and Stegemann, H., Enzym-elektrophorese in einschluß-polymerisaten des acrylamids.A. Amylasen, phosphorylasen, *Z. Naturforsch.*, 22, 949, 1967.

OTHER NAMES	Sucrose 6-glucosyltransferase
REACTION	Sucrose + $(1,6\text{-}\alpha\text{-}D\text{-glucosyl})_n$ = D-fructose + $(1,6\text{-}\alpha\text{-}D\text{-glucosyl})_{n+1}$
ENZYME SOURCE	Bacteria, green algae
SUBUNIT STRUCTURE	Heterotrimer[#] (bacteria)

METHOD 1

Visualization Scheme

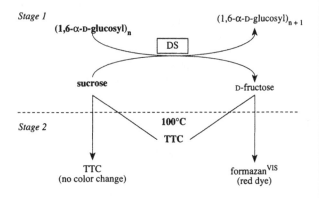

Staining Solution[1]

A. 0.2 *M* Sodium acetate buffer, pH 5.0   100 ml
   Sucrose   860 mg
   Dextran T10 (Pharmacia)   100 mg
B. 2,3,5-Triphenyltetrazolium chloride (TTC)   100 mg
   1 *N* NaOH   100 ml

Procedure

Incubate the electrophorized PAG in solution A for 30 min and dip into boiling solution B for 1 to 4 min until red-colored bands of DS appear. Too much heating causes the gel background to be stained pink. Immediately after the bands become well visible, wash and fix the gel in 7.5% acetic acid.

METHOD 2

Visualization Scheme

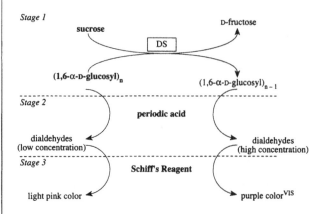

Staining Solution[1]

A. 0.2 *M* Sodium acetate buffer, pH 5.0   100 ml
   Sucrose   860 mg
   Dextran T10 (Pharmacia)   100 mg
   Merthiolate   10 mg
B. 1% Periodic acid (dissolved in ethanol) in 0.02 *M* sodium acetate
C. Schiff's reagent

Procedure

Incubate the electrophorized PAG in solution A for 15 h at 37°C. Treat the incubated gel with solution B, wash with distilled water, stain with Schiff's reagent, and wash with 7% acetic acid. The zones of DS activity develop as purple bands on a light pink background. Store the stained gel covered with Saran wrap.

Notes: When the gel is incubated in solution A, containing more dextran primer, the gel background is stained more intensely pink. Therefore, the addition of a minimum amount of a polysaccharide primer is recommended in this method.

The staining of proteins other than DS is a main disadvantage of this method. Artifactual staining of proteins is caused by the diffusion of the periodate-oxidized carbohydrate before and after staining with Schiff's reagent. The diffusion can be greatly slowed, and the staining artifact decreased, by following the stain by a cross-linking treatment of the carbohydrate–dye complex with cross-linking reagents (e.g., vinyl sulfone, formaldehyde, quinone, or 1,4-butanediol diglycidyl ether). This treatment can only be used after the initial destaining of the processed gel in 7.5% acetic acid. Protein staining artifacts can be prevented by using chymotrypsin (0.1% chymotrypsin in 50 m*M* Tris–HCl buffer (pH 7.8), containing 5 m*M* CaCl$_2$) to remove proteins from the gel at the stage after polysaccharide synthesis but before treating the gel with solutions B and C.[2]

The periodate-oxidized carbohydrate can also be stained using Alcian blue stain or the reductive dye-coupling procedure, which involves pararosaniline hydrochloride and NaBH$_3$CN.[2]

METHOD 3

Visualization Scheme

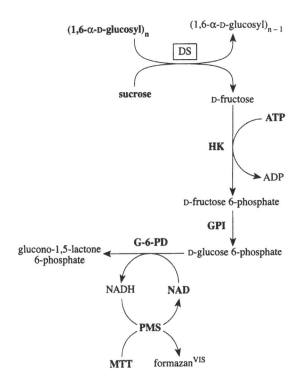

Staining Solution*

A. 0.005 M Sodium acetate buffer, pH 6.5   20 ml
    Sucrose   170 mg
    Dextran (low molecular weight)   50 mg
    ATP   30 mg
    NAD   20 mg
    MTT   10 mg
    PMS   1 mg
    Hexokinase (HK)   50 U
    Glucose-6-phosphate isomerase (GPI)   50 U
    Glucose-6-phosphate dehydrogenase (G-6-PD;
         NAD dependent; Sigma)   50 U
B. 2% Agarose solution (60°C)   20 ml

Procedure

Mix A and B components of the staining solution and pour the mixture over the surface of the gel. Incubate the gel at 37°C in the dark until dark blue bands appear. Fix the stained agarose overlay in 25% ethanol.

---

* New; recommended for use.

Notes: The main advantage of the method is the one-step staining of DS activity bands. This method is based on the ability of yeast HK to convert D-fructose to fructose-6-phosphate in the presence of ATP.

REFERENCES

1. Mukasa, H., Shimamura, A., and Tsumori, H., Direct activity stains for glycosidase and glucosyltransferase after isoelectric focusing in horizontal polyacrylamide gel layers, *Anal. Biochem.*, 123, 276, 1982.
2. Miller, A.W. and Robyt, J.F., Detection of dextransucrase and levansucrase on polyacrylamide gels by the periodic acid-Schiff stain: staining artifacts and their prevention, *Anal. Biochem.*, 156, 357, 1986.

2.4.1.7 — Sucrose Phosphorylase; SP

OTHER NAMES	Sucrose glucosyltransferase
REACTION	Sucrose + orthophosphate = D-fructose + α-D-glucose 1-phosphate
ENZYME SOURCE	Bacteria, fungi, green algae
SUBUNIT STRUCTURE	Monomer# (bacteria), dimer# (bacteria); see General Notes

METHOD 1

Visualization Scheme

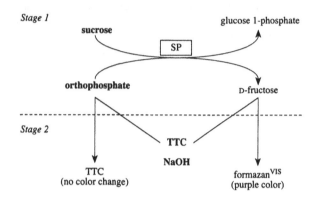

Staining Solution[1]

A. 3 mM Sodium phosphate buffer, pH 6.9
    200 mM Sucrose
B. 0.1% 2,3,5-Triphenyltetrazolium chloride (TTC)
    1 N NaOH

Procedure

Wash the electrophorized PAG in distilled water thoroughly and incubate in solution A at 30°C for 20 min. Wash the gel in distilled water again and place in solution B. Incubate the gel in this solution in the dark at room temperature until purple bands of SP activity appear on the pink background of the gel. Wash the stained gel in water and fix in 7.5% acetic acid.

Notes: This method is based on the capacity of D-fructose to reduce TTC at room temperature and alkaline conditions resulting in the formation of purple formazan.

METHOD 2

Visualization Scheme

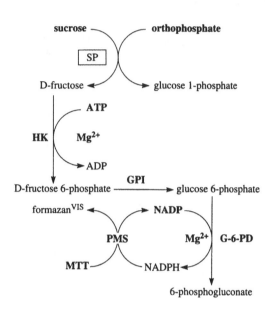

Staining Solution[2] (modified)

100 mM Phosphate buffer, pH 7.6
5 mg/ml Sucrose
2 U/ml Hexokinase (HK)
4 U/ml Glucose-6-phosphate isomerase (GPI)
1 U/ml Glucose-6-phosphate dehydrogenase (G-6-PD)
0.5 mg/ml NADP
1 mg/ml MgCl$_2$ (6H$_2$O)
1 mg/ml ATP
0.05 mg/ml PMS
0.5 mg/ml MTT

Procedure

Apply staining solution as a 1% agarose overlay and incubate the gel at 37°C in the dark until dark blue bands appear. Fix the stained gel or agarose overlay in 25% ethanol.

Notes: Activity bands of β-fructofuranosidase (see 3.2.1.26 — FF, Method 5) can also be detected by this method. Unlike SP, the bands of FF can also be developed in the staining solution lacking inorganic phosphate.

METHOD 3

Visualization Scheme

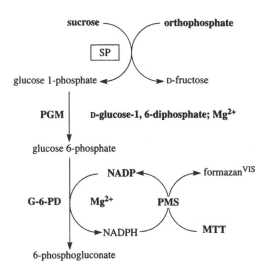

Staining Solution[2] (modified)

100 mM Phosphate buffer, pH 7.6
5 mg/ml Sucrose
10 μM D-Glucose-1,6-diphosphate
10 U/ml Phosphoglucomutase (PGM)
1 U/ml Glucose-6-phosphate dehydrogenase (G-6-PD)
0.5 mg/ml NADP
1 mg/ml $MgCl_2$ ($6H_2O$)
0.05 mg/ml PMS
0.5 mg/ml MTT

Procedure

Apply staining solution as a 1% agarose overlay and incubate the gel at 37°C in the dark until dark blue bands appear. Fix the stained gel or agarose overlay in 25% ethanol.

Notes: This method is specific for SP.

GENERAL NOTES

A dimeric SP was found in *Pseudomonas saccharophila*,[3] while the enzyme from *Leuconostoc mesenteroides* proved to be a monomer.[4]

REFERENCES

1. Gabriel, O. and Wang, S.-F., Determination of enzymatic activity in polyacrylamide gels: I. Enzymes catalyzing the conversion of nonreducing substrates to reducing products, *Anal. Biochem.*, 27, 545, 1969.
2. Mukasa, H., Tsumori, H., and Takeda, H., Renaturation and activity staining of glycosidases and glycosyltransferases in gels after sodium dodecyl sulfate–electrophoresis, *Electrophoresis*, 15, 911, 1994.
3. Silverstein, R., Voet, J., Reed, D., and Abeles, R.H., Purification and mechanism of action of sucrose phosphorylase, *J. Biol. Chem.*, 242, 1338, 1967.
4. Koga, T., Nakamura, K., Shirokane, Y., Mizusawa, K., Kitao, S., and Kikuchi, M., Purification and some properties of sucrose phosphorylase from *Leuconostoc mesenteroides*, *Agric. Biol. Chem.*, 55, 1805, 1991.

2.4.1.8 — Maltose Phosphorylase; MP

REACTION | Maltose + orthophosphate = D-glucose + β-D-glucose 1-phosphate
ENZYME SOURCE | Bacteria
SUBUNIT STRUCTURE | Monomer# (bacteria), dimer# (bacteria); see General Notes

METHOD

Visualization Scheme

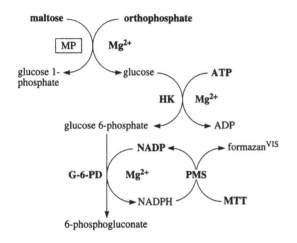

Staining Solution[1] (modified)

A. 0.2 *M* Potassium phosphate buffer, pH 7.2,
   containing 10 μM MgCl$_2$ — 14.5 ml
   40 m*M* NADP — 1 ml
   150 μM ATP — 1 ml
   0.2 U/μl Glucose-6-phosphate dehydrogenase
   (G-6-PD) — 0.5 ml
   0.2 U/μl Hexokinase (HK) — 0.5 ml
   100 m*M* Maltose — 2.5 ml
   PMS — 3 mg
   NBT — 60 mg
B. 1.5% Agarose solution (50°C) — 20 ml

Procedure

Mix A and B components of the staining solution and pour the mixture over the surface of the gel. Incubate the gel at 37°C in the dark until dark blue bands appear. Fix the stained gel or agarose overlay in 25% ethanol.

Notes: The specific activity of MP was found to be dependent on the carbon source used for cultivation of bacteria. However, the highest specific activity of the enzyme (0.18 U/mg protein) was observed for bacteria grown in media without any added sugar.[1]

OTHER METHODS

The product glucose 1-phosphate can be detected using linking enzymes phosphoglucomutase and glucose-6-phosphate dehydrogenase coupled with the PMS–MTT system (e.g., see 2.7.7.12 — GPUT).

GENERAL NOTES

The MP from *Lactococcus lactis* is a monomeric enzyme with a molecular mass of 75 kDa,[1] while the enzyme from *Lactobacillus brevis* is a dimer consisting of two similar subunits of 88 kDa each.[2] The dimeric subunit structure of MP was also detected in *Bacillus* sp.[3] A monomeric phosphorylase responsible for the phosphorolysis of maltose and the production of glucose and glucose 1-phosphate is found in the thermophile *Clostridium stercorarium*.[4]

REFERENCES

1. Nilsson, U. and Rädström, P., Genetic localization and regulation of the maltose phosphorylase gene, *malP*, in *Lactococcus lactis*, *Microbiology*, 147, 1565, 2001.
2. Hüwel, S., Haalck, L., Conrath, N., and Spener, F., Maltose phosphorylase from *Lactobacillus brevis*: purification, characterization, and application in a biosensor for *ortho*-phosphate, *Enz. Microb. Technol.*, 21, 413, 1997.
3. Inoue, Y., Ishii, K., Tomita. T., and Fukui, F., Purification and characterization of maltose phosphorylase from *Bacillus* sp. RK-1, *Biosci. Biotechnol. Biochem.*, 65, 2644, 2001.
4. Reichenbecher, M., Lottspeich, F., and Bronnenmeier, K., Purification and properties of a cellulose phosphorylase (CepA) and a cellodextrin phosphorylase (CepB) from the cellulolytic thermophile *Clostridium stercorarium*, *Eur. J. Biochem.*, 247, 262, 1997.

2.4.1.11 — Glycogen (Starch) Synthase; G(S)S

OTHER NAMES	UDPglucose–glycogen glucosyl-transferase (see also General Notes)
REACTION	UDPglucose + (1,4-α-D-glucosyl)$_n$ = UDP + (1,4-α-D-glucosyl)$_{n+1}$ (see General Notes)
ENZYME SOURCE	Fungi, plants, invertebrates, vertebrates
SUBUNIT STRUCTURE	Heterodimer[#] (invertebrates, vertebrates); see General Notes

METHOD 1

Visualization Scheme

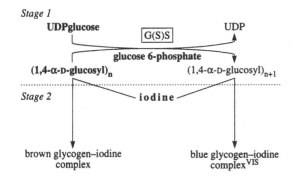

Staining Solution[1]

A. 50 mM Glycine–NaOH buffer, pH 8.6
10 mM UDPglucose
5 mM Dithiothreitol
20 mM EDTA
10 mM Glucose-6-phosphate
1% Rabbit liver glycogen

B. 0.1 M Acetate buffer, pH 4.7
0.26% Iodine
2.6% Potassium iodide

PROCEDURE

Incubate the electrophorized PAG overnight in the substrate solution (A) at 37°C. Rinse the gel with tap water and place in an iodine solution (B). The areas of G(S)S localization appear as blue bands on a brown background of the gel. The intensity of staining of G(S)S bands depends on the time of gel incubation in solution A (i.e., on the length of the synthesized polysaccharide molecules). Destain and store the developed gels in 7% acetic acid. Under this condition, gels retain color for several months. Upon restaining with iodine solution, gels recover the original intensity of the bands.

Notes: As low as 0.0075 U of G(S)S can be detected by the iodine method. Iodine staining of G(S)S activity bands is more intense when glucose-6-phosphate is included in the substrate solution (A).

METHOD 2

Visualization Scheme

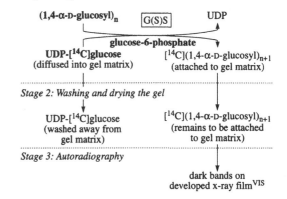

Reaction Mixture[1]

50 mM Glycine–NaOH buffer, pH 8.6
10 mM UDP-[^{14}C]glucose
5 mM Dithiothreitol
20 mM EDTA
10 mM Glucose-6-phosphate
1% Rabbit liver glycogen

Procedure

Stage 1. Incubate the electrophorized PAG overnight in the reaction mixture at 37°C for 10 to 24 h. During this stage, the growing glucan molecules grow into the gel matrix and become attached to the gel.

Stage 2. Rinse the gel with water to remove the radioactive UDP-[^{14}C]glucose. Dry the gel on Whatman 3MM chromatographic paper.

Stage 3. Autoradiograph the dry gel by exposing it to preflashed Kodak XR5 x-ray film at –80°C.

Dark bands on the developed x-ray film correspond to localization of G(S)S activity on the gel.

Notes: The sensitivity of in-gel detection of G(S)S activity is significantly improved by the use of labeled UDP-[^{14}C]glucose.

GENERAL NOTES

Both methods can be adapted to detect other sugar nucleotides: α-1,4-glucosyl transferases (e.g., see 2.4.1.21 — STS).[1]

The enzyme from animal tissues is a complex of a catalytic subunit and the protein glycogenin. For its activity the enzyme requires glucosylated glycogenin as a primer. The glucosylation of glycogenin is catalyzed by the glycogenin itself. Five molecules of glucose can be transferred to one molecule of glycogenin. The product, glucosylated glycogenin, acts as a primer for the reaction catalyzed by G(S)S. Thus, glycogenin is also known as a separate enzyme, glycogenin glucosyltransferase (EC 2.4.1.186).

2.4.1.11 — Glycogen (Starch) Synthase; G(S)S (continued)

The recommended name varies according to the source of the enzyme and the nature of its synthetic product.

Amylose or amylopectin could be also used as the exogenous primers. However, their water solubility is low, and solutions must be heated after thawing in order to solubilize completely the polysaccharides. In order to avoid heating (UDPglucose is heat labile), the use of glycogen as a primer is preferable.[1]

Animal G(S)S is known to exist in two forms, synthase D and synthase I. The more phosphorylated D form has little activity in the absence of the modifier, glucose-6-phosphate. The conversion of synthase D to synthase I is necessary to glycogen synthesis. There is another (intermediate) form of synthase in animals, the R form, which is derived from the dephosphorylation of synthase D and is the active species in the liver. These different forms of animal G(S)S display different catalytic properties, depending on a pH profile, the presence or absence of glucose-6-phosphate, and the concentration of UDPglucose.[2]

REFERENCES

1. Krisman, C.R. and Blumenfeld, M.L., A method for the direct measurement of glycogen synthase activity on gels after polyacrylamide gel electrophoresis, *Anal. Biochem.*, 154, 409, 1986.
2. Nuttall, F.Q. and Gannon, M.C., An improved assay for hepatic glycogen synthase in liver extracts with emphasis on synthase R, *Anal. Biochem.*, 178, 311, 1989.

2.4.1.16 — Chitin Synthase; CHS

OTHER NAMES	Chitin-UDP acetylglucosaminyl-transferase
REACTION	UDP-*N*-acetyl-D-glucosamine + [1,4-(*N*-acetyl-β-D-glucosaminyl)]$_n$ = UDP + [1,4-(*N*-acetyl-β-D-glucosaminyl)]$_{n+1}$
ENZYME SOURCE	Fungi, invertebrates
SUBUNIT STRUCTURE	Unknown or no data available

METHOD

Visualization Scheme

UDP-*N*-acetyl-D-glucosamine

CHS   phosphatidylserine; *N*-acetylglucosamine

UDP   chitin^{VIS}

Staining Solution[1]

30 mM Tris–HCl buffer, pH 7.5
32 mM *N*-Acetylglucosamine
4 mM Magnesium acetate
0.18 mg/ml Phosphatidylserine
1 mM UDP-*N*-acetyl-D-glucosamine

Procedure

Incubate the electrophorized PAG in a staining solution at 30°C for 12 h. The zones of CHS activity develop as opaque bands visible in daylight on a clean PAG background.

Notes: The enzyme catalyzes the synthesis of chitin directly from UDP-*N*-acetyl-D-glucosamine without the participation of a primer. The synthesis is markedly stimulated by *N*-acetylglucosamine. When the enzyme is solubilized, its properties are altered. Thus, the soluble enzyme requires a primer for activity, and it is not stimulated by *N*-acetylglucosamine. In the presence of phosphatidylserine the solubilized enzyme catalyzes the synthesis of chitin without the participation of a primer and is stimulated by *N*-acetylglucosamine.

GENERAL NOTES

The enzyme converts UDP-*N*-acetyl-D-glucosamine into UDP and chitin.[2]

REFERENCES

1. Kang, M.S., Elango, N., Mattia, E., Au-Young, J., Robbins, P.W., and Cabib, E., Isolation of chitin synthetase from *Saccharomyces cerevisiae*, *J. Biol. Chem.*, 259, 14966, 1984.
2. NC-IUBMB, *Enzyme Nomenclature*, Academic Press, San Diego, 1992, p. 204 (EC 2.4.1.16, Comments).

2.4.1.18 — 1,4-α-Glucan Branching Enzyme; GBE

OTHER NAMES	Branching enzyme, amylo-(1,4–1,6)-transglycosylase, Q-enzyme (see also General Notes)
REACTION	Transfers a segment of a 1,4-α-D-glucan chain to a primary hydroxyl group in a similar glucan chain (see General Notes)
ENZYME SOURCE	Bacteria, plants, invertebrates, vertebrates
SUBUNIT STRUCTURE	Unknown or no data available

METHOD

Visualization Scheme

Stage 1

Stage 2

Staining Solution[1] (modified)

A. 50 mM Tris–citrate buffer, pH 7.0
2 mM Ascorbic acid
B. 0.1 M Acetate buffer, pH 4.7
0.26% Iodine
2.6% Potassium iodide

Procedure

Incubate the electrophorized PAG containing 1% starch in solution A at room temperature for 4 h. Rinse the gel with distilled water and place in solution B for 15 min. The zones of GBE activity appear as sharp red bands on the blue stained background. Store the developed gels in 7% acetic acid.

Notes: GBE hydrolyzes α-(1 → 4)-linkages and forms α-(1 → 6)-linkages, thus converting amylose into amylopectin. This results in the shift of λ_{max} of the starch–iodine complex to shorter wavelengths (red color). The activity of α-amylase (see 3.2.1.1 — α-AMY, Method 1), visible as achromatic bands on a blue background of the gel, is also developed by this method. Similarly, activity bands of pullulanase (see 3.2.1.41 — DEG, Method 2, *Notes*) can also be developed on GBE zymograms. Pullulanase hydrolyzes the α-(1 → 6) bonds and leads to light blue bands on the blue background of the gel treated with iodine. The split-off chains are of a rather short length, and therefore are able to diffuse out of the PAG matrix.

GBE activity can also be detected with the naked eye as white bands visible after incubation of the gel in solution A, i.e., before iodine staining.

GENERAL NOTES

The enzyme hydrolyzes α-(1 → 4)-linkages following chain transfer to a primary hydroxyl group and the formation of α-(1 → 6)-linkages on suitable acceptor chains converts amylose into amylopectin. The recommended enzyme name varies depending on the nature of its synthetic product, glycogen or amylopectin, e.g., glycogen branching enzyme or amylopectin branching enzyme. The latter is commonly termed Q-enzyme.[1,2]

REFERENCES

1. Rammesmayer, G. and Praznik, W., Fast and sensitive simultaneous staining method of Q-enzyme, α-amylase, R-enzyme, phosphorylase and soluble starch synthase separated by starch: polyacrylamide gel electrophoresis, *J. Chromatogr.*, 623, 399, 1992.
2. NC-IUBMB, *Enzyme Nomenclature*, Academic Press, San Diego, 1992, p. 205 (EC 2.4.1.18, Comments).

2.4.1.20 — Cellobiose Phosphorylase; CBP

REACTION | Cellobiose + orthophosphate = α-D-glucose 1-phosphate + D-glucose
ENZYME SOURCE | Bacteria
SUBUNIT STRUCTURE | Monomer[#] (*Clostridium*)

Method 1

Visualization Scheme

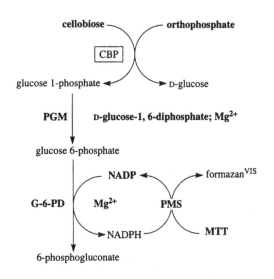

Staining Solution[1] (adapted, modified)

A. 100 mM Phosphate buffer, pH 7.5
   50 mM Cellobiose
   20 mM MgCl$_2$
   10 μM D-Glucose-1,6-diphosphate
   1 mM NADP
   0.5 U/ml Glucose-6-phosphate dehydrogenase (G-6-PD)
   0.5 U/ml Phosphoglucomutase (PGM)
   0.6 mM PMS
   2 mM MTT
B. 1.5% Agarose solution (50°C)

Procedure

Mix equal volumes of A and B components of the staining solution and pour the mixture over the surface of the gel. Incubate the gel at 37°C in the dark until dark blue bands appear. Fix the stained gel or agarose overlay in 25% ethanol.

Method 2

Visualization Scheme

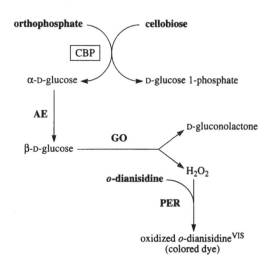

Staining Solution[1] (adapted, modified)

A. 100 mM Phosphate buffer, pH 7.5
   50 mM Cellobiose
   20 U/ml Glucose oxidase (GO)
   1.5 U/ml Peroxidase (PER)
   0.3 U/ml Aldose 1-epimerase (AE)
   0.3 mg/ml *o*-Dianisidine dihydrochloride (dissolved in a few drops of ethanol)
B. 1.5% Agarose solution (50°C)

Procedure

Mix equal volumes of A and B components of the staining solution and pour the mixture over the surface of the gel. Incubate the gel at 37°C until brown bands appear. Fix the stained gel or agarose overlay in 7% acetic acid.

Notes: Preparations of glucose oxidase and peroxidase should be catalase-free. If they are not, NaN$_3$ should be included in the stain to inhibit catalase activity.

2.4.1.20 — Cellobiose Phosphorylase; CBP (continued)

OTHER METHODS

The zones of D-glucose production can be visualized using auxiliary enzymes hexokinase and glucose-6-phosphate dehydrogenase coupled with the PMS–MTT system (e.g., see 3.2.1.28 — TREH, Method 1).

GENERAL NOTES

The enzyme is highly specific to cellobiose and cannot phosphorolyze any other glucosides or cellooligosaccharides.[2]

REFERENCES

1. Kitaoka, M., Aoyagi, C., and Hayashi, K., Colorimetric quantification of cellobiose employing cellobiose phosphorylase, *Anal. Biochem.*, 292, 163, 2001.
2. Alexander, J.K., Purification and specificity of cellobiose phosphorylase from *Clostridium thermocellum*, *J. Biol. Chem.*, 243, 2899, 1968.

2.4.1.21 — Starch Synthase; STS

OTHER NAMES	ADPglucose–starch glucosyltransferase (see also General Notes)
REACTION	ADPglucose + $(1,4$-α-D-glucosyl$)_n$ = ADP + $(1,4$-α-D-glucosyl$)_{n+1}$ (see *Notes*)
ENZYME SOURCE	Bacteria, fungi, plants
SUBUNIT STRUCTURE	Unknown or no data available

METHOD

Visualization Scheme

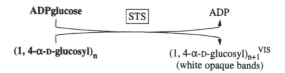

Staining Solution[1]

100 mM Glycine–NaOH buffer, pH 8.8
1 mM ADPglucose

Procedure

Rinse the electrophorized 7.5% PAG (containing 1% soluble starch) in water for 20 min and incubate in the staining solution at 37°C for 4 h. White bands of synthesized α-$(1 \rightarrow 4)$ chains are visible on the processed PAG.

Notes: White bands corresponding to phosphorylase activity (see 2.4.1.1 — PHOS) can also be developed by this method. It is suggested that phosphorylase can use ADPglucose as a substrate or ADPglucose can be metabolized to glucose 1-phosphate during the gel incubation procedure.[1] The mechanism of elongation of α-1,4-glucan chains by phosphorylase in the presence of ADPglucose remains unclear. To differentiate between STS and PHOS bands, an additional gel should be stained for PHOS activity (at 37°C for 1 h) using 50 mM Tris–HCl buffer (pH 6.2), containing 2.5 mM dithiothreitol, 10 mM glucose 1-phosphate, and 5 mM EDTA.[1]

OTHER METHODS

The product ADP can be detected using auxiliary enzymes pyruvate kinase and lactate dehydrogenase (e.g., see 2.7.1.30 — GLYCK, Method 3).

2.4.1.21 — Starch Synthase; STS (continued)

General Notes

The entry covers starch and glycogen synthases utilizing ADP-glucose. The recommended enzyme name varies according to the enzyme source and the nature of its synthetic product, e.g., starch synthase and bacterial glycogen synthase. The enzyme is similar to glycogen (starch) synthase (see 4.2.1.11 — G(S)S), but the mandatory nucleoside diphosphate sugar substrate is ADPglucose.[2]

References

1. Rammesmayer, G. and Praznik, W., Fast and sensitive simultaneous staining method of Q-enzyme, α-amylase, R-enzyme, phosphorylase and soluble starch synthase separated by starch: polyacrylamide gel electrophoresis, *J. Chromatogr.*, 623, 399, 1992.
2. NC-IUBMB, *Enzyme Nomenclature*, Academic Press, San Diego, 1992, p. 205 (EC 2.4.1.21, Comments).

2.4.1.90 — *N*-Acetyllactosamine Synthase; AGS

OTHER NAMES	UDPgalactose–*N*-acetylglucosamine β-D-galactosyltransferase, *N*-acetyl-galactosamine synthase, galactosyltransferase
REACTION	UDPgalactose + *N*-acetyl-D-glucosamine = UDP + *N*-acetyllactosamine
ENZYME SOURCE	Invertebrates, vertebrates
SUBUNITSTRUCTURE	Unknown or no data available

Method 1

Visualization Scheme

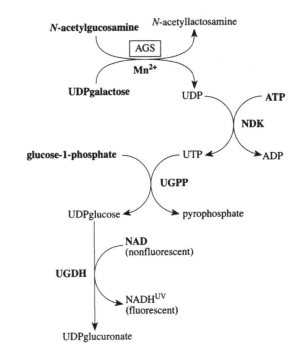

Staining Solution[1]

A. 0.1 *M* Sodium cacodylate buffer, pH 7.5   0.28 ml
    0.1 *M* MnCl$_2$ 0.5 ml
    25 U/ml Uridine-5'-diphosphoglucose
    Pyrophosphorylase (UGPP)   0.02 ml
    1000 U/ml Nucleoside-diphosphate
    Kinase (NDK)   0.02 ml
    2 U/ml UDPglucose dehydrogenase (UGDH)   0.02 ml
    200 mg/ml NAD   67 μl
    200 mg/ml ATP   0.05 ml
    200 mg/ml Glucose 1-phosphate   0.05 ml
    0.1 *M* UDPgalactose   0.02 ml
    5 *M N*-Acetylglucosamine   0.02 ml

B. 4% Agarose solution in 0.1 *M* sodium
    cacodylate buffer, pH 7.5 (60°C)

Procedure

Mix solution A with an equal volume of solution B (temperature of the mixture should be about 40°C) and pour over the surface of the gel. Incubate the gel at room temperature in a moist chamber for 4 to 15 h and view under long-wave UV light. The zones of AGS activity appear as light (fluorescent) bands on a dark (nonfluorescent) background. Record the zymogram or photograph using a yellow filter.

Notes: Inclusion of diaphorase and *p*-Iodonitrotetrazolium Violet in solution A allows one to develop AGS bands visible in daylight. This method, however, is less sensitive than the fluorescent one described above. Diaphorase cannot be replaced by PMS because the latter does not work as an electron acceptor at high Mn^{2+} concentrations.

Method 2

Visualization Scheme

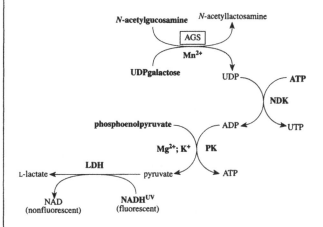

Staining Solution*

0.05 *M* Tris–HCl buffer, pH 7.5	5 ml
NADH	3 mg
ATP	5 mg
Phosphoenolpyruvate	10 mg
Nucleoside-diphosphate kinase (NDK)	10 U
Pyruvate kinase (PK)	20 U
Lactate dehydrogenase (LDH)	5 U
UDPgalactose	3 mg
N-Acetylglucosamine	10 mg
MnCl$_2$	10 mg
MgCl$_2$	10 mg
KCl	15 mg

Procedure

Apply the staining solution to the gel on a filter paper overlay and incubate at 37°C in a moist chamber until dark (nonfluorescent) bands visible in long-wave UV light appear on a light (fluorescent) background. Record the zymogram or photograph using a yellow filter.

Notes: Nucleoside-diphosphate kinase and ATP may be omitted from the staining solution because pyruvate kinase works sufficiently well with UDP.

---

* New; recommended for use.

OTHER METHODS

A. Bovine milk galactosyltransferase can be detected using an adaptation of the procedure with neoglyco-bovine serum albumin immobilized on nitrocellulose and radiolabeled UDPgalactose as substrates.[2] The enzyme should be transferred from the electrophoretic gel onto a nitrocellulose membrane containing immobilized neoglyco-bovine serum albumin, and the membrane should then be incubated in a buffered solution of radiolabeled UDPgalactose in the presence of manganese ions. After careful washing, the processed nitrocellulose membrane in 0.5 M EDTA, areas of radiolabeled neoglyco-bovine serum albumin, corresponding to localization of galactosyltransferase activity, can be developed by exposure of the membrane with x-ray film.

B. The enzyme from bovine milk can be transferred from the electrophoretic gel to a polyvinylidene difluoride (PVDF) membrane impregnated with substrate Lc$_3$Cer (GlcNAcβ1-3Galβ1-4Glcβ1-1Cer). After subsequent incubation of the membrane in a buffered solution containing UDPgalactose, areas of location of the enzyme can be detected using monoclonal antibodies specific to the product, nLc$_4$Cer, generated by AGS.[3] The method is based on the ability of the PVDF membrane to blot glycosphingolipids.[4] The method based on the use of a substrate-immobilized PVDF membrane for the detection of lipid-metabolizing enzymes was termed *Far Eastern blotting*.[5]

GENERAL NOTES

The AGS reaction is catalyzed by a component of EC 2.4.1.22, which is identical to EC 2.4.1.38, and by an enzyme from the Golgi apparatus of animal tissues. The EC 2.4.1.22 enzyme is a complex of two proteins A and B. In the absence of the B protein (α-lactalbumin), the enzyme catalizes the AGS reaction.[6]

REFERENCES

1. Pierce, M., Cummings, R.D., and Roth, S., The localization of galactosyltransferases in polyacrylamide gels by a coupled enzyme assay, *Anal. Biochem.*, 102, 441, 1980.
2. Parchment, R.E. and Shaper, J.H., Glycosyltransferases as probes for non-reducing terminal monosaccharide residues on nitrocellulose immobilized glycoproteins: the β(1–4)galactosyltransferase model, *Electrophoresis*, 8, 421, 1987.
3. Ishikawa, D., Kato, T., Handa, S., and Taki, T., New methods using polyvinylidene difluoride membranes to detect enzymes involved in glycosphingolipid metabolism, *Anal. Biochem.*, 231, 13, 1995.
4. Taki, T., Handa, S., and Ishikawa, D., Blotting of glycolipids and phospholipids from a high-performance thin-layer chromatogram to a polyvinylidene difluoride membrane, *Anal. Biochem.*, 221, 312, 1994.
5. Taki, T. and Ishikawa, D., TLC blotting: application to microscale analysis of lipids and as a new approach to lipid–protein interaction, *Anal. Biochem.*, 251, 135, 1997.
6. NC-IUBMB, *Enzyme Nomenclature*, Academic Press, San Diego, 1992, p. 205 (EC 2.4.1.22), p. 215 (EC 2.4.1.90).

2.4.2.1 — Purine-Nucleoside Phosphorylase; PNP

OTHER NAMES	Inosine phosphorylase, nucleoside phosphorylase
REACTION	Purine nucleoside + orthophosphate = purine + α-D-ribose 1-phosphate
ENZYME SOURCE	Bacteria, fungi, invertebrates, vertebrates
SUBUNIT STRUCTURE	Trimer (fungi, invertebrates, vertebrates)

METHOD

Visualization Scheme

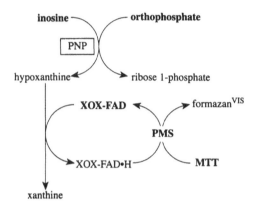

Staining Solution[1]

A. 0.05 M Phosphate buffer, pH 7.5   25 ml
    Inosine   5 mg
    MTT   5 mg
    PMS   5 mg
    Xanthine oxidase (XOX)   0.4 U
B. 2% Agar solution (60°C)   25 ml

Procedure

Mix A and B components of the staining solution and pour the mixture over the surface of the gel. Incubate the gel in the dark at 37°C until dark blue bands appear. Fix the stained agarose overlay in 25% ethanol.

Notes: The enzyme from animal species commonly exhibits stain artifacts consisting of one or two diffuse bands in the cathodal half of the gel. These bands do not represent endogenous enzyme and usually also develop on ADA (EC 3.5.4.4) and GDA (EC 3.5.4.3) zymograms.[2]

GENERAL NOTES

Specificity of this enzyme is not completely determined. It can also catalyze ribosyltransferase reactions specified by EC 2.4.2.5.[3]

REFERENCES

1. Edwards, Y.H., Hopkinson, D.A., and Harris, H., Inherited variants of human nucleoside phosphorylase, *Ann. Hum. Genet.*, 34, 395, 1971.
2. Richardson, B.J., Baverstock, P.R., and Adams, M., *Allozyme Electrophoresis: A Handbook for Animal Systematics and Population Studies*, Academic Press, Sydney, 1986, p. 204.
3. NC-IUBMB, *Enzyme Nomenclature*, Academic Press, San Diego, 1992, p. 236 (EC 2.4.2.1, Comments).

OTHER NAMES   Pyrumidine phosphorylase
REACTION   Uridine + orthophosphate = uracil + α-D-ribose 1-phosphate
ENZYME SOURCE   Bacteria, vertebrates
SUBUNIT STRUCTURE   Hexamer# (bacteria – *Escherichia coli*, *Salmonella typhimurium*)

METHOD

Visualization Scheme

Stage 1: Enzyme reaction

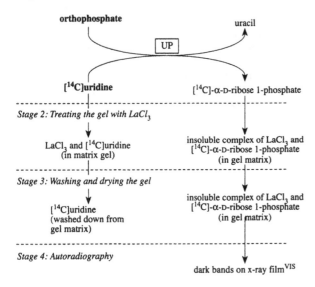

Reaction Mixture[1]

A. 0.2 *M* Potassium phosphate buffer, pH 7.5
   6 μCi [U-14C]Uridine (specific activity, 50 μCi/mmol) labeled on ribose

B. 0.1 *M* Tris–HCl buffer, pH 7.0
   0.1 *M* LaCl₃

Procedure

Stage 1. Incubate the electrophorized gel in solution A at 37°C for 2 h.
Stage 2. Place the gel in solution B and incubate at 4°C for 6 h.
Stage 3. Wash the gel for 12 h in deionized water. Dry the gel.
Stage 4. Expose the dry gel in tight contact with x-ray film (Kodak B-54) for 2 to 4 weeks. Develop exposed x-ray film.

The dark bands on the developed x-ray film correspond to localization of UP activity in the gel.

OTHER METHODS

The reaction catalyzed by UP is readily reversible. Thus, in principle, UP activity bands may be detected using histochemical methods of detection of orthophosphate (e.g., see 3.6.1.1 — PP), the product of the reverse UP reaction. This approach, however, has not yet been realized. Perhaps this is because D-ribose 1-phosphate, the substrate of the reverse UP reaction, is too expensive.

REFERENCES

1. Denney, R.M., Nichols, E.A., and Ruddle, F.H., Assignment of a gene for uridine phosphorylase to chromosome 7, *Cytogenet. Cell Genet.*, 22, 195, 1978.

OTHER NAMES	AMP pyrophosphorylase, transphosphoribosidase
REACTION	AMP + pyrophosphate = adenine + 5-phospho-α-D-ribose 1-diphosphate
ENZYME SOURCE	Bacteria, plants, invertebrates, vertebrates
SUBUNIT STRUCTURE	Dimer (vertebrates)

METHOD

Visualization Scheme

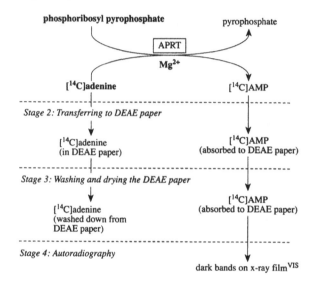

Stage 1: Enzyme reaction

phosphoribosyl pyrophosphate → pyrophosphate

APRT

Mg^{2+}

[^{14}C]adenine → [^{14}C]AMP

Stage 2: Transferring to DEAE paper

[^{14}C]adenine (in DEAE paper)

[^{14}C]AMP (absorbed to DEAE paper)

Stage 3: Washing and drying the DEAE paper

[^{14}C]adenine (washed down from DEAE paper)

[^{14}C]AMP (absorbed to DEAE paper)

Stage 4: Autoradiography

dark bands on x-ray film^{VIS}

Reaction Mixture[1]

55 mM Tris–HCl buffer, pH 7.4	10 ml
[^{14}C]Adenine (specific activity, 50 μCi/μmol; 50 μCi/ml) labeled on ribose	0.3 ml
0.2 M MgCl$_2$	2 ml
10 mM Phosphoribosyl pyrophosphate	1 ml

Procedure[2]

Stages 1 and 2. Apply a sheet of Whatman DE 81 (DEAE cellulose) paper to the surface of the electrophorized gel and pipette on the reaction mixture. Wrap the gel–DEAE paper combination in Saran wrap and incubate at 37°C for 1 h.

Stage 3. Remove the DEAE paper and wash on a Buchner funnel with 15 l of distilled water to remove the unreacted [^{14}C]adenine, leaving [^{14}C]AMP adsorbed to the DEAE paper at the sites of the enzyme reaction. Dry the DEAE paper.

Stage 4. Apply the DEAE paper to a sheet of x-ray film (Blue Brand, Kodak). Place the x-ray film–DEAE paper combination between two glass plates and wrap in a dark bag. Develop the x-ray film exposed for 4 days with a Kodak D 19 developer (Kodak Safelight 6B may be used).

The dark bands on the developed x-ray film correspond to localization of APRT activity on the gel.

Notes: Precipitation of [^{14}C]AMP with LaCl$_3$ directly in the gel matrix also may be used in visualization of APRT by the autoradiographic procedure.[3]

OTHER METHODS

A. A bioautographic method of visualization of APRT activity on electrophoretic gels was developed by Naylor and Klebe.[4] The principle of this method is the use of a microbial reagent to locate the enzyme after gel electrophoresis. The electrophorized gel is placed on an indicator agar seeded with mutant (pur A–) *E. coli*, which requires the product (adenine) of the APRT reaction. Bacteria proliferate to form visible bands at the sites of the enzymatic reaction. This method, however, is not appropriate for routine laboratory use and is recommended for visualization of enzymes that are difficult or impossible to detect with standard histochemical or other methods.

B. A general method for the localization of enzymes that produce pyrophosphate is available.[5] The method is based on insolubility of white calcium pyrophosphate precipitate, forming as a result of interaction between Ca^{2+} ions included in the reaction mixture and pyrophosphate molecules produced by an enzyme reaction. This method can also be applied to APRT electrophorized in clean gels (e.g., PAG or agarose gel). The zones of calcium pyrophosphate precipitation can be subsequently counterstained with Alizarin Red S. This procedure would be of advantage for starch or acetate cellulose gels.

C. The product of the reverse APRT reaction, AMP, can be detected histochemically using four linked reactions catalyzed by auxiliary enzymes 5'-nucleotidase, adenosine deaminase, purine-nucleoside phosphorylase, and xanthine oxidase coupled with the PMS–MTT system (see 2.7.4.3 — AK, Method 3).

D. The product AMP can also be detected in UV light using three linked reactions catalyzed by three auxiliary enzymes: adenylate kinase, pyruvate kinase, and lactate dehydrogenase (see 3.1.4.17 — CNPE, Method 1).

2.4.2.7 — Adenine Phosphoribosyltransferase; APRT (continued)

REFERENCES

1. Mowbray, S., Watson, B., and Harris, H., A search for electrophoretic variants of human adenine phosphoribosyl transferase, *Ann. Hum. Genet.*, 36, 153, 1972.
2. Harris, H. and Hopkinson, D.A., *Handbook of Enzyme Electrophoresis in Human Genetics*, North-Holland, Amsterdam, 1976 (loose-leaf, with supplements in 1977 and 1978).
3. Tischfield, J.A., Bernhard, H.P., and Ruddle, F.H., A new electrophoretic–autoradiographic method for visual detection of phosphotransferases, *Anal. Biochem.*, 53, 545, 1973.
4. Naylor, S.L. and Klebe, R.J., Bioautography: a general method for the visualization of isozymes, *Biochem. Genet.*, 15, 1193, 1977.
5. Nimmo, H.G. and Nimmo, G.A., A general method for the localization of enzymes that produce phosphate, pyrophosphate, or CO_2 after polyacrylamide gel electrophoresis, *Anal. Biochem.*, 121, 17, 1982.

2.4.2.8 — Hypoxanthine Phosphoribosyltransferase; HPRT

OTHER NAMES	Hypoxanthine–guanine phosphoribosyltransferase, IMP pyrophosphorylase, transphosphoribosidase, guanine phosphoribosyltransferase
REACTIONS	1. IMP + pyrophosphate = hypoxanthine + 5-phospho-α-D-ribose 1-diphosphate
	2. GMP + pyrophosphate = guanine + 5-phospho-α-D-ribose 1-diphosphate
ENZYME SOURCE	Bacteria, fungi, plants, invertebrates, vertebrates
SUBUNIT STRUCTURE	Tetramer (vertebrates); see General Notes

METHOD 1

Visualization Scheme

Stage 1: Enzyme reaction

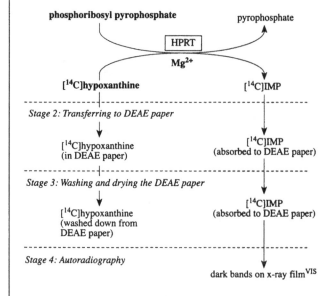

Stage 2: Transferring to DEAE paper

Stage 3: Washing and drying the DEAE paper

Stage 4: Autoradiography

Reaction Mixture[1]

55 mM Tris–HCl buffer, pH 7.4	10 ml
[^{14}C]Hypoxanthine (specific activity, 50 µCi/µmol; 50 µCi/ml)	0.3 ml
0.2 M MgCl$_2$	2 ml
10 mM Phosphoribosyl pyrophosphate	1 ml

Procedure

The procedure of HPRT detection by this method is similar to that described for APRT (see 2.4.2.7 — APRT).

Notes: [^{14}C]Hypoxanthine may be replaced by [^{14}C]guanine in the reaction mixture.[2]

Precipitation of [^{14}C]IMP with LaCl$_3$ directly in the gel matrix also may be used in visualization of HPRT by an autoradiographic procedure.[3]

METHOD 2

Visualization Scheme

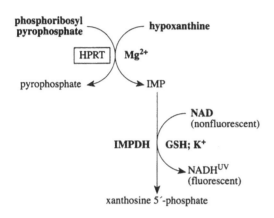

Staining Solution[4]

165 mM Tris–HCl buffer, pH 8.1
15 mM MgSO$_4$
300 mM KCl
6 mM Reduced glutathione (GSH)
3 mM Allopurinol
3 mM Hypoxanthine
2 mM Phosphoribosyl pyrophosphate
6 mM NAD
0.3 U/ml IMP dehydrogenase from *Aerobacter aerogenes* (IMPDH)

Procedure

Pour staining solution onto a clean glass plate (0.1 ml/10 cm^2 of gel to be stained). Place the electrophorized acetate cellulose sheet on the top of a glass plate with the porous side down (or starch gel with the cut surface down). Allow the staining solution to absorb completely into the gel matrix. Cover the gel with a thin plastic sheet to stop evaporation and incubate for 1 to 2 h at 37°C. View the gel under long-wave UV light for fluorescent bands. Record the zymogram or photograph using a yellow filter.

Notes: On prolonged incubation with staining solution, minor fluorescent bands in addition to the major HPRT bands can also appear. These bands also develop in a staining solution from which phosphoribosyl pyrophosphate and Mg^{2+} are lost. The appearance of some of these minor bands could be due to endogenous xanthine dehydrogenase or xanthine oxidase. Allopurinol is incorporated in a staining solution to inhibit these activities. Other minor bands can also sometimes develop in certain extracts used for electrophoresis. These bands can be eliminated

completely by heating the extracts for 2 min at 70°C. This pretreatment does not decrease the HPRT activity.[5]

The linking enzyme IMPDH is not commercially available; however, it can be relatively easily obtained from a guanine-requiring mutant of *Aerobacter aerogenes*.[6]

METHOD 3

Visualization Scheme

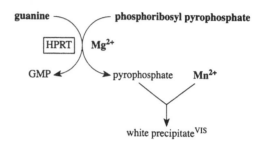

Staining Solution[7]

50 mM Tris–HCl buffer, pH 7.4
0.15 mM Guanine
1 mM Monothioglycerol
1 mM MgCl$_2$
1 mM MnCl$_2$
0.45 mM Phosphoribosyl pyrophosphate

Procedure

Wash the electrophorized PAG in 0.2 M Tris–HCl buffer (pH 7.4) for 15 min. Incubate the gel in staining solution at 37°C for about 45 to 60 min. The manganese pyrophosphate bands are visible due to light scattering. They are not apparent when the processed gel is placed directly on a dark background, but they are very distinct when the gel is held a few inches from a dark background and lighted indirectly. Submerge the stained gel in a solution of 0.01 M MnCl$_2$ in 0.02 M succinate (pH 6.0) for 30 to 60 min and store in 0.02 M succinate (pH 6.0).

Notes: The dense white band, which presumably arises from precipitation of anions that run at the dye-front, usually develops near the anodal end of the PAG.

The use of Ca^{2+} instead of Mn^{2+} ions in the staining solution is preferable because the white calcium pyrophosphate precipitate is clearly visible on a dark background.[8] The use of Ca^{2+} ions is also preferable because the zones of calcium pyrophosphate precipitation can be subsequently stained with Alizarin Red S. This procedure is of advantage for unclean gels (e.g., starch or acetate cellulose gels).

METHOD 4

Visualization Scheme

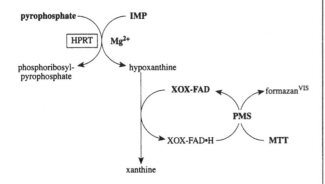

Staining Solution[9]

55 mM Tris–HCl buffer, pH 7.4, containing 8.3 mM

MgSO$_4$	8 ml
IMP (disodium salt)	50 mg
Xanthine oxidase (XOX)	0.1 U
22.5 mg/ml Na$_4$P$_2$O$_7$·10 H$_2$O	1 ml
2 mg/ml MTT	1 ml
0.4 mg/ml PMS	1 ml

Procedure

Pour staining solution over the surface of the gel, or apply it dropwise, and incubate the gel in the dark at 37°C until dark blue bands appear. Fix the stained gel in 25% ethanol.

GENERAL NOTES

Isozyme patterns of HPRT revealed at first in mammalian somatic cell hybrids suggested the dimeric subunit structure of the enzyme molecules.[9,10] However, isozyme patterns of HPRT scrutinized thoroughly in somatic cell hybrids and in germ cells from fetal ovaries of interspecific mouse hybrids provided strong evidence that the enzyme molecules are tetramers.[11]

REFERENCES

1. Watson, B., Gormley, I.P., Gardiner, S.E., Evans, H.J., and Harris, H., Reappearance of murine hypoxanthine guanine phosphoribosyl transferase activity in mouse A9 cells after attempted hybridization with human cell lines, *Exp. Cell Res.*, 75, 401, 1972.
2. Der Kaloustian, V.M., Byrne, R., Young, W.J., and Childs, B., An electrophoretic method for detecting hypoxanthine guanine phosphoribosyl transferase variants, *Biochem. Genet.*, 3, 299, 1963.
3. Tischfield, J.A., Bernhard, H.P., and Ruddle, F.H., A new electrophoretic–autoradiographic method for visual detection of phosphotransferases, *Anal. Biochem.*, 53, 545, 1973.
4. Van Diggelen, O.P. and Shin, S., A rapid fluorescent technique for electrophoretic identification of hypoxanthine phosphoribosyl transferase allozymes, *Biochem. Genet.*, 12, 375, 1974.
5. Van Diggelen, O.P., personal communication, 1977.
6. Hampton, A. and Nomura, A., Inosine 5'-phosphate dehydrogenase: site of inhibition by guanosine 5'-phosphate and of inactivation by 6-chloro and 6-mercapto purine ribonucleoside 5'-phosphates, *Biochemistry*, 3, 679, 1967.
7. Vasquez, B. and Bieber, A.L., Direct visualization of IMP-GMP: pyrophosphate phosphoribosyl transferase in polyacrylamide gels, *Anal. Biochem.*, 84, 504, 1978.
8. Chang, G.-G., Deng, R.-Y., and Pan, F., Direct localization and quantitation of aminoacyl-tRNA synthetase activity in polyacrylamide gel, *Anal. Biochem.*, 149, 474, 1985.
9. Van Someren, H., Van Henegouwen, H.B., Los, W., Wurzer-Figurelli, E., Doppert, B., Vervloet, M., and Meera Khan, P., Enzyme electrophoresis on cellulose acetate gel: II. Zymogram patterns in man–Chinese hamster somatic cell hybrids, *Humangenetik*, 25, 189, 1974.
10. Shows, T.B. and Brown, J.A., Human X-linked genes regionally mapped utilizing X-autosome translocations and somatic cell hybrids, *Proc. Natl. Acad. Sci. U.S.A.*, 72, 2125, 1975.
11. Bochkarev, M.N., Kulbakina, N.A., Zhdanova, N.S., Rubtsov, N.B., Zakian, S.M., and Serov, O.L., Evidence for tetrameric structure of mammalian hypoxanthine phosphoribosyltransferase, *Biochem. Genet.*, 25, 153, 1987.

OTHER NAMES	Poly(ADP-ribose) polymerase, poly(ADP) polymerase, poly(adenosine diphosphate ribose) polymerase, ADP-ribosyltransferase (polymerizing)
REACTION	NAD + (ADP-D-ribosyl)$_n$ – acceptor = nicotinamide + (ADP-D-ribosyl)$_{n+1}$ – acceptor
ENZYME SOURCE	Fungi, plants, invertebrates, vertebrates
SUBUNIT STRUCTURE	Unknown or no data available

METHOD

Visualization Scheme

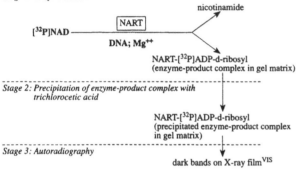

Stage 1: Enzyme Reaction

Stage 2: Precipitation of enzyme-product complex with trichlorocetic acid

Stage 3: Autoradiography

Reaction Mixture[1]

0.1 *M* Tris–HCl buffer, pH 8.0
10 m*M* MgCl$_2$
1 m*M* Dithiothreitol
1 m*M* NAD
10 μCi/ml [Adenin-2,8-^{32}P]NAD

Procedure

The method is developed for PAGs containing 1% SDS and 100 μg/ml calf thymus DNA activated by pancreatic deoxyribonuclease.

After completion of electrophoresis, wash the SDS-PAG in 1 1 of renaturing buffer (5 m*M* Tris–HCl, pH 8.0; 3 m*M* 2-mercaptoethanol) for 4 h with two changes at room temperature. Then wash the gel in 100 ml of renaturing buffer containing 6 *M* guanidine hydrochloride for 2 h with two changes at room temperature. Place the gel again in 1 1 of renaturing buffer and wash at 2°C for 16 to 18 h with several changes. After this, preincubate the gel in 50 ml of 100 m*M* Tris–HCl buffer (pH 8.0), 10 m*M* MgCl$_2$, and 1 m*M* dithiotreitol for 1 h. The processed gel is renatured and ready for NART detection by autoradiography.

Stage 1. Incubate the renatured gel in 10 ml of reaction mixture at 37°C for 8 to 12 h.

Stage 2. Wash the incubated gel five times in 1 1 of cold 5% (w/v) trichloroacetic acid. Dry the gel.

Stage 3. Autoradiograph the dry gel with x-ray film (Kodak AR).

The dark bands on the developed x-ray film correspond to localization of NART activity on the gel.

OTHER METHODS

A. An erasable blot technique is described below in which it is possible to detect both the enzyme (NART) activity and the enzyme protein.[2] After SDS-PAG electrophoresis, the gel is soaked at 37°C in 0.7 *M* 2-mercaptoethanol, 190 m*M* glycine, 25 m*M* Tris–HCl (pH 8.0), and 0.1% SDS. Proteins are electrotransferred at 4°C to a nitrocellulose membrane using 25 m*M* Tris, 192 m*M* glycine, and 20% (v/v) methanol. The blot is then soaked for 1 h in 10 ml of 50 m*M* Tris–HCl (pH 8.0), 100 m*M* NaCl, 1 m*M* dithiothreitol, 0.3% Tween 20 (renaturation buffer) containing 2 μg/ml DNase I–activated DNA, 20 μM Zn(II) acetate, and 2 m*M* MgCl$_2$. The blot is then incubated for 1 h in the same buffer containing 2 μCi/ml [α-^{32}P]NAD (New England Nuclear, Boston, Massachusetts) and washed four times for 15 min with 20 ml of renaturation buffer. The blot is then dried, and autoradiographed using Kodak X-AR film. In addition to the main activity band with a strong radioactive signal, several faint bands at higher and lower molecular weights are detected in preparations of HeLa and bovine kidney cell cultures. Using the erasable blot technique, the higher-molecular-weight signals and the lower-molecular-weight signals are shown to be due to automodified NART (specific radioactive signals) and to histones (nonspecific radioactive signals), respectively. The first step of this technique is a chemical erasure of nonspecific radioactive signals by washing the dried blot (processed as described above) in 10 ml of erasure solution (50 m*M* Tris–HCl (pH 8.0), 100 m*M* NaCl, 1 m*M* dithiothreitol, 2% SDS). The blot is erased by four successive washes of 10 ml of erasure solution for 15 min, dried, and autoradiographed. After this step all nonspecific signals are eliminated and only specific signals remain on the blot. The second step of the technique is an enzymatic erasure of covalently bound (specific) radioactive signals. The autoradiographed blot is soaked in a solution of phosphate-buffered saline containing 0.1% (v/v) Tween 20, and 5% nonfat dried milk (PBSTM) for 1 h, and then in a solution of 50 m*M* KPO$_4$ (pH 7.4), 50 m*M* KCl, 5 m*M* 2-mercaptoethanol, 10% (v/v) glycerol, and 0.1% Triton X-100 containing 5 U/ml of purified calf thymus poly(ADP-ribose) glycohydrolase (PARG) for 2 h. The blot is washed four times with 10 ml of renaturation buffer at 15-min intervals, dried, and autoradiographed as described above. The poly(ADP-ribose) covalently linked to NART is erased by PARG, and the specific radioactive signals are therefore identified. The blocking step using PBSTM is essential for the activity of PARG. When unblocked blots are used, the elimination of specific radioactive signals by PARG digestion requires 24 h instead of 2 h, providing evidence that PARG can be bound to the unblocked blot during the second (enzymatic digestion) step. Finally, after soaking the autoradiographed blot in phosphate-buffered saline for 15 min and then in PBSTM for 1 h at room temperature, the routine immunoblotting procedure can be performed to identify the enzyme protein.

A similar procedure (activity blotting technique) may be adapted to detect NART activity bands by blotting the enzyme protein on a gapped DNA-fixed nylon membrane and incubating it in a reaction mixture similar to that given in the method above.[3]

B. An immunoblotting procedure (for details, see Part II) may also be used to detect NART. Monoclonal antibodies specific to the calf thymus enzyme as well as antibodies specific to the human enzyme are available from Sigma.

GENERAL NOTES

The NART is a eukaryotic enzyme that transfers the ADP-D-ribosyl group of the NAD molecule to an acceptor carboxyl group on a histone or the enzyme protein itself. Further ADP-ribosyl groups are transferred to the 2'-position of the terminal adenosine moiety, building up an ADP-ribose homopolymer of different length (usually of 20 to 30 units). The polymer is synthesized in the presence of DNA and transferred to an acceptor protein.[4]

The enzyme is activated when cells are subjected to the action of DNA-damaging agents and is possibly involved in base excision repair of DNA (see 3.2.2.X' — PARG, General Notes).

The main disadvantage of the in-gel detection technique (see Method) is that it is time consuming and does not permit immunochemical and enzymological treatments desirable to estimate specificity of the technique. The use of the erasable blot technique (see Other Methods, A) allows for analysis of the specific and nonspecific radioactive signals on NART activity blots, and thus avoids this disadvantage.

REFERENCES

1. Scovassi, A.I., Stefanini, M., and Bertazzoni, U., Catalytic activities of human poly(ADP-ribose) polymerase from normal and mutagenized cells detected after sodium dodecyl sulfate–polyacrylamide gel electrophoresis, *Anal. Biochem.*, 259, 10973, 1984.
2. Desnoyers, S., Shah, G.M., Brochu, G., and Poirier, G.G., Erasable blot of poly(ADP-ribose) polymerase, *Anal. Biochem.*, 218, 470, 1994.
3. Seki, S., Akiyama, K., Watanabe, S., and Tsutsui, K., Activity gel and activity blotting methods for detecting DNA-modifying (repair) enzymes, *J. Chromatogr.*, 618, 147, 1993.
4. NC-IUBMB, *Enzyme Nomenclature*, Academic Press, San Diego, 1992, p. 240 (EC 2.4.2.30, Comments).

2.5.1.18 — Glutathione Transferase; GT

OTHER NAMES	Glutathione *S*-alkyltransferase, glutathione *S*-aryltransferase, glutathione *S*-transferase, *S*-(hydroxyalkyl) glutathione lyase, glutathione *S*-aralkyltransferase
REACTION	RX + glutathione–SH = HX + glutathione-*S*-*R* (see General Notes)
ENZYME SOURCE	Bacteria, fungi, plants, invertebrates, vertebrates
SUBUNIT STRUCTURE	Dimer (vertebrates)

METHOD 1

Visualization Scheme

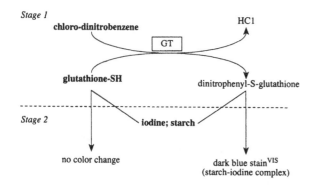

Staining Solution[1]

  A. 0.1 *M* Sodium phosphate buffer, pH 6.5   20 ml
        1-Chloro-2,4-dinitrobenzene (dissolved in
          0.8 ml of ethanol)   8 mg
        Reduced glutathione   14 mg
  B. 1.0% I_2 in 1.0% KI   0.9 ml
        H_2O 30 ml
  C. 2% Agar solution (60°C)   30 ml

Procedure

Saturate filter paper with solution A and place it on the cut surface of an electrophorized starch gel or PAG containing 0.04% (w/v) of soluble starch. Take care to avoid the formation of air bubbles between the paper overlay and gel. Incubate the overlayed gel at 37°C for 40 min. Remove the paper sheet. Mix solutions B and C and apply the mixture to the gel surface. An intense blue color appears immediately in areas where reduced glutathione has been conjugated to 1-chloro-2,4-dinitrobenzene by the action of GT. Record or photograph the zymogram.

METHOD 2

Visualization Scheme

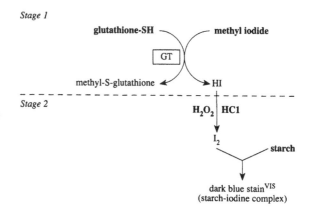

Staining Solution[2]

  A. 0.1 *M* Potassium phosphate buffer, pH 7.0
        7 m*M* Methyl iodide
        6 m*M* Reduced glutathione
  B. 0.3 *M* HCl
        0.3% H_2O_2

Procedure

Preincubate an electrophorized PAG containing 0.04% (w/v) soluble starch in two changes of 0.1 *M* potassium phosphate buffer (pH 7.0) for 10 min at room temperature with gentle agitation. Then place the gel in solution A and incubate at 37°C for periods of from 6 to 20 min with periodic agitation. At the end of the incubation period transfer the gel to solution B. Keep the gel under observation until blue bands indicating the local formation of iodine are observed. The bands appear abruptly following a lag period, the length of which depends on the amount of GT catalytic activity present. When the bands are well developed, record or photograph the zymogram quickly. After full development, the bands retain their definition for 2 to 3 h and then fade.

Notes: Two millimolar 1,3-dinitro-4-iodobenzene (IDNB) can be used as a substrate instead of methyl iodide. To increase the solubility of IDNB, 0.33 volume of glycerol should be added to 0.1 *M* potassium phosphate buffer (pH 7.0), containing 1.67 m*M* reduced glutathione.

METHOD 3

Visualization Scheme

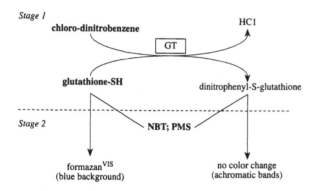

Staining Solution[3]

A. 0.1 M Potassium phosphate buffer, pH 6.5
    4.5 mM Reduced glutathione
    1 mM 1-Chloro-2,4-dinitrobenzene
    1 mM NBT
B. 0.1 M Tris–HCl buffer, pH 9.6
    3 mM PMS

Procedure

Incubate the electrophorized PAG in solution A at 37°C for 10 min under gentle agitation. Wash the gel with water and incubate in solution B at room temperature. Blue insoluble formazan appears on the gel surface in about 3 to 5 min, except in the GT areas. Wash the negatively stained gel with water and place in 1 M NaCl. Under these conditions the achromatic bands remain clearly visible for almost 1 month.

Notes: Artifactual achromatic bands, perhaps due to superoxide dismutase (see 1.15.1.1 — SOD), can be observed on gels stained using this method. These artifactual bands do not impair the method since they can be easily identified by comparing the test gel with a control gel incubated with all reagents except chloro-dinitrobenzene.

OTHER METHODS

The immunoblotting procedure (for details, see Part II) may be used to detect the enzyme. Monoclonal antibodies specific to *Shistosoma* GT are available from Sigma.

GENERAL NOTES

In the enzyme reaction R may be an aliphatic, aromatic, or heterocyclic group; X may be a sulfate, nitrile, or halide group.[4]

A considerable fraction of the enzyme can interact and move together with the dye, Bromophenol Blue. This GT retains its activity and therefore can be detected on electrophoretic gels at the dye-front. The amount of GT activity, at the dye-front, is dependent on the contact time of GT with the dye at room temperature.[5] A similar finding was reported for xanthine oxidase (see 1.1.3.22 — XOX, General Notes).

REFERENCES

1. Board, P.G., A method for the localization of glutathione *S*-transferase isozymes after starch gel electrophoresis, *Anal. Biochem.*, 105, 147, 1980.
2. Clark, A.G., A direct method for the visualization of glutathione *S*-transferase activity in polyacrylamide gels, *Anal. Biochem.*, 123, 147, 1982.
3. Ricci, G., Bello, M.L., Caccuri, A.M., Galiazzo, F., and Federici, G., Detection of glutathione transferase activity on polyacrylamide gels, *Anal. Biochem.*, 143, 226, 1984.
4. NC-IUBMB, *Enzyme Nomenclature*, Academic Press, San Diego, 1992, p. 246 (EC 2.5.1.18, Comments).
5. Özer, N., Erdemli, Ö., Sayek, I., and Özer, I., Resolution and kinetic characterization of glutathione *S*-transferase from human jejunal mucosa, *Biochem. Med. Methods Biol.*, 44, 142, 1990.

2.5.1.19 — 5-enol-Pyruvylshikimate-3-Phosphate Synthase; EPSPS

OTHER NAMES	3-Phosphoshikimate 1-carboxyvinyl-transferase (recommended name)
REACTION	Phosphoenolpyruvate + 3-phospho-shikimate = orthophosphate + 5-O-(1-carboxyvinyl)-3-phospho-shikimate
ENZYME SOURCE	Bacteria, fungi
SUBUNIT STRUCTURE	Unknown or no data available

METHOD

Visualization Scheme

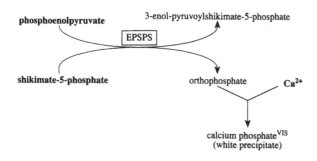

Staining Solution[1]

50 mM Glycine–KOH buffer, pH 10.0
1 mM Shikimate-5-phosphate
2 mM Phosphoenolpyruvate
10 mM Ca^{2+}

Procedure

Soak the electrophorized PAG in 50 mM glycine–KOH buffer (pH 10.0) for 20 to 30 min and incubate in a staining solution at room temperature until white bands of calcium phosphate precipitation become visible against a dark background. Photograph the stained gel by reflected light against a dark background. Store the stained gel in 50 mM glycine–KOH buffer (pH 10.0), containing 5 mM Ca^{2+}, either at 5°C or in the presence of an antibacterial agent at room temperature.

OTHER METHODS

The product orthophosphate may be detected using some other histochemical methods (e.g., see 3.6.1.1 — PP).

GENERAL NOTES

The enzyme is a domain in the *arom* multienzyme polypeptide of *Neurospora*.[2]

REFERENCES

1. Nimmo, H.G. and Nimmo, G.A., A general method for the localization of enzymes that produce phosphate, pyrophosphate, or CO$_2$ after polyacrylamide gel electrophoresis, *Anal. Biochem.*, 121, 17, 1982.
2. NC-IUBMB, *Enzyme Nomenclature*, Academic Press, San Diego, 1992, p. 563 (Nomenclature of Multienzymes, Symbolism).

OTHER NAMES	Aspartate transaminase (recommended name), aspartate aminotransferase, glutamic–aspartic transaminase, transaminase A
REACTION	L-Aspartate + 2-oxoglutarate = oxaloacetate + L-glutamate
ENZYME SOURCE	Bacteria, fungi, plants, protozoa, invertebrates, vertebrates
SUBUNIT STRUCTURE	Dimer (fungi, algae, plants, invertebrates, vertebrates)

METHOD 1

Visualization Scheme

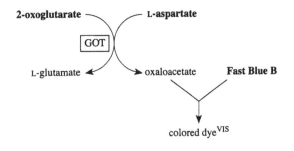

Staining Solution[1] (modified)

0.05 M Tris–HCl buffer, pH 7.5	100 ml
L-Aspartic acid (monosodium salt)	400 mg
2-Oxoglutaric acid (disodium salt)	180 mg
Fast Blue B salt	150 mg

Procedure

Incubate the gel in staining solution in the dark at 37°C until dark brown bands appear. Wash the gel in water and fix in 50% glycerol or in a water solution of reduced glutathione.

Notes: Some authors recommend including cofactor pyridoxal 5-phosphate in the staining solution. Usually, there is no need to do this because native GOT molecules contain sufficient amounts of tightly bound pyridoxal 5-phosphate molecules.

Fast Violet B, Fast Blue BB, and Fast Garnet GBC may be used instead of Fast Blue B.

Where enzyme activity is low, greater sensitivity of the method may be obtained using a postcoupling technique. When using this technique, apply all necessary ingredients except a diazo dye to the gel in the normal manner. Incubate the gel for a period of time (depending on the expected activity of the enzyme), and then add a diazo dye dissolved in a minimal volume of staining buffer.

METHOD 2

Visualization Scheme

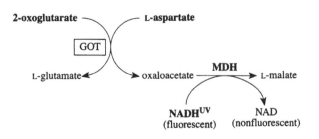

Staining Solution[2] (modified)

A. 0.1 M Sodium phosphate buffer, pH 7.5	25 ml
L-Aspartic acid (monosodium salt)	200 mg
2-Oxoglutaric acid (disodium salt)	90 mg
Malate dehydrogenase (MDH)	60 U
NADH	7 mg
B. 1.5% Agar solution (60°C)	25 ml

Procedure

Mix A and B components of the staining solution and pour the mixture over the surface of the gel. Incubate the gel at 37°C for 20 to 60 min and monitor under long-wave UV light. The zones of GOT activity appear as dark (nonfluorescent) bands on a light (fluorescent) background. Record the zymogram or photograph using a yellow filter.

Notes: When a zymogram that is visible in daylight is required, counterstain the processed gel with MTT–PMS solution. Achromatic bands of GOT appear almost immediately.

METHOD 3

Visualization Scheme

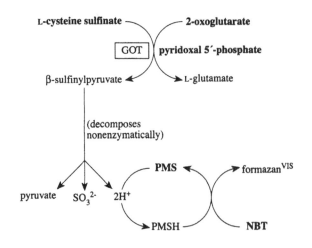

Staining Solution[3]

500 μM Tris–HCl buffer, pH 8.5
250 μM L-Cysteine sulfinate
25 μM 2-Oxoglutarate
0.5 μM Pyridoxal 5'-phosphate
0.4 mg/ml NBT
0.1 mg/ml PMS

Procedure

Incubate the gel in staining solution in the dark at 37°C until dark blue bands appear. Wash the gel in water and fix in 3% acetic acid or 25% ethanol.

Notes: This method is based on the cysteine sulfinate transamination activity of GOT. The method is highly sensitive. It is capable of detecting 0.3 mU per band of GOT activity. It is difficult to detect such a low level of activity on the gel by any of the other methods except Method 4 (see below).

A sensitive tetrazolium method is described in which the use of cysteine sulfinate as a substrate is combined with the use of glutamate dehydrogenase as a linking enzyme.[4] High sensitivity of the method is possibly due to the involvement of both the reaction products, β-sulfinylpyruvate and L-glutamate, in the formation of formazan.

METHOD 4

Visualization Scheme

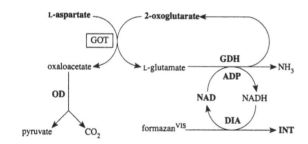

Staining Solution[5]

A. 180 mM Tris–HCl buffer, pH 8.1
   20 mM L-Aspartate
   10 mM 2-Oxoglutarate
   4 mM ADP
B. 80 mM NAD
C. 30 mM Iodonitrotetrazolium Violet (INT)
D. 100 mM Tris–HCl buffer, pH 7.5
   9.8 U/ml Diaphorase (DIA)
E. 1200 U/ml Glutamate dehydrogenase (GDH) in 50% glycerol
F. 125 U/ml Oxaloacetate decarboxylase (OD) in 100 mM Tris–HCl buffer, pH 7.5
G. 2% Agarose solution (45°C) in 100 mM Tris–HCl buffer, pH 7.5

Procedure

Mix 5 ml of solution A with 1 ml each of solutions B through F and incubate in the dark at 37°C for 15 min. Add this mixture to 10 ml of solution G and pour the mixture over the surface of the gel. Incubate the gel in the dark at 37°C until pink bands appear. Fix the gel in 5% acetic acid.

Notes: This method is suitable for quantitative detection of GOT isoenzymes. Product inhibition of GOT isozymes (especially mitochondrial isozyme) by oxaloacetate is prevented by oxaloacetate decarboxylase, which is included in the stain. GOT activity as low as 30 μU per band can be quantitated by incubating the gel overnight at room temperature in the dark. Pyruvate kinase (EC 2.7.1.40) has been reported to have decarboxylase activity and perhaps can be used instead of oxaloacetate decarboxylase to remove the excess of oxaloacetate.

This method is the most expensive in comparison with the other three described above.

GENERAL NOTES

The enzyme is a pyridoxal-phosphate protein. It also acts on L-tyrosine, L-phenylalanine, and L-tryptophan. GOT activity can be formed by controlled proteolysis from aromatic-amino acid transaminase (EC 2.6.1.57).[6]

REFERENCES

1. Decker, L.E. and Rau, E.M., Multiple forms of glutamic–oxalacetic transaminase in tissues, *Proc. Soc. Exp. Biol. Med.*, 112, 144, 1963.
2. Boyd, J.W., The extraction and purification of two isoenzymes of L-aspartate:2-oxoglutarate aminotransferase, *Biochim. Biophys. Acta*, 113, 302, 1966.
3. Yagi, T., Kagamiyama, H., and Nozaki, M., A sensitive method for the detection of aspartate: 2-oxoglutarate aminotransferase activity on polyacrylamide gels, *Anal. Biochem.*, 110, 146, 1981.
4. Jeremiah, S.J., Povey, S., Burley, M.W., Kielty, C., Lee, M., Spowart, G., Corney, G., and Cook, P.J.L., Mapping studies on human mitochondrial glutamate oxaloacetate transaminase, *Ann. Hum. Genet.*, 46, 145, 1982.
5. Nealon, D.A. and Rej, R., Quantitation of aspartate aminotransferase isoenzymes after electrophoretic separation, *Anal. Biochem.*, 161, 64, 1987.
6. NC-IUBMB, *Enzyme Nomenclature*, Academic Press, San Diego, 1992, p. 251 (EC 2.6.1.1, Comments).

2.6.1.2 — Glutamic–Pyruvic Transaminase; GPT

OTHER NAMES — Alanine transaminase (recommended name), alanine aminotransferase, glutamic–alanine transaminase, alanine transaminase

REACTION — L-Alanine + 2-oxoglutarate = pyruvate + L-glutamate

ENZYME SOURCE — Bacteria, fungi, plants, protozoa, invertebrates, vertebrates

SUBUNIT STRUCTURE — Dimer (fungi, invertebrates, vertebrates)

Method 1

Visualization Scheme

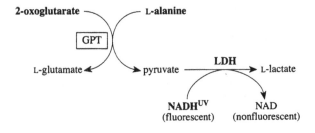

Staining Solution[1]

0.1 M Tris–HCl buffer, pH 7.6	10 ml
L-α-Alanine	178 mg
2-Oxoglutaric acid	13 mg
NADH	7 mg
Lactate dehydrogenase (LDH)	8 U

Procedure

Apply the staining solution to the gel on a filter paper overlay and, after incubation at 37°C for 30 to 60 min, view the gel under long-wave UV light. Dark (nonfluorescent) bands of GPT activity are visible on a light (fluorescent) background. Photograph the developed gel through a yellow filter. Nonfluorescent GPT bands may be made visible in daylight (achromatic bands on a blue background) after treating the processed gel with MTT–PMS solution.

Notes: When LDH suspension in $(NH_4)_2SO_4$ is used, glutamate dehydrogenase activity bands can also develop on a GPT zymogram obtained by this method. Thus, the use of NH_3-free preparations of LDH is recommended in this method.

Method 2

Visualization Scheme

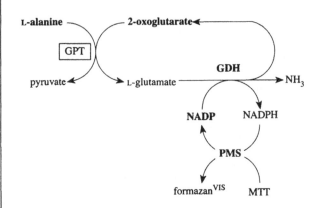

Staining Solution[2]

A. 0.2 M Tris–HCl buffer, pH 8.0	10 ml
400 mg/ml L-Alanine	5 ml
40 mg/ml 2-Oxoglutarate	2.5 ml
10 mg/ml NADP	0.5 ml
Glutamate dehydrogenase (GDH)	90 U
10 mg/ml MTT	0.5 ml
2.5 mg/ml PMS	0.5 ml
65 mg/ml KCN	0.5 ml
B. 0.15% Agar solution (60°C)	10 ml

Procedure

Mix A and B components of the staining solution, pour the mixture over the surface of the gel, and allow it to solidify. Incubate the gel in the dark at 37°C until dark blue bands appear. Fix the stained gel in 25% ethanol.

Notes: The linking enzyme, glutamate dehydrogenase (EC 1.4.1.3), also works well with NAD as a cofactor. The use of NAD instead of NADP is preferable because NAD is several times cheaper than NADP.

ADP may be included in the staining solution to stimulate the coupled reaction catalyzed by GDH (e.g., see 2.6.1.1 — GOT, Method 4).

2.6.1.2 — Glutamic–Pyruvic Transaminase; GPT (continued)

General Notes

The enzyme is a pyridoxal-phosphate protein.[3] However, usually there is no need to include this cofactor in the staining solution because native GPT molecules contain sufficient amounts of tightly bound pyridoxal 5-phosphate molecules.

References

1. Chen, S.-H. and Giblett, E.R., Polymorphism of soluble glutamic–pyruvic transaminase: a new genetic marker in man, *Science*, 173, 148, 1971.
2. Eicher, E.M. and Womack, J.E., Chromosomal location of soluble glutamic–pyruvic transaminase-1 (*Gpt-1*) in the mouse, *Biochem. Genet.*, 15, 1, 1977.
3. NC-IUBMB, *Enzyme Nomenclature*, Academic Press, San Diego, 1992, p. 251 (EC 2.6.1.2, Comments).

2.6.1.5 — Tyrosine Transaminase; TAT

OTHER NAMES	Tyrosine aminotransferase
REACTION	L-Tyrosine + 2-oxoglutarate = 4-hydroxyphenylpyruvate + L-glutamate
ENZYME SOURCE	Bacteria, protozoa, invertebrates, vertebrates
SUBUNIT STRUCTURE	Monomer (protozoa)

Method

Visualization Scheme

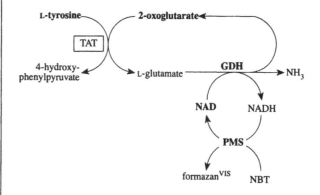

Staining Solution[1]

0.1 *M* Sodium phosphate buffer, pH 7.0	50 ml
NBT	30 mg
PMS	2 mg
NAD	50 mg
10 mg/ml 2-Oxoglutaric acid (neutralized)	10 ml
L-Tyrosine	100 mg
Glutamate dehydrogenase (GDH)	16 U
H_2O	30 ml

Procedure

Incubate the gel in staining solution in the dark at 37°C until dark blue bands appear. Wash the gel in water and fix in 25% ethanol.

Notes: ADP may be added to the staining solution to stimulate a coupled reaction catalyzed by GDH (e.g., see 2.6.1.1 — GOT, Method 4).

Other Methods

2-Hydroxyisocaproate dehydrogenase (EC 1.1.1.X) from *Lactobacillus delbrueckii* was successfully used as a linking enzyme to detect α-keto acid products generated by the forward TAT reaction.[2] The reaction catalyzed by the linking dehydrogenase is NADH dependent. Therefore, gel areas occupied by TAT and detected by the use of this linking enzyme should be observed in UV light as dark (nonfluorescent) bands on a light (fluorescent) background of the processed gel.

2.6.1.5 — Tyrosine Transaminase; TAT (continued)

GENERAL NOTES

The enzyme is a pyridoxal-phosphate protein. Native TAT molecules contain sufficient amounts of tightly bound pyridoxal 5-phosphate molecules, so usually there is no need to add the cofactor in the staining solution. The mitochondrial isozyme of TAT may be identical to EC 2.6.1.1 (see 2.6.1.1 — GOT). The three isozymic forms of plant TAT are interconverted by the cysteine endopeptidase bromelain.[3]

REFERENCES

1. Shaw, C.R. and Prasad, R., Starch gel electrophoresis of enzymes: a compilation of recipes, *Biochem. Genet.*, 4, 297, 1970.
2. Luong, T.N. and Kirsch, J.F., A continuous coupled spectrophotometric assay for tyrosine aminotransferase activity with aromatic and other nonpolar amino acids, *Anal. Biochem.*, 253, 46, 1997.
3. NC-IUBMB, *Enzyme Nomenclature*, Academic Press, San Diego, 1992, p. 252 (EC 2.6.1.5, Comments).

2.6.1.13 — Ornithine–Oxo-Acid Transaminase; OAT

OTHER NAMES	L-Ornithine aminotransferase, ornithine–oxo-acid aminotransferase
REACTION	L-Ornithine + 2-oxo-acid = L-glutamate 5-semialdehyde + L-amino acid
ENZYME SOURCE	Fungi, vertebrates
SUBUNIT STRUCTURE	Tetramer[a] (vertebrates)

METHOD 1

Visualization Scheme

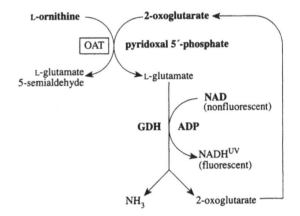

Staining Solution[1] (adapted)

A. 100 mM Tris–HCl buffer, pH 8.0
    0.1 mM Pyridoxal 5'-phosphate
    0.1 mM ADP
    2 mM NAD
    2 mM 2-Oxoglutarate
    10 mM L-Ornithine
    20 U/ml Glutamate dehydrogenase (GDH)
B. 2% Agarose solution (60°C)

Procedure

Mix equal volumes of A and B components of the staining solution and pour the mixture over the surface of the gel. Incubate the gel at 37°C for 30 min and view under long-wave UV light. Light (fluorescent) bands of OAT are visible on a dark (nonfluorescent) background of the gel. Record the zymogram or photograph using a yellow filter.

Notes: MTT and PMS may be added to the staining solution to obtain OAT zymograms visible in daylight.

METHOD 2

Visualization Scheme

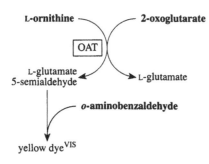

L-ornithine — 2-oxoglutarate

OAT

L-glutamate 5-semialdehyde ← → L-glutamate

— o-aminobenzaldehyde

yellow dye[VIS]

Staining Solution[2]

A. 0.1 M Potassium phosphate buffer, pH 8.15
   50 mM L-Ornithine
   25 mM 2-Oxoglutarate
   0.1% o-Aminobenzaldehyde
B. 3% Agar solution (60°C)

Procedure

Mix 19 ml of solution A with 5 ml of solution B and pour the mixture over the surface of the gel. Incubate the gel at 30°C until yellow bands appear. Record the zymogram or photograph by contact printing on a photographic paper placed under the stained agar overlay, with a photographic enlarger lamp as a light source.

GENERAL NOTES

OAT is a mitochondrial pyridoxal-phosphate enzyme protein. To prepare a mitochondrial extract for electrophoresis, the mitochondrial fraction should be isolated and suspended in an appropriate buffer at a concentration of 30% with respect to the original tissue weight. After sonication the mitochondrial suspension should be centrifuged at 100,000 g for 1 h. The supernatant may be used for electrophoresis.

REFERENCES

1. Akabayashi, A. and Kato, T., One-step and two-step fluorometric assay methods for general aminotransferases using glutamate dehydrogenase, *Anal. Biochem.*, 182, 129, 1989.
2. Yanagi, S., Tsutsumi, T., Saheki, S., Saheki, K., and Yamamoto, N., Novel and sensitive activity stains on polyacrylamide gel of serine and threonine dehydratase and ornithine aminotransferase, *Enzyme*, 28, 400, 1982.

OTHER NAMES	Glutamine–pyruvate aminotransferase, glutaminase II, glutamine transaminase L, glutamine–oxo-acid transaminase
REACTION	L-Glutamine + pyruvate = 2-oxoglutaramate + L-alanine
ENZYME SOURCE	Vertebrates
SUBUNIT STRUCTURE	Unknown or no data available

METHOD

Visualization Scheme

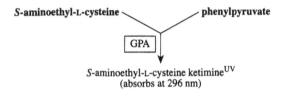

S-aminoethyl-L-cysteine ketimine^{UV}
(absorbs at 296 nm)

Reaction Mixture[1]

0.1 *M* Pyrophosphate buffer, pH 8.5
20 m*M* *S*-2-Aminoethyl-L-cysteine
0.4 m*M* Phenylpyruvate

Procedure

Incubate the electrophorized PAG in the reaction mixture at 20°C for 30 min. Wash the gel with water and scan at 296 nm using a scanning spectrophotometer. Sharp peaks on the photometric scan correspond to GPA bands on the PAG.

Notes: The method is based on the ability of GPA to act on *S*-aminoethyl-L-cysteine in the presence of phenylpyruvate.

The use of borate-containing buffers should be avoided because of the strong absorbance of the phenylpyruvate enol–borate complex at 300 nm.

Using an optical instrument constructed for two-dimensional spectroscopy of electrophoretic gels[2] and an interference filter with a peak transmission wavelength near 296 nm, a photocopy of the GPA zymogram can be made.

OTHER METHODS

A. Zymograms of GPA visible in daylight can be obtained using auxiliary alanine dehydrogenase (see 1.4.1.1 — ALADH) to detect the product L-alanine. The recommended staining solution should include L-glutamine, pyruvate, NAD, alanine dehydrogenase, MTT (or NBT), PMS, and pyrophosphate buffer (pH 8.5).

B. An auxiliary enzyme, alanopine dehydrogenase (see 1.5.1.17 — ALPD), can also be used to detect the product L-alanine. Staining solution recommended for the trial should include L-glutamine, pyruvate, NADH, and ALPD. Dark GPA bands will be visible in long-wave UV light on the fluorescent background of the gel. This method, however, has two obvious disadvantages: (1) the ALPD preparations are not yet commercially available; and (2) the control staining of an additional gel for lactate dehydrogenase is necessary. Nevertheless, when ALPD preparations are available, this method will be more appropriate for routine laboratory use than the spectrophotometric method detailed above.

GENERAL NOTES

The enzyme is a pyridoxal-phosphate protein. Native GPA molecules contain sufficient amounts of tightly bound pyridoxal 5-phosphate, so there is no need to add the cofactor in the staining solution. L-Methionine, L-cystathionine, and other sulfur compounds can act as donors; a number of keto acids, including glyoxylate, can act as acceptors.[3]

REFERENCES

1. Ricci, G., Caccuri, A.M., Bello, M.L., Solinas, S.P., and Nardini, M., Ketimine rings: useful detectors of enzymatic activities in solution and on polyacrylamide gel, *Anal. Biochem.*, 165, 356, 1987.
2. Klebe, R.J., Mancuso, M.G., Brown, C.R., and Teng, L., Two-dimensional spectroscopy of electrophoretic gels, *Biochem. Genet.*, 19, 655, 1981.
3. NC-IUBMB, *Enzyme Nomenclature*, Academic Press, San Diego, 1992, p. 253 (EC 2.6.1.15, Comments).

2.6.1.18 — β-Alanine–Pyruvate Transaminase; APT

REACTION   L-Alanine + 3-oxopropanoate = pyruvate + β-alanine

ENZYME SOURCE   Bacteria

SUBUNIT STRUCTURE   Unknown or no data available

METHOD

Visualization Scheme

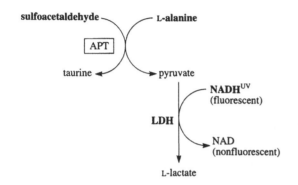

Staining Solution[1] (adapted)

0.1 *M* Potassium phosphate buffer, pH 8.0
2 μ*M* L-Alanine
10 m*M* Sulfoacetaldehyde
0.3 m*M* NADH
0.2 U/ml Lactate dehydrogenase (LDH)

Procedure

Overlay the gel with filter paper soaked in staining solution or apply staining solution as a 1% agarose overlay. Monitor the gel under UV light for dark (nonfluorescent) bands of APT visible on a light (fluorescent) background. Record the zymogram or photograph using a yellow filter.

Notes: When a zymogram that is visible in daylight is required, counterstain the processed gel with MTT–PMS solution to obtain achromatic bands of APT on a blue background. Fix the counterstained gel in 25% ethanol.

    An NADH-dependent reaction of the auxiliary enzyme tauropine dehydrogenase (see 1.5.1.23 — TAUDH) can also be used to detect taurine and pyruvate produced by APT. Unfortunately, this enzyme is not commercially available.

GENERAL NOTES

The enzyme is a pyridoxal-phosphate protein.[2] It is characterized by low substrate specificity for ω-amino carboxylic acids, mono- and diamines, and ω-amino sulfonates, including taurine. On the contrary, of the α-amino acids, only L-alanine is active in the enzyme reaction. Neither the D- nor the L-isomer of glutamate, aspartate, α-amino-*n*-butyrate, and phenylalanine could serve as an amino donor. Thus, the enzyme displays high amino acceptor specificity, i.e., α-keto acids other than pyruvate are inactive as amino acceptors.[1]

REFERENCES

1. Yonaha, K. and Toyama, S., Enzymatic determination of L-alanine with ω-amino acid:pyruvate aminotransferase, *Anal. Biochem.*, 101, 504, 1980.
2. NC-IUBMB, *Enzyme Nomenclature*, Academic Press, San Diego, 1992, p. 253 (EC 2.6.1.18, Comments).

OTHER NAMES	γ-Aminobutyric acid transaminase, β-alanine–oxoglutarate aminotransferase, β-alanine–oxoglutarate transaminase, 4-aminobutyrate aminotransferase
REACTION	4-Aminobutanoate + 2-oxoglutarate = succinate semialdehyde + L-glutamate
ENZYME SOURCE	Bacteria, vertebrates
SUBUNIT STRUCTURE	Dimer (vertebrates)

METHOD 1

Visualization Scheme

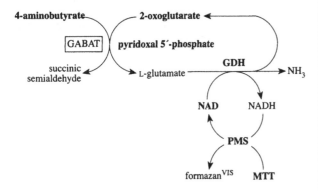

Staining Solution[1]

A.	0.5 M Tris–HCl buffer, pH 7.5	25 ml
	4-Aminobutyric acid	250 mg
	NAD	20 mg
	Glutamate dehydrogenase (GDH)	50 U
	0.1 M 2-Oxoglutaric acid (neutralized)	0.4 ml
	5 mg/ml MTT	1 ml
	5 mg/ml PMS	1 ml
B.	1.5% Agar solution (60°C)	25 ml

Procedure

Mix A and B components of the staining solution and pour the mixture over the surface of the gel. Incubate the gel in the dark at 37°C until dark blue bands appear. Fix the stained gel in 25% ethanol.

Notes: When the enzyme from mammalian tissues is detected, β-alanine may be used instead of 4-aminobutyric acid. The cofactor pyridoxal 5'-phosphate is recommended to be included in the starch gel after degassing (final concentration of 3 mM) and in the cathode buffer compartment (final concentration of 0.4 mM). A small amount of pyruvate may be added to the staining solution to inhibit lactate dehydrogenase activity.

METHOD 2

Visualization Scheme

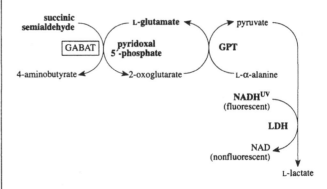

Staining Solution[1]

0.5 M Tris–HCl buffer, pH 7.5	10 ml
Succinic semialdehyde	0.1 ml
L-Glutamic acid	20 mg
D,L-α-Alanine	100 mg
Glutamic–pyruvic transaminase (GPT)	80 U
Lactate dehydrogenase (LDH)	150 U
NADH	20 mg

Procedure

Soak a piece of filter paper in a staining solution and apply to the gel surface. Wrap the gel in Saran wrap and incubate at 37°C for 30 to 60 min. View the gel under long-wave UV light. Dark (nonfluorescent) bands of GABAT activity are visible on a light (fluorescent) background. Record the zymogram or photograph using a yellow filter.

Notes: When a zymogram that is visible in daylight is required, counterstain the processed gel with MTT–PMS solution. Achromatic bands of GABAT activity appear on a blue background almost immediately.

The cofactor pyridoxal 5'-phosphate is recommended to be added to the starch gel after degassing (final concentration of 3 mM) and to the cathodal buffer (final concentration of 0.4 mM).

OTHER METHODS

Succinate-semialdehyde dehydrogenase coupled with the NADP–PMS–NBT system can be used to detect the product succinic semialdehyde) (see 4.1.1.15 — GDC).

REFERENCES

1. Jeremiah, S. and Povey, S., The biochemical genetics of human γ-aminobutyric acid transaminase, *Ann. Hum. Genet.*, 45, 231, 1981.

2.6.1.42 — Branched-Chain Amino Acid Transaminase; BCT

OTHER NAMES	Branched-chain L-amino acid amino-transferase, transaminase B
REACTION	L-Leucine + 2-oxoglutarate = 4-methyl-2-oxopentanoate + L-glutamate
ENZYME SOURCE	Bacteria, plants, invertebrates, vertebrates
SUBUNIT STRUCTURE	Hexamer[#] (bacteria), dimer[#] (bacteria, mammals)

METHOD 1

Visualization Scheme

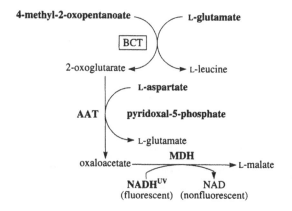

Staining Solution[1]

0.1 M Tris–HCl, pH 8.3
2.1 mM 4-Methyl-2-oxopentanoate
0.3 M L-Glutamate
0.2 M L-Aspartate
0.2 mM NADH
0.1 mM Pyridoxal 5-phosphate
8 U/ml Aspartate aminotransferase (AAT)
16 U/ml Malate dehydrogenase (MDH)

Procedure

Apply the staining solution to the gel surface on a filter paper overlay and incubate the gel in a moist chamber. After 10 to 30 min of incubation, monitor the gel under long-wave UV light for dark (nonfluorescent) bands visible on a light (fluorescent) background. Record the zymogram or photograph using a yellow filter.

Notes: Pyridoxal 5-phosphate is added to the staining solution as a cofactor of auxiliary enzyme AAT; it can be omitted, however, because native AAT molecules contain sufficient amounts of tightly bound pyridoxal 5-phosphate molecules.

Oxaloacetate generated by auxiliary AAT can be coupled with diazonium ions to generate colored dye visible in daylight (e.g., see 2.6.1.1 — GOT, Method 1). To carry out the coupling reaction, replace NADH and MDH in the staining solution by a diazonium salt (e.g., Fast Blue B, Fast Blue BB, Fast Garnet GBC).

METHOD 2

Visualization Scheme

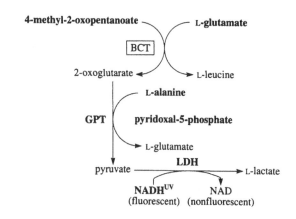

Staining Solution[1] (adapted)

0.1 M Tris–HCl, pH 8.3
2.1 mM 4-Methyl-2-oxopentanoate
0.3 M L-Glutamate
0.8 M L-Alanine
0.2 mM NADH
0.1 mM Pyridoxal 5-phosphate
8 U/ml Glutamic–pyruvic transaminase (GPT)
16 U/ml Lactate dehydrogenase (LDH)

Procedure

Apply the staining solution to the gel surface on a filter paper overlay and incubate the gel in a moist chamber. After 10 to 30 min of incubation, monitor the gel under long-wave UV light for dark (nonfluorescent) bands visible on a light (fluorescent) background. Record the zymogram or photograph using a yellow filter.

OTHER METHODS

A. A spectrophotometric method that uses L-leucine and 2-oxoglutarate as substrates and NAD-dependent D-2-hydroxyisocaproate dehydrogenase as the coupling enzyme can be adapted to detect BCT activity in electrophoretic gels.[2] This method has two limitations: (1) only L-leucine can be used in this method as a branched-chain substrate due to the specific properties of the auxiliary enzyme applied; and (2) D-2-hydroxyisocaproate dehydrogenase preparations are not commercially available.

B. A bioautographic procedure for visualization of BCT activity bands that was developed using *P. cerevisiae* ATCC strain 8081 requires L-leucine for growth and proliferation.[3] This method, however, is not as practical as Methods 1 and 2, given above.

2.6.1.42 — Branched-Chain Amino Acid Transaminase; BCT (continued)

GENERAL NOTES

It is recommended that coupling enzyme preparations containing sulfate ammonium be dialyzed prior to use. A wide variety of 2-oxo-acids can be applied with both methods, except for a few obvious exceptions, e.g., pyruvate in Method 2, which involves LDH.

The enzyme also acts on L-isoleucine and L-valine.[4]

REFERENCES

1. Schadewaldt, P. and Adelmeyer, F., Coupled enzymatic assay for estimation of branched-chain L-amino acid aminotransferase activity with 2-oxo acid substrates, *Anal. Biochem.*, 238, 65, 1996.
2. Schadewaldt, P., Hummel, W., Wendel, U., and Adelmeyer, F., Enzymatic method for determination of branched-chain amino acid aminotransferase activity, *Anal. Biochem.*, 230, 199, 1995.
3. Naylor, S.L. and Klebe, R.J., Bioautography: a general method for the visualization of isozymes, *Biochem. Genet.*, 15, 1193, 1977.
4. NC-IUBMB, *Enzyme Nomenclature*, Academic Press, San Diego, 1992, p. 256 (EC 2.6.1.42, Comments).

2.6.1.44 — Alanine–Glyoxylate Transaminase; AGTA

REACTION   L-Alanine + glyoxylate = pyruvate + glycine
ENZYME SOURCE   Vertebrates
SUBUNIT STRUCTURE   Dimer[#] (mammals)

METHOD

Visualization Scheme

Stage 1

Transferring of proteins (including AGTA) from SDS-PAG onto NC membrane

Stage 2

Treatment of NC membrane with streptavidin-ALP, formation of the biotinylated protein/streptavidin-ALP complexes, and elimination of nonspecific biotin signals by histochemical staining for ALP resulted in formation of a blue dye[VIS]

Stage 3

Treatment of NC membrane with biotinylated PTS1-receptor protein and formation of PTS1-containing protein (including AGTA)/biotinylated PTS1-receptor protein complexes.

Stage 4

Incubation of NC membrane with streptavidin-ALP and formation of PTS1-containing protein (including AGTA)/biotinylated PTS1-receptor protein/streptavidin-ALP complexes. Histochemical staining for ALP resulted in formation of a red dye[VIS']

Visualization Procedure[1]

Stage 1. Transfer proteins from the electrophoretic SDS-PAG onto a nitrocellulose (NC) membrane using the routine capillary blotting procedure (see Part II, Section 6).

Stage 2. Block the NC membrane containing blotted proteins for 1 h with 5% (w/v) dried milk powder (Nutricia) in Tris-buffered saline containing 0.05% (v/v) Tween 20 (TBST). Incubate the membrane for 1 h with streptavidin–alkaline phosphatase (ALP) diluted 1500 times in 1% (w/v) dried milk powder in TBST. Wash the membrane five times for 5 min with TBST and stain with naphthol-AS-GR-phosphate (0.2 mg/ml)/Fast Blue BN (0.35 mg/ml) in alkaline phosphatase buffer (10 mM Tris–HCL, pH 9.2; 100 mM NaCl; 5 mM MgCl$_2$).

This stage is required to eliminate any biotin signals on the membrane. In this way, endogenous biotinylated proteins are shielded by the precipitation of the water-insoluble blue dye formed as a result of the coupling of the diazonium salt (Fast Blue BN) with the naphthol-AS-GR produced by alkaline phosphatase.

Stage 3. Block the NC membrane for 30 min with 5% (w/v) dried milk powder in TBST, and incubate for 1 h in 1% (w/v) dried milk powder in TBST containing approximately 2 µg/ml bacterially expressed PTS1-receptor (purified or present in a crude bacterial lysate).

Stage 4. Wash the NC membrane 3 times with TBST, and incubate for 1 h with streptavidin–ALP diluted 5000 times in 1% (w/v) dried milk powder in TBST. Wash the membrane five times for 5 min with TBST and stain with naphthol-AS-TR-phosphate (0.05 mg/ml)/Fast Blue BN (0.2 mg/ml) in alkaline phosphatase buffer.

Red bands on the developed NC membrane correspond to localization of PST1-containing proteins (including AGTA) on the gel. For details concerning the identification of AGTA bands, see *Notes.*

Notes: The obtained Western blot simultaneously displays easily distinguishable blue and red bands. The blue bands indicate biotinylated peroxisomal proteins, and the red bands indicate PTS1-containing peroxisomal proteins (including AGTA).

The method is not specific for AGTA. It is developed to visualize proteins containing a peroxisomal targeting sequence (PTS1) consisting of the tripeptide serine–lysine–leucine. This sequence is specifically recognized by the PTS1-receptor protein responsible for the import of proteins containing PTS1 into peroxisomes. A bacterially expressed biotinylated C-terminal part of the PTS1-receptor protein is used to specifically recognize peroxisomal proteins. Streptavidin–ALP is then used to detect histochemically the peroxisomal protein–PTS1-receptor protein complexes. The method also detects some other abundant peroxisomal enzymes (urate oxidase, catalase, and palmitoyl-CoA oxidase). When 100 µg of proteins, present in purified rat liver peroxisomes, is subjected to electrophoresis in SDS-PAG, the restricted number of red bands are developed that correspond (in order of decreasing electrophoretic mobility) to palmitoyl-CoA oxidase (the 23-kDa subunit), urate oxidase, AGTA, catalase, and palmitoyl-CoA oxidase (the 70-kDa subunit). Thus, the method is suitable for detection and reliable identification of the enzyme from rat liver only. However, if the molecular weights of the enzyme polypeptides listed above are not changed dramatically during the divergent evolution of vertebrates, one can expect the same arrangement of these polypeptides on SDS-PAGs after electrophoresis of protein preparations obtained from liver peroxisomes of other vertebrates, or at least mammals.

OTHER METHODS

A. Areas of enzymatic production of pyruvate can be detected in long-wave UV light (dark bands on a fluorescent background) using auxiliary LDH as a linking enzyme (e.g., see 2.6.1.2 — GPT, Method 1).

B. Simultaneous detection of pyruvate and glycine produced by AGTA can be accomplished using auxiliary strombine dehydrogenase (see 1.5.1.22 — STRD) as a linking enzyme. This will result in dark (nonfluorescent) bands of AGTA visible on a light (fluorescent) background in long-wave UV light. STRD is not commercially available; however, preparations of this enzyme can be easily obtained in laboratory conditions.[2–4]

GENERAL NOTES

The enzyme is a pyridoxal-phosphate protein. Thus, the inclusion of pyridoxal-phosphate in the staining solutions supposed for the two methods outlined above (see Other Methods, A and B) may prove beneficial.

Two components are known for the animal enzyme. With one component, 2-oxobutanoate can replace glyoxylate. A second (liver) component also catalyzes the reaction of EC 2.6.1.51.[5]

REFERENCES

1. Fransen, M., Brees, C., Van Veldhoven, P.P., and Mannaerts, G.P., The after visualization of peroxisomal proteins containing a C-terminal targeting sequence on Western blot by using the biotinylated PTS1-receptor, *Anal. Biochem.*, 242, 26, 1996.
2. Storey, K.B., Miller, D.C., Plaxton, W.C., and Storey, J.J.M., Gas–liquid chromatography and enzymatic determination of alanopine and strombine in tissues of marine invertebrates, *Anal. Biochem.*, 125, 50, 1982.
3. Storey, K.B., Tissue-specific alanopine dehydrogenase and strombine dehydrogenase from the sea mouse, *Aphrodite aculeata* (Polychaeta), *J. Exp. Zool.*, 225, 369, 1983.
4. Nicchitta, C.V. and Ellington, W.R., Partial purification and characterization of a strombine dehydrogenase from the adductor muscle of the mussel *Modiolus squamosus*, *Comp. Biochem. Physiol.*, 77B, 233, 1984.
5. NC-IUBMB, *Enzyme Nomenclature*, Academic Press, San Diego, 1992, p. 257 (EC 2.6.1.44, Comments; EC 2.6.1.51, Comments).

2.6.1.64 — Glutamine Transaminase K; GTK

OTHER NAMES   Glutamine–phenylpyruvate transaminase (recommended name), glutamine–phenylpyruvate aminotransferase, glutamine transaminase from kidney

REACTION   L-Glutamine + phenylpyruvate = 2-oxoglutaramate + L-phenylalanine

ENZYME SOURCE   Vertebrates (mammals)

SUBUNIT STRUCTURE   Dimer# (mammals)

METHOD

Visualization Scheme

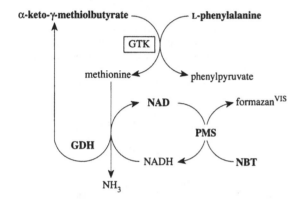

Staining Solution[1]

0.1 M Potassium phosphate buffer, pH 7.2
0.02 M L-Phenylalanine
0.005 M α-Keto-γ-methiolbutyrate
0.032 M NAD
0.1 mM PMS
1 mM NBT
12 U/ml Glutamate dehydrogenase from bovine liver (GDH)

Procedure

Wash the electrophorized PAG twice with distilled water and incubate in a staining solution in the dark at 37°C until dark blue bands appear. Fix the stained gel in 25% ethanol.

Notes: This method is based on the ability of an auxiliary enzyme, glutamate dehydrogenase (EC 1.4.1.3), to exhibit catalytic activity toward methionine.

The cytosolic enzyme from the rat kidney was shown to be identical with cytosolic cysteine conjugate β-lyase (see 4.4.1.13 — CCL) from the same source.[2]

GENERAL NOTES

The enzyme demonstrates broad substrate specificity. It is active with L-glutamine, L-methionine, L-phenylalanine, L-tyrosine, L-histidine, L-cystine, and the corresponding α-keto acids. GTK may be regarded as a fully reversible glutamine (methionine) aromatic amino acid aminotransferase. It is a pyridoxal-phosphate protein that has little activity on pyruvate and glyoxylate (cf. 2.6.1.15 — GPA).[3]

REFERENCES

1. Abraham, D.G. and Cooper, A.J.L., Glutamine transaminase K and cysteine S-conjugate β-lyase activity stains, *Anal. Biochem.*, 197, 421, 1991.
2. Cooper, A.J.L. and Anders, M.W., Glutamine transaminase K and cysteine conjugate β-lyase, *Ann. N.Y. Acad. Sci.*, 585, 118, 1990.
3. NC-IUBMB, *Enzyme Nomenclature*, Academic Press, San Diego, 1992, p. 259 (EC 2.6.1.64, Comments).

2.7.1.1 — Hexokinase; HK

OTHER NAMES	Glucokinase, hexokinase type IV, hexokinase D
REACTION	ATP + D-hexose = ADP + D-hexose 6-phosphate
ENZYME SOURCE	Bacteria, green algae, fungi, plants, protozoa, invertebrates, vertebrates
SUBUNIT STRUCTURE	Monomer (plants, invertebrates, vertebrates)

METHOD

Visualization Scheme

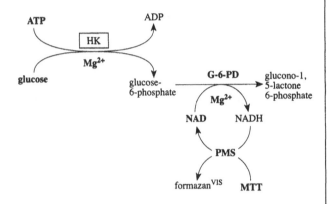

Staining Solution[1] (modified)

A. 0.1 M Tris–HCl buffer, pH 7.8 — 25 ml
   D-Glucose — 50 mg
   ATP (disodium salt) — 30 mg
   NAD — 15 mg
   MTT — 5 mg
   PMS — 1 mg
   MgCl$_2$·6H$_2$O — 10 mg
   Glucose-6-phosphate dehydrogenase (G-6-PD; NAD dependent; Sigma G5760 or G5885) — 20 U
B. 2% Agarose solution (60°C) — 25 ml

Procedure

Mix A and B components of the staining solution and pour the mixture over the surface of the gel. Incubate the gel in the dark at 37°C until dark blue bands appear. Fix the stained gel in 25% ethanol.

Notes: Final glucose concentrations of 1 and 100 mM are usually used to identify hexokinase and glucokinase (see General Notes) isozymes, respectively.[2,3]

When NADP-dependent G-6-PD is used as a linking enzyme, NADP should be substituted for NAD in the staining solution. However, the use of NAD-dependent G-6-PD results in cost savings because NAD is several times cheaper than NADP.

When high concentrations of glucose are used in the staining solution, the bands of glucose oxidase (see 1.1.3.4 — GO) and glucose dehydrogenase (see 1.1.1.47 — GD) can sometimes also be developed on hexokinase zymograms.

GENERAL NOTES

D-Mannose, D-fructose, sorbitol, and D-glucosamine also can act as acceptors; ITP and dATP can act as donors.

The enzyme from microorganisms and some invertebrates is known as glucokinase (EC 2.7.1.2). Glucokinase is activated by glucose at high concentrations. The HK isozyme from the mammalian liver has also sometimes been called glucokinase.[4]

REFERENCES

1. Eaton, G.M., Brewer, G.J., and Tashian, R.E., Hexokinase isozyme patterns of human erythrocytes and leukocytes, *Nature*, 212, 944, 1966.
2. Harris, H. and Hopkinson, D.A., *Handbook of Enzyme Electrophoresis in Human Genetics*, North-Holland, Amsterdam, 1976 (loose-leaf, with supplements in 1977 and 1978).
3. Richardson, B.J., Baverstock, P.R., and Adams, M., *Allozyme Electrophoresis: A Handbook for Animal Systematics and Population Studies*, Academic Press, Sydney, 1986, p. 198.
4. NC-IUBMB, *Enzyme Nomenclature*, Academic Press, San Diego, 1992, p. 262 (EC 2.7.1.1, Comments; EC 2.7.1.2, Comments).

REACTION	ATP + D-fructose = ADP + D-fructose-6-phosphate
ENZYME SOURCE	Plants, invertebrates
SUBUNIT STRUCTURE	Monomer (plants, invertebrates)

METHOD

Visualization Scheme

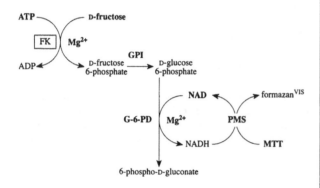

Staining Solution[1] (modified)

A. 0.1 M Tris–HCl buffer, pH 7.8 .... 25 ml
    D-Fructose .... 100 mg
    ATP .... 50 mg
    NAD .... 20 mg
    MTT .... 7 mg
    PMS .... 1 mg
    $MgCl_2 \cdot 6H_2O$ .... 10 mg
    Glucose-6-phosphate isomerase (GPI) .... 20 U
    Glucose-6-phosphate dehydrogenase (G-6-PD; NAD dependent; Sigma G5760 or G5885) .... 20 U
B. 2% Agarose solution (60°C) .... 25 ml

Procedure

Mix A and B components of the staining solution and pour the mixture over the surface of the gel. Incubate the gel in the dark at 37°C until dark blue bands appear. Fix the stained gel in 25% ethanol.

Notes: In mushrooms and invertebrates this enzyme is perhaps identical to hexokinase (2.7.1.1 — HK). Identical allozymic patterns were observed on FK and HK zymograms in mushrooms, nemerteans, sea urchins, sea stars, and insects.[2] Some FK isozymes, however, are specific for fructose and are not active with glucose. Such a fructose-specific isozyme was found, for example, in the nemertean *Tetrastemma nigrifrons*.[3] Two FK isozymes electrophoretically revealed in the plant *Arabidopsis thaliana* show a high specificity for fructose and do not stain when glucose or mannose is used as a substrate.[4]

REFERENCES

1. Jelnes, J.E., Identification of hexokinases and localization of a fructokinase and tetrazolium oxidase locus in *Drosophila melanogaster*, *Hereditas*, 67, 291, 1971.
2. Manchenko, G.P., The use of allozyme patterns in genetic identification of enzymes, *Isozyme Bull.*, 23, 102, 1990.
3. Manchenko, G.P. and Kulikova, V.I., Enzyme and colour variation in the hoplonemertean *Tetrastemma nigrifrons* from the Sea of Japan, *Hydrobiologia*, 337, 69, 1996.
4. Gonzali, S., Pistelli, L., De Bellis, L., and Alpi, A., Characterization of two *Arabidopsis thaliana* fructokinases, *Plant Sci.*, 160, 1107, 2001.

2.7.1.6 — Galactokinase; GALK

REACTION	ATP + D-galactose = ADP + α-D-galactose-1-phosphate
ENZYME SOURCE	Bacteria, fungi, protozoa, vertebrates
SUBUNIT STRUCTURE	Monomer (vertebrates)

Method 1

Visualization Scheme

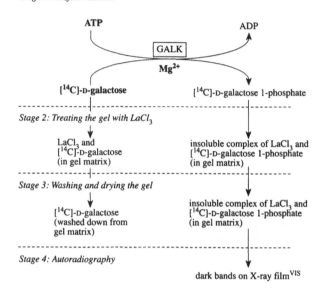

Stage 1: Enzyme reaction

Stage 2: Treating the gel with LaCl₃

Stage 3: Washing and drying the gel

Stage 4: Autoradiography

Reaction Mixture[1]

A. 0.2 M Tris–HCl buffer, pH 7.2
   3.61 mM ATP
   7.84 mM MgCl$_2$
B. [¹⁴C]-D-Galactose (specific activity, 45 to 50 mCi/mmol)
C. 0.1 M Tris–HCl buffer, pH 7.0
   0.1 M LaCl$_3$

Procedure

Stage 1. Add 50 μl of B to 15 ml of A. Incubate the gel in the resultant mixture at 37°C for 1 to 2 h.

Stage 2. After incubation place the gel in solution C and expose at 4°C for 6 h.

Stage 3. Wash the gel for 3 h in distilled water. Dry the gel.

Stage 4. Autoradiograph the dry gel with x-ray film (Kodak Blue Brand BB-54) for 1 to 2 weeks. Develop the exposed x-ray film.

The dark bands on the developed x-ray film correspond to localization of GALK activity in the gel.

Method 2

Visualization Scheme

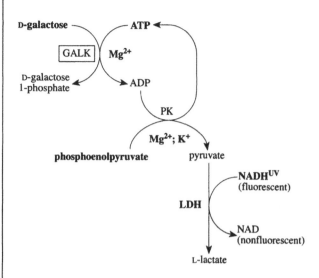

Staining Solution*

0.1 M Tris–HCl buffer, pH 7.5	5 ml
ATP	10 mg
D-Galactose	15 mg
KCl	1.5 mg
MgCl$_2$·6H$_2$O	2 mg
NADH	3.5 mg
Pyruvate kinase (PK)	40 U
Lactate dehydrogenase (LDH)	30 U
Phosphoenolpyruvate	10 mg

Procedure

Apply the staining solution to the gel surface on a filter paper overlay. Incubate the gel at 37°C for 30 to 60 min and view under long-wave UV light. Dark (nonfluorescent) bands of GALK are visible on a light (fluorescent) background. Record the zymogram or photograph using a yellow filter.

Notes: When a final zymogram that is visible in daylight is required, counterstain the processed gel with MTT–PMS solution. Achromatic bands of GALK activity appear on a blue background. Wash the negatively stained gel in water and fix in 25% ethanol.

Two additional gels should be stained for phosphoenolpyruvate phosphatase (see 3.1.3.18 — PGP, Method 2) and ATPase (see 3.6.1.3 — ATPASE, Method 2) activities, which can also be developed by this method.

---

* New; recommended for use.

227

METHOD 3

Visualization Scheme

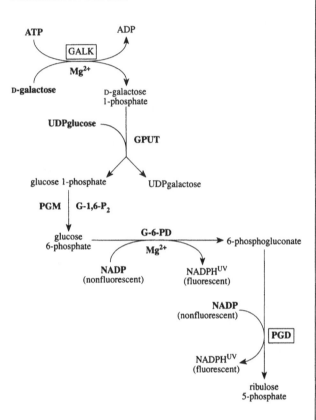

Staining Solution[2]

A. 1 *M* Tris–HCl buffer, pH 8.0   1.5 ml

     7.2 mg/ml D-Galactose   0.1 ml

     182 mg/ml ATP (disodium salt)   0.2 ml

     8 mg/ml UDPglucose (disodium salt)   0.75 ml

     1.4 mg/ml Glucose-1,6-diphosphate (G-1,6-P_2;

        tetracyclohexylammonium salt; $4H_2O$)   75 μl

     8 mg/ml $MgCl_2 \cdot 6H_2O$ 0.75 ml

     16 mg/ml NADP (disodium salt)   0.75 ml

     20 U/mg Galactose-1-phosphate uridyl

        transferase (GPUT)   0.2 mg

     2000 U/ml Phosphoglucomutase (PGM)   30 μl

     700 U/ml Glucose-6-phosphate dehydrogenase

        (G-6-PD)   60 μl

     120 U/ml Phosphogluconate dehydrogenase

        (PGD)   30 μl

     10 μl/ml 2-Mercaptoethanol   0.75 ml

B. 1% Agar solution (60°C)   5 ml

Procedure

Mix A and B components of the staining solution and pour the mixture over the surface of the gel. Incubate the gel at 37°C for 30 to 60 min and monitor under long-wave UV light. Light (fluorescent) bands of GALK activity appear on a dark (nonfluorescent) background of the gel. Record the zymogram or photograph using a yellow filter.

Notes: Phosphogluconate dehydrogenase is included in the staining solution to obtain additional reduction of NADP to NADPH and so enhance the sensitivity of the method.

When preparations with high activity of GALK are used for electrophoresis, phosphogluconate dehydrogenase may be omitted from the staining solution. This will allow the use of NAD-dependent glucose-6-phosphate dehydrogenase (Sigma G 5760 or G 5885) instead of the NADP-dependent form of the enzyme and NAD instead of NADP.

MTT and PMS can be added to staining solution A lacking 2-mercaptoethanol to detect positively stained activity bands of GALK visible in daylight.[3]

REFERENCES

1. Nichols, E.A., Elsevier, S.M., and Ruddle, F.H., A new electrophoretic technique for mouse, human and Chinese hamster galactokinase, *Cytogenet. Cell Genet.*, 13, 275, 1974.
2. Harris, H. and Hopkinson, D.A., *Handbook of Enzyme Electrophoresis in Human Genetics*, North-Holland, Amsterdam, 1976 (loose-leaf, with supplements in 1977 and 1978).
3. Stambolian, D., Galactokinase: technique for polyacrylamide gel isoelectric focusing, *Electrophoresis*, 7, 390, 1986.

2.7.1.7 — Mannokinase; MK

REACTION	ATP + D-mannose = ADP + D-mannose 6-phosphate
ENZYME SOURCE	Bacteria, invertebrates
SUBUNIT STRUCTURE	Unknown or no data available

METHOD

Visualization Scheme

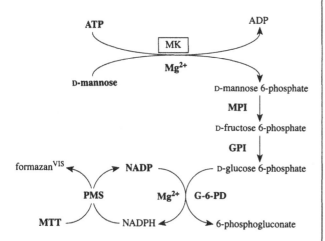

Staining Solution[1]

A. 60 mM Tris–HCl buffer, pH 8.0 — 25 ml
D-Mannose — 100 mg
ATP — 25 mg
NADP — 2.5 mg
MTT — 2.5 mg
PMS — 1.25 mg
Glucose-6-phosphate dehydrogenase (G-6-PD) — 0.7 U
Glucose-6-phosphate isomerase (GPI) — 7 U
Mannose-6-phosphate isomerase (MPI) — 0.4 U
B. 2% Agarose solution containing — 25 ml
25 mM MgCl$_2$ (60°C)

Procedure

Mix A and B components of the staining solution and pour the mixture over the surface of the gel. Incubate the gel in the dark at 37°C until dark blue bands appear. Fix the stained gel in 25% ethanol.

Notes: In *Drosophila melanogaster*, MK is identical to one of the three isozymes revealed on a hexokinase (2.7.1.1 — HK) zymogram.[1]

OTHER METHODS

Coupled reactions catalyzed by exogenous pyruvate kinase and lactate dehydrogenase can be used to detect the production of ADP by MK (e.g., see 2.7.1.6 — GALK, Method 2). Two additional gels should be stained for phosphoenolpyruvate phosphatase (see 3.1.3.18 — PGP, Method 2) and ATPase (see 3.6.1.3 — ATPASE, Method 2), for which activity bands can also be developed by this method.

REFERENCES

1. Jelnes, J.E., Identification of hexokinases and localization of a fructokinase and tetrazolium oxidase locus in *Drosophila melanogaster*, *Hereditas*, 67, 291, 1971.

2.7.1.11 — 6-Phosphofructokinase; PFK

OTHER NAMES	Phosphofructokinase, phosphohex-okinase, phosphofructokinase 1
REACTION	ATP + D-fructose-6-phosphate = ADP + D-fructose-1,6-bisphosphate
ENZYME SOURCE	Bacteria, fungi, plants, protozoa, invertebrates, vertebrates
SUBUNIT STRUCTURE	Monomer (fungi, vertebrates)

Method 1

Visualization Scheme

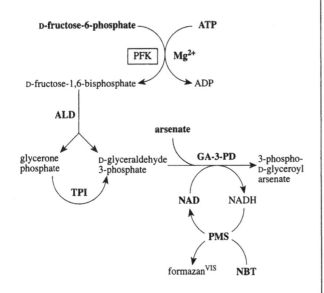

Staining Solution[1] (modified)

A.	100 mM Tris–HCl buffer, pH 8.3	5 ml
	NBT	4 mg
	PMS	0.24 mg
	NAD	8.6 mg
	MgCl$_2$ (anhydrous)	3.8 mg
	ATP	6.25 mg
	D-Fructose-6-phosphate (sodium salt)	3.5 mg
	EDTA	7.5 mg
	Triose-phosphate isomerase (TPI)	250 U
	Aldolase (ALD)	18 U
	Glyceraldehyde-3-phosphate dehydrogenase (GA-3-PD)	20 U
	Sodium arsenate	13 mg
B.	1% Agarose solution (60°C)	5 ml

Procedure

Mix A and B components of the staining solution and pour the mixture over the surface of the gel. Incubate the gel in the dark at 37°C until dark blue bands appear. Fix the stained gel in 25% ethanol.

Method 2

Visualization Scheme

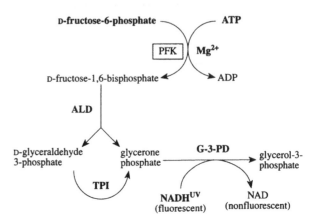

Staining Solution[2]

0.07 M Tris–HCl buffer, pH 8.1
2.8 mM MgCl$_2$
1.1 mM NADH
1.8 mM ATP
2.2 mM D-Fructose-6-phosphate
1.5 U/ml Glycerol-3-phosphate dehydrogenase (G-3-PD)
2 U/ml Aldolase (ALD)
111 U/ml Triose-phosphate isomerase (TPI)

Procedure

Apply the staining solution to the gel surface on a filter paper overlay. Incubate the gel at 37°C and monitor under long-wave UV light. Dark (nonfluorescent) bands of PFK activity are visible on a light (fluorescent) background. Record the zymogram or photograph using a yellow filter.

Notes: When a zymogram that is visible in daylight is required, counterstain the processed gel with MTT–PMS solution. Achromatic bands of PFK activity become readily visible on a blue background of the gel.

METHOD 3

Visualization Scheme

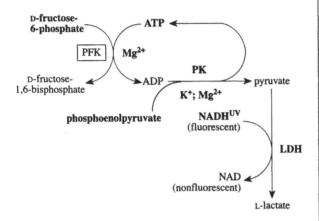

Staining Solution*

0.1 M Tris–HCl buffer, pH 8.3	5 ml
ATP (disodium salt)	10 mg
D-Fructose-6-phosphate	8 mg
Phosphoenolpyruvate	7 mg
NADH	3 mg
MgCl$_2$·6H$_2$O	8 mg
KCl	12 mg
Pyruvate kinase (PK)	18 U
Lactate dehydrogenase (LDH)	25 U

Procedure

Apply the staining solution to the gel surface on a filter paper overlay. Incubate the gel at 37°C and monitor under long-wave UV light. Dark (nonfluorescent) bands of PFK activity are visible on a light (fluorescent) background. Record the zymogram or photograph using a yellow filter. When a zymogram that is visible in daylight is required, counterstain the processed gel with MTT–PMS solution to obtain white bands of PFK activity on a blue background of the gel.

Notes: Stain a control gel in a staining solution lacking ATP, fructose-6-phosphate, and pyruvate kinase to identify possible bands of phosphoenolpyruvate phosphatase activity (see 3.1.3.18 — PGP, Method 2). An additional gel should also be stained for ATPase (see 3.6.1.3 — ATPASE, Method 2), for which bands also can be developed by this method.

OTHER METHODS

The immunoblotting procedure (for details, see Part II) can be used for visualization of PFK on electrophoretic gels using monoclonal antibodies against the human enzyme.[3]

GENERAL NOTES

The plant enzyme is inactivated *in vivo* by light and *in vitro* by dithiothrietol.[4]

REFERENCES

1. Niessner, H. and Beutler, E., Starch gel electrophoresis of phosphofructokinase in red cells, *Biochem. Med.*, 9, 73, 1974.
2. Anderson, J.E. and Giblett, E.R., Intraspecific red cell enzyme variation in the pigtailed macaque (*Macaca nemestrina, Biochem. Genet.*, 13, 189, 1975.
3. Vora, S., Wims, L.A., Durham, S., and Morrison, S.I., Production and characterization of monoclonal antibodies to the subunits of human phosphofructokinase: new tools for the immunochemical and genetic analyses of isozymes, *Blood*, 58, 823, 1981.
4. Kachru, R.B. and Anderson, L.E., Inactivation of pea leaf phosphofructokinase by light and dithiothreitol, *Plant Physiol.*, 55, 199, 1975.

* New; recommended for use.

REACTION ATP + adenosine = ADP + AMP
ENZYME SOURCE Fungi, protozoa, invertebrates, vertebrates
SUBUNIT STRUCTURE Monomer (vertebrates)

METHOD

Visualization Scheme

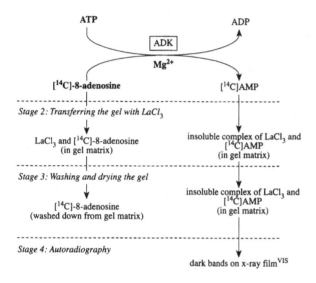

Stage 1: Enzyme reaction

ATP   ADP

ADK

Mg²⁺

[¹⁴C]-8-adenosine   [¹⁴C]AMP

Stage 2: Transferring the gel with LaCl₃

LaCl₃ and [¹⁴C]-8-adenosine   insoluble complex of LaCl₃ and
(in gel matrix)   [¹⁴C]AMP
                                      (in gel matrix)

Stage 3: Washing and drying the gel

[¹⁴C]-8-adenosine   insoluble complex of LaCl₃ and
(washed down from gel matrix)   [¹⁴C]AMP
                                      (in gel matrix)

Stage 4: Autoradiography

                                      dark bands on x-ray film^{VIS}

Reaction Mixture[1]

A. 20 mM Potassium phosphate buffer, pH 6.2   20 ml
    ATP (disodium salt)   4 mg
    MgCl₂ (anhydrous)   1 mg
    [¹⁴C]-8-Adenosine (40 to 60 μCi/mmol)   5 μCi
B. 0.1 M Tris–HCl buffer, pH 7.0
    0.1 M LaCl₃

Procedure

Stage 1. Incubate the gel in solution A at 37°C for 1 h.
Stage 2. Place the gel in solution B and incubate at 4°C for 6 h.
Stage 3. Wash the gel for 3 h with distilled water and dry.
Stage 4. Autoradiograph the dry gel with x-ray film for 2 weeks. The dark bands on the developed x-ray film correspond to localization of ADK activity in the gel.

Notes: The addition of dithiothreitol (about 0.2 mg/ml) to the electrophoretic starch gel is recommended during the last 30 sec of gel cooking.

OTHER METHODS

A. Areas of ADK localization and production of ADP can also be detected in long-wave UV light (dark bands on a fluorescent background) using linked reactions catalyzed by two auxiliary enzymes, pyruvate kinase and lactate dehydrogenase, in the presence of phosphoenolpyruvate and NADH. This method, however, requires a control staining of two additional gels for phosphoenolpyruvate phosphatase (see 3.1.3.18 — PGP, Method 2) and ATPase (see 3.6.1.3 — ATPASE, Method 2).

B. Areas of AMP production can be visualized as positively stained bands using four linking reactions catalyzed by auxiliary enzymes 5'-nucleotidase, adenosine deaminase, purine-nucleoside phosphorylase, and xanthine oxidase (e.g., see 2.7.4.3 — AK, Method 3).

GENERAL NOTES

The reaction catalyzed by ADK is essentially irreversible. Thus, the backward ADK reaction may not be used to develop some other practical methods for ADK detection.

REFERENCES

1. Klobutcher, L., Nichols, E., Kucherlapati, R., and Ruddle, F.H., Assignment of the gene for human adenosine kinase to chromosome 10 using a somatic cell hybrid clone panel, *Cytogenet. Cell Genet.*, 16, 171, 1976.

REACTION	ATP + thymidine = ADP + thymidine 5'-phosphate
ENZYME SOURCE	Bacteria, plants, invertebrates, vertebratesv
SUBUNIT STRUCTURE	See General Notes

METHOD

Visualization Scheme

Stage 1: Enzyme reaction

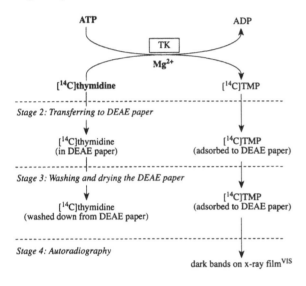

Stage 2: Transferring to DEAE paper

Stage 3: Washing and drying the DEAE paper

Stage 4: Autoradiography

dark bands on x-ray film[VIS]

Reaction Mixture[1]

0.01 M Tris–HCl buffer, pH 8.0
57 μM [^{14}C]-Thymidine (26.9 to 29.6 mCi/mmol)
5 mM ATP
5 mM MgCl$_2$

Procedure

Stages 1 and 2. Apply a sheet of Whatman DE 81 (DEAE cellulose) paper to the surface of the electrophorized gel and pipette on the reaction mixture. Wrap the gel–DEAE paper combination in Saran wrap and incubate at 37°C for 1.5 h.

Stage 3. Remove the DEAE paper and wash with 15 l of distilled water on a Buchner funnel to remove the unreacted [^{14}C]-thymidine, leaving [^{14}C]-TMP absorbed to the DEAE paper at the sites of the enzyme reaction. Dry the DEAE paper.

Stage 4. Apply the DEAE paper to a sheet of x-ray film (Blue Brand, Kodak). Place the x-ray film–DEAE paper combination between two glass plates wrapped in a dark bag. Develop the x-ray film exposed for 2 to 4 weeks with Kodak D 19 developer.

The dark bands on the developed x-ray film correspond to localization of TK activity on the gel.

Notes: The phosphorylated products of thymidine, iododeoxycytidine, and iododeoxyuridine are much more negatively charged than the substrates and bind firmly to DEAE paper. Thus, the paper binds labeled reaction products but not labeled substrates.

Using ^{125}iododeoxycytidine or ^{125}iododeoxyuridine as substrates, TK can also be visualized after electrophoresis, transferring to ion exchange DEAE paper and subsequent autoradiography.[2]

The Herpes virus and mammalian TK differ in substrate specificity. Halogenated deoxyuridine is efficiently phosphorylated by both mammalian and viral TK, whereas only viral TK efficiently utilizes halogenated deoxycytidine.

OTHER METHODS

Gel areas occupied by TK (areas of ADP production) can be detected using two auxiliary enzymes, pyruvate kinase and lactate dehydrogenase, as dark bands visible in long-wave UV light on a fluorescent background of the gel. This method, however, requires a control staining of two additional gels for phosphoenolpyruvate phosphatase (see 3.1.3.18 — PGP, Method 2) and ATPase (see 3.6.1.3 — ATPASE, Method 2), whose activity bands also can be developed by this method.

GENERAL NOTES

The mitochondrial form of mammalian TK is expressed in all cells and tissues. The active enzyme is a monomer of about 30 kDa.[3]

Human cytosolic TK is a tetramer (subunit molecular mass of about 24 kDa) in the presence of ATP, but a dimer in the presence of thymidine or without substrates.[4]

Native TK from broad bean (*Vicia faba* L.) seedlings presumably exists as a homotetramer of 30-kDa subunits and a monomer.[5]

REFERENCES

1. Migeon, B.R., Smith, S.W., and Leddy, C.L., The nature of thymidine kinase in the human–mouse hybrid cell, *Biochem. Genet.*, 3, 583, 1969.
2. Van Den Berg, K.J., Direct assay of thymidine kinase bound to ion-exchange paper for dot spotting and enzyme blotting analysis, *Anal. Biochem.*, 155, 149, 1986.
3. Jansson, O., Bohman, C., Munchpetersen, B., and Eriksson, S., Mammalian thymidine kinase-2: direct photoaffinity-labeling with [^{32}P]-dTTP of the enzyme from spleen, liver, heart and brain, *Eur. J. Biochem.*, 206, 485, 1992.
4. Munchpetersen, B., Tyrsted, G., and Cloos, L., Reversible ATP-dependent transition between two forms of human cytosolic thymidine kinase with different enzymatic properties, *J. Biol. Chem.*, 268, 15621, 1993.
5. Nosov, V.A., Thymidine kinase in higher-plant cells: 2. Isolation and purification of the enzyme from broad bean seedlings, *Russ. J. Plant Physiol.*, 42, 652, 1995.

OTHER NAMES	Triose kinase
REACTION	ATP + D-glyceraldehyde = ADP + D-glyceraldehyde-3-phosphate
ENZYME SOURCE	Vertebrates
SUBUNIT STRUCTURE	Dimer[a] (mammals)

METHOD

Visualization Scheme

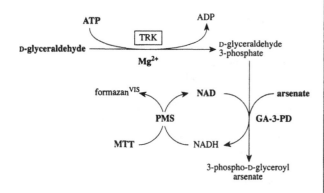

Staining Solution[1]

A.	0.2 M Tris–HCl buffer, pH 8.0	15 ml
	D,L-Glyceraldehyde	20 mg
	ATP	15 mg
	1.0 M MgCl$_2$	0.3 ml
	Sodium arsenate	50 mg
	0.5% NAD	0.5 ml
	0.5% MTT	0.5 ml
	0.5% PMS	0.5 ml
	Glyceraldehyde-3-phosphate dehydrogenase (GA-3-PD)	12 U
B.	2% Agar solution (60°C)	10 ml

Procedure

Mix A and B components of the staining solution and pour the mixture over the surface of the gel. Incubate the gel in the dark at 37°C until dark blue bands appear. Fix the stained gel in 25% ethanol.

Notes: In some species nonspecific staining of NAD-dependent dehydrogenases (e.g., aldehyde dehydrogenase) or aldehyde oxidase can occur with this method. Thus, interpretation of TRK zymograms should be made with caution.

OTHER METHODS

To avoid the problem of nonspecific stainings described above, the negative fluorescent method of detection of the product ADP may be used. This method involves two auxiliary enzymes, pyruvate kinase and lactate dehydrogenase (e.g., see 2.7.1.11 — PFK, Method 3). It should be taken into account, however, that this method also can detect aldehyde reductase (see 1.1.1.1 — ADH, Method 2) bands (due to the presence of glyceraldehyde and NADH in the stain), phosphoenolpyruvate phosphatase (see 3.1.3.18 — PGP, Method 2) bands (due to the presence of phosphoenolpyruvate and NADH), and ATPase (see 3.6.1.3 — ATPASE, Method 2) bands (due to the presence of ATP). Thus, at least three additional gels should be stained as controls to identify bands not caused by TRK activity.

REFERENCES

1. Aebersold, P.B., Personal communication, 1991.

2.7.1.30 — Glycerol Kinase; GLYCK

REACTION	ATP + glycerol = ADP + *sn*-glycerol-3-phosphate
ENZYME SOURCE	Bacteria, fungi, protozoa, invertebrates, vertebrates
SUBUNIT STRUCTURE	Dimer (invertebrates)

METHOD 1

Visualization Scheme

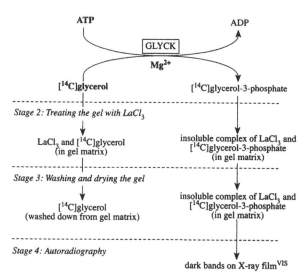

Stage 1: Enzyme reaction

Stage 2: Treating the gel with LaCl₃

Stage 3: Washing and drying the gel

Stage 4: Autoradiography

Reaction Mixture[1]

A. 49.5 m*M* Tris–384 m*M* glycine buffer, pH 8.4
    50 m*M* ATP
    50 m*M* MgCl₂
    7.4 μ*M* [¹⁴C]glycerol (specific activity, 134 mCi/mmol)
B. 0.1 *M* Tris–HCl buffer, pH 7.0
    0.1 *M* LaCl₃

Procedure

Stage 1. Incubate the gel in solution A at 37°C for 30 to 45 min.
Stage 2. Rinse the gel in water, place it in solution B, and incubate at 4°C for 6 h.
Stage 3. Wash the gel in deionized water for 12 h and dry.
Stage 4. Autoradiograph the dry gel with x-ray film for 2 to 6 days. Develop the exposed x-ray film.

The dark bands on the x-ray film correspond to localization of GLYCK activity in the gel.

METHOD 2

Visualization Scheme

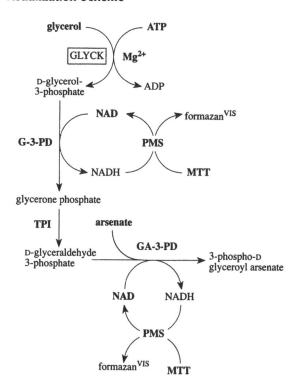

Staining Solution*

A.	0.1 *M* Tris–HCl buffer, pH 8.5	25 ml
	Glycerol	0.3 ml
	ATP (disodium salt)	20 mg
	MgCl₂·6H₂O	5 mg
	NAD	25 mg
	MTT	7 mg
	PMS	1 mg
	Sodium arsenate	50 mg
	Glycerol-3-phosphate dehydrogenase (G-3-PD)	90 U
	Triose-phosphate isomerase (TPI)	1000 U
	Glyceraldehyde-3-phosphate dehydrogenase (GA-3-PD)	90 U
B.	2% Agarose solution (60°C)	15 ml

Procedure

Mix A and B components of the staining solution and pour the mixture over the surface of the gel. Incubate the gel in the dark at 37°C until dark blue bands appear. Fix the stained gel in 25% ethanol.

Notes: Auxiliary enzymes triose-phosphate isomerase and glyceraldehyde-3-phosphate dehydrogenase are included to double the NAD-into-NADH conversion and formation of blue formazan. When preparations with high activity of GLYCK are used for electrophoresis, these enzymes and arsenate may be omitted from the staining solution.

235

2.7.1.30 — Glycerol Kinase; GLYCK (continued)

Method 3

Visualization Scheme

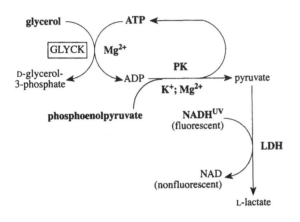

Staining Solution*

0.05 M Tris–HCl buffer, pH 8.5	10 ml
ATP (disodium salt)	10 mg
Glycerol	0.2 ml
Phosphoenolpyruvate (tricyclohexylammonium salt)	8 mg
NADH	4 mg
$MgCl_2 \cdot 6H_2O$	8 mg
KCl	20 mg
Pyruvate kinase (PK)	60 U
Lactate dehydrogenase (LDH)	50 U

Procedure

Apply the staining solution to the gel on a filter paper overlay. Incubate the gel at 37°C and monitor under long-wave UV light. Dark (nonfluorescent) bands of GLYCK activity are visible on a light (fluorescent) background. Record the zymogram or photograph using a yellow filter.

Notes: Stain two additional gels as controls for phosphoenolpyruvate phosphatase (see 3.1.3.18 — PGP, Method 2) and ATPase (see 3.6.1.3 — ATPASE, Method 2), which also can develop on GLYCK zymograms obtained by this method.

References

1. Tischfield, J.A., Bernhard, H.P., and Ruddle, F.H., A new electrophoretic–autoradiographic method for visual detection of phosphotransferases, *Anal. Biochem.*, 53, 545, 1973.

2.7.1.35 — Pyridoxine Kinase; PNK

OTHER NAMES	Pyridoxal kinase (recommended name)
REACTION	Pyridoxal + ATP = pyridoxal 5-phosphate + ADP (also see General Notes)
ENZYME SOURCE	Bacteria, fungi, vertebrates
SUBUNIT STRUCTURE	Unknown or no data available

Method

Visualization Scheme

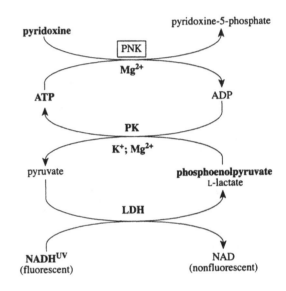

Staining Solution[1]

0.1 M Tris–HCl buffer, pH 8.0
0.2 mM Pyridoxine
0.5 mM ATP
0.2 mM NADH
5 mM Phosphoenolpyruvate
0.1 M KCl
0.01 M $MgCl_2$
7.8 U/ml Lactate dehydrogenase (LDH)
10.2 U/ml Pyruvate kinase (PK)

Procedure

Apply the staining solution to the gel surface on a filter paper overlay. Incubate the gel at 37°C and monitor under long-wave UV light. Dark (nonfluorescent) bands of PNK are visible on a light (fluorescent) background. Record the zymogram or photograph using a yellow filter.

---

* New; recommended for use.

2.7.1.35 — Pyridoxine Kinase; PNK (continued)

Notes: Faint additional bands of activity may also appear due to other kinases or phosphatases that catalyze the ATP-into-ADP conversion or formation of pyruvate from phosphoenolpyruvate. It is necessary, therefore, to carry out control stains on additional gels. The composition of a control stain for phosphoenolpyruvate phosphatase is given in 3.1.3.18 — PGP (Method 2) and for ATPase in 3.6.1.3 — ATPASE, Method 2.

GENERAL NOTES

Pyridoxine, pyridoxamine, and various derivatives can also act as acceptors.[2]

REFERENCES

1. Chern, C.J. and Beutler, E., Biochemical and electrophoretic studies of erythrocyte pyridoxine kinase in white and black Americans, *Am. J. Hum. Genet.*, 28, 9, 1976.
2. NC-IUBMB, *Enzyme Nomenclature*, Academic Press, San Diego, 1992, p. 266 (EC 2.7.1.35, Comments).

2.7.1.37 — Protein Kinase; PROTK

OTHER NAMES	Phosphorylase *b* kinase kinase, glycogen synthase *a* kinase, hydroxylalkyl-protein kinase, serine (threonine) protein kinase
REACTION	ATP + protein = ADP + phosphoprotein (see General Notes)
ENZYME SOURCE	Fungi, plants, protozoa, invertebrates, vertebrates
SUBUNIT STRUCTURE	Heterotetramer# (invertebrates), heterotrimer# (vertebrates); see General Notes

METHOD

Visualization Scheme

Stage 1: Enzyme reaction

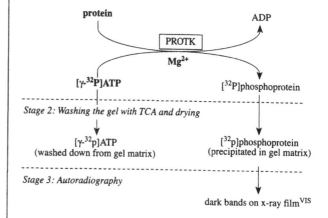

Stage 2: Washing the gel with TCA and drying

[γ-³²p]ATP
(washed down from gel matrix)

[³²p]phosphoprotein
(precipitated in gel matrix)

Stage 3: Autoradiography

dark bands on x-ray film^{VIS}

Reaction Mixture[1]

A. 10 m*M* Tris–HCl buffer, pH 6.75
   10 m*M* MgCl$_2$
   0.1 m*M* 2-Mercaptoethanol
   0.1 m*M* EDTA
   0.2 m*M* Ethylene glycol-bis(β-aminoethyl ether) *N*,*N*,*N'*,*N'*-tetraacetic acid

B. [γ-³²P]ATP (specific activity, 20 to 40 Ci/mmol; 1 mCi/ml)

C. 5 mg/ml Casein, or histone, in solution A containing 0.4 *M* NaCl

D. 5 mg/ml Protamine sulfate, or phosvitin, in solution A containing 0.2 *M* NaCl

E. 0.1 g/ml Trichloroacetic acid (TCA)
   5 m*M* Sodium pyrophosphate
   5 m*M* NaH$_2$PO$_4$

Procedure

Stage 1. Mix 150 μl of solution A with 15 μl of B and apply onto the "running" acetate cellulose strip (5.7×5.7 cm) with a fine brush. Prepare "substrate" acetate cellulose by incubation of another strip in solution C or D, depending on the specificity of the investigated PROTK. Press together the substrate and running strips between two glass slides for 30 min.

Stage 2. Fix the running strip in solution E. After extensive washing in this solution (five baths of 300 ml changed every 30 min), dry the strip.

Stage 3. Autoradiograph the dry running strip at room temperature with "no screen" film for 6 to 16 h. Develop the exposed x-ray film.

The dark bands on the film correspond to localization of PROTK activity on the running acetate cellulose gel.

Notes: The method is highly sensitive. Bands corresponding to PROTK activity of 0.25 pmol of phosphate transferred per minute at 30°C are easily detected after an overnight exposure. The method using acetate cellulose gel seems to be at least 50-fold more sensitive than methods using routine PAG.

A similar procedure was used to detect PROTK activity bands after electrophoresis of crude protein mixtures in SDS-PAGs containing 10 μg/ml bovine serum albumin. After incubating the gel in solution containing radiolabeled ATP, the excess ATP was removed by rinsing the gel briefly with water and incubating about 10 h in 600 ml of 40 mM HEPES (pH 7.4) containing 20 g of Dowex 2×8–50 anion exchange resin (chloride form). Labeled proteins were then fixed into the gel by a 1-h incubation in 250 ml of 10% isopropanol, 5% acetic acid, 1% sodium pyrophosphate. The processed gel was dried and exposed to x-ray film. Multiple PROTK bands were detected after electrophoresis of a mixture of human red blood cell proteins using this technique.[2]

OTHER METHODS

The immunoblotting procedure (for details, see Part II) may also be used to detect PROTK. Monoclonal antibodies reactive with human, rat, and mouse enzymes are available from Sigma.

GENERAL NOTES

Protein kinases represent a group of enzymes that are under review by NC-IUBMB.[3] There is evidence that PROTK from a particular source preferentially attacks the phosphoprotein from the same source. Some enzymes are activated by cyclic AMP or cyclic GMP, and some enzymes by neither. PROTKs can catalyze an autophosphorylation reaction that can be used to detect their activity bands on PAGs by autoradiography.[2] See also 2.7.1.123 — CDPK, 2.7.1.135 — TPK, 2.7.1.X — CAMPPK, and 2.7.1.X′ — TTK.

D. melanogaster PROTK (CK2) is a highly conserved protein kinase that is composed of catalytic (α) and regulatory (β) subunits associated as a $2\alpha2\beta$-heterotetramer. This enzyme interacts with a number of proteins. Most notable among these are Surf6, a nucleolar protein involved in RNA processing, and Spalt, a homeotic protein.[4]

The human AMP-activated protein kinase cascade plays an important role in the regulation of energy homeostasis within the cell. The enzyme is a heterotrimer composed of a catalytic subunit (α) and two regulatory subunits (β and γ). Comparative analysis of three different isoforms of the γ-subunit provides evidence that this subunit may participate directly in the binding of AMP within the complex.[5]

REFERENCES

1. Phan-Dinh-Tuy, F., Weber, A., Henry, J., Cottreau, D., and Kahn, A., Cellulose acetate electrophoresis of protein kinases: detection of the active forms using various substrates, *Anal. Biochem.*, 127, 73, 1982.

2. Geahlen, R.L., Anostario, M., Jr., Low, P.S., and Harrison, M.L., Detection of protein kinase activity in sodium dodecyl sulfate–polyacrylamide gels, *Anal. Biochem.*, 153, 151, 1986.

3. NC-IUBMB, *Enzyme Nomenclature*, Academic Press, San Diego, 1992, p. 266 (EC 2.7.1.37, Comments).

4. Trott, R.L., Kalive, M., Karandikar, U., Rummer, R., Bishop, C.P., and Bidwai, A.P., Identification and characterization of proteins that interact with *Drosophila melanogaster* protein kinase CK2, *Mol. Cell. Biochem.*, 227, 91, 2001.

5. Cheung, P.C.F., Salt, I.P., Davies, S.P., Hardie, D.G., and Carling, D., Characterization of AMP-activated protein kinase γ-subunit isoforms and their role in AMP binding, *Biochem. J.*, 346, 659, 2000.

2.7.1.40 — Pyruvate Kinase; PK

OTHER NAMES	Phosphoenolpyruvate kinase, phosphoenol transphosphorylase
REACTION	ATP + pyruvate = ADP + phosphoenolpyruvate
ENZYME SOURCE	Bacteria, fungi, green algae, plants, protozoa, invertebrates, vertebrates
SUBUNIT STRUCTURE	Tetramer (invertebrates, vertebrates)

METHOD 1

Visualization Scheme

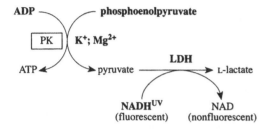

Staining Solution[1] (modified)

0.2 M Tris–HCl buffer, pH 8.0	10 ml
Phosphoenolpyruvate (monopotassium salt)	15 mg
ADP (disodium salt)	25 mg
NADH	7 mg
$MgCl_2 \cdot 6H_2O$	30 mg
KCl	40 mg
Lactate dehydrogenase (LDH)	50 U

Procedure

Apply the staining solution to the gel surface on a filter paper overlay. Incubate the gel at 37°C for 30 to 60 min and monitor under long-wave UV light. Dark (nonfluorescent) bands visible on a light (fluorescent) background correspond to localization of PK activity. Record the zymogram or photograph using a yellow filter.

Notes: In some organisms additional bands caused by phosphoenolpyruvate phosphatase (see 3.1.3.18 — PGP, Method 2) can also be developed on PK zymograms obtained by this method. Thus, the control staining of an additional gel in staining solution lacking ADP is required.

METHOD 2

Visualization Scheme

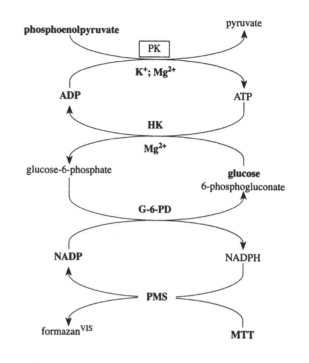

Staining Solution[2]

A. 0.1 M Tris–HCl buffer, pH 8.0
   0.1 M KCl
   10 mM $MgCl_2$
   5 mM Phosphoenolpyruvate
   3 mM ADP
   10 mM Glucose

B.	
NADP	10 mg
MTT	10 mg
PMS	1 mg
Hexokinase (HK)	16 U
Glucose-6-phosphate dehydrogenase (G-6-PD)	12 U

C. 1% Agarose solution (60°C)

Procedure

Add reagents B to 10 ml of solution A and then add 10 ml of solution C. Pour the resulting mixture over the surface of the gel and incubate in the dark at 37°C until dark blue bands appear. Fix the stained gel in 25% ethanol.

Notes: Additional bands caused by adenylate kinase activity can also develop on PK zymograms obtained by this method. Thus, an adenylate kinase control gel should also be available for comparison (see 2.7.4.3 — AK, Method 1). The use of AMP (final concentration of 2 mg/ml) in the PK stain may inhibit AK activity.

NAD-dependent G-6-PD is available (Sigma G 5760 or G 5885). When this form of the enzyme is used, NAD should be substituted for NADP in the staining solution. This substitution will result in cost savings because NAD is several times cheaper than NADP.

METHOD 3

Visualization Scheme

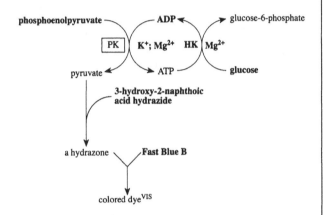

Staining Solution[3]

50 mM Tris–HCl buffer, pH 7.5	100 ml
1 M MgCl$_2$·6H$_2$O	1 ml
5 M KCl	3 ml
Phosphoenolpyruvate (trisodium salt; 5H$_2$O)	50 mg
ADP (sodium salt)	20 mg
Glucose	180 mg
3-Hydroxy-2-naphthoic acid hydrazide	50 mg
Fast Blue B salt	50 mg
Hexokinase (HK)	30 U

Procedure

Incubate the gel in staining solution in the dark at 36°C until colored bands appear. Rinse the stained gel in water and fix in 50% glycerol.

Notes: In this method, the carbonyl group of pyruvate condenses with the hydrazine group of 3-hydroxy-2-naphthoic acid hydrazide. Hydrazone molecules so formed are cross-linked with each other by the addition of a tetrazonium salt. The coupling reaction is directed by the 2-hydroxy group.

The role of glucose and HK is the regeneration of ADP.

OTHER METHODS

An immunoblotting procedure (for details, see Part II) using monoclonal antibodies specific to rabbit PK[4] can also be used for immunohistochemical localization of the enzyme protein on electrophoretic gels. This procedure, however, is not as suitable for routine laboratory use as the methods described above.

GENERAL NOTES

The enzyme from some mammalian species is activated by fructose-1,6-diphosphate (final concentration of 2 mg/ml).

REFERENCES

1. Shaw, C.R. and Prasad, R., Starch gel electrophoresis of enzymes: a compilation of recipes, *Biochem. Genet.*, 4, 297, 1970.
2. Chern, C.J. and Croce, C.M., Confirmation of the synteny of the human genes for mannose phosphate isomerase and pyruvate kinase and their assignment to chromosome 15, *Cytogenet. Cell Genet.*, 15, 299, 1975.
3. Vallejos, C.E., Enzyme activity staining, in *Isozymes in Plant Genetics and Breeding*, Part A, Tanksley, S.D. and Orton, T.J., Eds., Elsevier, Amsterdam, 1983, p. 469.
4. Hance, A.J., Lee, J., and Feitelson, M., The M1 and M2 isozymes of pyruvate kinase are the products of the same gene, *Biochem. Biophys. Res. Commun.*, 106, 492, 1982.

2.7.1.74 — Deoxycytidine Kinase; DCK

REACTION	Nucleoside triphosphate + deoxycytidine = nucleoside diphosphate + dCMP
ENZYME SOURCE	Bacteria, vertebrates
SUBUNIT STRUCTURE	Heterodimer# (bacteria; see General Notes), dimer# (vertebrates)

METHOD

Visualization Scheme

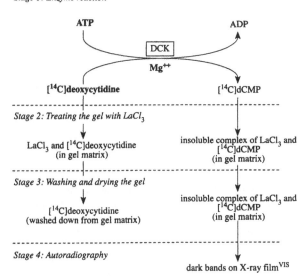

Stage 1: Enzyme reaction

ATP → ADP

DCK

Mg++

[14C]deoxycytidine   [14C]dCMP

Stage 2: Treating the gel with LaCl₃

LaCl₃ and [14C]deoxycytidine (in gel matrix)   insoluble complex of LaCl₃ and [14C]dCMP (in gel matrix)

Stage 3: Washing and drying the gel

[14C]deoxycytidine (washed down from gel matrix)   insoluble complex of LaCl₃ and [14C]dCMP (in gel matrix)

Stage 4: Autoradiography

dark bands on X-ray film^{VIS}

Reaction Mixture[1]

A. 20 mM Potassium phosphate buffer, pH 6.2   20 ml
   ATP (disodium salt)   4 mg
   MgCl₂ (anhydrous)   1 mg
   [14C]Deoxycytidine   5 µCi
B. 0.1 M Tris–HCl buffer, pH 7.0
   0.1 M LaCl₃

Procedure

Stage 1. Incubate the gel in solution A at 37°C for 1 h.

Stage 2. Remove solution A and incubate the gel in solution B at 4°C for 6 h.

Stage 3. Wash the gel for 3 h with distilled water. Dry the gel.

Stage 4. Autoradiograph the dry gel with x-ray film for 2 weeks.

The dark bands on the developed x-ray film correspond to localization of DCK activity on the gel.

OTHER METHODS

The areas of ADP production can be detected as dark bands visible in long-wave UV light using reactions catalyzed by two auxiliary enzymes, pyruvate kinase and lactate dehydrogenase (e.g., see 2.7.1.30 — GLYCK, Method 3). This method also detects phosphoenolpyruvate phosphatase and ATPase activity bands. Thus, control stainings of two additional gels for phosphoenolpyruvate phosphatase (see 3.1.3.18 — PGP, Method 2) and ATPase (see 3.6.1.3 — ATPASE, Method 2) are desirable.

GENERAL NOTES

The enzyme may be identical to deoxyadenosine kinase (see 2.7.1.76 — DAK) and deoxyguanosine kinase (see 2.7.1.113 — DGK).

All natural nucleoside triphosphates, except dCTP, can act as donors.[2]

Deoxynucleoside kinases required for growth of *Lactobacillus acidophilus* exist as heterodimeric pairs specific for deoxyadenosine and deoxycytidine or deoxyadenosine and deoxyguanosine.[3]

REFERENCES

1. Osborne, W.R.A. and Scott, C.R., Nucleoside kinases in B and T lymphoblastoid cells, *Isozyme Bull.*, 18, 68, 1985.
2. NC-IUBMB, *Enzyme Nomenclature*, Academic Press, San Diego, 1992, p. 270 (EC 2.7.1.74, Comments).
3. Ma, N., Ikeda, S., Guo, S.Y., Fieno, A., Park, I., Grimme, S., Ikeda, T., and Ives, D.H., Deoxycytidine kinase and deoxyguanosine kinase of *Lactobacillus acidophilus* R-26 are colinear products of a single gene, *Proc. Natl. Acad. Sci. U.S.A.*, 93, 14385, 1996.

2.7.1.76 — Deoxyadenosine Kinase; DAK

OTHER NAMES   Purine-deoxyribonucleoside kinase

REACTION   ATP + deoxyadenosine = ADP + dAMP

ENZYME SOURCE   Bacteria, vertebrates

SUBUNIT STRUCTURE   Heterodimer[a] (bacteria; see General Notes)

Method

Visualization Scheme

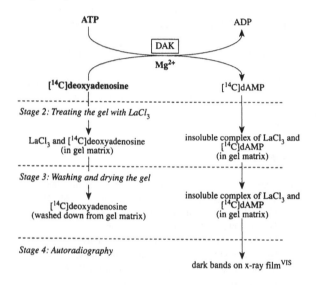

Stage 1: Enzyme reaction

Stage 2: Treating the gel with LaCl₃

Stage 3: Washing and drying the gel

Stage 4: Autoradiography

Reaction Mixture[1]

A.   20 mM Potassium phosphate buffer, pH 6.2   20 ml
   ATP (disodium salt)   4 mg
   $MgCl_2$ (anhydrous)   1 mg
   [^{14}C]Deoxyadenosine   5 μCi
B.   0.1 M Tris–HCl buffer, pH 7.0
   0.1 M $LaCl_3$

Procedure

Stage 1.   Incubate the gel in solution A at 37°C for 1 h.

Stage 2.   Remove solution A and incubate the gel in solution B at 4°C for 6 h.

Stage 3.   Wash the gel for 3 h with distilled water. Dry the gel.

Stage 4.   Autoradiograph the dry gel with x-ray film for 2 weeks.

The dark bands on the developed x-ray film correspond to localization of DAK activity in the gel.

Other Methods

Bands of DAK activity can be detected in long-wave UV light (dark bands on a fluorescent background) using the ADP-detecting system of two auxiliary enzymes, pyruvate kinase and lactate dehydrogenase (for example, see 2.7.1.30 — GLYCK, Method 3). This method, however, requires two additional gels to be stained for phosphoenolpyruvate phosphatase (see 3.1.3.18 — PGP, Method 2) and ATPase (see 3.6.1.3 — ATPASE, Method 2), for which activity bands also can develop on DAK zymograms obtained by this method.

General Notes

This enzyme may be identical to deoxycytidine kinase (see 2.7.1.74 — DCK) and deoxyguanosine kinase (see 2.7.1.113 — DGK).

Deoxyguanosine can also act as an acceptor.[2]

Deoxynucleoside kinases required for growth of *Lactobacillus acidophilus* exist as heterodimeric pairs specific for deoxyadenosine and deoxycytidine or deoxyadenosine and deoxyguanosine.[3]

References

1. Osborne, W.R.A. and Scott, C.R., Nucleoside kinases in B and T lymphoblastoid cells, *Isozyme Bull.*, 18, 68, 1985.
2. NC-IUBMB, *Enzyme Nomenclature*, Academic Press, San Diego, 1992, p. 270 (EC 2.7.1.76, Comments).
3. Ma, N., Ikeda, S., Guo, S.Y., Fieno, A., Park, I., Grimme, S., Ikeda, T., and Ives, D.H., Deoxycytidine kinase and deoxyguanosine kinase of *Lactobacillus acidophilus* R-26 are colinear products of a single gene, *Proc. Natl. Acad. Sci. U.S.A.*, 93, 14385, 1996.

2.7.1.90 — Pyrophosphate-Fructose-6-Phosphate 1-Phosphotransferase; PFPPT

OTHER NAMES	6-Phosphofructokinase (pyrophosphate), pyrophosphate-dependent phosphofructokinase
REACTION	Pyrophosphate + D-fructose-6-phosphate = orthophosphate + D-fructose-1,6-bisphosphate
ENZYME SOURCE	Bacteria, plants
SUBUNIT STRUCTURE	Dimer# (plants — pineapple leaves), heterotetramer# (plants — barley leaves); see General Notes

Method

Visualization Scheme

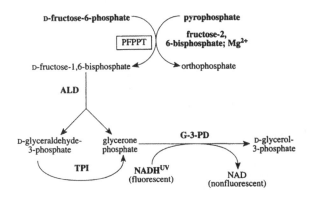

Staining Solution[1] (adapted)

A. 50 mM HEPES (*N*-2-hydroxyethylpiperazine-*N'*-2-ethanesulfonate) buffer, pH 8.0
    0.5 mM EDTA
    4 mM Mg(CH$_3$COO)$_2$
    1 mM NADH
    10 mM D-Fructose-6-phosphate
    2 mM Pyrophosphate
    0.2 U/ml Aldolase (ALD)
    0.2 U/ml Triose-phosphate isomerase (TPI)
    0.2 U/ml Glycerol-3-phosphate dehydrogenase (G-3-PD)
    100 nM D-Fructose-2,6-bisphosphate
B. 1.4% Agar solution (60°C)

Procedure

Mix equal volumes of A and B components of the staining solution and pour the mixture over the surface of the gel. Incubate the gel at 37°C and monitor under long-wave UV light. Dark (nonfluorescent) bands on a light (fluorescent) background indicate the sites of localization of PFPPT activity in the gel. Record the zymogram or photograph using a yellow filter.

Notes: The plant enzyme requires fructose-2,6-biphosphate for its activity. The bacterial enzyme is not activated by this cofactor.

Fructose-2,6-bisphosphate may be replaced by 6-phosphofructo-2-kinase (EC 2.7.1.105) and ATP. This substitution, however, causes additional development of 6-phosphofructokinase (see 2.7.1.11 — PFK) activity bands on PFPPT zymograms.

Other Methods

When a zymogram of PFPPT that is visible in daylight is required, G-3-PD and NADH in the staining solution given above should be replaced by glyceraldehyde-3-phosphate dehydrogenase, arsenate, NAD, PMS, and MTT (for example, see 4.1.2.13 — ALD, Method 1).

General Notes

Purified PFPPT from pineapple (*Ananas comosus*) leaves is represented by a single subunit of 61.5 kDa that is immunologically related to the potato tuber PFPPT α-subunit. The native form of pineapple PFPPT likely consists of a homodimer of 97.2 kDa, as determined by gel filtration. It is suggested that the β-subunit alone is sufficient to confer PFPPT with a high catalytic rate and the regulatory properties associated with activation by fructose-2,6-bisphosphate.[2]

The purified barley PFPPT consisted of two subunits, with apparent molecular masses of 65 (α) and 60 (β) kDa. Both the α- and the β-subunits are present in near stoichiometric amounts in all investigated tissues. In the absence of pyrophosphate, barley PFPPT elutes as a heterotetramer (2α2β), whereas it elutes as a heterooctamer (4α4β) in the presence of 20 mM pyrophosphate.[3]

References

1. Kora-Miura, Y., Fujii, S., Matsuda, M., Sato, Y., Kaku, K., and Kaneko, T., Electrophoretic determination of fructose-6-phosphate, 2-kinase, *Anal. Biochem.*, 170, 372, 1988.
2. Tripodi, K.E.J. and Podesta, F.E., Purification and structural and kinetic characterization of the pyrophosphate:fructose-6-phosphate 1-phosphotransferase from the Crassulacean acid metabolism plant, pineapple, *Plant Physiol.*, 113, 779, 1997.
3. Nielsen, T.H., Pyrophosphate-fructose-6-phosphate 1-phosphotransferase from barley seedlings: isolation, subunit composition and kinetic characterization, *Physiol. Plantarum*, 92, 311, 1994.

OTHER NAMES	Neomycin–kanamycin phospho-transferase, neomycin phospho-transferase
REACTION	ATP + kanamycin = ADP + kanamycin 3′-phosphate
ENZYME SOURCE	Bacteria
SUBUNIT STRUCTURE	Unknown or no data available

METHOD

Visualization Scheme

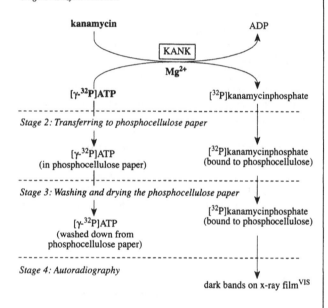

Stage 1: Enzyme reaction

Stage 2: Transferring to phosphocellulose paper

Stage 3: Washing and drying the phosphocellulose paper

Stage 4: Autoradiography

Reaction Mixture[1]

20 mM Tris–HCl buffer, pH 7.3
13 mM MgCl$_2$
100 mM NH$_4$Cl
0.5 mM Dithiothreitol
62.5 μg/ml Kanamycin
1.25 mM ATP
10 to 20 μCi/ml [γ-^{32}P]ATP

Procedure

Stage 1. Soak the electrophorized gel in reaction mixture at 4°C for 15 min in a plastic box to permit the reactants to diffuse into the gel. Then place the container in a 37°C incubator for 30 min to enable the reaction to occur.

Stage 2. Remove the reaction mixture and rinse the gel once with distilled water. Transfer the reaction products from the gel to prewetted Whatman P-81 phosphocellulose paper by capillary blotting overnight using distilled water as a transfer solution.

Stage 3. Wash the P-81 paper in 70°C distilled water twice and three times at room temperature, and then dry.

Stage 4. Apply the dry Whatman P-81 paper to a sheet of x-ray film and expose for 6 days.

The dark bands on the developed x-ray film correspond to localization of KANK activity on the gel.

GENERAL NOTES

This enzyme also acts on some other antibiotics (neomycin, paromycin, neamine, paromamine, vistamycin, and gentamycin A). KANK from *Pseudomonas aeruginosa* acts on bitirosin.[2]

REFERENCES

1. Fregien, N. and Davidson, N., Quantitative *in situ* gel electrophoretic assay for neomycin phosphotransferase activity in mammalian cell lysates, *Anal. Biochem.*, 148, 101, 1985.
2. NC-IUBMB, *Enzyme Nomenclature*, Academic Press, San Diego, 1992, p. 273 (EC 2.7.1.95, Comments).

OTHER NAMES — Phosphofructokinase 2

REACTION — ATP + D-fructose-6-phosphate = ADP + D-fructose-2,6-bisphosphate

ENZYME SOURCE — Invertebrates, vertebrates

SUBUNIT STRUCTURE — Dimer[#] (invertebrates, vertebrates); see General Notes

METHOD

Visualization Scheme

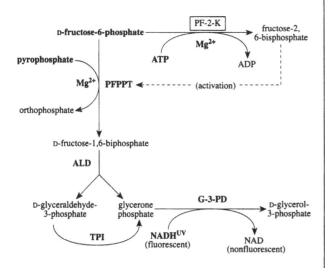

Staining Solution[1]

A. 50 mM HEPES (N-2-hydroxyethylpiperazine-N'-2-ethanesulfonate) buffer, pH 8.0
    0.5 mM EDTA
    4 mM Mg(CH₃COO)₂
    1 mM NADH
    10 mM ATP
    10 mM D-Fructose-6-phosphate
    2 mM Pyrophosphate
    0.2 U/ml Aldolase (ALD)
    0.2 U/ml Triose-phosphate isomerase (TPI)
    0.2 U/ml Glycerol-3-phosphate dehydrogenase (G-3-PD)
    0.1 U/ml Pyrophosphate–fructose-6-phosphate 1-phosphotransferase (PFPPT)
B. 1.4% Agar solution (60°C)

Procedure

Mix equal volumes of A and B components of the staining solution and pour the mixture over the surface of the gel. Incubate the gel at 37°C and monitor under long-wave UV light. Dark (nonfluorescent) bands on a light (fluorescent) background indicate the sites of localization of PF-2-K activity in the gel. Record the zymogram or photograph using a yellow filter.

Notes: 6-Phosphofructokinase (EC 2.7.1.11) activity bands can also develop on PF-2-K zymograms obtained by this method. However, PF-2-K bands do not develop when pyrophosphate is excluded from the staining solution. This difference may be used to differentiate activity bands of these two enzymes.

The visualization of PF-2-K activity on electrophoretic gels is based on the production of fructose-2,6-bisphosphate, coupled with the activation of auxiliary potato enzyme pyrophosphate-fructose-6-phosphate 1-phosphotransferase (see 2.7.1.90 — PFPPT). The product of the PFPPT reaction (fructose-1,6-bisphosphate) is then detected using three auxiliary enzymes: ALD, TPI, and G-3-PD.

OTHER METHODS

When a zymogram of PF-2-K that is visible in daylight is required, G-3-PD and NADH in the staining solution given above should be replaced by glyceraldehyde-3-phosphate dehydrogenase, arsenate, NAD, PMS, and MTT (e.g., see 4.1.2.13 — ALD, Method 1).

GENERAL NOTES

This enzyme is not identical with EC 2.7.1.11 (see 2.7.1.11 — PFK); it co-purifies with fructose-2,6-bisphosphate 2-phosphatase (EC 3.1.3.46).[2]

The PF-2-K isozyme from the rat testis is a homodimer of 55-kDa subunits arranged in a head-to-head fashion, with each monomer consisting of independent kinase and phosphatase domains.[3]

The enzyme from the mantle of the sea mussel *Mytilus galloprovincialis* is inhibited by citrate and phosphoenolpyruvate.[4]

REFERENCES

1. Kora-Miura, Y., Fujii, S., Matsuda, M., Sato, Y., Kaku, K., and Kaneko, T., Electrophoretic determination of fructose-6-phosphate, 2-kinase, *Anal. Biochem.*, 170, 372, 1988.
2. NC-IUBMB, *Enzyme Nomenclature*, Academic Press, San Diego, 1992, p. 274 (EC 2.7.1.105, Comments).
3. Hasemann, C.A., Istvan, E.S., Uyeda, K., and Deisenhofer, J., The crystal structure of the bifunctional enzyme 6-phosphofructo-2-kinase/fructose-2,6-bisphosphatase reveals distinct domain homologies, *Structure*, 4, 1017, 1996.
4. Vazquezillanes, M.D., Barcia, R., Ibarguren, I., Villamarin, J.J.A., and Ramosmartinez, J.J.I., Regulation of fructose-2,6-bisphosphate content in mantle tissue of the sea mussel *Mytilus galloprovincialis*: 1. Purification and properties of 6-phosphofructo-2-kinase, *Mar. Biol.*, 112, 277, 1992.

OTHER NAMES | Tyrosylprotein kinase, protein kinase (tyrosine), hydroxyaryl-protein kinase

REACTION | ATP + protein tyrosine = ADP + protein tyrosine phosphate

ENZYME SOURCE | Fungi, plants, invertebrates, vertebrates

SUBUNIT STRUCTURE | See General Notes

METHOD

Visualization Scheme

Stage 1: Enzyme reaction

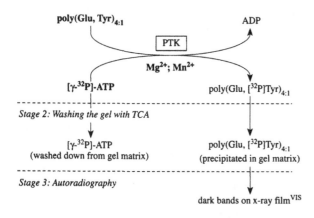

Stage 2: Washing the gel with TCA

Stage 3: Autoradiography

dark bands on x-ray film^{VIS}

Reaction Mixture[1]

A. 20 mM HEPES (N-[2-hydroxyethyl]piperazine-N'-[2-ethanesulfonate]) buffer, pH 7.4

5 mg/ml poly(Glu, Tyr)$_{4:1}$

33 μCi/ml [γ-^{32}P]ATP

50 μM ATP

10 mM MgCl$_2$

10 mM MnCl$_2$

B. 5% Trichloroacetic acid (TCA)

10 mM Sodium pyrophosphate

Procedure

Stage 1. Soak the electrophorized PAG (containing 10% glycerol, 0.1% Triton X-100) twice for 15 min in 20 mM HEPES buffer (pH 7.4) at 4°C. Remove the buffer and overlay the gel with solution A (35 μl/cm^2). Incubate the gel at 37°C for 30 min and then rinse the gel once with water.

Stage 2. Wash the gel several times (usually five 10-min washes) at room temperature on a shaker with solution B.

Stage 3. Dry the gel and autoradiograph it with Kodak X-Omat XK-1 film.

The dark bands on the developed x-ray film correspond to localization of PTK activity on the gel.

Notes: PTK acts on specific tyrosine-containing proteins associated with membrane vesicles. Nucleotides other than ATP may also act as phosphate donors.

If protein-tyrosine-phosphatase (EC 3.1.3.48), which may be present in preparations under analysis, comigrates with some PTK isozymes, it can decrease the intensity of development of these isozymes on autoradiographs. To inhibit this phosphatase activity, sodium vanadate (final concentration of 0.1 mM) should be included in solution A of the reaction mixture.

The identification of PTKs in SDS-PAG is difficult because they usually display no activities after protein renaturation by usual methods. This may be explained by their low degree or lack of autophosphorylation activity after renaturation. The procedure for detecting PTKs in SDS-PAG was improved by incorporation of the poly(Glu, Tyr) substrate (0.1 mg/ml) into the resolving gel. After electrophoresis, SDS was removed from PAG by two successive washes (100 ml, 60 min each) at room temperature in 20% isopropanol–50 mM imidazole, 28 mM iminodiacetic acid, and 10 mM β-mercaptoethanol (pH 7.0). To facilitate the unfolding of proteins within the gel matrix, PAG was then treated with a buffer containing 8 M guanidine HCl. Proteins were then renatured again by successive washes in 25 mM imidazole–14 mM iminodiacetic acid buffer (pH 7.0), containing 50 mM KCl, 10 mM β-mercaptoethanol, 0.04% Tween 20, and 10% sucrose. Proteins were then cross-linked within the gel matrix by 100 ml of 10% (v/v) glutaraldehyde treatment (30 min) and three washes with distilled water (250 ml, 20 min each). The glutaraldehyde-fixed PAG was then treated with 1 M KOH at 58°C for 2 h in order to decrease the relative proportions of labeled phosphoserine and phosphothreonine to labeled phosphotyrosine. Incubation of the PAG in hot alkali after glutaraldehyde cross-linking completely eliminated the activity of non-PTK enzymes.[2]

GENERAL NOTES

Nucleotides other than ATP may also act as donors. There are many eukaryotic genes coding for this type of enzyme, some of which are closely related.[3]

The human enzyme is autophosphorylated at Tyr residues. Autophosphorylation activates PTK approximately threefold. Polylysine (1 mM) activates the kinase on average approximately fourfold. Polylysine treatment concurrently results in both aggregation and activation of the enzyme. This suggests that only β-subunit domains participate in aggregation and activation without participation of α-subunit. Oligomers of the β-subunit domains such as tetramers and octamers are formed.[4]

2.7.1.112 — Protein-Tyrosine Kinase; PTK (continued)

REFERENCES

1. Glazer, R.I., Yu, G., and Knode, M.C., Analysis of tyrosine kinase activity in cell extracts using nondenaturing polyacrylamide gel electrophoresis, *Anal. Biochem.*, 164, 214, 1987.
2. Dupont, H., Audigier, S., and Chevalier, S., Screening of protein tyrosine kinases activated during neural induction in *Xenopus*, *Anal. Biochem.*, 237, 42, 1996.
3. NC-IUBMB, *Enzyme Nomenclature*, Academic Press, San Diego, 1992, p. 275 (EC 2.7.1.112, Comments).
4. Li, S.L., Yan, P.F., Paz, I.B., and Fujitayamaguchi, Y., Human insulin-receptor β-subunit transmembrane cytoplasmic domain expressed in a baculovirus expression system: purification, characterization, and polylysine effects on the protein tyrosine kinase-activity, *Biochemistry*, 31, 12455, 1992.

2.7.1.113 — Deoxyguanosine Kinase; DGK

REACTION	ATP + deoxyguanosine = ADP + dGMP
ENZYME SOURCE	Bacteria, vertebrates
SUBUNIT STRUCTURE	Heterodimer# (bacteria; see General Notes)

METHOD

Visualization Scheme

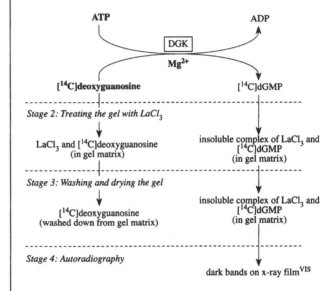

Stage 1: Enzyme reaction

Stage 2: Treating the gel with LaCl$_3$

Stage 3: Washing and drying the gel

Stage 4: Autoradiography

dark bands on x-ray film[VIS]

Reaction Mixture[1]

A. 20 mM Potassium phosphate buffer, pH 6.2 — 20 ml
ATP (disodium salt) — 4 mg
MgCl$_2$ (anhydrous) — 1 mg
[^{14}C]Deoxyguanosine — 5 μCi
B. 0.1 M Tris–HCl buffer, pH 7.0
0.1 M LaCl$_3$

Procedure

Stage 1. Incubate the gel in solution A at 37°C for 1 h.

Stage 2. Remove solution A and incubate the gel in solution B at 4°C for 6 h.

Stage 3. Wash the gel for 3 h with distilled water. Dry the gel.

Stage 4. Autoradiograph the dry gel with x-ray film for 2 weeks.

The dark bands on the developed x-ray film correspond to localization of DGK activity on the gel.

2.7.1.113 — Deoxyguanosine Kinase; DGK (continued)

OTHER METHODS

Areas of ADP production can be detected in UV light as dark bands on the fluorescent background of the gel using two auxiliary enzymes, pyruvate kinase and lactate dehydrogenase (for example, see 2.7.1.30 — GLYCK, Method 3). When using this method, stain two additional gels for phosphoenolpyruvate phosphatase (see 3.1.3.18 — PGP, Method 2) and ATPase (see 3.6.1.3 — ATPASE, Method 2), for which bands can also develop on DGK zymograms obtained by this method.

GENERAL NOTES

The enzyme may be identical to deoxyadenosine kinase (see 2.7.1.76 — DAK) and deoxycytidine kinase (see 2.7.1.74 — DCK).

Deoxyinosine can also act as an acceptor.[2]

Deoxynucleoside kinases required for growth of *Lactobacillus acidophilus* exist as heterodimeric pairs specific for deoxyadenosine and deoxycytidine or deoxyadenosine and deoxyguanosine.[3]

REFERENCES

1. Osborne, W.R.A. and Scott, C.R., Nucleoside kinases in B and T lymphoblastoid cells, *Isozyme Bull.*, 18, 68, 1985.
2. NC-IUBMB, *Enzyme Nomenclature*, Academic Press, San Diego, 1992, p. 275 (EC 2.7.1.113, Comments).
3. Ma, N., Ikeda, S., Guo, S.Y., Fieno, A., Park, I., Grimme, S., Ikeda, T., and Ives, D.H., Deoxycytidine kinase and deoxyguanosine kinase of *Lactobacillus acidophilus* R-26 are colinear products of a single gene, *Proc. Natl. Acad. Sci. U.S.A.*, 93, 14385, 1996.

2.7.1.119 — Hygromycin-B Kinase; HBK

OTHER NAMES	Hygromycin B phosphotransferase
REACTION	ATP + hygromycin B = ADP + 7″-O-phosphohygromycin
ENZYME SOURCE	Bacteria
SUBUNIT STRUCTURE	Unknown or no data available

METHOD

Visualization Scheme

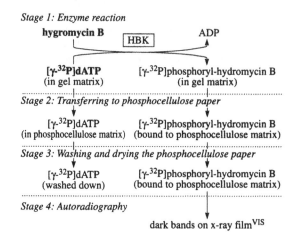

Reaction Mixture[1]

75 mM Tris–HCl, pH 7.4
10 mM MgSO$_4$
100 mM (NH$_4$)$_2$SO$_4$
1 mM Hygromycin B
20 mM ATP
50 μCi/ml [γ-^{32}P]ATP (3000 Ci/mmol)

Procedure

Stage 1. Cover the electrophorized 7.5% PAG with a thin layer (0.3 ml/cm²) of 0.7% agarose solution containing the reaction mixture. Incubate the gel for 2 h at 30°C to allow the reaction to proceed.

Stage 2. Transfer the phosphorylated hygromycin B to Whatman P-81 paper by the Southern blotting technique.

Stage 3. Wash the phosphocellulose paper several times with distilled water at 80°C, then at room temperature, and dry using an amplifying screen.

Stage 4. Apply dry Whatman P-81 paper to a sheet of MAFE RP-X1 film and expose for 6 to 24 h.

The dark bands on the developed x-ray film correspond to localization of HBK activity on the gel.

Notes: Detected bands are not sharp, probably due to diffusion of phosphoryl-hygromycin B in the agarose gel.

2.7.1.119 — Hygromycin-B Kinase; HBK (continued)

GENERAL NOTES

This enzyme phosphorylates the antibiotics hygromycin B, 1-*N*-hygromycin B, and desmomycin (but not hygromycin B$_2$) at the 7″-hydroxyl group in the destomic acid ring.[2]

REFERENCES

1. Zalacain, M., Pardo, J.M., and Jiménez, A., Purification and characterization of a hygromycin B phosphotransferase from *Streptomyces hygroscopicus*, *Eur. J. Biochem.*, 162, 419, 1987.
2. NC-IUBMB, *Enzyme Nomenclature*, Academic Press, San Diego, 1992, p. 276 (EC 2.7.1.119, Comments).

2.7.1.123 — Ca²⁺/Calmodulin-Dependent Protein Kinase; CDPK

OTHER NAMES	Microtubule-associated protein 2 kinase
REACTION	ATP + protein = ADP + *O*-phosphoprotein (see General Notes)
ENZYME SOURCE	Vertebrates
SUBUNIT STRUCTURE	Heterodimer* (vertebrates); however, see General Notes

METHOD

Visualization Scheme

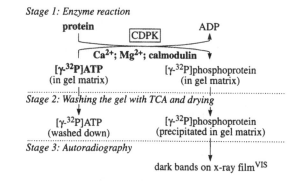

Stage 1: Enzyme reaction

protein   CDPK   ADP

Ca²⁺; Mg²⁺; calmodulin

[γ-³²P]ATP   [γ-³²P]phosphoprotein
(in gel matrix)   (in gel matrix)

Stage 2: Washing the gel with TCA and drying

[γ-³²P]ATP   [γ-³²P]phosphoprotein
(washed down)   (precipitated in gel matrix)

Stage 3: Autoradiography

dark bands on x-ray film^{VIS}

Reaction Mixture[1]

A. 40 m*M* HEPES-NaOH, pH 8.0 (see *Notes*)
0.1 m*M* Ethylene glycol *bis*(β-aminoethyl ether) *N,N′*-tetraacetic acid (EGTA)
5 m*M* Mg(CH$_3$COO)$_2$
0.15 m*M* CaCl$_2$
14 μg/ml Calmodulin
50 m*M* ATP
15 μCi/ml [γ-³²P]ATP (Amersham; 3000 Ci/mmol)
B. 5% (w/v) Trichloroacetic acid (TCA)
1% Sodium pyrophosphate

Procedure

After electrophoresis in an SDS-PAG containing 0.1 mg/ml microtubule-associated protein 2 (or other protein substrates added to the separation gel just prior to polymerization), wash the gel with two changes of 100 ml each of 20% 2-propanol in 50 m*M* Tris–HCl (pH 8.0) (buffer A) for 1 h, and then 250 ml of buffer A containing 5 m*M* 2-mercaptoethanol (buffer B) for 1 h at room temperature to remove the SDS. To denature the enzyme repeatedly (see General Notes), treat the washed gel first with two changes of 100 ml of 6 *M* guanidine HCl in buffer B at room temperature for 1 h, and then renature again with five changes of 250 ml each of buffer B containing 0.04% Tween 40 at 4°C for 16 h. Then preincubate the gel at 22°C for 30 min with 40 m*M* HEPES-NaOH (pH 8.0) containing 2 m*M* dithiothreitol, 0.1 m*M* EGTA, 5 m*M* Mg(CH$_3$COO)$_2$, 0.15 m*M* CaCl$_2$, and 14 μg/ml calmodulin. Finally, carry out the enzyme reaction

(Stage 1), the washing and drying of the gel (Stage 2), and the autoradiography (Stage 3) as indicated below:

Stage 1. Incubate the gel in solution A at 22°C for 1 h.

Stage 2. Wash the gel with solution B until the radioactivity of the solution becomes negligible (usually five changes of 500 ml each are necessary). Dry the gel on Whatman 3MM chromatographic paper.

Stage 3. Autoradiograph the dry gel with Fuji x-ray film at −80°C with an intensifying screen (Quanta III, DuPont).

The dark bands on the developed x-ray film correspond to localization of CDPK activity on the gel.

Notes: HEPES is an abbreviation of (*N*-[2-hydroxyethyl]piperazine *N*-[2-ethanesulfonic acid]).

Approximately 0.05 μg of the enzyme protein from the rat cerebral cortex could be detected on a gel containing no protein substrate as a result of autophosphorylation. When microtubule-associated protein 2 from the pig brain was included in the gel as a protein substrate, the sensitivity of the method was more than one order of magnitude higher.

Some additional radioactive bands besides α- and β-subunits of CDPK were also detected in the cytoskeletal fraction from the rat cerebral cortex when phosphorylation was carried out in the absence of Ca²⁺ and calmodulin. This provided evidence that there were several protein kinases that could phosphorylate microtubule-associated protein 2.

The method was shown to be useful for the detection of a variety of calmodulin-dependent protein kinases.

A new procedure for the detection of CDPK activity toward synthetic oligopeptides was developed.[2] Different synthetic oligopeptides were linked to high-molecular-weight amino acid polymers (e.g., poly-L-lysine) through their amino-terminal cysteinyl residue. These oligopeptide–polymer conjugates were efficiently retained in the SDS-PAG matrix during electrophoresis and served as substrates for different protein kinases, including CDPK. C-syntide-2 oligopeptide (CPLARTLS-VAGLPLKK) conjugated to poly(Lys) served as a substrate for two CDPKs (CaM-kinase II and CaM-kinase IV) from the rat cerebral cortex, as well as for the catalytic subunit of cAMP-dependent protein kinase from the bovine heart. The last enzyme, however, can be specifically detected using another oligopeptide–polymer conjugate (see 2.7.1.70 — CAMPPK). CaM-kinase II and CaM-kinase IV can be discriminated using CAMKAKS oligopeptide (CSQPSFQWRQPSLDVDVGD) conjugated to poly(Lys). CaM-kinase IV does not use this oligopeptide–polymer conjugate as a substrate, while CaM-kinase II does.[2]

Renaturation of CDPK after denaturation by SDS is markedly enhanced when the denatured enzyme protein is repeatedly denatured by concentrated guanidine HCl (or urea) and then renatured again.[2]

OTHER METHODS

The immunoblotting procedure (for details, see Part II) may be used to detect CDPK. Monoclonal and polyclonal antibodies specific to the enzyme from the chicken, mouse, human, and rat are available from Sigma.

GENERAL NOTES

The enzyme requires Ca²⁺ and calmodulin for activity. A wide range of proteins can act as an acceptor, including microtubule-associated protein 2, vimentin, synapsin, glycogen synthase, and myosin light chains.[3] The enzyme itself can also act as an acceptor, but more slowly.

The genetic basis of multiple CDPK isoforms and their functional differences, as well as the phylogenetic relationship between different CDPK enzyme proteins, is still poorly understood. For example, the rat CDPK IV is a monomeric multifunctional enzyme that is expressed only in subanatomical portions of the brain, T lymphocytes, and postmeiotic male germ cells. This form of CDPK may be involved both in preventing apoptosis during T cell development and in the early cascade of events that is required to activate the mature T cells in response to a mitogenic stimulus.[4]

REFERENCES

1. Kameshita, I. and Fujisawa, H., A sensitive method for detection of calmodulin-dependent protein kinase II activity in sodium dodecyl sulfate–polyacrylamide gel, *Anal. Biochem.*, 183, 139, 1989.

2. Kameshita, I. and Fujisawa, H., Detection of protein kinase activities toward oligopeptides in sodium dodecyl sulfate–polyacrylamide gel, *Anal. Biochem.*, 237, 198, 1996.

3. NC-IUBMB, *Enzyme Nomenclature*, Academic Press, San Diego, 1992, p. 277 (EC 2.7.1.123, Comments).

4. Means, A.R., Ribar, T.J., Kane, C.D., Hook, S.S., and Anderson, K.A., Regulation and properties of the rat Ca²⁺/calmodulin-dependent protein kinase IV gene and its protein products, *Recent Prog. Horm. Res.*, 52, 389, 1997.

REACTION ATP + *tau*-protein = ADP + *O*-phospho-*tau*-protein (see General Notes)

ENZYME SOURCE Vertebrates

SUBUNIT STRUCTURE Heterodimer[#] (mammals); see General Notes

METHOD

Visualization Scheme

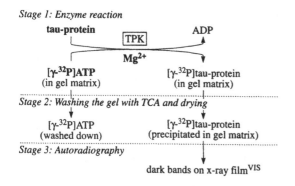

Stage 1: Enzyme reaction

tau-protein ──┤TPK├── ADP

Mg²⁺

[γ-³²P]ATP (in gel matrix) ──── [γ-³²P]tau-protein (in gel matrix)

Stage 2: Washing the gel with TCA and drying

[γ-³²P]ATP (washed down)   [γ-³²P]tau-protein (precipitated in gel matrix)

Stage 3: Autoradiography

dark bands on x-ray film[VIS]

Reaction Mixture[1,2]

A. 20 m*M* MES-NaOH, pH 6.8 (see *Notes*)
   1 m*M* Ethylene glycol *bis*(β-aminoethyl ether) *N,N*'-tetraacetic acid (EGTA)
   5 m*M* Mg(CH₃COO)₂
   50 m*M* ATP
   15 μCi/ml [γ-³²P]ATP (Amersham; 3000 Ci/mmol)
B. 5% (w/v) Trichloroacetic acid (TCA)
   1% Sodium pyrophosphate

Procedure

After electrophoresis in an SDS-PAG containing 0.4 mg/ml tau (added to the separation gel solution just prior to polymerization), wash the gel with two changes of 100 ml each of 20% 2-propanol in 50 m*M* Tris–HCl (pH 8.0) (buffer A) for 1 h, and then 250 ml of buffer A containing 5 m*M* 2-mercaptoethanol for 1 h at room temperature to remove the SDS. Then carry out the enzyme reaction (Stage 1), the washing and drying of the gel (Stage 2), and the autoradiography (Stage 3) as indicated below:

Stage 1. Incubate the gel in solution A of the reaction mixture at 22°C for 1 h.

Stage 2. Wash the gel with solution B until the radioactivity of the solution becomes negligible (usually five changes of 500 ml each are necessary). Dry the gel on Whatman 3MM chromatographic paper.

Stage 3. Autoradiograph the dry gel with Fuji x-ray film at −80°C with an intensifying screen (Quanta III, DuPont).

The dark bands on the developed x-ray film correspond to localization of TPK activity on the gel.

Notes: MES is an abbreviation of 2(*N*-morpholino)ethanesulfonic acid.

Additional radioactive bands belonging to *tau*-tubulin kinase (see 2.7.1.X' — TTK) can also be detected on gels developed using this method. However, TTK activity bands can be developed on gels containing 0.8 mg/ml β-tubulin, while TPK activity bands can not be developed on such gels. This difference may be used to differentiate between TPK and TTK activity bands.[1]

GENERAL NOTES

TPK requires tubulin for activity but has no ability to phosphorylate it. It is different from Ca²⁺/calmodulin-dependent protein kinase (see EC 2.7.1.123 — CDPK). Unlike some other protein kinases, TPK is not activated by Ca²⁺ and calmodulin (see EC 2.7.1.123 — CDPK) or cyclic nucleotides (e.g., see 2.7.1.X — CAMPPK).[3]

The enzyme from the rat brain is a complex composed of a catalytic subunit and a regulatory subunit. Expression of regulatory subunit pre-mRNA changes, coinciding with the developmental change of TPK activity, suggesting that its expression controls the phosphorylation of *tau*-protein by the TPK. It appears that the regulatory subunit serves as an activator of the catalytic subunit of TPK in neuronal cells.[4]

The explotable classification of protein kinases is imperfect and is now under review by NC-IUBMB (see 2.7.1.37 — PROTK, General Notes).

REFERENCES

1. Takahashi, M., Tomizawa, K., Sato, K., Ohtake, A., and Omori, A., A novel *tau*-tubulin kinase from bovine brain, *FEBS Lett.*, 372, 59, 1995.
2. Kameshita, I. and Fujisawa, H., A sensitive method for detection of calmodulin-dependent protein kinase II activity in sodium dodecyl sulfate–polyacrylamide gel, *Anal. Biochem.*, 183, 139, 1989.
3. NC-IUBMB, *Enzyme Nomenclature*, Academic Press, San Diego, 1992, p. 279 (EC 2.7.1.135, Comments).
4. Uchida, T., Ishiguro, K., Ohnuma, J., Takamatsu, M., Yonekura, S., and Imahori, K., Precursor of CDK5 activator, the 23-kDa subunit of *tau*-protein kinase-II: its sequence and developmental change in brain, *FEBS Lett.*, 355, 35, 1994.

REACTION   ATP + protein = ADP + O-phosphoprotein

ENZYME SOURCE   Vertebrates

SUBUNIT STRUCTURE Heterotetramer[#] (vertebrates); see General Notes

METHOD

Visualization Scheme

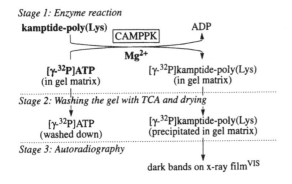

Stage 1: Enzyme reaction

Stage 2: Washing the gel with TCA and drying

Stage 3: Autoradiography

dark bands on x-ray film[VIS]

Reaction Mixture[1]

A. 40 mM HEPES-NaOH, pH 8.0
    0.1 mM Ethylene glycol bis(β-aminoethyl ether) N,N'-tetraacetic acid (EGTA)
    5 mM Mg(CH$_3$COO)$_2$
    50 mM ATP
    15 μCi/ml [γ-^{32}P]ATP (Amersham; 3000 Ci/mmol)

B. 5% (w/v) Trichloroacetic acid (TCA)
    1% Sodium pyrophosphate

Procedure

After electrophoresis in an SDS-PAG containing 4.5 μM Kamptide-poly(Lys) (added to the separation gel just prior to polymerization), wash the gel with two changes of 100 ml each of 20% 2-propanol in 50 mM Tris–HCl (pH 8.0) (buffer A) for 1 h, and then 250 ml of buffer A containing 5 mM 2-mercaptoethanol (buffer B) for 1 h at room temperature to remove the SDS. To denature the enzyme repeatedly (see General Notes), treat the washed gel first with two changes of 100 ml of 6 M guanidine HCl in buffer B at room temperature for 1 h, and then renaturate again with five changes of 250 ml each of buffer B containing 0.04% Tween 40 at 4°C for 16 h. Then preincubate the gel at 22°C for 30 min with 40 mM HEPES-NaOH (pH 8.0) containing 2 mM dithiothreitol, 0.1 mM EGTA, 5 mM Mg(CH$_3$COO)$_2$, and 0.15 mM cAMP. Finally, carry out the enzyme reaction (Stage 1), the washing and drying of the gel (Stage 2), and the autoradiography (Stage 3) as indicated below:

Stage 1. Incubate the gel in solution A at 22°C for 1 h.

Stage 2. Wash the gel with solution B until the radioactivity of the solution becomes negligible (usually five changes of 500 ml each are necessary). Dry the gel on Whatman 3MM chromatographic paper.

Stage 3. Autoradiograph the dry gel with Fuji x-ray film at –80°C with an intensifying screen (Quanta III, DuPont).

The dark bands on the developed x-ray film correspond to localization of CAMPPK activity on the gel.

Notes: About 2.5 pg of the catalytic subunit of CAMPPK from the bovine heart could be detected on the PAG containing 4.5 μM Kamptide-poly(Lys).

The method was shown to be useful for the detection of CAMPPK in crude extracts of the rat cerebellum, liver, spleen, kidney, heart, and PC12 cells. Ca^{2+}/calmodulin-dependent protein kinase (see 2.7.1.123 — CDPK) does not phosphorylate the Kamptide-poly(Lys), and therefore is not detected by this method.

Kamptide (CLRRWSVA) is a synthetic peptide in which a cysteine residue is added to the amino terminus of an analogue of Kemptide (LRRASLG), a specific peptide substrate for CAMPPK, to link to poly(Lys). The CAMKAS peptide (CSQPSFQWRQPSLDVDVGD) conjugated to poly(Lys) can also be used as substrate for CAMPPK. It should be taken into account, however, that CDPK (EC 2.7.1.123) also can phosphorylate this substrate.

OTHER METHODS

A. CAMPPK may be visualized on an SDS-containing PAG by autoradiography, using a method based on photoactivated covalent binding of radiolabeled 8-azido cAMP with the regulatory subunit of the enzyme in sample preparations before electrophoresis.[2]

B. The immunoblotting procedure (for details, see Part II) may be used to detect CAMPPK. Antibodies specific to the mouse enzyme are available from Sigma.

GENERAL NOTES

The changes in the electrophoretic mobility of cytosolic CAMPPK from murine erythroleukemic cells, observed when nonionic detergent Triton X-100 or Nonidet P-40 was included in the PAG, suggest that the enzyme contains a hydrophobic region interacting with these detergents. Thus, care must be taken when comparing and interpreting the banding patterns of CAMPPK obtained in routine gels and gels containing nonionic detergents.[3]

The enzyme molecules are heterodimers that can associate into heterotetramers consisting of two heterodimers. Each heterodimer consists of two different subunits, the regulatory (R) subunit and the catalytic (C) subunit. When both subunits are associated, the catalytic activity is inhibited. However, when cAMP binds to the regulatory subunit, the catalytic subunit is released and can then catalyze the phosphorylation reaction using various proteins as substrates. There is evidence that the mammalian enzyme functions as a heterodimer.[4] The heterodimeric structure of CAMPPK is consistent with the known physiological events required for cAMP-dependent activation of the kinase.[5]

Renaturation of oligomeric enzymes after denaturation by SDS is markedly enhanced when denatured oligomeric proteins are repeatedly denatured by concentrated guanidine HCl (or urea) and then renatured again.[6]

Many cAMP-dependent protein kinases are known. Classification of protein kinases is now under review by NC-IUBMB.[7] For example, protamine kinase (histone kinase) EC 2.7.1.70 is one of the cAMP-dependent protein kinases.

REFERENCES

1. Kameshita, I. and Fujisawa, H., Detection of protein kinase activities toward oligopeptides in sodium dodecyl sulfate–polyacrylamide gel, *Anal. Biochem.*, 237, 198, 1996.
2. Sato, M., Hiragun, A., and Mitsui, H., Differentiation-associated increase of cAMP-dependent type II protein kinase in murine preadipose cell line (ST 13), *Biochim. Biophys. Acta*, 844, 296, 1985.
3. Sprott, S., Hammond, K.D., and Savage, N., Electrophoretic separation of protein kinases: altered mobility with different crosslinking agents in the presence of certain detergents, *Electrophoresis*, 11, 29, 1990.
4. Nikolakaki, E., Fissentzidis, A., Giannakouros, T., and Georgatsos, J.J.G., Purification and characterization of a dimer form of the cAMP-dependent protein kinase from mouse liver cytosol, *Mol. Cell. Biochem.*, 197, 117, 1999.
5. Tung, C.S., Walsh, D.A., and Trewhella, J., A structural model of the catalytic subunit: regulatory subunit dimeric complex of the cAMP-dependent protein kinase, *J. Biol. Chem.*, 277, 12423, 2002.
6. Kameshita, I. and Fujisawa, H., A sensitive method for detection of calmodulin-dependent protein kinase II activity in sodium dodecyl sulfate–polyacrylamide gel, *Anal. Biochem.*, 183, 139, 1989.
7. NC-IUBMB, *Enzyme Nomenclature*, Academic Press, San Diego, 1992, p. 266 (EC 2.7.1.37, Comments), p. 270 (EC 2.7.1.70).

2.7.1.X' — Tau-Tubulin Kinase; TTK

REACTION   ATP + *tau*-protein = ADP + *O*-phos-
      pho-*tau*-protein (see General Notes)

ENZYME SOURCE   Vertebrates

SUBUNIT STRUCTURE   Monomer[#] (vertebrates)

Method

Visualization Scheme

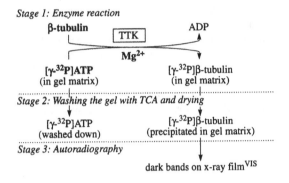

Stage 1: Enzyme reaction

Stage 2: Washing the gel with TCA and drying

Stage 3: Autoradiography

Reaction Mixture[1,2]

A.   20 mM MES-NaOH, pH 6.8 (see *Notes*)

     1 mM Ethylene glycol *bis*(β-aminoethyl ether)
     N,N'-tetraacetic acid (EGTA)

     5 mM Mg(CH$_3$COO)$_2$

     50 mM ATP

     15 μCi/ml [γ-^{32}P]ATP (Amersham; 3000 Ci/mmol)

B.   5% (w/v) Trichloracetic acid (TCA)

     1% Sodium pyrophosphate

Procedure

After electrophoresis in an SDS-PAG containing either 0.8
mg/ml β-tubulin or 0.4 mg/ml tau (added to the separation gel
solution just prior to polymerization), wash the gel with two
changes of 100 ml each of 20% 2-propanol in 50 mM Tris–HCl
(pH 8.0) (buffer A) for 1 h, and then 250 ml of buffer A containing 5 mM 2-mercaptoethanol for 1 h at room temperature to
remove the SDS. Then carry out the enzyme reaction (Stage 1),
the washing and drying of the gel (Stage 2), and the autoradiography (Stage 3) as indicated below:

Stage 1.   Incubate the gel in solution A of the reaction mixture
        at 22°C for 1 h.

Stage 2.   Wash the gel with solution B until the radioactivity of
        the solution becomes negligible (usually five changes
        of 500 ml each are necessary). Dry the gel on Whatman
        3MM chromatographic paper.

Stage 3.   Autoradiograph the dry gel with Fuji x-ray film at
        −80°C with an intensifying screen (Quanta III,
        DuPont).

The dark bands on the developed x-ray film correspond to
localization of TTK activity on the gel.

Notes: MES is an abbreviation of 2(*N*-morpholino)ethanesulfonic
acid.

General Notes

TTK also phosphorylates β-tubulin, microtubule-associated protein, and α-casein. However, it is different from Ca^{2+}/calmodu-
lin-dependent protein kinase (see EC 2.7.1.123 — CDPK)
because it is not activated by Ca^{2+} and calmodulin; it is not
activated by cyclic nucleotides.

When tau-containing gels are used, radioactive bands
belonging to tau-protein kinase (see 2.7.1.135 — TPK) can also
be developed on gels developed using the method described
above. However, unlike TPK, activity bands of TTK can also be
developed on gels containing β-tubulin, while TPK activity
bands can not be developed on such gels. This difference may
be used to differentiate between TTK and TPK activity bands.[1]

The explotable classification of protein kinases is now under
review by NC-IUBMB.[3]

References

1.   Takahashi, M., Tomizawa, K., Sato, K., Ohtake, A., and
    Omori, A., A novel tau-tubulin kinase from bovine brain,
    FEBS Lett., 372, 59, 1995.

2.   Kameshita, I. and Fujisawa, H., A sensitive method for detec-
    tion of calmodulin-dependent protein kinase II activity in
    sodium dodecyl sulfate–polyacrylamide gel, *Anal. Biochem.,*
    183, 139, 1989.

3.   NC-IUBMB, *Enzyme Nomenclature,* Academic Press, San
    Diego, 1992, p. 266 (EC 2.7.1.37, Comments).

2.7.2.3 — Phosphoglycerate Kinase; PGK

OTHER NAMES	3-Phosphoglycerate kinase, 3-phosphoglyceric phosphokinase
REACTION	ATP + 3-phospho-D-glycerate = ADP + 3-phospho-D-glyceroyl phosphate
ENZYME SOURCE	Bacteria, fungi, plants, protozoa, invertebrates, vertebrates
SUBUNIT STRUCTURE	Monomer (fungi, plants, invertebrates, vertebrates)

METHOD

Visualization Scheme

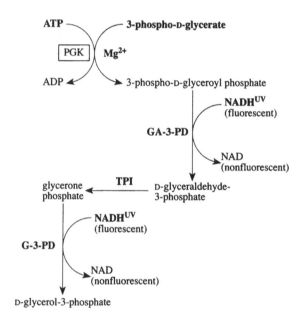

Staining Solution[1] (modified)

0.3 M Tris–HCl buffer, pH 8.0	5 ml
3-Phosphoglycerate (disodium salt)	20 mg
ATP (disodium salt; $3H_2O$)	30 mg
$MgCl_2 \cdot 6H_2O$	35 mg
Glyceraldehyde-3-phosphate dehydrogenase (GA-3-PD)	40 U
NADH	10 mg
Triose-phosphate isomerase (TPI)	100 U
Glycerol-3-phosphate dehydrogenase (G-3-PD)	10 U

Procedure

Apply the staining solution to the gel surface on a filter paper overlay. Incubate the gel at 37°C and monitor under long-wave UV light. Dark (nonfluorescent) bands visible on a light (fluorescent) background indicate the sites of PGK localization. Record the zymogram or photograph using a yellow filter.

Notes: Negatively stained bands of PGK visible in daylight can be developed by treating the processed gel with MTT–PMS solution.

Auxiliary enzymes TPI and G-3-PD are used to intensify the process of NADH-into-NAD conversion. These enzymes may be omitted from the staining solution when preparations with high PGK activity are analyzed.

OTHER METHODS

A. The reverse PGK reaction can be used to detect ATP production via two linked reactions catalyzed by two auxiliary enzymes, hexokinase and glucose-6-phosphate dehydrogenase (e.g., see 2.7.3.2 — CK, Method 2). However, this method is not specific since the adenylate kinase (AK) activity bands are stained together with PGK bands. Thus, an additional gel should be stained for AK as a control (see 2.7.4.3 — AK, Method 1).

B. Using the forward PGK reaction, the product ADP can be detected via two linked reactions catalyzed by auxiliary enzymes, pyruvate kinase and lactate dehydrogenase (e.g., see 2.7.1.11 — PFK, Method 3). This method reveals areas of PGK localization as dark (nonfluorescent) bands visible in long-wave UV light on the light (fluorescent) background of the gel. When using this method, two additional gels should be stained as controls to identify possible bands caused by phosphoenolpyruvate phosphatase (see 3.1.3.18 — PGP, Method 2) and ATPase (see 3.6.1.3 — ATPASE, Method 2).

REFERENCES

1. Beutler, E., Electrophoresis of phosphoglycerate kinase, *Biochem. Genet.*, 3, 189, 1969.

REACTION	ATP + creatine = ADP + phospho-creatine
ENZYME SOURCE	Invertebrates, vertebrates
SUBUNIT STRUCTURE	Dimer (cytoplasmic enzyme: tunicates, vertebrates), octamer[#] (mitochondrial enzyme: invertebrates, vertebrates); also see General Notes

Method 1

Visualization Scheme

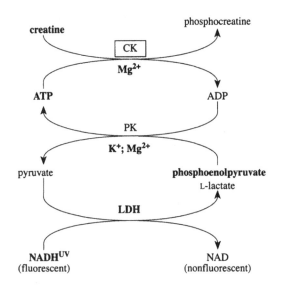

Staining Solution[1]

0.5 M Tris–HCl buffer, pH 8.0	10 ml
Creatine	30 mg
ATP (disodium salt; 3H$_2$O)	20 mg
Magnesium acetate (4H$_2$O)	40 mg
Potassium acetate	40 mg
Phosphoenolpyruvate (potassium salt)	15 mg
NADH	10 mg
Pyruvate kinase (PK)	2 U
Lactate dehydrogenase (LDH)	130 U

Procedure

Mix A and B components of the staining solution and pour the mixture over the surface of the gel. Incubate the gel in the dark until dark blue bands appear. Fix the stained gel in 25% ethanol.

Notes: When a zymogram that is visible in daylight is required, the processed gel should be counterstained with MTT–PMS solution, resulting in the appearance of achromatic bands on a blue background of the gel.

In some cases, additional bands of phosphatase activity can also be developed by this method. This is due to the ability of phosphatases from some sources to hydrolyze the phosphoenolpyruvate.[1,2] To identify these bands, an additional gel should be stained with staining solution lacking ATP, PK, and creatine (for details, see 3.1.3.18 — PGP, Method 2).

Method 2

Visualization Scheme

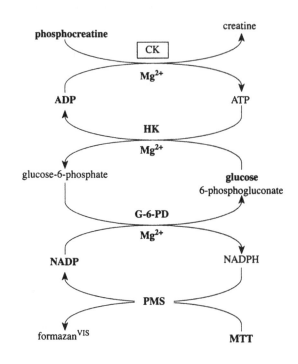

Staining Solution[3] (modified)

A.	0.1 M Tris–HCl buffer, pH 8.0	20 ml
	NADP	10 mg
	Phosphocreatine (disodium salt; 6H$_2$O)	25 mg
	ADP (disodium salt)	25 mg
	Glucose	60 mg
	MgCl$_2$·6H$_2$O	60 mg
	Glucose-6-phosphate dehydrogenase (G-6-PD)	20 U
	Hexokinase (HK)	20 U
	MTT	7 mg
	PMS	1 mg
B.	1.5% Agarose solution (60°C)	20 ml

Procedure

Apply the staining solution to the gel surface on a filter paper overlay. Incubate the gel at 37°C and monitor under long-wave UV light. Dark (nonfluorescent) bands visible on a light (fluorescent) background indicate the sites of CK localization. Record the zymogram or photograph using a yellow filter.

Notes: Bands of adenylate kinase (see 2.7.4.3 — AK) activity can also be developed with this method. Therefore, an additional gel should be stained for AK as a control. The addition of AMP (final concentration of 2 mg/ml) to the CK staining solution may be used to inhibit AK activity.

Preparations of NAD-dependent G-6-PD are now commercially available (Sigma G 5760 or G 5885). When this form of G-6-PD is used, NAD should be substituted for NADP in the staining solution. This substitution will result in cost savings because NAD is about five times less expensive than NADP.

P^1,P^5-di(adenosine-5′) pentaphosphate is a potent AK inhibitor. It can be used to prevent development of undesirable AK bands on phosphagen kinase zymograms.[4] It should be taken into account, however, that this compound is too expensive.

OTHER METHODS

An immunoblotting procedure (for details, see Part II) based on the utility of monoclonal antibodies specific to the hen enzyme[5] can also be used for visualization of the enzyme protein on electrophoretic gels. This procedure is not recommended for routine laboratory use; however, it may be of great value in special analyses of the enzyme.

GENERAL NOTES

Method 1 is more sensitive, but Method 2 is more convenient for routine laboratory use.

CKs constitute a group of different oligomeric isozymes with tissue-specific subcellular localization. Mitochondrial CK isozymes are octramers, while cytosolic CK isozymes of vertebrates are dimers. It was shown that the octamer is a primitive feature of CK.[6] The change from the primordial monomeric into the dimeric state was for a long time suspected to have physiological and evolutionary advantages. However, only recently was the physiological advantage of the dimeric CK structure proved by the evidence for cooperativeness between the two subunits of cytosolic CK. It was clearly demonstrated that cooperativeness between CK subunits results in an improvement of the catalytic properties of CK in the direction of ATP synthesis (reverse reaction).[7]

REFERENCES

1. Harris, H. and Hopkinson, D.A., *Handbook of Enzyme Electrophoresis in Human Genetics*, North-Holland, Amsterdam, 1976 (loose-leaf, with supplements in 1977 and 1978).
2. Manchenko, G.P., Allozymic variation and substrate specificity of phosphatase from sipunculid *Phascolosoma japonicum*, *Isozyme Bull.*, 21, 168, 1988.
3. Shaw, C.R. and Prasad, R., Starch gel electrophoresis of enzymes: a compilation of recipes, *Biochem. Genet.*, 4, 297, 1970.
4. Röhner, M., Bastrop, R., and Jürss, K., Colonization of Europe by two American genetic types or species of the genus *Marenzelleria* (Polychaeta: Spionidae): an electrophoretic analysis of allozymes, *Mar. Biol.*, 127, 277, 1996.
5. Morris, G.E. and Head, L.P., A monoclonal antibody against the skeletal muscle enzyme creatine kinase, *FEBS Lett.*, 145, 163, 1982.
6. Pineda, A.O., Jr. and Ellington, W.R., Structural and functional implications of the amino acid sequences of dimeric, cytoplasmic and octameric mitochondrial creatine kinases from a protostome invertebrate, *Eur. J. Biochem.*, 264, 67, 1999.
7. Hornemann, T., Rutishauser, D., and Wallimann, T., Why is creatine kinase a dimer? Evidence for cooperativity between the two subunits, *Biochim. Biophys. Acta*, 1480, 365, 2000.

REACTION	ATP + L-arginine = ADP + N-phospho-L-arginine
ENZYME SOURCE	Invertebrates
SUBUNIT STRUCTURE	Monomer (majority of protostome invertebrates), dimer (echinoderms), tetramer (polychaetes); also see General Notes

METHOD 1

Visualization Scheme

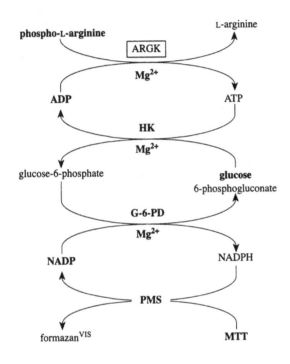

Staining Solution[1] (modified)

A.	0.1 M Tris–HCl buffer, pH 8.0	20 ml
	Phospho-L-arginine	20 mg
	ADP	30 mg
	Glucose	80 mg
	MgCl$_2$·6H$_2$O	8 mg
	NADP	20 mg
	PMS	1 mg
	MTT	8 mg
	Hexokinase (HK)	45 U
	Glucose-6-phosphate dehydrogenase (G-6-PD)	100 U
B.	1.5% Agarose solution (60°C)	20 ml

Procedure

Mix A and B components of the staining solution and pour the mixture over the surface of the gel. Incubate the gel in the dark at 37°C until dark blue bands appear. Fix the stained gel in 25% ethanol.

Notes: Bands of adenylate kinase (see 2.7.4.3 — AK) activity can also be developed with this method. Therefore, an additional gel should be stained for AK as a control. The addition of AMP (final concentration of 2 mg/ml) to the staining solution may be used to inhibit AK activity.

NAD-dependent G-6-PD is now commercially available (Sigma G 5760 or G 5885). When this form of the linking enzyme is used, NAD should be substituted for NADP in the staining solution. This replacement will result in cost savings because NAD is about five times less expensive than NADP.

P^1,P^5-di(adenosine-5′) pentaphosphate is a potent AK inhibitor. It can be used to prevent development of AK bands on ARGK zymograms.[2] It should be taken into account, however, that this compound is too expensive.

METHOD 2

Visualization Scheme

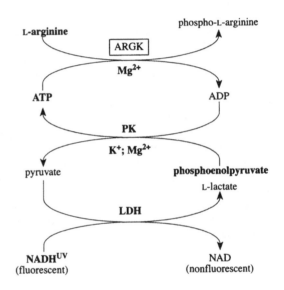

Staining Solution[3] (modified)

A.	0.1 M Tris–HCl buffer, pH 8.0	20 ml
	L-Arginine	20 mg
	ATP	30 mg
	MgCl$_2$·6H$_2$O	25 mg
	KCl	20 mg
	Phosphoenolpyruvate (tricyclohexylammonium salt)	30 mg
	NADH	15 mg
	Pyruvate kinase (PK)	90 U
	Lactate dehydrogenase (LDH)	150 U
B.	1.5% Agarose solution (60°C)	20 ml

Procedure

Mix A and B components of the staining solution and pour the mixture over the surface of the gel. Incubate the gel at 37°C and monitor under long-wave UV light. Dark (nonfluorescent) bands visible on a light (fluorescent) background correspond to localization of the enzyme on the gel. Record the zymogram or photograph using a yellow filter.

Notes: When a zymogram that is visible in daylight is required, counterstain the processed gel with MTT–PMS solution to reveal achromatic ARGK bands on a blue background of the gel.

Alkaline phosphatase (ALP; EC 3.1.3.1) and phosphoglycolate phosphatase (PGP; EC 3.1.3.18) activity bands can also be developed by this method. This is due to the ability of ALP and PGP from some animal species to hydrolyze the phosphoenolpyruvate.[4] To identify these bands, an additional gel should be stained in the staining solution lacking ATP, PK, and L-arginine (see also 3.1.3.18 — PGP, Method 2).

METHOD 3

Visualization Scheme

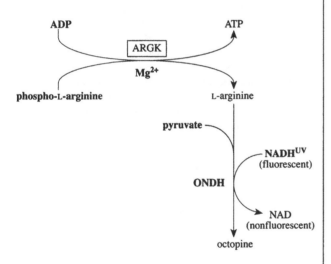

Staining Solution*

0.1 *M* Tris–HCl buffer, pH 8.0	10 ml
ADP	15 mg
Phospho-L-arginine	10 mg
MgCl$_2$·6H$_2$O	5 mg
Pyruvic acid (sodium salt)	10 mg
NADH	7 mg
Octopine dehydrogenase (ONDH)	20 U

Procedure

Apply the staining solution to the gel surface on a filter paper overlay. Incubate the gel in a moist chamber at 37°C for 30 min and monitor under long-wave UV light. Dark (nonfluorescent) bands visible on a light (fluorescent) background indicate the sites of ARGK localization. Record the zymogram or photograph using a yellow filter.

Notes: When a zymogram that is visible in daylight is required, counterstain the processed gel with MTT–PMS solution to obtain achromatic ARGK bands on a blue gel background.

Lactate dehydrogenase bands can also be developed by this method. Therefore, an additional gel should be stained for LDH activity as a control. In many invertebrate taxons, however, LDH activity is very low or not detectable at all.

METHOD 4

Visualization Scheme

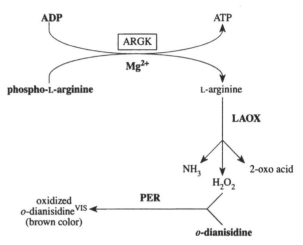

Staining Solution*

A.	0.1 *M* Tris–HCl buffer, pH 8.0	20 ml
	Phospho-L-arginine	20 mg
	ADP	30 mg
	Snake venom L-amino acid oxidase (LAOX)	1 U
	Peroxidase (PER)	200 U
	MgCl$_2$·6H$_2$O	16 mg
	o-Dianisidine (dihydrochloride)	10 mg
B.	2% Agarose solution (60°C)	20 ml

Procedure

Mix A and B components of the staining solution and pour the mixture over the surface of the gel. Incubate the gel at 37°C until colored bands of good intensity appear. Fix the stained gel and agarose overlay in 50% glycerol.

Notes: *o*-Dianisidine dihydrochloride is a potential carcinogen. 3-Amino-9-ethyl-carbazole is recommended for use in place of *o*-dianisidine. The use of *o*-dianisidine, however, gives better results.

METHOD 5

Visualization Scheme

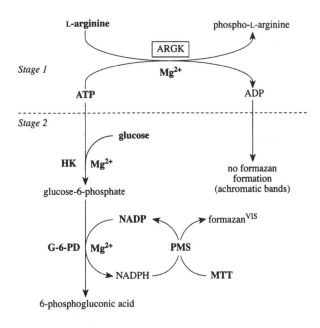

Staining Solution*

A. 0.05 *M* Tris–HCl buffer, pH 8.0   20 ml
    L-Arginine   20 mg
    ATP   30 mg
    $MgCl_2 \cdot 6H_2O$ 10 mg
B. 0.05 *M* Tris–HCl buffer, pH 8.0   20 ml
    Hexokinase (HK)   45 U
    Glucose-6-phosphate dehydrogenase (G-6-PD)   100 U
    NADP   15 mg
    Glucose   80 mg
    $MgCl_2 \cdot 6H_2O$ 16 mg
    MTT   8 mg
    PMS   1 mg
C. 1.5% Agarose solution (50°C)

Procedure

Mix solution A with 20 ml of solution C and pour the mixture over the surface of the gel. Incubate the gel at 37°C for 30 to 60 min. Remove the first agarose overlay. Mix solution B with 20 ml of solution C and pour the mixture over the surface of the preincubated gel. Incubate the gel covered with the second agarose overlay in the dark at 37°C until white bands appear on a blue background. Fix the stained agarose plate in 25% ethanol and dry on filter paper.

Notes: This negative stain may be explored in the absence of phospho-L-arginine.

---

* New; recommended for use.

GENERAL NOTES

Arginine kinase is a monomer in most invertebrates except echinoderms, where this enzyme is known to be a dimer. Subunits of the dimeric echinoderm ARGK are capable of *in vitro* hybridization with subunits of mammalian dimeric (cytosol) creatine kinase (see 2.7.3.2 — CK), thus providing evidence that these two enzymes are closely related. However, sudden emergence of the unusual dimeric ARGK in echinoderms still represents an unsolved problem. The solution of this problem was hindered by the long-standing point of view that only dimeric ARGK is present in echinoderms.

Two ARGK isozymes (ARGK-1 and ARGK-2) were found to be co-expressed in the sea star *Asterias amurensis*.[5] Activity of the ARGK-1 isozyme predominates in all the sea star tissues, while activity of the ARGK-2 isozyme is low in most tissues except ambulacra, where it is sufficient for electrophoretic analysis. Allozyme variation provides evidence that the ARGK-1 isozyme is a dimer (heterozygotes display three-banded phenotypes), while ARGK-2 is a monomer (heterozygotes are two-banded). Similar ARGK isozymes were also revealed in sea urchins.[6] It was concluded from these data that monomeric ARGK is characteristic of echinoderms and represents a missing link in the evolution of deuterostome phosphagen kinases. According to this view, phosphagen kinase evolution was progressing from monomeric ARGK in protostomes through gene duplication and coexistence of monomeric and dimeric ARGK isozymes in an ancestor common for echinoderms and chordates to dimeric CK.[5] This evolutionary scenario is supported by the presence of monomeric ARGK and dimeric CK in tunicates.[7]

Recently another course of phosphagen kinase evolution was suggested by Suzuki et al.,[8] who determined the cDNA-derived amino acid sequence of the dimeric echinoderm ARGK and compared it with those of other phosphagen kinases. These authors concluded that dimeric ARGK of echinoderms originated from a dimeric CK-like ancestral protein. It should be stressed, however, that this evolutionary way is based on the assumption that dimeric ARGK is the only molecular form of the enzyme in echinoderms. The co-expression of monomeric and dimeric ARGK isozymes in echinoderms and the presence of monomeric ARGK and dimeric CK in tunicates make this way not absolutely convincing and restore an old question about the origin of unusual dimeric ARGK in echinoderms.

REFERENCES

1. Fisher, S.E. and Whitt, G.S., Evolution of isozyme loci and their differential tissue expression: creatine kinase as a model system, *J. Mol. Evol.*, 12, 25, 1978.
2. Röhner, M., Bastrop, R., and Jürss, K., Colonization of Europe by two American genetic types or species of the genus *Marenzelleria* (Polychaeta: Spionidae): an electrophoretic analysis of allozymes, *Mar. Biol.*, 127, 277, 1996.
3. Bulnheim, H.-P. and Scholl, A., Genetic variation between geographic populations of the amphipods *Gammarus zaddachi* and *G. salinus*, *Mar. Biol.*, 64, 105, 1981.

2.7.3.3 — Arginine Kinase; ARGK (continued)

4. Manchenko, G.P., Allozymic variation and substrate specificity of phosphatase from sipunculid *Phascolosoma japonicum*, *Isozyme Bull.*, 21, 168, 1988.
5. Manchenko, G.P., Monomeric and dimeric arginine kinase isozymes are co-expressed in the sea star *Asterias amurensis*, *Gene Fam. Isozymes Bull.*, 33, 56, 2000.
6. Manchenko, G.P. and Yakovlev, S.N., Genetic divergence between three sea urchin species of the genus *Strongylocentrotus* from the Sea of Japan, *Biochem. Syst. Ecol.*, 29, 31, 2001.
7. Manchenko, G.P., Monomeric arginine kinase of echinoderms and tunicates: a missing link in the evolution of deuterostomean phosphagen kinases, paper presented in *Proceedings of International Symposium of Modern Achievements in Population, Evolutionary and Ecological Genetics*, Vladivostok, Russia, 1998, p. 11 (Abstract).
8. Suzuki, T., Kamidochi, M., Inoue, N., Kawamichi, H., Yazawa, Y., Furukohri, T., and Ellington, W.R., Arginine kinase evolved twice: evidence that echinoderm arginine kinase originated from creatine kinase, *Biochem. J.*, 340, 671, 1999.

2.7.4.3 — Adenylate Kinase; AK

OTHER NAMES	Myokinase
REACTION	ATP + AMP = ADP + ADP
ENZYME SOURCE	Bacteria, fungi, green algae, plants, protozoa, invertebrates, vertebrates
SUBUNIT STRUCTURE	Monomer (plants, invertebrates, vertebrates)

METHOD 1

Visualization Scheme

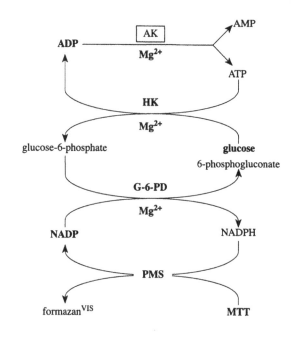

Staining Solution[1]

A.	0.2 *M* Tris–HCl buffer, pH 8.0	25 ml
	1 *M* MgCl$_2$	1 ml
	ADP (disodium salt)	25 mg
	Glucose	180 mg
	NADP	15 mg
	PMS	2 mg
	MTT	10 mg
	Hexokinase (HK)	25 U
	Glucose-6-phosphate dehydrogenase (G-6-PD)	15 U
B.	1.5% Agar solution (60°C)	25 ml

Procedure

Mix A and B components of the staining solution and pour the mixture over the surface of the gel. Incubate the gel in the dark at 37°C until dark blue bands appear. Fix the stained gel or agar overlay in 25% ethanol.

Notes: When a preparation of NAD-dependent G-6-PD is included in the staining solution, NAD should be substituted for NADP. The use of NAD-dependent G-6-PD (Sigma G 5760 or G 5885) in a stain results in cost savings because NAD is about five times less expensive than NADP. Furthermore, such a substitution eliminates the problem of enhancement of some AK bands, owing to the activity of comigrating endogenous phosphogluconate dehydrogenase (see 1.1.1.44 — PGD), of which bands can develop by Method 1 due to the production of 6-phosphogluconate by the linking enzyme, glucose-6-phosphate dehydrogenase, and the presence of NADP in the stain.

Method 2

Visualization Scheme

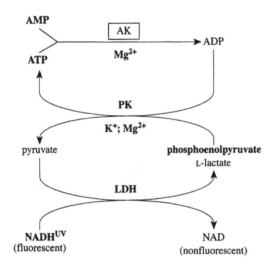

Staining Solution[2]

A. 0.5 M Tris–HCl buffer, pH 8.0 .... 10 ml
    AMP (disodium salt; 6H$_2$O) .... 25 mg
    ATP (disodium salt; 3H$_2$O) .... 25 mg
    Phosphoenolpyruvate (potassium salt) .... 25 mg
    NADH (disodium salt) .... 10 mg
    KCl .... 150 mg
    0.2 M MgCl$_2$ .... 2 ml
    Pyruvate kinase (PK; 400 U/ml) .... 0.1 ml
    Lactate dehydrogenase (LDH; 2750 U/ml) .... 0.1 ml
B. 2% Agar solution (60°C) .... 12 ml

Procedure

Mix A and B components of the staining solution and pour the mixture over the surface of the gel. Incubate the gel at 37°C and monitor under long-wave UV light. Dark (nonfluorescent) bands of AK activity are visible on a light (fluorescent) background of the gel. Record the zymogram or photograph using a yellow filter.

Notes: When preparations containing a phosphoenolpyruvate phosphatase activity are used for electrophoresis, additional bands can be observed on AK zymograms obtained by this method. These bands also develop in a staining solution lacking AMP, ATP, and PK (see 3.1.3.18 — PGP, Method 2), whereas AK bands do not.

Method 3

Visualization Scheme

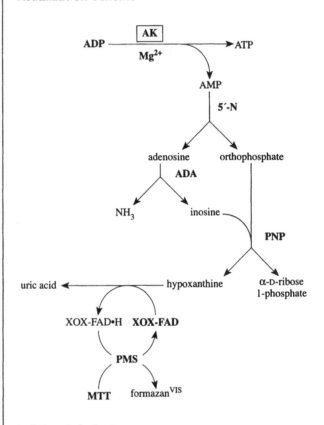

Staining Solution[3]

A. 0.2 M Tris–HCl buffer, pH 8.0 .... 1 ml
    ADP .... 10 mg
    MgCl$_2$ .... 10 mg
    PMS .... 2 mg
    MTT .... 8 mg
    5′-Nucleotidase (5′-N) .... 12 U
    Adenosine deaminase (ADA) .... 25 U
    Purine-nucleoside phosphorylase (PNP) .... 6 U
    Xanthine oxidase (XOX) .... 2 U
B. 1% Agar solution (60°C) .... 5 ml

Procedure

Mix A and B components of the staining solution and pour the mixture over the surface of the gel. Incubate the gel in the dark at 37°C until dark blue bands appear. Fix the stained gel in 25% ethanol.

2.7.4.3 — Adenylate Kinase; AK (continued)

Notes: This method is less practical than the other two described above; however, it demonstrates a general approach for visualizing enzymes releasing AMP.

GENERAL NOTES

In the forward reaction catalyzed by mammalian AK, AMP cannot be replaced by UMP, GMP, or CMP, using ATP as the second substrate. However, ATP can be replaced by UTP, ITP, CTP, or GTP, using AMP as the other substrate. The enzyme from the rabbit muscle is inactive toward IDP, GDP, and UDP in the backward reaction, while CDP is a substrate as well as ADP. The enzyme from Bakers yeast is specific for ADP as the nucleotide diphosphate. However, GTP or ITP can replace ATP, and GMP can replace AMP, but CTP, UTP, CDP, and UDP are inactive.

REFERENCES

1. Fildes, R.A. and Harris, H., Genetically determined variation of adenylate kinase in man, *Nature,* 209, 261, 1966.
2. Harris, H. and Hopkinson, D.A., *Handbook of Enzyme Electrophoresis in Human Genetics,* North-Holland, Amsterdam, 1976 (loose-leaf, with supplements in 1977 and 1978).
3 Friedrich, C.A., Chakravarti, S., and Ferrell, R.E., A general method for visualizing enzymes releasing adenosine or adenosine-5′-monophosphate, *Biochem. Genet.,* 22, 389, 1984.

2.7.4.4 — Nucleoside-Phosphate Kinase; NPK

OTHER NAMES	Nucleoside monophosphate kinase
REACTION	ATP + nucleoside monophosphate = ADP + nucleoside diphosphate
ENZYME SOURCE	Bacteria, fungi, vertebrates
SUBUNIT STRUCTURE	Monomer (vertebrates)

METHOD

Visualization Scheme

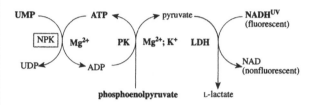

Staining Solution[1]

0.1 M Tris–HCl buffer, pH 7.8
4.4 mM UMP
3.6 mM ATP
2.1 mM Phosphoenolpyruvate
1 mM NADH
25 mM MgCl$_2$
100 mM K$_2$SO$_4$
20 U/ml Pyruvate kinase (PK)
50 U/ml Lactate dehydrogenase (LDH)

Procedure

Soak the filter paper in the staining solution and apply to the gel surface. Incubate the gel at 37°C and monitor under long-wave UV light. Dark (nonfluorescent) bands of NPK activity are visible on the light (fluorescent) background of the gel. Record the zymogram or photograph using a yellow filter.

Notes: When preparations containing a phosphoenolpyruvate phosphatase activity are electrophorized, additional bands can develop on NPK zymograms obtained by this method. These bands also develop in a staining solution lacking UMP, ATP, and PK (see 3.1.3.18 — PGP, Method 2), whereas NPK bands do not.

OTHER METHODS

The product of the reverse NPK reaction, ATP, can be detected using two linked reactions catalyzed by auxiliary enzymes hexokinase and glucose-6-phosphate dehydrogenase. This method, however, also develops adenylate kinase activity bands (see 2.7.4.3 — AK, Method 1).

2.7.4.4 — Nucleoside-Phosphate Kinase; NPK (continued)

GENERAL NOTES

Many nucleoside monophosphates can act as acceptors; other nucleoside triphosphates can act as donors.[2]

REFERENCES

1. Giblett, E.R., Anderson, J.J.E., Chen, S.-H., Teng, Y.-S., and Cohen, F., Uridine monophosphate kinase: a new genetic polymorphism with possible clinical implications, *Am. J. Hum. Genet.*, 26, 627, 1974.
2. NC-IUBMB, *Enzyme Nomenclature*, Academic Press, San Diego, 1992, p. 288 (EC 2.7.4.4, Comments).

2.7.4.6 — Nucleoside-Diphosphate Kinase; NDK

OTHER NAMES	Uridine-diphosphate kinase
REACTION	ATP + nucleoside diphosphate = ADP + nucleoside triphosphate
ENZYME SOURCE	Bacteria, fungi, plants, invertebrates, vertebrates
SUBUNIT STRUCTURE	Tetramer[#] (bacteria), hexamer[#] (fungi, plants, invertebrates, vertebrates)

METHOD 1

Visualization Scheme

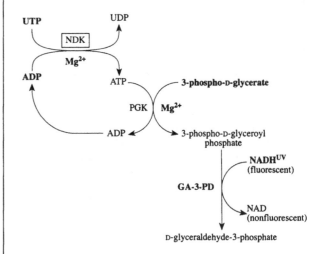

Staining Solution[1]

0.1 M Tris–HCl buffer, pH 7.8	10 ml
UTP	15 mg
ADP	4.9 mg
3-Phospho-D-glycerate (disodium salt)	8.6 mg
0.2 M KCl; 50 mM MgSO$_4$	0.2 ml
NADH	8 mg
Phosphoglycerate kinase (PGK)	60 U
Glyceraldehyde-3-phosphate dehydrogenase (GA-3-PD)	40 U

Procedure

Apply the filter paper soaked in the staining solution to the surface of the gel. Incubate the gel at 37°C in a moist chamber for 1 to 2 h and monitor under long-wave UV light. Dark (nonfluorescent) bands of NDK activity are visible on the light (fluorescent) background of the gel. Record the zymogram or photograph using a yellow filter.

2.7.4.6 — Nucleoside-Diphosphate Kinase; NDK (continued)

Notes: Adenylate kinase bands can also develop on NDK zymograms obtained by this method. Therefore, an additional gel should be stained for AK activity (see 2.7.4.3 — AK) as a control.

Additional auxiliary enzymes, triose-phosphate isomerase and glycerol-3-phosphate dehydrogenase, may be included in the staining solution to intensify the process of NADH-into-NAD conversion (for example, see 2.7.2.3 — PGK).

METHOD 2

Visualization Scheme

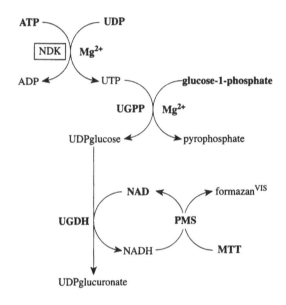

Staining Solution[2] (adapted)

A. 0.2 *M* Tris–HCl buffer, pH 8.0	6 ml
0.1 *M* MgCl$_2$	2 ml
200 mg/ml ATP	0.5 ml
200 mg/ml UDP	0.5 ml
200 mg/ml Glucose-1-phosphate	0.5 ml
200 mg/ml NAD	0.2 ml
25 U/ml UTP–glucose-1-phosphate uridylyltransferase (UGPP)	0.5 ml
2 U/ml UDPglucose dehydrogenase (UGDH)	0.2 ml
10 mg/ml PMS	0.1 ml
50 mg/ml MTT	0.1 ml
B. 2% Agarose solution (60°C)	10 ml

Procedure

Mix A and B components of the staining solution and pour the mixture over the surface of the gel. Incubate the gel in the dark at 37°C until dark blue bands appear. Fix the stained gel in 25% ethanol.

Notes: UDPglucose dehydrogenase preparation used in the staining solution should be free of NDK activity.

GENERAL NOTES

The enzyme is nonspecific with respect to its substrates. For example, the enzyme from human erythrocytes reacts with di- and triphosphate nucleotides containing either ribose or deoxyribose and any of the naturally occurring purine or pyrimidine bases. The enzyme does not catalyze an adenylate kinase type of reaction (see 2.7.4.3 — AK).

REFERENCES

1. Anderson, J.E., Teng, Y.-S., and Giblett, E.R., Stains for six enzymes potentially applicable to chromosomal assignment by cell hybridization, *Cytogenet. Cell Genet.*, 14, 465, 1975.
2. Pierce, M., Cummings, R.D., and Roth, S., The localization of galactosyltransferases in polyacrylamide gels by a coupled enzyme assay, *Anal. Biochem.*, 102, 441, 1980.

2.7.4.8 — Guanylate Kinase; GUK

OTHER NAMES   Deoxyguanylate kinase
REACTION   ATP + GMP = ADP + GDP
ENZYME SOURCE   Fungi, invertebrates, vertebrates
SUBUNIT STRUCTURE Monomer (vertebrates)

Method 1

Visualization Scheme

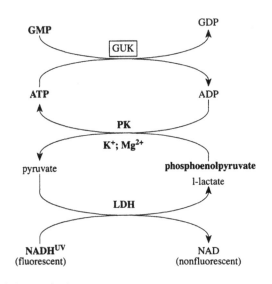

Staining Solution[1]

A. 0.2 M Tris–HCl buffer, pH 7.9
    1.5 mM ATP
    6 mM GMP
    4 mM Phosphoenolpyruvate
    0.6 mM NADH
    20 mM MgSO$_4$
    100 mM KCl
    1 mM CaCl$_2$
    2 U/ml Pyruvate kinase (PK)
    18 U/ml Lactate dehydrogenase (LDH)
B. 1% Agar solution (60°C)

Procedure

Mix equal volumes of A and B solutions and pour the mixture over the surface of the gel. Incubate the gel at 37°C and monitor under long-wave UV light. Dark (nonfluorescent) bands of GUK activity are visible on a light (fluorescent) background of the gel. Record the zymogram or photograph using a yellow filter.

Notes: When a zymogram that is visible in daylight is required, counterstain the processed gel with MTT–PMS solution. This will result in the appearance of achromatic GUK bands on a blue background of the gel.

   When preparations containing a phosphoenolpyruvate phosphatase activity are electrophorized, additional bands can be observed on GUK zymograms obtained by this method. These bands also develop in a staining solution lacking ATP, GMP, and PK (see 3.1.3.18 — PGP, Method 2), whereas GUK bands do not.

Method 2

Visualization Scheme

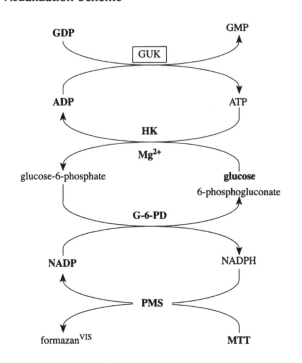

Staining Solution[1]

A. 0.2 M Tris–HCl buffer, pH 7.9
    1 mM ADP
    4 mM GDP
    20 mM Glucose
    0.8 mM NADP
    0.08% MTT
    0.04% PMS
    20 mM MgSO$_4$
    100 mM KCl
    0.2 U/ml Glucose-6-phosphate dehydrogenase
      (G-6-PD)
    0.4 U/ml Hexokinase (HK)
B. 1% Agar solution (60°C)

Procedure

Mix equal volumes of A and B solutions and pour the mixture over the surface of the gel. Incubate the gel in the dark at 37°C until dark blue bands appear. Fix the stained gel in 25% ethanol.

Notes: Adenylate kinase (AK) bands can also develop on GUK zymograms obtained by this method. Therefore, an additional gel should be stained for AK activity (see 2.7.4.3 — AK, Method 1) as a control.

References

1. Moon, E. and Christiansen, R.O., Guanylate kinase in man: multiple molecular forms, *Hum. Hered.*, 22, 18, 1972.

2.7.4.10 — Nucleoside-Triphosphate–Adenylate Kinase; NTAK

REACTION   AMP + nucleoside triphosphate = ADP + nucleoside diphosphate

ENZYME SOURCE Vertebrates

SUBUNIT STRUCTURE Monomer (vertebrates)

METHOD

Visualization Scheme

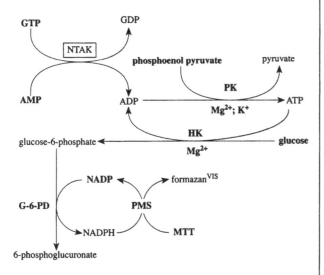

Staining Solution[1]

A.	0.3 M Tris–HCl buffer, pH 8.0	10 ml
	AMP (disodium salt; 6H$_2$O)	25 mg
	GTP	15 mg
	Phosphoenolpyruvate (potassium salt)	20 mg
	KCl	150 mg
	MgCl$_2$·6H$_2$O	40 mg
	NADP (disodium salt)	5 mg
	Glucose	40 mg
	Glucose-6-phosphate dehydrogenase (G-6-PD; 140 U/ml)	25 μl
	Hexokinase (HK; 280 U/ml)	25 μl
	Pyruvate kinase (PK; 400 U/ml)	50 μl
	MTT	5 mg
	PMS	0.5 mg
B.	1.5% Agar solution (60°C)	10 ml

Procedure

Mix A and B solutions and pour the mixture over the surface of the gel. Incubate the gel in the dark at 37°C until dark blue bands appear. Fix the stained gel in 25% ethanol.

Notes: This staining system also detects adenylate kinase activity bands, but NTAK activity is not detected by the ordinary adenylate kinase stains (see 2.7.4.3 — AK, Methods 1 to 3). These differences can be used to differentiate AK and NTAK bands.

It is possible to replace GTP with ITP in the staining solution. However, this will cause significant staining of the gel background because of the reaction of auxiliary HK with ITP and glucose, and the subsequent reaction of auxiliary G-6-PD with glucose-6-phosphate and NADP in the presence of PMS and MTT.

OTHER METHODS

NTAK activity can also be detected by negative fluorescent stain using pyruvate kinase and lactate dehydrogenase as linking enzymes. To prepare the necessary staining solution, substitute GTP for ATP in the negative fluorescent stain for adenylate kinase (see 2.7.4.3 — AK, Method 2).

REFERENCES

1. Wilson, D.E., Povey, S., and Harris, H., Adenylate kinases in man: evidence for a third locus, *Ann. Hum. Genet.*, 39, 305, 1976.

2.7.6.1 — Ribose-Phosphate Pyrophosphokinase; RPPPK

OTHER NAMES	5-Phosphoribosyl-1-pyrophosphate synthetase
REACTION	ATP + D-ribose 5-phosphate = AMP + 5-phospho-α-D-ribose 1-diphosphate
ENZYME SOURCE	Bacteria, vertebrates
SUBUNIT STRUCTURE	Unknown or no data available

METHOD 1

Visualization Scheme

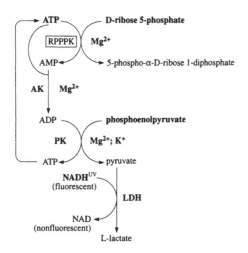

Staining Solution[1]

0.1 M Tris–HCl buffer, pH 7.8	5 ml
D-Ribose 5-phosphate (disodium salt)	12.4 mg
ATP	15 mg
Phosphoenolpyruvate	4.5 mg
NADH	3.5 mg
2 M KCl; 0.5 M MgSO₄	0.25 ml
Adenylate kinase (AK; Sigma)	50 μl
Pyruvate kinase (PK; Sigma)	50 μl
Lactate dehydrogenase (LDH; Sigma)	100 μl

Procedure

Soak the filter paper in the staining solution and apply it to the surface of the gel. Incubate the gel at 37°C for 2 to 3 h in a moist chamber and monitor under long-wave UV light. Dark (nonfluorescent) bands of RPPPK are visible on a light (fluorescent) background of the gel. Record the zymogram or photograph using a yellow filter.

Notes: When preparations containing a phosphoenolpyruvate phosphatase activity are electrophorized, additional bands can be observed on RPPPK zymograms obtained by this method. These bands also develop in a staining solution lacking ATP, D-ribose 5-phosphate, AK, and PK (see 3.1.3.18 — PGP, Method 2), whereas RPPPK activity bands do not.

METHOD 2

Visualization Scheme

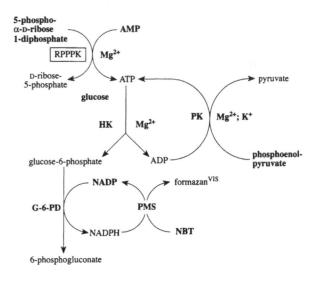

Staining Solution[2]

75 mM Potassium phosphate buffer, pH 7.4
1 mM 5-Phospho-α-D-ribose 1-diphosphate
0.5 mM AMP
1 mM Phosphoenolpyruvate
1 mM D-Glucose
0.4 mM NADP
10 mM MgCl₂
0.075 mg/ml PMS
0.75 mg/ml NBT
10 U/ml Pyruvate kinase (PK)
1 U/ml Glucose-6-phosphate dehydrogenase (G-6-PD)
2.25 U/ml Hexokinase (HK)
10% Glycerol

Procedure

Saturate the glass-supported Whatman 3MM chromatographic paper with staining solution and apply the gel directly to the chromatographic paper. Cover the gel with a second glass plate and wrap in aluminum foil to avoid nonspecific light-induced production of blue formazan. Incubate the gel at 37°C until dark blue bands appear. Fix the stained gel in 25% ethanol.

Notes: The addition of 3% Nonidet to a homogenization buffer and to electrode and gel buffers is recommended.

Pyruvate kinase and phosphoenolpyruvate are included in the stain to remove ADP, which inhibits RPPPK.

Intensely stained nonspecific bands that are not dependent on the 5-phospho-α-D-ribose 1-diphosphate also develop on RPPPK zymograms obtained using this method after electrophoresis of human tissue preparations. The nature of these bands is unknown.

2.7.6.1 — Ribose-Phosphate Pyrophosphokinase; RPPPK (continued)

OTHER METHODS

A. The product of the forward RPPPK reaction, AMP, can be detected using four reactions catalyzed by auxiliary enzymes 5′-nucleotidase, adenosine deaminase, purine-nucleoside phosphorylase, and xanthine oxidase (for example, see 2.7.4.3 — AK, Method 3).

B. Method 2, described above, may be modified by inclusion of lactate dehydrogenase and NADH in the staining solution in place of glucose-6-phosphate dehydrogenase, NADP, PMS, and NBT. This modification will result in the development of dark (nonfluorescent) bands of RPPPK activity visible in long-wave UV light on the light (fluorescent) background of the gel.

GENERAL NOTES

The forward reaction of RPPPK is favored. The enzyme is highly specific for its substrates.

REFERENCES

1. Anderson, J.E., Teng, Y.-S., and Giblett, E.R., Stains for six enzymes potentially applicable to chromosomal assignment by cell hybridization, *Cytogenet. Cell Genet.*, 14, 465, 1975.
2. Lebo, R.V. and Martin, D.W., Electrophoretic heterogeneity of 5-phosphoribosyl-1-pyrophosphate synthetase within and among humans, *Biochem. Genet.*, 16, 905, 1978.

2.7.7.4 — Sulfate Adenylyltransferase; SAT

OTHER NAMES	ATP-sulfurylase, sulfurylase
REACTION	ATP + sulfate (or molybdate) = pyrophosphate + adenylylsulfate (or adenylylmolybdate)
ENZYME SOURCE	Bacteria, fungi, plants
SUBUNIT STRUCTURE	Unknown or no data available

METHOD

Visualization Scheme

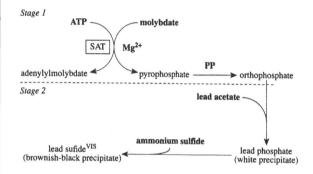

Staining Solution[1] (modified)

A. 0.1 M Tris–HCl buffer, pH 8.0 — 20 ml
ATP — 40 mg
Na_2MoO_4 — 80 mg
$MgCl_2 \cdot 6H_2O$ — 30 mg
Inorganic pyrophosphatase from yeast (PP) — 25 U
B. 2% Agarose solution (60°C) — 20 ml
C. 0.2% $Pb(CH_3COO)_2$ in 1% CH_3COOH
D. 5% Ammonium sulfide

Procedure

Mix A and B solutions and pour the mixture over the surface of the gel. Incubate the gel at 37°C for 30 to 40 min. Remove the agarose overlay and immerse the gel in solution C for 10 to 15 min. Discard solution C, rinse the gel in water, and treat with solution D. A brownish black precipitate of lead sulfide is generated in areas of SAT localization.

OTHER METHODS

An enzymatic method for detection of orthophosphate generated by auxiliary inorganic pyrophosphatase is available.[2] This method involves two additional linked reactions catalyzed by auxiliary enzymes purine-nucleoside phosphorylase and xanthine oxidase, coupled with the PMS–MTT system (for example, see 3.1.3.1 — ALP, Method 8). The enzymatic method may be preferable due to its sensitivity and generation of a nondiffusable formazan as the result of a one-step procedure.

2.7.7.4 — Sulfate Adenylyltransferase; SAT (continued)

REFERENCES

1. Skiring, G.W., Trudinger, P.A., and Shaw, W.H., Electrophoretic characterization of ATP:sulfate adenylyltransferase (ATP–sulfurylase) with use of polyacrylamide gels, *Anal. Biochem.*, 48, 259, 1972.
2. Klebe, R.J., Schloss, S., Mock, L., and Link, C.R., Visualization of isozymes which generate inorganic phosphate, *Biochem. Genet.*, 19, 921, 1981.

2.7.7.6 — DNA-Directed RNA Polymerase; DDRP

OTHER NAMES	RNA nucleotidyltransferase (DNA directed), RNA polymerase, RNA polymerase I, RNA polymerase II, RNA polymerase III
REACTION	Nucleoside triphosphate + RNA_n = pyrophosphate + RNA_{n+1} (see General Notes)
ENZYME SOURCE	Bacteria, fungi, plants, invertebrates, vertebrates
SUBUNIT STRUCTURE	Unknown or no data available

METHOD

Visualization Scheme

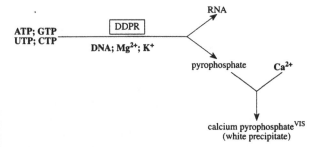

Staining Solution[1]

10 mM Tris–HCl buffer, pH 8.0
12 mM Mg^{2+}
0.1 mM EDTA
1 mM Dithiothreitol
0.2 M KCl
20 mM Ca^{2+}
0.8 mM ATP
0.8 mM CTP
0.8 mM GTP
0.8 mM UTP

Procedure

Soak an electrophorized PAG containing 1.6 mg/ml calf thymus DNA in 100 mM Tris–HCl buffer (pH 8.0) for 20 to 30 min and incubate in the staining solution at 37°C until white bands of calcium pyrophosphate precipitation appear. These bands are well visible when the gel is viewed against a dark background. When an activity stain of sufficient intensity is obtained, remove the gel from the staining solution and store in 50 mM glycine–KOH buffer, (pH 10.0) containing 5 mM Ca^{2+}, either at 5°C or at room temperature in the presence of an antibacterial agent.

Notes: The method is applicable for clean gels (PAG or agarose). When more opaque gels are used (e.g., acetate cellulose or starch gel), precipitated calcium pyrophosphate should be counterstained with Alizarin Red S.

OTHER METHODS

A. The product pyrophosphate can be detected enzymatically using three coupled reactions sequentially catalyzed by auxiliary enzymes bacterial pyrophosphate-fructose-6-phosphate 1-phosphotransferase, aldolase, and glycerol-3-phosphate dehydrogenase. The activity bands of DDRP developed by this method are visible in long-wave UV light as dark areas on a light (fluorescent) gel background. The substitution of glyceraldehyde-3-phosphate dehydrogenase for glycerol-3-phosphate dehydrogenase, and NAD, arsenate, PMS, and MTT for NADH results in the development of DDRP activity bands visible in daylight. The use of an additional auxiliary enzyme, triose-phosphate isomerase, accelerates reactions occurring in fluorogenic and chromogenic methods and enhances developing bands (for details, see 2.7.1.90 — PFPPT). Another enzymatic method for detection of pyrophosphate is available. It involves three auxiliary enzymes (inorganic pyrophosphatase, purine-nucleoside phosphorylase, and xanthine oxidase) coupled with a PMS–MTT system (for example, see 2.7.7.4 — SAT, Other Methods).

B. The enzyme can also be detected by autoradiography using a staining solution similar to that described above, but lacking Ca^{2+} and containing [^{14}C]ATP (final concentration of 1 μCi/ml) in addition to nonlabeled ATP.[2]

C. An immunoblotting procedure (for details, see Part II) based on the utility of monoclonal antibodies specific to the *Drosophila* and bovine enzyme[3–5] can also be used for immunohistochemical visualization of the enzyme protein on electrophoretic gels. This procedure is unsuitable for routine laboratory use in large-scale population genetic studies, but it may be of great value in special (e.g., biochemical, immunochemical, phylogenetic, genetic, etc.) analyses of DDRP.

GENERAL NOTES

The enzyme catalyzes the DNA template–directed extension of the 3'-end of an RNA strand one nucleotide at a time and can initiate an RNA chain *de novo*. Three forms of DDRP have been distinguished in eukaryotes on the basis of sensitivity to α-amanitin and the type of RNA synthesized.[6]

REFERENCES

1. Nimmo, H.G. and Nimmo, G.A., A general method for the localization of enzymes that produce phosphate, pyrophosphate, or CO_2 after polyacrylamide gel electrophoresis, *Anal. Biochem.*, 121, 17, 1982.
2. Uriel, J. and Lavialle, C., Autoradiographic method for characterization of DNA and RNA polymerases after gel electrophoresis, *Anal. Biochem.*, 42, 509, 1971.
3. Bona, M., Scheer, U., and Bautz, E.K., Antibodies to RNA polymerase-II(B) inhibit transcription in lampbrush chromosomes after microinjection into living amphibian oocytes, *J. Mol. Biol.*, 151, 81, 1981.
4. Christmann, J.L. and Dahmus, M.E., Monoclonal antibody specific for calf thymus RNA polymerases-IIO and polymerase-IIA, *J. Biol. Chem.*, 256, 1798, 1981.
5. Kramer, A. and Bautz, E.K., Immunological relatedness of subunits of RNA polymerase-II from insects and mammals, *Eur. J. Biochem.*, 117, 449, 1981.
6. NC-IUBMB, *Enzyme Nomenclature*, Academic Press, San Diego, 1992, p. 287 (EC 2.7.7.6, Comments).

2.7.7.7 — DNA-Directed DNA Polymerase; DDDP

OTHER NAMES   DNA nucleotidyltransferase (DNA directed); DNA polymerase I, II, or III; DNA polymerase α, β, or γ

REACTION   Deoxynucleoside triphosphate + DNA_n = pyrophosphate + DNA_{n+1}

ENZYME SOURCE   Bacteria, fungi, plants, invertebrates, vertebrates

SUBUNIT STRUCTURE   See General Notes

METHOD 1

Visualization Scheme

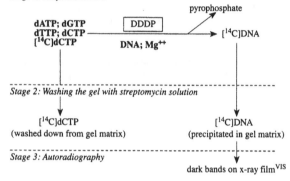

Stage 1: Enzyme Reaction

Stage 2: Washing the gel with streptomycin solution

[14C]dCTP (washed down from gel matrix)

[14C]DNA (precipitated in gel matrix)

Stage 3: Autoradiography

dark bands on x-ray film[VIS]

Reaction Mixture[1]

  A. 140 mM Tris–HCl buffer, pH 7.4
      6 mM $MgCl_2$
      2 mM 2-Mercaptoethanol
      0.06 mM dATP
      0.06 mM dGTP
      0.06 mM dTTP
      0.06 mM dCTP
      0.4 μCi/ml [14C]dCTP
      400 μg/ml Calf thymus DNA
  B. 2.4% Agar solution (42°C)
  C. 2% Streptomycin (water solution)

Procedure

Stage 1. Mix A and B components of the reaction mixture and pour the mixture over the surface of an electrophorized 6% PAG containing 0.8% agarose to form a 1-mm-thick reactive agar plate. Incubate the PAG–reactive plate combination at 37°C for 2 h.

Stage 2. Take off the reactive plate and wash it two times for 1 to 2 h in solution C. Dry the reactive plate on a sheet of filter paper.

Stage 3. Autoradiograph the dry reactive plate with x-ray film (Kodirex film, Kodak) for 1 to 2 weeks at room temperature.

The dark bands on the developed x-ray film correspond to localization of DDDP activity on the gel.

Notes: The DNA template (calf thymus DNA activated by DNase I, or poly[d(A-T)]) may be included directly into a running 10% PAG containing 0.1% SDS.[2]

METHOD 2

Visualization Scheme

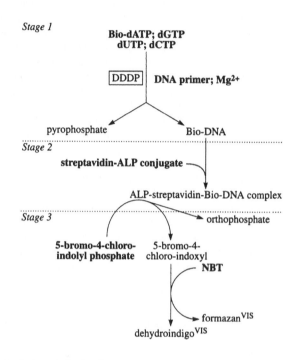

Stage 1

Bio-dATP; dGTP
dUTP; dCTP

DDDP   DNA primer; Mg^{2+}

pyrophosphate   Bio-DNA

Stage 2

streptavidin-ALP conjugate

ALP-streptavidin-Bio-DNA complex

Stage 3   orthophosphate

5-bromo-4-chloro-indolyl phosphate 5-bromo-4-chloro-indoxyl

NBT

formazan[VIS]

dehydroindigo[VIS]

Staining Solution[3]

  A. 50 mM Tris–HCl buffer, pH 8.0
      10 mM $Mg(CH_3COO)_2$
      0.4 mM EDTA
      10 mM Dithiothreitol
      400 μg/ml Bovine serum albumin (BSA)
      16% (v/v) Glycerol
      10 to 44 nM Biotin-labeled dATP (Bio-dATP)
      10 μM dCTP
      10 μM dGTP
      10 μM dTTP
  B. 0.1 M Tris–HCl, pH 7.5; 0.15 M NaCl (TN buffer)
      0.1 μg/ml Streptavidin–alkaline phosphatase (ALP) conjugate (this conjugate concentration corresponds to a dilution of about 1:5000 with respect to the commercial solution)
  C. 0.1 M Tris–HCl, pH 9.5; 5 mM $MgCl_2$; 0.1 M NaCl (ALP buffer)
      60 μg/ml 5-Bromo-4-chloro-3-indolyl phosphate
      100 μg/ml NBT

Procedure

Carry out electrophoresis in an SDS-PAG prepared following the discontinuous system described by Laemmli,[4] adapted to slab gels (7 × 8 × 0.05 cm) containing 21 μg/ml calf thymus DNA (activated by DNase I treatment) as a primer. After electrophoresis, wash the gel twice in 50 ml of 50 mM Tris–HCl (pH 7.5) for 15 min with gentle agitation at room temperature, and then incubate in 40 ml of renaturation buffer (50 mM Tris–HCl (pH 8.0), 10 mM 2-mercaptoethanol, 10 mM Mg(CH$_3$COO)$_2$, 400 μg/ml bovine serum albumin, 0.4 mM EDTA, and 16% (v/v) glycerol) for 3 h at room temperature with gentle agitation. Rinse the gel three times in 50 ml of Tris–HCl (pH 8.0) and incubate (without agitation) for 20 h at 37°C in 25 ml of solution A (see Staining Solution) to carry out the enzyme reaction. Rinse the gel briefly three times with several volumes of TN buffer (see solution B). Then block the gel for 1 h with 4% nonfat milk in TN buffer, rinse three times with TN buffer, and incubate for 10 to 15 h in solution B. Rinse the gel three times with ALP buffer (see solution C) for 20 min. Perform all the procedures following the enzyme reaction at room temperature with gentle agitation. Finally, place the gel in solution C to develop gel areas where the ALP–streptavidin–Bio-DNA complex is localized. These areas correspond to localization of DDDP activity. Carry out the gel staining and visualization of DDDP activity bands without agitation in the dark at room temperature for 1 to 4 days. Stop the ALP reaction by washing the gel three times with stop buffer (10 mM Tris–HCl, pH 8.0, and 1mM EDTA). Photograph the zymogram or dry between two cellophane papers and store in the dark for further use.

Notes: Biotin-labeled dATP in solution A can be substituted for digoxigenin-labeled dUTP (10 to 44 nM), and streptavidin–alkaline phosphatase conjugate in solution B for anti-digoxigenin antibody–alkaline phosphatase conjugate (0.75 U/ml).

It is necessary to eliminate the possibility that some bands could be due to endogenous ALP activity. This objective can be attained by performing control gel staining in the absence of the labeled deoxynucleotide.

This method allows the detection of DDDP activity bands in crude extracts; it is harmless and less expensive than the radioactive technique (see Method 1).

OTHER METHODS

A. The product pyrophosphate can be detected in clean gels (PAG or agarose) using the calcium pyrophosphate precipitation method.[5] The white bands of the precipitated calcium pyrophosphate are clearly visible when viewed against a dark background and can be photographed or scanned (e.g., see 2.7.7.6 — DDRP).

B. Pyrophosphate can be detected enzymatically using three coupled reactions catalyzed by the auxiliary enzymes inorganic pyrophosphatase, purine-nucleoside phosphorylase, and xanthine oxidase (see 3.6.1.1 — PP, Method 2). Another enzymatic method for detection of pyrophosphate is available. Two variants of this method exist that

allow observation of DDDP bands in long-wave UV light or daylight (see 2.7.7.6 — DDRP, Other Methods, A).

C. DNA polymerase can be detected using a novel polyacrylamide activity gel electrophoresis procedure.[6] In this procedure, the enzyme is initially resolved using either native PAG or SDS-PAG containing 5′-[^{32}P]-oligo(24-mer)DNA annealed with a single-stranded circular M13mp2 DNA (for more details, see 6.5.1.1 — DL(ATP), Method 2, *Notes*). The oligo(24-mer)DNA is complementary to positions 107 to 129 of M13mp2 DNA, but forms a 3′-terminal (T·C) mispair at position 106 (the 3′ to 5′ exonuclease activity of DNA polymerase exhibits a significant preference for mismatched over matched DNA substrates). Following electrophoresis in the first dimension (and protein renaturation, if required), the longitudinal strip of resolving PAG is used to carry out the *in situ* enzyme reaction using a reaction mixture containing 50 mM Tris–HCl (pH 7.5), 5 mM 2-mercaptoethanol, 16% (w/v) glycerol, 400 μg/ml bovine serum albumin, 50 mM KCl, 7 mM MgCl$_2$, and 100 μM dNTP, four (A, T, G, and C) deoxyribonucleoside triphosphates. The highly extended 5′-[^{32}P]DNA product is then resolved from the gel strip by a second dimension of electrophoresis through a 20% DNA sequencing PAG containing urea. After autoradiography, the position of DNA polymerase is detected via projection of a slowly migrating dark spot (indicating the position of a highly extended [^{32}P]DNA product on the autoradiogram) onto the starting strip of the substrate-containing PAG. The method is suitable to detect the 3′ to 5′ exonuclease activity of DNA polymerase (see General Notes). To detect exonuclease activity, the longitudinal strip of the substrate-containing PAG is incubated in the reaction mixture (see above) lacking deoxyribonucleoside triphosphates. After electrophoresis in the second dimension and autoradiography, the position of DNA polymerase is detected via projection of serial dark spots of increasing mobility (indicating positions of serial shortened [^{32}P]DNA products) onto the starting strip of the substrate-containing PAG.

GENERAL NOTES

The enzyme catalyzes the DNA template-directed extension of the 3′ end of a DNA strand one nucleotide at a time. It cannot initiate a DNA chain *de novo*.[7]

DNA polymerase I from *E. coli* is a monomeric protein with a molecular weight of 103 kDa that contains three distinct enzyme activities: polymerase, 5′ to 3′ exonuclease, and 3′ to 5′ exonuclease. DNA polymerase III holoenzyme from the same source contains two distinct activities: polymerase and 3′ to 5′ exonuclease. These activities are displayed by separate subunits, designated α and ε, with molecular weights of 140 and 27 kDa, respectively. All five classes (α, β, γ, δ, and ε) of eukaryotic DNA polymerase contain polymerase and 3′ to 5′ exonuclease activities.[8]

2.7.7.7 — DNA-Directed DNA Polymerase; DDDP (continued)

REFERENCES

1. Uriel, J. and Lavialle, C., Autoradiographic method for characterization of DNA and RNA polymerases after gel electrophoresis, *Anal. Biochem.*, 42, 509, 1971.
2. Karawya, E., Swack, J.A., and Wilson, S.H., Improved conditions for activity gel analysis of DNA polymerase catalytic polypeptides, *Anal. Biochem.*, 135, 318, 1983.
3. Venegas, J. and Solari, A., Colorimetric detection of DNA polymerase activity after sodium dodecyl sulfate polyacrylamide gel electrophoresis, *Anal. Biochem.*, 221, 57, 1994.
4. Laemmli, U.K., Cleavage of structural proteins during the assembly of the head of bacteriophage T4, *Nature*, 227, 680, 1970.
5. Nimmo, H.G. and Nimmo, G.A., A general method for the localization of enzymes that produce phosphate, pyrophosphate, or CO_2 after polyacrylamide gel electrophoresis, *Anal. Biochem.*, 121, 17, 1982.
6. Longley, M.J. and Mosbaugh, D.W., *In situ* detection of DNA-metabolizing enzymes following polyacrylamide gel electrophoresis, in *Methods in Enzymology*, Vol. 218, *Recombinant DNA*, Part I, Wu, R., Ed., Academic Press, New York, 1993, p. 587.
7. NC-IUBMB, *Enzyme Nomenclature*, Academic Press, San Diego, 1992, p. 288 (EC 2.7.7.7, Comments).
8. Longley, M.J. and Mosbaugh, D.W., Characterization of DNA metabolizing enzymes *in situ* following polyacrylamide gel electrophoresis, *Biochemistry*, 30, 2655, 1991.

2.7.7.8 — Polyribonucleotide Nucleotidyltransferase; PNT

OTHER NAMES	Polynucleotide phosphorylase, polyribonucleotide phosphorylase
REACTION	RNA_{n+1} + orthophosphate = RNA_n + nucleoside diphosphate
ENZYME SOURCE	Bacteria, fungi, plants, invertebrates, vertebrates
SUBUNIT STRUCTURE	Unknown or no data available

METHOD

Visualization Scheme

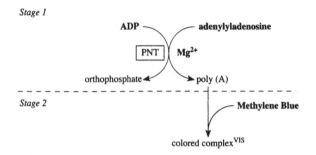

Staining Solution[1]

A. 75 mM Tris–HCl buffer, pH 9.0
   20 mM ADP
   5 mM $MgCl_2$
   0.2 mM EDTA
   0.2 mg/ml Adenylyladenosine (ApA)
B. 0.2% Methylene Blue
   0.4 M Sodium acetate buffer, pH 4.7

Procedure

Incubate the electrophorized gel at 37°C for 2 to 48 h in solution A. Rinse the gel with a solution of 7% acetic acid and transfer to solution B for 18 h. Destain the stained gel by diffusion with distilled water. Dark blue bands indicate the areas of PNT activity.

Notes: A 1% (w/v) solution of Acridine Orange or a 1.03% (w/v) solution of lanthanum chloride in 25% acetic acid can also be used to visualize the areas of poly(A) deposition on the gel. Gels stained with Acridine Orange should be destained electrophoretically with 7.5% acetic acid.

2.7.7.8 — Polyribonucleotide Nucleotidyltransferase; PNT (continued)

OTHER METHODS

A. The bands of PNT activity can be visualized by autoradiography using labeled substrate [^{14}C]ADP (final concentration of 1 μCi/ml) in a reaction mixture similar to that in solution A of the staining solution given above.[2] After an appropriate period of gel incubation in the reaction mixture, the poly(A) formed is precipitated in a gel matrix with streptomycin. The washed and dried gel is then autoradiographed with x-ray film.

B. Different methods of detection of orthophosphate can also be used for localization of PNT activity on electrophoretic gels (for details, see 3.1.3.1 — ALP, Methods 5, 6, 7, and 8; 3.1.3.2 — ACP, Methods 4 and 5).

GENERAL NOTES

ADP, IDP, GDP, UDP, and CDP can act as donors.[3]

REFERENCES

1. Killick, K.A., Polyribonucleotide phosphorylase from *Dictyostelium discoideum*, *Exp. Mycol.*, 4, 181, 1980.
2. Uriel, J. and Lavialle, C., Autoradiographic method for characterization of DNA and RNA polymerases after gel electrophoresis, *Anal. Biochem.*, 42, 509, 1971.
3. NC-IUBMB, *Enzyme Nomenclature*, Academic Press, San Diego, 1992, p. 288 (EC 2.7.7.8, Comments).

2.7.7.9 — UTP–Glucose-1-Phosphate Uridylyltransferase; UGPP

OTHER NAMES	UDPglucose pyrophosphorylase; glucose 1-phosphate uridylyltransferase
REACTION	UTP + α-D-glucose 1-phosphate = pyrophosphate + UDPglucose
ENZYME SOURCE	Bacteria, fungi, plants, invertebrates, vertebrates
SUBUNIT STRUCTURE	See General Notes

METHOD 1

Visualization Scheme

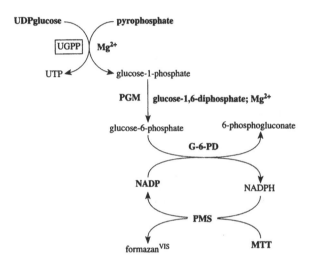

Staining Solution[1]

0.36 *M* Tris–HCl buffer, pH 8.0	0.8 ml
0.1 *M* MgCl$_2$	0.2 ml
UDPglucose (disodium salt)	6 mg
22.5 mg/ml Pyrophosphate (disodium salt; 10 H$_2$O)	0.2 ml
0.175 mg/ml Glucose-1,6-diphosphate (tetracyclohexylammonium salt; 4H$_2$O)	0.2 ml
4 mg/ml NADP (disodium salt)	0.2 ml
0.054 *M* EDTA, pH 7.0	0.2 ml
140 U/ml Glucose-6-phosphate dehydrogenase (G-6-PD)	5 μl
400 U/ml Phosphoglucomutase (PGM)	5 μl
2 mg/ml MTT	0.2 ml
0.4 mg/ml PMS	0.2 ml

Procedure

Apply the staining solution dropwise to the gel surface. Incubate the gel in the dark at 37°C until dark blue bands appear. Fix the stained gel in 25% ethanol.

METHOD 2

Visualization Scheme

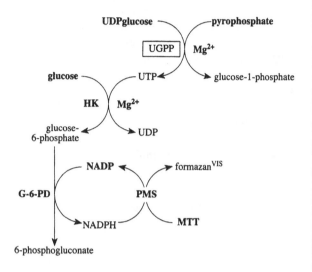

Staining Solution*

A.	0.4 M Tris–HCl buffer, pH 8.0	20 ml
	UDPglucose (disodium salt)	50 mg
	Pyrophosphate (disodium salt; 10 H_2O)	40 mg
	$MgCl_2 \cdot 6H_2O$	10 mg
	Glucose	80 mg
	Hexokinase from Bakers yeast (HK)	40 U
	Glucose-6-phosphate dehydrogenase (G-6-PD)	20 U
	NADP	10 mg
	MTT	6 mg
	PMS	1 mg
B.	2% Agarose solution (50°C)	20 ml

Procedure

Mix A and B solutions and pour the mixture over the surface of the gel. Incubate the gel in the dark at 37°C until dark blue bands appear. Fix the stained gel in 25% ethanol.

Notes: The method is based on the ability of yeast hexokinase to utilize UTP instead of ATP in its forward reaction. The hexokinase reaction rate with UTP is much lower than that with ATP but is quite enough to produce essential quantities of glucose-6-phosphate, which is then detected using auxiliary NADP-dependent glucose-6-phosphate dehydrogenase coupled with the PMS–MTT system.

METHOD 3

Visualization Scheme

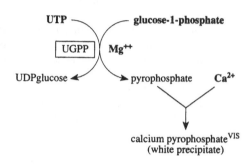

Staining Solution[2]

50 mM Glycine–KOH buffer, pH 9.0
0.5 mM Glucose 1-phosphate
8 mM Mg^{2+}
2 mM UTP
2 mM Ca^{2+}

Procedure

Following electrophoresis, soak PAG in 50 mM glycine–KOH buffer (pH 9.0) for 20 to 30 min and incubate in the staining solution at room temperature until white bands of calcium pyrophosphate precipitation appear. These bands are well visible when the gel is viewed against a dark background. When an activity stain of sufficient intensity is obtained, remove the gel from the staining solution and store in 50 mM glycine–KOH buffer (pH 10.0), containing 5 mM Ca^{2+}, either at 5°C or at room temperature in the presence of an antibacterial agent.

Notes: The method is developed for clean gels (PAG or agarose). When more opaque gels are used (e.g., acetate cellulose or starch gel), precipitated calcium pyrophosphate can be subsequently counterstained with Alizarin Red S.

OTHER METHODS

The product of the reverse UGPP reaction, UDPglucose, can be detected using a linked reaction catalyzed by the auxiliary enzyme NAD-dependent UDPglucose dehydrogenase coupled with the PMS–MTT system (see 1.1.1.22 — UGDH).

---

* New; recommended for use.

GENERAL NOTES

Although the enzyme from *Dictyostelium* is a homomer consisting of eight identical subunits, SDS-PAG electrophoresis coupled with the subsequent renaturation procedure proved suitable for the analysis of UGPP structural isoforms detected using the in-gel activity stain (Method 1).[3,4] The origin of these isoforms is uncertain.

REFERENCES

1. Van Someren, H., Van Henegouwen, H.B., Los, W., Wurzer-Figurelli, E., Doppert, B., Vervloet, M., and Meera Khan, P., Enzyme electrophoresis on cellulose acetate gel: II. Zymogram patterns in man–Chinese hamster somatic cell hybrids, *Humangenetik*, 25, 189, 1974.

2. Nimmo, H.G. and Nimmo, G.A., A general method for the localization of enzymes that produce phosphate, pyrophosphate, or CO_2 after polyacrylamide gel electrophoresis, *Anal. Biochem.*, 121, 17, 1982.

3. Manrow, R.E. and Dottin, R.P., Renaturation and localization of enzyme in polyacrylamide gels: studies with UDPglucose pyrophosphorylase of *Dictyostelium*, *Proc. Natl. Acad. Sci. U.S.A.*, 77, 730, 1980.

4. Manrow, R.E. and Dottin, R.P., Demonstration, by renaturation in O'Farrell gels, of heterogeneity in *Dictyostelium* uridine diphosphoglucose pyrophosphatase, *Anal. Biochem.*, 120, 181, 1982.

OTHER NAMES	UDPglucose–hexose-1-phosphate uridylyltransferase (recommended name), hexose-1-phosphate uridylyltransferase, uridyl transferase
REACTION	UDPglucose + α-D-galactose-1-phosphate = α-D-glucose 1-phosphate + UDPgalactose
ENZYME SOURCE	Bacteria, fungi, plants, protozoa, vertebrates
SUBUNIT STRUCTURE	Dimer (vertebrates)

METHOD

Visualization Scheme

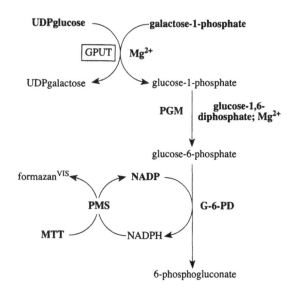

Staining Solution[1]

A. 0.3 *M* Tris–acetate buffer, pH 7.8 — 10 ml
Galactose-1-phosphate
 (dipotassium salt; 5H$_2$O) — 20 mg
UDPglucose (disodium salt) — 15 mg
NADP (disodium salt) — 10 mg
Glucose-1,6-diphosphate — 0.2 mg
0.2 *M* MgCl$_2$ — 1 ml
500 U/ml Phosphoglucomutase (PGM) — 20 μl
345 U/ml Glucose-6-phosphate dehydrogenase
 (G-6-PD) — 20 μl
2 mg/ml MTT — 1 ml
1 mg/ml PMS — 1 ml
B. 2% Agarose solution (50°C) — 15 ml

Procedure

Mix A and B components of the staining solution and pour the mixture over the surface of the gel. Incubate the gel in the dark at 37°C until dark blue bands appear. Fix the stained gel in 25% ethanol.

Notes: NAD-dependent glucose-6-phosphate dehydrogenase (Sigma G 5760 or G 5885) and NAD may be used instead of NADP-dependent glucose-6-phosphate dehydrogenase and NADP in the staining solution. This substitution will result in cost savings because NAD is about five times less expensive than NADP.

When NADP is used, 6-phosphogluconate dehydrogenase may be included in the staining solution to enhance NADP-into-NADPH conversion and thus to increase the sensitivity of the method.[2]

OTHER METHODS

The product of the backward GPUT reaction, UDPglucose, can be detected using the auxiliary enzyme UDPglucose dehydrogenase (see 1.1.1.22 — UGDH).

REFERENCES

1. Sparkes, M.C., Crist, M., and Sparkes, R.S., Improved technique for electrophoresis of human galactose-1-P-uridyl transferase (EC 2.7.7.12), *Hum. Genet.*, 40, 93, 1977.
2. Harris, H. and Hopkinson, D.A., *Handbook of Enzyme Electrophoresis in Human Genetics*, North-Holland, Amsterdam, 1976 (loose-leaf, with supplements in 1977 and 1978).

2.7.7.27 — Glucose 1-Phosphate Adenylyltransferase; GPAT

OTHER NAMES	ADPglucose pyrophosphorylase
REACTION	ATP + α-D-glucose 1-phosphate = pyrophosphate + ADPglucose
ENZYME SOURCE	Plants
SUBUNIT STRUCTURE	Heterotetramer[a] (plants)

METHOD

Visualization Scheme

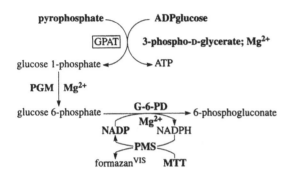

Staining Solution[1]

A. 2.5 μM HEPES buffer, pH 8.0
   240 μM $MgSO_4$
   60 μM Tetrasodium pyrophosphate
   65 μM 3-Phospho-D-glycerate
   24 μM ADPglucose

B. Glucose-6-phosphate dehydrogenase (G-6-PD)   10 U
   Phosphoglucomutase (PGM)   10 U
   NADP   6 mg
   PMS   0.6 mg
   MTT   1 mg

PROCEDURE

Mix 50 ml of A with B ingredients and apply to the surface of the gel. Incubate the gel in the dark at 37°C until dark blue bands appear. Fix the stained gel in 25% ethanol.

Notes: 3-Phosphoglycerate is added to activate GPAT. The addition of trace amounts of glucose-1,6-diphosphate can be necessary to activate the auxiliary enzyme, PGM.

OTHER METHODS

A. The product of the backward reaction, ATP, can be detected using two auxiliary enzymes (hexokinase and glucose-6-phosphate dehydrogenase) coupled with the PMS–MTT system (e.g., see 2.7.3.3 — ARGK, Method 1).

B. Activity bands of GPAT can be visualized using a staining procedure developed to detect the product of the forward reaction, ADPglucose.[2]

C. There are several methods suitable for detection of the product of the forward reaction, pyrophosphate (e.g., see 2.7.7.6 — DDRP).

REFERENCES

1. Hannah, L.C. and Nelson, O.E., Jr., Characterization of adenosine diphosphate glucose pyrophosphorylases from developing maize seeds, *Plant Physiol.*, 55, 297, 1975.

2. Weaver, S.H., Glover, D.V., and Tsai, C.Y., Nucleoside diphosphate glucose pyrophosphorylase isoenzymes of developing normal, *brittle-2*, and *shrunken-2* endosperms of *Zea mays* L., *Crop Sci.*, 12, 510, 1972.

OTHER NAMES | Terminal deoxyribonucleotidyl-transferase, terminal deoxynucleotidyl transferase, terminal addition enzyme

REACTION | Deoxynucleoside triphosphate + DNA_n = pyrophosphate + DNA_{n+1} (see also General Notes)

ENZYME SOURCE | Vertebrates

SUBUNIT STRUCTURE | Unknown or no data available

METHOD

VISUALIZATION SCHEME

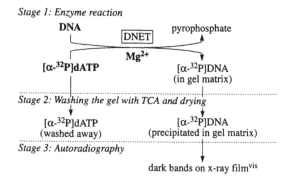

Stage 1: Enzyme reaction

Stage 2: Washing the gel with TCA and drying

Stage 3: Autoradiography

REACTION MIXTURE[1]

0.2 *M* Potassium cacodylate buffer, pH 7.2
1 m*M* 2-Mercaptoethanol
8 m*M* $MgCl_2$
0.05 m*M* [α-^{32}P]dATP at 0.25 µCi/nmol

PROCEDURE

The method is developed for a PAG containing 0.1% SDS and 15 mg/ml calf thymus DNA activated by pancreatic deoxyribonuclease.

After completion of electrophoresis, wash the gel with three changes of 2 l of 50 m*M* sodium phosphate (pH 7.4) containing 10 m*M* 2-mercaptoethanol for a total of 2 h at room temperature to remove the SDS. Equilibrate the gel with two changes of 500 ml of 6 *M* guanidine HCl (see *Notes*) in 50 m*M* sodium phosphate buffer (pH 7.4) and 10 m*M* 2-mercaptoethanol at room temperature for a total of 1 h. Remove the guanidine HCl by washing the gel with 50 m*M* sodium phosphate buffer (pH 7.4) containing 10 m*M* 2-mercaptoethanol and equilibrate the gel with 0.2 *M* potassium cacodylate buffer (pH 7.2), 1 m*M* 2-mercaptoethanol, and 8 m*M* $MgCl_2$.

Stage 1. Transfer the gel in the reaction mixture and incubate at 35°C for 17 h.

Stage 2. After incubation precipitate the radioactive DNA product on the gel with cold 5% trichloroacetic acid and 1% sodium pyrophosphate for 2 h, and then wash the gel with cold 5% trichloroacetic acid until no detectable radioactivity (<200 counts/min/ml) is found in the wash.

Stage 3. Autoradiograph the gel by exposing it to Kodak X-O-X AR5 film for 14 h at 4°C.

The dark bands on the developed x-ray film correspond to localization of DNET activity on the gel.

Notes: The repeated denaturation of the enzyme with 6 *M* guanidine HCl is necessary for its efficient renaturation after SDS-PAG electrophoresis.

GENERAL NOTES

The enzyme catalyzes the template-independent extension of the 3′ end of a DNA strand one nucleotide at a time and cannot initiate a chain *de novo*. The nucleoside may be ribo- or deoxyribo-.[2]

REFERENCES

1. Chang, L.M.S., Plevani, P., and Bollum, F.J., Proteolytic degradation of calf thymus terminal deoxynucleotidyl transferase, *J. Biol. Chem.*, 257, 5700, 1982.

2. NC-IUBMB, *Enzyme Nomenclature*, Academic Press, San Diego, 1992, p. 290 (EC 2.7.7.31, Comments).

2.7.7.48 — RNA-Directed RNA Polymerase; RNAP

OTHER NAMES   RNA nucleotidyltransferase (RNA directed), RNA-directed RNA replicase

REACTION   Nucleoside triphosphate + RNA_n = pyrophosphate + RNA_{n+1}

ENZYME SOURCE   Plants, vertebrates

SUBUNIT STRUCTURE   Unknown or no data available

METHOD

Visualization Scheme

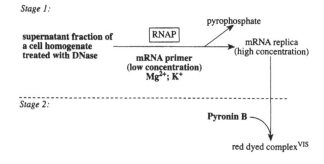

Stage 1:

Stage 2:

Staining Solution[1]

A. Supernatant fraction of a cell homogenate treated with DNase (see *Notes*)
B. 0.1 *M* Tris–maleate buffer, pH 8.0
   50 m*M* KCl
   10 m*M* $Mg(CH_3COO)_2$
   6 m*M* 2-Mercaptoethanol
C. 0.9% NaCl
D. 1% Pyronin B
   7.5% Acetic acid

Procedure

Following electrophoresis, incubate a PAG containing a mRNA primer (see *Notes*, below) for 5 min at 0°C in an A + B mixture (one part of A + nine parts of B), and then incubate at 37°C for 5 min. Rinse the incubated gel thoroughly in solution C and place in solution D for 5 min. Wash the stained gel in 7.5% acetic acid. Red bands indicate the RNAP activity areas on the gel.

Notes: To prepare mammalian preparations containing RNAP activity, peripheral leukocytes should be grown in a medium containing 1 to 10 μg/ml antigenic mRNA. Immunocompetitive leukocytes are then homogenized, and a supernatant fraction obtained after centrifugation at 24,000 rpm is used for electrophoresis.

Immunocompetitive leukocytes contain mRNA for synthesis of the antibody to antigenic mRNA. The antibody mRNA incorporated into PAG during the process of gel polymerization is used as a primer in the reaction catalyzed by RNAP.

The supernatant fraction of a cell homogenate treated with DNase (component A of the staining solution) is used as a source of the nucleoside triphosphates needed for RNA synthesis.

GENERAL NOTES

The enzyme catalyzes the RNA template–directed extension of the 3′ end of an RNA strand one nucleotide at a time; it can initiate a chain *de novo*.[2]

REFERENCES

1. Neuhoff, V., Schill, W.-B., and Jacherts, D., Nachweis einer RNA-abhängigen RNA-replicase aus immunologisch kompetenten zellen durch mikro-disk-elektrophorese, *Hoppe-Seyler's Z. Physiol. Chem.*, 351, 157, 1970.
2. NC-IUBMB, *Enzyme Nomenclature*, Academic Press, San Diego, 1992, p. 293 (EC 2.7.7.48, Comments).

OTHER NAMES	Reverse transcriptase, DNA nucleotidyltransferase (RNA directed), revertase
REACTION	Deoxynucleoside triphosphate + DNA_n = pyrophosphate + DNA_{n+1} (see also General Notes)
ENZYME SOURCE	Viruses, organisms containing retrotransposons
SUBUNIT STRUCTURE	Heterodimer^a (viruses)

METHOD

Visualization Scheme

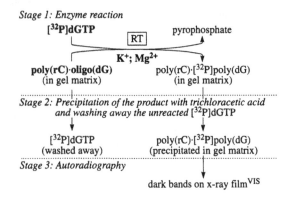

Stage 1: Enzyme reaction

Stage 2: Precipitation of the product with trichloracetic acid and washing away the unreacted [^{32}P]dGTP

Stage 3: Autoradiography

Reaction Mixture[1]

50 mM Tris–HCl buffer, pH 8.0
80 mM KCl
2.5 mM Dithiothreitol
4 mM MgCl$_2$
2.5 µCi/ml [^{32}P]dGTP (specific activity, 3000 µCi/mmol)

PROCEDURE

The method is developed for an SDS-PAG containing 8 µg/ml poly(rC)·oligo(dG) as the template primer.

After completion of electrophoresis, incubate the gel in renaturation buffer (50 mM Tris–HCl buffer, pH 8.0; 1 mM 2-mercaptoethanol; and 0.1 mM EDTA) for 30 min at 42°C and then for 2.5 h at 4°C with buffer changes every 30 min. Then treat the gel as described below:

Stage 1. Transfer the gel in the reaction mixture and incubate overnight at 42°C.

Stage 2. After incubation precipitate the radioactive product with cold 10% trichloroacetic acid containing 3% sodium pyrophosphate until no detectable radioactivity is found in the wash.

Stage 3. Dry and autoradiograph the gel.

The dark bands on the developed x-ray film correspond to localization of RT activity on the gel.

Notes: Under optimized conditions, the virus RT demonstrates highest activity with poly(rA)·oligo(dT) as the template primer. The activity with poly(rC)·oligo(dG) is about four times lower under the same conditions. However, the loss of poly(rA)·oligo(dT)-dependent activity of RT is more rapid during the storage of enzyme preparations. As a whole, the poly(rC)·oligo(dG)-dependent activity of RT is more heat stable and the use of this template primer in the activity gel analysis is therefore preferable.

Calf thymus high-molecular-mass DNA (Serva) activated by digestion with DNase I for 30 min, as well as poly(rA)·oligo(dT), can also be used as template primers for the in-gel detection of RT.[2]

GENERAL NOTES

RT catalyzes the RNA template-directed extension of the 3′ end of a DNA strand one deoxynucleotide at a time; it cannot initiate a chain *de novo*; it requires an RNA or DNA primer. DNA can also serve as a template.[2]

The enzyme from the immunodeficiency virus of the African green monkey is a heterodimer composed of a 64- and 50-kDa subunit. Using SDS-PAG electrophoresis and the in-gel detection method described above, it was demonstrated that the smaller RT subunit is enzymatically active, although to a lesser extent than the larger subunit.[1,3]

REFERENCES

1. Lüke, W., Hoefer, K., Moosmayer, D., Nickel, P., Hunsmann, G., and Jentsch, K.-D., Partial purification and characterization of the reverse transcriptase of the simian immunodeficiency virus TYO-7 isolated from an African green monkey, *Biochemistry*, 29, 1764, 1990.
2. NC-IUBMB, *Enzyme Nomenclature*, Academic Press, San Diego, 1992, p. 293 (EC 2.7.7.49, Comments).
3. Kraus, G., Behr, E., Baier, M., König, H., and Kurth, R., Simian immunodeficiency virus reverse transcriptase: purification and partial characterization, *Eur. J. Biochem.*, 192, 207, 1990.

2.8.1.1 — Thiosulfate Sulfurtransferase; TST

OTHER NAMES	Thiosulfate cyanide transsulfurase, thiosulfate thiotransferase, rhodanese
REACTION	Thiosulfate + cyanide = sulfite + thiocyanate
ENZYME SOURCE	Vertebrates
SUBUNIT STRUCTURE	Unknown or no data available

METHOD 1

Visualization Scheme

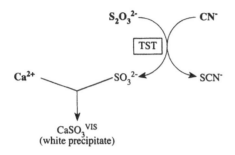

$S_2O_3^{2-}$ CN^-

TST

Ca^{2+} SO_3^{2-} SCN^-

$CaSO_3^{VIS}$
(white precipitate)

Staining Solution[1]

0.333 M Tris–acetate buffer, pH 8.5
0.1 M KCN
0.1 M $Na_2S_2O_3$
0.3 M $CaCl_2$

Procedure

Incubate the electrophorized PAG in the staining solution at 37°C until white opaque bands appear. These bands are well visible against a dark background.

Notes: Only freshly prepared staining solution should be used.

METHOD 2

Visualization Scheme

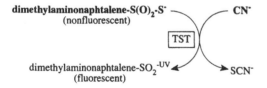

dimethylaminonaphtalene-S(O)$_2$-S⁻
(nonfluorescent) CN^-

TST

dimethylaminonaphtalene-SO$_2^{-UV}$
(fluorescent) SCN^-

Staining Solution[2]

A. 0.25 M Tris–acetate buffer, pH 8.2
    50 mM 5-Dimethylamino-1-naphthalene thiosulfonate
B. 0.25 M KCN

Procedure

Following electrophoresis, coat the top of the gel with solution A (0.7 ml per 8 × 10.5 cm of gel; because of background fluorescence, minimizing the volume of solution A is desirable). Spread solution evenly with a glass rod, and allow it to penetrate for 10 min. Then treat the gel with solution B (0.3 ml per 8 × 10.5 cm of gel). After 1 to 2 min, view the gel under long-wave UV light. Areas of TST activity are indicated by fluorescent bands (emission maximum of 500 to 510 nm; excitation at 325 nm). Record the zymogram or photograph using a yellow filter.

Notes: The synthesis of a fluorogenic substrate, 5-dimethyl-amino-1-naphthalene thiosulfonate anion, is carried out by stirring dansyl chloride in 10 volumes of an aqueous solution containing one analytical equivalent of Na_2S at 65°C for 2 to 3 h. The reaction mixture is then dried under reduced pressure in an atmosphere of nitrogen, and the crude product is extracted from the residue with hot ethanol. After crystallization at –20°C and recrystallization from ethanol, the thiosulfonate is contaminated with only trace amounts of the corresponding sulfinate or sulfonate and is suitable for use in the method described above.

METHOD 3

Visualization Scheme

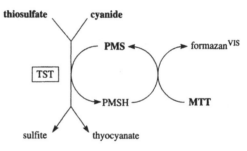

thiosulfate   cyanide

PMS   formazan^{VIS}

TST

PMSH   MTT

sulfite   thyocyanate

Staining Solution (recipe of R.D. Sage, given by Murphy et al.)[3]

0.2 M Tris–HCl, pH 8.0	50 ml
KCN	120 mg
$Na_2S_2O_3$	500 mg
5 mg/ml MTT	1 ml
5 mg/ml PMS	1 ml

Procedure

Incubate the gel in staining solution in the dark at room temperature until dark blue bands appear. Fix the stained gel in 25% ethanol.

Notes: The mechanism of PMS reduction is uncertain.

REFERENCES

1. Guilbault, G.G., Kuan, S.S., and Cochran, R., Procedure for rapid and sensitive detection of rhodanese separated by polyacrylamide gel electrophoresis, *Anal. Biochem.*, 43, 42, 1971.
2. Aird, B.A., Lane, J., and Westley, J., Methods for *in situ* visualization and assay of sulfurtransferases, *Anal. Biochem.*, 164, 554, 1987.
3. Murphy, R.W., Sites, J.J.W., Jr., Buth, D.G., and Haufler, C.H., Proteins: isozyme electrophoresis, in *Molecular Systematics*, 2nd ed., Hillis, D.M., Moritz, C., and Mable, B.K., Eds., Sinauer Associates, Sunderland, MA, 1996, p. 51.

OTHER NAMES — Glutathione-dependent thiosulfate reductase, sulfane reductase, thiosulfate reductase

REACTION — Thiosulfate + 2 glutathione = sulfite + oxidized glutathione + sulfide

ENZYME SOURCE — Fungi

SUBUNIT STRUCTURE — Unknown or no data available

Method

Visualization Scheme

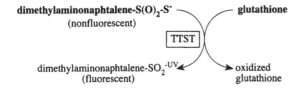

dimethylaminonaphtalene-S(O)$_2$-S$^-$
(nonfluorescent)

glutathione

TTST

dimethylaminonaphtalene-SO$_2^-$$^{-UV}$
(fluorescent)

oxidized glutathione

Staining Solution[1]

A. 0.25 M Tris–acetate buffer, pH 8.2
   50 mM 5-Dimethylamino-1-naphthalene thiosulfonate

B. 80 mM Glutathione, reduced, pH 9.0

Procedure

Coat the top of the gel with solution A (0.7 ml per 8 × 10.5 cm of gel). Spread solution evenly with a glass rod and allow it to penetrate for 10 min. Then treat the gel with solution B (0.3 ml per 8 × 10.5 cm of gel). After 1 to 2 min, view the gel under long-wave UV light. Areas of TTST activity are indicated by fluorescent bands (emission maximum of 500 to 510 nm; excitation at 325 nm). Record the zymogram or photograph using a yellow filter.

Notes: The fluorogenic substrate 5-dimethylamino-1-naphthalene thiosulfonate is not yet commercially available, but is available by a simple one-step synthesis from dansyl chloride (see 2.8.1.1 — TST, Method 2, *Notes*).

General Notes

The primary product of reaction is glutathione hydrodisulfide, which further reacts with glutathione to give oxidized glutathione and sulfide. L-Cysteine can also act as an acceptor.[2]

References

1. Aird, B.A., Lane, J., and Westley, J., Methods for *in situ* visualization and assay of sulfurtransferases, *Anal. Biochem.*, 164, 554, 1987.
2. NC-IUBMB, *Enzyme Nomenclature*, Academic Press, San Diego, 1992, p. 298 (EC 2.8.1.3, Comments).

OTHER NAMES	Nonspecific esterases; carboxylic ester hydrolases
REACTION	Hydrolyze ester bonds of various carboxylic esters; wide substrate specificity
ENZYME SOURCE	Bacteria, fungi, plants, protozoa, invertebrates, vertebrates
SUBUNIT STRUCTURE	Monomer (fungi, plants, invertebrates, vertebrates), dimer (fungi, plants, invertebrates, vertebrates), tetramer (vertebrates)

METHOD 1

Visualization Scheme

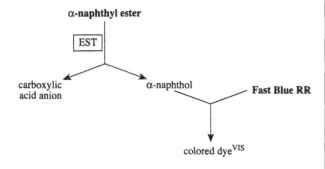

α-naphthyl ester

EST

carboxylic acid anion   α-naphthol   **Fast Blue RR**

colored dye[VIS]

Staining Solution[1] (modified)

0.05 M Phosphate buffer, pH 7.2	100 ml
α-Naphthyl acetate (dissolved in 1 ml of acetone)	10 mg
Fast Blue RR	50 mg

Procedure

Incubate the gel in staining solution in the dark at 37°C until dark gray or black bands appear. Wash the stained gel with water and fix in 3% acetic acid.

Notes: Esters of α-naphthols with acetate, propionate, and butyrate are the most commonly used artificial substrates for nonspecific esterases. Certain esterase isozymes have specificity for β-naphthyl acetate, which gives a reddish dye with Fast Blue RR salt. Thus, both α- and β-naphthyls may be included in the staining solution in some cases. Other dye couplers, such as Fast Blue BB, Fast Garnet GBC, and Fast Red TR, can also be used instead of Fast Blue RR. To reduce nonspecific background staining of the gel, 2 to 5 ml of 4% formaldehyde should be added to the staining solution.

METHOD 2

Visualization Scheme

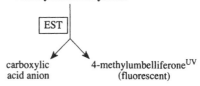

4-methylumbelliferyl ester

EST

carboxylic acid anion   4-methylumbelliferone[UV] (fluorescent)

Staining Solution[2]

50 mM Sodium acetate buffer, pH 5.2	10 ml
4-Methylumbelliferyl acetate (dissolved in a few drops of acetone)	4 mg

Procedure

Apply the staining solution to the gel surface on a filter paper overlay. Incubate the gel at room temperature and view under long-wave UV light. Fluorescent bands of EST activity usually appear after a few minutes. Record the zymogram or photograph using a yellow filter.

Notes: Prolonged incubation of the gel with a filter paper overlay results in a growing complication of banding patterns observed on zymograms of human erythrocyte esterase detected using 4-methylumbelliferyl acetate (4-MUA) as substrate. This is due to the presence of other EST isozymes capable of 4-MUA hydrolysis.[3] The same is true for 4-MUA esterases from other sources. Highly polymorphic dimeric 4-MUA esterase of human red cells was termed esterase D (EST-D).[4] This enzyme is included in the enzyme list under a separate number (EC 3.1.1.56).[5] 4-MUA esterases are widely distributed among plant and animal (both invertebrate and vertebrate) species; however, their homology to human EST-D in most cases remains uncertain. It was found that EST-D from human erythrocytes is identical to S-formylglutathione hydrolase (see 3.1.2.12 — FGH).[6]

4-Methylumbelliferyl butyrate and propionate also may be used as fluorogenic esterase substrates.

METHOD 3

Visualization Scheme

fluorescein diacetate

EST

acetate → ← fluorescein^{UV}
(fluorescent)

Staining Solution[7]

0.1 M Phosphate buffer, pH 6.5	100 ml
Fluorescein diacetate (dissolved in a few drops of acetone)	10 mg

Procedure

Incubate the gel in staining solution at 37°C for 30 min and inspect under long-wave UV light for yellow fluorescent bands. When the bands are well developed, record the zymogram or photograph using a yellow filter.

Notes: The visualization of yeast EST activity by using fluorescein diacetate as a substrate was shown to be a rapid zymogram technique with high resolution and sensitivity in both native and SDS-PAG electrophoresis.[8]

METHOD 4

Visualization Scheme

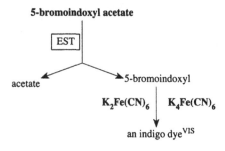

5-bromoindoxyl acetate

EST

acetate → ← 5-bromoindoxyl

$K_2Fe(CN)_6$ | $K_4Fe(CN)_6$

an indigo dye^{VIS}

Staining Solution[9]

0.15 M Sodium phosphate buffer, pH 6.5	90 ml
1% 5-Bromoindoxyl acetate in dimethyl sulfoxide	4 ml
10 mM K$_4$Fe(CN)$_6 \cdot$3H$_2$O	5 ml
10 mM K$_3$Fe(CN)$_6$	5 ml

Procedure

Incubate the gel in staining solution until colored bands appear. Wash the gel with water and fix in 3% acetic acid.

Notes: Potassium ferricyanide is used to accelerate oxidation of bromoindoxyl to insoluble indigo dye. Potassium ferrocyanide is added to prevent further oxidation of the dye.

The chromogenic substrate 5-bromo-4-chloro-3-indolyl acetate coupled with NBT (or MTT) can also be used to visualize EST activity, resulting in the formation of both the dehydroindigo and formazan (e.g., see 3.1.3.1 — ALP, Method 2).

GENERAL NOTES

All the methods described above can also detect carbonic anhydrase activity bands (see 4.2.1.1 — CA). Mammalian CA is inhibited by 1 mM acetazolamide. This specific inhibitor is used to identify CA bands on EST zymograms.

The esterases are a complex of enzymes capable of hydrolyzing ester bonds. Nonspecific esterases detected using artificial substrates include five different enzymes: carboxylesterase or ali-esterase (EC 3.1.1.1), arylesterase or A(aromatic) esterase (EC 3.1.1.2), acetylesterase (EST 3.1.1.6), acetylcholinesterase or true cholinesterase (EC 3.1.1.7), and cholinesterase or pseudocholinesterase (EC 3.1.1.8). Different artificial substrates and inhibitors can be used to identify and differentiate these esterases on the zymograms.[10–14] The validity of such an approach to esterase differentiation is, however, questionable.[15]

Two esterases can be identified using specific staining procedures. These are acetylcholinesterase (see 3.1.1.7 — ACHE) and cholinesterase (see 3.1.1.8 — CHE).

REFERENCES

1. Hunter, R.L. and Markert, C.L., Histochemical demonstration of enzymes separated by zone electrophoresis in starch gels, *Science*, 125, 1294, 1957.
2. Bargagna, M., Domenici, R., and Morali, A., Red cell esterase D polymorphism in the population of Tuscany, *Humangenetik*, 29, 251, 1975.
3. Nishigaki, I., Iton, T., and Ogasawara, N., Quantitative variations in polymorphic types of human red cell esterase D, *Ann. Hum. Genet.*, 47, 187, 1983.
4. Hopkinson, D.A., Mestriner, M.A., Cortner, J., and Harris, H., Esterase D: a new human polymorphism, *Ann. Hum. Genet.*, 37, 119, 1973.
5. NC-IUBMB, *Enzyme Nomenclature*, Academic Press, San Diego, 1992, p. 313 (EC 3.1.1.56).
6. Apeshiotis, F. and Bender, K., Evidence that S-formylglutathione hydrolase and esterase D polymorphisms are identical, *Hum. Genet.*, 74, 176, 1986.
7. Hopkinson, D.A., Coppock, J.S., Mühlemann, M.F., and Edwards, Y.H., The detection and differentiation of the products of the human carbonic anhydrase loci CA$_I$ and CA$_{II}$, using fluorogenic substrates, *Ann. Hum. Genet.*, 38, 155, 1974.
8. Lomolino, G., Lante, A., Crapisi, A., Spettoli, P., and Curioni, A., Detection of *Saccharomyces cerevisiae* carboxylesterase activity after native and sodium dodecyl sulfate electrophoresis by using fluorescein diacetate as substrate, *Electrophoresis*, 22, 1021, 2001.

9. Bender, K., Nagel, M., and Gunther, E., *Est*-6, a further polymorphic esterase in the rat, *Biochem. Genet.*, 20, 221, 1982.
10. Augustinson, K.-B., Electrophoretic separation and classification of blood plasma esterase, *Nature*, 181, 1786, 1958.
11. Holmes, R.S. and Masters, C.J., The developmental multiplicity and isoenzyme status of cavian esterases, *Biochim. Biophys. Acta*, 132, 379, 1967.
12. Holmes, R.S. and Masters, C.J., A comparative study of the multiplicity of mammalian esterases, *Biochim. Biophys. Acta*, 151, 147, 1968.
13. Holmes, R.S., Masters, C.J., and Webb, E.C., A comparative study of vertebrate esterase multiplicity, *Comp. Biochem. Physiol.*, 26, 837, 1968.
14. Tashian, R.E., The esterases and carbonic anhydrases in human erythrocytes, in *Biochemical Methods in Red Cell Genetics*, Yunis, J.J., Ed., Academic Press, New York, 1969, p. 307.
15. Choudhury, S.R., The nature of nonspecific esterase: a subunit concept, *J. Histochem. Cytochem.*, 20, 507, 1972.

3.1.1.3 — Triacylglycerol Lipase; TAGL

OTHER NAMES	Lipase, triglyceride lipase, tributyrase
REACTION	Triacylglycerol + H_2O = diacylglycerol + carboxylate
ENZYME SOURCE	Bacteria, fungi, plants, invertebrates, vertebrates
SUBUNIT STRUCTURE	Heterodimer[a] (fungi – *Aspergillus oryzae*); see General Notes

Method 1

Visualization Scheme

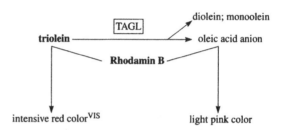

Staining Solution[1]

A.	2% Agar solution (60°C) in 0.1 M succinate buffer, pH 6.0	20 ml
B.	Rhodamin B	4 mg
	Triolein	0.5 g

Procedure

Add B components to solution A and mix at 60 to 65°C for 1 min using a high-speed homogenizer to prepare the emulsion. Form a 50-μm reactive agar plate between two glass plates heated to 40°C. Use a lower glass plate covered by silanized 100-μm polyester film and an upper glass plate covered by 100-μm polyester film treated with alkali.

Apply the electrophorized gel gently to the reactive agar plate formed and incubate at 40°C until light pink bands (fluorescent at 366 nm) appear on an intensive red gel background.

Notes: Olive oil can be used as a source of triglycerides and Sudan Black B as a lipid stain.[2]

METHOD 2

Visualization Scheme

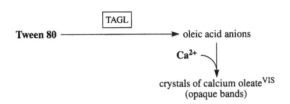

Staining Solution[3]

A. 1% Tween 80
0.02% CaCl$_2$
B. 0.5 mM Tris–HCl buffer, pH 9.2
4.2 mM NaCl

Procedure

Incubate the electrophorized PAG in a 1:1 mixture of A and B solutions at room temperature until opaque bands of calcium oleate appear. Photograph the stained gel on a black background.

Notes: The substrate Tween 80 (final concentration of 0.1%) can be included directly in the separating gel. In this case, the presence of Tween 80 is not necessary in the staining solution. In gels without Tween 80, very narrow bands are detected when the gel is incubated with the staining solution statically. Under these conditions the substrate does not penetrate into the gel and opaque bands are formed on the surface of the gel. These bands are lost when the gel is disturbed, but the bands appear again when the static conditions are restored.

This method can be adapted to starch gels via application of the staining solution as a 1% agar overlay. To observe opaque bands of TAGL activity, the agar overlay should be taken off from the gel surface after an appropriate period of time of incubation.

OTHER METHODS

A. Fluorogenic substrates 4-methylumbelliferyl oleate and monodecanoyl fluorescein can also be used to visualize TAGL activity bands on electrophoretic gels.[4]
B. The 2.5% substrate (tributiryn, triolein, diolein, or monoolein) can be included in the PAG before polymerization. After incubation of the electrophorized PAG in 50 mM Tris–HCl buffer (pH 7.4) containing 10 mM CaCl$_2$ for a few hours, the gel is immersed in boiling water for 2 min and, after cooling, stained with Nile Blue A solution (1 mg/ml in 0.4 N sulfuric acid, dissolved by heating) for 2 min. The stained gel is washed free of the stain until a pale pink background is obtained. Activity bands of TAGL appear as blue bands on a pink background, indicating sites where fatty acid anions are liberated. Care should be taken to minimize the gel washing when tributiryn-containing gels are used. This is because the butyric acid produced by TAGL is soluble in water.[5]

GENERAL NOTES

Banding patterns of TAGL from *Culex* IV instar larvae are the same on gels containing triolein and diolein, thus providing evidence that one and the same enzyme hydrolyzes triglyceride and diglyceride. This enzyme, however, is different from that hydrolyzing monoolein. The tributyrin-containing gel displays a banding pattern different from those obtained with monoolein and with diolein and triolein. These results provide evidence that in the insect several TAGL isozymes are expressed that demonstrate different substrate specificity.[5]

The purified enzyme from *Aspergillus oryzae* is formed from a glycoprotein and a monomeric protein with molecular masses of 25 and 29 kDa, respectively.[6]

REFERENCES

1. Höfelmann, M., Kittsteiner-Eberle, R., and Schreier, P., Ultrathin-layer agar gels: a novel print technique for ultrathin-layer isoelectric focusing of enzymes, *Anal. Biochem.*, 128, 217, 1983.
2. Jacobsen, T., Poulsen, O.M., and Hau, J., Enzyme activity electrophoresis and rocket immunoelectrophoresis for the qualitative and quantitative analysis of *Geotrichum candidum* lipase activity, *Electrophoresis*, 10, 49, 1989.
3. Nuero, O.M., García-Lepe, R., Lahoz, C., Santamaría, F., and Reyes, F., Detection of lipase activity on ultrathin-layer isoelectric focusing gels, *Anal. Biochem.*, 222, 503, 1994.
4. Cortner, J.A. and Swoboda, E., "Wolman's" disease: prenatal diagnosis; electrophoretic identification of the missing lysosomal acid lipase, *Am. J. Hum. Genet.*, 26, 23A, 1974.
5. Lakshmi, M.B. and Subrahmanyam, D., Separation and detection of lipases using polyacrylamide gel electrophoresis, *J. Chromatogr.*, 130, 441, 1977.
6. Toida, J., Arikawa, Y., Kondou, K., Fukuzawa, M., and Sekiguchi, J., Purification and characterization of triacylglycerol lipase from *Aspergillus oryzae*, *Biosci. Biotechnol. Biochem.*, 62, 759, 1998.

OTHER NAMES	Lecithinase A₂, phosphatidase, phosphatidolipase
REACTION	Phosphatidylcholine + H₂O = 1-acylglycerophosphocholine + carboxylate
ENZYME SOURCE	Invertebrates, vertebrates
SUBUNIT STRUCTURE	See General Notes

METHOD

Visualization Scheme

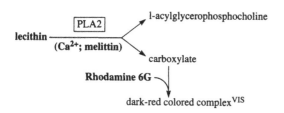

Staining Solution[1]

A. 100 mM Tris–HCl, pH 7.4
    20 mM CaCl₂
    5 μg/ml Melittin
B. 0.12% Rhodamine 6G

Procedure

The method is developed for a 9% PAG containing 5 mg/ml lecithin. To prepare the gel, polymerize a mixture of acrylamide (90 mg/ml) and N',N'-methylenebisacrylamide (2.4 mg/ml), diluted in an emulsion of lecithin (final concentration of 5 mg/ml), buffered with N,N,N',N'-tetramethylethylenediamine (0.05%, v/v) and Tris (45.4 mg/ml, adjusted to pH 8.9 with 1 N HCl). The emulsion of lecithin is prepared by evaporating a hexane solution of chromatographically purified lecithin *in vacuo*, and suspending the residue in water (1.5% emulsion) by repeatedly passing the suspension through a 22-gauge needle with a 5-ml syringe. Dilute the resulting suspension in a mixture of the remaining gel components, and sonicate the diluted suspension for 5 min. Initiate polymerization by the addition of ammonium persulfate (final concentration of 0.5 mg/ml).

After electrophoresis, incubate the lecithin-containing PAG overnight in a bath containing 100 to 200 ml of the 9:1 mixture of A and B components of the staining solution at 37°C with gentle shaking. Dark red bands indicating the presence of free unsaturated fatty acids are visible on a faint pink background of the lecithin-containing PAG. Longer incubation results in clearing of the lecithin emulsion in zones of the gel where PLA2 activity is localized. At the end of the incubation period, wash the gel with numerous changes of water to remove all the dye from cleared zones.

Photograph the developed PAG with reflected white light against a black background.

Notes: Some commercial preparations of lecithin contain impurities that inhibit polymerization of acrylamide. These impurities can be removed by activated charcoal treatment of a solution of the lecithin in hexane-chloroform.

Melittin (the bee venom peptide) enhances the PLA2 activity in the presence of calcium ions and displays no effect in their absence.

Detection of PLA2 activity is possible by the clearing of the lecithin-containing gel. Such a clearing results from local hydrolysis of the lecithin emulsion if large amounts of the enzyme are used. The sensitivity of the method is greatly increased by staining the lipids in the gel matrix with Rhodamine 6G. It is suggested that the presence of Rhodamine 6G causes a loss of free fatty acids (produced by the enzyme) from liposomes into the incubation solution, and thus prevents inhibition of PLA2 by the reaction product and permits the enzyme reaction to proceed to completion.

PLA2 requires calcium ions for activity. So, to prevent the hydrolysis of lecithin during electrophoresis, a chelating agent (2 mM EDTA) should be included in the electrode buffer. After electrophoresis, the enzyme is reactivated by calcium ions contained in the staining solution. The addition of EDTA directly to the lecithin-containing gel solution causes aggregation of lecithin suspension during the polymerization step.

GENERAL NOTES

The enzyme from scorpion *Pandinus imperator* is a heterodimer consisting of large (108 amino acid residues) and small (17 residues) subunits. Both subunits are encoded by the same mRNA. The precursor polypeptide includes sequences of large and small subunits separated only by a pentapeptide that is processed during maturation.[2]

The functional enzyme from the snake *Crotalus atrox* exists as a dimer, while that from the pig pancreas and the snake *Naja naja* is active as a monomer.[3]

REFERENCES

1. Shier, W.T. and Trotter, J.T., Phospholipase A2 electrophoretic variants and their detection after polyacrylamide gel electrophoresis, *Anal. Biochem.*, 87, 604, 1978.
2. Conde, R., Zamudio, F.Z., Becerril, B., and Possani, L.D., Phospholipin, a novel heterodimeric phospholipase A2 from *Pandinus imperator* scorpion venom, *FEBS Lett.*, 460, 447, 1999.
3. Ferreira, J.P.M., Sasisekharan, R., Louie, O., and Langer, R., A study on the functional subunits of phospholipase A₂ by enzyme immobilization, *Biochem. J.*, 303, 527, 1994.

3.1.1.7 — Acetylcholinesterase; ACHE

OTHER NAMES	True cholinesterase, cholinesterase I, cholinesterase
REACTION	Acetylcholine + H_2O = choline + acetate
ENZYME SOURCE	Vertebrates
SUBUNIT STRUCTURE	Tetramer[#] (vertebrates)

METHOD 1

Visualization Scheme

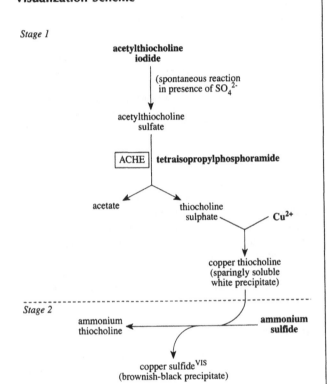

Stage 1

acetylthiocholine iodide

(spontaneous reaction in presence of SO_4^{2-}

acetylthiocholine sulfate

ACHE   tetraisopropylphosphoramide

acetate   thiocholine sulphate Cu^{2+}

copper thiocholine (sparingly soluble white precipitate)

Stage 2

ammonium thiocholine   **ammonium sulfide**

copper sulfide[VIS] (brownish-black precipitate)

Staining Solution[1]

A. 2.4 M Na$_2$SO$_4$ (see Procedure)
   6.9 mM Acetylthiocholine iodide
   6 mM CuSO$_4$
   22 mM Glycine
   75 mM Maleic acid
   150 mM NaOH
   25 mM MgCl$_2$
   0.01 mM Tetraisopropylphosphoramide (dissolved in 95% ethanol)
B. 40% Na$_2$SO$_4$
C. 10 mM (NH$_4$)$_2$S

Procedure

To prepare solution A, dissolve the Na_2SO_4 first, because the solution must be heated to effect solution. When Na_2SO_4 solution cools, add the other components. The pH of solution A should be 6.0.

Incubate the electrophorized gel in solution A at 37°C for 2 h. Rinse the gel in solution B and place in solution C. Brown bands that appear quite rapidly denote ACHE activity.

Notes: Tetraisopropylphosphoramide is used to inhibit cholinesterase (see 3.1.1.8 — CHE), which also can hydrolyze acetylthiocholine sulfate.

METHOD 2

Visualization Scheme

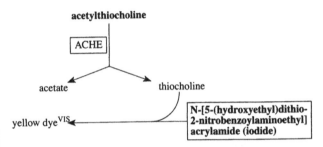

acetylthiocholine

ACHE

acetate   thiocholine

yellow dye[VIS]   N-[5-(hydroxyethyl)dithio-2-nitrobenzoylaminoethyl] acrylamide (iodide)

Staining Solution[2]

50 mM Phosphate buffer, pH 7.4
10 mM Acetylthiocholine
10 μM EDTA

Procedure

Incubate an electrophorized PAG prepared using N-[5-(hydroxy-ethyl)dithio-2-nitrobenzoylaminoethyl] acrylamide (iodide) in the staining solution until yellow bands appear.

Notes: To prevent development of cholinesterase (see 3.1.1.8 — CHE) activity bands, the electrophorized PAG should be initially preincubated in phosphate buffer (pH 7.4) containing 0.01 mM tetraisopropylphos-phoramide.

The method is based on the reduction of disulfide bonds of the chromogenic group (dithio-2-nitrobenzene) by thiocholine.

OTHER METHODS

An immunoblotting procedure (for details, see Part II) based on the utility of monoclonal antibodies specific to human ACHE[3] can also be used for immunohistochemical visualization of the enzyme protein on electrophoretic gels. This procedure is not appropriate for routine laboratory use, but may be of great value in special analyses of ACHE (e.g., biochemical, immunochemical, phyloge-netic, genetic, etc.). Monoclonal antibodies specific to the rat brain enzyme are now commercially available from Sigma.

3.1.1.7 — Acetylcholinesterase; ACHE (continued)

REFERENCES

1. Brewer, G.J., *An Introduction to Isozyme Techniques*, Academic Press, New York, 1970, p. 122.
2. Harris, R.B. and Wilson, I.B., Polyacrylamide gels which contain a novel mixed disulfide compound can be used to detect enzymes that catalyze thiol-producing reactions, *Anal. Biochem.*, 134, 126, 1983.
3. Fambrough, D.M., Engel, A.G., and Rosenberry, T.L., Acetylcholinesterase of human erythrocytes and neuromuscular junctions: homologies revealed by monoclonal antibodies, *Proc. Natl. Acad. Sci. U.S.A.*, 79, 1078, 1982.

3.1.1.8 — Cholinesterase; CHE

OTHER NAMES	Pseudocholinesterase, butyrylcholine esterase, nonspecific cholinesterase, choline esterase II (unspecific), benzoylcholinesterase
REACTION	Acylcholine + H_2O = choline + carboxylate
ENZYME SOURCE	Vertebrates
SUBUNIT STRUCTURE	Tetramer (vertebrates)

METHOD

Visualization Scheme

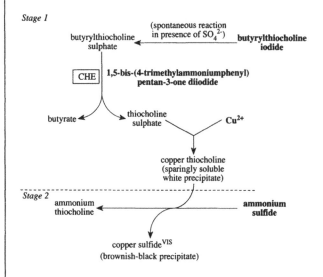

Staining Solution[1]

A. 2.4 *M* Na_2SO_4 (see Procedure)
6.9 m*M* Butyrylthiocholine iodide
6 m*M* $CuSO_4$
22 m*M* Glycine
75 m*M* Maleic acid
150 m*M* NaOH
25 m*M* $MgCl_2$
0.01 m*M* 1,5-Bis-(4-trimethylammoniumphenyl) pentan-3-one diiodide
B. 40% Na_2SO_4
C. 10 m*M* $(NH_4)_2S$

Procedure

To prepare solution A, dissolve the Na_2SO_4 first, because the solution must be heated to effect solution. When the Na_2SO_4 solution cools, add the other components. The pH of solution A should be 6.0.

Incubate the electrophorized gel in solution A at 37°C for 2 h. Rinse the gel in solution B and place in solution C. Brown bands that appear quite rapidly denote CHE activity.

3.1.1.8 — Cholinesterase; CHE (continued)

Notes: 1,5-Bis-(4-trimethylammoniumphenyl) pentan-3-one diiodide inhibits acetylcholinesterase (see 3.1.1.7 — ACHE), which also can hydrolyze butyrylthiocholine sulfate. Trimethyl(*p*-aminophenyl)-ammonium chloride can also be used for this purpose.

OTHER METHODS

A PAG prepared using *N*-[5-(hydroxyethyl)dithio-2-nitrobenzoylaminoethyl] acrylamide (iodide) can also be used to visualize CHE activity bands (for example, see 3.1.1.7 — ACHE, Method 2).

REFERENCES

1. Brewer, G.J., *An Introduction to Isozyme Techniques*, Academic Press, New York, 1970, p. 122.

3.1.1.11 — Pectinesterase; PE

OTHER NAMES	Pectin demethoxylase, pectin methoxylase, pectin methylesterase
REACTION	Pectin + n H_2O = n methanol + pectate
ENZYME SOURCE	Bacteria, fungi, plants
SUBUNIT STRUCTURE	Unknown or no data available

METHOD 1

Visualization Scheme

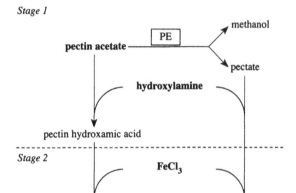

Staining Solution[1]

A. 2% Agar, 1% pectin solution (60°C) in 0.3 *M* acetate buffer, pH 4.5
B. 14% Hydroxylamine chloride in 60% ethanol
C. 14% NaOH in 60% ethanol
D. 25% HCl
E. 95% Ethanol
F. 2.5% $FeCl_3$ in 60% ethanol containing 0.1 *N* HCl

Procedure

Using solution A, form a 50-μm reactive agar–pectin plate between two glass plates heated to 40°C. Use a lower glass plate covered by silanized 100-μm polyester film and an upper glass plate covered by 100-μm polyester film treated with alkali. Apply an electrophorized gel to the reactive agar–pectin plate and incubate the gel and reactive agar–pectin plate combination at 40°C for 20 sec. Remove the gel and put the reactive plate in a 1:1 mixture of B and C solutions for 30 sec. Then treat the reactive plate for 30 sec in a 1:2 mixture of D and E solutions. Finally, place the reactive plate in solution F for 30 sec. Achromatic bands of PE activity appear on the reddish brown background of the gel. Fix the stained agar plate in 25% ethanol and dry on filter paper.

METHOD 2

Visualization Scheme

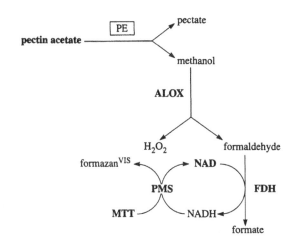

Staining Solution[2,3] (adapted)

A.	0.2 *M* Potassium phosphate buffer, pH 7.5	15 ml
	Alcohol oxidase (ALOX; EC 1.1.3.13;	
	from *Pichia*; Sigma)	80 U
	Formaldehyde dehydrogenase (FDH;	
	EC 1.2.1.46; Sigma)	5 U
	NAD	20 mg
	MTT	7 mg
	PMS	1 mg
B.	2% Agar, 10 mg/ml pectin solution in	
	0.2 *M* Potassium phosphate buffer,	
	pH 7.5 (60°C)	15 ml

Procedure

Mix A and B components of the staining solution and pour the mixture over the gel surface. Incubate the gel in the dark at room temperature until dark blue bands appear. Fix the stained gel in 25% ethanol.

Notes: This method is an adapted combination of a reduced fluorometric method developed to detect PE activity using auxiliary alcohol oxidase[2] and a histochemical method developed to detect trimethylamine-oxide aldolase (see 4.1.2.32 — TMAOA) using auxiliary formaldehyde dehydrogenase.[3] The activity of PEs detected by the method will increase with increasing methylation of a pectin preparation used as a PE substrate. The use of citrus pectins with a high degree of methyl esterification is recommended.

The method is suitable for detecting plant PEs that have alkaline pH optima in the range of pH 7.0 to 9.5. It is not applicable for bacterial and fungal PEs, which have acidic pH optima (pH 4.0 to 6.5). [2]

A chromogenic method suitable for detecting bacterial and fungal PEs can be developed using auxiliary peroxidase (see 1.11.1.7 — PER,

Method 1) to detect hydrogen peroxide generated by auxiliary alcohol oxidase. Both of these auxiliary enzymes work sufficiently well at pH 6.0.

METHOD 3

Visualization Scheme

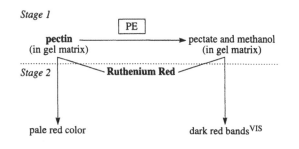

Staining Solution[4]

A.	100 m*M* Malic acid
B.	0.2% Ruthenium Red

Procedure

Incubate an electrophorized PAG containing 0.1% citrus pectin in 100 ml of solution A at room temperature for 90 min to cause a gradual pH change in the gel from the pH value of the gel buffer used to a pH of 3.0, allowing PE isozymes and other pectic enzymes (see *Notes*) to act on the pectin while passing through a suitable pH range. Following a brief rinse in distilled water, stain the gel in solution B for 30 to 120 min. Wash the stained gel with several changes of distilled water for 1 h to overnight. Dark red bands visible on a pale red background of the gel correspond to localization of PE.

Notes: The increased red staining in gel areas occupied by PE is explained by the increased number of stainable sites in the pectin produced by PE action and available to Ruthenium Red.

This method also detects activity bands caused by polygalacturonase (see 3.2.1.15 — PG, Method 2) and pectin lyase (see 4.2.2.10 — PECL). In pectin-containing PAGs processed as described above, PG produces colorless or pale zones and PL produces yellow zones visible on a pale red background of the gel. PE and PG were found to be useful for biochemical systematics of the genus *Penicillium*.[5]

A modification of this method is the three-step procedure that involves incubation of the gel in a pectin-containing buffered solution, washing the gel with distilled water to remove surface pectin, and staining the gel with Ruthenium Red to develop deep red bands of PE activity. The use of Ruthenium Red allows the visualizing of PE bands containing 0.3 mU of the enzyme activity and 0.125 ng of the enzyme protein. The absence of the need to include pectin in the gel presents several advantages when isoelectric focusing of PAGs with a wide range of pH values is used.[6]

3.1.1.11 — Pectinesterase; PE (continued)

REFERENCES

1. Höfelmann, M., Kittsteiner-Eberle, R., and Schreier, P., Ultrathin-layer agar gels: a novel print technique for ultrathin-layer isoelectric focusing of enzymes, *Anal. Biochem.*, 128, 217, 1983.
2. Wojciechowski, C.L. and Fall, R., A continuous fluorometric assay for pectin methylesterase, *Anal. Biochem.*, 237, 103, 1996.
3. Havemeister, W., Rehbein, H., Steinhart, H., Gonzales-Sotelo, C., Krogsgaard-Nielsen, M., and Jørgensen, B., Visualization of the enzyme trimethylamine oxide demethylase in isoelectric focusing gels by an enzyme-specific staining method, *Electrophoresis*, 20, 1934, 1999.
4. Cruickshank, R.H. and Wade, G.C., Detection of pectic enzymes in pectin-acrylamide gels, *Anal. Biochem.*, 107, 177, 1980.
5. Cruickshank, R.H. and Pitt, J.I., Identification of species in *Penicillium* subgenus *Penicillium* by enzyme electrophoresis, *Mycologia*, 79, 614, 1980.
6. Alonso, J., Rodriguez, M.T., and Canet, W., Detection of pectinesterase in polyacrylamide gels, *Electrophoresis*, 16, 39, 1995.

3.1.1.X — Poly(3-Hydroxybutyrate) Depolymerase; PHBD

REACTION | Endohydrolysis of carboxylic ester linkages in poly(3-hydroxybutyrate) (see General Notes)

ENZYME SOURCE | Bacteria, fungi

SUBUNIT STRUCTURE | Monomer[#] (bacteria)

METHOD

Visualization Scheme

poly(3-hydroxybutyrate) $\xrightarrow[\text{Ca}^{2+}]{\boxed{\text{PHBD}}}$ depolymerization products
(in gel matrix)   (diffuse from gel matrix)
opaque white background[vis]   translucent bands

Staining Solution[1]

100 mM Tris–HCl, pH 8.0
1 mM CaCl$_2$

Procedure

Incubate an electrophorized 3% PAG containing 2.5% (w/v) poly(3-hydroxybutyrate) (PHB; Zeneca Bioproducts, Billingham, England) in the staining solution at 30°C for 4 to 6 h. Translucent bands of PHBD activity are visible on an opaque background of the gel. Store the processed gel containing PHB in buffer or distilled water.

Notes: Sudan Red (traces) can be added to a PHB-containing PAG before polymerization. This will result in development of translucent bands visible on a red background of the gel.

The incorporation of 2.5% PHB imparts additional strength to the gel and facilitates its handling. The enzyme preparation fails to penetrate the gel on convenient electrophoresis and therefore should be pretreated with nonionic detergent Triton X-100 (final concentration of 0.5%). This facilitates enzyme migration into the gel without its activity being affected.

The method was used to detect extracellular PHBD after electrophoresis of an *Aspergillus* culture supernatant concentrated tenfold by reverse dialysis against sucrose, followed by 3 to 5 h of dialysis against glass distilled water.[1]

OTHER METHODS

Because depolymerization products generated by PHBD include 3-hydroxybutyrate, auxiliary enzyme 3-hydroxybutyrate dehydrogenase (see 1.1.1.30 — HBD) can be used to detect PHBD activity by the tetrazolium method. Preparations of HBD are commercially available. The pH optima of HBD and PHBD are within the same range (pH 7.0 to 8.0). This method, however, is expected to be much more expensive than that described above.

3.1.1.X — Poly(3-Hydroxybutyrate) Depolymerase; PHBD (continued)

GENERAL NOTES

The enzyme depolymerizing PHB may be identical with hydroxybutyrate–dimer hydrolase (EC 3.1.1.22), catalyzing the reaction (R)-3-$((R)$-3-hydroxybutanoyloxy)-butanoate + H_2O = 2 (R)-3-hydroxybutanoate.[2]

REFERENCES

1. Iyer, S., Joshi, A., and Desai, A., Rapid detection of a fungal poly(3-hydroxybutyrate) depolymerase using a new *in situ* gel activity staining technique, *Biotechnol. Tech.*, 11, 905, 1997.
2. NC-IUBMB, *Enzyme Nomenclature*, Academic Press, San Diego, 1992, p. 309 (EC 3.1.1.22).

3.1.2.1 — Acetyl-CoA Hydrolase; ACoAH

OTHER NAMES	Acetyl-CoA deacylase, acetyl-CoA acylase
REACTION	Acetyl-CoA + H_2O = CoA + acetate
ENZYME SOURCE	Plants, vertebrates
SUBUNIT STRUCTURE	Unknown or no data available

METHOD

Visualization Scheme

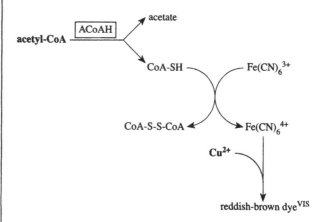

Staining Solution[1]

30 mM Phosphate buffer, pH 7.6	7.5 ml
100 mM Sodium phosphotartrate	2.5 ml
50 mM Cupric sulfate	2.5 ml
15 mM Potassium ferricyanide	12.5 ml
25 mM Magnesium chloride	25 ml
10 mM Acetyl-CoA	2.5 ml

Procedure

Incubate the gel in staining solution at 37°C until reddish brown bands appear.

3.1.2.1 — Acetyl-CoA Hydrolase; ACoAH (continued)

OTHER METHODS

A. A PAG prepared using *N*-[5-(hydroxyethyl)dithio-2-nitrobenzoylaminoethyl] acrylamide (iodide) can be used to visualize ACoAH activity bands. The method is based on the reduction of disulfide bonds of the chromogenic group (dithio-2-nitrobenzene) by thiol reagents, including CoA-SH produced by ACoAH (e.g., see 3.1.1.7 — ACHE, Method 2).

B. The product CoA-SH can be detected using the redox indicator 2,6-dichlorophenol indophenol coupled with the tetrazolium system (e.g., see 1.6.4.2 — GSR, Method 2).

C. The free thiol group of CoA-SH can also be detected using 5,5′-dithio-bis(2-nitrobenzoic acid) (e.g., see 1.6.4.2 — GSR, Method 3).

REFERENCES

1. Volk, M.J., Trelease, R.N., and Reeves, H.C., Determination of malate synthase activity in polyacrylamide gels, *Anal. Biochem.*, 58, 315, 1974.

3.1.2.6 — Hydroxyacylglutathione Hydrolase; HAGH

OTHER NAMES	Glyoxalase II
REACTION	*S*-(2-Hydroxyacyl)glutathione + H_2O = glutathione + 2-hydroxy carboxylate
ENZYME SOURCE	Vertebrates
SUBUNIT STRUCTURE	Monomer (vertebrates)

METHOD 1

Visualization Scheme

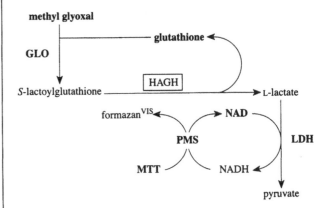

Staining Solution[1]

A.	0.1 *M* Tris–HCl buffer, pH 8.0	12.5 ml
	Methyl glyoxal	50 μl
	Glyoxalase I (GLO; 200 U/ml)	25 μl
	Lactate dehydrogenase (LDH; 2750 U/ml)	10 μl
	NAD	40 mg
	Glutathione (reduced form)	40 mg
	MTT	4 mg
	PMS	2 mg
B.	2% Agar solution (60°C)	12.5 ml

Procedure

Mix A and B components of the staining solution and pour the mixture over the gel surface. Incubate the gel in the dark at 37°C until dark blue bands appear. Fix the stained gel in 25% ethanol.

METHOD 2

Visualization Scheme

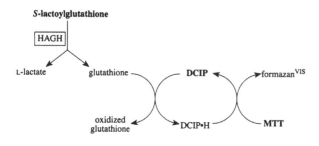

Staining Solution[2]

A.	0.2 M Tris–HCl buffer, pH 8.0	28.2 ml
	2,6-Dichlorophenol indophenol (DCIP)	6 mg
	MTT	24 mg
	50 mM S-Lactoylglutathione	1.8 ml
B.	1.8% Agar solution (50°C)	30 ml

Procedure

Mix A and B components of the staining solution and pour the mixture over the gel surface. Incubate the gel in the dark at 37°C until dark blue bands appear. Fix the stained gel in 25% ethanol.

GENERAL NOTES

The human red cell enzyme is able to catalyze the hydrolysis of S-formylglutathione and gives with this substrate 46% of the rate with S-lactoylglutathione (see also 3.1.2.12 — FGH, Method 1, *Notes*).

REFERENCES

1. Charlesworth, D., Starch gel electrophoresis of 4 enzymes from human red blood cells: glyceraldehyde-3-phosphate dehydrogenase, fructo-aldolase, glyoxalase II and sorbitol dehydrogenase, *Ann. Hum. Genet.*, 35, 477, 1972.
2. Uotila, L., Polymorphism of red cell S-formylglutathione hydrolase in a Finnish population, *Hum. Hered.*, 34, 273, 1984.

3.1.2.12 — *S*-Formylglutathione Hydrolase; FGH

REACTION   *S*-Formylglutathione + H$_2$O = glutathione + formate

ENZYME SOURCE   Plants, vertebrates

SUBUNIT STRUCTURE Dimer[#] (plants)

METHOD 1

Visualization Scheme

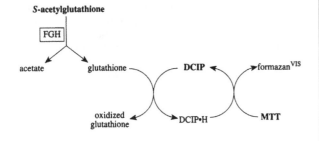

Staining Solution[1]

A. 0.2 *M* Tris–HCl buffer, pH 8.0   28.2 ml
   2,6-Dichlorophenol indophenol (DCIP)   6 mg
   MTT   24 mg
   50 m*M* *S*-acetylglutathione   1.8 ml
B. 1.8% Agar solution (50°C)   30 ml

Procedure

Mix A and B components of the staining solution and pour the mixture over the gel surface. Incubate the gel in the dark at 37°C until dark blue bands appear. Fix the stained gel in 25% ethanol.

Notes: The use of *S*-formylglutathione instead of *S*-acetylglutathione is preferable because it is a much more effective substrate of FGH than *S*-acetylglutathione (relative rates 100 and 0.5, respectively). However, *S*-formylglutathione of sufficient purity was not obtained by the author during development of this method. When using this substrate in the staining solution instead of *S*-acetylglutathione, the interpretation of FGH electrophoretic patterns should be made with caution because hydroxyacylglutathione hydrolase (see 3.1.2.6 — HAGH, General Notes) also is able to catalyze the hydrolysis of *S*-formylglutathione (with this substrate human red cell HAGH gives 46% of the rate with *S*-lactoylglutathione). Thus, when *S*-formylglutathione is used as a substrate in this method, a control staining for HAGH activity should be made for comparison.

METHOD 2

Visualization Scheme

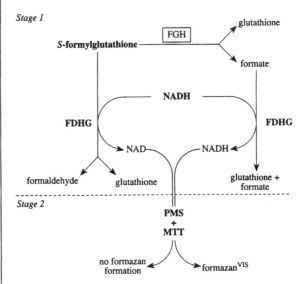

Staining Solution[1]

A. 0.1 *M* Sodium phosphate buffer, pH 6.0
   1 m*M* *S*-formylglutathione
   0.4 m*M* NADH
   0.01 U/ml Formaldehyde dehydrogenase (FDHG)
B. 0.1 *M* Sodium phosphate buffer, pH 7.5
   0.03 mg/ml PMS
   0.3 mg/ml NBT

Procedure

Incubate the gel in solution A at 23°C for 30 min and then transfer to solution B. Incubate the gel in solution B in the dark at 23°C until dark blue bands appear (30 to 60 min). Rinse the stained gel with water and fix in 25% ethanol.

Notes: Formaldehyde dehydrogenase is used to catalyze in its reverse reaction the oxidation of NADH to NAD in the presence of *S*-formylglutathione. At the location of FGH activity, however, *S*-formylglutathione is hydrolyzed, preventing there the oxidation of NADH, which is then visualized via the PMS–MTT system.

3.1.2.12 — *S*-Formylglutathione Hydrolase; FGH (continued)

GENERAL NOTES

FGH also hydrolyzes *S*-acetylglutathione (more slowly).[2] *S*-Formylglutathione is a much more effective substrate of FGH than *S*-acetylglutathione. However, Method 1 works well even with *S*-acetylglutathione. Using this method it is possible to detect a much lower amount of FGH than with Method 2. The use of Method 2 is recommended only to confirm that the staining results obtained with Method 1 are not affected by using *S*-acetylglutathione instead of *S*-formylglutathione.

Preparations of *S*-acetylglutathione and *S*-formylglutathione are not yet commercially available, but they can be synthesized and purified under laboratory conditions.[3–5]

The enzyme from human erythrocytes is identical to esterase D (see 3.1.1 ... — EST, Method 2, *Notes*).[6]

The plant enzyme displays thioesterase activity toward *S*-formylglutathione and carboxyesterase activity toward 4-methylumbelliferyl acetate.[7]

REFERENCES

1. Uotila, L., Polymorphism of red cell *S*-formylglutathione hydrolase in a Finnish population, *Hum. Hered.*, 34, 273, 1984.
2. NC-IUBMB, *Enzyme Nomenclature*, Academic Press, San Diego, 1992, p. 316 (EC 3.1.2.12, Comments).
3. Uotila, L., Preparation and assay of glutathione thiol esters: survey of human liver glutathione thiol esterases, *Biochemistry*, 12, 3938, 1973.
4. Uotila, L., Thioesters of glutathione, *Methods Enzymol.*, 77, 424, 1981.
5. Board, P.G. and Coggan, M., Genetic heterogeneity of *S*-formylglutathione hydrolase, *Ann. Hum. Genet.*, 50, 35, 1986.
6. Apeshiotis, F. and Bender, K., Evidence that S-formylglutathione hydrolase and esterase D polymorphisms are identical, *Hum. Genet.*, 74, 176, 1986.
7. Kordic, S., Cummins, I., and Edwards, R., Cloning and characterization of an S-formylglutathione hydrolase from *Arabidopsis thaliana*, *Arch. Biochem. Biophys.*, 399, 232, 2002.

3.1.3.1 — Alkaline Phosphatase; ALP

OTHER NAMES	Alkaline phosphomonoesterase, phosphomonoesterase, glycerophosphatase
REACTION	Orthophosphoric monoester + H_2O = alcohol + orthophosphate
ENZYME SOURCE	Bacteria, fungi, green algae, plants, protozoa, invertebrates, vertebrates
SUBUNIT STRUCTURE	Monomer (plants, invertebrates, vertebrates), dimer (plants, invertebrates, vertebrates)

METHOD 1

Visualization Scheme

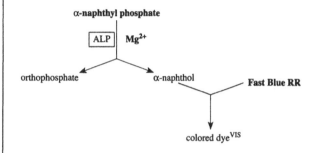

Staining Solution[1]

60 mM Borate buffer, pH 9.7	100 ml
α-Naphthyl phosphate	50 mg
Fast Blue RR	50 mg
10 mM MgCl$_2$	1 ml

Procedure

Incubate the gel in staining solution in the dark at 37°C until dark gray bands appear. Wash the stained gel with water and fix in 25% ethanol.

Notes: Other substrates (e.g., α-naphthyl acid phosphate and β-naphthyl phosphate) and other dye couplers (e.g., Fast Blue B) can also be used.

Using this method, nonspecifically stained light yellow bands of albumin, haptoglobin, and ceruloplasmin can also be developed on ALP zymograms after electrophoresis of blood serum preparations.

METHOD 2

Visualization Scheme

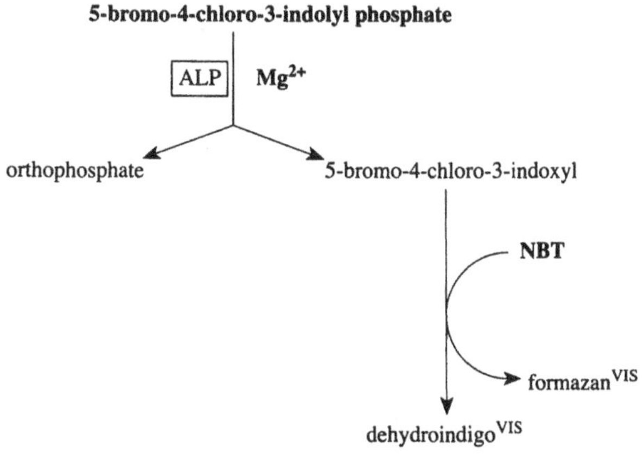

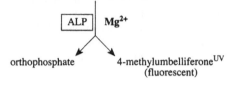

5-bromo-4-chloro-3-indolyl phosphate

ALP   Mg²⁺

orthophosphate   5-bromo-4-chloro-3-indoxyl

NBT

formazan^VIS

dehydroindigo^VIS

Staining Solution²

0.1 *M* Tris–HCl buffer, pH 9.0	95 ml
2 m*M* MgCl₂	5 ml
1% 5-Bromo-4-chloro-3-indolyl phosphate (*p*-toluidine salt, dissolved in dimethylformamide)	5 ml
NBT	10 mg

Procedure

Incubate the gel in staining solution in the dark at 37°C until dark blue bands appear. Wash the stained gel with water and fix in 3% acetic acid.

Notes: The hydroxyl group of the product 5-bromo-4-chloro-3-indoxyl tautomerizes, forming a ketone, and under alkaline conditions, dimerization occurs, forming a dehydroindigo. In the process of dimerizing, it releases hydrogen ions that reduce nitro blue tetrazolium to blue formazan. This one-step staining procedure gives distinct bands. Thus, the bands migrating close together can easily be distinguished from each other. The reaction causing formazan formation can be quantitated by reflective densitometry and appears to be a log linear relation over a wide concentration range of ALP.³

METHOD 3

Visualization Scheme

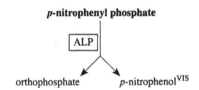

4-methylumbelliferyl phosphate

ALP   Mg²⁺

orthophosphate   4-methylumbelliferone^UV
(fluorescent)

Staining Solution⁴

A.	0.1 *M* Borate buffer, pH 9.7	10 ml
	4-Methylumbelliferyl phosphate	1 mg
	MgSO₄	50 mg
B.	2% Agar solution (55°C)	20 ml

Procedure

Mix A and B components of the staining solution and pour the mixture over the gel surface. Incubate the gel at 37°C and monitor under long-wave UV light. Fluorescent bands indicate localization of ALP activity in the gel. Record the zymogram or photograph using a yellow filter.

METHOD 4

Visualization Scheme

p-nitrophenyl phosphate

ALP

orthophosphate   *p*-nitrophenol^VIS

Staining Solution⁵

A.	0.89 *M* Diethanolamine	20 ml
	p-Nitrophenyl phosphate	4 mg
B.	2% Agarose solution (50°C)	20 ml

Procedure

Mix A and B components of the staining solution and pour the mixture over the gel surface. Incubate the gel at 37°C until yellow bands appear. Photograph the zymogram using a 436.8-nm interference filter.

METHOD 5

Visualization Scheme

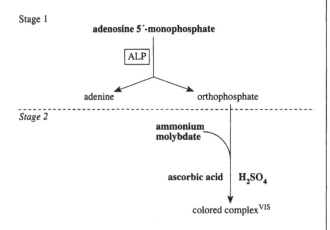

Stage 1

adenosine 5′-monophosphate

ALP

adenine   orthophosphate

Stage 2

ammonium molybdate

ascorbic acid | H₂SO₄

colored complex^VIS

Staining Solution[5]

A.	0.1 *M* Tris–HCl buffer, pH 8.0	20 ml
	AMP (sodium salt)	140 mg
B.	2% Agarose solution (60°C)	
C.	1.25% Ammonium molybdate	
	1 *M* H₂SO₄	
	10% Ascorbic acid	

Procedure

Mix equal volumes of A and B solutions and pour the mixture over the gel surface. Incubate the gel at 37°C for 30 min. Remove the first agarose overlay and cover the gel surface with a mixture consisting of equal volumes of B and C solutions. Blue bands appear almost immediately at sites of ALP activity. The bands are ephemeral; thus the zymogram should be recorded or photographed immediately.

Notes: Malachite Green is a basic dye that may be used coupled with ammonium molybdate to detect orthophosphate generated by alkaline phosphatase. Unlike the ammonium molybdate method, the Malachite Green–ammonium molybdate method results in permanently stained bands due to the formation of a stable malachite–phosphomolybdate complex of sharp blue-green color. Zymograms obtained by the Malachite Green–ammonium molybdate method may be preserved in 5% acetic acid containing 20% ethanol (e.g., see 3.1.3.2 — ACP, Method 4), or in 25% ethanol containing 2% glycerol (e.g., see 3.1.3.11 — FBP, Method 3).

METHOD 6

Visualization Scheme

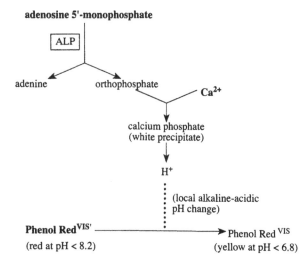

adenosine 5′-monophosphate

ALP

adenine   orthophosphate   Ca²⁺

calcium phosphate
(white precipitate)

H⁺

(local alkaline-acidic
pH change)

Phenol Red^VIS′ ⟶ Phenol Red ^VIS
(red at pH < 8.2)   (yellow at pH < 6.8)

Staining Solution[5]

A.	1% Phenol Red	0.25 ml
	AMP (sodium salt)	35 mg
	0.15 *M* NaCl	4.1 ml
	1.0 *M* CaCl₂	1 ml
B.	2% Agarose solution (50°C)	5 ml

Procedure

Wash an electrophorized acetate cellulose gel with water for 2 min to remove the gel buffer. Prepare solution A. Prior to the addition of CaCl₂, all other components of solution A should be adjusted to pH 8.0. Mix solutions A and B and pour the mixture over the porous side of the acetate cellulose gel. Incubate the gel at 37°C until yellow bands of ALP appear on a red background. Photograph the zymogram using a 570-nm interference filter.

Notes: Since changes in the absorption spectrum of Phenol Red occur between pH values 6.8 and 8.2, theoretically a change as small as 10^{-7} *M* hydrogen ion could be detected by this method.[6]

A white precipitate of calcium phosphate is visible against a dark background when a transparent gel (e.g., PAG) is used for electrophoresis.[7]

METHOD 7

Visualization Scheme

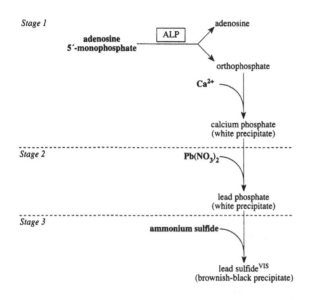

Staining Solution[5]

A. 90 m*M* Tris–2 m*M* EDTA–50 m*M* boric acid,
       pH 8.6   16 ml
       1 *M* CaCl₂   2 ml
       AMP (sodium salt)   140 mg
       1.5 *M* NaCl   1.6 ml
B. 2% Agarose solution (50°C)   20 ml
C. 0.1 *M* Tris–HCl buffer, pH 7.0
       3 m*M* Pb(NO₃)₂
D. 5% Ammonium sulfide

Procedure

Mix A and B components of the staining solution and pour the mixture over the gel surface. Incubate the gel at 37°C for 30 to 60 min. Remove the agarose overlay and wash the gel with water for 15 min. Place the gel in solution C for 30 min and then in solution D. Brownish black bands indicate the sites of ALP activity localization on the gel. Photograph the zymogram using a 436.8-nm interference filter.

METHOD 8

Visualization Scheme

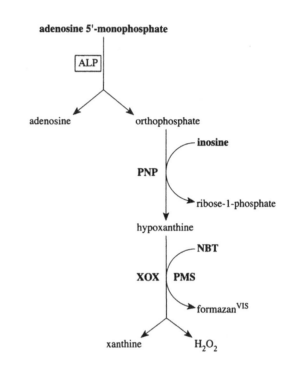

Staining Solution[5]

A. 0.1 *M* Tris–HCl buffer, pH 8.0   6 ml
       0.1 *M* EDTA   0.5 ml
       10 mg/ml NBT   1 ml
       2 mg/ml PMS   0.1 ml
       0.1 *M* Inosine   1.6 ml
       AMP (sodium salt)   140 mg
       0.75 U/ml Xanthine oxidase (XOX)   8 ml
       1.67 U/ml Purine-nucleoside phosphorylase
       (PNP)   0.2 ml
       H₂O   12 ml
B. 2% Agarose solution in 0.1 *M* Tris–HCl buffer,
       pH 8.0 (50°C)   20 ml

Procedure

Mix A and B components of the staining solution and pour the mixture over the gel surface. Incubate the gel in the dark at 37°C until dark blue bands appear. Fix the stained gel in 25% ethanol.

Notes: Alternative enzymatic methods for detection of orthophosphate are available that are based on coupled reactions involving either the glyceraldehyde-3-phosphate dehydrogenase[8] or the phosphorylase *a*[9] reactions. Both methods involve three enzymatic steps and are more complicated than the method presented above. Moreover, the coupled reactions of these alternative enzymatic methods are readily reversible and some of the substrates used are unstable.

METHOD 9

Visualization Scheme

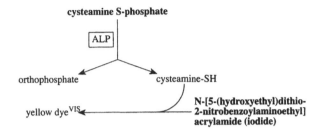

Staining Solution[10]

0.5 M Tris–HCl buffer, pH 8.0
5 mM Cysteamine S-phosphate

Procedure

Incubate an electrophorized PAG prepared using N-[5-(hydroxyethyl)dithio-2-nitrobenzoylaminoethyl] acrylamide (iodide) in the staining solution at 37°C until yellow bands appear.

METHOD 10

Visualization Scheme

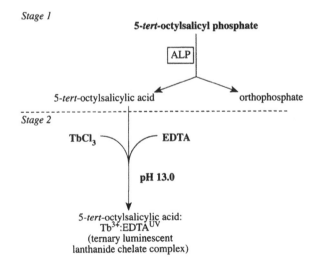

Staining Solution[11]

A. 0.1 M Tris–HCl buffer, pH 9.0
   0.1 M NaCl
   1 mM MgCl$_2$
   1 mM 5-$tert$-Octylsalicyl phosphate
B. 0.01 M HCl
   5 mM TbCl$_3$
   5 mM EDTA (tetrasodium salt)
C. 2.5 M Tris, pH 13.0

Procedure

Transfer proteins from an electrophorized gel onto a Nitran nylon membrane (0.45-μm pore size; Schleicher and Schull) using the routine blotting procedure. Incubate the membrane for 2 h in substrate solution A and blot with blotting paper on the non-protein-containing surface. Cover the membrane with developing solution (one part of solution B, one part of solution C, and three parts of deionized water) and observe luminescent ALP bands under 300- to 400-nm UV light, or photograph on Polaroid instant film with a TRP 100 time-resolved photographic camera (Kronem Systems, Inc., Mississauga, Ontario, Canada) using filters that pass 320- to 400-nm excitation and more than 515-nm emission wavelengths, a 440-μsec time delay, and a 4.1-msec measurement gate.

Notes: This method may be applied directly to the electrophoretic gel without transferring proteins onto the nylon membrane. In this situation the substrate and developing solutions should be applied as 1% agar or as filter paper overlays. The time of incubation of the gel with a substrate-containing overlay should be diminished with the purpose of reducing undesirable diffusion of the product 5-$tert$-octylsalicylic acid. The main disadvantage of nonmembrane-based detection of ALP by this method is that the detection should be carried out in a two-step procedure. This is because optimal formation of a luminescent lanthanide chelate complex occurs at pH 12.5, which is too high even for ALP.

The membrane-based method was used to detect biotinylated DNA in dot blot and Southern blot assays using streptavidin–ALP conjugate. High sensitivity that is comparable with radioisotopic detection was obtained with this method during quantitative assays.

Another substrate, 5-fluorosalicyl phosphate, may be used in nonmembrane-based detection of ALP via formation of a ternary luminescent lanthanide chelate complex (5-fluorosalicylic acid:Tb^{3+}:EDTA).[12]

OTHER METHODS

The immunoblotting procedure (for details, see Part II) may be used to detect ALP. Antibodies specific to the bacterial, calf intestinal, and human placental enzyme are available from Sigma.

GENERAL NOTES

The enzyme from some sources requires manganese ions for activity.[13]

Alkaline phosphatase zymograms should be compared to those of acid phosphatase (see 3.1.3.2 — ACP) because some taxa possess phosphatases capable of exhibiting activity under both acidic and alkaline conditions.[14]

There is phosphatase preferentially catalyzing the hydrolysis of β-glycerophosphate (glycerol-2-phosphate) (EC 3.1.3.19).[15] β-Glycerophosphate-specific alkaline (pH 8.0) phosphatase was found in cultured human limphoid cells and human–rodent hybrid cells. This phosphatase is a dimeric enzyme that demonstrates no hydrolyzing activity toward uridine 5′-monophosphate, cytidine 5′-monophosphate, deoxyinosine 5′-monophosphate, deoxyuridine 5′-monophosphate, p-nitrophenyl phosphate, 4-methylumbelliferyl phosphate, fructose-1,6-diphosphate, 3-phosphoglycerate, and 5′-phosphoribosyl-1-pyrophosphate and very slowly hydrolyzes α-glycerophosphate (glycerol-3-phosphate).[16]

All methods based on detection of liberated orthophosphate can only be used when electrophoresis is carried out in a phosphate-free buffer system.

ALP from some sources can hydrolyze phosphoenolpyruvate and cause the appearance of additional bands on the zymograms of some enzymes where pyruvate kinase and lactate dehydrogenase are used as linking enzymes to detect the product ADP (e.g., see 3.1.4.17 — CNPE, Method 1).

REFERENCES

1. Boyer, S.H., Alkaline phosphatase in human sera and placentae, *Science*, 134, 1002, 1961.
2. Dingjan, P.G., Postma, T., and Stroes, J.A.P., Quantitative differentiation of human-serum alkaline phosphatase isoenzymes with polyacrylamide disc gel electrophoresis, *Z. Klin. Chem. Klin. Biochem.*, 11, 167, 1973.
3. Blake, M.S., Johnston, K.H., Russell-Jones, G.J., and Gotschlich, E.C., A rapid, sensitive method for detection of alkaline phosphatase-conjugated anti-antibody in Western blots, *Anal. Biochem.*, 136, 175, 1984.
4. Benham, F.J., Cottell, P.C., Franks, L.M., and Wilson, P.D., Alkaline phosphatase activity in human bladder tumor cell lines, *J. Histochem. Cytochem.*, 25, 266, 1977.
5. Klebe, R.J., Schloss, S., Mock, L., and Link, G.R., Visualization of isozymes which generate inorganic phosphate, *Biochem. Genet.*, 19, 921, 1981.
6. Klebe, R.J., Mancuso, M.G., Brown, C.R., and Teng, L., Two-dimensional spectroscopy of electrophoretic gels, *Biochem. Genet.*, 19, 655, 1981.
7. Nimmo, H.G. and Nimmo, G.A., A general method for the localization of enzymes that produce phosphate, pyrophosphate, or CO_2 after polyacrylamide gel electrophoresis, *Anal. Biochem.*, 121, 17, 1982.
8. Cornell, N.W., Leadbetter, M.G., and Veech, R.L., Modifications in the enzymatic assay for inorganic phosphate, *Anal. Biochem.*, 95, 524, 1979.
9. Lowry, O.H., Schulz, D.W., and Passonneau, J.V., Effects of adenylic acid on the kinetics of muscle phosphorylase *a*, *J. Biol. Chem.*, 239, 1947, 1964.
10. Harris, R.B. and Wilson, I.B., Polyacrylamide gels which contain a novel mixed disulfide compound can be used to detect enzymes that catalyze thiol-producing reactions, *Anal. Biochem.*, 134, 126, 1983.
11. Templeton, E.F.G., Wong, H.E., Evangelista, R.A., Granger, T., and Pollak, A., Time-resolved fluorescence detection of enzyme-amplified lanthanide luminescence for nucleic acid hybridization assays, *Clin. Chem.*, 37, 1506, 1991.
12. Evangelista, R.A., Pollak, A., and Templeton, E.F.G., Enzyme-amplified lanthanide luminescence for enzyme detection in bioanalytical assays, *Anal. Biochem.*, 197, 213, 1991.
13. Wilcox, F.H., Hirschhorn, L., Taylor, B.A., Womack, J.E., and Roderick, T.H., Genetic variation in alkaline phosphatase of the house mouse (*Mus musculus*) with emphasis on a manganese-requiring isozyme, *Biochem. Genet.*, 17, 1093, 1979.
14. Richardson, B.J., Baverstock, P.R., and Adams, M., *Allozyme Electrophoresis: A Handbook for Animal Systematics and Population Studies*, Academic Press, Sydney, 1986, p. 170.
15. NC-IUBMB, *Enzyme Nomenclature*, Academic Press, San Diego, 1992, p. 320 (EC 3.1.3.19).
16. Wilson, D.E., Del Pizzo, R., Carritt, B., and Povey, S., Assignment of the human gene for β-glycerol phosphatase to chromosome 8, *Ann. Hum. Genet.*, 50, 217, 1986.

3.1.3.2 — Acid Phosphatase; ACP

OTHER NAMES	Acid phosphomonoesterase, phosphomonoesterase, glycerophosphatase
REACTION	Orthophosphoric monoester + H_2O = alcohol + orthophosphate
ENZYME SOURCE	Bacteria, fungi, green algae, plants, protozoa, invertebrates, vertebrates
SUBUNIT STRUCTURE	Monomer (plants, invertebrates, vertebrates), dimer (plants, invertebrates, vertebrates), tetramer (invertebrates)

METHOD 1

Visualization Scheme

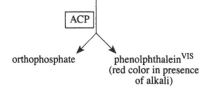

Staining Solution[1]

A. 0.1 M Citrate buffer, pH 5.5	25 ml
Phenolphthalein monophosphate (disodium salt)	50 mg
B. NH_4OH (concentrated solution)	5 ml

Procedure

Pour solution A on top of filter paper; overlay the gel and incubate at 37°C for 1.5 to 2 h. Remove filter paper and cover the gel with solution B. Red bands indicating localization of ACP activity on the gel appear after 1 to 2 min. Record or photograph the zymogram.

Notes: A method with phenolphthalein diphosphate as a chromogenic substrate is also available,[2] but it is less sensitive than that described above. The method with phenolphthalein monophosphate has a sensitivity comparable to the sensitivity of the method that uses 4-methylumbelliferyl phosphate (see Method 2, below).

METHOD 2

Visualization Scheme

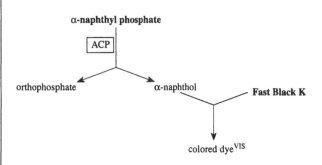

Staining Solution[1]

0.1 M Citrate buffer, pH 5.5	25 ml
4-Methylumbelliferyl phosphate	5 mg

Procedure

Apply the staining solution to the gel surface on a filter paper overlay and incubate the gel at 37°C for 15 to 30 min. Light (fluorescent) bands of ACP are visible under long-wave UV light. Record the zymogram or photograph using a yellow filter.

Notes: Zones of ACP that are very weak after prolonged incubation may be seen more clearly if the gel surface is made alkaline with ammonia to increase the level of fluorescence.[3]

METHOD 3

Visualization Scheme

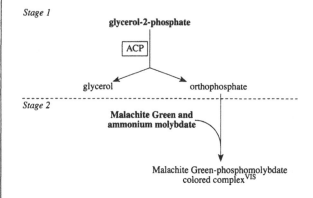

Staining Solution[4]

50 mM Acetate buffer, pH 5.0	100 ml
α-Naphthyl phosphate (disodium salt)	100 mg
Fast Black K salt	100 mg

Procedure

Incubate the gel in staining solution in the dark at 37°C until colored bands appear. Wash the stained gel in water and fix in 7% acetic acid.

Notes: Where the enzyme activity is weak, greater sensitivity may be obtained using a postcoupling technique. In this situation, all ingredients of the staining solution except Fast Black K (the possible inhibitor) are applied to the gel. The gel is incubated for a period of time (depending on the expected activity of the enzyme) and Fast Black K is then applied to the gel.

β-Naphthyl phosphate may be used as a substrate instead of α-naphthyl phosphate. Other dye couplers (e.g., Fast Blue B, Fast Blue BB, and Fast Garnet GBC) also may be used instead of Fast Black K.

Method 4

Visualization Scheme

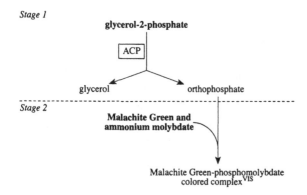

Staining Solution[5]

A.	100 m*M* Acetate buffer, pH 5.0	
	5 m*M* Glycerol-2-phosphate	
B.	4.2% Ammonium molybdate in 4 *N* HCl	30 ml
	0.045% Malachite Green (oxalate salt)	90 ml
	2% Sterox (a detergent diluent)	2.4 ml
C.	Solution B	60 ml
	H₂O	40 ml

Procedure

Preincubate an electrophorized gel in 100 m*M* acetate buffer (pH 5.0) for 30 min. Then incubate the gel in solution A (100 ml of solution A for 20 ml of gel volume) for 30 min at 37°C. Following incubation, rinse the gel quickly with water (usually 2 × 100 ml) to remove substrate and excess orthophosphate from the surface of the gel, and place the gel in solution C (see *Notes*). Development of sharp blue-green bands of ACP is usually complete in 10 to 20 min. Rinse the stained gel with several changes of water and store in 5% acetic acid containing 20% ethanol.

Notes: Prepare stock solution B as follows. Mix 30 ml of ammonium molybdate solution and 90 ml of Malachite Green solution for 20 min at room temperature. Then pass the resulting mixture through a Whatman No. 5 filter and add 2.4 ml of sterox solution. This stock solution may be prepared in advance and stored at 4°C, at which it is stable for about a week. Solution C is prepared using stock solution B and water immediately prior to use. Malachite Green hydrochloride can be used instead of Malachite Green oxalate salt.

This method is ideal for use in the assay of detergent-solubilized membrane-associated phosphatase activities. The staining procedure is easy to perform, results in a stabler colored product than does the routine acid molybdate method, and is free from interference by detergents. The Malachite Green technique may be used to detect orthophosphate at pH values 5.0 to 8.5.

The method can easily determine orthophosphate in a range of 0.5 to 10 nmol. However, it has two restrictions: (1) the use of citrate buffers prevents color development, and (2) the method can only be used with phosphate-free buffer systems.

Method 5

Visualization Scheme

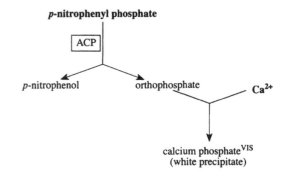

Staining Solution[6]

0.1 *M* Acetate buffer, pH 5.0
0.6 m*M* *p*-Nitrophenyl phosphate
0.2 *M* Ca²⁺

Procedure

Soak an electrophorized PAG in 0.1 *M* acetate buffer (pH 5.0) for 20 to 30 min and incubate in the staining solution at 37°C. View the gel against a dark background and, when white bands of calcium phosphate precipitation of sufficient intensity are obtained, remove the gel from staining solution and store in 50 m*M* glycine–KOH (pH 10.0), containing 5 m*M* Ca²⁺, either at 5°C or at room temperature in the presence of an antibacterial agent.

3.1.3.2 — Acid Phosphatase; ACP (continued)

Notes: The method can only be used when electrophoresis is carried out in a phosphate-free buffer system.

The precipitated calcium phosphate may be converted to brownish black lead sulfide (e.g., see 3.1.3.1 — ALP, Method 7) or subsequently stained with Alizarin Red S. The lead conversion stain and Alizarin Red S stain do not increase the sensitivity of the staining method for photographs or for scanning, although they would be of advantage for opaque gel systems such as starch or acetate cellulose.

Other Methods

Some other methods developed for alkaline phosphatase (see 3.1.3.1 — ALP) may be adapted for acid phosphatase.

General Notes

Acid phosphatase zymograms should be compared to those of alkaline phosphatase (see 3.1.3.1 — ALP) because some taxa possess phosphatases capable of exhibiting activity under both acidic and alkaline conditions.[7]

The enzyme from some sources requires magnesium or manganese ions or EDTA for activity.

References

1. Sparkes, M.C., Crist, M.L., and Sparkes, R.S., High sensitivity of phenolphthalein monophosphate in detecting acid phosphatase isoenzymes, *Anal. Biochem.*, 64, 316, 1975.
2. Hopkinson, D.A., Spencer, N., and Harris, H., Red cell acid phosphatase variants: a new human polymorphism, *Nature*, 199, 969, 1963.
3. Harris, H. and Hopkinson, D.A., *Handbook of Enzyme Electrophoresis in Human Genetics*, North-Holland, Amsterdam, 1976 (loose-leaf, with supplements in 1977 and 1978).
4. Shaw, C.R. and Prasad, R., Starch gel electrophoresis of enzymes: a compilation of recipes, *Biochem. Genet.*, 4, 297, 1970.
5. Zlotnick, G.W. and Gottlieb, M., A sensitive staining technique for the detection of phosphohydrolase activities after polyacrylamide gel electrophoresis, *Anal. Biochem.*, 153, 121, 1986.
6. Nimmo, H.G. and Nimmo, G.A., A general method for the localization of enzymes that produce phosphate, pyrophosphate, or CO_2 after polyacrylamide gel electrophoresis, *Anal. Biochem.*, 121, 17, 1982.
7. Richardson, B.J., Baverstock, P.R., and Adams, M., *Allozyme Electrophoresis: A Handbook for Animal Systematics and Population Studies*, Academic Press, Sydney, 1986, p. 162.

3.1.3.3 — Phosphoserine Phosphatase; PSP

REACTION	L(or D)-*O*-Phosphoserine + H_2O = L(or D)-serine + orthophosphate
ENZYME SOURCE	Bacteria, fungi, vertebrates
SUBUNIT STRUCTURE	Dimer (vertebrates)

Method

Visualization Scheme

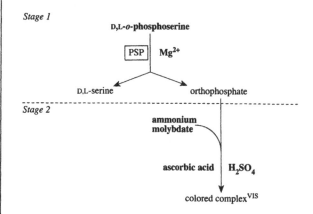

Staining Solution[1]

A.	0.1 *M* Tris–HCl buffer, pH 7.5	25 ml
	L-*O*-Phosphoserine	50 mg
	0.2 *M* MgCl$_2$	2 ml
B.	2% Agar solution (55°C)	
C.	2.5% Ammonium molybdate in 4 *N* H$_2$SO$_4$	25 ml
	Ascorbic acid	1.25 g

Procedure

Mix solution A with 25 ml of solution B and pour the mixture over the gel surface. Incubate the gel at 37°C for 1 to 4 h. Remove the agar overlay. Mix solution C with 25 ml of solution B and cover the preincubated gel with the resulting mixture. Dark blue bands due to the presence of orthophosphate appear at the sites of PSP activity. The bands are ephemeral; thus the zymogram should be recorded or photographed immediately.

Notes: The best resolution of PSP allozymes is achieved when 2-mercaptoethanol is incorporated in the gel. The method can only be used when electrophoresis is carried out in a phosphate-free buffer system.

Other Methods

Some alternative methods of detecting the product orthophosphate also may be used (see 3.1.3.1 — ALP, Methods 6 to 8; 3.1.3.2 — ACP, Methods 4 and 5).

References

1. Moro-Furlani, A.M., Turner, V.S., and Hopkinson, D.A., Genetical and biochemical studies on human phosphoserine phosphatase, *Ann. Hum. Genet.*, 43, 323, 1980.

REACTION	5′-Ribonucleotide + H_2O = ribonucleoside + orthophosphate
ENZYME SOURCE	Bacteria, fungi, plants, protozoa, invertebrates, vertebrates
SUBUNIT STRUCTURE	Tetramer[#] (fungi — yeast, vertebrates — membrane-bound form), heterodimer[#] (vertebrates — black rockfish white muscle), monomer[#] (vertebrates — rat brain cytosol, human seminal plasma), dimer[#] (vertebrates — bull seminal plasma)

METHOD 1

Visualization Scheme

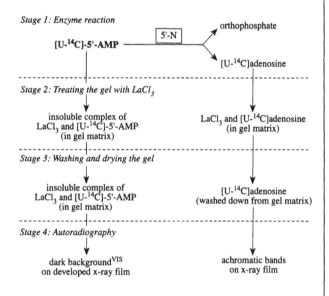

Reaction Mixture[1]

A. 56 mM Sodium phosphate buffer, pH 6.8
   1 μCi/ml [U-14C]-5′-AMP
B. 0.1 M Tris–HCl buffer, pH 7.0
   0.1 M LaCl₃

Procedure

Stage 1. Incubate the electrophorized gel in solution A at 37°C for 1 h.
Stage 2. Place the gel in solution B and incubate at 4°C for 6 h.
Stage 3. Wash the gel for 12 h in deionized water. Dry the gel.
Stage 4. Autoradiograph the dry gel with x-ray film (Kodak PR/R54) for 48 h. Develop the exposed x-ray film.

The white bands on a dark background of the developed x-ray film correspond to localization of 5′-N activity in the gel.

METHOD 2

Visualization Scheme

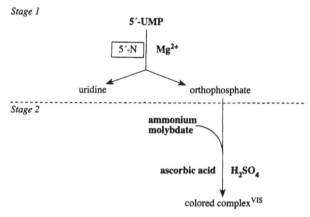

Staining Solution[2]

A. 0.1 M Tris–HCl buffer, pH 7.8 .... 10 ml
   5′-UMP .... 20 mg
   Glutathione (reduced form) .... 5 mg
   0.5 M MgSO₄ .... 2 ml
B. 2% Agar solution (60°C)
C. 2.5% Ammonium molybdate in 4 N H₂SO₄ .... 10 ml
   Ascorbic acid .... 1 g

Procedure

Mix solution A with an equal volume of solution B and pour the mixture over the gel surface. Incubate the gel at 37°C for 1 to 2 h. Remove the agar overlay containing the substrate. Mix solution C with an equal volume of solution B and pour the mixture over the gel surface. Dark blue bands appear at the sites of 5′-N activity. The bands are ephemeral; thus the zymogram should be recorded or photographed immediately.

Notes: The method can only be used when electrophoresis is carried out in a phosphate-free buffer system.

Malachite Green may be used coupled with ammonium molybdate to detect orthophosphate generated by 5′-N.[3] Unlike the ammonium molybdate method, the Malachite Green–ammonium molybdate method results in permanently stained bands due to the formation of a stable malachite–phosphomolybdate complex of sharp blue-green color. Zymograms obtained by this method may be preserved in 25% ethanol containing 2% glycerol (e.g., see 3.1.3.11 — FBP, Method 3).

OTHER METHODS

A. Several alternative methods for detecting the product orthophosphate are also available (see 3.1.3.1 — ALP, Methods 6 to 8; 3.1.3.2 — ACP, Methods 4 and 5).

B. An immunoblotting procedure (for details, see Part II) based on the utility of monoclonal antibodies specific to the rat enzyme[4] can be used for immunohistochemical visualization of the enzyme protein on electrophoretic gels.

C. Positive zymograms of 5′-N can be obtained using IMP as a substrate and two linking enzymes, purine-nucleoside phosphorylase and xanthine oxidase, coupled with the PMS–MTT system (for example, see 3.5.4.6 — AMPDA, Method 2).

D. Dark (nonfluorescent) bands of 5′-N visible in UV light on a light (fluorescent) background can be developed using AMP as a substrate and two coupled reactions catalyzed by auxiliary adenosine deaminase (3.5.4.4 — ADA, forward reaction) and glutamate dehydrogenase (1.4.1.2–4 — GDH, backward reaction). In this method 5′-N bands become visible due to NADH-to-NAD conversion. Negative zymograms of 5′-N visible in daylight may then be obtained by treating the fluorozymogram with the PMS–MTT mixture.

GENERAL NOTES

It is recommended that 0.75% amphoteric detergent (zwittergent-314) be included in the enzyme-containing preparations and electrophoretic gels. In the absence of the detergent the enzyme activity is very low.

Alkaline phosphatase (see 3.1.3.1 — ALP) can also act on some 5′-ribonucleotides (e.g., 5′-AMP). Thus, an additional gel should be stained for ALP activity for comparison using substrates other than 5′-ribonucleotides.

The enzyme displays wide specificity for 5′-nucleotides, including AMP, CMP, GMP, IMP, UMP, and the corresponding 2′-deoxynucleotide 5′-monophosphates. Many 5′-N isozymes are distinguishable by substrate and tissue specificity.[5] It is therefore stressed that this enzyme system deserves a thorough comparative study. In poeciliid fish, an isozyme originally referred to as an acid phosphatase appears to function physiologically as 5′-N, accepting several substrates.[5,6]

Two 5′-N isozymes (UMPH-1 and UMPH-2) are detected in rodents[7] and man.[8,9] One of these, UMPH-1, appears to be specific for pyrimidine 5′-nucleotides, while the other, UMPH-2, demonstrates a broader substrate specificity. For example, 4-methylumbelliferyl phosphate and β-naphthyl phosphate are hydrolyzed by the UMPH-2 but not by the UMPH-1 isozyme. Thus, activity bands of UMPH-2 are also developed on phosphatase zymograms.[7,8]

REFERENCES

1. Tucker-Pian, C., Bakay, B., and Nyhan, W.L., 5′-Nucleotidase: solubilization, radiochemical analysis, and electrophoresis, *Biochem. Genet.*, 17, 995, 1979.

2. Anderson, J.E., Teng, Y.-S., and Giblett, E.R., Stains for six enzymes potentially applicable to chromosomal assignment by cell hybridization, *Cytogenet. Cell Genet.*, 14, 465, 1975.

3. Queiroz-Claret, C. and Meunier, J.-C., Staining technique for phosphatases in polyacrylamide gels, *Anal. Biochem.*, 209, 228, 1993.

4. Bailyes, E.M., Newby, A.C., Siddle, K., and Luzio, J.P., Solubilization and purification of rat liver 5′-nucleotidase by use of zwitterionic detergent and monoclonal antibody immunoadsorbent, *Biochem. J.*, 203, 245, 1982.

5. Morizot, D.C. and Schmidt, M.E., Starch gel electrophoresis and histochemical visualization of proteins, in *Electrophoretic and Isoelectric Focusing Techniques in Fisheries Management*, Whitmore, D.H., Ed., CRC Press, Boca Raton, FL, 1990, p. 23.

6. Morizot, D.C. and Siciliano, M.J., Gene mapping in fishes and other vertebrates, in *Evolutionary Genetics of Fishes*, Turner, B.J., Ed., Plenum, New York, 1984, p. 173.

7. Swallow, D.M., Turner, V.S., and Hopkinson, D.A., Isozymes of rodent 5′-nucleotidase: evidence for two independent structural loci *Umph-1* and *Umph-2*, *Ann. Hum. Genet.*, 47, 9, 1983.

8. Wilson, D.E., Swallow, D.M., and Powey, S., Assignment of the human gene for uridine 5′-monophosphate phosphohydrolase (*UMPH2*) to the long arm of chromosome 17, *Ann. Hum. Genet.*, 50, 223, 1986.

9. Manco, L. and Amorim, A., Human erythrocyte pyrimidine 5′-nucleotidase isozymes: effect of sulfhydryl reagents and electrophoretic discrimination, *Electrophoresis*, 14, 1084, 1993.

REACTION	3'-Ribonucleotide + H$_2$O = ribonucleoside + orthophosphate
ENZYME SOURCE	Plants, protozoa
SUBUNIT STRUCTURE	Unknown or no data available

METHOD

Visualization Scheme

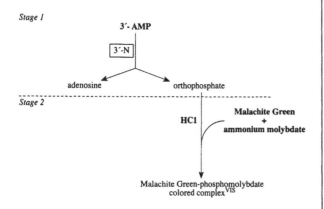

Stage 1

3'-AMP

3'-N

adenosine   orthophosphate

Stage 2

HCl   Malachite Green + ammonium molybdate

Malachite Green-phosphomolybdate colored complex^{VIS}

Staining Solution[1]

A. 100 mM Tris–HCl buffer, pH 8.5
    2.5 mM 3'-AMP
    0.05% CHAPSO (Sigma; a nondenaturing zwitterionic detergent)

B. 4.2% Ammonium molybdate in 4 N HCl .... 30 ml
    0.045% Malachite Green (oxalate salt) .... 90 ml
    2% Sterox (a detergent diluent) .... 2.4 ml

C. Solution B .... 60 ml
    H$_2$O .... 40 ml

Procedure

Incubate the gel in solution A (100 ml of solution A for 20 ml of gel volume) for 30 min at 37°C. Following incubation, rinse the gel quickly with water (usually 2 × 100 ml) to remove substrate and excess orthophosphate from the surface of the gel, and place the gel in solution C (see *Notes*). Color development is complete in 20 to 30 min. Rinse the stained gel with several changes of water and store in 5% acetic acid containing 20% ethanol.

Notes: Prepare stock solution B as follows. Mix 30 ml of ammonium molybdate solution and 90 ml of Malachite Green solution for 20 min at room temperature. Then pass the resulting mixture through a Whatman No. 5 filter and add 2.4 ml of sterox solution. This stock solution may be prepared in advance and stored at 4°C, at which it is stable for about a week. Solution C is prepared using stock solution B and water immediately prior to use. Malachite Green hydrochloride can be used instead of Malachite Green oxalate salt.

The method is ideal for use in the assay of detergent-solubilized membrane-associated 3'-N activity.

The staining procedure results in a stabler colored product than does the routine acid molybdate method.

The method can only be used with phosphate-free buffer systems. The use of citrate buffers prevents color development.

OTHER METHODS

A. Several alternative methods of detecting the product orthophosphate are also available (see 3.1.3.1 — ALP, Methods 6 to 8; 3.1.3.2 — ACP, Method 5).

B. Both products, adenosine and orthophosphate, can be detected using linked enzymatic reactions catalyzed by auxiliary enzymes adenosine deaminase, purine-nucleoside phosphorylase, and xanthine oxidase (e.g., see 3.5.4.4 — ADA, Method 1).

C. The product adenosine can be detected using two linked reactions catalyzed by auxiliary enzymes adenosine deaminase and glutamate dehydrogenase (see 3.5.4.4 — ADA, Method 2).

GENERAL NOTES

The enzyme demonstrates wide specificity for 3'-nucleotides.[2]

REFERENCES

1. Zlotnick, G.W. and Gottlieb, M., A sensitive staining technique for the detection of phosphohydrolase activities after polyacrylamide gel electrophoresis, *Anal. Biochem.*, 153, 121, 1986.

2. NC-IUBMB, *Enzyme Nomenclature*, Academic Press, San Diego, 1992, p. 318 (EC 3.1.3.6, Comments).

3.1.3.9 — Glucose-6-Phosphatase; G-6-PH

REACTION   D-Glucose-6-phosphate + H_2O = D-glucose + orthophosphate

ENZYME SOURCE   Invertebrates, vertebrates

SUBUNIT STRUCTURE   Dimer[a] (vertebrates; see General Notes)

Method

Visualization Scheme

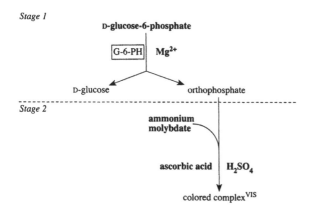

Stage 1

Stage 2

Staining Solution[1]

A.   0.1 M Tris–HCl buffer, pH 7.8   25 ml
     20 mM D-Glucose-6-phosphate   25 ml
     0.2 M $MgCl_2$ 2.5 ml
B.   2% Agar solution (60°C)
C.   2.5% Ammonium molybdate in 4 N H_2SO_4 50 ml
     Ascorbic acid   5 g

Procedure

Mix solution A with an equal volume of solution B and pour the mixture over the gel surface. Incubate the gel at 37°C for 30 min. Remove the first agar overlay. Mix solution C with an equal volume of solution B and pour the mixture over the gel surface. Dark blue bands of G-6-PH activity appear almost immediately. The bands are ephemeral; thus the zymogram should be recorded or photographed immediately.

Notes: The method can only be used with phosphate-free electrophoretic and staining buffers.

$MgCl_2$ may be omitted from the staining solution because the enzyme does not require any divalent ions for activity.

Other Methods

A.   Several alternative methods of detecting the product orthophosphate are also available (see 3.1.3.1 — ALP, Methods 6 to 8; 3.1.3.2 — ACP, Methods 4 and 5).

B.   The product glucose may be detected enzymatically using the linking enzyme glucose oxidase coupled with the MTT–PMS system or using two linking enzymes, glucose oxidase and peroxidase (see 1.1.3.4 — GO, Method 1 and Other Methods).

General Notes

The enzyme also catalyzes potent transphosphorylations from carbamoyl phosphate, hexose phosphates, pyrophosphate, phosphoenolpyruvate, and nucleoside di- and triphosphates, to D-glucose, D-mannose, 3-methyl-D-glucose, and 2-deoxy-D-glucose.[2] It may be identical to EC 3.1.3.1, EC 3.6.1.1, EC 3.9.1.1, EC 2.7.1.62, and EC 2.7.1.79.

The rat liver microsomal G-6-PH is suggested to be a monomer. No definitive evidence has been obtained for the assembly of microsomal monomers into dimers under functioning conditions.[3]

References

1.   Fisher, R.A., Turner, B.M., Dorkin, H.L., and Harris, H., Studies on human erythrocyte inorganic pyrophosphatase, *Ann. Hum. Genet.*, 37, 341, 1974.
2.   NC-IUBMB, *Enzyme Nomenclature*, Academic Press, San Diego, 1992, p. 319 (EC 3.1.3.9, Comments).
3.   Mithieux, G., Ajzannay, A., and Minassian, C., Identification of membrane-bound phosphoglucomutase and glucose-6-phosphatase by P[32]-labeling of rat-liver microsomal membrane-proteins with P[32] glucose-6-phosphate, *J. Biochem.*, 117, 908, 1995.

3.1.3.11 — Fructose-Bisphosphatase; FBP

OTHER NAMES — Hexosediphosphatase, fructose-1,6-diphosphatase

REACTION — D-Fructose-1,6-bisphosphate + H_2O = D-fructose-6-phosphate + orthophosphate

ENZYME SOURCE — Bacteria, fungi, plants, invertebrates, vertebrates

SUBUNIT STRUCTURE — Monomer (invertebrates), dimer (invertebrates), tetramer (vertebrates)

Method 1

Visualization Scheme

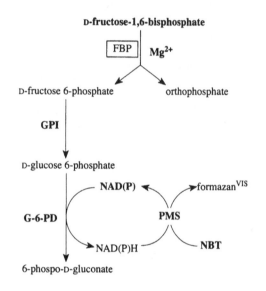

Staining Solution[1]

A.	0.1 M Tris–HCl buffer, pH 7.5	25 ml
	NADP	8 mg
	$MgSO_4 \cdot 7 H_2O$	50 mg
	2-Mercaptoethanol	1 μl
	PMS	1 mg
	NBT	10 mg
	D-Fructose-1,6-bisphosphate (sodium salt)	10 mg
	Glucose-6-phosphate isomerase (GPI)	20 U
	Glucose-6-phosphate dehydrogenase (G-6-PD)	20 U
B.	2% Agar solution (60°C)	25 ml

Procedure

Mix A and B components of the staining solution and pour the mixture over the gel surface. Incubate the gel in the dark at 37°C until dark blue bands appear. Fix the stained gel in 25% ethanol.

Notes: An NAD-dependent form of the auxiliary enzyme G-6-PD is available (Sigma G 5760 and G 5885). If NAD-dependent G-6-PD is used, NAD should be substituted for NADP in the staining solution. This substitution is beneficial because NAD is about five times less expensive than NADP.

Method 2

Visualization Scheme

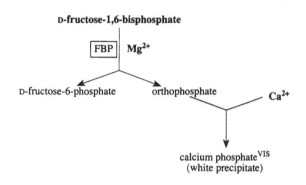

Staining Solution[2]

50 mM Glycine–KOH buffer, pH 10.0
10 mM Mg^{2+}
10 mM Ca^{2+}
10 mM KCl
5 mM D-Fructose-1,6-bisphosphate

Procedure

Soak an electrophorized PAG in 50 mM glycine–KOH buffer (pH 10.0) at 37°C for 20 to 30 min and incubate in the staining solution at 37°C. View the gel against a dark background and, when white bands of calcium phosphate precipitation of sufficient intensity are obtained, remove the gel from the staining solution and store in 50 mM glycine–KOH (pH 10.0), containing 5 mM Ca^{2+}, either at 5°C or at room temperature in the presence of an antibacterial agent.

Notes: The method may be used only when electrophoresis is carried out in a phosphate-free buffer system.

The precipitated calcium phosphate may be converted to brownish black lead sulfide (e.g., see 3.1.3.1 — ALP, Method 7) or counterstained with Alizarin Red S. These counterstains would be of advantage for opaque gel systems such as starch or acetate cellulose.

METHOD 3

Visualization Scheme

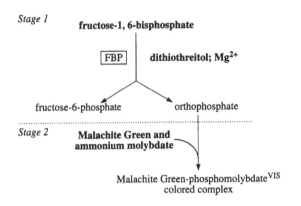

Stage 1   fructose-1, 6-bisphosphate

FBP   dithiothreitol; Mg^{2+}

fructose-6-phosphate   orthophosphate

Stage 2   **Malachite Green and ammonium molybdate**

Malachite Green-phosphomolybdate[VIS] colored complex

Staining Solution[3]

  A. 100 m*M* HEPES buffer, pH 8.1
      30 m*M* Dithiothreitol
      30 m*M* $MgCl_2$
  B. 100 m*M* HEPES buffer, pH 8.1
      30 m*M* Dithiothreitol
      30 m*M* $MgCl_2$
      1.2 m*M* Fructose-1,6-bisphosphate

C. H_2SO_4	60 ml
H_2O	300 ml
Malachite Green	440 mg
D. Solution C diluted four times with sulfuric acid:water	1:5
E. Solution D	10 ml
7.5% Ammonium molybdate	2.5 ml
11% Tween 20	0.2 ml

Procedure

After electrophoresis dip the gel in solution A for 10 min to activate the enzyme (see General Notes), and then incubate in 10 ml of solution B at room temperature with gentle agitation. Prepare solution C by slowly adding concentrated sulfuric acid to the deionized water. When the solution is at room temperature, add and dissolve Malachite Green. Prepare solution D just prior to use. Following 45 min of gel incubation, add 1.5 ml of solution E. Bands of blue-green color characteristic of the malachite–phosphomolybdate complex develop on the pale blue background of the gel within about 10 min. Fix the stained gel in 25% ethanol containing 2% glycerol.

Notes: Solution C may be stored several months at room temperature. The lower limit of orthophosphate detection by this method is in the order of 0.2 nmol.

    The reaction catalyzed by FBP in chloroplast stroma requires a reduced active state of the enzyme. The reduction of FBP *in vivo* occurs through the action of the ferredoxin–thioredoxin system. It can be obtained *in vitro* with dithiothreitol at pH 8.1 in the presence of Mg^{2+} ions. Under these conditions disulfide bridges are reduced into –SH groups and the enzyme becomes active. A reduction step (incubation of the gel in solution A) is therefore needed to shift chloroplast FBP into its active state.

OTHER METHODS

Some other methods of detecting the product orthophosphate are also available (see 3.1.3.1 — ALP, Methods 5 to 8).

REFERENCES

1. Shaw, C.R. and Prasad, R., Starch gel electrophoresis of enzymes: a compilation of recipes, *Biochem. Genet.*, 4, 297, 1970.
2. Nimmo, H.G. and Nimmo, G.A., A general method for the localization of enzymes that produce phosphate, pyrophosphate, or CO_2 after polyacrylamide gel electrophoresis, *Anal. Biochem.*, 121, 17, 1982.
3. Queiroz-Claret, C. and Meunier, J.-C., Staining technique for phosphatases in polyacrylamide gels, *Anal. Biochem.*, 209, 228, 1993.

3.1.3.13 — Bisphosphoglycerate Phosphatase; BPGP

REACTION   2,3-Bisphospho-D-glycerate + H_2O = 3-phospho-D-glycerate + orthophosphate (see also General Notes)

ENZYME SOURCE   Fungi, vertebrates

SUBUNIT STRUCTURE   Unknown or no data available

Method 1

Visualization Scheme

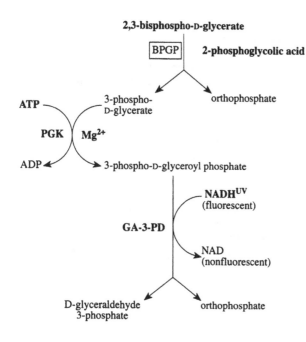

Staining Solution[1]

50 mM Triethanolamine–HCl buffer, pH 7.5

2 mM 2,3-Bisphospho-D-glycerate

3 mM ATP

1 mM MgCl₂

1 mM 2-Phosphoglycolic acid

0.25 mM NADH

1.6 U/ml Glyceraldehyde-3-phosphate dehydrogenase (GA-3-PD)

1.6 U/ml Phosphoglycerate kinase (PGK)

Procedure

Apply the staining solution to the gel surface on a filter paper overlay and incubate the gel at 37°C. Inspect the gel under long-wave UV light for dark (nonfluorescent) bands. Record the zymogram or photograph using a yellow filter.

Notes: When a zymogram that is visible in daylight is required, counterstain the processed gel with the PMS–MTT solution. Achromatic bands of BPGP appear on a blue background of the gel almost immediately.

Method 2

Visualization Scheme

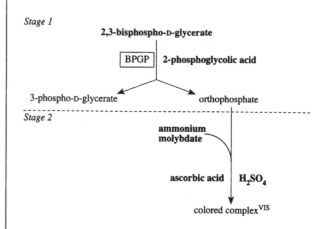

Staining Solution[2]

A. 0.2 M Histidine–HCl buffer, pH 6.5
     1 mM 2,3-Bisphospho-D-glycerate
     0.12 mM 2-Phosphoglycolate

B. 1.5% Agar solution (55°C)

C. 2% Agar solution (55°C)

D. 1.25% Ammonium molybdate in 1 M H_2SO_4 containing 0.15 M Ascorbic acid

Procedure

Mix equal volumes of A and B solutions, pour the mixture over the gel surface, and incubate at 37°C for 2 h. Remove the first agar overlay, mix equal volumes of solutions C and D, and pour the mixture over the gel surface. Dark blue bands appear within 20 min.

Notes: It should be taken into account that phosphoglycolate phosphatase (see 3.1.3.18 — PGP, Method 1) activity bands can also be developed by this method. Thus, control staining of an additional gel for PGP should be made. This problem can be avoided if a low concentration of 2-phosphoglycolic acid is used in the staining solution.

Other Methods

Several other methods may be used to detect the product orthophosphate (see 3.1.3.1 — ALP, Methods 6 to 8; 3.1.3.2 — ACP, Methods 4 and 5). All these methods may be used only when electrophoresis is carried out in a phosphate-free buffer system.

3.1.3.13 — Bisphosphoglycerate Phosphatase; BPGP (continued)

GENERAL NOTES

Comparative electrophoretic studies show that BPGP, phosphoglycerate mutase (see 5.4.2.1 — PGLM), and bisphosphoglycerate mutase (see 5.4.2.4 — BPGM) activities, at least in mammalian red cells, are determined by a single protein.[3]

2-Phosphoglycolate is used in both methods to activate the enzyme.

REFERENCES

1. Rosa, R., Gaillardon, J., and Rosa, J., Characterization of 2,3-diphosphoglycerate phosphatase activity: electrophoretic study, *Biochim. Biophys. Acta*, 293, 285, 1973.
2. Scott, E.M. and Wright, R.C., An alternate method for demonstration of bisphosphoglyceromutase (DPGM) on starch gels, *Am. J. Hum. Genet.*, 34, 1013, 1982.
3. Rosa, R., Audit, I., and Rosa, J., Evidence for three enzymatic activities in one electrophoretic band of 3-phosphoglycerate mutase from red cells, *Biochimie*, 57, 1059, 1975.

3.1.3.16 — Phosphoprotein Phosphatase; PPP

OTHER NAMES	Protein phosphatase-1, protein phosphatase-2A, protein phosphatase-2B, protein phosphatase-2C
REACTION	Phosphoprotein + H_2O = protein + orthophosphate
ENZYME SOURCE	Fungi, plants, invertebrates, vertebrates
SUBUNIT STRUCTURE	Heterotrimer[#] (eukaryotes); see General Notes

METHOD

Visualization Scheme

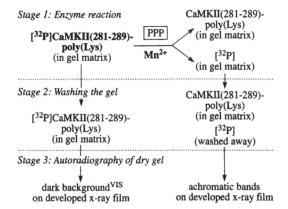

Reaction Mixture[1]

A. The substrate, CaMKII(281–289)-poly(Lys) peptide conjugate, obtained and phosphorylated with [^{32}P] as described below, is incorporated into the SDS-PAG mixture (2×10^5 cpm/ml) just prior to polymerization.

The [^{32}P]-labeled CaMKII(281–289)-poly(Lys) peptide conjugate is not commercially available and should be prepared as described by authors.[1] Briefly, CaMKII(281–289) peptide corresponds to the autophosphorylation site, MHRQETVDC, of calmodulin-dependent protein kinase II. It is synthesized using automated peptide synthesizer. Purified MHRQETVDC polypeptide is then linked to poly(Lys) (M_r 87,000; Sigma) by a heterobifunctional reagent, *N*-(ε-maleimidocaproyloxy)succinimide (Dojindo Laboratories). The resultant peptide conjugate is phosphorylated using [γ-^{32}P]ATP (Amersham) and an active 30-kDa proteolytic fragment of calmodulin-dependent protein kinase II.[2]

B. 20% Isopropanol in 50 m*M* Tris–HCl, pH 7.0

C. 50 m*M* Tris–HCl, pH 7.0, containing 5 m*M* 2-mercaptoethanol

D. 6 *M* Guanidine HCl (or 8 *M* urea)
   20 m*M* 2-Mercaptoethanol
   50 m*M* Tris–HCl, pH 7.0

E. 50 m*M* Tris–HCl, pH 7.0
   0.02% Tween 20
   20 m*M* 2-Mercaptoethanol
   1 m*M* $MnCl_2$

Procedure

To remove the SDS and to renaturate PPP, subject the SDS-PAG to a preliminary washing with two changes of 100 ml each of solution B (see the reaction mixture) for 60 min. Then rinse the gel with 200 ml of solution C for 10 min and treat with two changes of 50 ml of solution D for 60 min. The complete denaturation of the enzyme with 6 M guanidine HCl (or 8 M urea) is necessary for efficient renaturation of PPP. When such treatment is omitted, the intensity of activity bands of PPP revealed on the developed x-ray film becomes very weak.

Stages 1 and 2. Remove detergent (guanidine HCl or urea) quickly by vigorously shaking the gel in 100 ml of the renaturation solution E (with subsequent changes) for 10, 20, 30, 60, and 120 min. Such a treatment is sufficient for complete renaturation of the enzyme, the hydrolysis of [^{32}P], and the washing of it away from the gel matrix.

Stage 3. Dry the gel with a gel drier and expose it to Fuji x-ray film for several hours.

The white bands on the developed x-ray film correspond to localization of PPP activity in the gel.

Notes: Two other synthetic peptide conjugates, CPLARTLSVA-GLPLKK-poly(Lys) and CSQPSFQWRQPSLDVDVGD-poly(Lys), were used, but the results they give are not as good as those obtained with MHRQETVDC-poly(Lys).

OTHER METHODS

The immunoblotting procedure (for details, see Part II) may also be used to detect PPP. Monoclonal antibodies specific to the rabbit enzyme and antibodies reactive with the human enzyme are now available from Sigma.

GENERAL NOTES

Phosphoprotein phosphatase represents a group of enzymes removing the serine- or threonine-bound phosphate group from a wide range of phosphoproteins, including some enzymes phosphorylated by the action of a protein kinase. See also 3.1.3.48 — PTP.

The enzyme is a heterotrimer consisting of highly conserved structural (A) and catalytic (C) subunits. Diverse functions of PPP in the cell are determined by association of A and C subunits with a highly variable regulatory and targeting subunit (B) encoded by at least three different genes.[3]

The C subunit is reversibly methyl esterified by specific enzymes at a completely conserved C-terminal leucine residue. Methylation plays an essential role in promoting assembly of PPP, while demethylation has an opposing effect. Changes in methylation indirectly regulate PPP activity by controlling the binding of multiple regulatory B subunits to AC dimers.[4]

REFERENCES

1. Kameshita, I., Ishida, A., Okuno, S., and Fujisawa, H., Detection of protein phosphatase activities in sodium dodecyl sulfate–polyacrylamide gel using peptide substrates, *Anal. Biochem.*, 245, 149, 1997.
2. Ishida, A., Kitani, T., Okuno, S., and Fujisawa, H., Inactivation of Ca^{2+}/calmodulin-dependent protein kinase II by Ca^{2+}/calmodulin, *J. Biol. Chem.*, 115, 1075, 1994.
3. Li, X.H. and Virshup, D.M., Two conserved domains in regulatory B subunits mediate binding to the A subunit of protein phosphatase 2A, *Eur. J. Biochem.*, 269, 546, 2002.
4. Tolstykh, T., Lee, J., Vafai, S., and Stock, J.J.B., Carboxyl methylation regulates phosphoprotein phosphatase 2A by controlling the association of regulatory B subunits, *EMBO J.*, 19, 5682, 2000.

3.1.3.18 — Phosphoglycolate Phosphatase; PGP

REACTION	2-Phosphoglycolate + H_2O = glycolate + orthophosphate
ENZYME SOURCE	Plants, invertebrates, vertebrates
SUBUNIT STRUCTURE	Monomer (invertebrates), dimer (vertebrates)

METHOD 1

Visualization Scheme

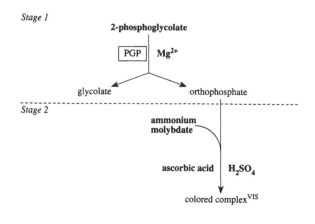

Staining Solution[1]

A. 0.1 M Tris–HCl buffer, pH 7.5   25 ml
    2-Phosphoglycolic acid   50 mg
    $MgSO_4\cdot7H_2O$ 10 mg
B. 2% Agar solution (60°C)
C. 2.5% Ammonium molybdate in 4 N H_2SO_4 25 ml
    Ascorbic acid   1.25 g

Procedure

Mix solution A with an equal volume of solution B and pour the mixture over the gel surface. Incubate the gel at 37°C for 1 to 2 h. Remove the agar overlay. Mix solution C with an equal volume of solution B and pour the mixture over the gel surface. Dark blue bands of PGP activity appear after a few minutes. Record the zymogram or photograph because the stain is ephemeral.

Notes: The method may be used only when electrophoresis is carried out in a phosphate-free buffer system.

METHOD 2

Visualization Scheme

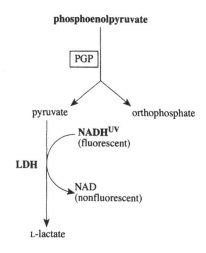

Staining Solution[1]

0.1 M Tris–HCl buffer, pH 7.5	5 ml
Phosphoenolpyruvate (potassium salt)	25 mg
NADH (disodium salt)	10 mg
2750 U/ml Lactate dehydrogenase (LDH)	50 μl

Procedure

Apply the staining solution to the gel surface on a filter paper overlay and monitor the gel under long-wave UV light. The PGP activity is seen as dark (nonfluorescent) bands on a light (fluorescent) background. When the bands are well developed, record the zymogram or photograph using a yellow filter.

Notes: This method is based on the phosphoenolpyruvate phosphatase activity of PGP from some sources (e.g., from human red cells).

OTHER METHODS

Several alternative methods of detection of the product orthophosphate are also available (see 3.1.3.1 — ALP, Methods 6 to 8; 3.1.3.2 — ACP, Methods 4 and 5). All these methods may be used only when electrophoresis is carried out in a phosphate-free buffer system.

GENERAL NOTES

PGP and alkaline phosphatase activities toward phosphoenolpyruvate can cause the development of additional bands on the zymograms of some other enzymes for which linked reactions catalyzed by pyruvate kinase and lactate dehydrogenase are used (for example, see 2.7.1.40 — PK, Method 1; 2.7.3.3 — ARGK, Method 2).

REFERENCES

1. Barker, R.F. and Hopkinson, D.A., Genetic polymorphism of human phosphoglycolate phosphatase (PGP), *Ann. Hum. Genet.*, 42, 143, 1978.

3.1.3.37 — Sedoheptulose-Bisphosphatase; SBP

REACTION   Sedoheptulose 1,7-bisphosphate +
                        H_2O = sedoheptulose 7-phosphate +
                        orthophosphate
ENZYME SOURCE   Plants
SUBUNIT STRUCTURE   Unknown or no data available

METHOD

Visualization Scheme

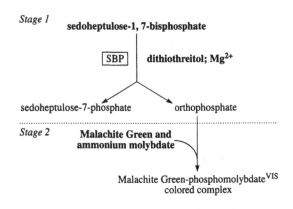

Stage 1   **sedoheptulose-1, 7-bisphosphate**

SBP   **dithiothreitol; Mg^{2+}**

sedoheptulose-7-phosphate   orthophosphate

Stage 2   **Malachite Green and
            ammonium molybdate**

Malachite Green-phosphomolybdate^{VIS}
colored complex

Staining Solution[1]

A. 100 mM HEPES buffer, pH 8.1
    30 mM Dithiothreitol
    30 mM $MgCl_2$
B. 100 mM HEPES buffer, pH 8.1
    30 mM Dithiothreitol
    30 mM $MgCl_2$
    1.2 mM Sedoheptulose 1,7-bisphosphate
C. H_2SO_4 60 ml
    H_2O 300 ml
    Malachite Green   440 mg
D. Solution C diluted four times with
       sulfuric acid:water 1:5
E. Solution D   10 ml
    7.5% Ammonium molybdate   2.5 ml
    11% Tween 20   0.2 ml

Procedure

After electrophoresis dip the gel in solution A for 10 min to activate the enzyme (see General Notes), and then incubate in 10 ml of solution B at room temperature with gentle agitation. Prepare solution C by slowly adding concentrated sulfuric acid to the deionized water. When the solution is at room temperature, add and dissolve Malachite Green. Prepare solution D just prior to use. Following 45 min of gel incubation, add 1.5 ml of solution E. Bands of blue-green color characteristic of the malachite–phosphomolybdate complex develop on the pale blue background of the gel within about 10 min. Fix the stained gel in 25% ethanol containing 2% glycerol.

Notes: Solution C may be stored several months at room temperature. The lower limit of orthophosphate detection by this method is in the order of 0.2 nmol.

GENERAL NOTES

The reaction catalyzed by SBP requires a reduced active state of the enzyme. The reduction of SBP *in vivo* occurs through the action of the ferredoxin–thioredoxin system. It can be obtained *in vitro* with dithiothreitol at pH 8.1 in the presence of Mg^{2+} ions. Under these conditions disulfide bridges are reduced into –SH groups and the enzyme becomes active. A reduction step (incubation of the gel in solution A) is therefore needed to shift the enzyme into its active state.

REFERENCES

1. Queiroz-Claret, C. and Meunier, J.-C., Staining technique for phosphatases in polyacrylamide gels, *Anal. Biochem.*, 209, 228, 1993.

OTHER NAMES	Phosphotyrosine phosphatase
REACTION	Protein tyrosine phosphate + H_2O = protein tyrosine + orthophosphate
ENZYME SOURCE	Fungi, plants, invertebrates, vertebrates
SUBUNIT STRUCTURE	Monomer# (vertebrates)

METHOD

Visualization Scheme

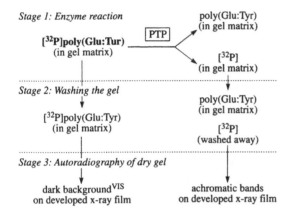

Stage 1: Enzyme reaction

[32P]poly(Glu:Tur) — PTP → poly(Glu:Tyr) (in gel matrix)
(in gel matrix) → [32P] (in gel matrix)

Stage 2: Washing the gel

[32P]poly(Glu:Tyr) → poly(Glu:Tyr) (in gel matrix)
(in gel matrix) → [32P] (washed away)

Stage 3: Autoradiography of dry gel

dark background VIS on developed x-ray film — achromatic bands on developed x-ray film

Reaction Mixture[1]

A. Substrate, poly(Glu:Tyr) (Sigma, catalog no. PO275; mol wt 20,000–50,000), phosphorylated with [32P] as described (see *Notes*), is incorporated into the SDS-PAG mixture prior to polymerization at approximately 10^5 cpm/ml

B. 50 mM Tris–HCl, pH 8.0, containing 20% isopropanol

C. 50 mM Tris–HCl, pH 8.0, containing 0.3% 2-mercapto-ethanol

D. 50 mM Tris–HCl, pH 8.0
6 M Guanidine HCl
1 mM EDTA

E. 50 mM Tris–HCl, pH 8.0
0.04% Tween 20
0.3% 2-Mercaptoethanol
1 mM EDTA

Procedure

To remove the SDS and to renaturate PTP, subject an SDS-PAG containing the [32P]-labeled substrate (see A in the reaction mixture) to a preliminary washing with 250 ml of solution B (see the reaction mixture) for either 1.5 h or overnight at room temperature. Then rinse the gel with 250 ml of solution C two times for 30 min each and treat with 250 ml of solution D for 90 min at room temperature. Such a treatment is sufficient for complete renaturation of the enzyme, the hydrolysis of [32P], and the washing of it away from the gel matrix.

Stages 1 and 2. Incubate the gel at room temperature in 250 ml of renaturation solution E three times (for 1 h each) with subsequent changes. Dry the gel.

Stage 3. Expose the dry gel to Xomat x-ray film (Kodak) with an intensifying screen at room temperature or –90°C for 24 h, or longer as required.

The white bands on the developed x-ray film correspond to localization of PTP activity in the gel.

Notes: The [32P]-labeled poly(Glu:Tyr) is not commercially available and should be prepared at laboratory conditions. Briefly, incubate 1 mg of poly(Glu:Tyr) with 50 µl of a 20% suspension of a recombinant tyrosine kinase catalytic domain, immobilized on agarose beads in 0.5 ml of kinase buffer (30 mM MgCl$_2$, 1 mM MnCl$_2$, 1 mM sodium orthovanadate, 1 mM ATP, 10 mM dithiothreitol, 0.05% Triton X-100, 50 mM imidazole (pH 7.2) to which 200 to 500 µCi of [γ-32P]ATP (3000 Ci/mmol) (New England Nuclear) has been added. Allow the kinase reaction to proceed with rotation at room temperature for 18 h. Terminate the reaction by sedimentation of the agarose beads and addition of an equal volume of 20% trichloroacetic acid. After 30 min on ice, sediment the precipitated poly(Glu:Tyr) at 12,000 g for 10 min at 4°C. Dissolve the pellet in 100 µl of 2 M Tris base, pass over a 15 × 0.7 cm G50 Sephadex column (Pharmacia) equilibrated in 50 mM imidazole (pH 7.2), and collect in 0.8-ml fractions. Measure the incorporation of 32P by scintillation counting. Specific activities of between 0.7 and 3×10^8 cpm/mg (80 to 130 nmol/mg phosphate) poly(Glu:Tyr) should be obtained.[1]

Using this method, PTP was detected at a level as low as 10 pg of enzyme protein loaded in a gel line.[1]

Significant renaturation of PTP activity was also detected when the guanidine step (treating the gel with solution D) was omitted. However, for many PTP isozymes optimal renaturation has been observed when this step was included.[1]

The method is specific for PTP. Activity bands of serine/threonine phosphoprotein phosphatase (see 3.1.3.16 — PPP) are not detected by this method.[1]

Substituting 33P for 32P should enhance the clarity of the bands of PTP activity detected by the method described above due to the lower energy of the β particles emitted by 33P.[1]

Epidermal growth factor receptor-kinase, c-src kinase, and focal adhesion kinase phosphorylated on tyrosine with 32PO$_4$ were incorporated into SDS-PAGs and tested as substrates for the in-gel detection of PTP in crude cell lysates obtained from different human cells. The revealed banding patterns of PTP proved similar for different cell types and for different substrates.[2]

It was found that the composition of the lysing buffer used to prepare PTP-containing extracts greatly influenced the activity and electrophoretic mobility of PTP from human acute promyelocytic leukemic cells. The best lysing buffer was 30 mM Tris–HCl (pH 6.8) containing

319

3.1.3.48 — Protein-Tyrosine-Phosphatase; PTP (continued)

150 mM NaCl, 0.5% Na-deoxycholate, 1% Nonidet P-40, and 0.1% SDS.[3]

Some other phosphotyrosyl proteins can also be used as PTP substrates (e.g., reduced carboxamidomethylated and maleylated lysozymes). However, phosphoseryl proteins (e.g., histone II phosphorylated by the cAMP-dependent protein kinase and phosphorylase b phosphorylated by the phosphorylase kinase) are not dephosphorylated by PTP from human platelet membranes.[4]

OTHER METHODS

The immunoblotting procedure (for details, see Part II) may also be used to detect PTP. Monoclonal antibodies reactive with the enzyme from the human, bovine, rat, and mouse, as well as monoclonal antibodies specific to the enzyme from human T cells, are now available from Sigma.

REFERENCES

1. Burridge, K. and Nelson, A., An in-gel assay for protein tyrosine phosphatase activity: detection of widespread distribution in cells and tissues, *Anal. Biochem.*, 232, 56, 1995.
2. Gates, R.E., Miller, J.L., and King, L.E., Jr., Activity and molecular weight of protein tyrosine phosphatases in cell lysates determined by renaturation after gel electrophoresis, *Anal. Biochem.*, 237, 208, 1996.
3. Calvert-Evers, J. and Hammond, K., The influence of lysis buffer composition on the expression and activity of protein tyrosine phosphatase, *Electrophoresis*, 21, 2944, 2000.
4. Dawicki, D.D. and Steiner, M., Identification of a protein-tyrosine phosphatase from human platelet membranes by an immobilon-based solid phase assay, *Anal. Biochem.*, 213, 245, 1993.

3.1.4.1 — Phosphodiesterase I; PDE-I

OTHER NAMES	5′-Exonuclease
REACTION	Hydrolytically removes 5′-nucleotides successively from the 3′-hydroxy termini of 3′-hydroxy-terminated oligonucleotides
ENZYME SOURCE	Bacteria, plants, protozoa, invertebrates, vertebrates
SUBUNIT STRUCTURE	Octamer[*] (plants — soybean leaves)

METHOD 1

Visualization Scheme

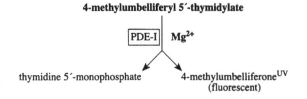

Staining Solution[1]

A. 50 mM Tris–HCl buffer, pH 9.0
   1 mg/ml 4-Methylumbelliferyl 5′-thymidylate
   1 mM MgCl$_2$
B. 1.8% Agarose solution (55°C)

Procedure

Mix equal volumes of A and B solutions and pour the mixture over the gel surface. Incubate the gel at 37°C and monitor under long-wave UV light. The PDE-I activity is seen as fluorescent bands. Record the zymogram or photograph using a yellow filter.

Notes: Another fluorogenic substrate, 4-methylumbelliferyl phenylphosphonate, also may be used for detection of PDE-I activity on electrophoretic gels.[2]

METHOD 2

Visualization Scheme

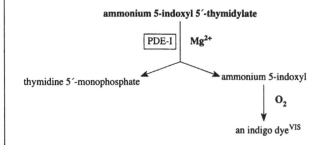

Staining Solution[1]

50 mM Tris–HCl buffer, pH 9.0
0.25 mg/ml Ammonium 5-indoxyl 5'-thymidylate
1 mM MgCl$_2$

Procedure

Incubate the gel in staining solution at 37°C. Colored bands at the sites of PDE-I activity appear after about 8 h of incubation.

Notes: Control staining for alkaline phosphatase (3.1.3.1 — ALP) should be made because some ALP isozymes also can hydrolyze ammonium 5-indoxyl 5'-thymidylate.

Method 3

Visualization Scheme

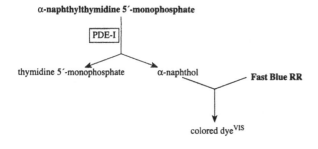

Staining Solution[3]

50 mM Tris–HCl buffer, pH 7.5	100 ml
Thymidine 5'-monophosphate α-naphthyl ester	10 mg
Fast Blue RR	100 mg

Procedure

Incubate the gel in staining solution in the dark at 37°C until colored bands appear. Wash the stained gel with water and fix in 50% glycerol or 3% acetic acid.

Notes: Nonspecifically stained bands were observed on PDE-I zymograms obtained using this method after electrophoresis of wheat leaf preparations. These bands were not observed after heating of preparations to 50°C before electrophoresis.[3]

Method 4

Visualization Scheme

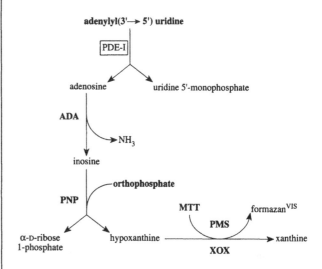

Staining Solution[4] (modified[5])

A.	10 mg/ml Adenylyl(3' → 5')uridine (in distilled water)	50 μl
	5 mg/ml PMS	200 μl
	5 mg/ml MTT	200 μl
	Purine-nucleoside phosphorylase (PNP; Boehringer)	10 μl
	Xanthine oxidase (XOX; Boehringer)	10 μl
	Adenosine deaminase (ADA; Sigma, type III)	20 μl
B.	1% Agarose solution in 50 mM sodium phosphate buffer, pH 8.0 (45°C)	9.5 ml

Procedure

Mix A and B components of the staining solution and pour the mixture over the gel surface. Incubate the gel in the dark at 37°C until dark blue bands appear. Fix the stained gel in 25% ethanol.

Notes: An additional gel should be stained in a staining solution lacking adenylyl(3' → 5')uridine to identify possible bands of "nothing dehydrogenase" (see 1.X.X.X — NDH).

Adenylyl(3' → 5')adenosine also may be used as a PDE-I substrate in this method. However, when it is used, bands of spleen exonuclease (see 3.1.16.1 — SE, Method 3) are also developed.

METHOD 5

Visualization Scheme

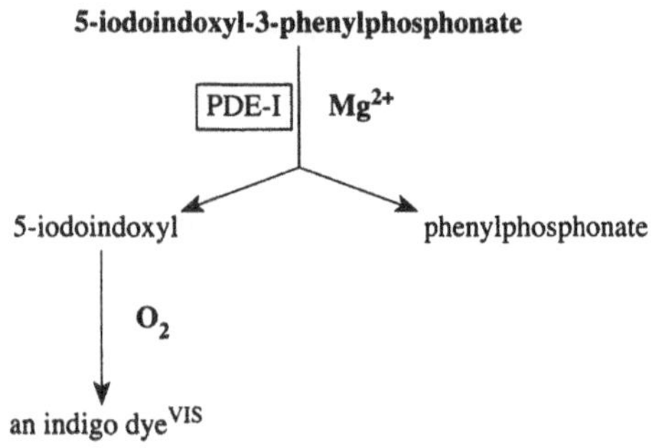

5-iodoindoxyl-3-phenylphosphonate

PDE-I | Mg²⁺

5-iodoindoxyl   phenylphosphonate

O₂

an indigo dye^VIS

Staining Solution[6]

50 mM Tris–HCl buffer, pH 9.0
1 mM MgCl₂
1 mg/ml 5-Iodoindoxyl-3-phenylphosphonate

Procedure

Incubate the gel in staining solution at 37°C until bluish-purplish bands appear (usually for about 10 h).

Notes: The substrate 5-iodoindoxyl-3-phenylphosphonate is specific for PDE-I and is not cleaved by alkaline phosphatase, as is the other chromogenic substrate, ammonium 5-indoxyl 5′-thymidylate (see above, Method 2, *Notes*).

OTHER METHODS

When adenylyl(3′ → 5′)uridine is used as a PDE-I substrate (see above, Method 4), the product adenosine also can be detected using two coupled reactions catalyzed by auxiliary enzymes adenosine deaminase and glutamate dehydrogenase (for details, see 3.5.4.4 — ADA, Method 2).

GENERAL NOTES

A 3′-phosphate terminus on the substrate inhibits hydrolysis.[7]

REFERENCES

1. Lo, K.W., Aoyagi, S., and Tsou, K.C., A fluorogenic method for the demonstration of human serum 5′-nucleotide phosphodiesterase isozymes after polyacrylamide gel electrophoresis, *Anal. Biochem.*, 117, 24, 1981.
2. Hawley, D.M., Crisp, M., and Hodes, M.E., The synthesis of 4-methylumbelliferyl phenylphosphonate and its use in an improved method for the zymogram analysis of phosphodiesterase I, *Anal. Biochem.*, 129, 522, 1983.
3. Wolf, G., Rimpau, J., and Lelley, T., Localization of structural and regulatory genes for phosphodiesterase in wheat (*Triticum aestivum*), *Genetics*, 86, 597, 1977.
4. Karn, R.C., Crisp, M., Yount, E.A., and Hodes, M.E., A positive zymogram method for ribonuclease, *Anal. Biochem.*, 96, 464, 1979.
5. Hodes, M.E. and Retz, J.E., A positive zymogram for distinguishing among RNase and phosphodiesterases I and II, *Anal. Biochem.*, 110, 150, 1981.
6. Gangyi, H., The synthesis of 5-iodoindoxyl-3-phenylphosphonate and its use in analysis of phosphodiesterase I, *Anal. Biochem.*, 185, 90, 1990.
7. NC-IUBMB, *Enzyme Nomenclature*, Academic Press, San Diego, 1992, p. 327 (EC 3.1.4.1, Comments).

3.1.4.3 — Phospholipase C; PLC

OTHER NAMES   Lipophosphodiesterase I, lecithinase C, *Clostridium welchii* α-toxin, *Clostridium oedematiens* β- and γ-toxins

REACTION   Phosphatidylcholine + H_2O = 1,2-diacylglycerol + choline phosphate

ENZYME SOURCE   Bacteria, protozoa, vertebrates

SUBUNIT STRUCTURE   Monomer[#] (bacteria – *Pseudomonas fluorescens*)

METHOD

Visualization Scheme

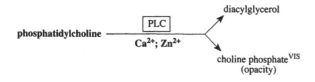

Staining Solution[1]

A. 20 mM Tris–HCl buffer, pH 7.4
0.15 M NaCl
1 mM CaCl$_2$
1 mM ZnCl$_2$

B. 3% Lecithin (or 20% egg yolk) emulsified in solution A using an ultrasonic disintegrator (amplitude of 8 μm for 1 to 2 min, 20°C)

C. 2% Agarose solution in solution A (56°C)

D. 1:1 (v/v) Mixture of solutions B and C

Procedure

Apply solution D to the PAG surface and incubate at 37°C in the humid chamber until opaque bands appear. Record the zymogram or photograph using Kodak photomicrography color film (PCF-36 2483).

Notes: The use of a decreased concentration of lecithin or egg yolk in the staining solution will result in opacity zones of decreased contrast.

OTHER METHODS

The immunoblotting procedure (for details, see Part II) may also be used to detect PLC. Antibodies reactive with the enzyme from the human, rat, and *Xenopus* are now available from Sigma.

REFERENCES

1. Smith, C.J. and Wadström, T., Isoelectric focusing in thin layer polyacrylamide gel combined with a zymogram method for detecting enzyme microheterogeneity: sample application, *Anal. Biochem.*, 65, 137, 1975.

3.1.4.12 — Sphingomyelin Phosphodiesterase; SPD

OTHER NAMES   Neutral sphingomyelinase, sphingomyelinase

REACTION   Sphingomyelin + H_2O = *N*-acyl-
sphingosine + choline phosphate

ENZYME SOURCE   Invertebrates, vertebrates

SUBUNIT STRUCTURE Unknown or no data available

METHOD

Visualization Scheme

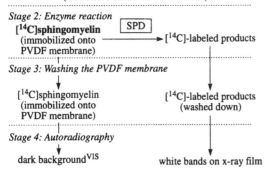

Stage 1: Blotting on PVDF membrane containing immobilized [^{14}C]sphingomyelin

SPD

(in gel matrix)

SPD

(transferred onto PVDF membrane)

Stage 2: Enzyme reaction

[^{14}C]**sphingomyelin**   SPD
(immobilized onto   ⟶ [^{14}C]-labeled products
PVDF membrane)

Stage 3: Washing the PVDF membrane

[^{14}C]sphingomyelin   [^{14}C]-labeled products
(immobilized onto   (washed down)
PVDF membrane)

Stage 4: Autoradiography

dark background^{VIS} white bands on x-ray film

Development by Autoradiography[1]

Stage 1. To prepare a substrate-coated polyvinylidene difluoride (PVDF) membrane, dip the membrane into a solution of [^{14}C]sphingomyelin (1 µCi/5 µmol) in
methanol and dry in air. Transfer the SPD molecules
from the electrophoretic gel to the PVDF membrane
via a routine blotting procedure.

Stages 2 and 3. Affix the gel to the PVDF membrane and incubate in 50 m*M* Tris–glycine buffer (pH 7.4) at 37°C
overnight.

Stage 4. Remove the PVDF membrane and expose to x-ray film
for 4 days to visualize the enzyme by autoradiography.

White bands on a dark background of the developed x-ray
film indicate areas of SPD localization in the electrophoretic gel.

GENERAL NOTES

The method using radiolabeled phospholipid substrates immobilized on a PVDF membrane may be adapted to detect activity
of some other phospholipases, such as phospholipase D or C.
Similarly, enzymes involved in the biosynthesis of phospholipids
(e.g., diacylglycerol, CDP-choline) and phosphocholine transferase may be detected through the incorporation of radiolabeled
choline onto a PVDF solid phase. This method may prove unsuitable for enzymes requiring high concentrations of detergents,
because under this condition phospholipid substrates may be
released from PVDF membranes. The method based on the use
of a substrate-immobilized PVDF membrane for the detection
of lipid-metabolizing enzymes was termed *Far Eastern blotting*.[2]

REFERENCES

1. Taki, T. and Chatterjee, S., An improved assay method for the
   measurement and detection of sphingomyelinase activity,
   Anal. Biochem., 224, 490, 1995.
2. Taki, T. and Ishikawa, D., TLC blotting: application to microscale analysis of lipids and as a new approach to lipid–protein
   interaction, *Anal. Biochem.*, 251, 135, 1997.

3.1.4.17 — 3',5'-Cyclic-Nucleotide Phosphodiesterase; CNPE

OTHER NAMES	3',5'-Cyclic AMP phosphodiesterase
REACTION	Nucleoside 3',5'-cyclic phosphate + H_2O = nucleoside 5'-phosphate
ENZYME SOURCE	Bacteria, fungi, invertebrates, vertebrates
SUBUNIT STRUCTURE	Monomer[a] (bacteria)

METHOD 1

Visualization Scheme

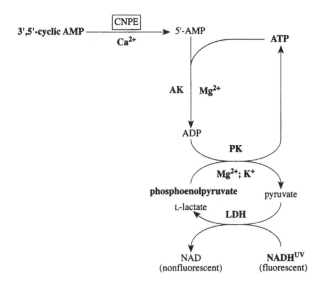

Staining Solution[1]

A. 0.1 M Tris–HCl buffer, pH 7.9	10 ml
3',5'-Cyclic AMP (monohydrate)	21 mg
$MgSO_4 \cdot 7H_2O$	40 mg
KCl	60 mg
$CaCl_2 \cdot 6H_2O$	15 mg
ATP (disodium salt; 3 H_2O)	7 mg
Phosphoenolpyruvate (potassium salt)	6 mg
NADH (disodium salt)	30 mg
Adenylate kinase (AK; 360 U/ml)	5 μl
Pyruvate kinase (PK; 400 U/ml)	5 μl
Lactate dehydrogenase (LDH; 2750 U/ml)	5 μl
B. 2% Agar solution (60°C)	5 ml

Procedure

Mix A and B components of the staining solution and pour the mixture over the gel surface. Incubate the gel at 37°C and monitor under long-wave UV light. The areas of CNPE activity are seen as dark (nonfluorescent) bands on a light (fluorescent) background. Record the zymogram or photograph using a yellow filter. When a zymogram that is visible in daylight is required, counterstain the processed gel with MTT–PMS solution to obtain white bands on a blue background.

Notes: The bands caused by phosphoenolpyruvate phosphatase activity of alkaline phosphatase (3.1.3.1 — ALP, General Notes) or phosphoglycolate phosphatase (see 3.1.3.18 — PGP, Method 2) can also be developed by this method.

METHOD 2

Visualization Scheme

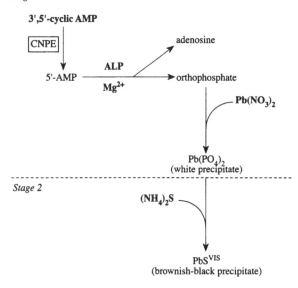

Staining Solution[2]

A. 0.32 mM Tris–maleate buffer, pH 7.0
    8 μM 3',5'-Cyclic AMP
    0.5 U/ml Alkaline phosphatase (ALP)
    12 mM $Pb(NO_3)_2$
    10 μM $MgSO_4$
B. 5% $(NH_4)_2S$

Procedure

Incubate the gel in solution A at 37°C for 2 h or until bands of white precipitation appear. Wash the gel in tap water for 1 h and place in solution B for 2 min. The black bands correspond to localization of CNPE activity in the gel. Wash the stained gel in 10 mM sodium acetate (pH 4.0) and store in 50% glycerol.

Notes: Solution A may be applied to the gel surface as a 1% agarose overlay. 3',5'-Cyclic GMP may be used as a CNPE substrate instead of 3',5'-cyclic AMP.[3] It should be kept in mind, however, that the bands of phosphodiesterase (EC 3.1.4.35), which are specific to 3',5'-cyclic GMP, can also be developed on CNPE zymograms obtained using cyclic GMP as a substrate.

Method 3

Visualization Scheme

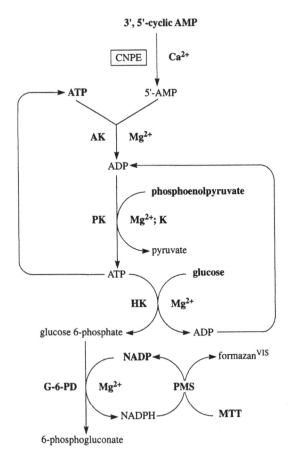

Staining Solution[4] (adapted)

A. 50 mM Tris–HCl buffer, pH 8.0
    100 mM 3′,5′-Cyclic AMP
    40 nM ATP
    1 U/ml Adenylate kinase (AK)
    2 U/ml Pyruvate kinase (PK)
    2 U/ml Hexokinase (HK)
    0.5 U/ml Glucose-6-phosphate dehydrogenase (G-6-PD)
    20 mM Phosphoenolpyruvate
    6 mg/ml Glucose
    1 mM NADP
    0.04 mg/ml PMS
    0.2 mg/ml MTT
    2 mM MgCl$_2$
    200 mM KCl
    0.06 mM CaCl$_2$

B. 1% Agarose solution (55°C)

Procedure

Mix equal volumes of A and B components of the staining solution and pour the mixture over the gel surface. Incubate the gel in the dark at 37°C until dark blue bands appear. Fix the stained gel in 25% ethanol.

Notes: Be careful with adding ATP to the staining solution. Only a trace quantity of ATP is needed to initiate ATP-ADP enzyme-cycling reactions used to amplify the 5′-AMP signal generated by CNPE activity. An excess of ATP in the staining solution results in undesirable staining of the gel background.

Concentrations of auxiliary enzymes in solution A can require some adjustments.

Other Methods

A. An immunoblotting procedure (for details, see Part II) based on the utility of monoclonal antibodies specific to bovine CNPE[5] can also be used for immunohistochemical visualization of the enzyme protein on electrophoretic gels. This procedure is not appropriate for routine laboratory use, but may be of great value in special analyses of CNPE.

B. The product 5′-AMP can be hydrolyzed by auxiliary alkaline phosphatase yielding orthophosphate and adenosine (see Method 2, above). The areas of adenosine production can then be detected by using additional auxiliary enzymes, adenosine deaminase, purine-nucleoside phosphorylase, and xanthine oxidase (e.g., see 3.1.4.1 — PDE-I, Method 4), or by using another set of auxiliary enzymes, including adenosine deaminase and glutamate dehydrogenase (see 3.5.4.4 — ADA, Method 2).

General Notes

The enzyme acts on 3′,5′-cyclic AMP, 3′,5′-cyclic dAMP, 3′,5′-cyclic IMP, 3′,5′-cyclic GMP, and 3′,5′-cyclic CMP.[6]

References

1. Monn, E. and Christiansen, R.O., Adenosine 3′,5′-monophosphate phosphodiesterase: multiple molecular forms, *Science*, 173, 540, 1971.
2. Goren, E.N., Hirsch, A.H., and Rosen, O.M., Activity stain for the detection of cyclic nucleotide phosphodiesterase separated by polyacrylamide gel electrophoresis and its application to the cyclic nucleotide phosphodiesterase of beef heart, *Anal. Biochem.*, 43, 156, 1971.
3. Nemoz, G., Prigent, A.-F., and Pacheco, H., Analysis of cyclic nucleotide phosphodiesterase by isoelectric focusing coupled to a specific activity stain, *Anal. Biochem.*, 133, 296, 1983.
4. Sugiyama, A. and Lurie, K.G., An enzymatic fluorometric assay for adenosine 3′,5′-monophosphate, *Anal. Biochem.*, 218, 20, 1994.
5. Hansen, R.S. and Beavo, J.A., Purification of two calcium calmodulin-dependent forms of cyclic nucleotide phosphodiesterase by using conformation specific monoclonal antibody chromatography, *Proc. Natl. Acad. Sci. U.S.A.*, 79, 2788, 1982.
6. NC-IUBMB, *Enzyme Nomenclature*, Academic Press, San Diego, 1992, p. 329 (EC 3.1.4.17, Comments).

OTHER NAMES   2′,3′-Cyclic-nucleotide 3′-phospho-diesterase (recommended name)

REACTION   Nucleoside 2′,3′-cyclic phosphate + H$_2$O = nucleoside 2′-phosphate

ENZYME SOURCE   Bacteria, vertebrates

SUBUNIT STRUCTURE   Unknown or no data available

METHOD

Visualization Scheme

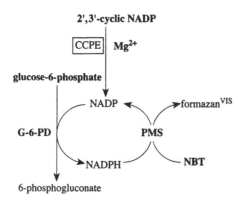

2′,3′-cyclic NADP

CCPE   Mg^{2+}

glucose-6-phosphate

NADP ← → formazan^{VIS}

G-6-PD   PMS

NADPH   NBT

6-phosphogluconate

Staining Solution[1]

A. 200 mM 4-Morpholine–ethanesulfonate buffer, pH 6.1
   60 mM MgCl$_2$
   0.2% Triton X-100
   0.1 mM 2′,3′-Cyclic NADP
   4 mg/ml Glucose-6-phosphate
   0.4 mg/ml NBT
   0.04 mg/ml PMS
   0.7 U/ml Glucose-6-phosphate dehydrogenase (G-6-PD)

B. 1% Agarose solution (55°C)

Procedure

Mix equal volumes of A and B components of the staining solution and pour the mixture over the gel surface. Incubate the gel in the dark at 37°C until dark blue bands appear. Fix the stained gel in 25% ethanol.

Notes: The enzyme is not water soluble; therefore, sample preparation and electrophoresis should be made in the presence of SDS or a mixture of Triton X-100, CHAPS, and urea. Thus, before staining, the gel should be made free of detergents by washing. The method described above was developed to detect the enzyme from myelin of the mammalian nervous system. Hydrolytic activity of the bacterial enzyme and mammalian enzyme from other tissues toward 2′,3′-cyclic NADP is very low.

GENERAL NOTES

The brain enzyme acts on 2′,3′-cyclic AMP more rapidly than on the CMP or UMP derivatives. The liver enzyme acts on 2′,3′-cyclic CMP more rapidly than on the purine derivatives. Just this latter enzyme has been called cyclic CMP phosphodiesterase.[2]

REFERENCES

1. Bradbury, J.M. and Thompson, R.J., Photoaffinity labeling of central-nervous-system myelin, *Biochem. J.*, 221, 361, 1984.
2. NC-IUBMB, *Enzyme Nomenclature*, Academic Press, San Diego, 1992, p. 330 (EC 3.1.4.37, Comments).

3.1.4.40 — CMP-Sialate Hydrolase; CMP-SH

OTHER NAMES | CMP-*N*-acylneuraminate phospho-diesterase (recommended name)

REACTION | CMP-*N*-acylneuraminate + H₂O = CMP + *N*-acylneuraminate

ENZYME SOURCE | Vertebrates

SUBUNIT STRUCTURE | Unknown or no data available

METHOD

Visualization Scheme

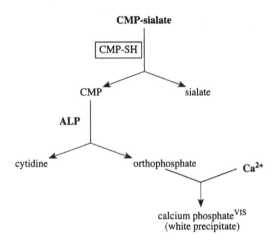

Staining Solution[1]

A. 0.15 *M* Tris–HCl buffer, pH 9.0
    5 m*M* CaCl₂
    0.1% Triton X-100
    3 U/ml Alkaline phosphatase (ALP)
B. 60 m*M* CMP–sialic acid

Procedure

Cover the surface of the electrophorized PAG dropwise with solution A (0.03 ml/cm²) and allow the solution to enter the gel. Then apply solution B in the same way and incubate the gel in a humid chamber at 37°C. View the gel against a dark background. White bands of calcium phosphate precipitation are visible after 15 to 120 min of incubation. Store the stained gel in 50 m*M* glycine–KOH (pH 10.0), containing 5 m*M* Ca²⁺, either at 5°C or at room temperature in the presence of an antibacterial agent. The gel can be photographed by reflected light against a dark background.

Notes: The precipitated calcium phosphate can be converted to brownish black lead sulfide (for details, see 3.1.3.1 — ALP, Method 7) or subsequently stained with Alizarin Red S.

The method may be used only when electrophoresis is carried out in a phosphate-free buffer system.

OTHER METHODS

Several other methods of detection of orthophosphate produced by auxiliary enzyme alkaline phosphatase are also available (see 3.1.3.1 — ALP, Methods 5, 6, and 8; 3.1.3.2 — ACP, Method 4).

REFERENCES

1. Van Dijk, W., Lasthuis, A.-M., Koppen, P.L., and Muilerman, H.G., A universal and rapid spectrophotometric assay of CMP–sialic acid hydrolase and nucleoside-diphosphosugar pyrophosphatase activities and detection in polyacrylamide gels, *Anal. Biochem.*, 117, 346, 1981.

3.1.4.X — Nonspecific Phosphodiesterase; NSPDE

REACTION	Hydrolyzes bis(p-nitrophenyl) phosphate
ENZYME SOURCE	Bacteria, invertebrates, vertebrates
SUBUNIT STRUCTURE	Unknown or no data available

METHOD

Visualization Scheme

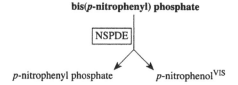

bis(p-nitrophenyl) phosphate

NSPDE

p-nitrophenyl phosphate   p-nitrophenol[VIS]

Staining Solution[1]

A. Bis(p-nitrophenyl) phosphate (disodium salt) 470 mg
   H_2O 125 ml
B. 0.5 M Sodium acetate buffer, pH 5.7 30 ml
   0.3% Tween 80 9 ml
   H_2O 15 ml
C. A mixture of solution A (4 ml) and
   solution B (5 ml)

Procedure

Cover the gel with Whatman No. 1 filter paper saturated with staining solution (solution C). Wrap the covered gel in Saran wrap and aluminum foil and incubate at 37°C for 2 h. Drain the gel surface and cover with 0.1 M NaOH. Yellow bands appear immediately at the sites of NSPDE action. The yellow color can be preserved for a few hours if the gel is enclosed in Saran wrap.

REFERENCES

1. Hodes, M.E., Crisp, M., and Gelb, E., Electrophoresis of acid phosphohydrolase isozymes on Cellogel, *Anal. Biochem.*, 80, 239, 1977.

3.1.4.X′ — NAD Phosphodiesterase; NADPDE

OTHER NAMES	Phosphodiesterase
REACTION	NAD + H_2O = adenosine + nicotinamide ribose + pyrophosphate
ENZYME SOURCE	Vertebrates
SUBUNIT STRUCTURE	Unknown or no data available

METHOD 1

Visualization Scheme

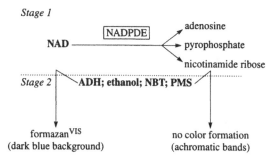

Stage 1

NAD — NADPDE → adenosine
→ pyrophosphate
→ nicotinamide ribose

Stage 2 **ADH; ethanol; NBT; PMS**

formazan[VIS]
(dark blue background)

no color formation
(achromatic bands)

Staining Solution[1]

A. 50 mM Phosphate buffer, pH 6.5
   0.65 mM NAD
B. 0.1 M Pyrophosphate buffer, pH 8.8 50 ml
   Ethanol 0.3 ml
   NBT 10 mg
   PMS 1 mg
   Alcohol dehydrogenase (ADH) 25 U

Procedure

Incubate an electrophorized gel in solution A at 37°C for 1 to 2 h. Rinse the gel with water and counterstain with solution B in the dark at 37°C to detect NAD. Achromatic bands visible on a blue background of the gel indicate areas occupied by NAD-degrading enzymes, including NADPDE. Wash the negatively stained gel with water and fix in 25% ethanol.

Notes: This method also detects activity bands of two other NAD-catabolizing enzymes, the NAD nucleotidase (see 3.2.2.5 — NADN) and the NAD pyrophosphatase (see 3.6.1.22 — NADPP). Methods specific for NADPDE are outlined below (see Other Methods).

METHOD 2

Visualization Scheme

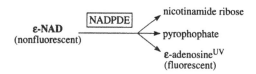

ε-NAD
(nonfluorescent)

NADPDE → nicotinamide ribose
→ pyrophophate
→ ε-adenosine[UV]
(fluorescent)

Staining Solution[2]

50 mM Tris–HCl buffer, pH 7.5
0.03% Lauryl dimethylamine N-oxide (LDAO)
150 μM ε-NAD (see *Notes*)

Procedure

The method is developed for SDS-PAG. So, after separation of proteins by SDS-PAG electrophoresis, wash the gel for 30 min in a solution containing 50 mM Tris–HCl buffer (pH 7.5) and 0.5% LDAO.

To visualize NAD-converting activities, incubate the gel for 15 min in the staining solution and place on a UV transilluminator to observe fluorescent bands. Photograph the developed gel using a 550-nm interference filter.

Notes: ε-NAD is a NAD derivative (1,N^6-etheno-NAD) containing a modified adenine ring whose fluorescence is greatly enhanced after its separation from the quenching nicotinamide moiety.

This method also detects activity bands of other NAD-catabolizing enzymes, the NAD nucleotidase (see 3.2.2.5 — NADN) and the NAD pyrophosphatase (see 3.6.1.22 — NADPP). After SDS-PAG electrophoresis of rat kidney and heart preparations, the fluorescent band of NADPDE (210,000 Da) migrates the slowest, the band of NADN (41,000 Da) migrates the fastest, and the band of NADPP (105,000 Da) occupies the intermediate position.[2] Because of a very large difference between molecular masses of the three NAD-degrading enzymes, the same position of their activity bands in the same tissues may be expected for all mammals.

Other Methods

A. The product adenosine can be detected using auxiliary enzymes adenosine deaminase, purine-nucleoside phosphorylase, and xanthine oxidase combined with PMS–MTT (see 3.1.4.1 — PDE-I, Method 4). This method will result in positively stained NADPDE bands visible in daylight.

B. The product adenosine can also be detected using auxiliary enzymes adenosine deaminase and glutamate dehydrogenase (see 3.5.4.4 — ADA, Method 2). This method will result in NADPDE activity bands visible in UV light as dark (nonfluorescent) bands on a light (fluorescent) background.

C. The product pyrophosphate can be detected by the fluorogenic enzymatic method, which involves three auxiliary enzymes, pyrophosphate–fructose-6-phosphate 1-phosphotransferase (the bacterial enzyme is preferable because it does not require fructose-2,6-bisphosphate as a cofactor), aldolase, and glycerol-3-phosphate dehydrogenase (see 2.7.1.90 — PFPPT).

D. Pyrophosphate can also be detected by the chromogenic enzymatic method, which involves three auxiliary enzymes, pyrophosphate–fructose-6-phosphate 1-phosphotransferase, aldolase, and glyceraldehyde-3-phosphate dehydrogenase, combined with PMS–MTT (see 2.7.1.90 — PFPPT, Other Methods).

General Notes

Only methods outlined in the Other Methods section may be considered as specific for NADPP. These methods are recommended for a trial.

References

1. Ravazzolo, R., Bruzzone, G., Garrè, C., and Ajmar, F., Electrophoretic demonstration and initial characterization of human red cell NAD(P)⁺ase, *Biochem. Genet.*, 14, 877, 1976.
2. Hagen, T.H. and Ziegler, M., Detection and identification of NAD-catabolizing activities in rat tissue homogenates, *Biochim. Biophys. Acta*, 1340, 7, 1997.

3.1.6.1 — Arylsulfatase; ARS

OTHER NAMES	Sulfatase
REACTION	Phenol sulfate + H_2O = phenol + sulfate
ENZYME SOURCE	Bacteria, fungi, green algae, plants, invertebrates, vertebrates
SUBUNIT STRUCTURE	Dimer (vertebrates)

METHOD 1

Visualization Scheme

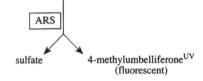

4-methylumbelliferyl sulfate

ARS

sulfate   4-methylumbelliferone[UV]
(fluorescent)

Staining Solution[1]

0.5 M Sodium acetate buffer, pH 5.6	2 ml
4-Methylumbelliferyl sulfate	8.8 mg

Procedure

Cover the gel surface with staining solution and incubate at 37°C in a humid chamber for 1 h. Then fix the gel in 10% formalin for 4 min and immerse in 0.25 M sodium carbonate–glycin buffer (pH 10.0) for 4 min. The sites of ARS activity appear as bright fluorescent bands when viewed under long-wave UV light.

Notes: Gels electrophorized in Tris–glycine buffer (pH 8.4 to 8.6) can display fluorescent bands caused by nonenzymatic cleavage of 4-methylumbelliferyl sulfate. To prevent this staining artifact, the use of acid electrophoretic buffers with a pH below 6.0 is recommended. Gels electrophorized in acid buffers do not display staining artifacts when stained using 4-methylumbelliferyl derivatives as fluorogenic substrates.[2]

METHOD 2

Visualization Scheme

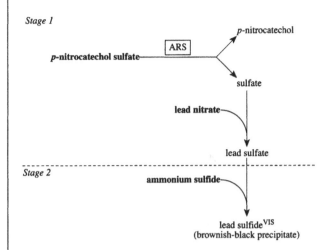

Stage 1

p-nitrocatechol

p-nitrocatechol sulfate — ARS

sulfate

lead nitrate

lead sulfate

Stage 2   ammonium sulfide

lead sulfide[VIS]
(brownish-black precipitate)

Staining Solution[3]

A.	50 mM Sodium acetate buffer, pH 5.5	100 ml
	p-Nitrocatechol sulfate	300 mg
	24% Lead nitrate	2 ml
B.	1% Ammonium sulfide	5 ml

Procedure

Incubate the gel in solution A for 1 to 2 h, and then add solution B and wait for dark bands to appear. Rinse the stained gel with 10 mM sodium acetate (pH 4.0) and store in 50% glycerol.

Notes: The brownish black lead sulfide precipitate will go into solution if the pH rises above 5.7.

REFERENCES

1. Rattazzi, M.C., Marks, J.S., and Davidson, R.G., Electrophoresis of arylsulfatase from normal individuals and patients with metachromatic leukodystrophy, *Am. J. Hum. Genet.*, 25, 310, 1973.
2. Chang, P.L., Ballantyne, S.R., and Davidson, R.G., Detection of arylsulfatase A activity after electrophoresis in polyacrylamide gels: problems and solutions, *Anal. Biochem.*, 97, 36, 1979.
3. Vallejos, C.E., Enzyme activity staining, in *Isozymes in Plant Genetics and Breeding*, Part A, Tanskley, S.D. and Orton, T.J., Eds., Elsevier, Amsterdam, 1983, p. 469.

OTHER NAMES | *E. coli* exonuclease III
REACTION | Exonucleolytic cleavage in the 3'- to 5'-direction to yield 5'-phospho-mononucleotides; preference for double-stranded DNA
ENZYME SOURCE | Bacteria
SUBUNIT STRUCTURE | Unknown or no data available

METHOD

Visualization Scheme

Stage 1: Electrophoretic separation of proteins in substrate-containing PAG (first dimension)

Stage 2: Vertical incision of the sample line of the first-dimension PAG and receipt of narrow (2 mm) substrate-containing gel strip

Stage 3: Enzyme reaction in the strip of the first-dimension PAG

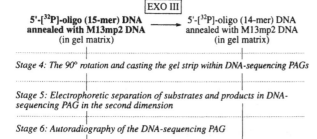

Stage 4: The 90° rotation and casting the gel strip within DNA-sequencing PAGs

Stage 5: Electrophoretic separation of substrates and products in DNA-sequencing PAG in the second dimension

Stage 6: Autoradiography of the DNA-sequencing PAG

slowly migrating dark line[VIS] across the DNA-sequencing PAG indicating the position of the substrate, 5'-labeled intact oligo (15-mer) DNA

quickly migrating dark spot[VIS'] indicating the product, 5'-labeled oligo (14-mer) DNA, on the DNA-sequencing PAG just opposite the position of EXO III on the strip of substrate-containing PAG

Substrate-Containing PAG[1,2]

Prepare denaturing (4% stacking and 10% resolving) SDS-PAG containing 5'-[^{32}P]-oligo(15-mer)DNA annealed with single-stranded circular M13mp2 DNA (see *Notes*).

Reaction Mixture[1,2]

50 mM Tris–HCl buffer, pH 8.0
5 mM 2-Mercaptoethanol
400 µg/ml Bovine serum albumin
15% (w/v) Glycerol
50 mM KCl
5 mM MgCl$_2$

Procedure

Stage 1. After electrophoresis, immerse the resolving substrate-containing SDS-PAG in 33 gel volumes of SDS extraction buffer: 10 mM Tris–HCl (pH 7.5), 5 mM 2-mercaptoethanol, 25% (v/v) 2-propanol. Gently agitate the gel on a gyratory shaker (60 rpm) at 25°C. Discard the buffer after 30 min and replace with the same volume of fresh SDS extraction buffer. Extract the SDS for an additional 30 min. Place the gel in

3 volumes of enzyme renaturation buffer (the reaction mixture lacking MgCl$_2$). After brief (30 sec) agitation, place the gel in 27 volumes of the enzyme renaturation buffer and incubate for 18 to 25 h with gentle agitation at 4°C.

Stage 2. Following renaturation of the enzyme within the SDS-PAG, cut each line of the substrate-containing PAG into 2-mm-wide longitudinal strips.

Stage 3. Place an individual strip into a 16 × 100 mm test tube containing 5 ml of the reaction mixture. Perform the enzyme reaction at 25°C for 30 min in a stoppered (Parafilm) tube that is placed on its side and gently shaken (60 rpm). Terminate the reaction by substituting the reaction mixture with an equal volume of ice-cold renaturation buffer containing 10 mM EDTA and gently shaking the tube for 30 min at 4°C.

Stages 4, 5, and 6. Following the *in situ* enzyme reaction, cast the processed strip of substrate-containing PAG within a 20% DNA sequencing PAG containing 8.3 M urea, avoiding formation of air bubbles. After polymerization, carry out electrophoresis in a second dimension, transfer the electrophorized 20% PAG to filter paper, dry, and autoradiograph.

Two dark areas are observed on the developed x-ray film. The slowly migrating dark area is a line across the 20% PAG indicating the position of 5'-labeled intact oligo(15-mer)DNA; two or more dark spots of faster mobility indicate the position of EXO III reaction products, the oligo(14-mer)DNA, the oligo(13-mer)DNA, etc. The position of EXO III is detected via projection of the dark spots onto the strip of substrate-containing PAG.

Notes: The oligo(15-mer)DNA is the 5'-GGCGATTAAGTTGGG-3'-oligonucleotide complementary to M13mp2 DNA (see below). The oligonucleotide is labeled at the 5'-end by T4 polynucleotide kinase with [γ-^{32}P]-ATP as described.[1,2] The labeled oligonucleotide (specific activity of 3.0 × 10^6 to 3.5 × 10^6 cpm/pmol of 5'-ends) is annealed with single-stranded M13mp2 DNA molecules (0.27 pmol of 5'-ends/µg of DNA) by heating to 70°C with slow cooling to 25°C.[3] M13mp2 DNA is a single-stranded circular DNA isolated from bacteriophage M13mp2 grown in *E. coli* strain JM107.[3] The substrate, [^{32}P]-labeled oligo(15-mer) (final concentration of about 8 ng/ml)/M13mp2 DNA (final concentration of about 2 µg/ml), is included in the resolving PAG before polymerization.[1,2]

The oligo(24-mer)DNA (5'-GTGCTGCAAGGCGATTAAGT-TGGT-3') can also be used as a substrate in place of the oligo(15-mer)DNA. In such a case, the product of the EXO III reaction will be represented by serial oligo(<24-mer)DNAs.

REFERENCES

1. Longley, M.J. and Mosbaugh, D.W., Characterization of DNA metabolizing enzymes *in situ* following polyacrylamide gel electrophoresis, *Biochemistry*, 30, 2655, 1991.
2. Longley, M.J. and Mosbaugh, D.W., *In situ* detection of DNA-metabolizing enzymes following polyacrylamide gel electrophoresis, *Methods Enzymol.*, 218, 587, 1993.
3. Kunkel, T.A., Roberts, J.D., and Zakour, R.A., Rapid and efficient site-specific mutagenesis without phenotypic selection, *Methods Enzymol.*, 154, 367, 1987.

3.1.16.1 — Spleen Exonuclease; SE

OTHER NAMES   3′-Exonuclease, spleen phosphodiesterase, phosphodiesterase II

REACTION   Exonucleolytic cleavage in the 5′- to
3′-direction to yield 3′-phospho-
mononucleotides

ENZYME SOURCE   Vertebrates

SUBUNIT STRUCTURE Heterotetramer($2\alpha2\beta$)[#] (mammals);
see General Notes

METHOD 1

Visualization Scheme

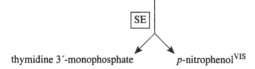

thymidine 3′-monophosphate-*p*-nitrophenyl

SE

thymidine 3′-monophosphate   *p*-nitrophenol[VIS]

Staining Solution[1]

A. Thymidine 3′-monophosphate–*p*-nitrophenyl
(ammonium salt)   276 mg
H_2O 1 ml

B. 0.5 *M* Ammonium acetate buffer, pH 5.7   16.7 ml
4 m*M* EDTA (sodium salt)   8.33 ml
1% Tween 80   0.05 ml
H_2O 14.67 ml

C. A mixture of solution B with 0.25 ml of
solution A

Procedure

Cover the gel with Whatman No. 1 filter paper saturated with
staining solution (solution C). Wrap the covered gel in Saran
wrap and aluminum foil and incubate at 37°C for 90 to 120 min.
Drain the gel surface and cover with 0.1 *M* NaOH. Yellow bands
appear immediately at the sites of SE action. The yellow color
can be preserved for a few hours if the gel is enclosed in Saran
wrap.

METHOD 2

Visualization Scheme

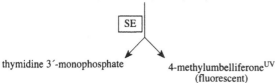

4-methylumbelliferyl 3′-thymidylate

SE

thymidine 3′-monophosphate   4-methylumbelliferone[UV]
(fluorescent)

Staining Solution[2]

2 m*M* water solution of 4-methylumbelliferyl 3′-thymidylate

Procedure

Incubate the gel electrophorized in an alkaline buffer system in
the staining solution at 37°C and monitor under long-wave UV
light. The areas of SE activity are seen as fluorescent bands.
Record the zymogram or photograph using a yellow filter.

METHOD 3

Visualization Scheme

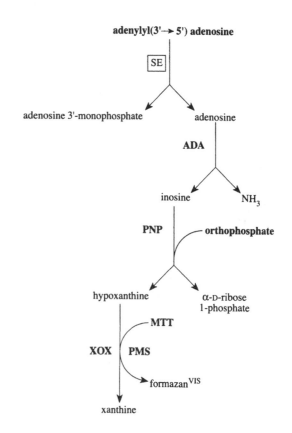

adenylyl(3′→ 5′) adenosine

SE

adenosine 3′-monophosphate   adenosine

ADA

inosine NH_3

PNP   orthophosphate

hypoxanthine   α-D-ribose
1-phosphate

MTT

XOX   PMS

formazan[VIS]

xanthine

Staining Solution[3]

A. 10 mg/ml Adenylyl(3′ → 5′)adenosine
(in distilled water)   50 μl
5 mg/ml PMS   200 μl
5 mg/ml MTT   200 μl
Purine-nucleoside phosphorylase
(PNP; Boehringer)   10 μl
Xanthine oxidase (XOX; Boehringer)   10 μl
Adenosine deaminase (ADA; Sigma, type III) 20 μl

B. 1% Agarose solution in 50 m*M* sodium
phosphate buffer, pH 8.0 (45°C)   9.5 ml

3.1.16.1 — Spleen Exonuclease; SE (continued)

Procedure

Mix A and B components of the staining solution and pour the mixture over the gel surface. Incubate the gel in the dark at 37°C until dark blue bands appear. Fix the stained gel in 25% ethanol.

Notes: Phosphodiesterase I (see 3.1.4.1 — PDE-I, Method 4) can also be developed on SE zymograms obtained by this method. However, PDE-I bands do not develop on SE zymograms obtained by this method when other dinucleoside monophosphates (UpA or CpA) are used as substrates. Moreover, only PDE-I bands develop when ApU is used as a substrate.

An additional gel should be stained in a staining solution lacking the dinucleoside monophosphate to identify "nothing dehydrogenase" (see 1.X.X.X — NDH) bands, which can also develop on SE zymograms obtained by this method.

GENERAL NOTES

The enzyme from beef spleen is a 160-kDa heterotetramer consisting of two types of subunits (α and β). Both monomeric subunits do not possess any exonuclease activity. The 80-kDa heterodimer shows only a low specific activity. Completely active is the 160-kDa $2\alpha2\beta$ heterotetramer.[4]

REFERENCES

1. Hodes, M.E., Crisp, M., and Gelb, E., Electrophoresis of acid phosphohydrolase isozymes on Cellogel, *Anal. Biochem.*, 80, 239, 1977.
2. Hawley, D.M., Tsou, K.C., and Hodes, M.E., Preparation, properties, and uses of two fluorogenic substrates for the detection of 5'- (venom) and 3'- (spleen) nucleotide phosphodiesterases, *Anal. Biochem.*, 117, 18, 1981.
3. Hodes, M.E. and Retz, J.E., A positive zymogram for distinguishing among RNase and phosphodiesterases I and II, *Anal. Biochem.*, 110, 150, 1981.
4. Mitkova, A.V., Stoynov, S.S., Bakalova, A.T., and Dolapchiev, L.B., Emergence of the active site of spleen exonuclease upon association of the two basic monomers of the tetrameric enzyme, *Int. J. Biochem Cell B*, 31, 1399, 1999.

3.1.21.1 — Deoxyribonuclease I; DNASE

OTHER NAMES	Pancreatic DNase, DNase, thymonuclease
REACTION	Endonucleolytic cleavage to 5'-phosphodinucleotide and 5'-phospho-oligonucleotide end-products
ENZYME SOURCE	Bacteria, fungi, plants, invertebrates, vertebrates
SUBUNIT STRUCTURE	Unknown or no data available

METHOD

Visualization Scheme

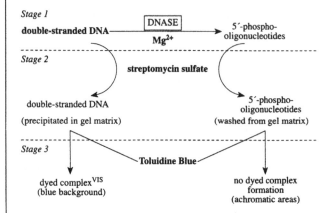

Staining Solution[1]

A. 20 m*M* Veronal–acetate buffer, pH 7.0
   0.1% Calf thymus DNA
   40 m*M* MgSO₄
B. 2% Agarose solution (50°C)
C. 10 m*M* Veronal–acetate buffer, pH 7.0
   1.5% Streptomycin sulfate
D. 0.125% Toluidine Blue
   0.5% Acetic acid
E. 2.5% Glycerol
   20% Methanol
   5% Acetic acid

Procedure

Mix equal volumes of A and B solutions and form a 1-mm-thick reactive agarose plate. Apply an electrophorized gel to the reactive agarose plate and incubate in a humid chamber at 40°C for 90 min. Wash the reactive agarose plate in solution C for 60 to 90 min and stain in solution D for 10 min. Wash the stained agarose plate in solution E. Areas of DNASE activity are seen as achromatic bands on a blue background.

Notes: Pyronin G may be used instead of Toluidine Blue, resulting in light red staining of a DNA-containing gel.[2] A Methyl Green:DNA dyed substrate complex also may be used to detect DNASE.[3] Native DNA may be specifically stained by the fluorogenic stain ethidium bromide.[4–6]

Incorporation of a DNA substrate directly into a running gel raises the sensitivity of negative stain methods considerably.[7] However, positively charged DNASE molecules could not be separated in electrophoretic gels containing DNA because they would bind to the negatively charged DNA. In this situation the use of SDS-containing PAG is recommended.[4] After denaturation with SDS and separation in an SDS-PAG containing DNA, many DNASE isozymes recover activity when SDS-PAGs are subsequently incubated in appropriate buffers without SDS.

The sensitivity with which DNASE activity can be detected after SDS-PAG electrophoresis varies widely, depending on the particular SDS preparation used for electrophoresis. Sensitivity of detection is greatly increased by using buffered 25% isopropanol, rather than buffer alone, to wash the SDS from the SDS-PAG after electrophoresis.[8]

The method described above is not specific. The bands of some other DNases with endonucleolytic activity also can be developed on DNASE zymograms.

Identification of DNASEs that display preference for double-stranded or single-stranded DNA can also be accomplished by electrophoresis in gels containing native and denatured DNA.[4]

Human urine DNase I contains a few sialic acid residues. Desialylation of the enzyme by sialidase treatment of the enzyme preparations before electrophoresis simplifies the enzyme pattern, the number of secondary isozyme bands being reduced.[6]

OTHER METHODS

A. A dried agarose film overlay (DAFO) method is developed that enhances the resolution of DNASE activity bands and sensitivity of the detection. The method is based on the staining of substrate-containing dry agarose film with ethidium bromide. Zones of DNA degradation are visible in UV light (312 nm) as dark bands on a light (fluorescent) background. Using this method it was possible to detect DNASE from human saliva samples of 2 to 5 µl.[9]

B. A fluorescence method is described that can be adapted to detect DNASE activity bands on electrophoretic gels. This method is based on the use of a quenched fluorophore–DNA covalent conjugate (fluorescein-labeled DNA) as a substrate and its subsequent dequenching due to degradation by DNASE.[10] The resulting fluorescence can be observed in long-wave UV light. The principle of quenching of fluorescent dyes upon conjugation to polymeric molecules is based on high local concentrations of self-quenched fluorophores on the polymeric backbone. Depolymerizing enzymes can reverse this state and generate a strong fluorescence signal.

C. The immunoblotting procedure (for details, see Part II) may also be used to detect DNASE. Antibodies specific to the enzyme from the bovine pancreas are available from Sigma.

GENERAL NOTES

Double-stranded DNA is a preferred substrate of DNASE.[11]

REFERENCES

1. Berges, J. and Uriel, J., Mise en evidence des activites ribo-nucleasiques et deoxyribonucleasiques apres electrophorese en acrylamide-agarose, *Biochimie*, 53, 303, 1971.

2. van Loon, L.C., Polynucleotide–acrylamide gel electrophoresis of soluble nucleases from tobacco leaves, *FEBS Lett.*, 51, 266, 1975.

3. Hodes, M.E., Crisp, M., and Gelb, E., Electrophoresis of acid phosphohydrolase isozymes on Cellogel, *Anal. Biochem.*, 80, 239, 1977.

4. Rosenthal, A.L. and Lacks, S.A., Nuclease detection in SDS–polyacrylamide gel electrophoresis, *Anal. Biochem.*, 80, 76, 1977.

5. Kim, H.S. and Liao, T.-H., Isoelectric focusing of multiple forms of DNase in thin layers of polyacrylamide gel and detection of enzymatic activity with a zymogram method following separation, *Anal. Biochem.*, 119, 96, 1982.

6. Yasuda, T., Mizuta, K., Ikehara, Y., and Kishi, K., Genetic analysis of human deoxyribonuclease I by immunoblotting and the zymogram method following isoelectric focusing, *Anal. Biochem.*, 183, 84, 1989.

7. Brown, T.L., Yet, M.-G., and Wold, F., Substrate-containing gel electrophoresis: sensitive detection of amylolytic, nucleolytic, and proteolytic enzymes, *Anal. Biochem.*, 122, 164, 1982.

8. Blank, A., Sugiyama, R.H., and Dekker, C.A., Activity staining of nucleolytic enzymes after sodium dodecyl sulfate–polyacrylamide gel electrophoresis: use of aqueous isopropanol to remove detergent from gels, *Anal. Biochem.*, 120, 267, 1982.

9. Tenjo, E., Sawazaki, K., Yasuda, T., Nadano, D., Takeshita, H., Iida, R., and Kishi, K., Salivary deoxyribonuclease I polymorphism separated by polyacrylamide gel-isoelectric focusing and detected by the dried agarose film overlay method, *Electrophoresis*, 14, 1042, 1993.

10. Trubetskoy, V.S., Hagstrom, J.E., and Budker, V.G., Self-quenched covalent fluorescent dye-nucleic acid conjugates as polymeric substrates for enzymatic nuclease assays, *Anal. Biochem.*, 300, 22, 2002.

11. NC-IUBMB, *Enzyme Nomenclature*, Academic Press, San Diego, 1992, p. 339 (EC 3.1.21.1, Comments).

OTHER NAMES Type II restriction enzyme, restriction endonuclease, restrictase

REACTION Endonucleolytic cleavage of DNA to give specific double-stranded fragments with terminal 5′-phosphates

ENZYME SOURCE Bacteria

SUBUNIT STRUCTURE Unknown or no data available

Method

Visualization Scheme

Stage 1: Electrophoretic separation of proteins in substrate-containing PAG (first dimension)

Stage 2: Vertical incision of the sample line of the first-dimension PAG and receipt of narrow (2 mm) substrate-containing gel strip

Stage 3: Enzyme reaction in the strip of the first-dimension PAG

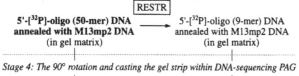

RESTR

5′-[^{32}P]-oligo (50-mer) DNA **annealed with M13mp2 DNA** (in gel matrix) → 5′-[^{32}P]-oligo (9-mer) DNA annealed with M13mp2 DNA (in gel matrix)

Stage 4: The 90° rotation and casting the gel strip within DNA-sequencing PAG

Stage 5: Electrophoretic separation of substrates and products in DNA-sequencing PAG in the second dimension

Stage 6: Autoradiography of the DNA-sequencing PAG

slowly migrating dark line^{VIS} across the DNA-sequencing PAG indicating the position of the substrate, 5′-labeled intact oligo(50-mer)DNA

quickly migrating dark spot$^{VIS'}$ indicating the product, 5′-labeled oligo(9-mer)DNA, on the DNA-sequencing PAG just opposite the position of RESTR on the strip of the substrate-containing PAG

Substrate-Containing PAG[1,2]

Prepare a native (2.5% stacking and 10% resolving) PAG or denaturing (4% stacking and 10% resolving) SDS-PAG containing 5′-[^{32}P]-oligo(50-mer)DNA annealed with single-stranded M13mp19 DNA (see *Notes*).

Reaction Mixture[1,2]

20 mM Tris–HCl buffer, pH 7.5
5 mM 2-Mercaptoethanol
400 μg/ml Bovine serum albumin
5% (w/v) Glycerol
8 mM MgCl$_2$

Procedure

Stage 1. After electrophoresis, immerse the resolving substrate-containing SDS-PAG in 33 gel volumes of SDS extraction buffer: 10 mM Tris–HCl (pH 7.5), 5 mM 2-mercaptoethanol, 25% (v/v) 2-propanol. Gently agitate the gel on a gyratory shaker (60 rpm) at 25°C. Discard the buffer after 30 min and replace with the same volume of fresh SDS extraction buffer. Extract SDS for an additional 30 min. Place the gel in 3 volumes of enzyme renaturation buffer (the reaction mixture lacking MgCl$_2$ and containing 50 mM NaCl). After brief (30 sec) agitation, place the gel in 27 volumes of the enzyme renaturation buffer and incubate for 18 to 25 h with gentle agitation at 4°C.

Stage 2. Following renaturation of the enzyme within an SDS-PAG or after native PAG electrophoresis, cut each line of the substrate-containing PAG into 2-mm-wide longitudinal strips.

Stage 3. Place an individual strip into a 16 × 100 mm test tube containing 5 ml of the reaction mixture. Perform the enzyme reaction at 25°C for 30 min in a stoppered (Parafilm) tube that is placed on its side and gently shaken (60 rpm). Terminate the reaction by substituting the reaction mixture with an equal volume of ice-cold renaturation buffer containing 10 mM EDTA and gently shaking the tube for 30 min at 4°C.

Stages 4, 5, and 6. Following the *in situ* enzyme reaction, cast the processed strip of substrate-containing PAG within a 20% DNA sequencing PAG containing 8.3 M urea, avoiding formation of air bubbles. After polymerization, carry out electrophoresis in a second dimension, transfer the electrophorized 20% PAG to filter paper, dry, and autoradiograph.

Two dark areas are observed on the developed x-ray film. The slowly migrating dark area is a line across the 20% PAG indicating the position of 5′-labeled intact oligo(50-mer)DNA. Dark spots (fast mobility) indicate the position(s) of RESTR reaction product(s), the 5′-labeled oligo(9-mer, 25-mer, 27-mer, or 36-mer)DNA (see *Notes*). The position of RESTR is detected via projection of the dark spot(s) onto the strip of substrate-containing PAG.

Notes: The oligo(50-mer)DNA is the 5′-CGGCCAGTGAATTC-GAGCTCGGTACCCGGGGATCCTCTAGAGTCGACCTG-3′ oligonucleotide complementary to M13mp19 DNA (see below). The oligonucleotide is labeled at the 5′ end by T4 polynucleotide kinase with [γ-^{32}P]-ATP as described.[1,2] The labeled oligonucleotide (specific activity of 3.0×10^6 to 3.5×10^6 cpm/pmol of 5′-ends) is annealed with single-stranded M13mp19 DNA molecules (0.27 pmol of 5′-ends/μg of DNA) by heating to 70°C with slow cooling to 25°C.[3] M13mp19 DNA is a single-stranded circular DNA isolated from bacteriophage M13mp19 grown in *E. coli* strain JM107.[3] The substrate, [^{32}P]-labeled 50-mer (final concentration of about 8 ng/ml)/M13mp19 DNA (final concentration of about 2 μg/ml), is included in the resolving PAG before polymerization.[1,2]

3.1.21.4 — Type II Site-Specific Deoxyribonuclease; RESTR (continued)

Restriction endonucleases from different sources recognize specific sequences in double-stranded DNA. The 5′-labeled oligo(50-mer)DNA used as a substrate in this method contains hexameric recognition sequences specific for *Eco*RI (G↓AATTC), *Kpn*I (GGTAC↓C), *Sma*I (CCC↓GGG), and *Xba*I (T↓CTAGA). The products of these restriction endonucleases will be the 5′-labeled oligo(9-mer)DNA, oligo(25-mer)DNA, oligo(27-mer)DNA, and oligo(36-mer)DNA, respectively.[1,2]

REFERENCES

1. Longley, M.J. and Mosbaugh, D.W., Characterization of DNA metabolizing enzymes *in situ* following polyacrylamide gel electrophoresis, *Biochemistry*, 30, 2655, 1991.
2. Longley, M.J. and Mosbaugh, D.W., *In situ* detection of DNA-metabolizing enzymes following polyacrylamide gel electrophoresis, *Methods Enzymol.*, 218, 587, 1993.
3. Kunkel, T.A., Roberts, J.D., and Zakour, R.A., Rapid and efficient site-specific mutagenesis without phenotypic selection, *Methods Enzymol.*, 154, 367, 1987.

3.1.26.3 — Ribonuclease III; RNASE III

OTHER NAMES	RNase O, RNase D
REACTION	Endonucleolytic cleavage to 5′-phosphomonoester (see General Notes)
ENZYME SOURCE	Bacteria, fungi
SUBUNIT STRUCTURE	Unknown or no data available

METHOD

Visualization Scheme

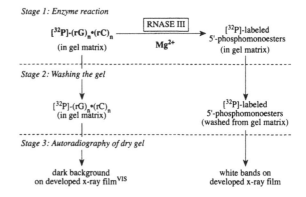

Reaction Mixture[1]

A. [^{32}P]-Labeled double-stranded $(rG)_n \cdot (rC)_n$ (300,000 to 400,000 cpm/20 ml of gel mixture) is incorporated into the running SDS-PAG mixture before polymerization

B. 40 m*M* Tris–HCl buffer, pH 8.0
0.1 m*M* EDTA
10 m*M* MgCl$_2$
0.1 m*M* Dithiothreitol
75 m*M* NaCl

C. Methanol:acetic acid:water 5:1:5 (v/v)

Procedure

Stages 1 and 2. Upon completion of electrophoresis, subject the gel containing double-stranded RNA (see A in the reaction mixture) to a preliminary washing at room temperature with two changes of 300 ml of water for 90 min, with gentle shaking. Then incubate the gel with 300 ml of reaction mixture (solution B). After 1 h of incubation at room temperature, replace the solution with 300 ml of a fresh solution of B and incubate the gel, with gentle shaking, at 37°C for a further 18 to 40 h.

Stage 3. After incubation, soak the gel in solution C for 30 min at room temperature, dry the gel, and subject it to autoradiography using Kodirex film (Kodak). Autoradiography is completed in 1 day.

The white bands on the developed x-ray film correspond to localization of RNASE III activity in the gel.

Notes: The sensitivity of the method can be increased by using buffered 25% isopropanol to wash the detergent from the gel after electrophoresis.[2]

Some nonspecific endoribonucleases degrading RNA-DNA hybrids (e.g., see 3.1.26.4 — CTRH) can also be developed on RNASE III autoradiograms obtained using this method.[1]

GENERAL NOTES

This ribonuclease cleaves a multimeric tRNA precursor at the spacer region; it is also involved in the processing of precursor rRNA, hnRNA, and early T_7-mRNA. Additionally, it cleaves double-stranded RNA.[3]

REFERENCES

1. Huet, J., Sentenac, A., and Fromageot, P., Detection of nucleases degrading double helical RNA and of nucleic acid–binding proteins following SDS–gel electrophoresis, *FEBS Lett.*, 94, 28, 1978.
2. Blank, A., Sugiyama, R.H., and Dekker, C.A., Activity staining of nucleolytic enzymes after sodium dodecyl sulfate–polyacrylamide gel electrophoresis: use of aqueous isopropanol to remove detergent from gels, *Anal. Biochem.*, 120, 267, 1982.
3. NC-IUBMB, *Enzyme Nomenclature*, Academic Press, San Diego, 1992, p. 342 (EC 3.1.26.3, Comments).

OTHER NAMES	Endoribonuclease H (calf thymus), RNase H
REACTION	Endonucleolytic cleavage to 5′-phosphomonoester (see General Notes)
ENZYME SOURCE	Bacteria, fungi, plants, vertebrates
SUBUNIT STRUCTURE	Unknown or no data available

Method

Visualization Scheme

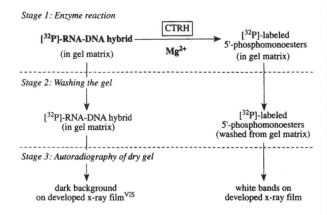

Stage 1: Enzyme reaction

[^{32}P]-RNA-DNA hybrid ———CTRH/Mg^{2+}———→ [^{32}P]-labeled 5′-phosphomonoesters
(in gel matrix)   (in gel matrix)

Stage 2: Washing the gel

[^{32}P]-RNA-DNA hybrid
(in gel matrix)

[^{32}P]-labeled 5′-phosphomonoesters (washed from gel matrix)

Stage 3: Autoradiography of dry gel

dark background on developed x-ray film^{VIS}

white bands on developed x-ray film

Reaction Mixture[1]

A. [^{32}P]-Labeled RNA-DNA hybrids (300,000 to 400,000 cpm/20 ml of gel mixture) are incorporated into the running SDS-PAG mixture before polymerization; synthetic [^{32}P]-labeled $(rA)_n \cdot (dT)_n$ and $(rG)_n \cdot (dC)_n$ are used as substrates

B. 40 mM Tris–HCl buffer, pH 8.0
   0.1 mM EDTA
   10 mM MgCl$_2$
   0.1 mM Dithiothreitol
   75 mM NaCl

C. Methanol:acetic acid:water 5:1:5 (v/v)

Procedure

Stages 1 and 2. Upon completion of electrophoresis, subject the gel containing RNA-DNA hybrids (see A in the reaction mixture) to a preliminary washing at room temperature with two changes of 300 ml of water for 90 min, with gentle shaking. Then incubate the gel with 300 ml of the reaction mixture (solution B). After 1 h of incubation at room temperature, replace the solution with 300 ml of a fresh solution of B and incubate the gel, with gentle shaking, at 37°C for a further 18 to 40 h.

Stage 3. After incubation, soak the gel in solution C for 30 min at room temperature, dry the gel, and subject it to autoradiography using Kodirex film (Kodak). Autoradiography is completed in 1 day.

The white bands on the developed x-ray film correspond to localization of CTRH activity in the gel.

Notes: The sensitivity of CTRH detection may be increased by using buffered 25% isopropanol to wash the detergent from the gel after electrophoresis (e.g., see 3.1.21.1 — DNASE, *Notes*).

General Notes

This ribonuclease acts on RNA-DNA hybrids.[2]

References

1. Huet, J., Sentenac, A., and Fromageot, P., Detection of nucleases degrading double helical RNA and of nucleic acid–binding proteins following SDS–gel electrophoresis, *FEBS Lett.*, 94, 28, 1978.

2. NC-IUBMB, *Enzyme Nomenclature*, Academic Press, San Diego, 1992, p. 342 (EC 3.1.26.4, Comments).

OTHER NAMES   Pancreatic ribonuclease (recommended name), pancreatic RNase, endoribonuclease I, RNase A, RNase I, RNase

REACTION   Endonucleolytic cleavage to 3′-phosphomononucleotides and 3′-phosphooligonucleotides ending in Cp or Up with 2′,3′-cyclic phosphate intermediates

ENZYME SOURCE   Bacteria, fungi, plants, invertebrates, vertebrates

SUBUNIT STRUCTURE   Monomer (vertebrates)

Method 1

Visualization Scheme

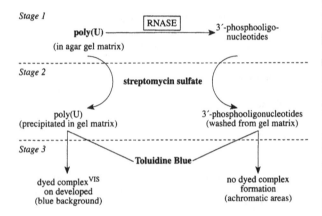

Stage 1

poly(U) ⟶ [RNASE] ⟶ 3′-phosphooligonucleotides
(in agar gel matrix)

Stage 2

streptomycin sulfate

poly(U)   3′-phosphooligonucleotides
(precipitated in gel matrix)   (washed from gel matrix)

Stage 3

Toluidine Blue

dyed complex^{VIS} no dyed complex
on developed   formation
(blue background)   (achromatic areas)

Staining Solution[1]

A. 50 mM Veronal–acetate buffer, pH 7.0
   10 mM EDTA
   0.01% Poly(U) or 1% yeast RNA
B. 2% Agarose solution (50°C)
C. 10 mM Veronal–acetate buffer, pH 7.0
   1.5% Streptomycin sulfate
D. 0.125% Toluidine Blue
   0.5% Acetic acid
E. 2.5% Alycerin
   20% Methanol
   5% Acetic acid

Procedure

Mix equal volumes of A and B solutions and form a 1-mm-thick reactive agarose plate. Apply an electrophorized gel to the reactive agarose plate and incubate in a humid chamber at 40°C for 90 min. Wash the reactive agarose plate in solution C for 60 to 90 min and stain in solution D for 10 min. Wash the stained agarose plate in solution E. Areas of RNASE activity are seen as achromatic bands on a blue background.

Notes: Pyronin B and Acridine Orange may be used instead of Toluidine Blue to stain RNA.

A substrate solution may be applied on Whatman No. 1 filter paper when electrophoresis is carried out in acetate cellulose gel. After incubation, the acetate cellulose gel is fixed in a cold mixture of 25% ethanol and 5% acetic acid for 30 min, stained in a mixture of 2% aqueous Pyronin B (10 ml), 0.4 M sodium acetate (45 ml), and 0.4 M acetic acid (45 ml) for 15 min, and destained in a mixture of equal volumes of 0.4 M sodium acetate and acetic acid. The gel is further destained in methanol:water:acetic acid (5:5:1) for 5 min and placed in the acetate–acetic acid destaining solution. The procedure results in a uniform cranberry stain except at the sites of RNASE action.[2]

An RNA substrate may be included directly into the running PAG before polymerization.[3] In this case the RNASE inhibitor (2 mM copper chloride) should also be added to the running gel. After electrophoresis the gel is preincubated in an appropriate buffer to remove the RNASE inhibitor and then placed in an appropriate buffer for a period sufficient for the enzyme action and stained with Toluidine Blue as described above.

A ribosomal RNA-containing SDS-PAG can be stained using the fluorogenic dye ethidium bromide. RNASE activity results in dark (nonfluorescent) bands visible in UV light on a light (fluorescent) background of the gel.[4] Sensitivity of RNASE detection in RNA-containing SDS-PAG is greatly increased by using buffered 25% isopropanol, rather than buffer alone, to wash detergent from the electrophorized PAG.[5]

It was found that the optimal RNA concentration in a 1% agarose reactive plate is 2 mg/ml. Fixation of an incubated agarose reactive plate in 0.5 N HCl (30 min) and subsequent washing in 0.5% acetic acid before staining with Toluidine Blue gives better results than the procedure described above.[6]

The method is not specific. The bands of some other RNASEs with endonucleolytic activity can also be developed on ribonuclease I zymograms.

RNA substrate may be applied to the electrophoretic gel as a dried agarose film overlay.[7]

METHOD 2

Visualization Scheme

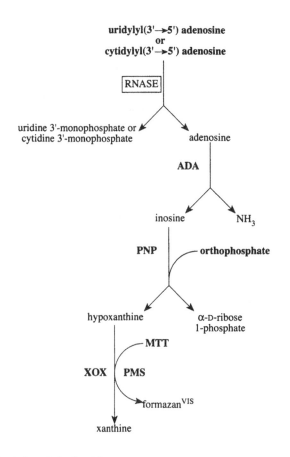

uridylyl(3'→5') adenosine
or
cytidylyl(3'→5') adenosine

RNASE

uridine 3'-monophosphate or
cytidine 3'-monophosphate   adenosine

ADA

inosine   NH$_3$

PNP   orthophosphate

hypoxanthine   α-D-ribose
1-phosphate

MTT

XOX   PMS

formazan^{VIS}

xanthine

Staining Solution[8,9]

A. 10 mg/ml Uridylyl(3' → 5')adenosine, or
    cytidylyl(3' → 5')adenosine, in distilled water   50 μl
    5 mg/ml PMS   200 μl
    5 mg/ml MTT   200 μl
    Purine-nucleoside phosphorylase
    (PNP; Boehringer)   10 μl
    Xanthine oxidase (XOX; Boehringer)   10 μl
    Adenosine deaminase (ADA; Sigma, type III)   20 μl
B. 1% Agarose solution in 50 m*M* sodium
    phosphate buffer, pH 8.0 (45°C)   9.5 ml

Procedure

Mix A and B components of the staining solution and pour the mixture over the gel surface. Incubate the gel in the dark at 37°C until dark blue bands appear. Fix the stained gel in 25% ethanol.

Notes: Phosphodiesterase II (see 3.1.16.1 — SE) can also be developed on RNASE zymograms obtained by this method. However, phosphodiesterase II bands are also developed by this method when adenylyl(3' → 5')adenosine is used as a substrate, whereas RNASE

activity bands are not. These differences in substrate specificity may be used for identification of RNASE and phosphodiesterase II bands on the zymograms obtained by the method described above.

A control gel should also be stained in a staining solution lacking the dinucleoside monophosphates and the auxiliary enzymes to identify "nothing dehydrogenase" (see 1.X.X.X — NDH) bands, which also can develop in the presence of PMS and MTT.

OTHER METHODS

A. When UpA or CpA are used as RNASE substrates, the product adenosine can also be detected using two auxiliary enzymes: adenosine deaminase (to convert adenosine into inosine and ammonia) and glutamate dehydrogenase (to convert ammonia and exogenous 2-oxoglutarate and NADH into L-glutamate and NAD). Gel areas where conversion of NADH into NAD occurs are visible in long-wave UV light as dark bands on the fluorescent background of the gel (for details, see 3.5.4.4 — ADA, Method 2). An additional gel should be stained for phosphodiesterase II as described above (see Method 2, *Notes*).

B. An auxiliary enzyme, 3'-nucleotidase, may be used to detect 3'-phosphomononucleotides produced by RNASE (for details, see 3.1.3.6 — 3'-N).

C. A fluorescence method is described that can be adapted to detect RNASE activity bands on electrophoretic gels. This method is based on the use of quenched fluorophore–RNA covalent conjugate (fluorescein-labeled ribosomal RNA) as a substrate and its subsequent dequenching due to degradation by RNASE. The resulting fluorescence can be observed in long-wave UV light. The principle of quenching of fluorescent dyes upon conjugation to polymeric molecules is based on high local concentrations of self-quenched fluorophores on the polymeric backbone. Depolymerizing enzymes can reverse this state and generate a strong fluorescence signal.[10]

D. A dry agarose film containing ethidium bromide and RNA substrate was used to detect activity of basic RNASEs in an electrophorized PAG. After incubation of the PAG–agarose film sandwich, RNASE activity bands are visible in UV light as dark (nonfluorescent) bands on a light (fluorescent) background.[11]

E. The RNASE activity areas can be directly visualized on an electrophoretic SDS-PAG containing 0.3 mg/ml poly(C) by placing the gel (after removal of SDS) on top of a 20 × 20 cm Silicagel TLC plate with F$_{254}$ fluorescent indicator (Merck), covered with Saran wrap using a 256-nm UV lamp. RNASE activity areas are visible as light bands on a dark background. The gel can be photographed with a Polaroid MP4 Land Camera provided with a yellow filter (Cokin 1A). Once the activity picture is taken, the gel can be further stained with Coomassie Blue or, for better contrast, with the negative activity staining with Toluidine Blue (see Method 1) or silver nitrate. The method has several advantages as it combines, by means of different staining procedures, high resolving power, sensitivity, and specificity with a rapid, reproducible, and simultaneous analysis of purity of RNASE samples on the same gel. Less than 1 pg of RNASE can be detected by this method in less than 2 h after completion of electrophoresis.[12]

General Notes

Proteins, such as albumin, can penetrate the RNA-containing overlay and give false RNASE bands on negative zymograms obtained using Method 1.[8] This has not been observed, and it is not expected to occur in the positive zymogram Method 2 or in Other Methods (see above), since these methods rely on the production of adenosine or 3′-phosphomononucleotide for the staining reaction to occur.

References

1. Berges, J. and Uriel, J., Mise en evidence des activites ribonucleasiques et deoxyribonucleasiques apres electrophorese en acrylamide–agarose, *Biochimie*, 53, 303, 1971.
2. Hodes, M.E., Crisp, M., and Gelb, E., Electrophoresis of acid phosphohydrolase isozymes on Cellogel, *Anal. Biochem.*, 80, 239, 1977.
3. Randles, J.W., Ribonuclease isozymes in Chinese cabbage, systematically infected with turnip yellow mosaic virus, *Virology*, 36, 556, 1968.
4. Rosenthal, A.L. and Lacks, S.A., Nuclease detection in SDS–polyacrylamide gel electrophoresis, *Anal. Biochem.*, 80, 76, 1977.
5. Blank, A., Sugiyama, R.H., and Dekker, C.A., Activity staining of nucleolytic enzymes after sodium dodecyl sulfate–polyacrylamide gel electrophoresis: use of aqueous isopropanol to remove detergent from gels, *Anal. Biochem.*, 120, 267, 1982.
6. Thomas, J.M. and Hodes, M.E., Improved method for ribonuclease zymogram, *Anal. Biochem.*, 113, 343, 1981.
7. Yasuda, T., Takeshita, H., and Kishi, K., Activity staining for detection of ribonucleases using dried agarose film overlay method after isoelectric focusing, *Methods Enzymol.*, 341, 94, 2001.
8. Karn, R.C., Crisp, M., Yount, E.A., and Hodes, M.E., A positive zymogram method for ribonuclease, *Anal. Biochem.*, 96, 464, 1979.
9. Hodes, M.E. and Retz, J.E., A positive zymogram for distinguishing among RNase and phosphodiesterases I and II, *Anal. Biochem.*, 110, 150, 1981.
10. Trubetskoy, V.S., Hagstrom, J.E., and Budker, V.G., Self-quenched covalent fluorescent dye-nucleic acid conjugates as polymeric substrates for enzymatic nuclease assays, *Anal. Biochem.*, 300, 22, 2002.
11. Nadano, D., Yasuda, T., Sawazaki, K., Takeshita, H., and Kishi, K., pH gradient electrophoresis of basic ribonucleases in sealed slab polyacrylamide gels: detection and inhibition of enzyme activity in the gel, *Electrophoresis*, 17, 104, 1996.
12. Bravo, J., Fernandez, E., Ribo, M., de Llorens, R., and Cuchillo, C.M., A versatile negative-staining ribonuclease zymogram, *Anal. Biochem.*, 219, 82, 1994.

3.2.1.1 — α-Amylase; α-AMY

OTHER NAMES   Glycogenase

REACTION   Endohydrolysis of 1,4-α-D-gluco-sidic linkages in polysaccharides containing three or more 1,4-α-linked D-glucose units

ENZYME SOURCE   Bacteria, fungi, plants, invertebrates, vertebrates

SUBUNIT STRUCTURE   Monomer (algae, plants, invertebrates, vertebrates)

Method 1

Visualization Scheme

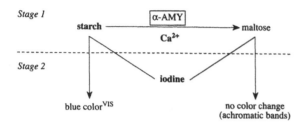

Staining Solution[1]

A. 50 mM Sodium acetate buffer, pH 5.6   100 ml
    1 M CaCl$_2$ 2 ml
B. 10 mM I$_2$
    14 mM KI

Procedure

Incubate an electrophorized starch gel (or PAG containing 0.5% soluble starch) in solution A at 37 to 50°C for 1 h. Discard solution A, wash the gel with distilled water, and stain in an appropriate volume of solution B. The areas of enzyme activity develop on a dark blue background of the gel as light blue or translucent bands, depending on incubation time and the enzyme activity. Discard solution B and rinse the stained gel with water. Record or photograph the zymogram as quickly as possible.

Notes: PAG and acetate cellulose gels can be negatively stained for α-AMY activity by a two-step procedure using an agar overlay[2] or ultrathin agarose plate[3] containing soluble starch and KI–iodine solution.

Dyed substrates Amylopectin Azure[4] and Amylose Azure[5] also may be used to obtain the negative α-AMY zymograms by a one-step procedure. These substrates should be applied to the gel surface as 1% agar overlays.

Amylase-sensitive test paper prepared using Procion Red MX2B-amylopectin is recommended for rapid detection of α-AMY activity on polyacrylamide gels.[6,7] This method gives negative α-AMY zymograms and is sensitive enough to detect the low levels of α-AMY found in urine and serum, and could thus be used to detect α-AMY isozymes in clinical samples.

It should be kept in mind that some negatively stained bands on α-AMY zymograms obtained using the methods described above may be caused by phosphorolytic activity of phosphorylase (see 2.4.1.1 — PHOS, Method 3) when electrophoresis or staining procedures are carried out in the presence of phosphate-containing buffers.[8,9] Calcium chloride (final concentration of 0.02 M) inhibits phosphorolytic digestion of the starch by phosphorylase.

No reliable differences can be detected between α-AMY and β-amylase (β-AMY; EC 3.2.1.2) activities with the negative staining procedures given above. However, plant α- and β-AMY activities can be discerned when they are present in the same gel. α-AMY tolerates a short exposure to 70°C and is insensitive to mercury, copper, and silver. It is activated by calcium and inactivated by low pH (below 3.6). β-AMY is heat labile, sensitive to mercury, copper, and silver, and is not activated by calcium. It tolerates acidic pH (below 3.6).[2] When a starch–iodine system is used to obtain plant amylase zymograms, α-AMY activity areas are seen as light or light bluish bands on a solid blue background, while β-AMY activity areas are seen as reddish bands on a solid blue background.[10]

When the starch–iodine method is used to detect α-AMY from animal serum or some plant tissues, additional negatively stained bands not associated with amylase activity also can be developed. It was established that some plant albumins are responsible for these false activity bands.[11] This is due to the ability of some proteins to preclude the formation of the starch–iodine complex. The most probable reason for the inhibitory effect of albumins on the starch–iodine reaction is the presence of free sulfhydryl groups in albumin molecules. These groups reduce iodine presented in a staining solution into iodide so that no starch–iodine colored complex can be formed in gel areas occupied by (–SH)-rich protein molecules.

The use of an SDS-PAG containing a copolymerized substrate was shown to be preferable for separating and detecting low-molecular-mass α-AMY isozymes insensitive to SDS. The main advantage of this technique is identification of amylase isozymes and determination of their molecular mass in the same gel. However, when PAGs were prepared at polyacrylamide concentrations lower than 10%, there were some difficulties in retaining starch in the gel matrix during electrophoresis. This resulted in a loss of background uniformity of gels when they were stained and in a limitation of resolving high-molecular-mass isozymes.[12]

Achromatic activity bands of α-AMY are developed on starch-containing PAGs stained for the glucan branching enzyme (see 2.4.1.18 — GBE, *Notes*) and pullulanase (see 3.2.1.41 — DEG, Method 2, *Notes*).[13]

METHOD 2

Visualization Scheme

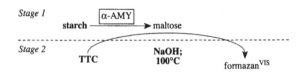

Staining Solution[14]

A. 0.2 *M* Sodium acetate buffer, pH 5.0   100 ml
    Soluble starch (dissolved in the buffer by
    heating)   0.2 g
B. 2,3,5-Triphenyltetrazolium chloride (TTC)   0.1 g
    1 *N* NaOH   100 ml

Procedure

Incubate an electrophorized PAG in solution A at 37°C for 30 min, keeping the gel on a glass plate, surface down. Heat solution B using a water bath. Immediately after the staining solution B starts to boil, dip the gel into the solution by keeping the gel surface down and heat for 3 to 4 min until red-colored bands appear. Immediately after heating, the gel should be washed with 7.5% acetic acid.

Notes: To reduce nonspecific staining of the gel background, treat the gel with 0.1 *M* iodoacetamide for 5 min before staining with solution B.

The use of a thin (0.8 mm) PAG is preferable for the detection of α-AMY activity by this method.

OTHER METHODS

A. Mammalian glycosylated α-amylases can generate glucose (see General Notes, below) and thus can be detected using linking enzymes hexokinase and glucose-6-phosphate dehydrogenase coupled with the PMS–MTT system (e.g., see 3.2.1.26 — FF, Method 5).
B. The immunoblotting procedure (for details, see Part II) may be used to detect α-AMY. Monoclonal antibodies specific to the enzyme from human saliva are available from Sigma.

GENERAL NOTES

Glycosylated α-amylase from the rat liver has a strong affinity to glycogen and is capable of glucose formation from maltotriose and maltopentose oligosaccharides.[15] Similar properties are described for glycosylated α-amylase from human saliva.[16]

REFERENCES

1. Siepmann, R. and Stegemann, H., Enzym-elektrophorese in einschluß-polymerisaten des acrylamids. A. Amylasen, phosphorylasen, *Z. Naturforsch.*, 22b, 949, 1967.
2. Frydenberg, O. and Nielsen, G., Amylase isozymes in germinating barley seeds, *Hereditas*, 54, 123, 1966.
3. Höffelmann, M., Kittsteiner-Eberle, R., and Schreier, P., Ultrathin-layer agar gels: a novel print technique for ultrathin-layer isoelectric focusing of enzymes, *Anal. Biochem.*, 128, 217, 1983.
4. Schiwara, H.-W., Detection of isoamylases with amylopectin azure as a substrate, *Z. Klin. Chem. Klin. Biochem.*, 11, 319, 1973.
5. Rinderknecht, H., Wilding, P., and Haverback, B.J., A new method for the determination of α-amylase, *Experientia*, 23, 805, 1967.
6. Whitehead, P.H. and Kipps, A.E., A test paper for detecting saliva stains, *J. Forensic Sci. Soc.*, 15, 39, 1975.
7. Burdett, P.E., Kipps, A.E., and Whitehead, P.H., A rapid technique for the detection of amylase isoenzymes using an enzyme sensitive "test-paper," *Anal. Biochem.*, 72, 315, 1976.
8. Brewer, G.J., *An Introduction to Isozyme Techniques*, Academic Press, New York, 1970, p. 133.
9. Vallejos, C.E., Enzyme activity staining, in *Isozymes in Plant Genetics and Breeding*, Part A, Tanskley, S.D. and Orton, T.J., Eds., Elsevier, Amsterdam, 1983, p. 469.
10. Chao, S.F. and Scandalios, J.G., Identification and genetic control of starch-degrading enzymes in maize endosperm, *Biochem. Genet.*, 3, 537, 1969.
11. Zimniak-Przybylska, Z. and Przybylska, J., Interference of *Pisum* seed albumins with detecting amylase activity on electropherograms: an apparent relationship between protein patterns and amylase zymograms, *Genet. Pol.*, 17, 133, 1976.
12. Martinez, T.F., Alarcon, F.J., Diaz-Lopez, M., and Moyano, F.J., Improved detection of amylase activity by sodium dodecyl sulfate–polyacrylamide gel electrophoresis with copolymerized starch, *Electrophoresis*, 21, 2940, 2000.
13. Rammesmayer, G. and Praznik, W., Fast and sensitive simultaneous staining method of Q-enzyme, α-amylase, R-enzyme, phosphorylase and soluble starch synthase separated by starch–polyacrylamide gel electrophoresis, *J. Chromatogr.*, 623, 399, 1992.
14. Mukasa, H., Shimamura, A., and Tsumori, H., Direct activity stains for glycosidase and glucosyltransferase after isoelectric focusing in horizontal polyacrylamide gel layers, *Anal. Biochem.*, 123, 276, 1982.
15. Koyama, I., Komine, S., Hokari, S., Yakushijin, M. Matsunaga, T., and Komoda, T., Expression of α-amylase gene in rat liver: liver-specific amylase has a high affinity to glycogen, *Electrophoresis*, 22, 12, 2001.
16. Koyama, I., Komine, S., Yakushijin, M.S., Hokari, S., and Komoda, T., Glycosylated salivary α-amylases are capable of maltotriose hydrolysis and glucose formation, *Comp. Biochem. Physiol.*, 126B, 553, 2000.

3.2.1.3 — Glucan 1,4-α-Glucosidase; GAMY

OTHER NAMES	Glucoamylase, amyloglucosidase, γ-amylase, lysosomal α-glucosidase, acid maltase, exo-1,4-α-glucosidase
REACTION	Hydrolysis of terminal 1,4-linked α-D-glucose residues successively from nonreducing ends of the chains with release of β-D-glucose
ENZYME SOURCE	Bacteria
SUBUNIT STRUCTURE	Unknown or no data available

METHOD

Visualization Scheme

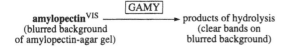

amylopectin^{VIS} ⟶ [GAMY] ⟶ products of hydrolysis
(blurred background   (clear bands on
of amylopectin-agar gel) blurred background)

Staining Solution[1]

100 m*M* Acetate buffer, pH 5.0
0.2% Amylopectin
2% Agar

Procedure

Dissolve amylopectin–agar suspension in a microwave oven, cool to 60°C, and pour the solution on a horizontal glass plate of an appropriate size to form the amylopectin–agar plate. After electrophoresis in native PAG, soak the gel twice for 10 min in 50 m*M* acetate buffer (pH 5.0) containing 25% (v/v) 2-propanol. Then immerse the gel twice in fresh acetate buffer (containing no 2-propanol) for 10 min; blot the gel dry and overlay on top of the amylopectin–agar plate. Place the gel sandwich into a box, seal the box with Parafilm, and incubate at 37°C for 4 h. Zones of glucoamylase location are visible as colorless bands.

Notes: Activity bands of isoamylase (see 3.2.1.68 — IA), α-dextrin endo-1,6-α-glucosidase (see 3.2.1.41 — DEG), and α-amylase (see 3.2.1.1 — α-AMY) are also developed by this method. However, when the processed amylopectin–agar plate is treated with iodine vapor (see 3.2.1.68 — IA), activity bands of IA and DEG become bluish purple, while bands of GAMY and α-AMY remain colorless.

The reliable discrimination between GAMY and α-AMY bands represents a problem that still remains unsolved.

REFERENCES

1. González, R.D., A simple and sensitive method for detection of pullulanase and isoamylase activities in polyacrylamide gels, *Biotechnol. Tech.*, 8, 659, 1994.

3.2.1.4 — Cellulase; CEL

OTHER NAMES	Endo-1,4-β-glucanase
REACTION	Endohydrolysis of 1,4-β-D-glucosidic linkages in cellulose, lichenin, and cereal β-D-glucans
ENZYME SOURCE	Bacteria, fungi, plants, invertebrates
SUBUNIT STRUCTURE	Monomer[#] (bacteria), heterodimer[#] (bacteria)

METHOD 1

Visualization Scheme

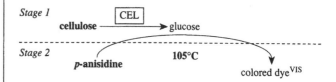

Stage 1 cellulose ⟶ [CEL] ⟶ glucose

Stage 2 *p*-anisidine   105°C   colored dye^{VIS}

Staining Solution[1]

A. 0.1 *M* Acetate buffer, pH 5.0
    0.4% Carboxymethyl cellulose
B. 1% *p*-Anisidine hydrochloride (ethanol solution)

Procedure

Spray the gel with solution A, cover with a sheet of Whatman No. 1 chromatographic paper, and incubate at 37°C for 10 to 20 min. After incubation dry the paper at 105°C, spray with solution B, and dry again. The areas of enzyme localization and liberation of reducing sugar are seen as positively stained bands on the paper print.

Notes: The inclusion of inert polymers into substrate solution A is desirable to raise the viscosity. The areas of enzymatically produced reducing sugars can also be positively stained using triphenyl tetrazolium chloride.[2] For example, see 3.2.1.1 — α-AMY, Method 2.

METHOD 2

Visualization Scheme

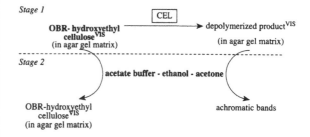

Stage 1 OBR-hydroxyethyl cellulose^{VIS} ⟶ [CEL] ⟶ depolymerized product^{VIS}
(in agar gel matrix)   (in agar gel matrix)

Stage 2 acetate buffer - ethanol - acetone

OBR-hydroxyethyl cellulose^{VIS} (in agar gel matrix)   achromatic bands

Staining Solution[3]

   A. Ostazin Brilliant Red H-3B
        (OBR)–hydroxyethyl cellulose   150 mg
        H_2O 10 ml
   B. 3% Agar solution in 0.2 *M* acetate buffer,
        pH 4.5 (60°C)   20 ml
   C. 0.05 *M* Acetate buffer (pH 5.4)–96%
        ethanol–acetone 1:2:1 (v/v)

Procedure

Mix A and B solutions and pour the mixture between two polyester sheets mounted on glass plates and separated by plastic spacer bars (0.75 mm). Lay an electrophorized gel on an agar gel containing a dyed substrate, which is preheated to about 40°C over a hot plate, and incubate at room temperature until the first most active enzyme zones become visible. Separate the agar layer from the electrophoretic gel and dip into solution C. The achromatic zones of CEL activity develop as a result of solubilization of the depolymerized colored substrate. The duration of the washing in solution C (3 to 20 h) depends on the extent of substrate depolymerization.

A part of the enzyme-released dyed fragments of OBR–hydroxyethyl cellulose stains areas occupied by the enzyme on the electrophoretic gel, but with a lower color contrast and poor sharpness in comparison with achromatic bands developed on a destained agar replica.

Notes: A carboxymethyl cellulose-containing agar replica preincubated with an electrophorized gel may be negatively stained for CEL activity with Congo Red.[2,4,5] It was found that carboxymethyl cellulose is able to adhere to PAG. This allows use of direct incubation of the electrophoretic gel in a solution of 1% carboxymethyl cellulose in 0.1 *M* citrate buffer (pH 4.8) at 50°C for a period as short as 15 min, and then visualization of negatively stained CEL activity bands by treating the gel with 0.1% Congo Red at room temperature for 10 min with constant shaking.[6] An electrophorized PAG containing carboxymethyl cellulose and preincubated in an appropriate buffer may then be negatively stained for CEL activity with iodine solution.[7] The inclusion of the substrate into the running PAG is preferable because the separated CEL isozymes do not need to diffuse out of the gel, and therefore isozymes with slight differences in mobility can be readily distinguished. To prevent enzymatic hydrolysis of the substrate during electrophoresis, the use of a denaturing SDS-PAG is recommended.

OBR–hydroxyethyl cellulose is commercially available (Sigma), but it can be easily prepared under laboratory conditions using hydroxyethyl cellulose and Ostazin Brilliant Red H-3B.[8]

Method 3

Visualization Scheme

4-methylumbelliferyl-β-D-cellobioside

CEL

D-cellobiose   4-methylumbelliferone[UV]
                         (fluorescent)

Staining Solution[9]

0.1 *M* Succinate buffer, pH 5.8
1 m*M* 4-Methylumbelliferyl β-D-cellobioside

Procedure

Apply the staining solution to the gel surface on a filter paper overlay and incubate the gel at 37°C in a humid chamber. Light (fluorescent) bands of CEL activity are visible under long-wave UV light. Record the zymogram or photograph using a yellow filter.

Notes: The method is based on cellobiohydrolase activity of CEL from some sources (e.g., *Clostridium*).

Method 4

Visualization Scheme

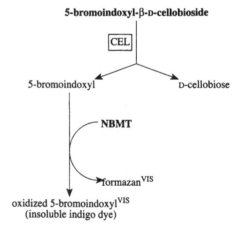

5-bromoindoxyl-β-D-cellobioside

CEL

5-bromoindoxyl   D-cellobiose

NBMT

formazan[VIS]

oxidized 5-bromoindoxyl[VIS]
(insoluble indigo dye)

Staining Solution[10]

A.	5-Bromoindoxyl-β-D-cellobioside	5 mg
	Nitro blue monotetrazolium chloride (NBMT)	20 mg
	Dimethylformamide	0.5 ml
B.	0.1 *M* Sodium acetate buffer, pH 5.0	20 ml

Procedure

Mix A and B components of the staining solution and incubate the gel in the resulting mixture at 40°C in the dark until dark blue bands appear. The time of development is 2 to 15 h, depending on the concentration of the active enzyme in the gel.

Notes: The staining solution can be used many times.

The product 5-bromoindoxyl can also be coupled with diazonium salts (e.g., Fast Blue B or Fast Red TR). However, there is a high background level with these azo dyes.

METHOD 5

Visualization Scheme

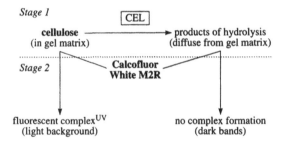

Stage 1

cellulose ⟶ products of hydrolysis
(in gel matrix) (diffuse from gel matrix)

Stage 2 Calcofluor White M2R

fluorescent complex[UV] no complex formation
(light background) (dark bands)

Staining Solution[11]

0.5 *M* Tris–HCl, pH 8.9
0.01% Calcofluor White M2R

Procedure

Stain an electrophorized 15% PAG containing 0.2% (w/v) carboxymethyl cellulose (CM-cellulose) (low viscosity; Sigma C-8758; boiled prior to incorporation into gel) in the staining solution for 5 min at 37°C. Wash the stained gel in distilled water for at least 30 min at room temperature to destain zones of CM-cellulose lysis. Observe dark (nonfluorescent) bands of CEL on a light (fluorescent) background in long-wave UV light.

Notes: The method is based on the use of fluorochrome Calcofluor White M2R, specific for β-1,4-linkages.

REFERENCES

1. Eriksson, K.-E. and Petterson, B., Zymogram technique for detection of carbohydrates, *Anal. Biochem.*, 56, 618, 1973.
2. Bartley, T.D., Murphy-Holland, K., and Eveleigh, D.E., A method for the detection and differentiation of cellulase components in polyacrylamide gels, *Anal. Biochem.*, 140, 157, 1984.
3. Biely, P., Marković, O., and Mislovičová, D., Sensitive detection of endo-1,4-β-glucanases and endo-1,4-β-xylanases in gels, *Anal. Biochem.*, 144, 147, 1985.
4. Beguin, P., Detection of cellulase activity in polyacrylamide gels using Congo Red–stain agar replicas, *Anal. Biochem.*, 131, 333, 1983.
5. Bertheau, Y., Madgidi-Hervan, E., Kotoujansky, A., Nguyen-The, C., Andro, T., and Coleno, A., Detection of depolymerase isoenzymes after electrophoresis or electrofocusing, or in titration curves, *Anal. Biochem.*, 139, 383, 1984.
6. Mathew, R. and Rao, K.K., Activity staining of endoglucanases in polyacrylamide gels, *Anal. Biochem.*, 206, 50, 1992.
7. Goren, R. and Huberman, M., A simple and sensitive staining method for the detection of cellulase isozymes in polyacrylamide gels, *Anal. Biochem.*, 75, 1, 1976.
8. Biely, P., Mislovičová, D., and Toman, R., Soluble chromogenic substrates for the assay of endo-1,4-β-xylanases and endo-1,4-β-glucanases, *Anal. Biochem.*, 144, 142, 1985.
9. Schwarz, W.H., Bronnenmeier, K., Gräbnitz, F., and Staudenbauer, W.L., Activity staining of cellulases in polyacrylamide gels containing mixed linkage β-glucans, *Anal. Biochem.*, 164, 72, 1987.
10. Chernoglazov, V.M., Ermolova, O.V., Vozny, Y.V., and Klyosov, A.A., A method for detection of cellulases in polyacrylamide gels using 5-bromoindoxyl-β-D-cellobioside: high sensitivity and resolution, *Anal. Biochem.*, 182, 250, 1989.
11. Côté, F., Ouakfaoui, S.E., and Asselin, A., Detection of β-1,3 and β-1,4-glucans after native and denaturing polyacrylamide gel electrophoresis, *Electrophoresis*, 12, 69, 1991.

OTHER NAMES   Endo-1,3-β-glucanase, laminarinase

REACTION   Endohydrolysis of 1,3- or 1,4-linkages in β-D-glucans when the glucose residue whose reducing group is involved in the linkage to be hydrolyzed is itself substituted at C-3

ENZYME SOURCE   Bacteria, plants, invertebrates

SUBUNIT STRUCTURE   Monomer[a] (bacteria)

METHOD

Visualization Scheme

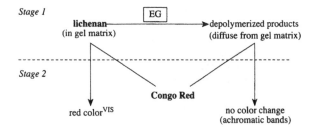

Staining Solution[1]

A. 0.1 *M* Succinate buffer, pH 5.8
   10 m*M* Dithiothreitol
B. 1 mg/ml Congo Red in 0.1 *M* Tris–HCl buffer, pH 8.0
C. 1 *M* NaCl

Procedure

Wash an electrophorized SDS-PAG containing 0.1% lichenan (a mixed β-1,4- and β-1,3-linked glucan) five times for at least 30 min with cold solution A to remove SDS. Then place gel in 0.1 *M* succinate buffer (pH 5.8) and incubate for 30 to 60 min at 37 to 60°C. After incubation, rinse the gel in water and place in solution B for 10 min at room temperature. Destain the gel in solution C for another 10 min. Achromatic or light yellowish activity bands are visible on a deep red background of the gel.

Notes: Barley β-glucan may also be used as an EG substrate instead of lichenan.

Both mixed linkage β-glucans are suitable for the simultaneous detection of β-glucanases with specificities either for β-1,4- (see 3.2.1.4 — CEL), β-1,3- (3.2.1.39 — LAM), or β-1,3-1,4- (3.2.1.6 — EG) linkages. Therefore, it is advisable to perform parallel runs with substrate gels containing either carboxymethyl cellulose (substrate for CEL) or laminarin (substrate for LAM).

The substrate (either barley β-glucan or lichenan) can be incorporated into an ultrathin (0.4 mm) 7% PAG. After electrophoresis or electrofocusing in a native PAG, the gel is dried with cold air, laid onto the substrate-containing PAG, incubated at 37°C for 1 h, and stained with Congo Red.[2]

REFERENCES

1. Schwarz, W.H., Bronnenmeier, K., Gräbnitz, F., and Staudenbauer, W.L., Activity staining of cellulases in polyacrylamide gels containing mixed linkage β-glucans, *Anal. Biochem.*, 164, 72, 1987.
2. Menteur, S., Jestin, L., Risacher, T., and Branlard, G., Visualization of barley β-glucan degrading isozymes after gel isoelectric focusing, *Electrophoresis*, 16, 1019, 1995.

3.2.1.8 — Endo-1,4-β-Xylanase; EX

REACTION	Endohydrolysis of 1,4-β-D-xylosidic linkages in xylans
ENZYME SOURCE	Bacteria, fungi, plants, invertebrates
SUBUNIT STRUCTURE	Unknown or no data available

METHOD 1

Visualization Scheme

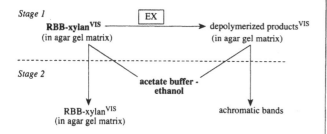

Staining Solution[1]

A. Remazol Brilliant Blue R (RBB)–xylan   150 mg
    H₂O   10 ml
B. 3% Agar solution in 0.2 M acetate buffer,
    pH 4.6 (60°C)   20 ml
C. 0.05 M Acetate buffer
    (pH 5.4)–96% ethanol 1:2 (v/v)

Procedure

Prepare solution A by heating at 60 to 70°C and mix with solution B. Pour the mixture between two polyester sheets mounted on glass plates and separated by plastic spacer bars (0.75 mm) to form a 0.75-mm-thick reactive agar plate. Lay an electrophorized gel on an agar gel containing a dyed substrate, which is preheated to about 40°C over a hot plate, and incubate at room temperature until the enzyme zones become clearly visible against a white light. Then dip agar replica into solution C for 3 to 20 h. The enzyme-degraded substrate zones are destained and appear as pale blue or almost colorless areas on a blue background of the agar replica.

Notes: The dyed substrate RBB–xylan is commercially available from Sigma Chemical Company, but it can be easily prepared in the laboratory.[2]

METHOD 2

Visualization Scheme

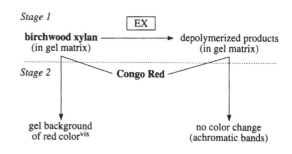

Staining Solution[3]

A. 50 mM Sodium acetate buffer, pH 5.5
B. 1 mg/ml Congo Red in 0.1 M Tris–HCl buffer, pH 8.0
C. 1 M NaCl

Procedure

Wash an electrophorized SDS-PAG containing 0.1% birchwood xylan in 2.5% Triton X-100 for 60 min to remove SDS, and then incubate in solution A for 30 sec to 10 min at 85 to 90°C. Stop the reaction by incubation of the gel at 4°C. Stain the gel in solution B for 20 min at room temperature with gentle shaking. Destain the gel in solution C until achromatic or light yellowish activity bands are visible on a deep red background of the gel. The addition of 1 N HCl to solution C causes the background to turn dark blue.

Notes: This method was used to detect xylanolytic enzymes of thermophilic bacteria.[3,4]

REFERENCES

1. Biely, P., Marković, O., and Mislovičová, D., Sensitive detection of endo-1,4-β-glucanases and endo-1,4-β-xylanases in gels, *Anal. Biochem.*, 144, 147, 1985.
2. Biely, P., Mislovičová, D., and Toman, R., Soluble chromogenic substrates for the assay of endo-1,4-β-xylanases and endo-1,4-β-glucanases, *Anal. Biochem.*, 144, 142, 1985.
3. Sunna, A. and Antranikian, G., Growth and production of xylanolytic enzymes by the extreme thermophilic anaerobic bacterium *Thermotoga thermarum*, *Appl. Microbiol. Biotechnol.*, 45, 671, 1996.
4. Sunna, A., Prowe, S.G., Stoffregen, T., and Antranikian, G., Characterization of the xylanases from the new isolated thermophilic xylan-degrading *Bacillus thermoleovorans* strain K-3d and *Bacillus flavothermus* strain LB3A, *FEMS Microbiol. Lett.*, 148, 209, 1997.

OTHER NAMES | Limit dextrinase, isomaltase, sucrase-isomaltase

REACTION | Hydrolysis of 1,6-β-D-glucosidic linkages in isomaltose and dextrins produced from starch and glycogen by α-amylase

ENZYME SOURCE | Bacteria, fungi, plants, invertebrates, vertebrates

SUBUNIT STRUCTURE | Monomer[#] (bacteria, vertebrates)

METHOD 1

Visualization Scheme

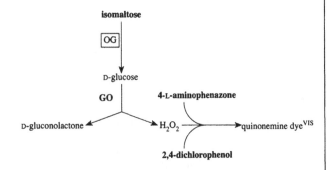

Staining Solution[1]

A. 2 *M* Sodium phosphate buffer, pH 6.0
   25 m*M* Isomaltose
B. 2,4-Dichlorophenol (sulfonated)
C. 10 U/ml Glucose oxidase (GO)
   10 U/ml Peroxidase (PER)
   0.5 mg/ml 4-L-Aminophenazone
D. 1.6% Agarose solution (60°C)

Procedure

Mix 0.75 ml of A with 0.5 ml of B, 2.5 ml of C, and 6.25 ml of D. Pour the mixture over the gel surface and incubate at 37°C until colored bands appear. Record or photograph the zymogram because the stain is not permanent.

Notes: Preparations of glucose oxidase and peroxidase should be catalase-free. If they are not, NaN₃ should be included in the stain to inhibit catalase activity.

Other methods may be used to detect hydrogen peroxide via the linked peroxidase reaction (see 1.11.1.7 — PER).

METHOD 2

Visualization Scheme

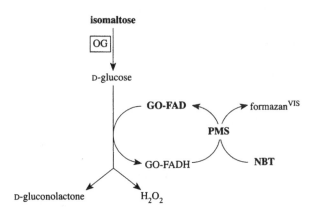

Staining Solution[1]

A. 2 *M* Sodium phosphate buffer, pH 6.0 .... 1 ml
   H₂O .... 4 ml
   PMS .... 2 mg
   NBT .... 5 mg
   Isomaltose .... 43 mg
   Glucose oxidase (GO) .... 100 U
B. 2% Agarose solution (60°C) .... 5 ml

Procedure

Mix A and B components of the staining solution and pour the mixture over the gel surface. Incubate the gel at 37°C in the dark until dark blue bands appear. Fix the stained gel in 25% ethanol or 5% acetic acid.

OTHER METHODS

A. The product D-glucose may also be detected using linking enzymes hexokinase and NAD(P)-dependent glucose-6-phosphate dehydrogenase coupled with PMS–MTT (see 3.2.1.26 — FF, Method 5).
B. An immunoblotting procedure (for details, see Part II) based on the utility of monoclonal antibodies specific to the rat enzyme[2] can be used for immunohistochemical visualization of the enzyme protein on electrophoretic gels. This procedure is not practical; however, it may be indispensable in special immunogenetic analyses of OG.

REFERENCES

1. Finlayson, S.D., Moore, P.A., Johnston, J.J.R., and Berry, D.R., Two staining methods for selectively detecting isomaltase and maltase activity in electrophoretic gels, *Anal. Biochem.*, 186, 233, 1990.
2. Hauri, H.P., Quaroni, A., and Isselbacher, K.J., Monoclonal antibodies to sucrase-isomaltase: probes for the study of postnatal development and biogenesis of the intestinal microvillus membrane, *Proc. Natl. Acad. Sci. U.S.A.*, 77, 6629, 1980.

3.2.1.11 — Dextranase; DEX

REACTION	Endohydrolysis of 1,6-α-D-gluco-sidic linkages in dextran
ENZYME SOURCE	Bacteria, fungi, plants, invertebrates, vertebrates
SUBUNIT STRUCTURE	Monomer# (fungi)

METHOD 1

Visualization Scheme

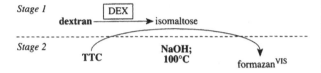

Staining Solution[1]

A. 0.2 M Sodium acetate buffer, pH 5.0   100 ml
    Dextran T 10 (Pharmacia)   0.1 g
B. 2,3,5-Triphenyltetrazolium chloride (TTC)   0.1 g
    1 N NaOH   100 ml

Procedure

Incubate an electrophorized PAG at 37°C for 30 min, keeping the gel on a glass plate, surface down, in a beaker of an appropriate size containing solution A. Heat solution B in another beaker in a water bath. Immediately after the staining solution B starts to boil, dip the gel into the solution by keeping the gel surface down and heat for 3 to 4 min until red-colored bands appear. Wash the stained gel with 7.5% acetic acid. Too much heating causes the gel background to be stained pink.

Notes: The use of a thin (0.8 mm) PAG is preferable for the detection of DEX activity by this method.

To reduce nonspecific staining of the gel background, treat the gel with 0.1 M iodoacetamine for 5 min before staining with solution B.

METHOD 2

Visualization Scheme

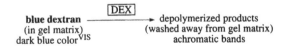

Detecting Mixture[2]

200 mM Sodium citrate buffer, pH 6.0
1% Triton X-100

Procedure

Place an electrophorized SDS-PAG containing 0.25% blue dextran in the detecting mixture and shake gently overnight at 37°C. DEX is thus located as achromatic bands on a blue background of the gel.

Notes: This method coupled with its modification was used to study dextranase inhibitors and properties of the DEX inhibitor complex of *Streptococcus sobrinus* (see General Notes).

OTHER METHODS

Two linking enzymes, isomaltase and glucose oxidase, may be used coupled with the PMS–NBT system to detect the product isomaltose (see 3.2.1.10 — OG, Method 2).

GENERAL NOTES

To detect DEX inhibitors, the electrophorized SDS-PAG containing blue dextran (see Method 2) is renatured by incubation in 0.2 M sodium citrate buffer (pH 6.0) containing 1% Triton X-100 overnight. The renatured gel is then incubated for up to 24 h in the renaturation buffer containing 1% Triton X-100 and 0.5 U/ml exogenous DEX under study. The activity of DEX inhibitors is revealed as dark blue bands of blue dextran in gel areas occupied by inhibitor species where the depolymerizing activity of DEX is arrested.[2]

REFERENCES

1. Mukasa, H., Shimamura, A., and Tsumori, H., Direct activity stains for glycosidase and glucosyltransferase after isoelectric focusing in horizontal polyacrylamide gel layers, *Anal. Biochem.*, 123, 276, 1982.
2. Wellington, J.E., Shaw, J.M., and Walker, G.J, Dissociation and electrophoretic separation of dextranase and dextranase inhibitor from a tightly bound enzyme-inhibitor complex of *Streptococcus sobrinus*, *Electrophoresis*, 14, 613, 1993.

OTHER NAMES   Endochitinase, chitodextrinase, 1,4-
                     β-poly-*N*-acetylglucosaminidase,
                     poly-β-glucosaminidase

REACTION   Random hydrolysis of *N*-acetyl-β-D-
                     glucosaminide 1,4-β-linkages in
                     chitin and chitodextrins

ENZYME SOURCE   Bacteria, diatoms, fungi, plants, pro-
                     tozoa, invertebrates, tunicates

SUBUNIT STRUCTURE   Dimer# (bacteria)

Method 1

Visualization Scheme

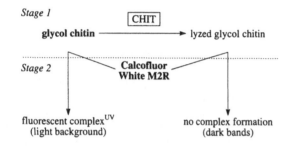

Stage 1

glycol chitin ⟶ lyzed glycol chitin

- -

Stage 2   Calcofluor
                  White M2R

fluorescent complex^UV   no complex formation
(light background)   (dark bands)

Substrate-Containing PAG[1]

7.5% Polyacrylamide gel solution
100 m*M* Sodium acetate buffer, pH 5.0
0.1% Glycol chitin (see *Notes*)

Staining Procedure

Preincubate the electrophorized native PAG in 200 ml of 150 m*M* sodium acetate buffer (pH 5.0) for 5 min. Then put the gel on a clean glass plate and cover with the substrate-containing PAG. Eliminate the liquid between the gels and the glass plate by gently sliding a test tube over the surface of the substrate-containing gel. Incubate the gels at 37°C for 1 h in a plastic container under a moist condition. Following incubation, transfer the gels in a freshly prepared solution of 0.015% Calcofluor White M2R in 0.5 *M* Tris–HCl (pH 8.9) for 5 min. Remove the gels and wash in distilled water for 1 to 2 h at room temperature to destain zones of chitin lysis. Observe dark (nonfluorescent) bands of CHIT on a light (fluorescent) background in a long-wave UV light.

Notes: Photograph the developed gel just following the staining procedure using Polaroid type 55 film and UV haze and 02 orange filters. Use an exposure time of 1 to 5 min with a 127-nm lens at *f* 4.7.

After electrophoresis in an SDS-PAG containing 0.01% (w/v) glycol chitin, wash the gel for 2 h at 37°C with shaking in 100 m*M* sodium acetate buffer (pH 5.0) containing 1% Triton X-100 (purified through a mixed-bed resin deionizing column AG 501-X8). Then stain the renatured gel with Calcofluor White M2R, destain, and photograph as described for the native PAG.

Glycol chitin can be obtained by acetylation of commercially available glycol chitosan. In brief, dissolve 5 g of glycol chitosan in 100 ml of 10% acetic acid by grinding in a mortar. Allow the resulted viscous solution to stand overnight at room temperature. Slowly add methanol (450 ml) and vacuum the filter through a Whatman No. 4 filter paper. Transfer the filtrate into a beaker and add 7.5 ml of acetic anhydride with magnetic stirring. Allow the resulting gel to stand for 30 min at room temperature and cut into small pieces. Discard the liquid extruding from the gel pieces. Transfer gel pieces to a Waring Blender cover with methanol, and homogenize for 4 min at top speed. Centrifuge the suspension at 27,000 *g* for 15 min at 4°C. Resuspend the gelatinous pellet in about 1 volume of methanol, homogenize, and centrifuge as in the preceding step. Resuspend the pellet in 500 ml of distilled water containing 0.02% sodium azide and homogenize again for 4 min. This is the 1% stock solution of glycol chitin ready for use in the substrate-containing PAG.[1]

It was found that this method has several drawbacks.[2] First, it lacks the ability to resolve closely migrating CHIT isoforms. Second, it has a narrow dynamic range. Practically, this makes it difficult to visualize simultaneously all of the CHIT isoforms present in a complex mixture. Finally, it requires the use of expensive, instant photographic film as a means of data storage.[2] While the fluorescent staining of glycol chitin reveals some chitinolytic zones, improved resolution and sensitivity are achieved with a modification of the silver staining procedure commonly used to detect proteins and nucleic acids (see Method 3, below).

Method 2

Visualization Scheme

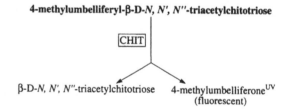

4-methylumbelliferyl-β-D-*N*, *N'*, *N''*-triacetylchitotriose

CHIT

β-D-*N*, *N'*, *N''*-triacetylchitotriose   4-methylumbelliferone^UV
                                                   (fluorescent)

Staining Solution[3]

100 m*M* Potassium phosphate buffer, pH 6.7
   (or 100 m*M* sodium acetate, pH 4.8)
0.75% Agarose
0.3 mg/ml 4-Methylumbelliferyl β-D-*N*,*N'*,*N''*-triacetylchito-
   triose

Procedure

Dissolve agarose–substrate suspension in a microwave oven, cool to 42°C, and apply to the gel surface. Incubate the gel at room temperature and monitor under long-wave UV light. The areas of CHIT activity are seen as fluorescent bands.

Notes: Staining solution can be applied to the gel surface using an AC membrane overlay. Fluorescence of CHIT activity bands can be enhanced by spraying the gel with ammonia after removal of the agarose overlay.

A native PAG, starch gel, or SDS-PAG can be used to study CHIT by this method. When SDS-PAG is used, soak the electrophorized gel in 25% isopropanol for 15 min and then rinse in water. Since isopropanol made wetting of the gel difficult, pour the substrate–agarose solution onto a glass plate, and then place the gel face down in the molten agarose.

Activity bands of lysozyme (see 3.2.1.17 — LZ) can also be detected by this method when bacterial preparations are used for electrophoresis.

4-Methylumbelliferyl β-D-N,N′-diacetylchitobiose can also be used as a fluorogenic substrate to detect CHIT activity bands.[4,5]

METHOD 3

Visualization Scheme

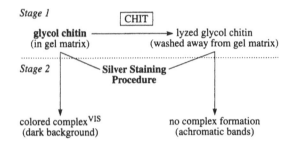

Stage 1

glycol chitin ⟶ lyzed glycol chitin
(in gel matrix)   (washed away from gel matrix)

Stage 2 —— **Silver Staining Procedure**

colored complex[VIS]   no complex formation
(dark background)   (achromatic bands)

Staining Solution[2,6]

A. 50% Methanol (v/v)
   12% Acetic acid (v/v)
   0.0185% (w/v) Formaldehyde
B. 40% (v/v) Ethanol
C. 30% (v/v) Ethanol
D. 0.2 g/l $Na_2S_2O_3 \cdot 5H_2O$ (freshly prepared)
E. 2 g/l $AgNO_3$
   0.75 ml/l 37% Formaldehyde
F. 60 g/l Na_2CO_3
   0.5 ml/l 37% Formaldehyde
   4 mg/ml $Na_2S_2O_3 \cdot 5\ H_2O$
G. 50% Methanol (v/v)
   12% Acetic acid (v/v)

Procedure

Immerse an electrophorized substrate-containing SDS-PAG (see *Notes*) in 0.1 *M* sodium acetate buffer (pH 5.0) containing 1% (v/v) deionized Triton X-100, place on a shaker, and incubate overnight at room temperature. Fix the gel in solution A for 1 h, and then sequentially wash in solutions B and C for 10 min each. Treat the gel with solution D for 1 min, and rinse in distilled water three times (each for 20 sec). Impregnate the gel with solution E for 20 min and rinse in distilled water two times (each for 20 sec). The times of pretreatment, the subsequent rinsing, and the rinsing of the impregnated gel should be observed exactly in order to ensure reproducible image development. Dilute the developer (solution F) with water 1:4 (v/v) to prevent staining of proteins within the lytic zones and to provide better control

over development. Development time can be determined by visual inspection (usually it is 1 to 2 min). Rinse the processed gel with water for 10 sec and place in solution G for 10 min. After development is stopped, wash the gel in 30% methanol for 20 min, then 10% methanol for 20 min, and store in 10% methanol at 4°C until drying. Dry the stained gel between two sheets of cellophane or a sheet of cellophane and filter paper.

Notes: The method is developed for an SDS-PAG[7] containing 0.01% (w/v) glycol chitin.[2] Glycol chitin can be obtained by acetylation of commercially available glycol chitosan as described[1] (see Method 1, *Notes*).

Nondenaturing PAGs containing glycol chitin can be used to directly detect a combination of SDS-tolerant and SDS-intolerant chitinolytic enzymes.

This method is more precise and sensitive than Method 1.

Chitinolytic activity of the egg white lysozyme is readily detected by this method.

OTHER METHODS

The use of carboxymethyl–chitin–Remazol Brilliant Violet as a substrate suitable for in-gel detection of chitinase activity bands was suggested.[8]

GENERAL NOTES

The inclusion of 2-mercaptoethanol in sample buffers is not recommended for CHIT activity gels.[1]

The insect enzyme is synthesized as an inactive precursor that is activated by limited proteolysis. Such a conclusion was inferred from results obtained through SDS-PAG electrophoresis coupled with the immunoblotting procedure using an antibody against *Bombyx mori* CHIT.[9]

The activity of endochitinase (3.2.1.14 — CHIT) can be distinguished from the activity of exochitinase, which catalyzes a progressive release of diacetylchitobiose units from the non-reducing end of chitin chains (see 3.2.1.X — CCB). Unlike endochitinase, exochitinase can not generate fluorescent 4-methylumbelliferone from 4-methylumbelliferyl β-D-N,N′,N″-triacetylchitotriose.[10]

Some chitinases display the activity defined in lysozyme (see 3.2.1.17 — LZ).[11]

REFERENCES

1. Trudel, J. and Asselin, A., Detection of chitinase activity after polyacrylamide gel electrophoresis, *Anal. Biochem.*, 178, 362, 1989.
2. Marek, S.M., Roberts, C.A., Beuselinck, P.R., and Karr, A.L., Silver stain detection of chitinolytic enzymes after polyacrylamide gel electrophoresis, *Anal. Biochem.*, 230, 184, 1995.
3. Tronsmo, A. and Harman, G.E., Detection and quantification of N-acetyl-β-D-glucosaminidase, chitobiosidase, and endochitinase in solutions and on gels, *Anal. Biochem.*, 208, 74, 1993.

4. Chernin, L., Ismailov, Z., Haran, S., and Chet, I., Chitinolytic *Enterobacter agglomerans* antagonistic to fungal plant pathogens, *Appl. Environ. Microbiol.*, 61, 1720, 1995.

5. Zhang, Z., Yuen, G.Y., Sarath, G., and Penheiter A.R., Chitinases from the plant disease biocontrol agent, *Stenotrophomonas maltophilia* C3, *Phytopathology*, 91, 204, 2001.

6. Blum, H., Beier, H., and Gross, H.J., Improved silver staining of plant proteins, RNA and DNA in polyacrylamide gels, *Electrophoresis*, 8, 93, 1987.

7. Laemmli, U.K., Cleavage of structural proteins during the assembly of the head of bacteriophage T4, *Nature*, 227, 680, 1970.

8. Kalix, S. and Buchenauer, H., Direct detection of β-1,3-glucanase in plant extracts by polyacrylamide gel electrophoresis, *Electrophoresis*, 16, 1016, 1995.

9. Koga, D., Fujimoto, H., Funakoshi, T., Utsumi, T., and Ide, A., Appearance of chitinolytic enzymes in integument of *Bombyx mori* during the larval–pupal transformation: evidence for zymogenic forms, *Insect Biochem.*, 19, 123, 1989.

10. McBride, J.D., Stubberfield, C.R., and Hayes, D.J., Electrophoretic detection of chitinase isoenzymes using the PhastSystem, *Electrophoresis*, 14, 165, 1993.

11. NC-IUBMB, *Enzyme Nomenclature*, Academic Press, San Diego, 1992, p. 348 (EC 3.2.1.14, Comments).

3.2.1.15 — Polygalacturonase; PG

OTHER NAMES	Pectin depolymerase, pectinase
REACTION	Random hydrolysis of 1,4-α-D-galactosiduronic linkages in pectate and other galacturonans
ENZYME SOURCE	Bacteria, fungi, plants
SUBUNIT STRUCTURE	Heterodimer[e] (plants – *Lycopersicon esculentum*); see General Notes

METHOD 1

Visualization Scheme

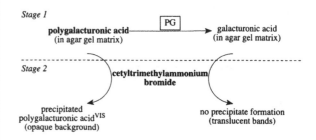

Staining Solution[1]

A. 0.1 *M* Citrate buffer, pH 5.0
   1.5% Agar
   0.5% Polygalacturonic acid (sodium salt)

B. 1% Cetyltrimethylammonium bromide

Procedure

Prepare solution A by heating in a boiling water bath; cool to 50°C and pour over the surface of an electrophorized PAG preincubated in 0.1 *M* citrate buffer (pH 5.0) for 20 min. Incubate the overlayed gel at 37°C for 2 to 3 h. Remove the agar overlay and place it in solution B. The areas of PG activity appear as translucent bands on an opaque background of the reactive agar plate.

Notes: Bands of polygalacturonate lyase (see 4.2.2.2 — PGL) and galacturan 1,4-α-galacturonidase (see 3.2.1.67 — GG) activities also can develop on PG zymograms obtained by this method. The main difference in substrate specificity of the three enzymes possessing polygalacturonase activity is the ability of PG to hydrolyze pectin (PGL and GG do not act on this polysaccharide). This difference may be used to identify PG activity bands.

METHOD 2

Visualization Scheme

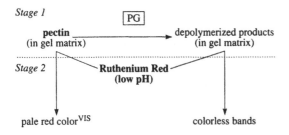

Stage 1

PG

pectin ————————————→ depolymerized products
(in gel matrix)   (in gel matrix)

Stage 2   **Ruthenium Red**
                     (low pH)

pale red color[VIS]   colorless bands

Staining Solution[2]

A. 100 m*M* Malic acid
B. 0.2% Ruthenium Red

Procedure

Incubate an electrophorized PAG containing 0.1% citrus pectin in 100 ml of solution A at room temperature for 90 min to cause a gradual pH change in the gel from the pH value of the gel buffer to a pH of 3.0, allowing PG isozymes and other pectic enzymes (see *Notes*) to act on the pectin while passing through a suitable pH range. Following a brief rinse in distilled water, stain the gel in solution B for 30 to 120 min. Wash the stained gel with several changes of distilled water for 1 h to overnight. Colorless or pale bands visible on a pale red background of the gel correspond to localization of PG.

Notes: PG hydrolyzes the pectin chains to units not receptive to staining or else to oligouronides that are leached from the gel more readily than is the undegraded pectin.

This method also detects activity bands caused by pectinesterase (see 3.1.1.11 — PE, Method 3) and pectin lyase (see 4.2.2.10 — PECL). In pectin-containing PAGs processed as described above, PE produces zones of dark red color and PECL produces yellow zones visible on a pale red background of the gel. PG and PE were found to be useful for biochemical systematics of the genus *Penicillium*.[3]

GENERAL NOTES

To detect PG inhibitors, the electrophorized PAG is first overlayed with sodium polygalacturonate and then treated with exogenous PG under study. The activity of PG inhibitors is revealed after staining the processed gel with Ruthenium Red. Colored bands of polygalacturonate are detected in gel areas occupied by inhibitor species where hydrolytic activity of PG is arrested.[4]

The PG-1 isozyme from tomato fruit is a heterodimer consisting of catalytic and noncatalytic subunits, whereas the PG-2 isozyme consists of only a catalytic subunit. A noncatalytic subunit is present in fruit of all developmental stages, but is absent in vegetative tissues.[5]

REFERENCES

1. Nguyen-The, C., Bertheau, Y., and Coleno, A., Étude des isoenzymes de polygalacturonases, d'endoglucanases de *Rhizopus* spp. et *Mucor* spp. et différenciation d'isolats dans le sud-est de la France, *Can. J. Bot.*, 62, 2670, 1984.
2. Cruickshank, R.H. and Wade, G.C., Detection of pectic enzymes in pectin-acrylamide gels, *Anal. Biochem.*, 107, 177, 1980.
3. Cruickshank, R.H. and Pitt, J.I., Identification of species in *Penicillium* subgenus *Penicillium* by enzyme electrophoresis, *Mycologia*, 79, 614, 1980.
4. Favaron, F., Gel detection of *Allium porrum* polygalacturonase-inhibiting protein reveals a high number of isoforms, *Physiol. Mol. Plant P.*, 58, 239, 2001.
5. Moore, T. and Bennett, A.B., Tomato fruit polygalacturonase isozyme: 1. Characterization of the β-subunit and its state of assembly *in vivo*, *Plant Physiol.*, 106, 1464,1994.

OTHER NAMES   Muramidase
REACTION   Hydrolysis of 1,4-β-linkages between *N*-acetylmuramic acid and *N*-acetyl-D-glucosamine residues in a peptidoglycan and between *N*-acetyl-D-glucosamine residues in chitodextrins
ENZYME SOURCE   Plants, vertebrates
SUBUNIT STRUCTURE Monomer (vertebrates)

METHOD

Visualization Scheme

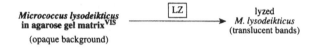

Micrococcus lysodeikticus in agarose gel matrix[VIS] (opaque background) → lyzed *M. lysodeikticus* (translucent bands)

Staining Solution[1]

A. 2% Agarose solution in 0.06 *M* phosphate buffer, pH 6.2 (55°C)

B. 0.12% Suspension of *Micrococcus lysodeikticus* in 0.06 *M* phosphate buffer, pH 6.2

Procedure

Mix equal volumes of solution A and suspension B and pour the mixture over the gel surface. Incubate the gel at 37°C until translucent bands appear. Fix the developed agarose overlay in 10% acetic acid.

Notes: Some chitinases (see EC 3.2.1.14 — CHIT, Methods 2 and 3) and autolysins (see 3.4.24.38 — AUTOL) also display the activity defined in LZ.

The use of an SDS-PAG containing 0.2% of autoclaved *M. lysodeikticus* cells revealed several peptidoglycan hydrolase activities in *Leuconostoc* species after renaturation of separated proteins. Renaturation was carried out by incubation of the SDS-PAG in 50 m*M* potassium phosphate buffer (pH 6.5) containing 0.1% Triton X-100 with gentle agitation at 37°C for 16 h. The renatured PAG was stained in 0.01% KOH containing 0.1% Methylene Blue for 2 h and destained in distilled water to obtain more contrast zymograms. One of these hydrolytic activities was possibly due to *N*-acetylmuramoyl-L-alanine amidase (EC 3.5.1.28).[2]

OTHER METHODS

An SDS-PAG containing 0.01% (w/v) glycol chitin can be used to detect LZ activity bands.[3] After renaturation and incubation, the gel is stained in solution of 0.015% Calcofluor White M2R in 0.5 *M* Tris–HCl (pH 8.9) and then destained in distilled water. Dark (nonfluorescent) bands of LZ are visible on a light (fluorescent) background in long-wave UV light. This method detects 10 ng of purified hen egg white lysozyme. The method can be improved by staining the glycol chitin using adaptation of the silver staining procedure[4] (e.g., see 3.2.1.14 — CHIT, Method 3).

REFERENCES

1. Azen, E.A., Genetic polymorphism of basic proteins from parotid saliva, *Science*, 176, 673, 1972.
2. Cibik, R. and Chapot-Chartier, M.-P., Autolysis of dairy leuconostocs and detection of peptidoglycan hydrolases by renaturing SDS-PAGE, *J. Appl. Microbiol.*, 89, 862, 2000.
3. Trudel, J. and Asselin, A., Detection of chitinase activity after polyacrylamide gel electrophoresis, *Anal. Biochem.*, 178, 362, 1989.
4. Marek, S.M., Roberts, C.A., Beuselinck, P.R., and Karr, A.L., Silver stain detection of chitinolytic enzymes after polyacrylamide gel electrophoresis, *Anal. Biochem.*, 230, 184, 1995.

OTHER NAMES   Exo-α-sialidase (recommended name), neuraminidase

REACTION   Hydrolysis of 2,3-, 2,6-, and 2,8-glycosidic linkages joining terminal nonreducing *N*- or *O*-acylneuraminyl residues to galactose, *N*-acetylhexosamine, or *N*- or *O*-acylated neuraminyl residues in oligosaccharides, glycoproteins, glycolipids, or colominic acid

ENZYME SOURCE   Cells infected by some viruses, bacteria, protozoa, invertebrates, vertebrates

SUBUNIT STRUCTURE   Tetramer# (colifage 63D), tetramer of trimers (α, β, γ)# (protozoa – *Tritrichomonas mobilensis*)

METHOD

Visualization Scheme

2′-(4-methylumbelliferyl)-α-D-N-acetylneuraminic acid

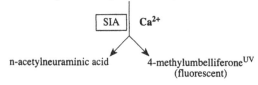

n-acetylneuraminic acid   4-methylumbelliferone^{UV} (fluorescent)

Staining Solution[1]

0.2 *M* Acetate buffer, pH 5.0
5 m*M* CaCl$_2$
0.5 m*M* 2′-(4-Methylumbelliferyl)-α-D-*N*-acetylneuraminic acid

Procedure

Preincubate an electrophorized gel in 0.2 *M* acetate buffer (pH 5.0) containing 5 m*M* CaCl$_2$. Apply the staining solution to the gel surface with filter paper and incubate the gel at room temperature for 15 to 30 min in a humid chamber. The areas of SIA activity are seen in long-wave UV light as fluorescent bands. Fluorescence of SIA activity bands can be enhanced by spraying the processed gel with ammonia. Record the zymogram or photograph using a yellow filter.

Notes: In case of low sialidase activity, treat the processed gel for 2 min with 0.133 *M* glycine buffer (pH 10.0) containing 60 m*M* NaCl and 40 m*M* Na$_2$CO$_3$ to intensify the fluorescence.

OTHER METHODS

A. Methoxyphenyl-α-D-*N*-acetylneuraminic acid may be used as a chromogenic substrate for SIA. After enzymatic action the methoxyphenol released should be coupled with a diazonium salt, yielding an insoluble dye.

B. An immunoblotting procedure (for details, see Part II) based on the utility of monoclonal antibodies specific to the influenza-A virus enzyme[2] also can be used for immunohistochemical visualization of the enzyme protein on electrophoretic gels. This procedure is unsuitable for routine laboratory use, but may be of value in special analyses of SIA.

C. A fluorogenic compound, 4-trifluoromethylumbelliferyl-α-D-*N*-acetylneuraminic acid (CF$_3$MU-Neu5Ac), was chemically synthesized and tested as a substrate for viral, bacterial, and eukaryotic sialidases. It proved to be a very sensitive sialidase substrate suitable for the detection of eukaryotic sialidases, which are expressed poorly and show low specific activity.[3] At neutral pH, the CF$_3$MU fluorescence was more than ten times stronger than 4-methylumbelliferone released after hydrolytic cleavage of another fluorogenic substrate, 4-methylumbelliferyl-α-D-*N*-acetylneuraminic acid.

D. Sialidase resolved in a native PAG and transferred to a polyvinylidene difluoride (PVDF) membrane impregnated with IV^{3}NeuAcαnLc$_4$Cer can be detected using monoclonal antibodies specific to the product, nLc$_4$Cer, generated by sialidase.[4] The method is based on the ability of the PVDF membrane to blot glycosphingolipids.[5] The method based on the use of a substrate-immobilized PVDF membrane for the detection of lipid-metabolizing enzymes was termed *Far Eastern blotting*.[6]

GENERAL NOTES

The advantages of the fluorescent method are the commercial availability of the substrate and the visualization of the product after a single reaction step.

The bacterial enzyme releases essentially all of the sialic acid from a variety of substrates. The enzyme from some bacteria (e.g., *Vibrio cholerae*) requires Ca^{2+} for activity. The mammalian enzyme is less efficient in that it releases only 20 to 30% of the sialic acid from most substrates.

REFERENCES

1. Berg, W., Gutschker-Gdaniec, G., and Schauer, R., Fluorescent staining of sialidases in polyacrylamide gel electrophoresis and ultrathin-layer isoelectric focusing, *Anal. Biochem.*, 145, 339, 1985.

2. Webster, R.G., Kendal, A.P., and Gerhard, W., Analysis of antigenic drift in recently isolated influenza-A (HINI) viruses using monoclonal antibody preparations, *Virology*, 96, 258, 1979.

3. Engstler, M., Talhouk, J.J.W., Smith, R.E., and Schauer, R., Chemical synthesis of 4-trifluoromethylumbelliferyl-α-D-*N*-acetylneuraminic acid glycoside and its use for the fluorometric detection of poorly expressed natural and recombinant sialidases, *Anal. Biochem.*, 250, 176, 1997.

4. Ishikawa, D., Kato, T., Handa, S., and Taki, T., New methods using polyvinylidene difluoride membranes to detect enzymes involved in glycosphingolipid metabolism, *Anal. Biochem.*, 231, 13, 1995.

5. Taki, T., Handa, S., and Ishikawa, D., Blotting of glycolipids and phospholipids from a high-performance thin-layer chromatogram to a polyvinylidene difluoride membrane, *Anal. Biochem.*, 221, 312, 1994.

6. Taki, T. and Ishikawa, D., TLC blotting: application to microscale analysis of lipids and as a new approach to lipid–protein interaction, *Anal. Biochem.*, 251, 135, 1997.

3.2.1.20 — α-Glucosidase; α-GLU

OTHER NAMES	Maltase, glucoinvertase, glucosidosucrase, maltase-glucoamylase
REACTION	Hydrolysis of terminal, nonreducing 1,4-linked α-D-glucose residues with release of α-D-glucose
ENZYME SOURCE	Bacteria, fungi, plants, invertebrates, vertebrates
SUBUNIT STRUCTURE	Monomer (vertebrates)

METHOD 1

Visualization Scheme

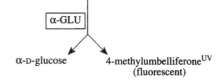

Staining Solution[1]

0.1 *M* Citrate buffer, pH 4.0	20 ml
4-Methylumbelliferyl α-D-glucoside	10 mg

Procedure

Apply the staining solution to the gel surface with filter paper and incubate the gel at 37°C in a humid chamber. Remove the filter paper after 30 to 60 min and view the gel under long-wave UV light. Areas of α-GLU activity are seen as fluorescent bands. Record the zymogram or photograph using a yellow filter.

Notes: Fluorescence of α-GLU activity bands can be enhanced by spraying the gel with ammonia after removal of the filter paper overlay.

METHOD 2

Visualization Scheme

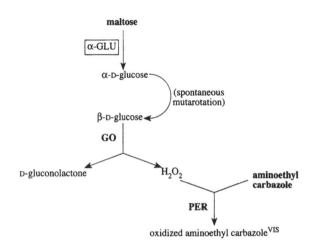

Staining Solution[1]

A. 0.1 *M* Citrate buffer, pH 5.0	10 ml
Maltose (optional)	50 mg
1000 U/ml Glucose oxidase (GO)	50 μl
2500 U/ml Peroxidase (PER)	50 μl
25 mg/ml 3-Amino-9-ethyl carbazole	
(dissolved in acetone)	2 ml
B. 2% Agar solution (60°C)	12 ml

Procedure

Mix A and B components of the staining solution and pour the mixture over the gel surface. Incubate the gel at 37°C until reddish brown bands appear. Fix the stained agar overlay in 7% acetic acid and dry on a filter paper sheet or on a glass plate of appropriate size.

Notes: o-Dianisidine hydrochloride may be used instead of aminoethyl carbazole. It should be remembered, however, that this redox dye is carcinogenic.

The product D-glucose may be detected using glucose oxidase coupled with PMS–NBT (e.g., see 3.2.1.10 — OG, Method 2).

METHOD 3

Visualization Scheme

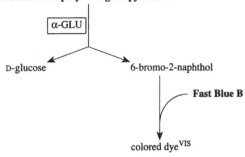

Staining Solution[2]

A.	50 m*M* Potassium phosphate buffer, pH 6.9	
	1 m*M* EDTA	
B.	50 m*M* Potassium phosphate buffer, pH 6.9	10 ml
	6-Bromo-2-naphthyl α-D-glucopyranoside	
	(dissolved in minimal volume of	
	dimethyl sulfoxide)	5 mg
C.	50 m*M* Potassium phosphate buffer, pH 6.9	2 ml
	Fast Blue B	20 mg

Procedure

Soak an electrophorized gel in solution A for 5 min and place in solution B at room temperature for 15 min. Then add solution C. Colored bands of α-GLU activity appear after 3 to 5 min. Fix the stained gel in 8% acetic acid–25% ethanol.

OTHER METHODS

A. An immunoblotting procedure (for details, see Part II) based on the utility of monoclonal antibodies specific to the human enzyme[3] can be used for immunohistochemical visualization of the enzyme protein on electrophoretic gels. This procedure is unsuitable for routine laboratory use but may be of great value in special analyses of α-GLU.

B. The product glucose can be detected using linking enzymes hexokinase and NAD(P)-dependent glucose-6-phosphate dehydrogenase coupled with the PMS–MTT system (see 2.7.1.1 — HK).

3.2.1.20 — α-Glucosidase; α-GLU (continued)

GENERAL NOTES

Sucrose α-glucosidase (see 3.2.1.48 — SG) activity bands also can be detected by all the methods described above. In contrast to α-GLU, this enzyme hydrolyzes isomaltose. This difference may be used to differentiate SG and α-GLU bands.

The enzyme from yeast can not be renatured after SDS-PAG electrophoresis.[4]

REFERENCES

1. Harris, H. and Hopkinson, D.A., *Handbook of Enzyme Electrophoresis in Human Genetics*, North-Holland, Amsterdam, 1976 (loose-leaf, with supplements in 1977 and 1978).
2. Spielman, L.L. and Mowshowitz, D.B., A specific stain for α-glucosidases in isoelectric focusing gels, *Anal. Biochem.*, 120, 66, 1982.
3. Hilkens, J., Tager, J.M., Buijs, F., Brouwer-Kelder, B., Van Thienen, G.M., Tegelaers, F.P., and Hilgers, J., Monoclonal antibodies against human acid α-glucosidase, *Biochim. Biophys. Acta*, 678, 7, 1981.
4. Mukasa, H., Tsumori, H., and Takeda, H., Renaturation and activity staining of glycosidases and glycosyltransferases in gels after sodium dodecyl sulfate–electrophoresis, *Electrophoresis*, 15, 911, 1994.

3.2.1.21 — β-Glucosidase; β-GLU

OTHER NAMES	Gentiobiase, cellobiase, amygdalase, linamarase
REACTION	Hydrolysis of terminal, nonreducing β-D-glucose residues with release of β-D-glucose
ENZYME SOURCE	Bacteria, fungi, red algae, plants, invertebrates, vertebrates
SUBUNIT STRUCTURE	Monomer (invertebrates — insects), dimer (plants)

METHOD 1

Visualization Scheme

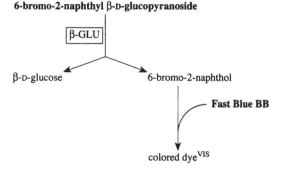

6-bromo-2-naphthyl β-D-glucopyranoside

β-GLU

β-D-glucose   6-bromo-2-naphthol

Fast Blue BB

colored dye[VIS]

Staining Solution[1]

50 mM Phosphate buffer, pH 6.5	70 ml
Polyvinylpyrrolidone	1.6 g
6-Bromo-2-naphthyl β-D-glucopyranoside (dissolved in 10 ml of acetone)	50 mg
Fast Blue BB	100 mg

Procedure

Incubate the gel in staining solution in the dark at 37°C for 2 to 3 h or until blue bands appear. Rinse the gel with water and fix in 50% glycerol.

Notes: This method was successfully applied to native and SDS-containing PAGs during electrophoretic study of multiple isoforms of the enzyme (linamarase) from flax (*Linum usitatissimum*) seeds.[2]

α-Naphthyl β-D-glucopyranoside and Fast Red B can also be used to detect β-GLU activity bands in native and denaturing PAGs.[3]

METHOD 2

Visualization Scheme

4-methylumbelliferyl-β-D-glucoside

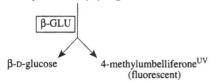

β-D-glucose   4-methylumbelliferone[UV]
                     (fluorescent)

Staining Solution*

0.1 M Phosphate buffer, pH 5.0	20 ml
4-Methylumbelliferyl β-D-glucoside	10 mg

Procedure

Lay a piece of filter paper saturated with the staining solution on top of the gel. Incubate the gel for 30 to 60 min at 37°C. Remove the filter paper and view the gel under long-wave UV light. Areas of β-GLU activity are seen as fluorescent bands. Record the zymogram or photograph using a yellow filter.

Notes: The fluorescence of β-GLU activity bands can be enhanced by spraying the gel with ammonia.

OTHER METHODS

A. Two linked reactions catalyzed by auxiliary enzymes glucose oxidase and peroxidase, coupled with redox dyes aminoethyl carbazole or *o*-dianisidine hydrochloride, can be used to detect the product β-D-glucose (e.g., see 3.2.1.20 — α-GLU, Method 2).

B. The product D-glucose also can be detected using auxiliary enzyme glucose oxidase coupled with the PMS–NBT system (e.g., see 3.2.1.10 — OG, Method 2).

GENERAL NOTES

The enzyme from sweet almonds cannot be renatured after SDS-PAG electrophoresis.[4]

REFERENCES

1. Stuber, C.W., Goodman, M.M., and Johnson, F.M., Genetic control and racial variation of β-glucosidase isozymes in maize (*Zea mays* L.), *Biochem. Genet.*, 15, 383, 1977.
2. Fieldes, M.A. and Gerhardt, K.E., An examination of the β-glucosidase (linamarase) banding pattern in flax seedlings using Ferguson plots and sodium dodecyl sulfate–polyacrylamide gel electrophoresis, *Electrophoresis*, 15, 654, 1994.
3. Vargić, T. and Mrša, V., Detection of exo-β-1,3-glucanase activity in polyacrylamide gels after electrophoresis under denaturing or nondenaturing conditions, *Electrophoresis*, 15, 903, 1994.
4. Mukasa, H., Tsumori, H., and Takeda, H., Renaturation and activity staining of glycosidases and glycosyltransferases in gels after sodium dodecyl sulfate electrophoresis, *Electrophoresis*, 15, 911, 1994.

---

* Adapted from 3.2.1.20 — α-GLU, Method 1.

3.2.1.22 — α-Galactosidase; α-GAL

OTHER NAMES	Melibiase
REACTION	Hydrolysis of terminal, nonreducing α-D-galactose residues in α-D-galactosides, including galactose oligosaccharides, galactomannans, and galactolipids
ENZYME SOURCE	Bacteria, fungi, plants, vertebrates
SUBUNIT STRUCTURE	Dimer (vertebrates)

METHOD 1

Visualization Scheme

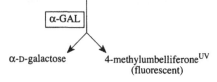

4-methylumbelliferyl-α-D-galactoside

α-GAL

α-D-galactose   4-methylumbelliferone[UV] (fluorescent)

Staining Solution[1] (modified)

0.2 M Phosphate–citrate buffer, pH 5.0	10 ml
4-Methylumbelliferyl α-D-galactoside	10 mg

Procedure

Lay a piece of filter paper saturated with the staining solution on top of the gel. Incubate the gel for 30 to 60 min at 37°C. Remove the filter paper and view the gel under long-wave UV-light.

Notes: Spray the gel with ammonia to enhance fluorescence.

METHOD 2

Visualization Scheme

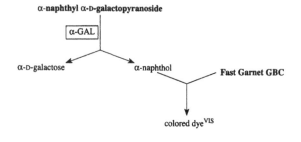

α-naphthyl α-D-galactopyranoside

α-GAL

α-D-galactose   α-naphthol   **Fast Garnet GBC**

colored dye[VIS]

Staining Solution[2]

0.1 M Sodium acetate buffer, pH 5.0	100 ml
1% α-Naphthyl α-D-galactopyranoside (in acetone)	3 ml
Fast Garnet GBC	100 mg

Procedure

Incubate the gel in staining solution in the dark at 30°C until colored bands appear. Rinse the gel with water and fix in 50% glycerol.

Notes: Fast Blue B or Fast Blue RR can be used instead of Fast Garnet GBC. 6-Bromo-2-naphthyl α-D-galactopyranoside may also be used as a substrate.

GENERAL NOTES

Phosphate, citrate, acetate, and phosphate–citrate buffers (pH 4.0 to 7.0) are usually used to detect α-GAL from different sources.

The enzyme appears to be inactivated by storage in liquid nitrogen.[3]

REFERENCES

1. Beutler, E. and Kuhl, W., Biochemical and electrophoretic studies of α-galactosidase in normal man, in patients with Fabry's disease and in *Equidae*, *Am. J. Hum. Genet.*, 24, 237, 1972.
2. Vallejos, C.E., Enzyme activity staining, in *Isozymes in Plant Genetics and Breeding*, Part A, Tanskley, S.D. and Orton, T.J., Eds., Elsevier, Amsterdam, 1983, p. 469.
3. Morizot, D.C. and Schmidt, M.E., Starch gel electrophoresis and histochemical visualization of proteins, in *Electrophoretic and Isoelectric Focusing Techniques in Fisheries Management*, Whitmore, D.H., Ed., CRC Press, Boca Raton, FL, 1990, p. 23.

3.2.1.23 — β-Galactosidase; β-GAL

OTHER NAMES	Lactase
REACTION	Hydrolysis of terminal nonreducing β-D-galactose residues in β-D-galactosides
ENZYME SOURCE	Bacteria, fungi, plants, invertebrates, vertebrates
SUBUNIT STRUCTURE	Monomer (invertebrates, vertebrates), dimer (plants, vertebrates)

METHOD 1

Visualization Scheme

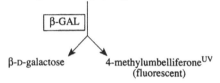

4-methylumbelliferyl β-D-galactoside

β-GAL

β-D-galactose   4-methylumbelliferone[UV] (fluorescent)

Staining Solution[1]

A. 0.5 *M* Acetate buffer, pH 5.0
    3.5 mg/ml 4-Methylumbelliferyl β-D-galactoside
B. 1 *M* Carbonate–bicarbonate buffer, pH 10.0

Procedure

Apply staining solution A to the gel surface with filter paper and incubate the gel at 37°C for 30 min. Fluorescent bands of β-GAL activity are seen under long-wave UV light after spraying the gel with solution B. Record the zymogram or photograph using a yellow filter.

METHOD 2

Visualization Scheme

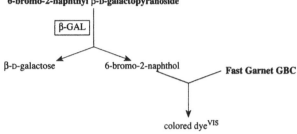

6-bromo-2-naphthyl β-D-galactopyranoside

β-GAL

β-D-galactose   6-bromo-2-naphthol   **Fast Garnet GBC**

colored dye[VIS]

Staining Solution[2]

0.1 *M* Acetate buffer, pH 3.6
1 mg/ml 6-Bromo-2-naphthyl β-D-galactopyranoside
1 mg/ml Fast Garnet GBC

Procedure

Incubate the gel in staining solution in the dark at 37°C until colored bands appear. Wash the gel with water and fix in 50% glycerol.

METHOD 3

Visualization Scheme

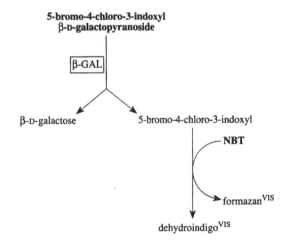

5-bromo-4-chloro-3-indoxyl β-D-galactopyranoside

β-GAL

β-D-galactose   5-bromo-4-chloro-3-indoxyl

NBT

formazan[VIS]

dehydroindigo[VIS]

Staining Solution[3]

0.1 *M* Sodium phosphate buffer, pH 7.5	95 ml
1% 5-Bromo-4-chloro-3-indolyl β-D-galactopyranoside (in dimethylformamide)	5 ml
NBT	10 mg

Procedure

Incubate the gel in staining solution in the dark at 37°C until dark blue bands appear. Wash the gel with water and fix in 25% ethanol.

METHOD 4

Visualization Scheme

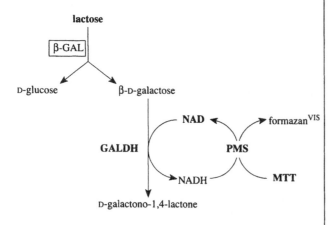

Staining Solution[4]

A.	0.1 M Sodium phosphate buffer, pH 7.0	25 ml
	Lactose (glucose-free; Sigma)	2 g
	NAD	25 mg
	Galactose dehydrogenase (GALDH; Boehringer)	150 μg
	MTT	5 mg
	PMS	5 mg
B.	2% Agar solution (60°C)	25 ml

Procedure

Mix A and B components of the staining solution and pour the mixture over the gel surface. Incubate the gel in the dark at 37°C until dark blue bands appear. Fix the stained gel in 25% ethanol.

Notes: Lactase (EC 3.2.1.108) activity bands can also develop on β-GAL zymograms obtained by this method. This enzyme from intestinal mucosa is isolated as a complex that also catalyzes the reaction of glycosylceramidase (EC 3.2.1.62).

METHOD 5

Visualization Scheme

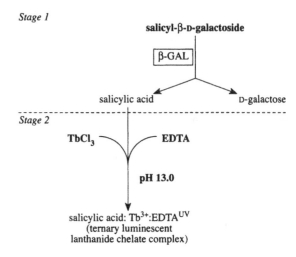

Staining Solution[5] (adapted)

A.	0.1 M Formate buffer, pH 4.0 (see *Notes*)	
	0.5 mM Salicyl-β-D-galactoside (see *Notes*)	
	1 mM MgCl$_2$	
	1 mM Dithiothreitol	
B.	0.01 M HCl	
	5 mM TbCl$_3$	
	5 mM EDTA (tetrasodium salt)	
C.	2.5 M Tris, pH 13.0	

Procedure

Apply substrate solution A to the gel surface with a filter paper or 1% agarose overlay. Incubate the gel at 37°C for 30 min. Remove the first application and apply the next one containing developing solution (one part of solution B, one part of solution C, and three parts of deionized water). Observe luminescent β-GAL bands under 300- to 340-nm UV light and record the zymogram or photograph on Polaroid instant film with a TRP 100 time-resolved photographic camera (Kronem Systems, Inc., Mississauga, Ontario, Canada), using a filter combination that provides excitation in the range of 320 to 400 nm and measuring emission more than 515 nm, with a measurement time delay and gate of 440 and 4.1 msec, respectively.

Notes: The substrate, salicyl-β-D-galactoside, is not yet commercially available. However, it can be synthesized under laboratory conditions according to the following procedure.[5] To a stirred solution of methyl salicylate (0.304 g in 5 ml of absolute ethanol), add 0.96 ml of 15% sodium ethoxide solution in ethanol under argon. Remove volatile components from the formed precipitate (methyl salicylate sodium salt) on a rotary evaporator. Dry the precipitate by addition of distilled pyridine followed by vacuum evaporation. Add 10 ml of 2 mM acetobromo-α-D-galactose in dimethylformamide to 10 ml of 2 mM methyl salicylate sodium salt in dry distilled dimethylformamide with stirring under argon. Allow the reaction to proceed (with stirring under argon) at room temperature for 9 days. Filter the resulting mixture through a sintered glass funnel and concentrate the filtrate on a rotary evaporator to produce a solid residue. Dissolve the product in dichloromethane–methanol, and then chromatograph on a flash silica gel column with 98% dichloromethane–2% triethylamine as the eluting solvent. Pool fractions containing the product with $R_f = 0.60$ and concentrate to produce about 500 mg of salicyl tetra-*O*-acetyl-β-D-galactoside methyl ester, which should then be recrystallized from ethyl acetate–toluene (containing a small amount of hexane) to produce 83 mg of pure product. Dissolve pure salicyl tetra-*O*-acetyl-β-D-galactoside methyl ester (4.82 mg) by stirring in 2 ml of 0.1 M KOH in methanol with heating in an oil bath at 70°C for 3 h to prepare a stock solution of 5 mM salicyl-β-D-galactoside potassium salt.

This substrate is unsuitable for the enzyme with a pH optimum near 7.0. This is probably due to ionization of the carboxyl group, which leaves a negative charge on the substrate molecule and reduces its suitability as a substrate.

Other Methods

An immunoblotting procedure (for details, see Part II) based on the utility of monoclonal antibodies specific to the *E. coli* enzyme[6] can also be used for immunohistochemical visualization of the enzyme protein on electrophoretic gels. This method is unsuitable for routine laboratory use, but may be useful in special analyses of β-GAL. Monoclonal antibodies specific to the enzyme from *E. coli* are now available from Sigma.

General Notes

The enzyme from some sources (e.g., mouse and human) also hydrolyzes β-D-fucosides and β-D-glucosides.[4,7]

References

1. Grzeschik, K.H., Grzeschik, A.M., Benoff, S., Romeo, G., Siniscalco, M., Van Someren, H., Meera Khan, P., Westerveld, A., and Bootsma, D., X-linkage of human α-galactosidase, *Nature New Biol.*, 240, 48, 1972.
2. Seyedyazdani, R., Floderus, Y., and Lundin, L.-G., Molecular nature of β-galactosidase from different tissues in two strains of the house mouse, *Biochem. Genet.*, 13, 733, 1975.
3. Shows, T.B., Scrafford-Wolff, L.R., Brown, J.J.A., and Masler, M.H., G$_{M1}$-Gangliosidosis: chromosome 3 assignment of the β-galactosidase-A gene (βGAL$_A$), *Somat. Cell Genet.*, 5, 147, 1979.
4. Ho, M.W., Povey, S., and Swallow, D., Lactase polymorphism in adult British natives: estimating allele frequences by enzyme assays in autopsy samples, *Am. J. Hum. Genet.*, 34, 650, 1982.
5. Evangelista, R.A., Pollak, A., and Templeton, E.F.G., Enzyme-amplified lanthanide luminescence for enzyme detection in bioanalytical assays, *Anal. Biochem.*, 197, 213, 1991.
6. Duncan, R.J., Hewitt, J., and Weston, P.D., Inactivation of β-galactosidase by monoclonal antibodies, *Biochem. J.*, 205, 219, 1982.
7. Seyedyazdani, R. and Lundin, L.-G., Genetic relationship between β-galactosidase and β-fucosidase in the mouse, *Biochem. Genet.*, 12, 441, 1974.

REACTION	Hydrolysis of terminal, nonreducing α-D-mannose residues in α-D-mannosides
ENZYME SOURCE	Fungi, plants, protozoa, invertebrates, vertebrates
SUBUNIT STRUCTURE	Dimer[#] (protozoa; neutral form of α-MAN), heterotetramer[#] (protozoa; acid form of α-MAN), heterodimer[#] (vertebrates; α-MAN from porcine caudal epididymal fluid), tetramer[#] (plants; vertebrates — hen oviduct)

METHOD 1

Visualization Scheme

4-methylumbelliferyl α-D-mannopyranoside

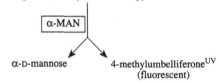

α-D-mannose   4-methylumbelliferone[UV] (fluorescent)

Staining Solution[1]

0.05 M Citrate–phosphate buffer, pH 4.0 to 6.0
2 mg/ml 4-Methylumbelliferyl α-D-mannopyranoside

Procedure

Lay a piece of filter paper saturated with staining solution on top of the gel and incubate at 37°C for 45 min. Remove the filter paper overlay and sprinkle 7.4 N NH$_4$OH over the gel surface. View the gel under long-wave UV light and note zones of enzyme activity as bands of fluorescence. Record the zymogram or photograph through a yellow filter.

METHOD 2

Visualization Scheme

p-nitrophenyl α-D-mannopyranoside

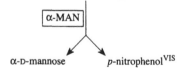

α-D-mannose   p-nitrophenol[VIS]

Staining Solution[2]

25 mM Citrate buffer, pH 4.5
5 mM p-Nitrophenyl α-D-mannopyranoside

Procedure

Preincubate an electrophorized gel in 0.2 M acetate buffer (pH 4.9) at room temperature for 30 min and then incubate at 37°C in a staining solution until yellow bands appear. Photograph the zymogram using a 436.8-nm interference filter.

Notes: To enhance the band coloration caused by p-nitrophenol, drain the gel surface and cover with 0.1 M NaOH. To preserve the yellow color for a few hours, enclose the processed gel in Saran wrap.

METHOD 3

Visualization Scheme

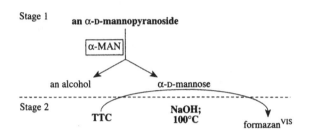

Stage 1   **an α-D-mannopyranoside**

α-MAN

an alcohol   α-D-mannose

Stage 2   TTC   NaOH; 100°C   formazan[VIS]

Staining Solution[2]

A. 25 mM Citrate buffer, pH 4.5
    5 mM α-D-Mannopyranoside (e.g., benzyl α-D-mannopyranoside, methyl α-D-mannopyranoside, p-nitrophenyl α-D-mannopyranoside)
B. 0.1 M Iodoacetamide
C. 0.1% 2,3,5-Triphenyltetrazolium chloride (TTC)
    0.5 N NaOH

Procedure

Incubate an electrophorized PAG in solution A at room temperature for 30 to 45 min. Rinse the gel with water and place in solution B for 5 min. Again, rinse the gel with water thoroughly and place in solution C heated in a boiling water bath. Violet bands on a pink background appear after 1 to 2 min. Wash the stained gel with water and fix in 7.5% acetic acid.

Notes: Treatment of the gel with iodoacetamide is needed to prevent nonspecific reduction of TTC by some nonenzymatic proteins and the PAG itself.

GENERAL NOTES

The enzyme from jack beans, unless heated before electrophoresis, can be renatured after SDS-PAG electrophoresis by removing the SDS with the aid of Triton X-100.[3]

REFERENCES

1. Poenaru, L. and Dreyfus, J.-C., Electrophoretic heterogeneity of human α-mannosidase, *Biochim. Biophys. Acta*, 303, 171, 1973.
2. Gabriel, O. and Wang, S.-F., Determination of enzymatic activity in polyacrylamide gels: I. Enzymes catalyzing the conversion of nonreducing substrates to reducing products, *Anal. Biochem.*, 27, 545, 1969.
3. Mukasa, H., Tsumori, H., and Takeda, H., Renaturation and activity staining of glycosidases and glycosyltransferases in gels after sodium dodecyl sulfate electrophoresis, *Electrophoresis*, 15, 911, 1994.

3.2.1.25 — β-Mannosidase; β-MAN

OTHER NAMES	Mannanase, mannase
REACTION	Hydrolysis of terminal, nonreducing β-D-mannose residues in β-D-mannosides
ENZYME SOURCE	Invertebrates (snails)
SUBUNIT STRUCTURE	See General Notes

METHOD 1

Visualization Scheme

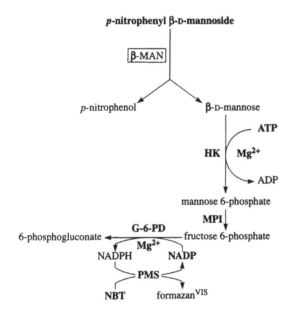

Staining Solution[1] (modified)

250 mM Triethanolamine buffer, pH 7.6
0.1 mg/ml p-Nitrophenyl β-D-mannoside
2.0 U/ml Hexokinase (HK)
1.5 U/ml Mannose-6-phosphate isomerase (MPI)
4 U/ml Glucose-6-phosphate isomerase (GPI)
1 U/ml Glucose-6-phosphate dehydrogenase (G-6-PD)
2.0 mg/ml ATP
0.5 mg/ml NADP
1 mg/ml MgCl$_2$ (6H$_2$O)
0.05 mg/ml PMS
0.5 mg/ml NBT

Procedure

Preincubate an electrophorized gel in 0.1 M acetate buffer (pH 4.9) at room temperature for 30 min, and then incubate at 37°C in the staining solution until dark blue bands appear. Wash the stained gel in water and fix in 25% ethanol.

Notes: The main problem of this method consists in the difference between the pH optimum (acidic) of β-MAN and the pH optimum (alkaline) of auxiliary enzymes used in the stain. It is important, therefore, to equilibrate the gel initially in a buffer with an acidic pH (4.0 to 5.0) before staining at an alkaline pH (7.6). The ionic strength of the acid buffer should be lower than that of the staining solution to ensure the gradual acid-to-alkaline change of the pH value in the gel during its incubation in the staining solution.

METHOD 2

Visualization Scheme

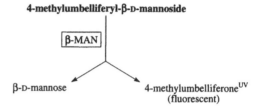

Staining Solution (see *Notes*)

0.05 M Citrate–phosphate buffer, pH 4.0 to 6.0
2 mg/ml 4-Methylumbelliferyl β-D-mannoside

Procedure

Lay a piece of filter paper saturated with staining solution on top of the gel and incubate at 37°C for 15 to 45 min. Remove the filter paper overlay and sprinkle 7.4 N NH$_4$OH over the gel surface. View the gel under long-wave UV light and note zones of enzyme activity as bands of fluorescence. Record the zymogram or photograph through a yellow filter.

Notes: This staining solution is the same as that used to detect α-mannosidase (see α-MAN — 3.2.1.24, Method 1), except 4-methylumbelliferyl β-D-mannoside is substituted for 4-methyl-umbelliferyl α-D-mannoside.

GENERAL NOTES

The enzyme from a snail can not be renatured after SDS-PAG electrophoresis.[1] This suggests that the enzyme molecules are oligomers.

REFERENCES

1. Mukasa, H., Tsumori, H., and Takeda, H., Renaturation and activity staining of glycosidases and glycosyltransferases in gels after sodium dodecyl sulfate electrophoresis, *Electrophoresis*, 15, 911, 1994.

OTHER NAMES	Invertase, saccharase
REACTION	Hydrolysis of terminal nonreducing β-D-fructofuranoside residues in β-D-fructofuranosides
ENZYME SOURCE	Bacteria, fungi, plants
SUBUNIT STRUCTURE	Monomer (plants)

METHOD 1

Visualization Scheme

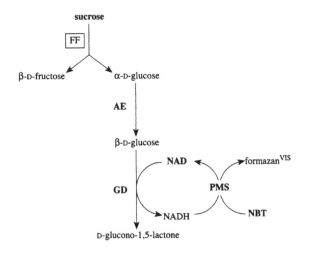

Staining Solution[1]

0.12 M Sodium phosphate buffer, pH 7.6
1.3 mM NAD
1.2 mM Sucrose
0.4 mM PMS
0.6 mM Iodonitrotetrazolium chloride (INT)
0.15 M NaCl
2 U/ml Aldose 1-epimerase (AE)
5 U/ml Glucose dehydrogenase (GD)

Procedure

Incubate the gel in staining solution in the dark at 37°C until dark brownish red bands appear. Wash the stained gel with water and fix in 25% ethanol.

Notes: Aldose 1-epimerase may be omitted from the staining solution because conversion of α-D-glucose into β-D-glucose can occur as a result of spontaneous mutarotation, although more slowly.

METHOD 2

Visualization Scheme

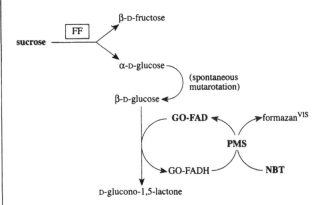

Staining Solution[2]

85 mM Citrate–phosphate buffer, pH 6.0
100 mM Sucrose
0.33 mg/ml PMS
0.33 mg/ml NBT
10 U/ml Glucose oxidase (GO)

Procedure

Incubate the gel in the dark at 37°C in staining solution filtered through glass wool until dark blue bands appear. Wash the stained gel with water and fix in 25% ethanol.

Notes: The staining solution should be prepared, filtered, and used in complete darkness.

The main problem concerning the use of glucose oxidase is the presence of varying amounts of FF in most commercial preparations of this enzyme. Only glucose oxidase preparations free of FF activity should be used.

METHOD 3

Visualization Scheme

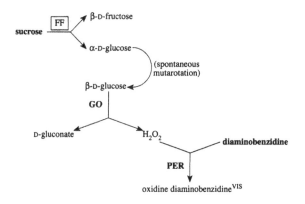

Staining Solution²

85 mM Citrate–phosphate buffer, pH 6.5
100 mM Sucrose
10 U/ml Glucose oxidase (GO)
10 U/ml Peroxidase (PER)
0.30 mg/ml 3,3'-Diaminobenzidine

Procedure

Incubate the gel at 37°C in staining solution filtered through glass wool until colored bands appear. Wash the stained gel with water and fix in 5% acetic acid.

Notes: Only glucose oxidase preparations free of FF activity should be used.

Because diaminobenzidine is a carcinogen, its solutions should be handled with extreme care.

METHOD 4

Visualization Scheme

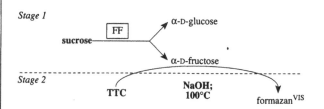

Staining Solution³

A. 0.2 M Acetate buffer, pH 5.0   100 ml
    Sucrose   860 mg
B. 2,3,5-Triphenyltetrazolium chloride (TTC)   100 mg
    1 N NaOH   100 ml

Procedure

Incubate an electrophorized PAG in solution A at 37°C for 30 min and place in solution B heated in a boiling water bath for 1 to 2 h. Red bands appear in gel areas containing β-D-fructose produced by FF. Fix the stained gel in 7.5% acetic acid.

Notes: To prevent possible nonspecific reduction of triphenyltetrazolium by some proteins and by the PAG itself, it is recommended that the gel be treated with 0.1 M iodoacetamide solution for 5 min before being stained in solution B.

METHOD 5

Visualization Scheme

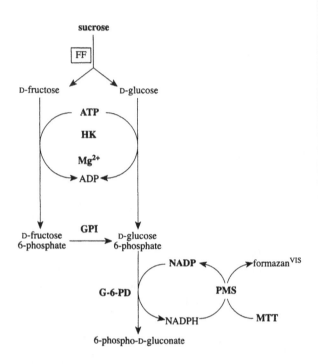

Staining Solution*

A.	0.1 *M* Tris–HCl buffer, pH 7.0	20 ml
	Sucrose	300 mg
	ATP	40 mg
	NADP	20 mg
	MTT	10 mg
	PMS	1 mg
	MgCl$_2$·6H$_2$O	40 mg
	Hexokinase (HK)	100 U
	Glucose-6-phosphate isomerase (GPI)	80 U
	Glucose-6-phosphate dehydrogenase (G-6-PD)	50 U
B.	1.5% Agarose solution (55°C)	20 ml

Procedure

Mix A and B components of the staining solution and pour the mixture over the gel surface. Incubate the gel in the dark at 37°C until dark blue bands appear. Fix the stained gel in 25% ethanol.

Notes: The formazan formation may be doubled by addition of 6-phosphogluconate dehydrogenase to the staining solution. This modified method is supposed to be the most sensitive one among other methods described above.

---

* New; recommended for use.

GENERAL NOTES

Bands of sucrose α-glucosidase (see 3.2.1.48 — SG) activity can also develop on FF zymograms obtained using the methods described above. This enzyme isolated from intestinal mucosa also displays activity toward isomaltose, while FF does not. This difference in substrate specificity may be used to differentiate the bands of activity caused by these two enzymes hydrolyzing sucrose.

The enzyme from yeast can be renatured after SDS-PAG electrophoresis unless heated before electrophoresis.[4]

REFERENCES

1. Babczinski, P., Fractionation of yeast invertase isozymes and determination of enzymatic activity in sodium dodecyl sulfate–polyacrylamide gels, *Anal. Biochem.*, 105, 328, 1980.
2. Faye, L., A new enzymatic staining method for the detection of radish β-fructosidase in gel electrophoresis, *Anal. Biochem.*, 112, 90, 1981.
3. Mukasa, H., Shimamura, A., and Tsumori, H., Direct activity stains for glycosidase and glucosyltransferase after isoelectric focusing in horizontal polyacrylamide gel layers, *Anal. Biochem.*, 123, 276, 1982.
4. Mukasa, H., Tsumori, H., and Takeda, H., Renaturation and activity staining of glycosidases and glycosyltransferases in gels after sodium dodecyl sulfate electrophoresis, *Electrophoresis*, 15, 911, 1994.

3.2.1.28 — α,α-Trehalase; TREH

REACTION   α,α-Trehalose + H$_2$O = 2 D-glucose

ENZYME SOURCE   Fungi, plants, invertebrates, vertebrates

SUBUNIT STRUCTURE   Monomer (invertebrates), dimer (invertebrates)

METHOD 1

Visualization Scheme

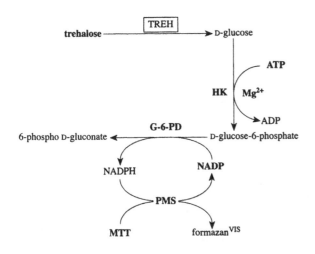

Staining Solution[1]

A. 0.1 M Tris–HCl buffer, pH 7.0
     6 mM MgCl$_2$
     1 mM ATP
     20 mM α,α-Trehalose
     0.2 mg/ml MTT
     0.1 mg/ml PMS
     0.4 mM NADP
     1.4 U/ml Hexokinase (HK)
     1.2 U/ml Glucose-6-phosphate dehydrogenase (G-6-PD)

B. 1.5% Agarose solution (55°C)

Procedure

Mix equal volumes of solutions A and B and pour the mixture over the gel surface. Incubate the gel in the dark at 37°C until dark blue bands appear. Fix the stained gel in 25% ethanol.

METHOD 2

Visualization Scheme

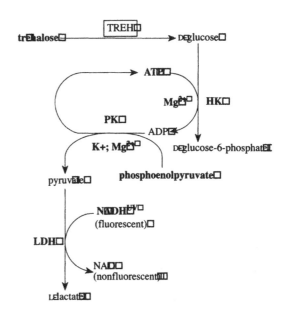

Staining Solution*

A. 0.2 M Tris–HCl buffer, pH 7.0
     30 mM α,α-Trehalose
     10 mM MgCl$_2$
     0.1 M KCl
     1 mM ATP
     0.7 mg/ml NADH
     1.5 mg/ml Phosphoenolpyruvate (tricyclohexylammonium salt)
     1.4 U/ml Hexokinase (HK)
     4.5 U/ml Pyruvate kinase (PK)
     7.5 U/ml Lactate dehydrogenase (LDH)

B. 2% Agarose solution (55°C)

Procedure

Mix equal volumes of solutions A and B and pour the mixture over the gel surface. Incubate the gel at 37°C and view under long-wave UV light. The areas of TREH activity are seen as dark bands on a fluorescent background. Record the zymogram or photograph using a yellow filter.

Notes: When a zymogram that is visible in daylight is required, counterstain the processed gel with MTT–PMS solution. This will result in the development of achromatic bands visible on a blue background of the gel. Fix the negative zymogram in 25% ethanol.

* New; recommended for use.

METHOD 3

Visualization Scheme

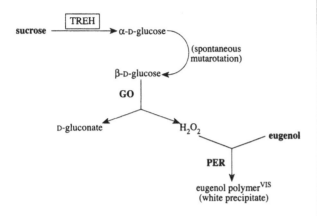

Staining Solution[2]

50 m*M* Sodium citrate buffer, pH 5.5
25 m*M* α,α-Trehalose
6 U/ml Glucose oxidase (GO)
0.9 U/ml Peroxidase (PER)
0.8 m*M* Eugenol

Procedure

Incubate an electrophorized PAG in a staining solution at 37°C until white bands of eugenol polymer deposition appear. Place the stained gel in 7.5% acetic acid in 5% methanol for clearing of the gel background. Although sufficient clearing occurs within 4 h (at 23°C), the gel may be exposed to acetic acid–methanol solution overnight to further reduce the background.

Notes: Chromogenic peroxidase substrates (e.g., diaminobenzidine, *o*-dianizidine, aminoethyl carbazole) may be used instead of eugenol (e.g., see 3.2.1.26 — FF, Method 3). The use of chromogenic peroxidase substrates will allow application of this method to unclear starch and acetate cellulose gels.

Because glucose oxidase used in this method as an auxiliary enzyme is a FAD-containing flavoprotein, PMS–MTT may be used in the stain instead of peroxidase and eugenol. This substitution will result in the appearance of dark blue bands of TREH activity on a white background of the gel (e.g., see 3.2.1.26 — FF, Method 2).

Aldose 1-epimerase (see 5.1.3.3 — AE) may be added to the staining solution to accelerate spontaneous mutarotation of α-D-glucose to β-D-glucose (e.g., see 3.2.1.26 — FF, Method 1).

Glucose dehydrogenase (EC 1.1.1.47) coupled with PMS–MTT may be used instead of glucose oxidase to detect the product, β-D-glucose (e.g., see 3.2.1.26 — FF, Method 1).

METHOD 4

Visualization Scheme

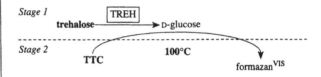

Staining Solution[3]

A. 0.5 *M* Citrate buffer, pH 5.6
B. 30 m*M* Citrate buffer, pH 5.6
   0.1 *M* α,α-Trehalose
C. 0.1 *M* Iodoacetamide
D. 0.2% 2,3,5-Triphenyltetrazolium chloride (TTC)

Procedure

Soak an electrophorized PAG in solution A for 5 min and incubate in solution B at 30°C for 20 min. Then wash the gel with distilled water and immerse in solution C for 5 min. Rinse the gel again with water, immerse in solution D (in the dark), and place in a boiling water bath for 4 min. Violet bands appear in gel areas where TREH activity is localized. Fix the stained gel in 7.5% acetic acid.

Notes: Treatment of the gel with iodoacetamide is needed to prevent nonspecific reduction of TTC by some nonenzymatic proteins and the PAG itself.

REFERENCES

1. Burton, R.S. and La Spada, A., Trehalase polymorphism in *Drosophila melanogaster*, *Biochem. Genet.*, 24, 715, 1986.
2. Killick, K.A. and Wang, L.-W., The localization of trehalase in polyacrylamide gels with eugenol by coupled enzyme assay, *Anal. Biochem.*, 106, 367, 1980.
3. Oliver, M.J., Huber, R.E., and Williamson, J.H., Genetic and biochemical aspects of trehalase from *Drosophila melanogaster*, *Biochem. Genet.*, 16, 927, 1978.

3.2.1.31 — β-Glucuronidase; β-GUS

REACTION	β-D-Glucuronoside + H$_2$O = alcohol + D-glucuronate
ENZYME SOURCE	Bacteria, invertebrates, vertebrates
SUBUNIT STRUCTURE	Tetramer (vertebrates)

METHOD 1

Visualization Scheme

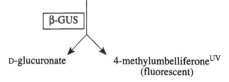

4-methylumbelliferyl β-D-glucuronide

β-GUS

D-glucuronate   4-methylumbelliferone[UV]
(fluorescent)

Staining Solution[1]

0.5 M Tris–citrate buffer, pH 4.5	10 ml
4-Methylumbelliferyl β-D-glucuronide	2 mg

Procedure

Apply the staining solution to the gel surface with filter paper and incubate at 37°C for about 1.5 h. The areas of β-GUS activity are seen under long-wave UV light as fluorescent bands. Record the zymogram or photograph using a yellow filter.

METHOD 2

Visualization Scheme

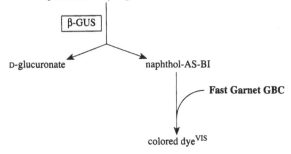

naphthol-AS-BI β-D-glucuronide

β-GUS

D-glucuronate   naphthol-AS-BI

Fast Garnet GBC

colored dye[VIS]

Staining Solution[2]

A.	0.2 M Sodium acetate–HCl buffer, pH 5.0	15 ml
	Naphthol-AS-BI β-D-glucuronide	10 mg
	Fast Garnet GBC	10 mg
B.	2% Agar solution (60°C)	10 ml

Procedure

Mix A and B solutions and pour the mixture over the gel surface. Incubate the gel at 37°C in the dark until colored bands of sufficient intensity appear. Wash the stained gel in water and fix in 7% acetic acid.

Notes: 6-Bromo-2-naphthyl β-D-glucuronide coupled with Fast Blue B can also be used to obtain the positive zymograms of β-GUS.

METHOD 3

Visualization Scheme

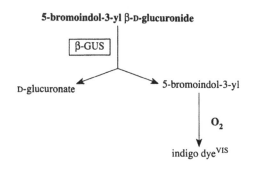

5-bromoindol-3-yl β-D-glucuronide

β-GUS

D-glucuronate   5-bromoindol-3-yl

O$_2$

indigo dye[VIS]

Staining Solution[3]

0.4 M Acetate buffer, pH 4.5	100 ml
5-Bromoindol-3-yl β-D-glucuronide	45 mg

Procedure

Incubate the gel in staining solution at 37°C until dark blue bands appear. Fix the stained gel in 10% trichloroacetic acid for 8 h and store in 7% acetic acid.

Notes: The indigogenic substrate 5-bromo-4-chloro-3-indolyl β-D-glucuronide may be used coupled with NBT (e.g., see 3.1.3.1 — ALP, Method 2). This combination will give simultaneous formation of blue dehydroindigo dye and blue formazan in gel areas where β-GUS activity is localized.

METHOD 4

Visualization Scheme

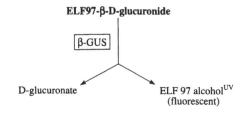

ELF97-β-D-glucuronide

β-GUS

D-glucuronate   ELF 97 alcohol[UV]
(fluorescent)

Staining Solution[4]

50 mM Sodium phosphate buffer, pH 7.0
25 μM ELF 97-β-D-glucuronide (Molecular Probes, Eugene, OR)

3.2.1.31 — β-Glucuronidase; β-GUS (continued)

Procedure

Incubate an electrophorized SDS-PAG in two 10-min changes of 0.1% Triton X-100 in 50 mM sodium phosphate buffer (pH 7.0) to remove SDS and then in staining solution at room temperature for 60 min to overnight. Observe green fluorescent bands of β-GUS using a midrange UV transilluminator operating at 302 nm, such as the Foto/UV 450 transilluminator (Fotodyne Inc., Hartford, WI) or the UVP transilluminator/Polaroid MP4+ camera system (UVP, Upland, CA).

Notes: This method can be combined with the SYPRO Ruby dye method of detection of total proteins after SDS-PAG electrophoresis. This combination permits convenient dichromatic detection of β-GUS (due to green fluorescence of ELF 97 alcohol) and general proteins (red fluorescence due to direct interaction of SYPRO Ruby dye with basic amino acids in proteins after removal of SDS from the gel).[4]

REFERENCES

1. Harris, H. and Hopkinson, D.A., *Handbook of Enzyme Electrophoresis in Human Genetics*, North-Holland, Amsterdam, 1976 (loose-leaf, with supplements in 1977 and 1978).
2. Aebersold, P.B., Winans, G.A., Teel, D.J., Milner, G.B., and Utter, F.M., *Manual for Starch Gel Electrophoresis: A Method for the Detection of Genetic Variation*, NOAA Technical Report NMFS 61, U.S. Department of Commerce, National Marine Fisheries Service, Seattle, WA, 1987.
3. Yoshida, K., Iino, N., Koga, I., and Kato, K., Demonstration of β-glucuronidase in disc electrophoresis by means of 5-bromoindol-3-yl-β-D-glucuronide, *Anal. Biochem.*, 58, 77, 1974.
4. Kemper, C., Steinberg, T.H., Jones, L., and Patton, W.F., Simultaneous, two-color fluorescence detection of total protein profiles and β-glucuronidase activity in polyacrylamide gel, *Electrophoresis*, 22, 970, 2001.

3.2.1.32 — Xylan Endo-1,3-β-Xylosidase; XYL

OTHER NAMES	Xylanase, endo-1,3-β-xylanase
REACTION	Random hydrolysis of 1,3-β-D-xylosidic linkages in 1,3-β-D-xylans
ENZYME SOURCE	Bacteria
SUBUNIT STRUCTURE	Unknown or no data available

METHOD

Visualization Scheme

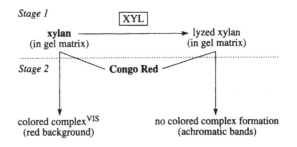

Staining Procedure[1]

The method is developed for an SDS-PAG containing 0.1% xylan. Dissolve xylan at a concentration of 1% in 0.1 N NaOH. Incorporate xylan into the SDS-PAG by adding a 1/10 volume of the prepared 1% alkaline stock solution to the separating gel solution prepared according to Laemmli.[2] Neutralize the mixture with 6 N HCl, before the addition of ammonium persulfate and TEMED. The xylan, although it becomes insoluble after neutralization, remains a uniform suspension in the gel.

After electrophoresis, wash the gel by five successive 30-min washes in 250 ml of cold 0.1 M phosphate buffer (pH 7.0) with slight shaking (the first two washes contain 25% isopropanol). Then submerse the gel in 0.1 M phosphate buffer (pH 7.0) for 30 to 60 min at 55°C to promote the hydrolysis of the substrate by XYL. Stain the gel with 1 mg/ml solution of Congo Red for 10 to 30 min and destain in 1 M NaCl for another 10 to 30 min.

Clear activity bands of XYL are visible on a deep red background.

Notes: The identification of depolymerizing enzymes that are fractionated by SDS-PAG electrophoresis and that have water-insoluble substrates usually is achieved through the agar replica method.[3] This method, however, suffers from pure resolution due to the slow diffusion of renatured enzymes out of electrophoretic gels. The method described above results in the development of sharp activity bands and allows differentiation of XYL isozymes with slight differences in molecular weight.

3.2.1.32 — Xylan Endo-1,3-β-Xylosidase; XYL (continued)

REFERENCES

1. Chen, P. and Buller, C.S., Activity staining of xylanases in polyacrylamide gels containing xylan, *Anal. Biochem.*, 226, 186, 1995.
2. Laemmli, U.K., Cleavage of structural proteins during the assembly of the head of bacteriophage T4, *Nature*, 227, 680, 1970.
3. Gabriel, O. and Gersten, D.M., Staining for enzymatic activity after gel electrophoresis, *Anal. Biochem.*, 203, 1, 1992.

3.2.1.35 — Hyaluronoglucosaminidase; HYAL

OTHER NAMES	Hyaluronidase
REACTION	Random hydrolysis of 1,4-linkages between N-acetyl-β-D-glucosamine and D-glucuronate residues in hyaluronate
ENZYME SOURCE	Bacteria, invertebrates, vertebrates
SUBUNIT STRUCTURE	Unknown or no data available

METHOD

Visualization Scheme

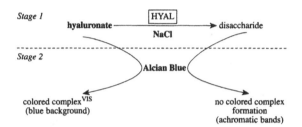

Staining Solution[1]

A. 50 mM Phosphate–citrate buffer, pH 5.0
    1 mg/ml Hyaluronic acid
    0.15 M NaCl
B. 1% Alcian Blue in 7% acetic acid

Procedure

Join an electrophorized acetate cellulose strip with a second acetate cellulose strip soaked in solution A, avoiding the appearance of air bubbles between the strips. Place the strips between two glass plates and incubate at 37°C for 30 min. Dry the substrate-containing strip and stain in solution B for 1 min. Wash the stained acetate cellulose strip with water until achromatic bands appear on a blue background.

Notes: Hyaluronic acid may be incorporated directly into the running PAG prior to polymerization. After incubation in an appropriate buffer solution, the gel may be negatively stained with "Stains-all"[2] or Alcian Blue.[3]

Two other enzymes possessing hyaluronidase activity can develop on HYAL zymograms obtained by this method. These are hyaluronoglucuronidase (EC 3.2.1.36) and hyaluronate lyase (EC 4.2.2.1). Chondroitinase (EC 4.2.2.4) activity bands can also be developed by this method when an alkaline buffer is used in the stain.

An electrophoretic PAG can be stained for HYAL activity using a substrate-containing gel consisting of 1% agarose, 1% bovine serum albumin, and 0.8 mg/ml hyaluronic acid. After electrophoresis, wash the gel in 0.3 M phosphate buffer (pH 5.3) for 10 to 15 min, place directly on the substrate-containing gel, and incubate at 37°C for 48 h. After incubation, remove the gel and place the substrate-containing gel replica in 2 M acetic acid. Gel areas where the hyaluronic acid is digested appear as clear zones on an opaque background of the substrate-containing gel plate. The zymogram can be photographed against a dark background. The developed zymogram is stable indefinitely when stored in distilled water or acetic acid.[4]

3.2.1.35 — Hyaluronoglucosaminidase; HYAL (continued)

General Notes

To detect HYAL inhibitors, an electrophorized SDS-PAG containing 1 mg/ml hyaluronic acid is incubated in 3% Triton X-100 in 50 mM HEPES (pH 7.4) for 1 h with agitation in order to remove the SDS. The gel is rinsed twice with 50 mM HEPES (pH 7.4) and transferred to the appropriate buffer containing 0.5 U/ml HYAL under study. After 16 h incubation of the gel at 37°C with agitation, the activity of HYAL inhibitors is revealed by staining the processed gel with a solution of 0.5% Alcian Blue in 3% acetic acid for 1 h and subsequent destaining in 7% acetic acid, changing the destaining solution once every hour until positively stained bands of HYAL inhibitors appear (usually 2 to 3 h). Colored bands of hyaluronic acid are detected in gel areas occupied by inhibitor species where hydrolytic activity of HYAL is arrested.[5]

References

1. Herd, J.K., Tschida, J., and Motycka, L., The detection of hyaluronidase on electrophoresis membranes, *Anal. Biochem.*, 61, 133, 1974.
2. Fiszer-Szafarz, B., Hyaluronidase polymorphism detected by polyacrylamide gel electrophoresis: application to hyaluronidases from bacteria, slime molds, bee, and snake venoms, bovine testes, rat liver lysosomes, and human serum, *Anal. Biochem.*, 143, 76, 1984.
3. Mio, K. and Stern, R., Reverse hyaluronan substrate gel zymography procedure for the detection of hyaluronidase inhibitors, *Glycoconjugate J.*, 17, 761, 2000.
4. Steiner, B. and Cruce, D., A zymographic assay for detection of hyaluronidase activity on polyacrylamide gels and its application to enzymatic activity found in bacteria, *Anal. Biochem.*, 200, 405, 1992.

3.2.1.37 — Xylan 1,4-β-Xylosidase; XX

OTHER NAMES	Xilobiase, β-xylosidase, exo-1,4-β-xylosidase
REACTION	Hydrolysis of 1,4-β-D-xylans so as to remove successive D-xylose residues from the nonreducing termini
ENZYME SOURCE	Bacteria, fungi, plants
SUBUNIT STRUCTURE	Unknown or no data available

Method

Visualization Scheme

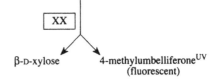

4-methylumbelliferyl β-D-xyloside

β-D-xylose   4-methylumbelliferone[UV] (fluorescent)

Staining Solution[1]

A.	50 mM Sodium acetate buffer, pH 5.5	100 ml
B.	50 mM Sodium acetate buffer, pH 5.5	10 ml
	1% 4-Methylumbelliferyl β-D-xyloside (dissolved in 50% acetone)	0.3 ml

Procedure

Soak an electrophorized gel in solution A for 30 to 45 min. Discard solution A and place a piece of filter paper saturated with solution B on top of the gel. Incubate the gel for 30 min to 2 h and view under long-wave UV light for fluorescent bands. Record the zymogram or photograph using a yellow filter. The resolution of the bands will be lost in a short period of time.

References

1. Vallejos, C.E., Enzyme activity staining, in *Isozymes in Plant Genetics and Breeding*, Part A, Tanskley, S.D. and Orton, T.J., Eds., Elsevier, Amsterdam, 1983, p. 469.

3.2.1.39 — Glucan Endo-1,3-β-D-Glucosidase; LAM

OTHER NAMES   Laminarinase, endo-1,3-β-glucanase

REACTION   Hydrolysis of 1,3-β-D-glucosidic linkages in 1,3-β-D-glucans

ENZYME SOURCE   Bacteria, fungi, algae, plants, invertebrates, vertebrates (fishes)

SUBUNIT STRUCTURE   Unknown or no data available

METHOD 1

Visualization Scheme

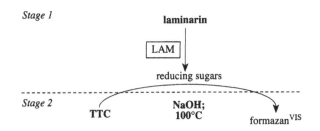

Staining Solution[1]

A.   0.05 M Potassium acetate buffer, pH 5.0   75 ml

B.   Laminarin   1 g

     H₂O   75 ml

C.   1.0 M NaOH   100 ml

     2,3,5-Triphenyltetrazolium chloride (TTC)   150 mg

Procedure

Wash an electrophorized PAG with distilled water three times and preincubate in 0.05 M potassium acetate buffer (pH 5.0) for 5 min with slow shaking. Before mixing A and B components of the staining solution, dissolve laminarin in water by heating in a boiling water bath. Incubate the gel in A + B mixture at 40°C for 30 min. Wash the gel three times with distilled water and place onto a glass tray containing solution C. Keep the tray in a boiling water bath until red bands appear (about 10 min).

Notes: To reduce the pink background, put the gel into 7.5% acetic acid as soon as the bands of interest clearly appear. To avoid breaking the gel, it should be placed in a mixture containing 3% glycerol, 40% methanol, 10% acetic acid, and 47% water (v/v). This mixture, however, destains the gel slowly.

A control gel should be incubated under the same conditions, except that laminarin is omitted in order to identify nonspecifically stained bands.

METHOD 2

Visualization Scheme

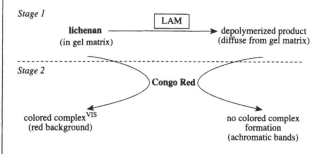

Staining Solution[2]

A.   0.1 M Succinate buffer, pH 5.8

     10 mM Dithiothreitol

B.   1 mg/ml Congo Red in 0.1 M Tris–HCl buffer, pH 8.0

C.   1 M NaCl

Procedure

Wash an electrophorized SDS-PAG containing 0.1% lichenan (a mixed β-1,4- and β-1,3-linked glucan) five times for at least 30 min with cold solution A to remove detergent. Then submerge the gel in 0.1 M succinate buffer (pH 5.8) and incubate for 30 to 60 min at 37 to 60°C. After incubation rinse the gel in water and place in solution B for 10 min at room temperature. Destain the gel in solution C for another 10 min. Achromatic or light yellowish activity bands are visible on a deep red background of the gel.

Notes: Barley β-glucan may also be used as a LAM substrate instead of lichenan. Both mixed linkage β-glucans are suitable for the simultaneous detection of β-glucanases with specificities for either β-1,4- (see 3.2.1.4 — CEL), β-1,3-, 1,4- (see 3.2.1.6 — EG), or β-1,3- (3.2.1.39 — LAM) linkages. It is therefore advisable to perform parallel runs with substrate gels containing either carboxymethyl cellulose (substrate for CEL) or laminarin (substrate for LAM).

METHOD 3

Visualization Scheme

CM-curdlan-RBB $\xrightarrow{\boxed{\text{LAM}}}$ depolymerized CM-curdlan-RBB
(in gel matrix)   (washed away from gel matrix)
blue background[VIS]   clear bands

Visualization Procedure[3]

CM–curdlan–RBB is a carboxymethyl (CM) polysaccharide linked with the dye Remazol Brilliant Blue (RBB) (Loewe Biochemica, Otterfing, Germany). It is used as a substrate for LAM and included in a native PAG before polymerization (final concentration of 1.33 mg/ml in separating gel). After electrophoresis, incubate the gel in 100 mM acetate buffer (pH 5.0) at 30°C until clear zones of LAM activity appear on a blue background of the gel. Dry the stained gel.

METHOD 4

Visualization Scheme

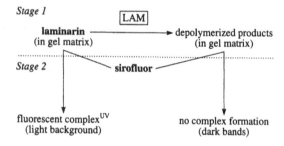

Stage 1

laminarin $\xrightarrow{\boxed{\text{LAM}}}$ depolymerized products
(in gel matrix)   (in gel matrix)

Stage 2   **sirofluor**

fluorescent complex[UV]   no complex formation
(light background)   (dark bands)

Staining Solution[4]

0.1 M Glycine–NaOH buffer, pH 11.5
35 mg/ml Sirofluor (see *Notes*)

Procedure

Incubate an electrophorized 15% PAG containing 0.2% (w/v) laminarin (Sigma L-9634) in 50 mM sodium acetate buffer (pH 5.0) at 37°C for 45 min. Stain the gel in the staining solution for at least 15 min at 37°C. No destaining step is required. Observe dark (nonfluorescent) bands of LAM on a light (fluorescent) background in long-wave UV light (however, see *Notes*, below).

Notes: The method is based on the use of the fluorochrome sirofluor, specific for β-1,3-linkages. Sirofluor is present in commercially available Aniline Blue and can be purified in a laboratory using the procedure described below.

Pass the solution of Aniline Blue (1.82 g in 35 ml of 5% triethylamine) through a dry cellulose column (Bio-Rad Cellex N-1, 50 g). Develop the column (25 × 200 mm) with 65 ml of 1% triethylamine.

Extract the yellow fluorescent moiety separated from Aniline Blue with 200 ml of boiling water, and then filter through Whatman No. 1 paper. Lyophilize the eluate, resuspend in 1.5 ml of 15 triethylamine, and pass through another dry cellulose column. Extract, filter, and flash-evaporate the fluorochrome to a final volume of 30 ml. Add 10 g of Dowex-50X (H⁺ form) resin (Sigma). Mix the mixture for 10 min, neutralize with NaOH, and filter and lyophilize as above. The resulting powder (sirofluor) is ready for use.

Commercial Aniline Blue can also be used to detect negatively stained LAM activity bands visible in long-wave UV light;[5,6] however, the use of purified sirofluor is preferable.

Two types of bands corresponding to endo-1,3-β-glucanase activity can be detected by this method: (1) dark nonfluorescent bands, corresponding to extensive degradation of laminarin, and (2) zones of fluorescence higher than the background level, corresponding to partial enzymatic action on the substrate. It is suggested that this enhanced fluorescence is related to β-1,6-linked groups of the main polysaccharide chain. Dark nonfluorescent band can appear within a high-fluorescence band after prolonged incubation of the electrophoretic gel in 50 mM sodium acetate buffer (pH 5.0) before staining.

Alkali-soluble *Saccharomyces cerevisiae* β-1,3-glucan can also be used as a LAM substrate incorporated in the PAG (0.6 mg/ml) before polymerization. Following electrophoresis and incubation in 10 mM citrate–phosphate buffer (pH 5.0), the gel may be stained with Aniline Blue (0.025% in 150 mM potassium phosphate buffer, pH 8.6)[6] or sirofluor (see procedure above) to observe dark bands of LAM in UV light.

GENERAL NOTES

When compared with laminarin-embedded PAGs, the alkali-soluble yeast glucan-embedded PAGs allow a more precise detection of activity bands of endo-1,3-β-glucanases from bacteria, fungi, and plants. Another advantage with the use of the yeast glucan as a substrate is the capacity to analyze endo-1,3-β-glucanases after SDS-PAG electrophoresis.[6]

A procedure was developed for double activity staining of laminarinase and peroxidase (see 1.11.1.7 — PER, Method 1) in the same PAG using amino-ethyl-carbazole for peroxidase staining and laminarin and triphenyltetrazolium for laminarinase. This procedure saves time and sample material.[7]

REFERENCES

1. Pan, S.-Q., Ye, X.-S., and Kuć, J., Direct detection of β-1,3-glucanase isozymes on polyacrylamide electrophoresis and isoelectrofocusing gels, *Anal. Biochem.*, 182, 136, 1989.
2. Schwarz, W.H., Bronnenmeier, K., Gräbnitz, F., and Staudenbauer, W.L., Activity staining of cellulases in polyacrylamide gels containing mixed linkage β-glucans, *Anal. Biochem.*, 164, 72, 1987.
3. Kalix, S. and Buchenauer, H., Direct detection of β-1,3-glucanase in plant extracts by polyacrylamide gel electrophoresis, *Electrophoresis*, 16, 1016, 1995.
4. Côté, F., Ouakfaoui, S.E., and Asselin, A., Detection of β-1,3 and β-1,4-glucans after native and denaturing polyacrylamide gel electrophoresis, *Electrophoresis*, 12, 69, 1991.

3.2.1.39 — Glucan Endo-1,3-β-D-Glucosidase; LAM (continued)

5. Côté, F., Letarte, J., Grenier, J., Trudel, J., and Asselin, A., Detection of β-1,3-glucanase activity after native polyacrylamide gel electrophoresis: application to tobacco pathogenesis-related proteins, *Electrophoresis*, 10, 527, 1989.
6. Grenier, J. and Asselin, A., Detection of β-1,3-glucanase activity in gels containing alkali-soluble yeast glucan, *Anal. Biochem.*, 212, 301, 1993.
7. Shimoni, M., A method for activity staining of peroxidase and β-1,3-glucanase isozymes in polyacrylamide electrophoresis gels, *Anal. Biochem.*, 220, 36, 1994.

3.2.1.41 — α-Dextrin Endo-1,6-α-Glucosidase; DEG

OTHER NAMES	Pullulanase, limit dextranase, debranching enzyme, amylopectin 6-glucanohydrolase, "R-enzyme"
REACTION	Hydrolysis of 1,6-α-D-glucosidic linkages in pullulan, amylopectin, and glycogen, and in the α- and β-amylase limit dextrins of amylopectin and glycogen
ENZYME SOURCE	Bacteria, fungi, plants
SUBUNIT STRUCTURE	Unknown or no data available

METHOD 1

Visualization Scheme

Reactive Red-pullulan^{VIS} [DEG] → Reactive Red-maltotriose
(in gel matrix) (diffuses from gel matrix)
red background light bands

Staining Solution[1]

A. 0.2 *M* Acetate buffer, pH 5.0
   5% Reactive Red–pullulan
B. 4.2% Agar solution (60°C)

Procedure

Mix equal volumes of solutions A and B and pour the mixture over the surface of an electrophorized gel soaked in 0.2 *M* acetate buffer (pH 5.0) for 10 to 20 min. Incubate the gel at 37°C in a humid chamber until light bands appear on a red background. Fix the stained gel in 3% acetic acid.

Notes: Reactive Red–pullulan is obtained with the dyestuff covalently bound to pullulan.[2]

The specific dye-conjugate substrate, red pullulan (RP), is now commercially available from Megazyme (North Rocks, NSW, Australia). The method was developed that used the substrate RP (10 mg/ml) incorporated in the resolving native or SDS-PAG.[3] It is specific and allows detection of about 1.2 AU (arbitrary absorbance unit; the AU value is the value of the absorbance at 510 nm after 20 min of reaction at 40°C × 100). The incorporation of RP directly in electrophoretic gels is of advantage because it increases the sharpness of DEG activity bands and allows the detection of DEG activity under native conditions or in the presence of SDS.

METHOD 2

Visualization Scheme

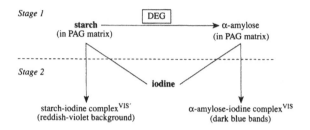

Staining Solution[4]

A. 0.1 M Sodium citrate buffer, pH 5.0
B. 14 mM Potassium iodide
   10 mM Iodine

Procedure

Incubate an electrophorized PAG containing 0.1% starch in solution A for 5 h and place in solution B. Dark blue bands of DEG activity appear on a reddish violet background. Record or photograph the zymogram.

Notes: Development of DEG activity bands is caused by hydrolysis of the α-1,6-branch linkages of the starch by the enzyme action, resulting in a greater capacity of the unit chain to adopt the helical configuration for iodine complex formation.

Amylase activity bands (light bands) also develop on DEG zymograms obtained by this method.

Bands of DEG activity can also develop on phosphorylase zymograms obtained using the backward reaction of the enzyme (see 2.4.1.1 — PHOS, Method 1).

DEG activity bands can be detected by incubating an electrophoretic PAG in the substrate-containing solution (1% soluble starch, 0.5% β-limited dextrin, 1% amylopectin in 50 mM Tris-HCl buffer, pH 9.0) at 30°C for 30 min, rinsing with water, and subsequently staining with iodine solution (1% J_2, 10% KJ in 50% methanol).[5] Utilization of the substrate in a solution form is fast and sensitive (about 1 mU of DEG per band can be detected). However, the use of the substrate incorporated into the PAG matrix may be preferable because it results in more sharp activity bands.

Activity bands of DEG can also be developed on a starch-containing PAG stained for the activity of the glucan branching enzyme (see 2.4.1.18 — GBE, *Notes*). It is suggested that light blue bands of DEG developed on a blue background of the processed gel stained with iodine are due to the diffusion of unit chains of short length from the starch-containing PAG.[6]

OTHER METHODS

A simple pullulan–agar replica method was developed. It detects zones occupied by DEG as colorless bands visible on the pullulan–agar gel preincubated with an electrophoretic PAG and then treated with ethanol. The sensitivity of the method is such that 0.0015 U of DEG activity can be detected easily.[7] The use of this method in combination with the amylopectin–agar replica method enabled differentiation between activity bands caused by DEG, isoamylase, and glucoamylase (see 3.2.1.3 — GAMY, *Notes*; 3.2.1.68 — IA, *Notes*).

GENERAL NOTES

Four different pullulanases are known. The first one (pullulan hydrolase type I, also known as neopullulanase) attacks α-1,4-glycosidic linkages in pullulan-forming panose. The second one (pullulan hydrolase type II, also known as isopullulanase) attacks α-1,4-glycosidic linkages in pullulan-forming isopanose. The third one (pullulanase type I) specifically hydrolyzes the α-1,6-glycosidic linkages in pullulan-forming maltotriose. The fourth one (pullulanase type II) attacks α-1,6-glycosidic linkages in pullulan and α-1,4-glycosidic linkages in other polysaccharides. Of these, the only debranching enzyme is pullulanase type I.[8]

REFERENCES

1. Yang, S.-S. and Coleman, R.D., Detection of pullulanase in polyacrylamide gels using pullulan-reactive red agar plates, *Anal. Biochem.*, 160, 480, 1987.
2. Rinderknecht, M., Wilding, P., and Haverback, B.J., A new method for the determination of α-amylase, *Experientia*, 23, 805, 1967.
3. Furegon, L., Curioni, A., and Peruffo, A.D.B., Direct detection of pullulanase activity in electrophoretic polyacrylamide gels, *Anal. Biochem.*, 221, 200, 1994.
4. Gerbrandy, S.J. and Verleur, J.D., Phosphorylase isoenzymes: localization and occurrence in different plant organs in relation to starch metabolism, *Phytochemistry*, 10, 261, 1971.
5. Kim, C.-H., Specific detection of pullulanase type I in polyacrylamide gels, *FEMS Microbiol. Lett.*, 116, 327, 1994.
6. Rammesmayer, G. and Praznik, W., Fast and sensitive simultaneous staining method of Q-enzyme, α-amylase, R-enzyme, phosphorylase and soluble starch synthase separated by starch–polyacrylamide gel electrophoresis, *J. Chromatogr.*, 623, 399, 1992.
7. González, R.D., A simple and sensitive method for detection of pullulanase and isoamylase activities in polyacrylamide gels, *Biotechnol. Tech.*, 8, 659, 1994.
8. Kim., C.-H., Choi, H.I., and Lee, D.S., Pullulanases of alkaline and broad pH range from a newly isolated alkalophilic *Bacillus* sp. S-1 and a *Micrococcus* sp. Y-1, *J. Ind. Microbiol.*, 12, 48, 1993.

3.2.1.48 — Sucrose α-Glucosidase; SG

OTHER NAMES	Sucrose α-glucohydrolase, sucrase, sucrase-isomaltase
REACTION	Hydrolysis of sucrose and maltose by an α-D-glucosidase–type action
ENZYME SOURCE	Invertebrates, vertebrates
SUBUNIT STRUCTURE	Tetramer (invertebrates); also see General Notes

METHOD 1

Visualization Scheme

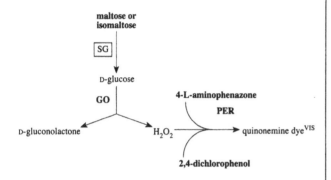

Staining Solution[1]

A. 2 M Sodium phosphate buffer, pH 6.0
    25 mM Maltose or isomaltose
B. 2,4-Dichlorophenol (sulfonated)
C. 10 U/ml Glucose oxidase (GO)
    10 U/ml Peroxidase (PER)
    0.5 mg/ml 4-L-Aminophenazone
D. 1.6% Agarose solution (60°C)

Procedure

Mix 0.75 ml of A with 0.5 ml of B, 2.5 ml of C, and 6.25 ml of D. Pour the resulting mixture over the gel surface and incubate at 37°C until colored bands appear. Record or photograph the zymogram because the stain is not permanent.

Notes: Preparations of glucose oxidase and peroxidase should be catalase-free. If they are not, NaN$_3$ should be included in the staining solution to inhibit catalase activity.

Other methods may be used to detect hydrogen peroxide via peroxidase (see 1.11.1.7 — PER).

METHOD 2

Visualization Scheme

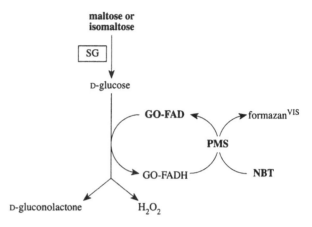

Staining Solution[1]

A.	2 M Sodium phosphate buffer, pH 6.0	1 ml
	H$_2$O	4 ml
	PMS	2 mg
	NBT	5 mg
	Maltose or isomaltose	43 mg
	Glucose oxidase (GO)	100 U
B.	2% Agarose solution (60°C)	5 ml

Procedure

Mix A and B solutions and pour the mixture over the gel surface. Incubate the gel at 37°C in the dark until dark blue bands appear. Fix the stained gel in 25% ethanol or 5% acetic acid.

OTHER METHODS

A. The product D-glucose can also be detected using linking enzymes hexokinase and NAD(P)-dependent glucose-6-phosphate dehydrogenase coupled with PMS–MTT (e.g., see 2.7.1.1 — HK).
B. An immunoblotting procedure (for details, see Part II) based on the utility of monoclonal antibodies specific to the rat enzyme[2] can be used for immunohistochemical visualization of the enzyme protein on electrophoretic gels.

GENERAL NOTES

This enzyme is isolated from intestinal mucosa as a single polypeptide chain also hydrolyzing isomaltose by an oligo-1,6-glucosidase–type reaction (see 3.2.1.10 — OG).[3]

α-Glucosidase (see 3.2.1.20 — α-GLU) can also be detected by Methods 1 and 2 described above when maltose is used as a substrate. However, in contrast to SG, α-GLU is not able to hydrolyze isomaltose. This difference may be used to differentiate SG and α-GLU activities.

3.2.1.48 — Sucrose α-Glucosidase; SG (continued)

When isomaltose is used as a substrate, the oligo-1,6-glucosidase (see 3.2.1.10 — OG) activity bands can also develop on SG zymograms. In contrast to SG, OG is not able to hydrolyze maltose. This distinction may be used to discriminate between these enzymes.

REFERENCES

1. Finlayson, S.D., Moore, P.A., Johnston, J.R., and Berry, D.R., Two staining methods for selectively detecting isomaltase and maltase activity in electrophoretic gels, *Anal. Biochem.*, 186, 233, 1990.
2. Hauri, H.P., Quaroni, A., and Isselbacher, K.J., Monoclonal antibodies to sucrase-isomaltase-probes for the study of postnatal development and biogenesis of the intestinal microvillus membrane, *Proc. Natl. Acad. Sci. U.S.A.*, 77, 6629, 1980.
3. NC-IUBMB, *Enzyme Nomenclature*, Academic Press, San Diego, 1992, p. 352 (EC 3.2.1.48, Comments).

3.2.1.51 — α-L-Fucosidase; α-FUC

REACTION	α-L-Fucoside + H_2O = alcohol + L-fucoside
ENZYME SOURCE	Bacteria, plants, invertebrates, vertebrates
SUBUNIT STRUCTURE	Monomer (vertebrates)

METHOD

Visualization Scheme

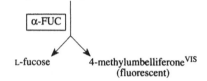

4-methylumbelliferyl α-L-fucoside

α-FUC

L-fucose   4-methylumbelliferone[VIS] (fluorescent)

Staining Solution[1]

A. 0.1 *M* Citrate–phosphate buffer, pH 4.8
   0.2 mg/ml 4-Methylumbelliferyl α-L-fucoside
B. 85 m*M* Glycine–carbonate buffer, pH 10.0

Procedure

Apply solution A to the gel surface on a filter paper overlay. Incubate the gel in a humid chamber at 37°C for 30 to 60 min and then place in solution B. Fluorescent bands of α-FUC activity are seen under long-wave UV light. Record the zymogram or photograph using a yellow filter.

REFERENCES

1. Turner, B.M., Beratis, N.G., Turner, V.S., and Hirschhorn, K., Isozyme of human α-L-fucosidase detectable by starch gel electrophoresis, *Clin. Chim. Acta*, 57, 29, 1974.

3.2.1.52 — Hexosaminidase; HEX

OTHER NAMES β-*N*-Acetylhexosaminidase (recommended name), *N*-acetyl-β-glucosaminidase (see General Notes)

REACTION Hydrolysis of terminal nonreducing *N*-acetyl-D-hexosamine residues in *N*-acetyl-β-D-hexosaminides

ENZYME SOURCE Bacteria, plants, invertebrates, vertebrates

SUBUNIT STRUCTURE Unknown or no data available

Method 1

Visualization Scheme

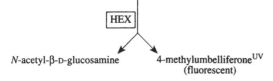

4-methylumbelliferyl *N*-acetyl-β-D-glucosaminide

HEX

N-acetyl-β-D-glucosamine   4-methylumbelliferone[UV] (fluorescent)

Staining Solution[1]

A. 1 *M* Sodium citrate buffer, pH 4.0 (or 4.5)
3 m*M* 4-Methylumbelliferyl *N*-Acetyl-β-D-glucosaminide
B. 2% Agarose solution (60°C)

Procedure

Mix equal volumes of A and B components of the staining solution and pour the mixture over the gel surface. Incubate the gel at 37°C and monitor under long-wave UV light. The areas of HEX activity appear as fluorescent bands. Spray the processed gel with ammonia to enhance fluorescence. Record the zymogram or photograph using a yellow filter.

Notes: Staining solution (solution A) may be applied to the gel surface with a filter paper overlay.

When an SDS-PAG is used to resolve HEX isozymes, the enzyme activity should be restored before staining by washing the gel in solution containing Triton X-100. The substrate, 4-methylumbelliferyl *N*-acetyl-β-D-glucosaminide, may be incorporated directly into PAG.[2]

Method 2

Visualization Scheme

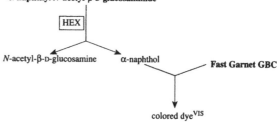

α-naphthyl *N*-acetyl-β-D-glucosaminide

HEX

N-acetyl-β-D-glucosamine   α-naphthol   Fast Garnet GBC

colored dye[VIS]

Staining Solution[3]

0.1 *M* Citrate buffer, pH 4.5 to 6.5
1 mg/ml α-Naphthyl *N*-acetyl-β-D-glucosaminide (dissolved in the buffer by heating)
1 mg/ml Fast Garnet GBC

Procedure

Incubate the gel in staining solution in the dark at 37°C until pink bands appear. Wash the stained gel with water and fix in 50% glycerol.

Notes: Naphthol-A*S*-BI 2-acetamido-2-deoxy-β-D-glucopyranoside (dissolved in ethanol) may be used in the staining solution instead of α-naphthyl *N*-acetyl-β-D-glucosaminide.

Method 3

Visualization Scheme

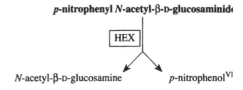

p-nitrophenyl *N*-acetyl-β-D-glucosaminide

HEX

N-acetyl-β-D-glucosamine   *p*-nitrophenol[VI]

Staining Solution[4]

25 m*M* Citrate buffer, pH 5.5
5 m*M* *p*-Nitrophenyl *N*-acetyl-β-D-glucosaminide

Procedure

Rinse an electrophorized gel with water and soak in 0.2 *M* acetate buffer (pH 4.9) for 30 min. Incubate the gel in staining solution at room temperature until yellow bands appear. Photograph the zymogram using a 436.8-nm interference filter.

Notes: To enhance the band coloration caused by *p*-nitrophenol, drain the gel surface and cover with 0.1 *M* NaOH. To preserve the yellow color for a few hours, enclose the processed gel in Saran wrap.

3.2.1.52 — Hexosaminidase; HEX (continued)

General Notes

This enzyme is also known as *N*-acetyl-β-glucosaminidase (EC 3.2.1.30; now a deleted entry).[5]

References

1. Gilbert, F., Kucherlapati, R., Creagan, R.P., Murnane, M.J., Darlington, G.J., and Ruddle, F.H., Tay-Sachs' and Sandhoff's diseases: the assignment of genes for hexosaminidase A and B to individual human chromosomes, *Proc. Natl. Acad. Sci. U.S.A.*, 72, 263, 1975.
2. Lincoln, S.P., Fermor, T.R., and Wood, D.A., A novel method for detecting acetylglucosaminidase activity on polyacrylamide gels, *J. Microbiol. Methods*, 20, 79, 1994.
3. Swallow, D.M., Evans, L., Saha, N., and Harris, H., Characterization and tissue distribution of *N*-acetyl hexosaminidase C: suggestive evidence for a separate hexosaminidase locus, *Ann. Hum. Genet.*, 40, 55, 1976.
4. Gabriel, O. and Wang, S.-F., Determination of enzymatic activity in polyacrylamide gels: I. Enzymes catalyzing the conversation of nonreducing substrates to reducing products, *Anal. Biochem.*, 27, 545, 1969.
5. NC-IUBMB, *Enzyme Nomenclature*, Academic Press, San Diego, 1992, p. 350 (EC 3.2.1.30).

3.2.1.53 — β-*N*-Acetylgalactosaminidase; AGA

OTHER NAMES	β-Galactosaminidase
REACTION	Hydrolysis of terminal nonreducing *N*-acetyl-D-galactosamine residues in *N*-acetyl-β-D-galactosaminides
ENZYME SOURCE	Invertebrates, vertebrates
SUBUNIT STRUCTURE	Unknown or no data available

Method

Visualization Scheme

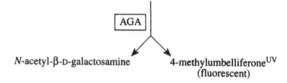

4-methylumbelliferyl *N*-acetyl-β-D-galactosaminide

AGA

N-acetyl-β-D-galactosamine   4-methylumbelliferone[UV]
(fluorescent)

Staining Solution[1]

A. 0.1 *M* Phosphate–citrate buffer, pH 9.5   15 ml
   4-Methylumbelliferyl *N*-acetyl-β-D-galactosaminide
   (dissolved in 0.25 ml of dimethyl sulfoxide)   5 mg
B. 2% Agar solution (60°C)   10 ml

Procedure

Mix A and B solutions and pour the mixture over the gel surface. Incubate the gel at 37°C until fluorescent bands visible under long-wave UV light appear. Record the zymogram or photograph using a yellow filter.

References

1. Aebersold, P.B., Winans, G.A., Teel, D.J., Milner, G.B., and Utter, F.M., *Manual for Starch Gel Electrophoresis: A Method for the Detection of Genetic Variation*, NOAA Technical Report NMFS 61, U.S. Department of Commerce, National Marine Fisheries Service, Seattle, WA, 1987.

3.2.1.55 — α-N-Arabinofuranosidase; AF

OTHER NAMES	Arabinosidase, α-arabinosidase
REACTION	Hydrolysis of terminal nonreducing α-L-arabinofuranoside residues in α-L-arabinosides
ENZYME SOURCE	Bacteria, fungi, protozoa, plants, lower vertebrates (blue shark)
SUBUNIT STRUCTURE	Unknown or no data available

METHOD 1

Visualization Scheme

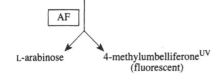

4-methylumbelliferyl α-L-arabinoside

L-arabinose   4-methylumbelliferone[UV]
(fluorescent)

Staining Solution[1]

0.1 M Phosphate–citrate buffer, pH 4.0	5 ml
4-Methylumbelliferyl α-L-arabinoside	10 mg

Procedure

Apply the staining solution to the gel surface on a filter paper overlay. Incubate the gel at 37°C for 30 to 60 min, and then spray with ammonia solution. Fluorescent bands of AF activity are seen under long-wave UV light. Record the zymogram or photograph using a yellow filter.

METHOD 2

Visualization Scheme

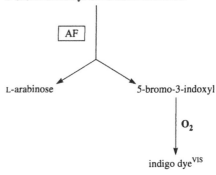

5-bromo-3-indolyl-α-L-arabinofuranoside

L-arabinose   5-bromo-3-indoxyl

O₂

indigo dye[VIS]

Staining Solution[2] (adapted)

0.1 M Phosphate–citrate buffer, pH 4.0
3 µl/ml 2% 5-Bromo-3-indolyl-α-L-arabinofuranoside
(dissolved in dimethylformamide; see also *Notes*)

Procedure

Incubate the gel in staining solution at 37°C until dark blue bands appear. Fix the stained gel in 10% trichloroacetic acid for several hours and store in 7% acetic acid.

Notes: The chromogenic substrate 5-bromo-3-indolyl-α-L-arabinofuranoside can be synthesized under laboratory conditions as described.[2]

GENERAL NOTES

Some β-galactosidases (see 3.2.1.23 — β-GAL) also hydrolyze α-L-arabinosides.[3]

REFERENCES

1. Eitner, B.J., α-L-Arabinofuranosidase expression in the blue shark, *Prionace glauca, Isozyme Bull.*, 22, 49, 1989.
2. Berlin, W. and Sauer, B., *In situ* color detection of α-L-arabinofuranosidase, a "no-background" reporter gene, with 5-bromo-3-indolyl-α-L-arabinofuranoside, *Anal. Biochem.*, 243, 171, 1996.
3. NC-IUBMB, *Enzyme Nomenclature*, Academic Press, San Diego, 1992, p. 353 (EC 3.2.1.55, Comments).

OTHER NAMES	Poly(galacturonate) hydrolase, exopolygalacturonase
REACTION	$(1,4\text{-}\alpha\text{-D-Galacturonide})_n + H_2O = (1,4\text{-}\alpha\text{-D-galacturonide})_{n-1} + \text{D-galacturonate}$
ENZYME SOURCE	Fungi, plants
SUBUNIT STRUCTURE	Unknown or no data available

METHOD

Visualization Scheme

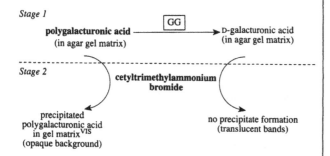

Stage 1

polygalacturonic acid ———[GG]———→ D-galacturonic acid
(in agar gel matrix)   (in agar gel matrix)

Stage 2

cetyltrimethylammonium bromide

precipitated polygalacturonic acid in gel matrix[VIS] (opaque background)   no precipitate formation (translucent bands)

Staining Solution[1]

A. 0.1 *M* Citrate buffer, pH 5.0
    1% Polygalacturonic acid
B. 3% Agarose solution (60°C)
C. 1% Water solution of cetyltrimethylammonium bromide

Procedure

Mix solutions A and B and pour the mixture over the surface of an electrophorized PAG. Incubate the PAG at 37°C for 1 to 4 h. Place the PAG and a substrate-containing agarose plate in solution C. Translucent bands of GG activity appear on the opaque background of the PAG and agarose plate. The bands with high GG activity are more clearly visible on the agarose plate, while the bands with low GG activity are more distinct on the PAG.

Notes: Two other enzymes with polygalacturonase activity can also be developed on GG zymograms obtained by this method. These are polygalacturonase (see 3.2.1.15 — PG) and polygalacturonate lyase (see 4.2.2.2 — PGL). The main difference in substrate specificity between enzymes with polygalacturonase activity is the ability of PG to hydrolyze pectin; PGL and GG do not act on this polysaccharide. Moreover, PGL displays maximal activity in alkaline conditions, while both GG and PG display optimal activity in acid conditions. These differences can be used to identify GG activity bands.

REFERENCES

1. Bertheau, Y., Madgidi-Hervan, E., Kotoujansky, A., Nguyen-The, C., Andro, T., and Coleno, A., Detection of depolymerase isoenzymes after electrophoresis or electrofocusing, or in titration curves, *Anal. Biochem.*, 139, 383, 1984.

OTHER NAMES	Debranching enzyme
REACTION	Hydrolysis of 1,6-α-D-glucosidic branch linkages in glycogen, amylopectin, and their β-limited dextrins
ENZYME SOURCE	Bacteria, algae, plants
SUBUNIT STRUCTURE	Tetrameric and hexameric forms[#] (plants)

METHOD

Visualization Scheme

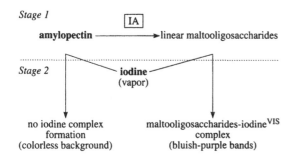

Stage 1

amylopectin [IA] → linear maltooligosaccharides

Stage 2 iodine (vapor)

no iodine complex formation (colorless background) — maltooligosaccharides-iodine[VIS] complex (bluish-purple bands)

Staining Solution[1]

A. 100 mM Acetate buffer, pH 5.0
    0.2% Amylopectin
    2% Agar
B. Iodine (few crystals)

Procedure

Dissolve amylopectin–agar suspension A in a microwave oven, cool to 60°C, and pour the solution on a horizontal glass plate of an appropriate size to form the amylopectin–agar plate. After electrophoresis in a native PAG, soak the gel twice for 10 min in 50 mM acetate buffer (pH 5.0) containing 25% (v/v) 2-propanol. Then immerse the gel twice in fresh acetate buffer (containing no 2-propanol) for 10 min; blot the gel dry and overlay it on top of the amylopectin–agar plate. Place the gel sandwich in a box, close it tightly, and incubate at 37°C for 4 h. The amylopectin debranching activity develops as bluish purple bands when the processed amylopectin–agar plate is treated with iodine vapor generated by a few crystals of iodine (component B of the staining solution).

Notes: In the amylopectin–agar replica plates, both isoamylase and α-dextrin endo-1,6-α-glucosidase (see 3.2.1.41 — DEG) produce bluish purple bands. However, isoamylase differs from DEG by its inability to attack pullulan.[2] A different specificity toward pullulan can be used to discriminate between these two debranching enzymes. Activity bands specific to DEG can be developed using a pullulan–agar replica plate as a control. Prepare the pullulan–agar plate (using 0.5% pullulan instead of amylopectin) and incubate with an electrophoretic PAG as described above for isoamylase. After incubation, pour a few milliliters of ethanol on the pullulan–agar plate surface to precipitate pullulan; zones of pullulan hydrolysis become visible as colorless bands on a blurred background. The Reactive Red–pullulan method can also be used for specific detection of DEG activity bands (see 3.2.1.41 — DEG, Method 1).

Activity bands of glucoamylase (see 3.2.1.3 — GAMY) and α-amylase (see 3.2.1.1 — α-AMY) are also developed by the IA method given above. However, unlike IA and DEG, activity bands of GAMY and α-AMY are colorless.

The sensitivity of the method is such that 0.0004 U of IA can be detected easily.

The method is applicable to renatured SDS-PAG and thus can be used for estimating the molecular weight of the enzyme.[3]

GENERAL NOTES

All starch and glycogen debranching enzymes from plants and various bacteria can be classified into two distinct types, an isoamylase type and a pullulanase type.[4]

REFERENCES

1. Gonzalez, R.D., A simple and sensitive method for detection of pullulanase and isoamylase activities in polyacrylamide gels, *Biotechnol. Tech.*, 8, 659, 1994.
2. NC-IUBMB, *Enzyme Nomenclature*, Academic Press, San Diego, 1992, p. 355 (EC 3.2.1.68, Comments).
3. Lim, W.J., Park, S.R., Cho, S.J., Kim, M.K., Ryu, S.K., Hong, S.Y., Seo, W.T., Kim, H., and Yun, H.D., Cloning and characterization of an intracellular isoamylase gene from *Pectobacterium chrysanthemi* PY35, *Biochem. Biophys. Res. Commun.*, 287, 348, 2001.
4. Fujita, N., Kubo, A., Francisco, P.B., Nakakita, M., Harada, K., Minaka, N., and Nakamura, Y., Purification, characterization, and cDNA structure of isoamylase from developing endosperm of rice, *Planta*, 208, 283, 1999.

3.2.1.75 — Glucan Endo-1,6-β-Glucosidase; GEG

OTHER NAMES   Endo-1,6-β-glucanase
REACTION   Random hydrolysis of 1,6-β-link-
                      ages in 1,6-β-D-glucans
ENZYME SOURCE   Fungi
SUBUNIT STRUCTURE   Unknown or no data available

Method 1

Visualization Scheme

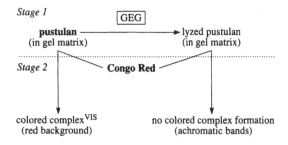

Staining Solution[1]

A. 50 mM Potassium acetate buffer, pH 5.5
   0.1% Pustulan
   0.85% Agarose
B. 1% Congo Red
C. 1 M NaCl
D. 50 mM Potassium acetate buffer, pH 5.5

Procedure

The method is developed for SDS-PAG. Renature proteins by soaking the electrophorized SDS-PAG in 50 ml of 50 mM Tris–HCl buffer (pH 8.2) containing 1% casein, 2 mM EDTA, and 0.1% sodium azide. Then incubate the gel for three periods of 20 min in this buffer at 4°C with gentle shaking. Wash the gel in 50 ml of 50 mM potassium acetate buffer (pH 5.5) at 4°C and incubate for 10 min in this buffer. Prepare the substrate-containing agarose gel of 0.5-mm thickness by boiling components A of the staining solution. Overlay the replica gel onto the renatured PAG and incubate the gel sandwich at 25°C for a period of time from 2 h to overnight, depending on the activity level of the sample to be assayed (see *Notes*). Stain the PAG with Coomassie for proteins and the replica gel with Congo Red for GEG by placing the agarose gel in solution B for 30 min. Wash the stained replica gel in solution C to remove the unbound dye. The areas of GEG activity are seen as achromatic bands on a red background of the replica gel. To obtain more contrast, immerse the agarose gel in solution D for at least 10 min. After this, achromatic bands of GEG remain clear on a dark blue background of the gel.

Notes: The same effect may be produced by placing the replica gel in 5% acetic acid.[2]

In order to obtain activity bands as sharp as possible, the contact time between electrophoretic and replica gels should be set between 2

and 4 h for samples with high GEG activity (about 10 to 25 mU per band). Bands with less activity could be detected with overnight contact, but this will result in very diffuse bands.

Method 2

Visualization Scheme

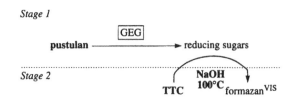

Staining Solution[1]

A. 50 mM Potassium acetate buffer, pH 5.5
B. 50 mM Potassium acetate buffer, pH 5.5
   1% Pustulan
C. 0.15% 2,3,5-Triphenyltetrazolium chloride (TTC)
   1 M NaOH

Procedure

The method is developed for an isoelectrofocusing PAG. Rinse the electrophorized PAG with distilled water and incubate with slow shaking for 5 min at room temperature in solution A. Following this, incubate the gel at 37°C for 1.5 h in solution B prepared by autoclaving and subsequent cooling. Wash the gel extensively with distilled water and place in a glass tray with solution C. Microwave the tray at maximum power. Shake the gel every 30 sec until dark red activity bands are developed on a transparent background.

Notes: This method is a modification of the procedure developed for the detection of activity bands of laminarinase (see 3.2.1.39 — LAM, Method 1).[3]

As little as 1 mU of GEG activity can be detected by this method.

General Notes

The enzyme acts on lutean, pustulan, and 1,6-oligo-β-D-glucosides.[4]

The maximal activity of GEG (100%) is detected against pustulan (a linear β-1,6-glucan) and yeast cell wall glucan (70%), which is a β-1,3:β-1,6 glucan (4:1). Approximately 20% of the maximal activity is detected against laminarin, which is mainly composed of β-1,3-glucan with β-1,6-glycosidic linkages as branches at the ratio 7:1 of linkage types. No GEG activity is detected against pachyman, which is a linear β-1,3-glucan, or against chitin and chitosan, or CM-cellulose, soluble starch, and dextran.[5]

3.2.1.75 — Glucan Endo-1,6-β-Glucosidase; GEG (continued)

REFERENCES

1. Soler, A., de la Cruz, J., and Llobell, A., Detection of β-1,6-glucanase isozymes from *Trichoderma* strains in sodium dodecyl sulfate–polyacrylamide gel electrophoresis and isoelectrofocusing gels, *J. Microbiol. Methods*, 35, 245, 1999.
2. Beguin, P., Detection of cellulase activity in polyacrylamide gels using Congo red–stain agar replicas, *Anal. Biochem.*, 131, 333, 1983.
3. Pan, S.-Q., Ye, X.-S., and Kuc, J., Direct detection of β-1,3-glucanase isozymes on polyacrylamide electrophoresis and isoelectrofocusing gels, *Anal. Biochem.*, 182, 136, 1989.
4. NC-IUBMB, *Enzyme Nomenclature*, Academic Press, San Diego, 1992, p. 356 (EC 3.2.1.75, Comments).
5. de la Cruz, J. and Llobell, A., Purification and properties of a basic endo-β-1,6-glucanase (BGN16.1) from the antagonistic fungus *Trichoderma harzianum*, *Eur. J. Biochem.*, 265, 145, 1999.

3.2.1.81 — Agarase; AG

REACTION	Endo-hydrolysis of 1,3-β-D-galactosidic linkages in agarose, giving the tetramer as the predominant product
ENZYME SOURCE	Bacteria
SUBUNIT STRUCTURE	Unknown or no data available

METHOD

Visualization Scheme

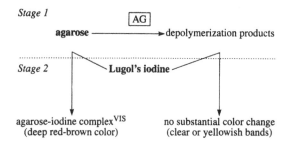

Staining Solution[1]

A. 0.5% Agarose solution in 20 mM Tris–HCl buffer, pH 7.0 (50°C)
B. Lugol's iodine

Procedure

Wash an electrophorized native or SDS-PAG in distilled water and bath for 15 min in 20 mM Tris–HCl buffer (pH 7.0). Pour solution A over the surface of the gel and incubate at 37°C for 1 h. Flood the gel with solution B to visualize the bands due to agarolytic activity.

Notes: The enzyme is heat stabile and renaturable after SDS-PAG electrophoresis. The inclusion of agarose into the nondenaturing PAG results in a clear-cut retardation of the AG mobility.

GENERAL NOTES

The enzyme also acts on porfyran.[2]

REFERENCES

1. Ghadi, S.C., Muraleedharan, U.D., Jawaid, S., Screening for agarolytic bacteria and development of a novel method for *in situ* detection of agarase, *J. Mar. Biotechnol.*, 5, 194, 1997.
2. NC-IUBMB, *Enzyme Nomenclature*, Academic Press, San Diego, 1992, p. 357 (EC 3.2.1.81, Comments).

3.2.1.91 — Cellulose 1,4-β-Cellobiosidase; CC

OTHER NAMES   Exo-cellobiohydrolase

REACTION   Hydrolysis of 1,4-β-D-glucosidic linkages in cellulose and cellotetraose, releasing cellobiose from the nonreducing ends of the chains

ENZYME SOURCE   Bacteria, fungi

SUBUNIT STRUCTURE   Unknown or no data available

METHOD

Visualization Scheme

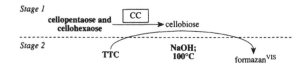

Staining Solution[1]

A. 0.1 M Phosphate buffer, pH 5.0 to 6.5
   0.25% Cellopentaose and cellohexaose mixture
B. 0.1 M Iodoacetamide
C. 0.1% 2,3,5-Triphenyltetrazolium chloride (TTC)
   0.5 M NaOH

Procedure

Wash an electrophorized PAG for 15 min in two changes of 0.1 M phosphate buffer (pH 5.0 to 6.5) with gentle shaking. Rinse the gel with distilled water and place in solution A for 15 min at 50°C. Rinse the gel again with distilled water and place in solution B for 5 min. Then rinse the gel again and place in a 2-l beaker containing solution C. Heat the solution over a gas burner with gentle agitation for several minutes until red bands are evident. The heating must be terminated before the appearance of a general red background. Rinse the gel immediately with distilled water and soak in several changes of 7.5% acetic acid. The heating and initial storage of the stained gel must be carried out in the dark.

Notes: The bands of cellulase activity (see 3.2.1.4 — CEL) can also develop on CC zymograms obtained by this method. However, CC activity bands do not develop on CEL negative zymograms obtained using carboxymethyl cellulose as a substrate and Congo Red staining of the poly-β-1,4-glucopyranoside (see 3.2.1.4 — CEL, Method 2, *Notes*). This difference can be used to differentiate the activity bands caused by CEL and CC. Specific detection of the product cellobiose can be performed using cellobiose-specific auxiliary enzymes (see below).

OTHER METHODS

The product cellobiose can be detected via chromogenic reactions catalyzed by auxiliary enzymes cellobiose dehydrogenase (quinone, EC 1.1.5.1) and cellobiose dehydrogenase (acceptor, EC 1.1.99.18) coupled with the PMS–MTT and DCIP–MTT systems, respectively.

REFERENCES

1. Bartley, T.D., Murphy-Holland, K., and Eveleigh, D.E., A method for the detection and differentiation of cellulase components in polyacrylamide gels, *Anal. Biochem.*, 140, 157, 1984.

3.2.1.96 — Mannosyl-Glycoprotein Endo-β-N-Acetylglucosamidase; MGEA

OTHER NAMES   *N,N'*-Diacetylchitobiosyl β-*N*-acetylglucosaminidase, endo-β-*N*-acetylglucosaminidase

REACTION   Endohydrolysis of the *N,N'*-diacetylchitobiosyl unit in high-mannose glycopeptides and glycoproteins containing the -[Man(GlcNAc)$_2$]Asn structure. One *N*-acetyl-D-glucosamine residue remains attached to the protein; the rest of the oligosaccharide is released intact

ENZYME SOURCE   Bacteria

SUBUNIT STRUCTURE   Unknown or no data available

METHOD

Visualization Scheme

Micrococcus luteus   lyzed *M. luteus*
(in gel matrix) ——→ (in gel matrix)
opaque background^{VIS} translucent bands

Detecting Mixture[1]

0.5 mg/ml suspension of spray-dried *Micrococcus luteus* ATCC 4698 cells in an SDS-PAG solution (bacteria suspended in the gel solution prior to the addition of ammonium persulfate and polymerization).

Procedure

After electrophoresis, wash the gel in distilled water (250 ml per gel) for 30 min at room temperature. Renaturate the enzyme at 37°C in 250 ml of 0.1 *M* phosphate buffer (pH 7.0) with gentle shaking. Periodically place the gel on an immuno viewer-MU (Jookoo Sangyo Co., Tokyo, Japan) and inspect in indirect light. When transparent bands are well developed on a cloudy gel background, photograph the gel to document observation.

Notes: Incubation of the gel for 3 h is sufficient for purified MGEA from *Staphylococcus aureus* (about 500 pg per line) to regain activity and to produce a visible transparent band. After 24 h of incubation, 160 pg per line gave a reasonably clear band.

Intrinsic bacteriolytic enzymes of *M. luteus*, used as a substrate for staphylococcal MGEA, displayed no detectable activity up to 48 h incubation. However, when *S. aureus* cells were used as a substrate, heat treatment (30 min incubation in boiling water) of cells was required to avoid interference in the renaturation by intrinsic bacteriolytic enzymes.

This method may be extended to the analysis of other bacteriolytic enzymes.

GENERAL NOTES

This hydrolytic activity is represented by a group of related enzymes.[2]

REFERENCES

1. Sugai, M., Komatsuzawa, H., Tomita, S., Akiyama, T., Miyake, Y., and Suginaka, H., Detection of a staphylococcal endo-β-*N*-acetylglucosaminidase using polyacrylamide gels, *J. Microbiol. Methods*, 13, 11, 1991.
2. NC-IUBMB, *Enzyme Nomenclature*, Academic Press, San Diego, 1992, p. 359 (EC 3.2.1.96, Comments).

3.2.1.X — Chitin 1,4-β-Chitobiosidase; CCB

OTHER NAMES   Chitobiosidase
REACTION   Hydrolytic release of chitobioside from chitin
ENZYME SOURCE   Bacteria, fungi
SUBUNIT STRUCTURE   Unknown or no data available

METHOD

Visualization Scheme

4-methylumbelliferyl-β-D-N, N'-diacetylchitobioside

$\boxed{\text{CCB}}$

β-D-N, N'-diacetylchitobioside   4-methylumbelliferone[UV]

Staining Solution[1]

100 mM Potassium phosphate buffer, pH 6.7
   (or 100 mM sodium acetate, pH 4.8)
0.75% Agarose
0.3 mg/ml 4-Methylumbelliferyl β-D-N,N'-diacetylchitobioside

Procedure

Dissolve agarose–substrate suspension in a microwave oven, cool to 42°C, and apply to the gel surface. Incubate the gel at room temperature and monitor under long-wave UV light. The areas of CCB activity are seen as fluorescent bands.

Notes: The fluorescence of CCB activity bands can be enhanced by spraying the gel with ammonia after removal of the agarose overlay.

A native PAG, starch gel, or SDS-PAG can be used to study CCB by this method. When an SDS-PAG is used, soak the electrophorized gel in 25% isopropanol for 15 min to reactivate the enzyme, and then rinse the gel in water before staining. Since isopropanol made wetting of the gel difficult, pour the substrate–agarose solution onto a glass plate, and then place the gel face down in the molten agarose.

4-Methylumbelliferyl-N-acetyl-D-glucosaminide was also used as a fluorogenic substrate to detect N-acetyl-D-glucosaminidase-type chitinase activity in the SDS-PAG.[2]

REFERENCES

1. Tronsmo, A. and Harman, G.E., Detection and quantification of N-acetyl-β-D-glucosaminidase, chitobiosidase, and endochitinase in solutions and on gels, *Anal. Biochem.*, 208, 74, 1993.
2. Chen, K.-S., Lee, K.-K., and Chen, H.-C., A rapid method for detection of N-acetylglucosaminidase-type chitinase activity in crossed immunoelectrophoresis and sodium dodecyl sulfate–polyacrylamide gel electrophoresis gels using 4-methylumbelliferyl-N-acetyl-D-glucosaminide as substrate, *Electrophoresis*, 15, 662, 1994.

3.2.2.5 — NAD Nucleosidase; NADN

OTHER NAMES	NADase, DPNase, DPN hydrolase, NAD glycohydrolase
REACTION	NAD + H₂O = nicotinamide + ADPribose
ENZYME SOURCE	Bacteria, fungi, invertebrates, vertebrates
SUBUNIT STRUCTURE	Unknown or no data available

METHOD 1

Visualization Scheme

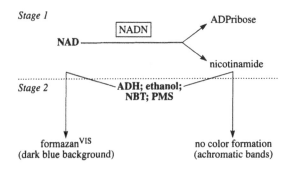

Staining Solution[1]

A. 50 mM Phosphate buffer, pH 6.5
   0.65 mM NAD

B. 0.1 M Pyrophosphate buffer, pH 8.8	50 ml
Ethanol	0.3 ml
NBT	10 mg
PMS	1 mg
Alcohol dehydrogenase (ADH)	25 U

Procedure

Incubate an electrophorized gel in solution A at 37°C for 1 to 2 h. Rinse the gel with water and counterstain with solution B in the dark at 37°C to detect NAD. Achromatic bands visible on a blue background of the gel indicate areas occupied by NADN. Wash the negatively stained gel with water and fix in 25% ethanol.

Notes: This method gives good results with cellulose acetate gel but can also be applied to PAG and starch gel.

The method also detects some other NAD-catabolizing activities (see General Notes).

METHOD 2

Visualization Scheme

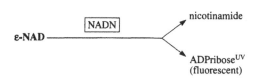

Staining Solution[2]

50 mM Tris–HCl buffer, pH 7.5
0.03% Lauryl dimethylamine N-oxide (LDAO)
150 μM ε-NAD (see *Notes*)

Procedure

The method is developed for an SDS-PAG. So, after separation of proteins by SDS-PAG electrophoresis, wash the gel for 30 min in a solution containing 50 mM Tris–HCl buffer (pH 7.5) and 0.5% LDAO.

To visualize NAD-converting activities, incubate the gel for 15 min in the staining solution and place on a UV transilluminator to observe fluorescent bands. Photograph the developed gel using a 550-nm interference filter.

Notes: ε-NAD is a NAD derivative (1,N^6-etheno-NAD) containing a modified adenine ring whose fluorescence is greatly enhanced after its separation from the quenching nicotinamide moiety. The method also detects some other NAD-catabolizing activities (see General Notes).

METHOD 3

Visualization Scheme

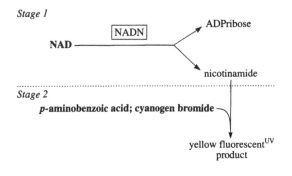

Staining Solution[3]

A. 0.1 M Phosphate buffer, pH 6.5
   12 mM NAD

B. 2% (w/v) p-Aminobenzoic acid

3.2.2.5 — NAD Nucleosidase; NADN (continued)

Procedure

Incubate an electrophorized gel in solution A for 1 to 2 h at 37°C. Place the preincubated gel in an atmosphere of cyanogen bromide in a sealed container and spray with solution B. Deep yellow bands, brightly fluorescent under long-wave UV light, indicate gel areas occupied by NADN activity. Record the zymogram immediately. The bands are visible for a few minutes and then fade.

Notes: This method generates very diffuse NADN activity bands and therefore is not as good as Methods 1 and 2 (however, see General Notes).

GENERAL NOTES

Only method 3 specifically detects activity bands of NADN. Methods 1 and 2 also detect activity bands of two other NAD-catabolizing enzymes, the NAD pyrophosphatase (see 3.6.1.22 — NADPP) and the NAD phosphodiesterase (see 3.1.4.X′ — NADPDE). However, using Method 2, the only fluorescent band was detected in the rat spleen, which was identified as an NADN activity band.[2] So, Method 2 may be considered a specific one for the rat spleen NADN and very likely for the spleen NADN of other mammalian species.

REFERENCES

1. Ravazzolo, R., Bruzzone, G., Garrè, C., and Ajmar, F., Electrophoretic demonstration and initial characterization of human red cell NAD(P)⁺ase, *Biochem. Genet.*, 14, 877, 1976.
2. Hagen, T.H. and Ziegler, M., Detection and identification of NAD-catabolizing activities in rat tissue homogenates, *Biochim. Biophys. Acta*, 1340, 7, 1997.
3. Flechner, I., Hirshorn, S., and Bekierkunst, A., A method of localization of soluble NAD-glycohydrolase from mycobacterial extracts after electrophoresis on cellulose acetate, *Life Sci.*, 7, 1327, 1968.

3.2.2.6 — NAD(P) Nucleosidase; NAD(P)N

OTHER NAMES	NAD(P) glycohydrolase
REACTION	NAD(P) + H₂O = nicotinamide + ADPribose(P)
ENZYME SOURCE	Bacteria, fungi, invertebrates, vertebrates
SUBUNIT STRUCTURE	Unknown or no data available

METHOD 1

Visualization Scheme

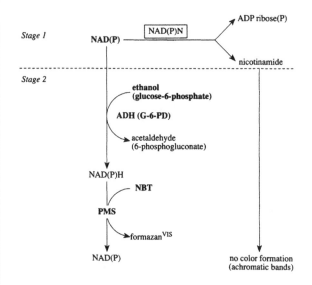

Staining Solution[1]

A. 50 mM Phosphate buffer, pH 6.5
    0.65 mM NAD or NADP

B. 0.1 M Pyrophosphate buffer, pH 8.8	50 ml
Ethanol	0.3 ml
NBT	10 mg
PMS	1 mg
Alcohol dehydrogenase (ADH)	25 U

C. 0.1 M Tris–HCl buffer, pH 8.0	50 ml
Glucose-6-phosphate (disodium salt)	10 mg
NBT	10 mg
PMS	1 mg
Glucose-6-phosphate dehydrogenase (G-6-PD)	25 U

Procedure

Incubate an electrophorized gel in solution A at 37°C for 1 to 2 h. Rinse the gel with water and counterstain with solution B or C in the dark at 37°C to detect NAD or NADP, respectively. Achromatic bands visible on a blue background of the gel indicate areas occupied by NAD(P)N. Wash the negatively stained gel with water and fix in 25% ethanol.

Notes: This method gives good results with a cellulose acetate gel but can also be applied to PAG and starch gel.

3.2.2.6 — NAD(P) Nucleosidase; NAD(P)N (continued)

METHOD 2

Visualization Scheme

Stage 1

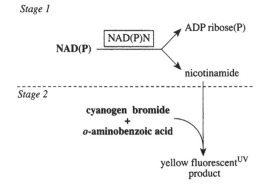

Stage 2

Staining Solution[2]

A. 0.1 M Phosphate buffer, pH 6.5
   12 mM NAD or NADP
B. 2% (w/v) p-Aminobenzoic acid

Procedure

Incubate an electrophorized gel in solution A for 1 to 2 h at 37°C. Place the preincubated gel in an atmosphere of cyanogen bromide in a sealed container and spray with solution B. Deep yellow bands, brightly fluorescent under long-wave UV light, indicate gel areas occupied by NAD(P)N activity. Record the the zymogram immediately. The bands are visible for a few minutes and then fade.

Notes: This method is not as practical as Method 1. It generates very diffuse NAD(P)N activity bands.

GENERAL NOTES

The use of NAD- and NADP-containing staining solutions to stain separate gels can help to discriminate between NAD(P)N and the related enzyme (see 3.2.2.5 — NADN) specific to NAD.

REFERENCES

1. Ravazzolo, R., Bruzzone, G., Garrè, C., and Ajmar, F., Electrophoretic demonstration and initial characterization of human red cell NAD(P)+ase, *Biochem. Genet.*, 14, 877, 1976.
2. Flechner, I., Hirshorn, S., and Bekierkunst, A., A method of localization of soluble NAD–glycohydrolase from mycobacterial extracts after electrophoresis on cellulose acetate, *Life Sci.*, 7, 1327, 1968.

3.2.2.X — Uracil-DNA Glycosylase; UDG

OTHER NAMES	Uracil *N*-glycosylase
REACTION	Catalyzes the removal of uracil residues from both single- and double-stranded DNA (but not RNA); leaves the DNA sugar-phosphodiester backbone intact
ENZYME SOURCE	Bacteria
SUBUNIT STRUCTURE	Unknown or no data available

METHOD

Visualization Scheme

Stage 1: Electrophoretic separation of proteins in substrate-containing PAG (first dimension)

Stage 2: Vertical incision of the sample line of the first-dimension PAG and receipt of narrow (2 mm) substrate-containing gel strip

Stage 3: Enzyme reaction in the strip of the first-dimension PAG and generation of apyrimidinic (AP) site at residue 15

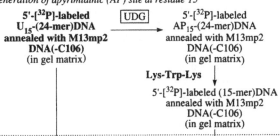

Stage 4: The 90° rotation and casting the gel strip within DNA-sequencing PAG

Stage 5: Electrophoretic separation of substrate and product in DNA-sequencing PAG in the second dimension

Stage 6: Autoradiography of the DNA-sequencing PAG

slowly migrating dark line[VIS] across the DNA-sequencing PAG indicating the position of the substrate, 5'-labeled intact U15-(24-mer)DNA

quickly migrating dark spot[VIS] indicating the product, 5'-labeled oligo(15-mer)DNA, on the DNA-sequencing PAG just opposite the position of UDG on the strip of the substrate-containing PAG

Substrate-Containing PAG[1,2]

Prepare a denaturing (4% stacking and 10% resolving) SDS-PAG containing 5'-[^{32}P]-labeled U_{15}-(24-mer)DNA annealed with single-stranded M13mp2(-C106) DNA (see *Notes*).

Reaction Mixture[1,2]

A. 70 mM HEPES–KOH buffer, pH 7.5
   5 mM 2-Mercaptoethanol
   1 mM EDTA
   400 μg/ml Bovine serum albumin
   15% (w/v) Glycerol
B. 1 mM Sodium cacodilate, pH 6.5
   0.1 mM EDTA
   1 mM NaCl
   100 μM Lys-Trp-Lys

Procedure

Stage 1. After electrophoresis, immerse the resolving substrate-containing SDS-PAG in 33 gel volumes of SDS extraction buffer: 10 mM Tris–HCl (pH 7.5), 5 mM 2-mercaptoethanol, and 25% (v/v) 2-propanol. Gently agitate the gel on a gyratory shaker (60 rpm) at 25°C. Discard the buffer after 30 min and replace with the same volume of fresh SDS extraction buffer. Extract SDS for an additional 30 min. Place the gel in 3 volumes of enzyme renaturation buffer (solution A of the reaction mixture). After brief (30 sec) agitation, place the gel in 27 volumes of the enzyme renaturation buffer and incubate for 18 to 25 h with gentle agitation at 4°C and then at 25°C for 30 min.

Stage 2. Following renaturation of the enzyme within the SDS-PAG, cut each line of the substrate-containing PAG into 2-mm-wide longitudinal strips.

Stage 3. Place an individual strip into a 16 × 100 mm test tube containing 5 ml of solution B of the reaction mixture. Incubate the gel strip at 25°C for 30 min in a stoppered (Parafilm) tube that is placed on its side and gently shaken (60 rpm). The tripeptide Lys-Trp-Lys diffuses into the gel, binds DNA at an apyrimidinic (AP$_{15}$) site produced by UDG, and promotes breakage of the phosphodiester bond specifically on the 3′-side of the AP site, like AP lyase (EC 4.2.99.18). Following this β-elimination reaction, a 5′-[^{32}P]-labeled oligo(15-mer)DNA is generated. Terminate the reaction by substituting the solution B with an equal volume of ice-cold renaturation buffer containing 10 mM EDTA and gently shaking the tube for 30 min at 4°C.

Stages 4, 5, and 6. Following the *in situ* enzyme reaction, cast the processed strip of substrate-containing PAG within a 20% DNA sequencing PAG containing 8.3 M urea, avoiding formation of air bubbles. After polymerization, carry out electrophoresis in a second dimension, transfer the electrophorized 20% PAG to filter paper, dry, and autoradiograph.

Two dark areas are observed on developed x-ray film. The slowly migrating dark area is a line across the 20% PAG indicating the position of intact 5′-[^{32}P]-labeled U$_{15}$-(24-mer)DNA. The fast migrating dark area is a spot indicating the position of 5′-[^{32}P]-labeled oligo(15-mer)DNA. The position of UDG is detected via projection of the dark spot onto the strip of substrate-containing PAG.

Notes: The U$_{15}$-(24-mer)DNA is a 5′-GTGCTGCAAGGC-GAUTAAGTTGGT-3′ oligonucleotide complementary to M13mp2 DNA(-C106) (see below). The oligonucleotide is labeled at the 5′ end by T4 polynucleotide kinase with [γ-^{32}P]-ATP as described.[1,2] The labeled oligonucleotide (specific activity of 3.0 × 10^6 to 3.5 × 10^6 cpm/pmol of 5′ ends) is annealed with single-stranded circular M13mp2 DNA(-C106) molecules (0.27 pmol of 5′ ends/μg of DNA) by heating to 70°C, with slow cooling to 25°C. M13mp2 DNA(-C106) is a DNA isolated from bacteriophage M13mp2(-C106) grown in *E. coli* strain JM107.[3] The substrate, 5′-[^{32}P]-labeled U$_{15}$-(24-mer)DNA (final concentration of about 8 ng/ml)/M13mp2 DNA(-C106) (final concentration of about 2 μg/ml), is included in the resolving SDS-PAG before polymerization.[1,2]

REFERENCES

1. Longley, M.J. and Mosbaugh, D.W., Characterization of DNA metabolizing enzymes *in situ* following polyacrylamide gel electrophoresis, *Biochemistry*, 30, 2655, 1991.

2. Longley, M.J. and Mosbaugh, D.W., *In situ* detection of DNA-metabolizing enzymes following polyacrylamide gel electrophoresis, *Methods Enzymol.*, 218, 587, 1993.

3. Kunkel, T.A., Roberts, J.D., and Zakour, R.A., Rapid and efficient site-specific mutagenesis without phenotypic selection, *Methods Enzymol.*, 154, 367, 1987.

3.2.2.X′ — Poly(ADP-Ribose) Glycohydrolase; PARG

REACTION	Hydrolyzes poly(ADP-ribose) polymer bound covalently at its reducing end to acceptor proteins such as histones or NAD ADP-ribosyltransferase (EC 2.4.2.30) to produce free ADP-ribose moiety (see also General Notes)
ENZYME SOURCE	Fungi, plants, invertebrates, vertebrates
SUBUNIT STRUCTURE	Unknown or no data available

METHOD

Visualization Scheme

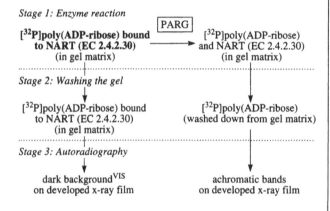

Stage 1: Enzyme reaction

Stage 2: Washing the gel

Stage 3: Autoradiography

Reaction Mixture[1]

50 mM Sodium phosphate buffer, pH 7.5
50 mM NaCl
10% Glycerol
1% Triton X-100
10 mM 2-Mercaptoethanol

Procedure

Stages 1 and 2. Incubate an electrophorized SDS-PAG containing 250 nM [P^{32}]poly(ADP-ribose) bound to NAD ADP-ribosyltransferase (see General Notes) in 5 volumes of renaturation buffer (the reaction mixture) at room temperature for 24 h. Change buffer five times. Incubate the gel for an additional 3 h at 37°C.

Stage 3. Autoradiograph the processed gel overnight at room temperature using Kodak XAR-5 film.

Achromatic bands on the developed x-ray film correspond to localization of PARG activity in the gel.

Notes: To prepare P^{32}-automodified NAD ADP-ribosyltransferase, incubate 20 U of the enzyme for 30 min at 30°C in 0.9 ml of 50 mM Tris–HCl (pH 8.0), 8 mM MgCl$_2$, 10 mM dithiothreitol, 10% (v/v) glycerol, 25 µl/ml activated DNA, 1 mM NAD, 75 µCi [P^{32}]NAD, and 10% ethanol. Preincubate the reaction mixture for 5 min at 30°C before addition of the enzyme. After 30 min incubation at 30°C, preincubate the reaction mixture with 100 µl of 3 M sodium acetate (pH 5.2) and 700 µl of isopropanol. Collect the pellet by centrifugation at 13,000 g for 5 min at 4°C. Wash the pellet three times with 70% ethanol, dry, and dissolve in 1 ml of 10 mM Tris–HCl buffer (pH 8.0) containing 1 mM EDTA. This preparation is included in the SDS-PAG before polymerization and used as a radiolabeled substrate of PARG. The use of a free P^{32}-polymer in the gel as a PARG substrate results in its migration during electrophoresis and is therefore unsuitable for the zymogram technique.

The procedure is sensitive enough to detect 0.5 mU of the enzyme per band.

GENERAL NOTES

Poly(ADP-ribosyl)ation is an important posttranslational modification in eukaryotic cells. Two nuclear enzymes take part in regulation of the metabolism of poly(ADP-ribose). NAD ADP-ribosyltransferase (see 2.4.2.30 — NART) synthesizes from NAD a homopolymer of ADP-ribose consisting of repeating ADP-ribose units linked by ribosyl-ribose glycosidic bonds with infrequent branchings. This poly(ADP-ribose) polymer is bound covalently at its reducing end to acceptor proteins (mostly histones), including NART itself. The turnover of the poly(ADP-ribose) polymer is an important event in cellular response to DNA damaging agents. During repair of the damaged DNA, poly(ADP-ribosyl)ation modulates the chromatin structure and the DNA binding capacity of NART is reduced by automodification, resulting in decreased competition with nuclear enzymes involved in DNA reparation.[1]

REFERENCES

1. Brochu, G., Shah, G.M., and Poirier, G.G., Purification of poly(ADP-ribose) glycohydrolase and detection of its isoforms by a zymogram following one- or two-dimensional electrophoresis, *Anal. Biochem.*, 218, 265, 1994.

3.2.3.1 — Thioglucosidase; TG

OTHER NAMES	Myrosinase, sinigrinase, sinigrase
REACTION	Thioglucoside + H$_2$O = thiol + sugar
ENZYME SOURCE	Plants, vertebrates
SUBUNIT STRUCTURE	Dimer[#] (plants)

METHOD

Visualization Scheme

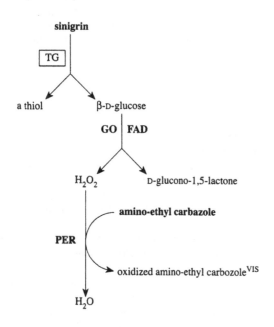

Staining Solution[1]

A. 0.2 *M* Acetate buffer, pH 5.5   50 ml
    Sinigrin   0.1 g
    FAD   2 mg
    Glucose oxidase (GO)   30 U
    Peroxidase (PER)   5 mg
    3-Amino-9-ethyl carbazole (dissolved in
     2 ml of formamide)   10 mg
B. 1.5% Agar solution (60°C)   50 ml

Procedure

Mix A and B components of the staining solution and pour the mixture over the gel surface. Incubate the gel at 37°C in the dark until reddish brown bands appear. Wash the gel with water and fix in 7% acetic acid.

Notes: The PMS–NBT system can be used coupled with linking enzyme glucose oxidase to produce blue formazan in gel areas occupied by TG activity (e.g., see 3.2.1.26 — FF, Method 2). Other chromogenic substrates may be used for peroxidase in place of amino-ethyl carbazole (see 1.11.1.7 — PER, Method 1, *Notes*).

Glucose oxidase used as a linking enzyme is a flavoprotein containing tightly bound FAD molecules. Thus, FAD may be omitted from the staining solution.

OTHER METHODS

A. The linking enzyme β-D-glucose dehydrogenase coupled with the PMS–INT system may be used to detect the product β-D-glucose (e.g., see 3.2.1.26 — FF, Method 1).
B. The product β-D-glucose may be detected using two reactions catalyzed by auxiliary enzymes hexokinase and glucose-6-phosphate dehydrogenase, coupled with the PMS–MTT system (e.g., see 3.2.1.28 — TREH, Method 1).

Both these methods, however, should be applied in two-step procedures because of differences in pH optima between TG (maximal activity at acid conditions) and β-D-glucose dehydrogenase, hexokinase, and glucose-6-phosphate dehydrogenase (maximal activities at alkaline conditions).

REFERENCES

1. Vaughan, J.G., Gordon, E., and Robinson, D., The identification of myrosinase after the electrophoresis of *Brassica* and *Sinapis* seed proteins, *Phytochemistry*, 7, 1345, 1968.

3.3.1.1 — Adenosylhomocysteinase; AHC

OTHER NAMES   *S*-Adenosylhomocysteine hydrolase

REACTION   *S*-Adenosyl-L-homocysteine + H_2O
                            = adenosine + L-homocysteine

ENZYME SOURCE   Vertebrates

SUBUNIT STRUCTURE Unknown or no data available

METHOD

Visualization Scheme

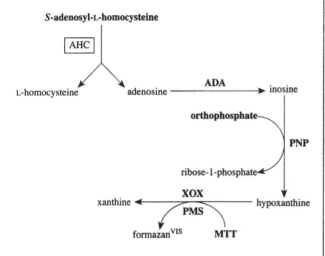

Staining Solution[1] (adapted)

A. 0.1 *M* Phosphate buffer, pH 7.5
     0.476 *µM S*-Adenosyl-L-homocysteine
     1.33 U/ml Adenosine deaminase (ADA)
     0.033 U/ml Purine-nucleoside phosphorylase (PNP)
     0.033 U/ml Xanthine oxidase (XOX)
     0.1 mg/ml MTT
     0.02 mg/ml PMS

B. 1.5% Agarose solution (55°C)

Procedure

Mix equal volumes of A and B solutions and pour the mixture over the gel surface. Incubate the gel at 37°C in the dark until dark blue bands appear. Fix the stained gel in 25% ethanol.

Notes: This method demonstrates a general approach for visualizing enzymes releasing adenosine.[2]

OTHER METHODS

A. An immunoblotting procedure (for details, see Part II) based on the utility of monoclonal antibodies specific to the human enzyme[3] can be used for immunohistochemical visualization of the enzyme protein on electrophoretic gels. This procedure may be of great value in special analyses of AHC, but it is unsuitable for routine use in population genetics studies.

B. A method exists that is based on the strong absorbance at 296 nm of a ketimine ring, which is formed by the reaction of the enzymatic product L-homocysteine with 3-bromopyruvate.[4] The AHC activity bands can be detected by gel scanning at 296 nm (one-dimensional spectroscopy procedure). A special optic device constructed for two-dimensional spectroscopy of electrophoretic gels[5] can be used to photograph AHC zymograms obtained by this method.

REFERENCES

1. Corbo, R.M., Scacchi, R., Palmarino, R., Lucarelli, P., Carapella-De Luca, E., and Businco, L., Detection of heterozygotes in three Italian families with adenosindeaminase deficiency and severe combined immunodeficiency, *Neonatalogica*, 2, 144, 1987.

2. Friedrich, C.A., Chakravarti, S., and Ferrell, R.E., A general method for visualizing enzymes releasing adenosine or adenosine-5′-monophosphate, *Biochem. Genet.*, 22, 389, 1984.

3. Hershfield, M.S. and Francke, U., The human genes for *S*-adenosylhomocysteine hydrolase and adenosine deaminase are syntenic on chromosome 20, *Science*, 216, 739, 1982.

4. Ricci, G., Caccuri, A.M., Lo Bello, M., Solinas, S.P., and Nardini, M., Ketimine rings: useful detectors of enzymatic activities in solution and on polyacrylamide gel, *Anal. Biochem.*, 165, 356, 1987.

5. Klebe, R.J., Mancuso, M.G., Brown, C.R., and Teng, L., Two-dimensional spectroscopy of electrophoretic gels, *Biochem. Genet.*, 19, 655, 1981.

OTHER NAMES — Leucyl aminopeptidase (recommended name), cytosol aminopeptidase, leucyl peptidase, peptidase S, naphthylamidase

REACTION — Release of an N-terminal amino acid, Xaa┼Xbb-, in which Xaa is preferably Leu, but may be other amino acids, including Pro, although not Arg or Lys, and Xbb may be Pro; amino acid amides and methyl esters are also readily hydrolyzed, but rates on arylamides are exceedingly low

ENZYME SOURCE — Bacteria, fungi, plants, protozoa, invertebrates, vertebrates

SUBUNIT STRUCTURE — Monomer (fungi, plants, invertebrates, vertebrates)

METHOD 1

Visualization Scheme

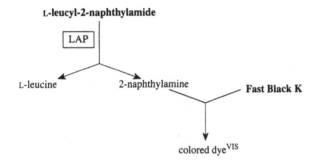

Staining Solution[1]

0.1 M Phosphate buffer, pH 5.8	100 ml
L-Leucyl-2-naphthylamide	40 mg
Fast Black K	60 mg

Procedure

Incubate the gel in filtered staining solution in the dark at 37°C until dark blue bands appear. Wash the stained gel with water and fix in 7% acetic acid.

Notes: The staining of the gel should be carried out with extreme caution because the product 2-naphthylamine is believed to be a carcinogen that causes bladder tumors.

Aminopeptidase activity bands detected using L-leucyl-2-naphthylamide in sera of pregnant women proved to be due to alanine aminopeptidase (see 3.4.11.2 — AAP) and cystyl-aminopeptidase (see 3.4.11.3 — CAP) activities.[2]

METHOD 2

Visualization Scheme

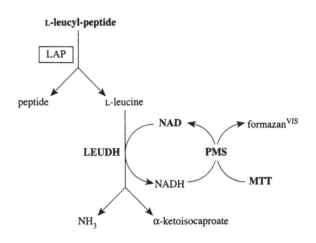

Staining Solution[3] (adapted)

100 mM Tricine–NaOH buffer, pH 8.3
2 mg/ml L-Leucyl–L-leucyl–L-leucine (or L-leucyl–L-valine, or L-leucyl–glycyl–glycine)
10 U/ml Leucine dehydrogenase (LEUDH)
2 mg/ml NAD
0.4 mg/ml MTT
0.04 mg/ml PMS

Procedure

Apply the staining solution to the gel dropwise. Incubate the gel in a humid chamber in the dark at 37°C until dark blue bands appear. Fix the stained gel in 25% ethanol.

Notes: This method is too expensive to be used in large-scale population assays.

The bacterial enzyme, leucine dehydrogenase, displays optimal activity in alkaline conditions, while LAP from many sources is most active at pH values below 7.0. Thus, a two-step procedure of LAP detection by this method may sometimes be desirable.

REFERENCES

1. Brewer, G.J., *An Introduction to Isozyme Techniques*, Academic Press, New York, 1970, p. 100.
2. Sanderink, G.-J.C.M., Artur, Y., Galteau, M.-M., Wellman-Bednawska, M., and Siest, G., Multiple forms of serum aminopeptidases separated by micro two-dimensional electrophoresis under nondenaturing conditions, *Electrophoresis*, 7, 471, 1986.
3. Takamiya, S., Ohshima, T., Tanizawa, K., and Soda, K., A spectrophotometric method for the determination of aminopeptidase activity with leucine dehydrogenase, *Anal. Biochem.*, 130, 266, 1983.

OTHER NAMES Membrane alanyl aminopeptidase (recommended name), microsomal aminopeptidase, aminopeptidase M, aminopeptidase N, particle-bound aminopeptidase, amino-oligopeptidase, membrane aminopeptidase I, pseudo-leucine aminopeptidase, peptidase E

REACTION Release of an N-terminal amino acid, Xaa┼Xbb-, from a peptide, amide, or arylamide; Xaa is preferably Ala, but may be most amino acids, including Pro (slow action); when a terminal hydrophobic residue is followed by a prolyl residue, the two may be released as an intact Xaa-Pro dipeptide

ENZYME SOURCE Plants, invertebrates, vertebrates

SUBUNIT STRUCTURE Monomer (plants, invertebrates)

METHOD

Visualization Scheme

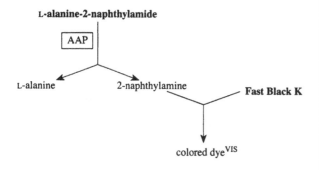

L-alanine-2-naphthylamide

AAP

L-alanine   2-naphthylamine   Fast Black K

colored dye^{VIS}

Staining Solution[1] (modified)

0.1 M Phosphate buffer, pH 5.8	100 ml
L-Alanine-2-naphthylamide hydrochloride	30 mg
Fast Black K	50 mg

Procedure

Incubate the gel in filtered staining solution in the dark at 37°C until dark blue bands appear. Wash the stained gel with water and fix in 7% acetic acid.

Notes: The staining of the gel should be carried out with extreme caution because the product 2-naphthylamine is believed to be a carcinogen that causes bladder tumors.

The enzyme splits α-amino acids (preferentially alanine, not proline) from peptides, amides, and 4-nitroanilides.

Manwell and Baker reported that the addition of $CoCl_2$ (final concentration of 2 mM) to the staining solution speeds up the staining of AAP from sheep sera.[2] It should be stressed, however, that neutral arylamidase detected in sheep sera is the cytosol alanyl aminopeptidase (EC 3.4.11.14), which can use the same substrate as AAP but requires Co^{2+} ions for activity.[3]

Fast Garnet GBC can be used in the staining solution instead of Fast Black K.[4]

OTHER METHODS

The product L-alanine can be detected using auxiliary NAD-dependent L-alanine dehydrogenase coupled with PMS–MTT. A two-step procedure of detection should be used because of the difference in pH optima between AAP (about pH 6.0) and L-alanine dehydrogenase (about pH 10.0).

REFERENCES

1. Okada, Y., Kawamura, K., Kawashima, A., and Mori, M., Hystochemistry and biology of leucine aminopeptidases, *Acta Histochem. Cytochem.*, 8, 265, 1975.
2. Manwell, C. and Baker, C.M.A., "Leucine aminopeptidase" (neutral arylamidase) in sheep sera: improved resolution with gradient gel electrophoresis, *Comp. Biochem. Physiol.*, 84B, 601, 1985.
3. NC-IUBMB, *Enzyme Nomenclature*, Academic Press, San Diego, 1992, p. 376 (EC 3.4.11.14, Comments).
4. Sanderink, G.-J.C.M., Artur, Y., Galteau, M.-M., Wellman-Bednawska, M., and Siest, G., Multiple forms of serum aminopeptidases separated by micro two-dimensional electrophoresis under nondenaturing conditions, *Electrophoresis*, 7, 471, 1986.

OTHER NAMES	Cystinyl aminopeptidase (recommended name), oxytocinase, cystine aminopeptidase
REACTION	Release of an N-terminal amino acid, Cys┼Xaa-, in which the half-cystine residue is involved in a disulfide loop, notably in oxytocin or vasopressin; hydrolysis rates on a range of aminoacyl arylamides usually exceed that for the cystinyl derivative
ENZYME SOURCE	Vertebrates
SUBUNIT STRUCTURE	Unknown or no data available

Method 1

Visualization Scheme

L-cystine di-2-naphthylamide

CAP

L-cystine   2-naphthylamine   **Fast Black K**

colored dye^{VIS}

Staining Solution[1]

0.2 M Tris–maleate buffer, pH 6.0	90 ml
L-Cystine di-2-naphthylamide (dissolved in 10 ml of dimethylformamide)	30 mg
Fast Black K	30 mg

Procedure

Incubate the gel in filtered staining solution in the dark until dark blue bands appear. Wash the stained gel with water and fix in 7% acetic acid.

Notes: Because L-cystine di-2-naphthylamine is a potential carcinogen, solutions of this substrate should be handled with extreme caution.

Method 2

Visualization Scheme

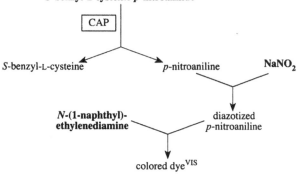

S-benzyl-L-cysteine-p-nitroanilide

CAP

S-benzyl-L-cysteine   p-nitroaniline   NaNO$_2$

N-(1-naphthyl)-ethylenediamine   diazotized p-nitroaniline

colored dye^{VIS}

Staining Solution[2]

A.	S-Benzyl-L-cysteine-p-nitroanilide	26 mg
	2-Methoxyethanol	20 ml
B.	0.1 M Citrate–phosphate buffer, pH 6.0	50 ml
	NaNO$_2$	50 mg
	N-(1-Naphthyl)-ethylenediamine dihydrochloride	10 mg
C.	12.5% Trichloroacetic acid	100 ml

Procedure

Prepare solution B before use. Use solution A stored at 4°C in a brown bottle. Mix one part of solution A with three parts of solution B. Incubate the gel in the resulting mixture at 37°C until pink bands appear. Wash the gel 1 min with water to remove excess NaNO$_2$ and substrate. Place the gel in solution C. The intensity of the pink bands becomes maximal 5 to 6 min after treatment with solution C and remains constant for 1 h.

Notes: The method is based on the enzymatic release of p-nitroaniline, which is diazotized with sodium nitrite and subsequently coupled with a chromogen, N-(1-naphthyl)-ethylenediamine. Less than 0.5 mU per band can be detected by this method. The use of S-benzyl-L-cysteine-p-nitroanilide and L-leucine-p-nitroanilide as CAP substrates results in the same banding pattern of CAP from human seminal plasma.[3]

A pink azo dye formed by this method does not precipitate in the gel matrix, and thus stable zymograms may not be obtained. The enzymoblotting method was developed to overcome this problem.[4] This method is based on transferring proteins from an electrophoretic gel to a nitrocellulose (NC) membrane, and subsequent staining of membrane-bound enzymes using a procedure very similar to that described above. The enzymoblotting method of detection of enzymes producing p-nitroaniline allows one to obtain zymograms that are stable for 16 months when stored between two plastic sheets, sealed with adhesive tape, at –18°C.

3.4.11.3 — Cystyl-Aminopeptidase; CAP (continued)

GENERAL NOTES

Both methods are specific for CAP, but Method 2 is more sensitive. It is also preferable because no carcinogens are used at any stage of staining.

REFERENCES

1. Kleiner, H. and Brouet-Yager, M., Separation of L-cystinyl-di-2-naphthylamide hydrolase ("oxytocinase") isoenzymes by acrylamide gel electrophoresis of human pregnancy, *Clin. Chim. Acta*, 40, 177, 1972.
2. Van Buul, T. and Van Oudheusden, A.P.M., A specific detection method for multiple forms of cystine aminopeptidase (oxytocinase-isoenzymes) after polyacrylamide gel electrophoresis, *J. Clin. Chem. Clin. Biochem.*, 16, 187, 1978.
3. Roy, A.C., Saha, N., Tan, S.M., Kamarul, F.Z., and Ratnam, S.S., A new technique for detecting oxytocinase activity in electrophoresis gels, *Electrophoresis*, 13, 396, 1992.
4. Ohlsson, B.G., Weström, B.R., and Karlsson, B.W., Enzymoblotting: a method for localizing proteinases and their zymogens using *para*-nitroanilide substrates after agarose gel electrophoresis and transfer to nitrocellulose, *Anal. Biochem.*, 152, 239, 1986.

3.4.11.6 — Arginine Aminopeptidase; ARAP

OTHER NAMES	Arginyl aminopeptidase (recommended name), aminopeptidase B, Cl$^-$-activated arginine aminopeptidase, cytosol aminopeptidase IV
REACTION	Release of N-terminal arginine (and to a lesser extent lysine), preferentially from a dipeptide or tripeptide
ENZYME SOURCE	Plants, invertebrates, vertebrates
SUBUNIT STRUCTURE	Monomer (plants)

METHOD

Visualization Scheme

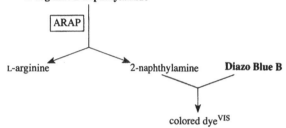

Staining Solution[1]

0.1 *M* Phosphate buffer, pH 6.6	100 ml
L-Arginine-2-naphthylamide	100 mg
Diazo Blue B	100 mg

Procedure

Incubate the gel in filtered staining solution in the dark at 37°C until colored bands appear. Rinse the stained gel with water and fix in 5% acetic acid.

Notes: L-Lysine-2-naphthylamide may be used in place of L-arginine-2-naphthylamide and Fast Black K salt in place of Diazo Blue B.

OTHER METHODS

L-Arginine *p*-nitroanilide and L-lysine *p*-nitroanilide can be used as ARAP substrates. Enzymatically generated *p*-nitroaniline, which is too faintly colored, can then be diazotized and coupled with *N*-(1-naphthyl)-ethylenediamine to produce a readily visible red azo dye. This method may prove more sensitive than that described above, especially when applied to the enzyme transferred from an electrophoretic gel to a nitrocellulose membrane (e.g., see 3.4.21.4 — T, Method 3).

GENERAL NOTES

Cytosolic enzyme from mammalian tissues is activated by Cl$^-$ ions (Cl$^-$-activated arginine aminopeptidase).[2]

3.4.11.6 — Arginine Aminopeptidase; ARAP (continued)

REFERENCES

1. Okada, Y., Kawamura, K., Kawashima, A., and Mori, M., Hystochemistry and biology of leucine aminopeptidases, *Acta Histochem. Cytochem.*, 8, 265, 1975.
2. NC-IUBMB, *Enzyme Nomenclature*, Academic Press, San Diego, 1992, p. 375 (EC 3.4.11.6, Comments).

3.4.11 or 13 ... — Peptidases; PEP

OTHER NAMES	Exopeptidases, dipeptidases, tripeptidases, aminopeptidases
REACTIONS	1. Dipeptide + H_2O = L-amino acids
	2. Tripeptide + H_2O = L-amino acid + dipeptide
ENZYME SOURCE	Bacteria, fungi, algae, plants, protozoa, invertebrates, vertebrates
SUBUNIT STRUCTURE	Monomer (fungi, invertebrates, vertebrates), dimer (fungi, invertebrates, vertebrates), tetramer (invertebrates), hexamer (vertebrates)

METHOD 1

Visualization Scheme

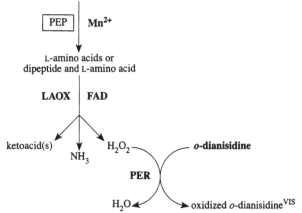

Staining Solution[1]

A. 0.2 *M* Phosphate buffer, pH 7.5 — 25 ml
 o-Dianisidine dihydrochloride — 5 mg
 Snake venom L-amino acid oxidase (LAOX) — 2 to 3 U
 Peroxidase (PER) — 200 U
 0.1 *M* MnCl$_2$ — 0.5 ml
 Dipeptide (or tripeptide); see *Notes* — 20 mg
B. 2% Agar solution (55°C) — 25 ml

Procedure

Mix A and B components of the staining solution and pour the mixture over the gel surface. Incubate the gel at 37°C until brown bands appear. Fix the stained gel in 50% glycerol.

Notes: *o*-Dianisidine dihydrochloride is a potential carcinogen. Solutions of this chromatogen should be handled with caution. An alternative chromatogen, 3-amino-9-ethyl carbazole, may be used in place of *o*-dianisidine dihydrochloride. However, *o*-dianisidine dihydrochloride usually gives better results. Do not use *o*-dianisidine (without diHCl) or benzydine as acceptors in the coupled peroxidase reaction, as these are extremely carcinogenic.

The staining system in Method 1 depends on the release from the peptide of an L-amino acid that is sensitive to the action of snake venom L-amino acid oxidase. In practice this requires the liberation from the peptide of the following L-amino acids: isoleucine, leucine, methionine, phenylalanine, tryptophan, and tyrosine. For example, most of these L-amino acids are effectively oxidized by L-amino acid oxidases from the snakes *Agkistrodon caliginosus* and *Bothrops atrox*. Therefore, this method could be applied to detect peptidases that hydrolyze peptides consisting of the L-amino acids, which may be oxidized by an L-amino acid oxidase used. A number of di- and tripeptides are available for use in PEP stains.[2,3] Crude snake venom (about 0.1 to 0.2 mg/ml) is usually used in the staining solution as the source of L-amino acid oxidase.

METHOD 2

Visualization Scheme

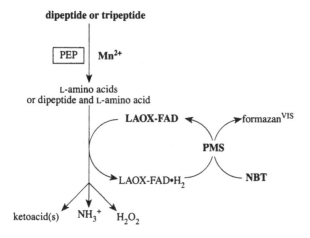

Staining Solution[4]

50 mM Tris–HCl buffer, pH 8.0	10 ml
L-Amino acid oxidase (from *Agkistrodon caliginosus*)	2 U
Dipeptide (or tripeptide); see *Notes*	10 mg
NBT	5 mg
PMS	1 mg

Procedure

Incubate the gel in staining solution in the dark at 37°C until dark blue bands appear. Wash the stained gel with water and fix in 25% ethanol.

Notes: The use of NBT gives better results with acetate cellulose and starch gels. For PAGs, the use of INT is preferable.

L-Amino acid oxidase from the snake *Agkistrodon caliginosus* is specific to L-phenylalanine, L-tryptophan, L-leucine, L-tyrosine, and L-isoleucine. Therefore, when LAOX from this source is used, the method could be applied to peptidases that hydrolyze peptides consisting of the L-amino acids mentioned above. Other preparations of L-amino acid oxidase with differing substrate specificities are commercially available.

METHOD 3

Visualization Scheme

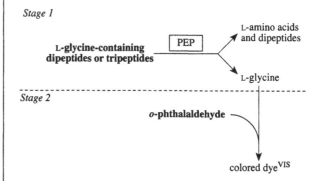

Staining Solution[5]

A. 2 mg/ml Glycyl-L-leucine, pH 7.8 (adjusted with 0.1 N NaOH)
B. 1:1 (w/v) *o*-Phthalaldehyde in 48% ethanol

Procedure

Apply solution A to the gel surface on a filter paper overlay. Incubate the gel at 37°C for 1 h. Remove the filter paper and cover the gel with solution B. Areas of PEP activity appear as dark green bands.

Notes: This method can be applied to detect only those peptidases that hydrolyze L-glycine-containing peptides and produce L-glycine.

METHOD 4

Visualization Scheme

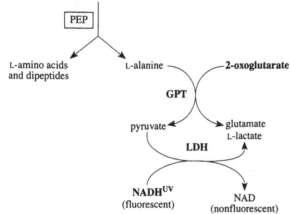

L-alanine-containing
dipeptides or tripeptides

PEP

L-amino acids
and dipeptides

L-alanine — 2-oxoglutarate

GPT

pyruvate ← → glutamate
L-lactate

LDH

NADH^{UV}
(fluorescent)

NAD
(nonfluorescent)

Staining Solution[6]

A. 0.1 *M* Tris–HCl buffer, pH 7.5   25 ml
    L-Alanine-containing di- or tripeptide
    (e.g., Ala-Gly, Ala-Ala-Ala)   15 mg
    Glutamic–pyruvic transaminase (GPT)   3.2 U
    Lactate dehydrogenase (LDH)   27.5 U
    NADH (disodium salt)   10 mg
    2-Oxoglutarate (disodium salt)   20 mg
B. 2% Agar solution (55°C)   25 ml

Procedure

Mix A and B components of the staining solution and pour the mixture over the gel surface. Incubate the gel at 37°C and monitor under long-wave UV light. Areas of PEP activity are visible as dark (nonfluorescent) bands on the fluorescent background of the gel. Record the zymogram or photograph using a yellow filter.

Notes: This method can be applied to detect only those peptidases that hydrolyze L-alanine-containing peptides and produce L-alanine.

When a final zymogram that is visible in daylight is required, the processed gel can be counterstained with the PMS–MTT solution. This will result in the appearance of achromatic bands on a blue background.

The product L-alanine can be detected using auxiliary NAD-dependent L-alanine dehydrogenase coupled with the PMS–MTT system.

GENERAL NOTES

This group of peptide hydrolases is represented by exopeptidases that act only near the ends of polypeptide chains. Those acting at a free N terminus liberate a single amino acid residue (aminopeptidases, EC 3.4.11 ...). Other exopeptidases are specific for dipeptides (dipeptidases, EC 3.4.13 ...) or tripeptides. The nomenclature of peptidases is troublesome.[7]

REFERENCES

1. Lewis, W.H.P. and Harris, H., Human red cell peptidases, *Nature*, 215, 351, 1967.
2. Lewis, W.H.P. and Truslove, G.M., Electrophoretic heterogeneity of mouse erythrocyte peptidases, *Biochem. Genet.*, 3, 493, 1969.
3. Harris, H. and Hopkinson, D.A., *Handbook of Enzyme Electrophoresis in Human Genetics*, North-Holland, Amsterdam, 1976 (loose-leaf, with supplements in 1977 and 1978).
4. Sugiura, M., Ito, Y., Hirano, K., and Sawaki, S., Detection of dipeptidase and tripeptidase activities on polyacrylamide gel and cellulose acetate gel by the reduction of tetrazolium salts, *Anal. Biochem.*, 81, 481, 1977.
5. Kühnl, P., Anneken, K., and Spielmann, V., PEP A^9, a new, unstable variant in the peptidase A system, *Hum. Genet.*, 47, 187, 1979.
6. Rapley, S., Lewis, W.H.P., and Harris, H., Tissue distribution, substrate specificities and molecular sizes of human peptidases determined by separate gene loci, *Ann. Hum. Genet.*, 34, 307, 1971.
7. NC-IUBMB, *Enzyme Nomenclature*, Academic Press, San Diego, 1992, p. 371.

3.4.13.9 — Proline Dipeptidase; PDP

OTHER NAMES	X-Pro dipeptidase (recommended name), prolidase, iminodipeptidase, peptidase D, γ-peptidase
REACTION	Hydrolysis of Xaa+Pro dipeptides; also acts on aminoacyl-hydroxyproline analogues; no action on Pro+Pro
ENZYME SOURCE	Bacteria, fungi, invertebrates, vertebrates
SUBUNIT STRUCTURE	Dimer (fungi, invertebrates, vertebrates)

METHOD

Visualization Scheme

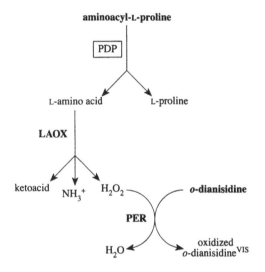

Staining Solution[1]

A. 0.2 M Phosphate buffer, pH 7.5	25 ml
o-Dianisidine dihydrochloride	5 mg
L-Amino acid oxidase (LAOX)	8 U
Peroxidase (PER)	200 U
Leucyl-L-proline or phenylalanyl-proline or glycyl-L-proline	20 mg
B. 2% Agar solution (55°C)	25 ml

Procedure

Mix A and B solutions and pour the mixture over the gel surface. Incubate the gel at 37°C until brown bands appear. Fix the stained gel in 50% glycerol.

Notes: Because o-dianisidine dihydrochloride is a potential carcinogen, solutions containing this chromatogen should be handled with caution. Other chromogenic and fluorogenic peroxidase substrates may be used in place of o-dianisidine dihydrochloride (see 1.11.1.7 — PER, Method 1, *Notes*).

OTHER METHODS

A. The PMS–MTT system can replace the peroxidase system. The FAD that is tightly bound to L-amino acid oxidase is reduced when the enzyme catalyzes the oxidation of L-amino acid. Then MTT is reduced by FADH in the presence of PMS (see 3.4.11 or 13 ... — PEP, Method 2).

B. o-Phthalaldehyde can be used to detect the product L-glycine when glycyl-L-proline is used as a substrate for PDP (see 3.4.11 or 13 ... — PEP, Method 3).

C. Two linked reactions catalyzed by glutamate–pyruvate transaminase and lactate dehydrogenase can be used to detect the product L-alanine when alanyl-L-proline is used as a substrate for PDP (see 3.4.11 or 13 ... — PEP, Method 4).

GENERAL NOTES

PDP is an Mn^{2+}-activated enzyme, cytosolic from most animal tissues.[2]

REFERENCES

1. Lewis, W.H.P. and Harris, H., Human red cell peptidases, *Nature*, 215, 351, 1967.
2. NC-IUBMB, *Enzyme Nomenclature*, Academic Press, San Diego, 1992, p. 378 (EC 3.4.13.9, Comments).

3.4.14.5 — Dipeptidyl-Peptidase IV; DPP

OTHER NAMES | Dipeptidyl aminopeptidase IV, Xaa-Pro-dipeptidyl-aminopeptidase, Gly-Pro naphthylamidase, postproline dipeptidyl aminopeptidase IV

REACTION | Release of an N-terminal dipeptide, Xaa-Xbb+Xcc, from a polypeptide, preferentially when Xbb is Pro, provided Xcc is neither Pro nor hydroxyproline

ENZYME SOURCE | Vertebrates

SUBUNIT STRUCTURE | Trimer* (porcine seminal plasma)

METHOD

Visualization Scheme

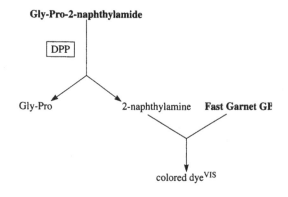

Staining Solution[1]

0.2 *M* Tris–HCl buffer, pH 7.8
0.5 m*M* Gly-Pro-2-naphthylamide
1.25 mg/ml Fast Garnet GBC

Procedure

Incubate the gel in staining solution in the dark at 37°C until colored bands appear. Rinse the stained gel with water and fix in 5% acetic acid.

REFERENCES

1. Yoshimoto, T. and Walter, R., Post-proline dipeptidyl aminopeptidase (dipeptidyl aminopeptidase IV) from lamb kidney: purification and some enzymatic properties, *Biochim. Biophys. Acta*, 485, 391, 1977.

3.4.16.1 — Serine Carboxypeptidase; SCP

OTHER NAMES | Serine-type carboxypeptidase (recommended name); see also General Notes

REACTION | Release of a C-terminal amino acid with broad specificity

ENZYME SOURCE | Bacteria, fungi, plants, invertebrates, vertebrates

SUBUNIT STRUCTURE | Unknown or no data available

METHOD

Visualization Scheme

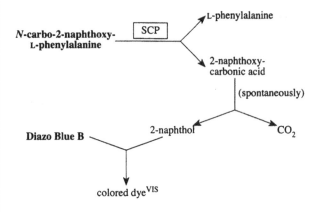

Staining Solution[1]

0.1 *M* Tris–HCl buffer, pH 7.0	150 ml
N-Carbo-2-naphthoxy-D,L-phenylalanine	60 mg
Diazo Blue B	45 mg
25% CaCl$_2$	3 drops
10% ZnCl$_2$	2 drops

Procedure

Incubate the gel in staining solution in the dark at 37°C until colored bands appear. Wash the stained gel with water and fix in 50% glycerol.

Notes: The enzyme represents a group of carboxypeptidases with broad specificity and usually with an optimum pH of 4.5 to 6.0. It releases different C-terminal amino acids, including L-proline, L-arginine, and L-lysine.

Different *N*-carbobenzoxy amino acids can be used as substrates for SCP. Releasing C-terminal L-amino acids can then be detected using coupled reactions catalyzed by L-amino acid oxidase and peroxidase (e.g., see 3.4.11 or 13 ... — PEP, Methods 1 and 2).

3.4.16.1 — Serine Carboxypeptidase; SCP (continued)

GENERAL NOTES

This enzyme activity represents a group of related carboxypeptidases of broad specificity, inhibited by diisopropyl fluorophosphate, and sensitive to thiol-blocking reagents. It is not homologous with serine-type endopeptidases (EC 3.4.21 ...). Sources include *Penicillium* (S_1, S_2, P), yeast (carboxypeptidase Y, yeast protease C, carboxypeptidase C), plants (carboxypeptidase C, serine carboxypeptidase II, phaseolin), and mammals (lisosomal carboxypeptidase A, cathepsin A).[2]

REFERENCES

1. Schopf, T.J., Population genetics of bryozoans, in *Biology of Bryozoans*, Woollacott, R.M. and Zimmer, R.L., Eds., Academic Press, New York, 1977, p. 459.
2. NC-IUBMB, *Enzyme Nomenclature*, Academic Press, San Diego, 1992, p. 382 (EC 3.4.16.1, Comments).

3.4.17.1 — Carboxypeptidase A; CPA

OTHER NAMES	Carboxypolypeptidase, pancreatic carboxypeptidase A, tissue carboxypeptidase A
REACTION	Release of a C-terminal amino acid, but little or no action with -Asp, -Glu, -Arg, -Lys, or -Pro
ENZYME SOURCE	Bacteria, invertebrates, vertebrates
SUBUNIT STRUCTURE	Dimer[d] (*Pseudomonas*)

METHOD

Visualization Scheme

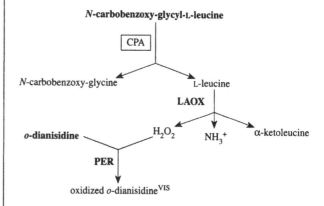

Staining Solution[1,2]

A.	0.2 *M* Tris–HCl buffer, pH 7.5	30 ml
	N-Carbobenzoxy-glycyl-L-leucine	20 mg
	Snake venom L-amino acid oxidase (LAOX)	2 to 3 U
	Peroxidase (PER; Sigma, type II)	2 to 3 mg
	o-Dianisidine dihydrochloride	10 mg
B.	2% Agar solution (55°C)	30 ml

Procedure

Mix A and B components of the staining solution and pour the mixture over the gel surface. Incubate the gel at 37°C until brown bands appear. Fix the stained gel in 50% glycerol.

Notes: Other *N*-carbobenzoxy blocked dipeptides are commercially available and may be used as substrates for CPA. C-terminal L-amino acids should be leucine, isoleucine, phenylalanine, tyrosine, methionine, and tryptophan when a linked reaction catalyzed by L-amino acid oxidase is used to detect CPA.

The MTT–PMS system may be used in the staining solution instead of the peroxidase–*o*-dianisidine system (e.g., see 3.4.11 or 13 ... — PEP, Method 2).

Solutions containing *o*-dianisidine dihydrochloride should be handled with caution because this chromatogen is a potential carcinogen.

3.4.17.1 — Carboxypeptidase A; CPA (continued)

OTHER METHODS

An immunoblotting procedure (for details, see Part II) may be used to detect CPA. Monoclonal antibodies specific to the enzyme from the bovine pancreas are available from Sigma.

REFERENCES

1. Baker, J.E., Isolation and properties of digestive carboxypeptidases from midguts of larvae of the black carpet beetle *Attagenus megatoma*, *Insect Biochem.*, 11, 583, 1981.
2. Baker, J.E., Application of capillary thin layer isoelectric focusing in polyacrylamide gel to the study of alkaline proteinases in stored-product insects, *Comp. Biochem. Physiol.*, 71B, 501, 1982.

3.4.19.3 — 5-Oxoprolyl-Peptidase; OPP

OTHER NAMES	Pyroglutamyl-peptidase I (recommended name), pyrrolidone-carboxylate peptidase, pyroglutamyl aminopeptidase, pyroglutamate aminopeptidase
REACTION	Release of an N-terminal pyroglutamyl group from a polypeptide, provided the next residue is not proline
ENZYME SOURCE	Bacteria, vertebrates
SUBUNIT STRUCTURE	Monomer[#]

METHOD

Visualization Scheme

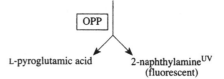

L-pyroglutamic acid 2-naphthylamide

OPP

L-pyroglutamic acid   2-naphthylamine[UV] (fluorescent)

Staining Solution[1]

0.1 *M* Potassium phosphate buffer, pH 7.0
2 m*M* L-Pyroglutamic acid-2-naphthylamide (dissolved in ethanol)
10 m*M* 2-Mercaptoethanol
10 m*M* EDTA

Procedure

Incubate the gel in staining solution at 30°C for 1 h.

Observe fluorescent bands of OPP activity under UV light. Record the zymogram or photograph using a yellow filter.

Notes: The product 2-naphthylamine can be coupled with Fast Blue B (or other diazonium salts), resulting in the formation of colored dye visible in daylight (e.g., see 3.4.21.1 — CT, Method 2). The postcoupling procedure may prove beneficial.

Because 2-naphthylamine derivatives are potential carcinogens, the OPP staining solutions should be handled with extreme caution.

L-Pyroglutamyl-7-amino-4-methylcoumarin can also be used as a sensitive fluorogenic substrate of OPP.[2]

OTHER METHODS

The substrate L-pyroglutamic acid-*p*-nitroanilide can also be used to visualize OPP activity bands on electrophoretic gels. Enzymatically generated *p*-nitroaniline, which is too faintly colored, can then be diazotized and coupled with *N*-(1-naphthyl)-ethylenediamine to produce a readily visible red azo dye. This method may prove more sensitive than that described above, especially when applied to the enzyme transferred from an electrophoretic gel to a nitrocellulose membrane (e.g., see 3.4.21.4 — T, Method 3).

GENERAL NOTES

Two classes of pyroglutamyl aminopeptidase are known.[2] The first class includes bacterial and animal (type I) cytosolic enzymes with similar biochemical characteristics (just EC 3.4.19.3 — OPP). This enzyme is present in all mammalian tissues tested, except blood, and demonstrates broad substrate specificity. It cleaves *N*-terminal pGlu from a range of biologically active peptides, including thyrotropin-releasing hormone, luliberin, bombesin, and neurotensin. The second class is represented by animal type II pyroglutamyl aminopeptidase (EC 3.4.19.6), which is an ectoenzyme (i.e., an integral membrane enzyme protein with an extracellulary localized active site). This enzyme is represented by two isozymes. The first one is localized in the central nervous system (with the highest level of activity in the brain), while the second one is localized in the liver. The serum enzyme is a secreted form of liver isozyme. The type II enzyme takes part in inactivation of the neuronally released thyrotropin-releasing hormone and is thought to be the first neuropeptide-specific peptidase to be characterized.

The use of specific inhibitors[3] is recommended to discriminate between type I (inhibited by thiol-blocking reagents) and type II (inhibited by metal chelators) pyroglutamyl aminopeptidases, which can be present together in some mammalian tissue preparations contaminated with serum.

REFERENCES

1. Sullivan, J.J., Muchnicky, E.E., Davidson, B.E., and Jago, G.R., Purification and properties of the pyrrolidonecarboxylate peptidase of *Streptococcus falcium*, *Aust. J. Biol. Sci.*, 30, 543, 1977.
2. Gallager, S., O'Leary, R.M., and O'Connor, B., The development of two fluorometric assays for the determination of pyroglutamyl aminopeptidase type-II activity, *Anal. Biochem.*, 250, 1, 1997.
3. NC-IUBMB, *Enzyme Nomenclature*, Academic Press, San Diego, 1992, p. 386 (EC 3.4.19.3), p. 387 (EC 3.4.19.6).

OTHER NAMES   Chymotrypsin A, chymotrypsin B

REACTION   Preferential cleavage: Tyr†, Trp†, Phe†, Leu†

ENZYME SOURCE   Invertebrates, vertebrates

SUBUNIT STRUCTURE Monomer[a] (vertebrates)

METHOD 1

Visualization Scheme

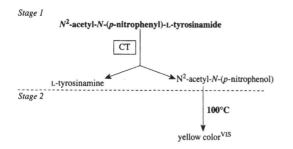

Staining Solution[1]

50 mM Tris–HCl buffer, pH 8.0

5 mM N^2-Acetyl-N-(p-nitrophenyl)-L-tyrosinamide (dissolved in dimethylformamide)

Procedure

Apply the staining solution to the gel surface on a chromatographic paper (Whatman 3MM) overlay and incubate at 40°C for 10 to 15 min. Remove the paper overlay and dry it at 100°C. Yellow bands indicate areas of CT activity.

METHOD 2

Visualization Scheme

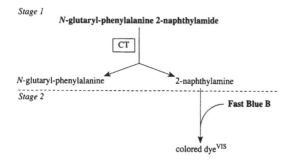

Staining Solution[2]

A. 50 mM Phosphate buffer, pH 8.0

    0.5 mg/ml N-Glutaryl-phenylalanine-2-naphthylamide (Merck; dissolved in dimethylformamide)

B. 50 mM Phosphate buffer, pH 8.0

    1 mg/ml Fast Blue B (or 2 mg/ml Fast Garnet GBC)

Procedure

Incubate the gel in solution A at 37°C for 30 min, wash briefly with water to remove excess naphthylamine, and place in solution B. After appearance of the orange-red bands, transfer the gel into water. The color is stable to light, and the gel may be stored at room temperature for months.

Notes: The postcoupling procedure used for the detection of CT activity has the advantage that the enzyme bands in the gel are rather distinct and bright in color. Furthermore, by using this procedure, a possible inhibitory effect of the diazonium salt on CT activity is avoided. The substrate (amino acid naphthylamide) lends itself well to the postcoupling procedure, since the chromogen (2-naphthylamine) released from the substrate has a great affinity for the protein, and thus no significant diffusion occurs.

Because the substrate is a 2-naphthylamine derivative and therefore a potential carcinogen, solutions of this substrate should be handled with extreme caution.

METHOD 3

Visualization Scheme

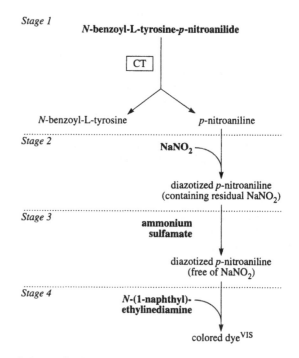

Staining Solution[3]

A. 0.2 M Tris–HCl buffer, pH 7.8

    0.5 mg/ml N-Benzoyl-L-tyrosine-p-nitroanilide (dissolved in dimethyl sulfoxide)

B. 0.1% $NaNO_2$

    1.0 M HCl

C. 0.5% Ammonium sulfamate

    1.0 M HCl

D. 0.05% N-(1-Naphthyl)-ethylenediamine

    47.5% Ethanol

Procedure

Immediately after electrophoresis of active CT, cover the gel by a nitrocellulose (NC) membrane. Perform blotting by covering the NC membrane with sheets of filter paper and paper towels under pressure (1 kg) for 30 min at room temperature (this time period is optimal for an agarose gel). After blotting remove the NC membrane, dry lightly on air, and stain as described below.

For analysis of chymotrypsinogen, use the same procedure, but before staining, preincubate the NC membrane for 1 h at room temperature with the activating enzyme enteropeptidase or trypsin (1.0 or 0.1 mg/ml, respectively), dissolved in 0.2 M Tris–HCl buffer (pH 7.8) containing 50 mM CaCl$_2$. After activation, incubate the NC membrane at room temperature in solution A for 60 min. Then diazotize the product p-nitroaniline as follows: put the NC membrane into solution B for 5 min; after another 5 min of incubation in solution C, put the membrane in solution D.

When the developing red-colored bands are of a maximal intensity (usually after 1 to 5 min), take up the NC membrane and allow solution D to drip off. Put the moist NC membrane between two plastic sheets (overhead film; air bubbles should be avoided), seal with adhesive tape, and store at −18°C.

Notes: N-succinyl-L-phenylalanine-p-nitroanilide may be used in solution A instead of N-benzoyl-L-tyrosine-p-nitroanilide.

A red azo dye formed by this method does not precipitate in the gel. The transfer of the enzyme from the electrophoretic gel to a NC membrane is used to avoid this disadvantage.

If the NC membrane is incubated too long in substrate solution A, the product p-nitroaniline is produced in such amounts that it is released from proteinase zones on the NC membrane, giving a yellow color to the substrate solution and causing undesirable background staining after diazotization and coupling with N-(1-naphthyl)-ethylenediamine. Therefore, incubation in solution A must be stopped just after the yellow bands have appeared.

A similar method was used to detect CT activity after SDS-PAG electrophoresis.[4] N-Acetyltyrosine p-nitroanilide (ATNA; predissolved in N,N-dimethylformamide) was used as a specific CT substrate. The electrophorized gel was immersed and shaken in 25% (v/v) isopropanol in 10 mM Tris–HCl buffer (pH 7.9) for 10 min twice to remove the SDS, and finally equilibrated with 100 mM Tris–HCl buffer (pH 7.9) for 15 min. After removal of the SDS, the gel was incubated with a 1% agarose overlay containing 0.5 mM ATNA or incubated directly in 0.5 mM ATNA solution. After incubation, an agarose overlay or PAG was immersed into a 50-ml solution of 1 N HCl for 3 min to terminate the enzyme reaction, and then 40 ml of 0.2% (w/v) water solution of sodium nitrite was added with gentle shaking and allowed to settle for 10 min. The excess sodium nitrite was destroyed by adding 40 ml of 0.5% (w/v) ammonium amidosulfonate and gently shaking the gel for 5 min. Finally, the diazotized gel was immersed into 1% N-(1-naphthyl)-ethylenediamine solution in 0.5 N HCl and gently agitated until CT activity bands appeared (usually after 5 to 15 min).

GENERAL NOTES

Chymotrypsins A and B are formed from chymotrypsinogens A and B.[5]

REFERENCES

1. Gertler, A., Tencer, Y., and Tinman, G., Simultaneous detection of trypsin and chymotripsin with 4-nitroanilide substrates on cellulose acetate electropherograms, *Anal. Biochem.*, 54, 270, 1973.
2. Hagenmaier, H.E., Polyacrylamide gel electrophoresis as a method for differentiation of gut proteinases which catalyze the hydrolysis of amino acid naphthylamides, *Anal. Biochem.*, 63, 579, 1975.
3. Ohlsson, B.G., Weström, B.R., and Karlsson, B.W., Enzymoblotting: a method for localizing proteinases and their zymogens using *para*-nitroanilide substrates after agarose gel electrophoresis and transfer to nitrocellulose, *Anal. Biochem.*, 152, 239, 1986.
4. Hou, W.-C., Chen, H.-J., Chen, T.-E., and Lin, Y.-H., Detection of protease activities using specific aminoacyl or peptidyl p-nitroanilides after sodium dodecyl sulfate–polyacrylamide gel electrophoresis and its applications, *Electrophoresis*, 20, 486, 1999.
5. NC-IUBMB, *Enzyme Nomenclature*, Academic Press, San Diego, 1992, p. 388 (EC 3.4.21.1, Comments).

OTHER NAMES	α-Trypsin, β-trypsin
REACTION	Preferential cleavage: Arg†, Lys†
ENZYME SOURCE	Bacteria, invertebrates, vertebrates
SUBUNIT STRUCTURE	Monomer# (vertebrates)

METHOD 1

Visualization Scheme

Stage 1

N^2-benzoyl-N-(p-nitrophenyl)-L-argininamide

T

L-argininamide N^2-benzoyl-N-(p-nitrophenol)

Stage 2

100°C

yellow color^{VIS}

Staining Solution[1]

50 mM Tris–HCl buffer, pH 8.0
5 mM N^2-Benzoyl-N-(p-nitrophenyl)-L-argininamide
(dissolved in dimethylformamide)

Procedure

Apply the staining solution to the gel surface on a chromatographic paper (Whatman 3MM) overlay and incubate at 40°C for 10 to 15 min. Remove the paper overlay and dry it at 100°C. Yellow bands indicate areas of T activity.

METHOD 2

Visualization Scheme

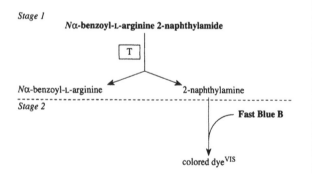

Stage 1

$N\alpha$-benzoyl-L-arginine 2-naphthylamide

T

$N\alpha$-benzoyl-L-arginine   2-naphthylamine

Stage 2

Fast Blue B

colored dye^{VIS}

Staining Solution[2]

A. 50 mM Phosphate buffer, pH 8.0
0.5 mg/ml $N\alpha$-Benzoyl-D,L-arginine-2-naphthylamide
(dissolved in dimethylformamide)
B. 50 mM Phosphate buffer, pH 8.0
1 mg/ml Fast Blue B (or 2 mg/ml Fast Garnet GBC)

Procedure

Incubate the gel in solution A at 37°C for 30 min, wash briefly with water to remove excess naphthylamine, and place in solution B. After appearance of the orange-red bands, transfer the gel into water. The color is stable to light, and the stained gel may be stored at room temperature for months.

Notes: The postcoupling procedure used for the detection of T activity has the advantage that the enzyme bands in the gel are rather distinct and bright in color. Furthermore, by using this procedure, a possible inhibitory effect of the diazonium salt on the T activity is avoided. The substrate (amino acid naphthylamide) lends itself well to the postcoupling procedure, since the chromogen (2-naphthylamine) released from the substrate has a great affinity for the protein, and thus no significant diffusion occurs.

Because the substrate is a 2-naphthylamine derivative and therefore a potential carcinogen, solutions of this substrate should be handled with extreme caution.

METHOD 3

Visualization Scheme

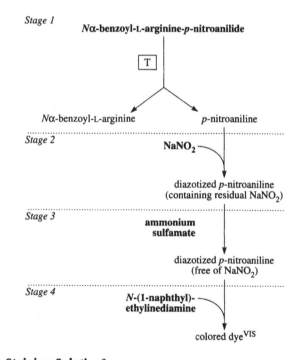

Stage 1

$N\alpha$-benzoyl-L-arginine-p-nitroanilide

T

$N\alpha$-benzoyl-L-arginine p-nitroaniline

Stage 2

NaNO$_2$

diazotized p-nitroaniline
(containing residual NaNO$_2$)

Stage 3

**ammonium
sulfamate**

diazotized p-nitroaniline
(free of NaNO$_2$)

Stage 4

**N-(1-naphthyl)-
ethylinediamine**

colored dye^{VIS}

Staining Solution[3]

A. 0.2 M Tris–HCl buffer, pH 7.8
0.5 mg/ml $N\alpha$-Benzoyl-D,L-arginine-p-nitroanilide
(Sigma; dissolved in buffer at 95°C)
B. 0.1% NaNO$_2$
1.0 M HCl
C. 0.5% Ammonium sulfamate
1.0 M HCl
D. 0.05% N-(1-Naphthyl)-ethylenediamine (Sigma)
47.5% Ethanol

Procedure

Immediately after electrophoresis of active T, cover the gel with a nitrocellulose (NC) membrane. Perform the blotting by covering the NC membrane with sheets of filter paper and paper towels under pressure (1 kg) for 30 min at room temperature (this time period is optimal for an agarose gel). After blotting, remove the NC membrane, dry lightly in air, and stain as described below.

For analysis of trypsinogen, use the same procedure, but before staining, preincubate the NC membrane for 1 h at room temperature with the activating enzyme, enteropeptidase (1.0 mg/ml), dissolved in 0.2 M Tris–HCl buffer (pH 7.8) containing 50 mM CaCl$_2$.

After staining, incubate the NC membrane at room temperature in solution A for 60 min. Then diazotize the product p-nitroaniline as follows: put the NC membrane into solution B for 5 min; after another 5 min of incubation in solution C, put the membrane in solution D.

When the developing red-colored bands are of a maximal intensity (usually after 1 to 5 min), take up the NC membrane and allow solution D to drip off. Put the moist NC membrane between two plastic sheets (overhead film; air bubbles should be avoided), seal with adhesive tape, and store at −18°C.

Notes: A red azo dye formed by this method does not precipitate in the gel. The transfer of the enzyme from the electrophoretic gel to a NC membrane is used to avoid this disadvantage.

If the NC membrane is incubated too long in substrate solution A, the product p-nitroaniline is produced in such amounts that it is released from proteinase zones on the NC membrane, giving a yellow color to the substrate solution and causing undesirable background staining after diazotation and coupling with N-(1-naphthyl)-ethylenediamine. Therefore, incubation in solution A must be stopped just after the yellow bands have appeared.

A similar method was used to detect T activity after SDS-PAG electrophoresis.[4] An electrophorized gel was immersed and shaken in 25% (v/v) isopropanol in 10 mM Tris–HCl buffer (pH 7.9) for 10 min twice to remove the SDS, and finally equilibrated with 100 mM Tris–HCl buffer (pH 7.9) for 15 min. The gel was then incubated with a 1% agarose overlay containing 0.5 mM N-benzoylarginine p-nitroanilide (BANA) or incubated directly in 0.5 mM BANA solution. After incubation, the agarose overlay or PAG was immersed into a 50-ml solution of 1 N HCl for 3 min to terminate the enzyme reaction, and then 40 ml of 0.2% (w/v) water solution of sodium nitrite was added with gentle shaking and allowed to settle for 10 min. The excess sodium nitrite was destroyed by adding 40 ml of 0.5% (w/v) ammonium amidosulfonate and gently shaking the gel for 5 min. Finally, the diazotized gel was immersed into 1% N-(1-naphthyl)-ethylenediamine solution in 0.5 N HCl and gently agitated until T activity bands appeared (usually after 5 to 15 min).

OTHER METHODS

A. Trypsin can also be detected via the hydrogen ion formed during the tryptic hydrolysis of the substrate protamine sulfate by means of the color change of the pH indicator Phenol Red. In order to remove the electrophoretic buffer, which can interfere with the pH change, the electrophoretic gel should be washed with 1 mM borate (pH 8.0) for 2 to 5 min prior to being placed in contact with a substrate-containing agarose plate. Substrate–agarose consists of 0.01% Phenol Red, 20 mM CaCl$_2$, 1 mM borate (pH 8.0), 10 mg/ml protamine sulfate, and 1% agarose. Sites of trypsin activity are indicated by a color change.[5]

B. Trypsin can be detected by bioautography (for more details, see Part II) by placing an electrophoretic gel in contact with an indicator agar containing 1.5% Bacto agar, 1 mg/ml protamine sulfate in *E. coli* minimal medium (see Appendix A-1), and 0.01 O.D./ml exponentially growing Arg H$^-$ arginine-required *E. coli*. Bands of growing *E. coli* are observed 3 h after initiation of the assay at locations where L-arginine is generated by trypsin in the electrophoretic gel.[5]

C. Two-dimensional spectroscopy of electrophoretic gels (for more details, see Part II) permits detection of trypsin activity bands by purely optical means.[5] In this procedure trypsin is detected using substrate α-N-benzoyl-L-arginine ethyl ester (BAEE), α-N-benzoyl-L-arginine-p-nitroanilide (BAPNA), or α-N-benzoyl-L-arginine 7-amido-4-methyl-coumarin (BAAMC). All reactions are carried out by placing an electrophoretic gel in contact with a 1% substrate–agarose plate containing 0.1 M Tris–HCl buffer (pH 8.0), 20 mM CaCl$_2$, and 5 mM BAEE (or 2.5 mM BAPNA, or 0.05 mM BAAMC). When BAAMC is used as a substrate, the product is detected due to the fluorescence excited at 313 nm and emitted at 420 nm. When BAEE and BAPNA are used as substrates, the product formation is detected due to the absorption at 253 and 392 nm, respectively. A special optical device permits one to take a photograph of the developed agarose overlay.

D. The enzyme can be labeled before SDS-PAG electrophoresis by biotinylated aprotinin[6] (T-specific inhibitor) and then detected on a nitrocellulose blot using an avidin–alkaline phosphatase conjugate and histochemical procedure of detection of alkaline phosphatase activity (see 3.1.3.1 — ALP, Method 2). The method proved highly sensitive and could detect as little as 0.2 ng (5 fmol) of trypsin. The method is not absolutely specific and therefore is applicable to the purified T preparations only. Biotinylated serine proteinase-specific inhibitors may prove to be useful probes suitable for detection (see 3.4.21.5 — THR, Others Methods, B) and identification of certain serine proteinases when used in combination with proteinase-specific chloromethyl ketone inhibitors.[7-9]

GENERAL NOTES

Two-dimensional spectroscopy and bioautography methods are not widely used because they are too complex. It should be pointed out, however, that the bioautographic method is one of the most sensitive and detects about 0.04 U of T activity toward BAEE per band.

The enzyme is isolated from bacteria (*Streptomyces griseus* trypsin), insects (cocconase), and the pancreases of many mammals and from lower vertebrates, including crayfish. Cattle β-trypsin is formed from trypsinogen by cleavage of one peptide bond. Further peptide bond cleavages produce α-trypsin and other T isoforms.[10]

REFERENCES

1. Gertler, A., Tencer, Y., and Tinman, G., Simultaneous detection of trypsin and chymotripsin with 4-nitroanilide substrates on cellulose acetate electropherograms, *Anal. Biochem.*, 54, 270, 1973.
2. Hagenmaier, H.E., Polyacrylamide gel electrophoresis as a method for differentiation of gut proteinases which catalyze the hydrolysis of amino acid naphthylamides, *Anal. Biochem.*, 63, 579, 1975.
3. Ohlsson, B.G., Weström, B.R., and Karlsson, B.W., Enzymo-blotting: a method for localizing proteinases and their zymogens using *para*-nitroanilide substrates after agarose gel electrophoresis and transfer to nitrocellulose, *Anal. Biochem.*, 152, 239, 1986.
4. Hou, W.-C., Chen, H.-J., Chen, T.-E., and Lin, Y.-H., Detection of protease activities using specific aminoacyl or peptidyl *p*-nitroanilides after sodium dodecyl sulfate–polyacrylamide gel electrophoresis and its applications, *Electrophoresis*, 20, 486, 1999.
5. Klebe, R.J., Mancuso, M.G., Brown, C.R., and Teng, L., Two-dimensional spectroscopy of electrophoretic gels, *Biochem. Genet.*, 19, 655, 1981.
6. Melrose, J., Ghosh, P., and Patel, M., Biotinylated aprotinin: a versatile probe for the detection of serine proteinases on Western blots, *Int. J. Biochem. Cell Biol.*, 27, 891,1995.
7. Powers, J.C., Gupron, B.F., Harley, A.D., Nishino, N., and Whitley, R.J., Specificity of porcine pancreatic elastase, human leucocyte elastase and cathepsin G: inhibition with chloromethyl ketones, *Biochim. Biophys. Acta*, 485, 156, 1977.
8. Kettner, C. and Shaw, E., Inactivation of trypsin-like enzymes with peptides of arginine chloromethyl ketones, *Methods Enzymol.*, 80, 826, 1981.
9. Kay, G., Bailie, R., Halliday, I.M., Nelson, J., and Walker, B., The synthesis, kinetic characterization and application of biotinylated aminoacylchloromethanes for the detection of chymotrypsin and trypsin-like serine proteinases, *Biochem. J.*, 283, 455, 1992.
10. NC-IUBMB, *Enzyme Nomenclature*, Academic Press, San Diego, 1992, p. 388 (EC 3.4.21.4, Comments).

3.4.21.5 — Thrombin; THR

OTHER NAMES	Fibrinogenase
REACTION	Selective cleavage of Arg┼Gly bonds in fibrinogen to form fibrin and release fibrinopeptides A and B
ENZYME SOURCE	Vertebrates
SUBUNIT STRUCTURE	Monomer[a] (mammals)

METHOD

Visualization Scheme

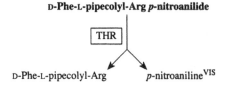

D-Phe-L-pipecolyl-Arg *p*-nitroanilide

THR

D-Phe-L-pipecolyl-Arg   *p*-nitroaniline[VIS]

Staining Solution[1]

A. 0.1 *M* Sodium barbital buffer, pH 7.75
   0.2 *M* NaCl
   3.4 m*M* CaCl$_2$
   1.4 m*M* MgCl$_2$
   2 m*M* D-Phe-L-pipecolyl-Arg-*p*-nitroanilide
B. 3% Agarose solution (60°C)

Procedure

Mix equal volumes of solutions A and B and pour the mixture over the gel surface. Incubate the gel at 37°C until yellow bands appear.

Notes: The product *p*-nitroaniline is usually too faint to be directly visualized in the gel and is registrated spectrophotometrically at 405 nm. When *p*-nitroaniline is diazotized and coupled with naphthylethylenedi-amine, a readily visible red azo dye is formed. Unfortunately, this dye does not precipitate in the gel. However, by using a method based on the transfer of enzymes from the electrophoretic gel to an immobilizing matrix of nitrocellulose, this disadvantage can be overcome.[2] An example application of this method is given above (see 3.4.21.4 — T, Method 3).

OTHER METHODS

Thrombin can be labeled by biotinylated THR-specific inhibitor D-Phe-Pro-Arg-CH$_2$Cl prior to electrophoresis and then detected after electrophoretic separation in SDS-PAG using an avi-din–alkaline phosphatase conjugate and histochemical detection of alkaline phosphatase activity (see 3.1.3.1 — ALP, Method 2) on nitrocellulose blots.[3] Sensitivity of the method depends on the length of a spacer connecting biotin with the THR-specific inhibitor. A 7- to 14-atom spacer is needed for sensitive (about 10 ng/band) detection of THR. The method is not absolutely specific and therefore is applicable to purified THR preparations only. Biotinylated serine proteinase-specific inhibitors may prove to be useful probes suitable for detection and identification of certain serine proteinases when used in combination with proteinase-specific chloromethyl ketone inhibitors.[4–6]

GENERAL NOTES

THR is formed from prothrombin.[7]

REFERENCES

1. Wagner, O.F., Bergmann, I., and Binder, B.R., Chromogenic substrate autography: a method for detection, characterization, and quantitative measurement of serine proteases after sodium dodecyl sulfate–polyacrylamide gel electrophoresis or isoelectric focusing in polyacrylamide gels, *Anal. Biochem.*, 151, 7, 1985.
2. Ohlsson, B.G., Weström, B.R., and Karlsson, B.W., Enzymoblotting: a method for localizing proteinases and their zymogens using *para*-nitroanilide substrates after agarose gel electrophoresis and transfer to nitrocellulose, *Anal. Biochem.*, 152, 239, 1986.
3. Anderson, P.J. and Bock, P.E., Biotin derivatives of D-Phe-Pro-Arg-CH₂Cl for active-site-specific labeling of thrombin and other serine proteases, *Anal. Biochem.*, 296, 254, 2001.
4. Powers, J.C., Gupron, B.F., Harley, A.D., Nishino, N., and Whitley, R.J., Specificity of porcine pancreatic elastase, human leucocyte elastase and cathepsin G: inhibition with chloromethyl ketones, *Biochim. Biophys. Acta*, 485, 156, 1977.
5. Kettner, C. and Shaw, E., Inactivation of trypsin-like enzymes with peptides of arginine chloromethyl ketones, *Methods Enzymol.*, 80, 826, 1981.
6. Kay, G., Bailie, R., Halliday, I.M., Nelson, J., and Walker, B., The synthesis, kinetic characterization and application of biotinylated aminoacylchloromethanes for the detection of chymotrypsin and trypsin-like serine proteinases, *Biochem. J.*, 283, 455, 1992.
7. NC-IUBMB, *Enzyme Nomenclature*, Academic Press, San Diego, 1992, p. 389 (EC 3.4.21.5, Comments).

OTHER NAMES   Acrosomal proteinase
REACTION   Preferential cleavage: Arg†, Lys†
ENZYME SOURCE   Invertebrates, vertebrates
SUBUNIT STRUCTURE   Monomer# (vertebrates)

METHOD

Visualization Scheme

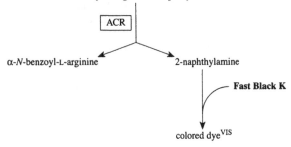

α-*N*-benzoyl-*L*-arginine 2-naphthylamide

ACR

α-*N*-benzoyl-*L*-arginine   2-naphthylamine

Fast Black K

colored dye[VIS]

Staining Solution[1]

A. 0.1 *M* Phosphate buffer, pH 7.5
    0.9 mg/ml Fast Black K
B. 0.1 *M* Phosphate buffer, pH 7.5
    1.25 mg/ml α-*N*-Benzoyl-*D,L*-arginine-2-naphthylamide
    (dissolved in dimethyl sulfoxide)
    0.85 mg/ml Fast Black K

Procedure

Wash an electrophorized gel in filtered solution A at 37°C for 3 to 5 min and then transfer to filtered solution B. Incubate the gel in solution B at 37°C in the dark until blue bands appear. Wash the stained gel in water and fix in 7% acetic acid.

Notes: Fast Garnet GBC and Fast Blue B can also be used as coupling dyes. The most stable dye is obtained, however, with Fast Black K.

Because the substrate is a 2-naphthylamine derivative and therefore a potential carcinogen, solutions of this substrate should be handled with extreme caution.

The method described above is not specific for ACR. However, this trypsin-like proteinase occurs only in spermatozoa (in acrosomes) and thus can be easily identified based on its tissue specificity.

OTHER METHODS

α-*N*-Benzoyl-*L*-arginine-*p*-nitroanilide can also be used as a chromogenic substrate for ACR. Enzymatically released *p*-nitroaniline, which is too faintly colored, can then be diazotized and coupled with *N*-(1-naphthyl)-ethylenediamine to produce a readily visible red azo dye. This method may prove more sensitive than that described above, especially when applied to the enzyme transferred from the electrophoretic gel to a nitrocellulose membrane (e.g., see 3.4.21.4 — T, Method 3).

GENERAL NOTES

ACR occurs in spermatozoa. The active enzyme is formed from proacrosin by limited proteolysis.[2]

REFERENCES

1. Garner, D.L., Improved zymographic detection of bovine acrosin, *Anal. Biochem.*, 67, 688, 1975.
2. NC-IUBMB, *Enzyme Nomenclature*, Academic Press, San Diego, 1992, p. 389 (EC 3.4.21.10, Comments).

OTHER NAMES	Serum kallikrein, kininogenin, kininogenase, arginine esterase
REACTION	Selective cleavage of some Arg† and Lys† bonds, including Lys†Arg and Arg†Ser in (human) kininogen to release bradykinin
ENZYME SOURCE	Vertebrates
SUBUNIT STRUCTURE	Unknown or no data available

METHOD 1

Visualization Scheme

N-α-benzoyl-L-arginine ethyl ester

PKK

N-α-benzoyl-L-arginine → ethanol

NAD → formazan^{VIS}

ADH   PMS

NADH   NBT

acetaldehyde

Staining Solution[1]

80 mM Tris–EDTA–borate buffer, pH 9.2	150 ml
N-α-Benzoyl-L-arginine ethyl ester	250 mg
NAD	100 mg
PMS	1 mg
NBT	30 mg
KCN	1 mg
Alcohol dehydrogenase (ADH; suitable for determination of ethanol; Sigma)	1500 U
CaCl$_2$	300 mg

Procedure

Incubate the gel in staining solution in the dark at 37°C until dark blue bands appear. Fix the stained gel in 7% acetic acid.

Notes: Bands of trypsin activity can also be developed by this method. These bands, however, can be easily identified by adding a soybean trypsin inhibitor in the staining solution.

METHOD 2

Visualization Scheme

H-D-Pro-Phe-Arg *p*-nitroanilide

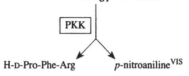

PKK

H-D-Pro-Phe-Arg   *p*-nitroaniline^{VIS}

Staining Solution[2]

A. 0.1 *M* Sodium barbital buffer, pH 7.75
    0.2 *M* NaCl
    3.4 m*M* CaCl$_2$
    1.4 m*M* MgCl$_2$
    2.4 m*M* H-D-Pro-Phe-Arg-*p*-nitroanilide
B. 3% Agarose solution (60°C)

Procedure

Mix equal volumes of A and B solutions and pour the mixture onto a prewarmed glass plate to give a 2-mm-thick gel. Place the electrophorized SDS-PAG (washed in an excess volume of 2.5% Triton X-100 for 50 min and rinsed with distilled water for 15 min) on top of a substrate-containing agarose gel and incubate at 37°C in a moist chamber for up to 12 h. Areas of PKK activity appear as yellow bands in the agarose gel. Record the zymogram or scan on a densitometer at 405 nm.

Notes: The chromogenic substrate H-D-Val-Leu-Arg-*p*-nitroanilide can also be used to detect tissue kallikrein (EC 3.4.21.35).

The product *p*-nitroaniline is usually too faint colored, but when it is diazotized and coupled with *N*-(1-naphthyl)-ethylenediamine, a readily visible red azo dye is formed. However, this dye does not precipitate in the gel. To overcome this disadvantage, a method that is based on transferring proteins from the electrophoretic gel to a nitrocellulose membrane and subsequent staining of the membrane-bound enzyme was developed[3] (an example application of this method is 3.4.21.4 — T, Method 3).

The method described above was used to visualize PKK after electrophoresis of purified enzyme preparations of plasma and tissue (urine) kallikreins from the human and mouse. Its specificity was not tested in detail.

GENERAL NOTES

The enzyme activates coagulation factors XII and VII and plasminogen. PKK is formed from prokallikrein by factor XIIa.[4]

3.4.21.34 — Plasma Kallikrein; PKK (continued)

REFERENCES

1. Fujimoto, Y., Moriya, H., Yamaguchi, K., and Moriwaki, C., Detection of arginine esterase of various kallikrein preparations on gellified electrophoretic media, *J. Biochem. (Tokyo)*, 71, 751, 1972.
2. Wagner, O.F., Bergmann, I., and Binder, B.R., Chromogenic substrate autography: a method for detection, characterization, and quantitative measurement of serine proteases after sodium dodecyl sulfate–polyacrylamide gel electrophoresis or isoelectric focusing in polyacrylamide gels, *Anal. Biochem.*, 151, 7, 1985.
3. Ohlsson, B.G., Weström, B.R., and Karlsson, B.W., Enzymoblotting: a method for localizing proteinases and their zymogens using *para*-nitroanilide substrates after agarose gel electrophoresis and transfer to nitrocellulose, *Anal. Biochem.*, 152, 239, 1986.
4. NC-IUBMB, *Enzyme Nomenclature*, Academic Press, San Diego, 1992, p. 391 (EC 3.4.21.34, Comments).

3.4.21.36 — Elastase; EL

OTHER NAMES	Pancreatic elastase (recommended name), pancreatopeptidase E, pancreatic elastase I
REACTION	Hydrolysis of proteins, including elastin; preferential cleavage: Arg†
ENZYME SOURCE	Bacteria, fungi, vertebrates
SUBUNIT STRUCTURE	Monomer[#] (mammals)

METHOD 1

Visualization Scheme

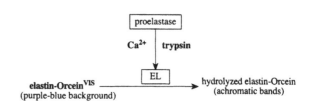

Substrate-Containing Reactive Plate[1]

To prepare a substrate-containing reactive agarose plate, grind elastin–Orcein crystals (100 mg) in a mortar with 1 ml of 20 mM Tris–HCl buffer (pH 8.8). Wash the particles twice with the same buffer by centrifugation until a colorless supernatant is obtained to remove contaminating dye-bound polypeptides. Suspend the pelleted particles in 1 ml of the buffer and mix with 49 ml of 2% agarose solution in 20 mM Tris–HCl buffer (pH 8.8, 60°C). Pour the mixture onto a glass plate of an appropriate size. After solidification, the substrate-containing agarose reactive plate is ready for use.

Procedure

Activate proelastase in an electrophorized gel by incubating the gel for 30 min at 37°C in 20 mM Tris–HCl (pH 8.2), 20 mM CaCl$_2$, and 50 μg/ml trypsin. After incubation wash the gel twice with 20 mM Tris-HCl buffer (pH 8.8) to remove excess trypsin. Then blot the gel with filter paper and lay onto a substrate-containing reactive plate. Incubate the electrophoretic gel–reactive plate combination at 37°C for 2 h or until colorless bands appear on the purple-blue background of the reactive agarose plate. Fix the stained reactive plate in 3% acetic acid.

Notes: Without activation of the proenzyme with trypsin, no EL activity could be detected.

Only a small part of the substrate complex usually is broken down as a result of EL activity so that it is worthwhile to regain the remaining part of the particles from the agarose gel. To achieve this, an agarose gel containing elastin–Orcein should be liquefied at 70°C by heating in a water bath. The pH of the solution should be lowered to 3 by addition of HCl. The color of the particles changes reversibly from purple-blue at pH 8.8 to pink at pH 3. Under this acid condition, an agarose solution stays liquid at room temperature, and the elastin–Orcein particles can easily be spun down by centrifugation for 2 min at 5000 g. The isolated elastin–Orcein should then be suspended in buffer, washed twice by centrifugation, and stored in a refrigerator.

Elastin–Congo Red can also be used as a colored substrate complex to obtain the negative EL zymograms.

METHOD 2

Visualization Scheme

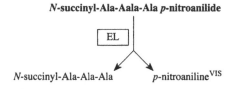

N-succinyl-Ala-Aala-Ala p-nitroanilide

EL

N-succinyl-Ala-Ala-Ala   p-nitroaniline[VIS]

Substrate-Containing Reactive Plate[2]

Prepare 0.125 M stock solution of N-succinyl-L-Ala-L-Ala-L-Ala-p-nitroanilide (SAPNA) in N-methylpirrolidone by heating at 60°C for 10 min and stirring. Store this solution at 4°C in a dark bottle. Prepare the 1.5 mM substrate solution just before use by dilution of the stock solution with 0.2 M Tris–HCl (pH 8.0). Suspend the agar in 0.2 M Tris–HCl, pH 8.0 (1.5% w/v), heat to 100°C to liquefy, and cool to 60°C. Add 1.5 mM substrate solution to the agar solution (10 μl/ml) and stir for 3 min at 60°C. Pour the resulting mixture onto a preheated (45°C) glass plate of an appropriate size and allow to solidify. Use a substrate-containing reactive plate immediately or store in an airtight bag at 4°C.

Procedure

Treat an electrophorized PAG with trypsin to convert proelastase into elastase as described in Method 1. Apply the reactive plate on top of the PAG, ensuring that no air bubbles are trapped between the two slabs. Incubate the PAG–reactive plate combination at 37°C in a humid box. Bands of a yellow reaction product (p-nitroaniline) appear almost immediately. When the bands are well developed, record, photograph, or scan the zymogram on a densitometer at 405 nm immediately, since the yellow bands increase progressively in intensity.

Notes: The product p-nitroaniline can be diazotized and coupled with N-(1-naphthyl)-ethylenediamine, resulting in a readily visible red azo dye. However, this dye does not precipitate in the gel. To overcome this disadvantage, the method of nitrocellulose enzyme blotting may be used[3] (for example, see 3.4.21.4 — T, Method 3).

GENERAL NOTES

Method 1 is highly specific for EL; however, the considerable time necessary for developing the zymogram in this procedure enhances the potential risk that enzyme bands placed closely together could mix with each other by diffusion. This method has a detection limit of 0.5 μg of EL per band after incubation at 37°C overnight. The same amount of EL can be detected with Method 2 within a few minutes. It should be taken into account, however, that Method 2 can also detect some elastase-like esterases cleaving SAPNA but not elastin. Such esterases may occur in certain biological fluids, e.g., human synovial fluid.

The enzyme is a homologue of chymotrypsin. Mammalian EL is formed by activation of pancreatic proelastase by trypsin.[4]

REFERENCES

1. Dijkhof, J. and Poort, C., Visualization of proelastase in polyacrylamide gels, *Anal. Biochem.*, 83, 315, 1977.
2. Gardi, C. and Lungarella, G., Detection of elastase activity with a zymogram method after isoelectric focusing in polyacrylamide gel, *Anal. Biochem.*, 140, 472, 1984.
3. Ohlsson, B.G., Weström, B.R., and Karlsson, B.W., Enzymoblotting: a method for localizing proteinases and their zymogens using *para*-nitroanilide substrates after agarose gel electrophoresis and transfer to nitrocellulose, *Anal. Biochem.*, 152, 239, 1986.
4. NC-IUBMB, *Enzyme Nomenclature*, Academic Press, San Diego, 1992, p. 392 (EC 3.4.21.36, Comments).

3.4.21.40 — Esteroprotease; EP (deleted from the Enzyme List;[1] see General Notes)

OTHER NAMES	Submandibular proteinase A, esteroprotease A, trypsin-like esteroprotease, arginine-esteroprotease
REACTION	Hydrolysis of proteins by cleavage of arginyl bonds; hydrolysis of ester bonds in acetyl-methionine and tosyl-arginine esters
ENZYME SOURCE	Vertebrates (mammalian submandibular gland)
SUBUNIT STRUCTURE	Unknown or no data available

METHOD 1

Visualization Scheme

N-acetyl-L-methionine α-naphthyl ester

EP

N-acetyl-L-methionine → α-naphthol → Fast Red TR

↓

colored dye^{VIS}

Staining Solution[2]

66.6 m*M* Phosphate buffer, pH 6.5

0.75 m*M* *N*-Acetyl-L-methionine α-naphthyl ester (dissolved in dimethyl sulfoxide)

1.5 mg/ml Fast Red TR

Procedure

Dissolve the substrate in a minimal volume of dimethyl sulfoxide, and then dissolve the dye in the buffer. Add the substrate solution to the dye solution under vigorous stirring. Incubate the gel in the resulting mixture at room temperature until red-brown bands appear. Fix the stained gel in 50% glycerol or 3% acetic acid.

METHOD 2

Visualization Scheme

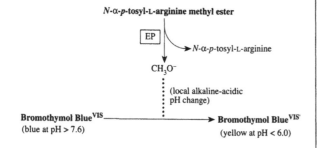

N-α-*p*-tosyl-L-arginine methyl ester

EP

→ *N*-α-*p*-tosyl-L-arginine

CH₃O⁻

⋮ (local alkaline-acidic pH change)

Bromothymol Blue^{VIS} ———— → Bromothymol Blue$^{VIS'}$

(blue at pH > 7.6)　　　　　　　(yellow at pH < 6.0)

Staining Solution[3]

A.	50 m*M* Tris–EDTA–borate buffer, pH 8.6	10 ml
	Bromothymol Blue	100 mg
B.	*N*-α-*p*-Tosyl-L-arginine methyl ester	100 mg
	H₂O	10 ml

Procedure

Place an electrophorized acetate cellulose gel horizontally and cover with solution A. After 1 min blot the gel and cover it with solution B. The areas of EP activity appear as yellow bands on a blue background of the gel. Record or photograph the zymogram.

METHOD 3

Visualization Scheme

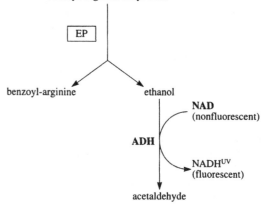

benzoyl-arginine-ethyl ester

EP

benzoyl-arginine ← → ethanol

NAD (nonfluorescent)

ADH

NADH^{UV} (fluorescent)

↓

acetaldehyde

Staining Solution[4]

Buffer, pH 8.7, containing 154 m*M* semicarbazide, 154 m*M* sodium pyrophosphate, 40 m*M* glycine

10 m*M* Benzoyl-arginine-ethyl ester

10 m*M* NAD

40 U/ml Alcohol dehydrogenase

Procedure

Immerse an acetate cellulose membrane (6 × 8 cm) in 2.5 ml of the staining solution and apply to the surface of the gel, avoiding the formation of air bubbles. Incubate the gel at 37°C and monitor under long-wave UV light for fluorescent bands of EP activity. Photograph the gel using a yellow filter.

Notes: Acetyl-phenylalanyl-arginine-ethyl ester can also be used as an EP substrate in place of benzoyl-arginine-ethyl ester.

To obtain a positively stained EP zymogram, counterstain the processed gel with PMS–MTT solution.

Method 4

Visualization Scheme

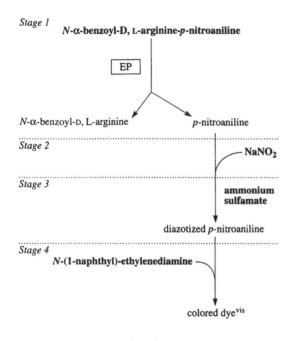

Stage 1

N-α-benzoyl-D, L-arginine-p-nitroaniline

EP

N-α-benzoyl-D, L-arginine   *p*-nitroaniline

Stage 2

— NaNO₂

Stage 3

ammonium sulfamate

diazotized *p*-nitroaniline

Stage 4

N-(1-naphthyl)-ethylenediamine

colored dye^vis

Staining Solution[4] (modified)

A. 50 mM Sodium phosphate buffer, pH 7.6
B. 10 mg N-α-Benzoyl-D,L-arginine-p-nitroanilide (dissolved in 0.5 ml of dimethyl sulfoxide by shaking and heating in a water bath at 80°C)
C. 0.1% NaNO₂
    1.0 M HCl
D. 0.5% Ammonium sulfamate
    1.0 M HCl
E. 0.05% N-(1-Naphthyl)-ethylenediamine
    47.5% Ethanol

Procedure

Add 2.5 ml of solution A to solution B. Soak an acetate cellulose membrane (6 × 8 cm) in the mixture and apply to the surface of the gel, avoiding the formation of air bubbles. Incubate the gel for 30 to 60 min at 37°C to reveal yellow bands of *p*-nitroaniline, resulting from the hydrolysis of the substrate by EP. Then diazotize the released *p*-nitroaniline. First, put the membrane into solution C for 5 min. Next, incubate the membrane in solution D for another 5 min to destroy the excess sodium nitrite. Finally, put the membrane in solution E. When the developing reddish violet bands are of a maximal intensity (usually after 1 to 5 min), take up the membrane and allow solution E to drip off. Put the moist cellulose acetate membrane between two plastic sheets (overhead film; air bubbles should be avoided), seal with adhesive tape, and store at −18°C.

Method 5

Visualization Scheme

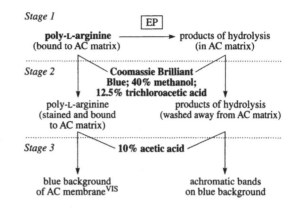

Stage 1

EP

poly-L-arginine ⟶ products of hydrolysis
(bound to AC matrix)   (in AC matrix)

Stage 2   **Coomassie Brilliant Blue; 40% methanol; 12.5% trichloroacetic acid**

poly-L-arginine   products of hydrolysis
(stained and bound   (washed away from AC matrix)
to AC matrix)

Stage 3   **10% acetic acid**

blue background   achromatic bands
of AC membrane^VIS   on blue background

Staining Solution[4] (modified)

A. 1% Poly-L-arginine in 50 mM sodium phosphate buffer, pH 7.6
B. 0.01% Coomassie Brilliant Blue R-250
    40% Methanol
    12.5% Trichloroacetic acid
C. 10% Acetic acid

Procedure

Soak an acetate cellulose (AC) membrane (6 × 8 cm) in 2.5 ml of solution A and apply to the surface of the gel, avoiding the formation of air bubbles. Incubate the gel for 30 to 60 min at 37°C. Remove the membrane from the gel and stain with solution B for 20 min. Destain the membrane by washing in solution C. Areas of EP activity are detected as white bands on a blue background of the processed AC membrane.

General Notes

The methods described above are not specific for EP. Additional bands caused by esteroprotease E develop on the zymograms obtained by Method 1. Method 2 develops activity bands of both EP and esteroprotease, called "tamase."[5] Activity bands caused by different esteroproteases can be identified by using both methods in parallel, based on different substrate specificities of these esteroproteases, as indicated in the table below:

Substrates	Esteroproteases		
	EP	Esteroprotease E	Tamase
N-Acetyl-L-methionine α-naphthyl ester	+	+	−
N-α-Tosyl-L-arginine methyl ester	+	−	+

423

3.4.21.40 — Esteroprotease; EP (deleted from the Enzyme List;[1] see General Notes) (continued)

Esteroproteases from mammalian submandibular glands displayed limited substrate specificity. The nomenclature of EPs has not been developed because the substrate specificities of individual EPs for a long time remained not sufficiently well defined. For example, 21 esteroproteases from the mouse submandibular gland were classified into four groups based on their substrate specificities toward benzoyl-arginine-ethyl ester, N-α-benzoyl-D,L-arginine-p-nitroanilide, and poly-L-lysine.[4] These groups were further subdivided into eight subgroups by their sensitivity to inhibitors and androgen dependence, providing evidence that submaxillary gland esteroproteases represent a complex and heterogenous group of proteinases. As a result, this enzyme entry was deleted from the Enzyme List published in 1992.[1] Nevertheless, the methods listed above are expected to be useful in further studies of esteroproteases from mammalian submaxillary glands.

REFERENCES

1. NC-IUBMB, *Enzyme Nomenclature*, Academic Press, San Diego, 1992, p. 393 (EC 3.4.21.40; deleted entry: Submandibular Proteinase A).
2. Schaller, E. and Von Deimling, O., Methionine-α-naphthyl ester, a useful chromogenic substrate for esteroproteases of the mouse submandibular gland, *Anal. Biochem.*, 93, 251, 1979.
3. Skow, L.C., Genetic variation at a locus (TAM-1) for submaxillary gland protease in the mouse and its location on chromosome 7, *Genetics*, 90, 713, 1978.
4. Kurosawa, N. and Ogita, Z.-I., Classification of mouse submaxillary gland esteroproteases by their substrate specificities, *Electrophoresis*, 10, 189, 1989.
5. Otto, J. and Von Deimling, O., Prt-4 and Prt-5: new constituents of a gene cluster on chromosome 7 coding for esteroproteases in the submandibular gland of the house mouse (*Mus musculus*), *Biochem. Genet.*, 19, 431, 1981.

3.4.21.73 — Urokinase; UK

OTHER NAMES	u-Plasminogen activator (recommended name), urinary plasminogen activator, cellular plasminogen activator
REACTION	Preferential cleavage of Arg+Val bonds in plasminogen and conversion of plasminogen into plasmin
ENZYME SOURCE	Vertebrates
SUBUNIT STRUCTURE	Unknown or no data available

METHOD 1

Visualization Scheme

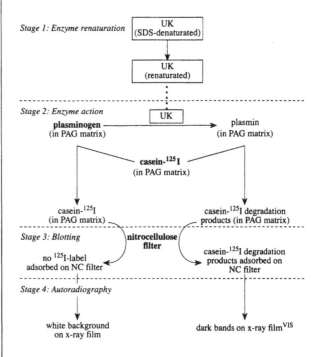

Substrate-Containing PAG[1]

Electrophoresis of UK is carried out according to Laemmli[2] in an SDS-PAG containing 0.08% casein-[125]I (specific activity of 6 to 7 × 10⁴ cpm/μg) and 46 μg/ml plasminogen.

Procedure

Stage 1. Shake the electrophorized SDS-PAG gently in cold 2.5% Triton X-100 for 60 min at 4°C, rinse thoroughly with water, and shake again in 0.1 Tris–HCl (pH 8.1) for 10 min.

Stage 2. Incubate the washed PAG for 2 h at 37°C in a humid chamber.

Stage 3. Wash a nitrocellulose filter (0.45 μm; Schleicher and Schull) in water and then in 0.1 *M* Tris–HCl (pH 8.1). Remove excess liquid by blotting on Whatman 3MM paper, and then lay the filter in a dish. Dip the incubated

PAG in 0.1 M Tris–HCl (pH 8.1) for about 5 sec, mount on a nitrocellulose filter, and blot for 60 min at 37°C in a humid chamber. Shake the blotted filter in phosphate-buffered saline solution for 60 min at room temperature and remove excess liquid by blotting on Whatman 3MM paper.

Stage 4. Dry the blotted and washed nitrocellulose filter, cover with Saran wrap, and expose to x-ray film (Kodak X-omat RP-2) at –70°C for 5 to 48 h.

The dark bands on the developed x-ray film correspond to the position of UK in the PAG.

Notes: The washed and incubated PAG can also be stained with Coomassie Brilliant Blue to develop achromatic bands of UK activity on a blue gel background. However, autoradiographic detection is about ten times more sensitive than the Coomassie staining procedure.

The sensitivity and accuracy of autoradiographic detection are higher when SDS and an iodinated substrate of high quality are used.

Autoradiographic detection is especially advantageous for crude samples containing proteins comigrating with UK. Such proteins develop on the Coomassie-stained PAG as dark bands that interfere with or completely mask achromatic zones of UK.

METHOD 2

Visualization Scheme

pyroglutamyl-Gly-Arg *p*-nitroanilide

UK

pyroglutamyl-Gly-Arg   *p*-nitroaniline[VIS]

Staining Solution[3]

A. 0.1 M Sodium barbital buffer, pH 7.75
    0.2 M NaCl
    3.4 mM CaCl$_2$
    1.4 mM MgCl$_2$
    0.9 mM Pyroglutamyl-Gly-Arg-*p*-nitroanilide
B. 3% Agarose solution (60°C)

Procedure

Mix equal volumes of A and B solutions and pour the mixture onto a prewarmed glass plate to give a 2-mm-thick gel. Place an electrophorized SDS-PAG (washed as described in Method 1) on top of a substrate-containing agarose gel and incubate at 37°C in a moist chamber for up to 12 h. Areas of UK activity appear as yellow bands in the agarose gel. Record the zymogram or scan on a densitometer at 405 nm.

Notes: This method was used to visualize UK after electrophoresis of purified enzyme preparations and, perhaps, is not as specific as Method 1.

The chromogenic substrate H-D-Ile-Pro-Arg-*p*-nitroanilide may also be used.

The product *p*-nitroaniline is usually too faint colored, but when it is diazotized and coupled with *N*-(1-naphthyl)-ethylenediamine, a readily visible red azo dye is formed. This dye, however, does not precipitate in the gel. To overcome this disadvantage, a method may be used that is based on transferring proteins from the electrophoretic gel to a nitrocellulose membrane[4] (an example application of this method is 3.4.21.4 — T, Method 3).

OTHER METHODS

An immunoblotting procedure (for details, see Part II) based on the utility of monoclonal antibodies specific to the human enzyme[5] can also be used for immunohistochemical visualization of the enzyme protein on electrophoretic gels. This procedure is too complex for routine laboratory use, but it may be of great value in special analyses of UK.

GENERAL NOTES

The enzyme was formerly classified as EC 3.4.21.31.

REFERENCES

1. Miskin, R. and Soreq, H., Sensitive autoradiographic quantification of electrophoretically separated proteases, *Anal. Biochem.*, 118, 252, 1981.
2. Laemmli, U.K., Cleavage of structural proteins during the assembly of the head of bacteriophage T4, *Nature*, 227, 680, 1970.
3. Wagner, O.F., Bergmann, I., and Binder, B.R., Chromogenic substrate autography: a method for detection, characterization, and quantitative measurement of serine proteases after sodium dodecyl sulfate–polyacrylamide gel electrophoresis or isoelectric focusing in polyacrylamide gels, *Anal. Biochem.*, 151, 7, 1985.
4. Ohlsson, B.G., Weström, B.R., and Karlsson, B.W., Enzymoblotting: a method for localizing proteinases and their zymogens using *para*-nitroanilide substrates after agarose gel electrophoresis and transfer to nitrocellulose, *Anal. Biochem.*, 152, 239, 1986.
5. Kaltoft, K., Nielsen, L.S., Zeuthen, J., and Dan, K., Monoclonal antibody that specifically inhibits a human MR 52000 plasminogen activating enzyme, *Proc. Natl. Acad. Sci. U.S.A.*, 79, 3720, 1982.

OTHER NAMES   Endopeptidases, proteases

REACTIONS   Attack all denatured and many native proteins and catalyze endohydrolysis of peptide bonds; do not have a substrate specificity in the ordinary sense. Many proteinases have a very close but not entirely identical specificity. The products resulting from their action on the same protein as the substrate differ quantitatively and qualitatively

ENZYME SOURCE   Bacteria, fungi, protozoa, plants, invertebrates, vertebrates

SUBUNIT STRUCTURES Mostly monomers[#]

GENERAL PRINCIPLES OF DETECTION

Many methods are developed for the detection of PROT activity on electrophoretic gels. The majority of them are based on incorporation of specific proteins (usually denatured) directly into separating gels or agarose gels that are held in contact with electrophoretic gels. After incubation, substrate-containing gels are treated with protein-precipitating agents (e.g., trichloroacetic acid or an acetic acid–methanol–water mixture) and stained for general proteins with Amidoblack or Coomassie Brilliant Blue dyes. The most frequently used PROT substrates are casein,[1–3] hemoglobin,[1,3,4] bovine serum albumin,[5,6] gelatin,[7,8] and keratin.[9] The methods of PROT detection that use substrate-containing PAGs are also known as zymographic methods. Various modifications of the zymographic technique are known, including two-dimensional zymography.[10] Zymography is a simple, sensitive, quantifiable, and functional assay for analyzing PROT. However, comparative quantitation of zymograms is difficult because staining and destaining of processed substrate-containing gels are separate procedures that are difficult to reproduce. The use of a single-step zymographic procedure allows one to overcome this disadvantage. This procedure utilizes 0.01% PhastBlue (Pharmacia) in the destaining solution, 1:3:6 glacial acetic acid:methanol:distilled water, and thus merges the traditionally separated staining and destaining steps into a single procedure.[11] The single-step staining method overcomes the problem of over-staining, leading to the loss of sensitivity, and the problem of overdestaining, which can bleach the bands and make them unsuitable for quantitation. The inclusion of a substrate protein in SDS-PAG can reduce the migration rate of PROT, making questionable the reliability of determination of PROT molecular weights. This disadvantage of substrate-containing PAG electrophoresis can be eliminated. The use of 0.5% casein solution as a PROT substrate is shown to be quite sensitive to detect 5 ng of trypsin per band.[12] A method of visualization of PROT with gelatinolytic activity on an SDS-PAG is described that uses electrotransfer of proteins from an SDS-PAG lacking the protein substrate onto an NC membrane preincubated with 0.3% gelatin solution. PROTs are reactivated during the transfer and hydrolyze gelatin. After general protein staining, PROT activity bands appear as white areas on a dark background.[13] A very similar procedure is described that uses a protein substrate immobilized on an acetate cellulose membrane pretreated with methacryloxypropyltrimethoxysilane.[14] When a print is taken with such a substrate-containing membrane, the gel is not contaminated by the immobilized substrate protein and can be further stained for general proteins. A simple two-step procedure enabling both the detection of PROT without interference by gelatin and the visual discrimination of PROTs that are sensitive and insensitive to specific inhibitors is developed as a complement to the standard gelatin-containing SDS-PAG electrophoresis.[15] The proteins are first resolved by SDS-PAG electrophoresis and then electrotransferred into a thin (0.75 mm) accompanying PAG containing 0.1% gelatin. PROT activity bands are developed by a gelatin proteolysis step in the substrate-containing PAG in the presence or absence of specific inhibitors, allowing the assessment of PROT classes. Commercial photographic sheet films containing a thin, uniform layer of gelatin were shown to be a very convenient and highly sensitive form of substrate for PROT.[16] Before use, a sheet film should be cleared with a nonhardening fixing solution in total darkness, washed thoroughly, and dried.

Proteinases can also be detected by incubating electrophoretic gels with colored substrates such as cytochrome c,[17] elastin–Orcein,[18] or Azocoll.[19] A simple and rapid procedure for visualization of PROT was developed that avoids the copolymerization of the substrate with the gel and does not require subsequent staining of the substrate-containing gel for general proteins. This procedure utilizes soluble azocasein as a substrate. An electrophoretic gel is incubated in a buffered 1% azocasein solution, immersed in 12.5% trichloroacetic acid, and treated with 1 N NaOH. Clear zones on an orange background of the gel indicate proteolytic activity.[20] The sensitivity of this method is at the submicrogram protein level for trypsin. All these methods are negative, i.e., they develop achromatic PROT bands on a colored background. The exception is the so-called "caseogram" method, which produces opaque PROT bands, well visible in sidelight, in an overlay gel containing skim milk.[3] Many PROTs have been electrophoretically studied using the detection methods listed above: chymotrypsins (EC 3.4.21.1–2), trypsin (EC 3.4.21.4), thrombin (EC 3.4.21.5), plasmin (EC 3.4.21.7), urokinase (EC 3.4.21.73), elastases (EC 3.4.21.36–37), cathepsin B (EC 3.4.22.1), papain (EC 3.4.22.2), ficin (EC 3.4.22.3), bromelain (EC 3.4.22.4), pepsins (EC 3.4.23.1–2), chymosin (EC 3.4.23.4), cathepsin D (EC 3.4.23.5), microbial aspartic proteases (all EC 3.4.23.6), collagenase (EC 3.4.24.3), thermolysin (EC 3.4.24.4), euphorbain (EC 3.4.99.7), keratin hydrolase (EC 3.4.99 ...), and others.

Another approach in detecting PROT activities is the use of chromogenic peptide substrates containing the terminal groups 2-naphthylamide, 1- or 2-naphthyl esters, ethyl or methyl esters, and p-nitroanilide. Naphthylamine and naphthols liberated as a result of PROT action are then coupled with a suitable diazonium salt (e.g., Fast Black K, Fast Blue B, Fast Garnet GBC, Fast Red TR) to generate an insoluble azo dye. The use of derivatives of 4-methoxy-2-naphthylamine, which complexes much more quickly with diazonium salts than does 2-naphthylamine, reduces the problem of diffusion during color development.

Besides that, when a postcoupling technique is used, the progress of enzyme action toward those substrates can be assessed from time to time by viewing the gel under long-wave UV light, where the blue bands of fluorescent reaction product, 1-methoxy-3-naphthylamine, are visible.[21] Peptide substrates containing the terminal groups 2-naphthylamide and a naphthyl ester are not always as specific and sensitive as substrates containing the *p*-nitroanilide group.[22] The enzymatically released yellow-colored *p*-nitroaniline is observed visually or registered spectrophotometrically at 405 nm. However, *p*-nitroaniline is usually too faintly colored to be readily observed on the gel, but when diazotized and coupled with naphthylethylenediamine, a clearly visible red azo dye is formed. This dye does not precipitate in a gel matrix and preservation of zymograms is not possible. However, when the "enzymoblotting" method (based on transferring enzymes from the electrophoretic gel to a nitrocellulose membrane) is used, the colored reaction products may be bound to nitrocellulose, thus increasing both the sensitivity and the resolution of the method and providing zymograms that are stable for a long time when stored frozen.[22] An important advantage of this method is that a single substrate can be used for both PROT activity staining on electrophoretic gels and spectrophotometric assay on soluble fractions during detection and purification processes to reduce specificity and quantification problems.[23] Peptides containing the ethyl ester terminal group are used as substrates for esteroproteases. Ethanol liberated as a result of enzyme action is then detected using alcohol dehydrogenase and the PMS–MTT system.[24] Substrates containing the ethyl ester terminal group liberate the CH_3O^- acidic ion, which causes a local pH change in the gel areas where esteroprotease activities are localized. The alkaline–acidic pH change is then detected via the indicator dye Bromothymol Blue.[25]

Peptides containing the 4-methylcoumarin terminal group can also be used as fluorogenic substrates for PROT detection. A peptide derivative, *N*-(*o*-aminobenzoyl-Ala-Ala-Phe-Phe-Ala-Ala)-*N'*-2,4-dinitrophenyl ethylenediamine, containing a fluorescent *o*-aminobenzoyl moiety as well as *N*-2,4-dinitrophenyl ethylenediamine (the group that causes fluorescence quenching) was synthesized and shown to be a useful fluorogenic substrate suitable for detection of human gastricsin (EC 3.4.23.3), cathepsin D (EC 3.4.23.5), and HIV proteinase (EC 3.4.23.16).[26] These aspartic proteinases hydrolyze the Phe-Phe peptide bond in the substrate and separate the fluorescent and quenching moieties. This leads to a considerable increase in the fluorescence intensity of *o*-aminobenzoyl residue. Fluorogenic PROT substrates containing either the 4-methyl coumarinyl-7-amide or 4-trifluoromethyl coumarinyl-7-amide group are also used to detect proteinase activities.[27] Neurolysin (EC 3.4.24.16) can be specifically detected using fluorogenic substrates derived from the neurotensin pELYENKPRRPYIL (pGlu-Leu-Tyr-Glu-Asn-Lys-Pro-Arg-Arg-Pro-Tyr-Ile-Leu) sequence by the introduction of *N*-2,4-dinitrophenyl ethylenediamine (EDDnp) at the C-terminal end of the sequence and by the substitution of the pyroglutamic (pE) residue at the N terminus for *o*-aminobenzoic acid (Abz). The two most selective fluorogenic substrates for neurolysin are Abz-LYENKPRRP-Q-EDDnp and Abz-NKPRRP-Q-EDDnp, whose

cleavage bond is R–R. These substrates can be used to discriminate between three closely related members of the metalloproteinase M3 family: the neurolysin, the thimet oligopeptidase (EC 3.4.24.15), and the neprilysin (EC 3.4.24.11).[28] Oligopeptide-7-amino-4-trifluoromethylcoumarin derivatives, immobilized on cellulose diacetate overlay membranes, can be used as substrates suitable to detect fluorescent PROT activity bands visible in long-wave UV light.[29] When appropriate substrates are available, this (enzyme overlay membrane) technique may be applied for the rapid detection and characterization of any proteinases. The quenched conjugates of BODIPY dye-labeled casein are very suitable for use as fluorogenic PROT substrates that are hydrolyzed to low-molecular-weight highly fluorescent fragments during PROT reaction. These substrates can be successfully used in the gel detection of serine proteinases, acid proteinases, sulfhydryl proteinases, and metalloproteinases.[30] This method allows detection of as few as about 2×10^{-7} U of proteinase K.

A number of chromogenic substrates suitable for PROT activity detection on electrophoretic gels are now commercially available. Some of them are listed below:

Chromogenic Substrates	Proteinases
N-Acetyl-L-methionine α-naphthyl ester	Submandibular esteroprotease A
N²-Acetyl-*N*-(*p*-nitrophenyl)-L-tyrosinamide[a]	Chymotrypsin
Asp-Glu-Val-Asp-*p*-nitroanilide	Caspase-3
N-α-Benzoyl-D,L-arginine-β-naphthylamide	Trypsin
N-α-Benzoyl-L-arginine ethyl ester	Plasma kallikrein
N-α-Benzoyl-D,L-arginine-*p*-nitroanilide	Trypsin
N²-Benzoyl-*N*-(*p*-nitrophenyl)-L-argininamide[a]	Trypsin
N-Benzoyl-Phe-Val-Arg-*p*-nitroanilide	Thrombin, trypsin, reptilase
N-Benzoyl-Pro-Phe-Arg-*p*-nitroanilide	Plasma kallikrein, thrombin-like proteinase (*Agkistrodon contortrix*)
N-Benzoyl-L-tyrosine-*p*-nitroanilide	Chymotrypsin
N-CBZ-Ala-Arg-Arg-4-methoxy β-naphthylamide[b]	Cathepsin B
N-CBZ-Gly-Gly-Arg-β-naphthylamide	Human serum trypsin
N-CBZ-Gly-Gly-Leu-*p*-nitroanilide	Subtilisin
N-CBZ-Leu-Leu-Glu-β-naphthylamide	Cation-sensitive proteinase
N-CBZ-Pro-Phe-His-Leu-Leu-Val-Tyr-Ser-β-naphthylamide	Renin
pGlu-Gly-Arg-*p*-nitroanilide[c]	Urokinase
pGlu-Phe-Leu-*p*-nitroanilide	Papain, ficin, bromelain
N-Glutaryl-phenylalanine-β-naphthylamide	Chymotrypsin
H-D-Ile-Pro-Arg-*p*-nitroanilide	Human epidermal keratin hydrolase
D-Phe-L-pipecolyl-Arg-*p*-nitroanilide	Thrombin
H-D-Pro-Phe-Arg-*p*-nitroanilide	Plasma kallikrein
N-Succinyl-Ala-Ala-Ala-*p*-nitroanilide	Elastase

Chromogenic Substrates (continued)	Proteinases
N-Succinyl-Ala-Ala-Pro-Leu-p-nitroanilide	Elastase
N-Succinyl-Ala-Ala-Pro-Phe-p-nitroanilide	Chymotrypsin, human leukocyte cathepsin G
N-Succinyl-phenylalanine-p-nitroanilide	Chymotrypsin
N-α-p-Tosyl-L-arginine methyl ester	Tamase, submandibular esteroprotease A
N-p-Tosyl-Gly-Pro-Arg-p-nitroanilide	Thrombin
N-p-Tosyl-Gly-Pro-Lys-p-nitroanilide	Plasmin
H-D-Val-Leu-Arg-p-nitroanilide	Tissue kallikrein

[a] Yellow p-nitrophenol is liberated from the product after treatment of the gel at high temperature (100°C).

[b] CBZ = carbobenzoxy.

[c] pGlu = pyroglutamyl.

Usually it is not possible to attribute the bands developed using chromogenic peptide substrates to any one proteinase with certainty except when purified enzymes are electrophorized. In some cases, however, specific PROTs can be identified using certain chromogenic substrates coupled with specific inhibitors and taking into account such information as pH optimum, tissue specificity, the preference to cleave peptide bonds only at certain few amino acids, existence of a proenzyme, and differences in proenzyme-activating agents (see General Notes).

The detection methods that may be considered as semispecific for certain proteinases are given separately (see 3.4.21.1 — CT; 3.4.21.4 — T; 3.4.21.5 — THR; 3.4.21.10 — ACR; 3.4.21.73 — UK; 3.4.21.34 — PKK; 3.4.21.36 — EL; 3.4.21.40 — EP; 3.4.23.1–3 — P).

The only way to specifically detect certain proteinases on electrophoretic gels is to use specific antibodies and labeled antiantibodies, i.e., to use the immunoblotting procedure (for details, see Part II). For example, this procedure was used to detect human uropepsinogen (see 3.4.23.1–3 — P, Other Methods, A). Monoclonal and polyclonal antibodies against human matrix metalloproteinases (EC 3.4.24 ...) and cathepsins (EC 3.4.22 ...), and human placenta and bovine skeletal muscle calpains (EC 3.4.22.17) are now commercially available from Sigma.

A procedure of two-dimensional PAG electrophoresis can be used to identify PROT based on electrophoretic analysis of peptide banding patterns generated by different endoproteinases due to proteolysis of the same protein substrate. According to this procedure, a narrow vertical strip of the resolving native PAG containing proteinases is placed horizontally together with a narrow strip containing a substrate protein. After preelectrophoresis, needed to bring together proteinases and a substrate protein, and the subsequent preincubation, needed to accomplish the hydrolysis, electrophoresis is carried out in a second dimension in the presence of SDS to separate the hydrolysis products. The location of endoproteinase activity in the native PAG strip is revealed by the disappearance of a substrate protein and the appearance of proteolytic products in the SDS-PAG. Comparison of banding patterns of peptide fragments produced by endoproteinases permits identification of proteinases generating similar or different products.[31]

GENERAL NOTES

Unlike peptidases (see 3.4.11 or 13 ... — PEP), which are exopeptidases, proteinases (EC 3.4.21–24 ...) are endopeptidases. The proteinases of sub-subclasses EC 3.4.21 ... (serine proteinases) have an active center histidine and serine involved in the catalytic process; those of EC 3.4.22 ... (cysteine, or thiol, proteinases) have a cysteine in the active center; those of EC 3.4.23 ... have a pH optimum below 5, due to the involvement of an aspartic acid residue in the catalytic process (aspartic, or acidic, proteinases); and those of EC 3.4.24 ... are metalloproteins using a metal ion (normally Zn^{2+}) in the catalytic mechanism (metalloproteinases). A number of proteinases are also known that cannot yet be assigned to any of the sub-subclasses listed above; these are listed in the sub-subclass EC 3.4.99.... Some of these proteinases may be eventually transferred to the sub-subclasses listed above, whereas others may prove to use novel catalytic mechanisms.[32]

Many PROTs exist in proenzyme forms and are converted into active enzymes only under certain conditions. For example, plasminogen is converted into plasmin by urokinase; activation of proelastase is under trypsin control; pepsinogen is converted into pepsin in acid conditions (0.1 M HCl). In normal tissues, the active forms of many PROTs are under control of tissue proteinase inhibitors. Examples are keratin hydrolase and urokinase.[9] Zymogen forms of some matrix metalloproteases (e.g., gelatinase A, EC 3.4.24.24) are activated by denaturing in the presence of SDS and subsequent renaturing in the presence of Triton X-100.[33] When a protein substrate is incorporated directly into the separating gel, the use of SDS-PAG is recommended to denature PROT and to prevent their action during the electrophoresis. After completion of electrophoresis, the SDS should be removed by washing the gel in 2.5% Triton X-100 to renature the PROT. Subsequent equilibration of the gel in an appropriate buffer is also needed for optimal PROT activity. Staining of aspartic (or acid) PROT (EC 3.4.23 ...) is carried out at a pH below 5 (usually at pH 2.5 to 3.5, but sometimes even at pH 1.5; e.g., see 3.4.23.1–3 — P).

Combination of the substrate-containing SDS-PAG electrophoresis with the use of class-specific irreversible proteinase inhibitors can be used to discriminate between different proteinases, e.g., plant cysteine and serine proteinases.[34] Cysteine (or thyol) PROTs require thyol reagents (e.g., cysteine) for their stability. So, cysteine is usually included in extraction and staining solutions when these proteinases are analyzed.

A method termed *reverse zymography* was designed to study PROT inhibitors using the electrophoretic technique. This method detects the presence of PROT inhibitors directly in the gel after their electrophoretic separation.[35] Generally speaking, the term *zymography* relates to any technique used to obtain a zymogram — i.e., the electrophoretic gel histochemically stained to visualize activity bands of a certain enzyme.[36] However, this term is commonly used to designate methods based on the use of electrophoretic gels containing a substrate protein (e.g., gelatin) and used to develop negatively stained PROT bands. To prevent the enzyme–substrate interaction, electrophoresis is carried out in denaturing conditions in the presence of SDS.[37]

Following electrophoresis, SDS is removed by incubating the gel in an appropriate buffer (usually containing Triton X-100). This allows separated PROTs to renature and autoactivate. Their activities are revealed by an absence of protein staining in gel areas occupied by PROTs where the substrate has been digested. Reverse zymography is designed to detect the presence of proteinase inhibitors rather than proteinases. It uses SDS-PAGs containing a substrate protein and a proteinase under study. After renaturation, the activity of proteinase inhibitors is revealed by the presence of positive protein staining in gel areas occupied by inhibitory species where activity of a renatured proteinase is inhibited and proteolysis arrested.[35] A modification of this method involves the use of SDS-PAG with a subsequent immersion of an electrophorized and renatured gel in an appropriate proteinase solution, and then in a protein substrate solution.[38] Further modification of the method consists in the use of native PAG lacking SDS.[39] Fluorogenic proteinase-specific substrates can be used to detect areas of localization of corresponding proteinase inhibitors in the electrophoretic gel using the "inhibitor overlay membrane technique."[27] This technique results in the appearance of dark (nonfluorescent) inhibitor bands on a light (fluorescent) background. An important advantage of the reverse zymography technique is that it allows the study of enzyme inhibitors in crude preparations, the determination of a number of inhibitory isoforms and their localization, the estimation of molecular weights of inhibitors using SDS-PAG electrophoresis, the investigation of properties of enzyme–inhibitor complexes, etc.

It is difficult to detect some proteinases (e.g., matrilysin and collagenases) at low levels in conventional casein or gelatin zymography. Heparin can be used to enhance the zymographic assays for these proteinases. With the addition of heparin to the enzyme sample, matrilysin and collagenases can be detected in transferrin and gelatin zymography at levels of 30 pg and 0.2 ng, respectively.[40] Exact mechanisms of proteinase activity enhancement are unknown. It is suggested, however, that heparin can enhance proteinase activity by (1) inducing a conformational change that increases activity, (2) facilitating refolding after SDS denaturation, (3) reducing autolysis, or (4) helping to anchor proteinases in the gel during long periods of incubation.[40] There are other cases in which heparin enhances proteinase activity. Good examples are antithrombin III[41] and human mast cell tryptase.[42]

REFERENCES

1. Andary, T.J. and Dabich, D., A sensitive polyacrylamide disc gel method for detection of proteinases, *Anal. Biochem.*, 57, 457, 1974.
2. Höfelmann, M., Kittsteiner-Eberle, R., and Schreier, P., Ultrathin-layer agar gels: a novel print technique for ultrathin-layer isoelectric focusing of enzymes, *Anal. Biochem.*, 128, 217, 1983.
3. Foltmann, B., Szecsi, P.B., and Tarasova, N.I., Detection of proteases by clotting of casein after gel electrophoresis, *Anal. Biochem.*, 146, 353, 1985.
4. Kaminski, E. and Bushuk, W., Detection of multiple forms of proteolytic enzymes by starch–gel electrophoresis, *Can. J. Biochem.*, 46, 1317, 1968.
5. Hanley, W.B., Boyer, S.H., and Naughton, M.A., Electrophoretic and functional heterogeneity of pepsinogen in several species, *Nature*, 209, 996, 1966.
6. Herd, J.K. and Motycka, L., Detection of proteolytic enzymes in agar electrophoresis, *Anal. Biochem.*, 53, 514, 1973.
7. Foissy, H.A., A method for demonstrating bacterial proteolytic isoactivities after electrophoresis in acrylamide gels, *J. Appl. Bacteriol.*, 37, 133, 1974.
8. Every, D., Quantitative measurement of protease activities in slab polyacrylamide gel electrophoretograms, *Anal. Biochem.*, 116, 519, 1981.
9. Hibino, T., Purification and characterization of keratin hydrolase in psoriatic epidermis: application of keratin–agarose plate and keratin–polyacrylamide enzymography methods, *Anal. Biochem.*, 147, 342, 1985.
10. Ong, K.L. and Chang, F.N., Analysis of proteins from different phase variants of the entomopathogenic bacteria *Photorhabdus luminescens* by two-dimensional zymography, *Electrophoresis*, 18, 834, 1997.
11. Leber, T.M. and Balkwill, F.R., Zymography: a single-step staining method for quantitation of proteolytic activity on substrate gels, *Anal. Biochem.*, 249, 24, 1997.
12. Lundy, F.T., Magee, A.C., Blair, I.S., and McDowell, D.A., A new method for the detection of proteolytic activity in *Pseudomonas lundensis* after sodium dodecyl sulfate–polyacrylamide gel electrophoresis, 16, 43, 1995.
13. Moos, J., Detection of gelatinolytic enzyme activities after sodium dodecyl sulfate–electrophoresis and protein blotting, *Electrophoresis*, 12, 444, 1991.
14. Kujat, R., Rehydratable ultrasensitive cellulose acetate substrate film for protease detection in ultrathin-layer isoelectric focusing not interfering with subsequent silver staining of focused proteins in the gel, *Electrophoresis*, 8, 298, 1987.
15. Visal-Shah, S., Vrain, T.C., Yelle, S., Nguyen-Quoc, B., and Michaud, D., An electroblotting, two-step procedure for the detection of proteinases and the study of proteinase/inhibitor complexes in gelatin-containing polyacrylamide gels, *Electrophoresis*, 22, 2646, 2001.
16. Burger, W.C. and Schroeder, R.L., A sensitive method for detecting endopeptidases in electrofocused thin-layer gels, *Anal. Biochem.*, 71, 384, 1976.
17. Ward, C.W., Detection of proteolytic enzymes in polyacrylamide gels, *Anal. Biochem.*, 74, 242, 1976.
18. Dijkhof, J. and Poort, C., Visualization of proelastase in polyacrylamide gels, *Anal. Biochem.*, 83, 315, 1977.
19. Lynn, K.R. and Clevette-Radford, N.A., Staining for protease activity on polyacrylamide gels, *Anal. Biochem.*, 117, 280, 1981.
20. Peyronel, D.V. and Cantera, A.M.B., A simple and rapid technique for postelectrophoretic detection of proteases using azocasein, *Electrophoresis*, 16, 1894, 1995.
21. Mort, J.S. and Leduc, M., A simple, economical method for staining gels for cathepsin B–like activity, *Anal. Biochem.*, 119, 148, 1982.

22. Ohlsson, B.G., Weström, B.R., and Karlsson, B.W., Enzymoblotting: a method for localizing proteinases and their zymogens using *para*-nitroanilide substrates after agarose gel electrophoresis and transfer to nitrocellulose, *Anal. Biochem.*, 152, 239, 1986.

23. Hou, W.-C., Chen, H.-J., Chen, T.-E., and Lin, Y.-H., Detection of protease activities using specific aminoacyl or peptidyl *p*-nitroanilides after sodium dodecyl sulfate–polyacrylamide gel electrophoresis and its application, *Electrophoresis*, 20, 486, 1999.

24. Fujimoto, Y., Morija, H., Yamaguchi, K., and Moriwaki, C., Detection of arginine esterase of various kallikrein preparations on gellified electrophoretic media, *J. Biochem. (Tokyo)*, 71, 751, 1972.

25. Skow, L.C., Genetic variation at a locus (TAM-1) for submaxillary gland protease in the mouse and its location on chromosome 7, *Genetics*, 90, 713, 1978.

26. Filippova, I.Y., Lysogorskaya, E.N., Anisimova, V.V., Suvorov, L.I., Oksenoit, E.S., and Stepanov, V.M., Fluorogenic peptide substrates for assay of aspartyl proteinases, *Anal. Biochem.*, 234, 113, 1996.

27. Weder, J.K.P., Haußner, K., and Bokor, M.V., Use of fluorogenic substrates to visualize trypsin and chymotrypsin inhibitors after electrophoresis, *Electrophoresis*, 14, 220, 1993.

28. Oliveira, V., Campos, M., Hemerly, J.P., Ferro, E.S., Camargo, A.C.M., Juliano, M.A., and Juliano, L., Selective neurotensin-derived internally quenched fluorogenic substrates for neurolysin (EC 3.4.24.16): comparison with thimet oligopeptidase (EC 3.4.24.15) and neprilysin (EC 3.4.24.11), *Anal. Biochem.*, 292, 257, 2001.

29. Shori, D.K., Proctor, G.B., and Garrett, J.R., Development and application of a method for the detection, elution, and characterization of rat submandibular proteinases separated on isoelectric focusing gels reveals male/female differences, *Anal. Biochem.*, 211, 123, 1993.

30. Jones, L.J., Upson, R.H., Haugland, R.P., Panchuk-Voloshina, N., Zhou, M., and Haugland, R.P., Quenched BODIPY dye-labeled casein substrates for the assay of protease activity by direct fluorescence measurement, *Anal. Biochem.*, 251, 144, 1997.

31. Petersen, G.R. and Van Etten, J.L., Detection of endoproteinases in polyacrylamide gels, *Electrophoresis*, 4, 433, 1983.

32. NC-IUBMB, *Enzyme Nomenclature*, Academic Press, San Diego, 1992, p. 371.

33. Kleiner, D.E. and Stetler-Stevenson, W.G., Quantitative zymography: detection of picogram quantities of gelatinase, *Anal. Biochem.*, 218, 325, 1994.

34. Michaud, D., Faye, L., and Yelle, S., Electrophoretic analysis of plant cysteine and serine proteinases using gelatin-containing polyacrylamide gels and class-specific proteinase inhibitors, *Electrophoresis*, 14, 94, 1993.

35. Oliver, G.W., Leferson, J.D., Stetler-Stevenson, W.G., and Kleiner, D.E., Quantitative reverse zymography: analysis of picogram amounts of metalloproteinase inhibitors using gelatinase A and B reverse zymograms, *Anal. Biochem.*, 244, 161, 1997.

36. Hunter, R.L. and Markert, C.L., Histochemical demonstration of enzymes separated by zone electrophoresis in starch gels, *Science*, 125, 1294, 1957.

37. Laemmli, U.K., Cleavage of structural proteins during the assembly of the head of bacteriophage T4, *Nature*, 227, 680, 1970.

38. Garcia-Carreño, F.L., Dimes, L.E., and Haard, N.F., Substrate-gel electrophoresis for composition and molecular weight of proteinases or proteinaceous proteinase inhibitors, *Anal. Biochem.*, 214, 65, 1993.

39. Felicioli, R., Garzelli, B., Vaccari, L., Melfi, D., and Balestreri, E., Activity staining of protein inhibitors of proteases on gelatin-containing polyacrylamide gel electrophoresis, *Anal. Biochem.*, 244, 176, 1997.

40. Yu, W.-H. and Woessner, J.F., Jr., Heparin-enhanced zymographic detection of matrilysin and collagenases, *Anal. Biochem.*, 293, 38, 2001.

41. Evans, D.L., Marshall, C.J., Christey, P.B., and Carrell, R.W., Heparin binding site, conformational change, and activation of antithrombin, *Biochemistry*, 31, 12629, 1992.

42. Hunt, J.E., Stevens, R.L., Austen, K.F., Zhang, J., Xia, Z., and Ghildyal, N., Natural disruption of the mouse mast cell protease 7 gene, *J. Biol. Chem.*, 271, 2851, 1995.

OTHER NAMES	Bean endopeptidase, vicilin peptidohydrolase, phaseolin
REACTION	Hydrolysis of proteins, such as azocasein; preferential cleavage: Asn┼ in small molecule substrates such as Boc-Asn┼OPhNO$_2$
ENZYME SOURCE	Plants, invertebrates, vertebrates
SUBUNIT STRUCTURE	Unknown or no data available

METHOD

Visualization Scheme

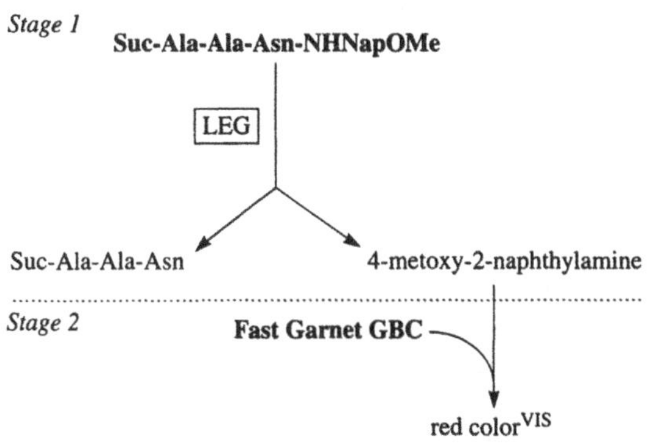

Stage 1

Suc-Ala-Ala-Asn-NHNapOMe

LEG

Suc-Ala-Ala-Asn   4-metoxy-2-naphthylamine

Stage 2   **Fast Garnet GBC**

red color[VIS]

Staining Solution[1]

A. 39.5 mM Citric acid–121 mM Na$_2$HPO$_4$, pH 5.8
   1 mM Dithiothreitol
   1 mM EDTA
   100 μM Suc-Ala-Ala-Asn-NHNapOMe
B. 3 mg/ml Fast Garnet GBC in the citrate–phosphate buffer (see above)

Procedure

Incubate electrophorized PAG in solution A for 10 min at room temperature and then add solution B. Red bands indicate areas of LEG activity.

Notes: The substrate, Suc-Ala-Ala-Asn-NHNapOMe (Suc-Ala-Ala-Asn-4-methoxy-2-naphthylamid), is not commercially available and should be synthesized as described by authors.

The enzyme activity toward Suc-Ala-Ala-Asn-NHNapOMe is almost completely (99%) inhibited by 100 nM egg white crystalline and is not inhibited by 10 μM *trans*-epoxysuccinyl-L-leucylamido(4-guanidino)butane (E-64). These LEG properties differ from those of cathepsin B, which cleaved Suc-Ala-Ala-Asn-NHNapOMe at a rate that is less than 0.2% of that observed with LEG. Moreover, cathepsin B activity is inhibited by E-64. These properties are characteristic of LEG from the pig kidney. It would be advisable, however, to include controls with E-64 when studying the enzyme from other sources, to exclude possible interference from cathepsins.

Because Suc-Ala-Ala-Asn-NHNapOMe is a 2-naphthylamine derivative and therefore a potential carcinogen, solutions of this substrate should be handled with caution.

GENERAL NOTES

The plant enzyme is thought to be involved in the hydrolysis of stored seed proteins.[2]

REFERENCES

1. Johansen, H.T., Knight, C.G., and Barrett, A.J., Colorimetric and fluorimetric microplate assays for legumain and a staining reaction for detection of the enzyme after electrophoresis, *Anal. Biochem.*, 273, 278, 1999.
2. NC-IUBMB, *Enzyme Nomenclature*, Academic Press, San Diego, 1992, p. 403 (EC 3.4.22.34, Comments).

REACTION   Cleaves substrates after aspartate residues. Is capable of cleaving poly(ADP-ribose) polymerase (PARP). The tetrapeptide Asp-Glu-Val-Asp of the PARP is identified as the consensus cleavage site. The cleavage of PARP occurs at the onset of apoptosis, a form of programmed cell death that plays a fundamental role in normal biological processes as well as several disease states

ENZYME SOURCE   Vertebrates

SUBUNIT STRUCTURE See General Notes

METHOD 1

Visualization Scheme

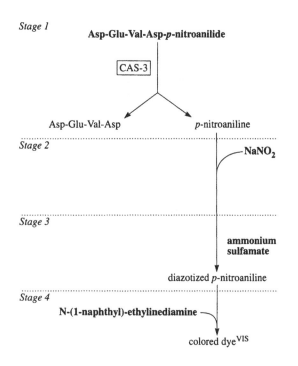

Stage 1

Asp-Glu-Val-Asp-*p*-nitroanilide

CAS-3

Asp-Glu-Val-Asp   *p*-nitroaniline

Stage 2

NaNO$_2$

Stage 3

ammonium sulfamate

diazotized *p*-nitroaniline

Stage 4

N-(1-naphthyl)-ethylinediamine

colored dye^{VIS}

Staining Solution[1] (adapted)

A. 50 m*M* Sodium phosphate buffer, pH 7.0
   0.5 mg/ml Asp-Glu-Val-Asp-*p*-nitroanilide (available from CLONTECH Laboratories, Inc., Palo Alto, CA)
B. 0.1% NaNO$_2$
   1.0 *M* HCl
C. 0.5% Ammonium sulfamate
   1.0 *M* HCl
D. 0.05% *N*-(1-Naphthyl)-ethylenediamine
   47.5% Ethanol

Procedure

Immediately after electrophoresis, transfer proteins to a nitrocellulose (NC) membrane using a routine blotting procedure (e.g., see 3.4.21.1 — CT, Method 3, Procedure). After blotting, incubate the NC membrane in solution A at 37°C for 60 min. Then diazotize the released *p*-nitroaniline as follows: put the NC membrane into solution B for 5 min; after another 5 min of incubation in solution C (to destroy the excess sodium nitrite), put the membrane in solution D.

When the developing red-colored bands are of a maximal intensity (usually after 1 to 5 min), take up the NC membrane and allow solution D to drip off. Put the moist NC membrane between two plastic sheets (overhead film; air bubbles should be avoided), seal with adhesive tape, and store at –18°C.

Notes: This method is a combination of the colorimetric method developed to detect *in vivo* CAS-3 activity[1] and the enzymoblotting method developed to detect proteinase activities using *p*-nitroanilide substrates.[2]

The CAS-3 activity was found to be four- to sevenfold higher in the human 32D cell cultures induced by cycloheximide, actinomycin D, and some other apoptotic stimuli.[1]

METHOD 2

Visualization Scheme

Asp-Glu-Val-Asp-7-amino-4-trifluoromethylcoumarin

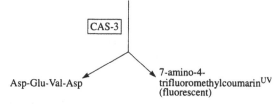

CAS-3

Asp-Glu-Val-Asp   7-amino-4-trifluoromethylcoumarin^{UV} (fluorescent)

Staining Solution[3] (modified)

20 m*M* HEPES buffer, pH 7.2, containing 10% glycerol, 1 m*M* EDTA, 10 m*M* dithiothreitol, 100 m*M* NaCl, 0.1% Chaps
100 μ*M* Asp-Glu-Val-Asp-7-amino-4-trifluoromethylcoumarin (Enzyme System Products)

Procedure

Cut the cellulose diacetate membrane to appropriate size, wet with the staining solution, gently blot to remove excess liquid, and apply to the gel surface. Incubate the gel at 37°C in a humid chamber. Monitor the progress of the proteinase reaction by transillumination at 365 nm. When sufficient fluorescence is developed (usually within 15 min), carefully remove the membrane from the gel, immerse for 5 min in 10% glycerol, and allow to air dry. The fluorescent banding pattern is captured and can be analyzed with a UVP Inc. Data Acquisition system using ImageStore 7500 software. The gel can then be stained for general proteins.

3.4.22.X — Caspase-3; CAS-3 (continued)

Notes: The lowest quantity of CAS-3, approximately 30 fmol, can be detected as a bright, well-defined band of enzymatic activity. The sensitivity of Method 2 is sufficient to study the activation of CAS-3 during apoptosis.[3]

OTHER METHODS

The immunoblotting procedure (for details, see Part II) may be used to detect CAS-3. Antibodies specific to the human enzyme are available from Sigma.

GENERAL NOTES

The human caspase family consists of about ten cysteine proteinases that possess the unusual ability to cleave substrates after aspartate residues. This is vital to their role in apoptosis.[1]

Procaspase-3 is a homodimeric protein.[4] The proteolytic conversion of proCAS-3 to active CAS-3 is a committed step in caspase-dependent forms of apoptosis.[5] The active enzyme is present as a part of aposome and microaposome protein complexes.[6]

REFERENCES

1. Gurtu, V., Kain, S.R., and Zhang, G., Fluorometric and colorimetric detection of caspase activity associated with apoptosis, *Anal. Biochem.*, 251, 98, 1997.
2. Ohlsson, B.G., Weström, B.R., and Karlsson, B.W., Enzymoblotting: a method for localizing proteinases and their zymogens using *para*-nitroanilide substrates after agarose gel electrophoresis and transfer to nitrocellulose, *Anal. Biochem.*, 152, 239, 1986.
3. Breithaupt, T.B., Shires, A.L., Voehringer, D.W., Herzenberg, L.A., and Herzenberg, L.A., Isoelectric focusing and enzyme overlay membrane analysis of caspase 3 activation, *Anal. Biochem.*, 292, 313, 2001.
4. Bose, K. and Clark, A.C., Dimeric procaspase-3 unfolds via a four-state equilibrium process, *Biochemistry*, 40, 14236, 2001.
5. Stennicke, H.R., Jurgensmeier, J.M., Shin, H., Deveraux, Q., Wolf, B.B., Yang, X., Zhou, Q., Ellerby, H.M., Ellerby, L.M., Bredesen, D., Green, D.R., Reed, J.C., Froelich, C.J., and Salvesen, G.S., Procaspase 3 is the major physiologic target of caspase 8, *J. Biol. Chem.*, 273, 27084, 1998.
6. Cain, K., Brown, D.G., Langlais, C., and Cohen, G.M., Caspase activation involves the formation of the aposome, a large (similar to 700 kDa) caspase-activating complex, *J. Biol. Chem.*, 274, 22686, 1999.

3.4.23.1–3 — Pepsin; P

OTHER NAMES	Pepsin A (EC 3.4.23.1), pepsin B (EC 3.4.23.2), pepsin C or gastricsin (EC 3.4.23.3)
REACTIONS	Pepsin A: Preferential cleavage: hydrophobic, preferably aromatic, residues in P1 and P1′ positions
	Pepsin B: Degradation of gelatin; little activity on hemoglobin
	Pepsin C: More restricted specificity than pepsin A, but shows preferential cleavage at Tyr+ bonds; high activity on hemoglobin
ENZYME SOURCE	Vertebrates
SUBUNIT STRUCTURE	Monomer (vertebrates)

METHOD

Visualization Scheme

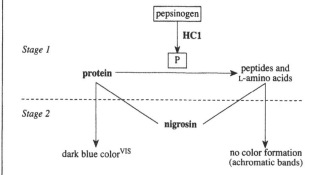

Staining Solution[1]

A. 0.06 *N* HCl, pH 1.4
0.65% Bovine hemoglobin
B. 10% Acetic acid in 50% methanol
C. 0.2 mg/ml Nigrosin in solution B

Procedure

Immerse the gel in solution A for 10 min, and then place in a humid chamber and incubate at 37°C for 1 h. Immerse the incubated gel in solution B for 18 h to fix the undigested protein and to wash away products of degradation. Stain the gel in solution C and wash in solution B to remove unbound Nigrosin. The gel areas occupied by P are indicated by achromatic bands visible on a dark blue background.

Notes: Other proteins, e.g., bovine serum albumin, can be used in place of hemoglobin, and Amidoblack can be used in place of Nigrosin.[2]

Some proteinases with pepsin-like activity (e.g., cathepsin D) can also be detected by this method. The bands caused by other than pepsin proteinase activities can, in some cases, be identified by omitting the stage of treating the gel with HCl.

3.4.23.1–3 — Pepsin; P (continued)

OTHER METHODS

A. The immunoblotting procedure was used to detect human uropepsinogen after PAG electrophoresis using polyclonal rabbit anti-uropepsinogen antibodies and commercial preparation of peroxidase-labeled goat anti-rabbit immunoglobulin.[3] This method is highly specific; however, its application is very limited.

B. A zymographic method is developed in which pepsin(ogen) activity bands are detected as opaque bands visible in an overlay gel of agarose containing skim milk as a substrate. This type of zymogram, also known as caseogram or caseogram print, is sensitive enough to detect 35 pg of pepsin(ogen) per band.[4]

REFERENCES

1. Harris, H. and Hopkinson, D.A., *Handbook of Enzyme Electrophoresis in Human Genetics*, North-Holland, Amsterdam, 1976 (loose-leaf, with supplements in 1977 and 1978).
2. Hanley, W.B., Boyer, S.H., and Naughton, M.A., Electrophoretic and functional heterogeneity of pepsinogen in several species, *Nature*, 209, 996, 1966.
3. Kishi, K. and Yasuda, T., Newly characterized genetic polymorphism of uropepsinogen group A (PGA) using both isoelectric focusing and immunoblotting, *Hum. Genet.*, 75, 209, 1987.
4. Till, O., Baumann, E., and Linss, W., Zymography with caseogram prints: quantification of pepsinogen, *Anal. Biochem.*, 292, 22, 2001.

3.4.24.38 — Autolysin; AUTOL

OTHER NAMES	*Chlamidomonas* cell wall degrading protease, lysin, peptidoglycan hydrolase
REACTION	Cleavage of the proline- and hydroxyproline-rich proteins of the *Chlamidomonas* cell wall (see also General Notes)
ENZYME SOURCE	Bacteria
SUBUNIT STRUCTURE	See General Notes

METHOD

Visualization Scheme

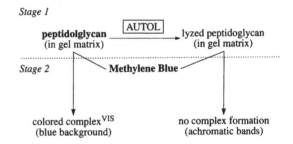

Staining Solution[1]

A. 25 m*M* Tris–HCl buffer, pH 6.9
   10 m*M* MgCl$_2$
   0.1% Triton X-100
B. 0.1% KOH
   0.01% Methylene Blue

Procedure

Carry out electrophoresis using a 10% SDS-PAG containing *o*-acetylated peptidoglycan (see *Notes*). Following electrophoresis, wash the gel in deionized water for 30 min at room temperature and then place in solution A (renaturation buffer). After 30 min, transfer the gel in fresh solution A and incubate at 37°C for 18 h. Visualize lytic bands caused by AUTOL as clear bands against a blue background by staining the processed gel in solution B for 3 h and subsequently destaining it in deionized water.

Notes: To prepare *o*-acetylated peptidoglycan, the freeze-dried cells of *Micrococcus luteus* were chemically acetylated as previously described.[2] In brief, suspend 40 mg of dried cells in 20 ml of acetic anhydride and 1.0 ml of pyridine. Stir gently for 1 h at 37°C and then harvest cells by ultracentrifugation (100,000 *g* at 4°C). Wash the cells with cold 100 m*M* sodium phosphate buffer (pH 6.0). Repeat washing and harvesting procedures four times. A homogenous suspension of *o*-acetylated *M. luteus* cells is prepared by ultrasonication of 12.5 mg of cells in 5 ml of 2 *M* Bis–Tris buffer (pH 6.8) for a combined total of 2 min using a Heat Systems Sonicator XL ultrasonic processor. This suspension (0.5 ml) is incorporated into resolving SDS-PAG solution (4 ml) just before polymerization.

The use of *o*-acetylated peptidoglycan as a substrate allows discrimination between AUTOL and lysozyme. Lysozyme (see 3.2.1.17 — LZ) is also able to lyse bacterial cells. However, this peptidoglycan hydrolase is inhibited by *o*-acetylated peptidoglycan, while AUTOL is not. This difference can be used to discriminate between these two peptidoglycan hydrolases.[1]

General Notes

The enzyme is a peptidoglycan hydrolase. Peptidoglycan is a heteropolymer of alternating *N*-acetylglucosamine and *N*-acetylmuramic acid, with associated peptides comprising the rigid cell wall layer of most bacteria. Autolysins produced by bacteria are thought to participate in the biosynthesis of peptidoglycan by providing sites for the incorporation of new material. Peptidoglycan is also the substrate of lysozyme produced by a host organism as a defense against bacterial invasion. The *o*-acetylation of the C-6 hydroxyl group of muramyl residues inhibits the activity of lysozyme.[3]

The enzyme exhibits a monomer–dimer association equilibrium, through the COOH-terminal part of the AUTOL polypeptide. Dimerization is regulated by choline interaction and involves the preferential binding of two molecules of choline per AUTOL dimer.[4]

References

1. Strating, H. and Clarke, A.J., Differentiation of bacterial autolysins by zymogram analysis, *Anal. Biochem.*, 291, 149, 2001.
2. Brumfitt, W., Wardlaw, A.C., and Park, J.T., Development of lysozyme resistance in *Micrococcus lysodeikticus* and its association with an increased *o*-acetyl content of the cell wall, *Nature*, 181, 1783, 1958.
3. Clarke, A.J. and Dupont, C., *o*-Acetylated peptidoglycan: its occurrence, pathobiological significance, and biosynthesis, *Can. J. Microbiol.*, 38, 85, 1992.
4. Usobiaga, P., Medrano, F.J., Gasset, M., Garcia, J.L., Saiz, J.L., Rivas, G., Laynez, J., and Menendez, M., Structural organization of the major autolysin from *Streptococcus pneumoniae*, *J. Biol. Chem.*, 271, 6832, 1996.

3.5.1.1 — Asparaginase; ASP

OTHER NAMES	Asparaginase II
REACTION	L-Asparagine + H_2O = L-aspartate + NH_3
ENZYME SOURCE	Bacteria, green alga, fungi, plants, invertebrates, vertebrates
SUBUNIT STRUCTURE	Tetramer[#] (bacteria), dimer[#] (fungi), monomer[#] (plants)

METHOD

Visualization Scheme

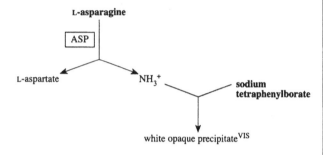

Staining Solution[1]

A. 0.2 *M* Tris–phosphate buffer, pH 8.0   20 ml
    L-Asparagine   52 mg
    Sodium tetraphenylborate   70 mg
B. 2% Agar solution   20 ml

Procedure

Mix A and B solutions and pour the mixture over the surface of an electrophorized PAG. Incubate the gel at 37°C in a humid chamber until white opaque bands appear.

Notes: All solutions used for ASP electrophoresis and detection should be made using bidistilled water lacking ammonia.

OTHER METHODS

A. The areas of ammonia production can be detected using the backward reaction of a linking enzyme, glutamate dehydrogenase (e.g., see 3.5.4.4 — ADA, Method 2). Dark (nonfluorescent) bands of the enzyme activity are observed on a light (fluorescent) background in long-wave UV light.

B. The local pH change due to the production of ammonia can be detected using the pH indicator dye Phenol Violet (e.g., see 3.5.4.4 — ADA, Method 3), the NBT (or MTT)–dithiothreitol system (e.g., see 3.5.3.1 — ARG, Method 1), or neutral $AgNO_3$ solution containing photographic developers (e.g., see 3.5.1.5 — UR, Method 2).

C. The areas of L-aspartate production can be detected using two linking enzymes, glutamic–oxaloacetic transaminase and malate dehydrogenase (e.g., see 2.6.1.1 — GOT, Method 2). Dark (nonfluorescent) bands of ASP activity are visible in long-wave UV light on the light (fluorescent) background of the gel.

D. Positive tetrazolium staining of the areas of L-aspartate production can be achieved by using two linking enzymes, glutamic–oxaloacetic transaminase and glutamate dehydrogenase (e.g., see 2.6.1.1 — GOT, Method 4).

REFERENCES

1. Pajdak, E. and Pajdak, W., A simple and sensitive method for detection of L-asparaginase by polyacrylamide gel electrophoresis, *Anal. Biochem.*, 50, 317, 1972.

3.5.1.2 — Glutaminase; GLUT

REACTION	L-Glutamine + H$_2$O = L-glutamate + NH$_3$
ENZYME SOURCE	Bacteria, fungi, invertebrates, vertebrates
SUBUNIT STRUCTURE	Heterodimer$^\#$ (fungi – *Debaryomyces*), tetramer$^\#$ (bacteria – *Rhizobium etli*)

METHOD

Visualization Scheme

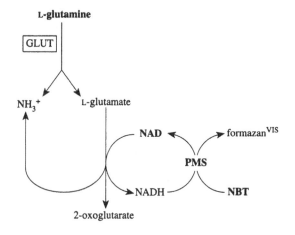

Staining Solution[1]

0.1 *M* Potassium phosphate buffer, pH 7.1
15 m*M* L-Glutamine
20 U/ml Glutamate dehydrogenase (GDH)
2 mg/ml NAD
2 mg/ml NBT
0.04 mg/ml PMS

Procedure

Incubate the gel in staining solution in the dark at 37°C until dark blue bands appear. Fix the stained gel in 25% ethanol.

Notes: Use L-glutamine that is free of L-glutamate, or purify crude L-glutamine using a Dowex 1-C1 column.

The use of native PAGs containing nondenaturant detergents 3-(3-cholamidopropyl)-dimethylammonio-1-propane sulfonate (CHARS) or Triton X-100 is recommended to resolve and to detect GLUT activity bands by the NBT–PMS method similar to that described above. Because GLUT is a mitochondrial protein, the use of crude Triton X-100 extracts of mitochondria for electrophoresis is preferable.[2]

OTHER METHODS

A. The local pH change due to the production of ammonia can be detected using the pH indicator dye Phenol Violet (e.g., see 3.5.4.4 — ADA, Method 3), the NBT (or MTT)–dithiothreitol system (e.g., see 3.5.3.1 — ARG, Method 1), or neutral AgNO$_3$ solution containing photographic developers (e.g., see 3.5.1.5 — UR, Method 2).

B. Tetraphenylborate can also be used to detect ammonia ions produced by GLUT (see 3.5.1.1 — ASP). This method is applicable only to transparent electrophoretic gels.

C. The product L-glutamate can be detected using three auxiliary enzymes, the L-glutamate oxidase (LGO; EC 1.4.3.11), the alanine transaminse (GPT; EC 2.6.1.2), and the horseradish peroxidase (PER; EC 1.11.1.7).[3] This method involves two different enzymes coupled in the opposite direction, so that the substrate (L-glutamate) of one of them (LGO) is the product of the other (GPT) and vice versa. Under such conditions, the reaction turns out with no consumption of recycling substrate, while other products of enzymatic reactions (in particular hydrogen peroxide) are accumulated with each turn of the cycle and are readily detected by colorometric PER reaction. The use of this method is of special value for detecting GLUT after electrophoresis of preparations containing low concentrations of the enzyme.

GENERAL NOTES

Mammalian GLUT is associated with inner mitochondrial membranes, is activated by phosphate, and is the first enzyme in glutamine catabolism. In the absence of phosphate ions, the enzyme from mammalian mitochondria displays no activity.[2]

REFERENCES

1. Davis, J.N. and Prusiner, S., Stain for glutaminase activity, *Anal. Biochem.*, 54, 272, 1973.
2. Aledo, J.C., Gomezbiedma, S., Segura, J.A., Molina, M., Decastro, I.N., Marquez, J., Native polyacrylamide-gel electrophoresis of membrane-proteins: glutaminase detection after *in situ* specific activity staining, *Electrophoresis*, 14, 88, 1993.
3. Valero, E. and Garcia-Carmona, F., A continuous spectrophotometric method based on enzymatic cycling for determining L-glutamate, *Anal. Biochem.*, 259, 265, 1998.

REACTION	Urea + H_2O = CO_2 + 2 NH_3
ENZYME SOURCE	Bacteria, fungi, plants, vertebrates
SUBUNIT STRUCTURE	Heterohexamer($3\alpha3\beta$)[#] (bacteria – a gram positive coccoid), dimer[#] (fungi – *Schizosaccharomyces pombe*), hexamer[#] (plants)

Method 1

Visualization Scheme

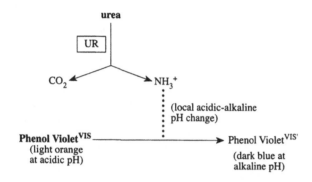

Staining Solution[1]

Urea	5 g
Agar	1 g
Phenol Violet (saturated water solution)	10 ml
H_2O	90 ml

Procedure

Dissolve the components of the staining solution by heating in a boiling water bath, cool the solution to 45°C, and pour over the gel surface. Incubate the gel at 37°C until dark blue bands appear on a light orange background of the gel. Record or photograph the zymogram.

Notes: Electrode and gel buffers used for UR electrophoresis should be of minimal ionic strength.

Method 2

Visualization Scheme

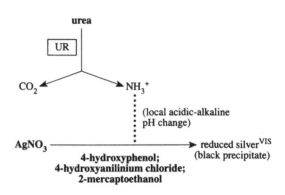

Staining Solution[2]

A. 10 mM 2-Mercaptoethanol
B. 10 mM 2-Mercaptoethanol
   250 mM Urea
C. 0.2 mg/ml 4-Hydroxyphenol
   0.2 mg/ml 4-Hydroxyanilinium chloride (freshly prepared and adjusted to pH 8.0 ± 0.1 by 0.25 M NaOH)
D. 2 mg/ml $AgNO_3$

Procedure

After electrophoresis, immerse the PAG sequentially in 100 ml of solution A (30 min: 10 min × three changes), solution B (3 min), and solution C (2 min) under continuous shaking. Wash the gel quickly (15 sec) with 100 ml of glass-distilled water to eliminate solution C from both the gel surface and the container, and immerse in 100 ml of solution D. The UR activity bands become visible at this last step, the faintest ones requiring about 5 min. After this period of time, only the background increases. This undesirable process can be prevented by immersing the gel in 5% (v/v) acetic acid. However, the developed bands show a tendency to bleach in this solution. Thus, it is recommended that the stained and fixed gel be washed with abundant glass-distilled water and then dried or photographed.

Notes: Using this method, 0.015 U of UR per band can be detected.

Method 3

Visualization Scheme

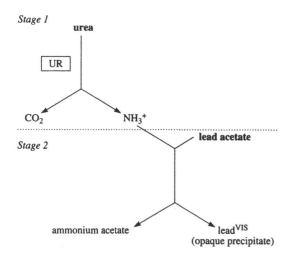

Stage 1

urea

UR

CO_2 NH_3^+

Stage 2

lead acetate

ammonium acetate   lead^{VIS}
(opaque precipitate)

Staining Solution[3]

A. 50 mM Sodium acetate buffer, pH 5.0
   1 mM EDTA
B. 20 mM Sodium acetate buffer, pH 5.0
   1 mM EDTA
C. 1.5% (w/v) Urea
   1 mM EDTA
D. 0.1 M Lead acetate

Procedure

Preincubate an electrophoretic PAG in solution A for 30 min at room temperature. Transfer the gel to solution B for an additional 30 min. Then place the gel in solution C. After 5 to 20 min of incubation, depending on the enzyme activity in preparations under question, transfer the gel into solution D. The regions of the gel containing UR activity are visible as white opaque bands.

Notes: The use of Cresol Red (a pH-sensitive indicator yellow at pH 7.1 and crimson at pH 8.8) resulted in development of crimson UR bands on a yellow background of the gel. However, if the reaction of UR was continued too long, it resulted in distortion of stained activity bands due to the diffusion of ammonia generated by UR during urea hydrolysis. The lead acetate (a highly potent inhibitor of UR) was initially employed to terminate the Cresol Red reaction. Further, lead acetate was used alone to detect UR activity as opaque bands on a transparent background of PAG. Because of the highly insoluble nature of lead, lead precipitate bands are easily visible and enable detection of less than 0.6 U of UR activity. UR in amounts greater than 2.5 U showed color development within 5 to 10 sec. If the reaction is not terminated after an appropriate period of gel incubation, diffusion of ammonia will result in the formation of broad distorted UR bands. It is essential therefore to first determine the optimum time of incubation in solution C before transferring the gel into the lead acetate solution (D) in order to prevent excess diffusion of ammonia.

Method 4

Visualization Scheme

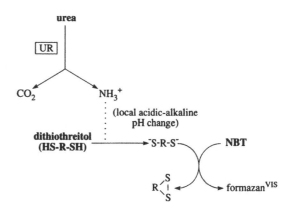

urea

UR

CO_2 NH_3^+
(local acidic-alkaline pH change)

dithiothreitol
(HS-R-SH) $^-$S-R-S$^-$ NBT

R$\langle$ S | S   formazan^{VIS}

Staining Solution[4]

27.4 mM Citrate buffer, pH 6.0
79 mM Urea
0.038% NBT
210 μM Dithiothreitol

Procedure

Equilibrate an electrophorized gel in 50 mM citrate buffer (pH 6.0) for 1 to 2 h and incubate in the staining solution at 37°C before blue bands of UR activity appear.

Notes: The exchange of 27.4 mM citrate buffer (pH 6.0) for 5 μM acetate buffer (pH 6.0) allows the development of localized alkalinity by much less UR activity.[5] The improved tetrazolium method is initiated by washing the electrophorized PAG five times for 10 min in the acetate buffer under continuous shaking. The last wash is performed in distilled water for 10 min and the gel placed in a staining solution similar to that given above. Dithiothreitol is added to the staining solution just before use, and the gel and staining solution are sealed in a plastic bag after trapped air bubbles have been removed. The gel is incubated at 37°C without shaking and the reaction stopped by a 5-min incubation in 20 mM hydrochloric acid and subsequent repeated washes in distilled water. The improved tetrazolium method detects as little as 25 μU of jack bean UR.

Other Methods

A. The areas of ammonia production can also be detected using the backward reaction of a linking enzyme, glutamate dehydrogenase (e.g., see 3.5.4.4 — ADA, Method 2). Dark (nonfluorescent) bands of enzyme activity are observed on a light (fluorescent) background in long-wave UV light.
B. The areas of ammonia production can also be detected using tetraphenylborate (see 3.5.1.1 — ASP). This method is applicable only to transparent electrophoretic gels.

C. An immunoblotting procedure (for details, see Part II) may also be used to detect UR. Monoclonal antibodies against the enzyme from the jack bean are now available from Sigma.

GENERAL NOTES

The comparison of different methods developed for the detection of UR reaction products showed that the improved tetrazolium method (see Method 4, *Notes*) is the most sensitive. As little as 25 mU of the enzyme can be detected by this method within 2 h following PAG electrophoresis.

UR is the only plant enzyme containing nickel.[6] The incorporation of nickel ions into apoUR is essential for UR activity. In bacteria, this process depends on a number of accessory proteins.[7] It was demonstrated that the overall structure of the UR activation complex is conserved between bacteria and plants.[8]

Plant UR is a hexamer. At least one free-SH cysteine residue per monomer is required for catalytic activity of UR. Activity-protecting agents (sodium sulfite or 2-mercaptoethanol) are usually used to avoid the inactivation of the enzyme due to oxidation of free-SH groups. However, the enzyme preparations maintained in solutions containing 2-mercaptoethanol exhibit slow inactivation. Because of this, the use of sodium sulfite is preferable. It was found, however, that this reagent causes a 10% increase in the anodic electrophoretic mobility of the native UR.[9]

REFERENCES

1. Daly, M.P. and Tully, E.R., Detection of ammonia-producing enzymes after electrophoresis: a screening procedure for adenosine deaminase in blood, *Biochem. Soc. Trans.*, 5, 1756, 1977.
2. Martin de Llano, J.J., Garcia-Segura, J.M., and Gavilanes, J.G., Selective silver staining of urease activity in polyacrylamide gels, *Anal. Biochem.*, 177, 37, 1989.
3. Shaik-M, M.B., Guy, A.L., and Pancholy, S.K., An improved method for the detection and preservation of urease activity in polyacrylamide gels, *Anal. Biochem.*, 103, 140, 1980.
4. Fishbein, W.N., The structural basis for the catalytic complexity of urease: interacting and interconvertible molecular species (with a note on isozyme classes), *Ann. N. Y. Acad. Sci.*, 147, 857, 1969.
5. Witte, C.-P. and Medina-Escobar, N., In-gel detection of urease with nitro blue tetrazolium and quantification of the enzyme from different crop plants using the indophenol reaction, *Anal. Biochem.*, 290, 102, 2001.
6. Gerendás, J., Polacco, J.C., Freyermuth, S.K., and Sattelmacher, B., Significance of nickel for plant growth and metabolism, *J. Plant Nutr. Soil Sci.*, 162, 241, 1999.
7. Mobley, H.L.T., Island, M.D., and Hausinger R.P., Molecular biology of microbial ureases, *Microbiol. Rev.*, 59, 451, 1995.
8. Witte, C.-P., Isidore, E., Tiller, S.A., Davies, H.V., and Taylor, M.A., Functional characterisation of urease accessory protein G (ureG) from potato, *Plant Mol. Biol.*, 45, 169, 2001.
9. Martin de Llano, J.J. and Gavilanes, J.G., Increased electrophoretic mobility of sodium sulfite–treated jack bean urease, *Electrophoresis*, 13, 300, 1992.

3.5.1.11 — Penicillin Amidase; PA

OTHER NAMES   Penicillin acylase, penicillin G acylase

REACTION   Penicillin + H_2O = carboxylate + 6-aminopenicillanate

ENZYME SOURCE   Bacteria

SUBUNIT STRUCTURE   Multichain heteromer[#] (see General Notes)

METHOD

Visualization Scheme

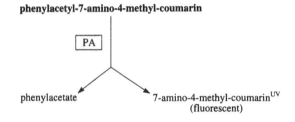

phenylacetyl-7-amino-4-methyl-coumarin

PA

phenylacetate   7-amino-4-methyl-coumarin[UV] (fluorescent)

Staining Solution[1]

100 mM Sodium phosphate buffer, pH 7
50 µM Phenylacetyl-7-amino-4-methyl-coumarin (see *Notes*)

Procedure

Soak the gel in the staining solution for 5 min and inspect under long-wave UV light for fluorescent bands of PA activity. Record the zymogram or photograph using a yellow filter.

Notes: The fluorogenic substrate, phenylacetyl-7-amino-4-methyl-coumarin, can be synthesized at laboratory conditions according to the following protocol. Dissolve 1 g of 7-amino-4-methyl-coumarin (AMC) in 11.42 ml of dioxan:1-methyl-2-pyrrolidon (1:1 v/v) at room temperature. Dissolve 132.3 µl of phenylacetylchloride in 868 µl of dioxan. Very slowly add 1 ml of the AMC solution to the phenylacetylchloride solution. To this mixture add slowly 20 µl of triethylamine. After adding another 1 ml of the AMC solution, add 100 µl of triethylamine and stir the mixture at room temperature for 10 min. Add the mixture to ice-cold water, filter off the insoluble product, and wash again with ice-cold water. Recrystallize the precipitate from hot ethanol/chloroform/1-methyl-2-pyrrolidon (5% from each). Wash crystals with cold ethanol and dry *in vacuo* to leave a white solid. Dissolve the phenylacetyl-7-amino-4-methyl-coumarin crystals in dimethyl sulfoxide (1 mg/ml, 2.93 mM) and then dilute in 100 mM sodium phosphate buffer (pH 7.4) to a final concentration of 50 µM.

The fluorescence of enzymatically generated 7-amino-4-methyl-coumarin is excited at 383 nm and emitted at 455 nm.

GENERAL NOTES

The enzyme is a multichain heteromeric protein decomposed into its subunits under denaturing conditions. After separation in an SDS-PAG, the PA molecules cannot be renatured to their native state.

REFERENCES

1. Ninkovic, M., Riester, D., Wirsching, F., Dietrich, R., Schweinhorst, A., Fluorogenic assay for penicillin G acylase activity, *Anal. Biochem.*, 292, 228, 2001.

OTHER NAMES	Dehydropeptidase II, histozyme, hippuricase, benzamidase, acylase I
REACTION	N-Acyl-L-amino acid + H_2O = fatty acid anion + L-amino acid
ENZYME SOURCE	Bacteria, fungi, vertebrates
SUBUNIT STRUCTURE	Dimer (vertebrates)

METHOD

Visualization Scheme

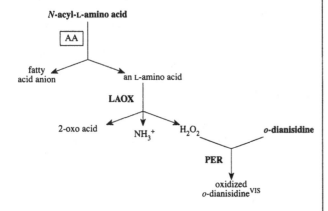

Staining Solution[1]

0.3 M Sodium phosphate buffer, pH 7.5	5 ml
N-Acetyl-L-methionine (or N-formyl-L-methionine)	10 mg
L-Amino acid oxidase (LAOX; crude snake venom from *Agkistrodon piscivorus* or *Crotalus adamanteus*)	3 mg
Peroxidase (PER)	3 mg
o-Dianisidine dihydrochloride	5 mg
12.5 mM $MnCl_2$	0.2 ml

Procedure

Apply the staining solution to the surface of an electrophorized acetate cellulose or starch gel and incubate in a humid chamber at 37°C until brown bands appear. Fix the stained gel in 7% acetic acid.

Notes: When an acetate cellulose gel is used for AA electrophoresis, only about 10 μg of protein (2 to 5 μl of cell extract) is needed to detect the enzyme activity bands by this method.

OTHER METHODS

A. A bioautographic procedure can also be used to visualize AA activity bands on electrophoretic gels.[2] The principle of bioautography is the visualization of an enzyme by a zone of bacterial growth that results when an auxotrophic bacterium is supplied with a product of the enzyme (see Part II for details). This procedure has its own limitations (e.g., to produce optimal banding patterns, conditions must be maintained that do not interfere with either bacterial growth or enzyme activity) and is not as practical as the method described above.

B. A procedure for spectrophotometric detection of AA activity bands on an electrophoretic PAG has also been developed.[3] The method is based on the use of N-acetyl-L-cysteine as a substrate and on the strong absorbance at 296 nm of a ketimine ring that is formed by the reaction of the enzymatic product (L-cysteine) with 3-bromopyruvate. The procedure allows one to visualize up to about 1 to 10 mU of the enzyme, but it also is not practical because it requires the use of a scanning spectrophotometer or a special optic device used for two-dimensional spectroscopy of electrophoretic gels.[4]

REFERENCES

1. Qavi, H. and Kit, S., Electrophoretic patterns of aminoacylase-1 (ACY-1) isozymes in vertebrate cells and histochemical procedure for detecting ACY-1 activity, *Biochem. Genet.*, 18, 669, 1980.
2. Naylor, S.L., Shows, T.B., and Klebe, R.J., Bioautographic visualization of aminoacylase-1: assignment of the structural gene ACY-1 to chromosome 3 in man, *Somat. Cell Genet.*, 5, 11, 1979.
3. Ricci, G., Caccuri, A.M., Lo Bello, M., Solinas, S.P., and Nardini, M., Ketimine rings: useful detectors of enzymatic activities in solution and on polyacrylamide gel, *Anal. Biochem.*, 165, 356, 1987.
4. Klebe, R.J., Mancuso, M.G., Brown, C.R., and Teng, L., Two-dimensional spectroscopy of electrophoretic gels, *Biochem. Genet.*, 19, 655, 1981.

3.5.1.27 — *N*-Formylmethionylaminoacyl-tRNA Deformylase; FMDF

OTHER NAMES	Peptide deformylase
REACTION	*N*-formyl-L-methionylaminoacyl-tRNA + H$_2$O = formate + L-methionylaminoacyl-tRNA (see also General Notes)
ENZYME SOURCE	Bacteria
SUBUNIT STRUCTURE	Unknown or no data available

METHOD

Visualization Scheme

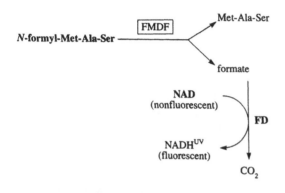

Staining Solution[1] (adapted)

50 m*M* Potassium phosphate buffer, pH 7.5
2 m*M* *N*-Formyl-Met-Ala-Ser (Bachem)
12 m*M* NAD
1.2 U/ml Formate dehydrogenase (FD)

Procedure

Apply the staining solution to the gel surface as a filter paper overlay. Incubate the gel at 37°C and monitor under long-wave UV light. Fluorescent bands of FMDF are visible on a dark background of the gel. Photograph the developed gel using yellow filter.

Notes: The processed gel can be counterstained using PMS–MTT solution to obtain dark blue bands of FMDF visible in daylight.

GENERAL NOTES

In eubacteria, protein synthesis starts with *N*-formylmethionine. This is because a formyl group is added onto Met-tRNA^{fMet} prior to its binding to the ribosome. The subsequent removal of the *N*-formyl group from nascent polypeptides is catalyzed by FMDF (also known as peptide deformylase) according to the reaction

$$N\text{-formyl-Met-X} + H_2O \rightarrow \text{formate} + \text{Met-X}$$

where X stands for at least one amino acid.[1]

REFERENCES

1. Lazennec, C. and Meinnel, T., Formate dehydrogenase-coupled spectrophotometric assay of peptide deformylase, *Anal. Biochem.*, 244, 180, 1997.

REACTION   Chitin + H_2O = chitosan + acetate
ENZYME SOURCE   Fungi, plants
SUBUNIT STRUCTURE Monomer# (fungi)

METHOD 1

Visualization Scheme

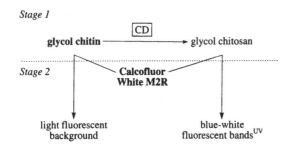

Stage 1

glycol chitin ⟶ [CD] ⟶ glycol chitosan

Stage 2 — Calcofluor White M2R

light fluorescent background   blue-white fluorescent bands[UV]

Substrate-Containing PAG[1]

7.5% Polyacrylamide gel solution
50 mM HEPES-KOH buffer, pH 7.0
0.1% Glycol chitin (see General Notes)

Visualization Procedure[1]

Preincubate the electrophorized PAG in 200 ml of 50 mM HEPES-KOH buffer (pH 7.0) for 5 min with gentle shaking. Then put the PAG on a clean glass plate and cover with a PAG overlay containing glycol chitin. Eliminate the liquid between the gels and the glass plate by gently sliding a test tube over the surface of the overlay gel. Incubate gels at 37°C for 4 to 5 h in a plastic container under moist conditions. Transfer the gels in a freshly prepared solution of 0.01% Calcofluor White M2R in 0.5 M Tris–HCl (pH 8.9) for 5 min. Remove the gels and wash in distilled water for at least 2 h at room temperature. Observe zones of CD activity as bands of blue-white fluorescence visible in long-wave UV light on a light fluorescent background.

METHOD 2

Visualization Scheme

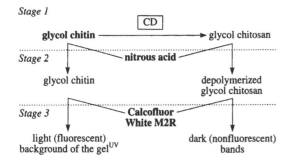

Stage 1

glycol chitin ⟶ [CD] ⟶ glycol chitosan

Stage 2 — nitrous acid

glycol chitin   depolymerized glycol chitosan

Stage 3 — Calcofluor White M2R

light (fluorescent) background of the gel[UV]   dark (nonfluorescent) bands

Substrate-Containing PAG[1]

7.5% Polyacrylamide gel solution
50 mM HEPES-KOH buffer, pH 7.0
0.1% Glycol chitin (see General Notes)

Visualization Procedure[1]

Preincubate the electrophorized PAG in 200 ml of 50 mM HEPES-KOH buffer (pH 7.0) for 5 min with gentle shaking.

Stage 1. Put the PAG on a clean glass plate and cover with a 7.5% PAG overlay containing glycol chitin. Eliminate the liquid between the gels and the glass plate by gently sliding a test tube over the surface of the overlay gel. Incubate the gels at 37°C for 4 to 5 h in a plastic container under moist conditions.

Stage 2. Expose the gels to nitrous acid generated by mixing, just before use, 72 ml of 5.5 M NaNo$_2$ with 28 ml of 1 N H$_2$SO$_4$. After 10 min in nitrous acid, wash the gels in distilled water for 5 min.

Stage 3. Place the gels in a freshly prepared solution of 0.01% Calcofluor White M2R in 0.5 M Tris–HCl (pH 9.0) for 5 min. Wash the gels in distilled water for at least 2 h with several changes.

Dark nonfluorescent bands of CD activity are visible in long-wave UV light on a light fluorescent background of the gel.

GENERAL NOTES

Photograph the developed gel just following the staining procedure using Polaroid type 55 film and UV haze and 02 orange filters. Use an exposure time of from 1 to 5 min with a 127-nm lens at f 4.7.[2]

Glycol chitin is synthesized from commercially available glycol chitosan as described for chitinase (see 3.2.1.14 — CHIT, Method 1).[2]

After SDS-PAG electrophoresis, incubate the gel overnight at 37°C with gentle reciprocal shaking in 50 mM HEPES-KOH buffer (pH 7.0) containing 1% Triton X-100 (purified through a mixed-bed resin deionizing column AG 501-X8). CD activity is then visualized as for native gels.

REFERENCES

1. Trudel, J. and Asselin, A., Detection of chitin deacetylase activity after polyacrylamide gel electrophoresis, *Anal. Biochem.*, 189, 249, 1990.
2. Trudel, J. and Asselin, A., Detection of chitinase activity after polyacrylamide gel electrophoresis, *Anal. Biochem.*, 178, 362, 1989.

3.5.1.X — Pantetheinase; PANT

REACTION	Pantetheine + H_2O = cysteamine + pantothenate; also acts on a variety of pantetheine derivatives
ENZYME SOURCE	Vertebrates (mammals)
SUBUNIT STRUCTURE	Unknown or no data available

Method

Visualization Scheme

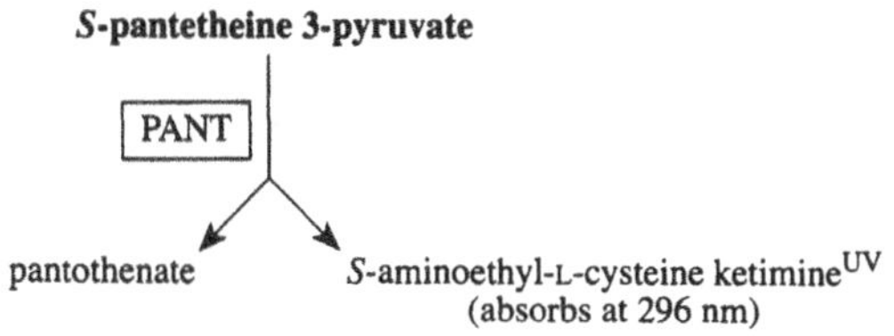

S-pantetheine 3-pyruvate

PANT

pantothenate   S-aminoethyl-L-cysteine ketimine[UV]
(absorbs at 296 nm)

Staining Solution[1]

0.1 M Phosphate buffer, pH 8.0
4 mM S-Pantetheine 3-pyruvate

Procedure

Incubate an electrophorized PAG in the staining solution at 20°C for 30 min, wash with water, and scan at 296 nm using a scanning spectrophotometer (e.g., Beckman gel scanner apparatus).

Notes: The method is semiquantitative. After a 30-min incubation the intensity of the peak observed on a photometric scan is proportional to the amount of enzyme applied on the gel, in the range of 0.001 to 0.01 U.

Two-dimensional spectrograms of processed PAGs can also be obtained using a special optical device constructed for two-dimensional spectroscopy of electrophoretic gels.[2]

References

1. Ricci, G., Caccuri, A.M., Lo Bello, M., Solinas, S.P., and Nardini, M., Ketimine rings: useful detectors of enzymatic activities in solution and on polyacrylamide gel, *Anal. Biochem.*, 165, 356, 1987.
2. Klebe, R.J., Mancuso, M.G., Brown, C.R., and Teng, L., Two-dimensional spectroscopy of electrophoretic gels, *Biochem. Genet.*, 19, 655, 1981.

3.5.2.2 — Dihydropyrimidinase; HYD

OTHER NAMES   Hydantoinase (in bacteria)

REACTIONS
1. Hydantoin + H_2O = hydantoate
2. 5,6-Dihydrouracil + H_2O = 3-ureidopropanoate

ENZYME SOURCE   Bacteria (hydantoinase), mammals (dihydropyrimidinase)

SUBUNIT STRUCTURE   Dimer$^{#}$ (bacteria), tetramer$^{#}$ (bacteria)

METHOD

Visualization Scheme

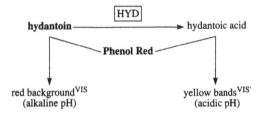

Staining Solution[1]

A. 40 mM Tris–HCl, pH 8.0
    2% Hydantoin
    0.01% Phenol Red
B. 2% Agarose solution (60°C)

Procedure

Mix equal volumes of A and B components of the staining solution and pour the mixture over the surface of the gel. Incubate the gel at 60°C until bright yellow bands of HYD activity appear on a red background of the gel. Record or photograph the zymogram (see *Notes*).

Notes: Use dihydrouracil as a substrate to detect the enzyme from animal tissues.

Photograph the zymogram using a 570-nm interference filter.

Drawing on the gel (using a felt-tip pen) in the position of the developed activity bands will allow permanent positions to be noted after the bands have faded.

Since changes in the absorption spectrum of Phenol Red occur between pH values 6.8 and 8.2, theoretically a change as small as 10^{-7} M hydrogen ion could be detected by this method.[2]

GENERAL NOTES

Bacterial hydantoinase catalyzes the hydrolysis of a variety of hydantoins and is suggested to be a bacterial counterpart of the animal dihydropyrimidinase, which is involved in the catabolic degradation of pyrimidines. Bacterial enzyme also displays considerable activity toward dihydrouracil (about one third of that toward hydantoin). A close evolutionary relationship was proposed between bacterial hydantoinase, animal dihydropyrimidinase, allantoinase (see 3.5.2.5 — ALL), and dihydroorotase (see 3.5.2.3 — DHO) on the basis of their structural and functional similarities. All these enzymes hydrolyze the cyclic amide bond (–CO–NH–) in either five- or six-membered rings. These enzymes are also known to catalyze the reverse reactions.[3]

REFERENCES

1. Kim, G.-J., Lee, S.-G., Park, J.-H., and Kim, H.-S., Direct detection of the hydantoinase activity on solid agar plates and electrophoretic acrylamide gels, *Biotechnol. Tech.*, 11, 511, 1997.
2. Klebe, R.J., Mancuso, M.G., Brown, C.R., and Teng, L., Two-dimensional spectroscopy of electrophoretic gels, *Biochem. Genet.*, 19, 655, 1981.
3. Kim, G.-J. and Kim, H.-S., Identification of the structural similarity in the functionally related amidohydrolases acting on the cyclic amide ring, *Biochem. J.*, 330, 295, 1998.

3.5.2.3 — Dihydroorotase; DHO

OTHER NAMES   Carbamoylaspartic dehydrase
REACTION (S)-Dihydroorotate H_2O =
                      N-carbamoyl-L-aspartate
ENZYME SOURCE   Bacteria, mammals
SUBUNIT STRUCTURE   Dimer[a] (bacteria)

METHOD

Visualization Scheme

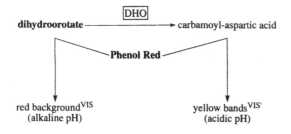

Staining Solution[1] (adapted)

A. 40 mM Tris–HCl, pH 8.0
   2% Dihydroorotate
   0.01% Phenol Red
B. 2% Agarose solution (60°C)

Procedure

Mix equal volumes of A and B components of the staining solution and pour the mixture over the surface of the gel. Incubate the gel at 60°C until bright yellow bands of DHO activity appear on a red background of the gel. Record or photograph the zymogram (see *Notes*).

Notes: Photograph the zymogram using a 570-nm interference filter.

Drawing on the gel (using a felt-tip pen) in the position of the developed activity bands will allow permanent positions to be noted after the bands have faded.

Since changes in the absorption spectrum of Phenol Red occur between pH values 6.8 and 8.2, theoretically a change as small as 10^{-7} M hydrogen ion could be detected by this method.[2]

GENERAL NOTES

A close evolutionary relationship was proposed between dihydroorotase, hydantoinase (see 3.5.2.2 — HYD), and allantoinase (see 3.5.2.5 — ALL) on the basis of their structural and functional similarities. These amidohydrolases hydrolyze the cyclic amide bond (–CO–NH–) in either five- or six-membered rings and also catalyze the reverse reactions.[3]

REFERENCES

1. Kim, G.-J., Lee, S.-G., Park, J.-H., and Kim, H.-S., Direct detection of the hydantoinase activity on solid agar plates and electrophoretic acrylamide gels, *Biotechnol. Tech.*, 11, 511, 1997.
2. Klebe, R.J., Mancuso, M.G., Brown, C.R., and Teng, L., Two-dimensional spectroscopy of electrophoretic gels, *Biochem. Genet.*, 19, 655, 1981.
3. Kim, G.-J. and Kim, H.-S., Identification of the structural similarity in the functionally related amidohydrolases acting on the cyclic amide ring, *Biochem. J.*, 330, 295, 1998.

3.5.2.5 — Allantoinase; ALL

REACTION   Allantoin + H_2O = allantoate
ENZYME SOURCE   Bacteria, fungi, vertebrates
SUBUNIT STRUCTURE   Tetramer[#] (bacteria)

METHOD

Visualization Scheme

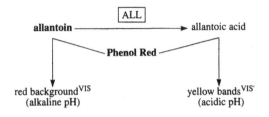

red background[VIS]   yellow bands[VIS']
(alkaline pH)   (acidic pH)

Staining Solution[1] (adapted)

    A.   40 mM Tris–HCl, pH 8.0
        2% Allantoin
        0.01% Phenol Red
    B.   2% Agarose solution (60°C)

Procedure

Mix equal volumes of A and B components of the staining solution and pour the mixture over the surface of the gel. Incubate the gel at 60°C until bright yellow bands of ALL activity appear on a red background of the gel. Record or photograph the zymogram (see *Notes*).

Notes: Photograph the zymogram using a 570-nm interference filter.

Drawing on the gel (using a felt-tip pen) in the position of the developed activity bands will allow permanent positions to be noted after the bands have faded.

Since changes in the absorption spectrum of Phenol Red occur between pH values 6.8 and 8.2, theoretically a change as small as 10^{-7} M hydrogen ion could be detected by this method.[2]

GENERAL NOTES

A close evolutionary relationship was proposed between allantoinase, bacterial hydantoinase and animal dihydropyrimidinase (see 3.5.2.2 — HYD), and dihydroorotase (see 3.5.2.3 — DHO) on the basis of their structural and functional similarities. These amidohydrolases hydrolyze the cyclic amide bond (–CO–NH–) in either five- or six-membered rings and also catalyze the reverse reactions.[3]

REFERENCES

1. Kim, G.-J., Lee, S.-G., Park, J.-H., and Kim, H.-S., Direct detection of the hydantoinase activity on solid agar plates and electrophoretic acrylamide gels, *Biotechnol. Tech.*, 11, 511, 1997.
2. Klebe, R.J., Mancuso, M.G., Brown, C.R., and Teng, L., Two-dimensional spectroscopy of electrophoretic gels, *Biochem. Genet.*, 19, 655, 1981.
3. Kim, G.-J. and Kim, H.-S., Identification of the structural similarity in the functionally related amidohydrolases acting on the cyclic amide ring, *Biochem. J.*, 330, 295, 1998.

3.5.2.6 — Penicillinase; PEN

OTHER NAMES	β-Lactamase (recommended name), cephalosporinase
REACTION	β-Lactam + H$_2$O = substituted β-amino acid
ENZYME SOURCE	Bacteria
SUBUNIT STRUCTURE	Unknown or no data available

METHOD

Visualization Scheme

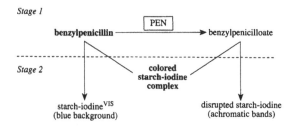

Staining Solution[1]

A. 10 mM Potassium phosphate buffer, pH 7.0
   0.5 M NaCl
   100 mg/ml Benzylpenicillin
B. 0.25 M I$_2$
   1.25 M KI
C. 2% Starch soluble for iodometry (Fisher Scientific); dissolved in distilled water by heating and stirring

Procedure

Add 1 part of solution B to 100 parts of freshly prepared solution C. Dip a strip of Whatman 3MM paper in starch–iodine solution and hang to dry overnight. Store the prepared strips in a dark, cool, and dry place before use.

Place an electrophorized PAG on a glass plate of an appropriate size and cover with a strip of Whatman 3MM paper soaked in solution A. Apply a second glass plate, avoiding formation of air bubbles, press down firmly and evenly, and incubate at room temperature for 8 to 10 min. Remove the top glass plate and the first paper application. Cover the gel with starch–iodine paper saturated with solution A and press down firmly and evenly with another glass plate. Invert the gel with glass plates and observe white bands of PEN activity on a dark blue background. Mark the position of achromatic bands and the "origin" on a glass plate by a marker.

Notes: This method is based on the ability of penicilloic acid, produced by PEN, to reduce I$_2$ and hence to decolorize the starch–iodine complex.

OTHER METHODS

An immunoblotting procedure (for details, see Part II) based on the utility of antibodies specific to the *E. coli* enzyme and ^{125}I-protein A as a radiolabel for anti-antibodies is also available.[1] This procedure, however, is more time-consuming and labor-intensive than that described above.

GENERAL NOTES

This activity is represented by a group of enzymes of varying specificities hydrolyzing β-lactams. Some of these enzymes act more rapidly on penicillins, some more rapidly on cephalosporins.[2]

REFERENCES

1. Tai, P.C., Zyk, N., and Citri, N., *In situ* detection of β-lactamase activity in sodium dodecyl sulfate–polyacrylamide gels, *Anal. Biochem.*, 144, 199, 1985.
2. NC-IUBMB, *Enzyme Nomenclature*, Academic Press, San Diego, 1992, p. 432 (EC 3.5.2.6, Comments).

OTHER NAMES   L-α-Aminocaprolactam hydrolase, L-lysinamidase

REACTION   L-Lysine 1,6-lactam + H_2O = L-lysine

ENZYME SOURCE   Bacteria

SUBUNIT STRUCTURE Unknown or no data available

METHOD

Visualization Scheme

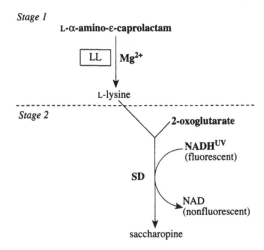

Stage 1

L-α-amino-ε-caprolactam

LL | Mg^{2+}

L-lysine

Stage 2

2-oxoglutarate

NADH^{UV} (fluorescent)

SD

NAD (nonfluorescent)

saccharopine

Staining Solution[1] (adapted)

A. 0.05 *M* Tris–HCl buffer, pH 9.0
   0.3 m*M* $MgCl_2$
   2.5 m*M* L-α-Amino-ε-caprolactam

B. 2% Agar solution (60°C)

C. 0.1 *M* Potassium phosphate buffer, pH 6.8
   0.3 m*M* NADH
   2.5 m*M* 2-Oxoglutarate
   1 U/ml Saccharopine dehydrogenase (NAD, L-lysine forming) (SD)

Procedure

Mix equal volumes of A and B solutions and pour the mixture over the gel surface. Incubate the gel at 37°C for 30 to 60 min. Remove the agar overlay. Apply solution C to the gel surface on a filter paper overlay and incubate for 20 to 30 min at 37°C in a moist chamber. View the gel under long-wave UV light. Dark (nonfluorescent) bands of LL activity are visible on a light (fluorescent) background. Record the zymogram or photograph using a yellow filter.

Notes: Lysine lactamase shows maximum reactivity at a pH of about 9.0, and the rate of substrate hydrolysis declines markedly at pH values below 8.0. Saccharopine dehydrogenase has a narrow pH optimum at 6.0 to 7.0. The use of a two-step staining procedure allows one to avoid the problem of different pH optima of the two enzymes.

REFERENCES

1. Laber, B. and Amrhein, N., A spectrophotometric assay for *meso*-diaminopimelate decarboxylase and L-α-amino-ε-caprolactam hydrolase, *Anal. Biochem.*, 181, 297, 1989.

3.5.3.1 — Arginase; ARG

OTHER NAMES	Arginine amidinase, canavanase
REACTION	L-Arginine + H_2O = L-ornithine + urea
ENZYME SOURCE	Bacteria, fungi, plants, invertebrates, vertebrates
SUBUNIT STRUCTURE	Hexamer[#] (bacteria), tetramer[#] (fungi, plants), trimer[#] (mammals)

METHOD 1

Visualization Scheme

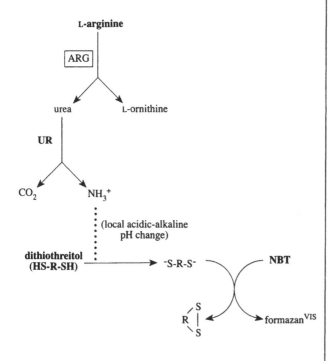

Staining Solution[1]

A.	2 mg/ml Urease (UR; Sigma, type VI)	1 ml
	0.85 M L-Arginine, pH 6.8	2 ml
	0.1 M Dithiothreitol	0.6 ml
	1.3% NBT	0.3 ml
B.	2% Agar solution (45°C)	15 ml

Procedure

Mix A and B components of the staining solution and pour the mixture over the gel surface. Incubate the gel at 37°C until dark blue bands appear. Wash the stained gel with water and fix in 25% ethanol.

Notes: All solutions used for ARG electrophoresis and detection should be made using bidistilled water lacking ammonia. Electrode and gel buffers used for ARG electrophoresis should be of minimal ionic strength.

METHOD 2

Visualization Scheme

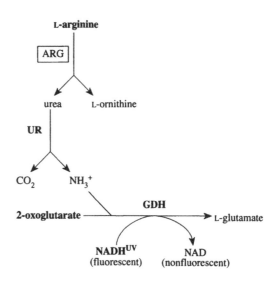

Staining Solution[2]

0.1 M Tris–HCl buffer, pH 7.6	5 ml
1.0 M L-Arginine, pH 7.6	1 ml
Urease (UR; Sigma, type VI)	20 U
2-Oxoglutarate	25 mg
NADH	10 mg
Glutamate dehydrogenase (GDH; Sigma; 500 U/ml)	50 μl

Procedure

Apply the staining solution to the gel surface on a filter paper overlay and incubate at 37°C until dark (nonfluorescent) bands visible under long-wave UV light on a light (fluorescent) background appear.

OTHER METHODS

The areas of ammonia production can also be detected using the pH indicator dye Phenol Violet (e.g., see 3.5.4.4 — ADA, Method 3), sodium tetraphenylborate (e.g., see 3.5.1.1 — ASP), and neutral $AgNO_3$ solution containing photographic developers (e.g., see 3.5.1.5 — UR, Method 2).

GENERAL NOTES

The enzyme activity (at least in mammals) is inhibited by citrate and borate. Thus, electrophoretic and staining buffers containing these substances are unsuitable for ARG electrophoresis and detection.

The enzyme from some sources (e.g., bacteria and plants) is activated by manganese ions.

3.5.3.1 — Arginase; ARG (continued)

REFERENCES

1. Farron, F., Arginase isozymes and their detection by catalytic staining in starch gel, *Anal. Biochem.*, 53, 264, 1973.
2. Nelson, R.L., Povey, S., Hopkinson, D.A., and Harris, H., The detection after electrophoresis of enzymes involved in ammonia metabolism using L-glutamate dehydrogenase as a linking enzyme, *Biochem. Genet.*, 15, 1023, 1977.

3.5.4.3 — Guanine Deaminase; GDA

OTHER NAMES	Guanase, guanine aminase
REACTION	Guanine + H_2O = xanthine + NH_3
ENZYME SOURCE	Invertebrates, vertebrates
SUBUNIT STRUCTURE	Dimer (vertebrates)

METHOD

Visualization Scheme

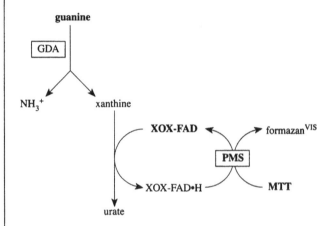

Staining Solution[1]

A. 0.2 *M* Tris–HCl buffer, pH 7.6   20 ml
     1 mg/ml Guanine (prepared by dissolving 50 mg
       of guanine in 10 ml of warm 0.1 *N* NaOH and
       made up to 50 ml with water)   3 ml
     4 U/ml Xanthine oxidase (XOX)   25 μl
     MTT   7.5 mg
     PMS   2.5 mg
B. 2% Agar solution (60°C)   25 ml

Procedure

Mix A and B components of the staining solution and pour the mixture over the gel surface. Incubate the overlayed gel in the dark at 37°C until dark blue bands appear. Fix the zymogram in 25% ethanol.

OTHER METHODS

A. The areas of ammonia production can be detected using the backward reaction of a linking enzyme, glutamate dehydrogenase (e.g., see 3.5.3.1 — ARG, Method 2). Dark (nonfluorescent) bands of GDA activity are observed on a light (fluorescent) background of the gel in long-wave UV light.
B. The areas of ammonia production can also be detected using sodium tetraphenylborate, which precipitates in the presence of ammonia ions, resulting in the formation of white opaque bands visible in a PAG (see 3.5.1.1 — ASP).

3.5.4.3 — Guanine Deaminase; GDA (continued)

C. The local pH change due to the production of ammonia can be detected using the pH indicator dye Phenol Violet (e.g., see 3.5.4.4 — ADA, Method 3), the NBT (or MTT)–dithiothreitol system (e.g., see 3.5.3.1 — ARG, Method 1), and neutral AgNO$_3$ solution containing photographic developers (e.g., see 3.5.1.5 — UR, Method 2).

REFERENCES

1. Harris, H. and Hopkinson, D.A., *Handbook of Enzyme Electrophoresis in Human Genetics*, North-Holland, Amsterdam, 1976 (loose-leaf, with supplements in 1977 and 1978).

3.5.4.4 — Adenosine Deaminase; ADA

REACTION	Adenosine + H$_2$O = inosine + NH$_3$
ENZYME SOURCE	Bacteria, fungi, protozoa, invertebrates, vertebrates
SUBUNIT STRUCTURE	Monomer (invertebrates, vertebrates)

Method 1

Visualization Scheme

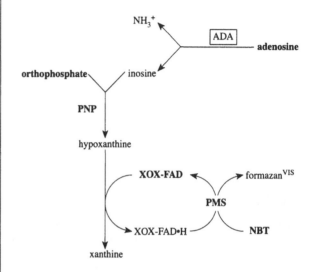

Staining Solution[1]

A.	0.05 *M* Phosphate buffer, pH 7.5	25 ml
	Adenosine	20 mg
	NBT	5 mg
	PMS	5 mg
	Xanthine oxidase (XOX)	0.08 U
	Purine-nucleoside phosphorylase (PNP)	0.08 U
B.	2% Agar solution (60°C)	25 ml

Procedure

Mix A and B components of the staining solution and pour the mixture over the gel surface. Incubate the gel in the dark at 37°C until dark blue bands appear. Fix the stained gel in 25% ethanol.

Method 2

Visualization Scheme

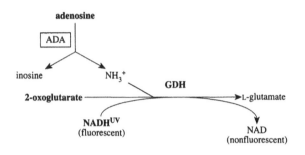

Staining Solution[2]

0.1 M Tris–HCl buffer, pH 7.6	5 ml
Adenosine	30 mg
2-Oxoglutarate	25 mg
NADH	10 mg
500 U/ml Glutamate dehydrogenase (GDH)	50 μl

Procedure

Apply the staining solution to the gel surface on a filter paper overlay and incubate at 37°C until dark (nonfluorescent) bands visible in long-wave UV light appear on a light (fluorescent) background. Photograph the developed gel using a yellow filter.

Notes: To make ADA bands visible in daylight, cover the processed gel with a second filter paper overlay containing PMS and MTT (or NBT). Achromatic bands corresponding to nonfluorescent areas visible in UV light appear almost immediately. The negative zymogram may be stored in 25% ethanol or 3% acetic acid.

Method 3

Visualization Scheme

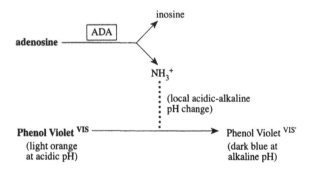

Staining Solution[3]

Adenosine	1 g
Phenol Violet (saturated water solution)	10 ml
Agar	1 g

Procedure

Dissolve components of the staining solution at 100°C and cool to 45°C. Pour the mixture over the gel surface and incubate until dark blue bands appear on a light orange background.

Notes: All solutions used for ADA electrophoresis and detection should be made using bidistilled water lacking ammonia. Electrode and gel buffers should be of minimal ionic strength.

Other Methods

A. The areas of ammonia production can also be detected using sodium tetraphenylborate, which precipitates in the presence of ammonia ions, resulting in the formation of white opaque bands visible in PAG (e.g., see 3.5.1.1 — ASP).

B. The local pH change due to the production of ammonia can be detected using the NBT (MTT)–dithiothreitol system (e.g., see 3.5.3.1 — ARG, Method 1) and neutral AgNO$_3$ solution containing photographic developers (e.g., see 3.5.1.5 — UR, Method 2).

C. A bioautographic procedure (for details, see Part II) for ADA location in electrophoretic gels has also been developed.[4] However, this procedure is more complex and less practical than any one of those described above.

References

1. Spencer, N., Hopkinson, D.A., and Harris, H., Adenosine deaminase polymorphism in man, *Ann. Hum. Genet.*, 32, 9, 1968.
2. Nelson, R.L., Povey, S., Hopkinson, D.A., and Harris, H., The detection after electrophoresis of enzymes involved in ammonia metabolism using L-glutamate dehydrogenase as a linking enzyme, *Biochem. Genet.*, 15, 1023, 1977.
3. Daly, M.P. and Tully, E.R., Detection of ammonia-producing enzymes after electrophoresis: a screening procedure for adenosine deaminase in blood, *Biochem. Soc. Trans.*, 5, 1756, 1977.
4. Naylor, S.L., Bioautographic visualization of enzymes, in *Isozymes: Current Topics in Biological and Medical Research*, Vol. 4, Rattazzi, M.C., Scandalios, J.M., and Whitt, G.S., Eds., Alan R. Liss, New York, 1980, p. 69.

3.5.4.5 — Cytidine Deaminase; CDA

REACTION   Cytidine + H_2O = uridine + NH_3
ENZYME SOURCE   Bacteria, vertebrates
SUBUNIT STRUCTURE   Tetramer (vertebrates)

METHOD 1

Visualization Scheme

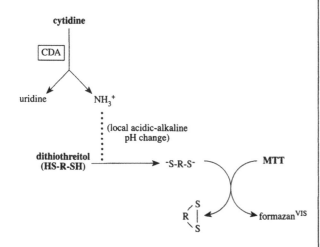

Staining Solution[1]

A. Cytidine	15 mg
0.3% Dithiothreitol	1 ml
5% MTT (or NBT)	0.3 ml
H_2O	10 ml
B. 2% Agar solution (60°C)	10 ml

Procedure

Mix A and B components of the staining solution and pour the mixture over the gel surface. Incubate the gel at 37°C until dark blue bands appear. Fix the stained gel in 25% ethanol.

Notes: This method requires that the electrophoretic gel be only minimally buffered, since staining is dependent on the ability of thiol groups in dithiothreitol to reduce a tetrazolium dye nonenzymatically when there is a rise in pH. All solutions used for CDA electrophoresis and detection should be made using bidistilled water lacking ammonia.

METHOD 2

Visualization Scheme

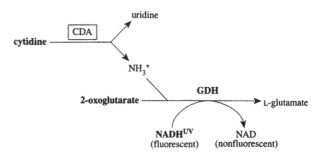

Staining Solution[2]

0.1 *M* Tris–HCl buffer, pH 7.6	5 ml
Cytidine	40 mg
Dithiothreitol	10 mg
ADP (disodium salt)	10 mg
2-Oxoglutarate	25 mg
NADH	10 mg
500 U/ml Glutamate dehydrogenase (GDH)	50 μl

Procedure

Apply the staining solution to the gel surface on a filter paper overlay and incubate at 37°C until dark (nonfluorescent) bands visible in long-wave UV light appear on a light (fluorescent) background. Record or photograph the developed gel using a yellow filter.

Notes: To make CDA bands visible in daylight, cover the processed gel with filter paper soaked in PMS–MTT solution. Achromatic CDA bands appear on a blue background of the gel almost immediately. Fix the negatively stained gel in 25% ethanol.

OTHER METHODS

A. The areas of ammonia production can be detected using sodium tetraphenylborate, which precipitates in the presence of ammonia ions, resulting in the formation of white opaque bands visible in PAG (see 3.5.1.1 — ASP).

B. The local pH change due to the production of ammonia can also be detected using pH indicator dye Phenol Violet (e.g., see 3.5.4.4 — ADA, Method 3) and neutral AgNO$_3$ solution containing photographic developers (e.g., see 3.5.1.5 — UR, Method 2).

REFERENCES

1. Teng, Y.-S., Anderson, J.E., and Giblett, E.R., Cytidine deaminase: a new genetic polymorphism demonstrated in human granulocytes, *Am. J. Hum. Genet.*, 27, 492, 1975.
2. Nelson, R.L., Povey, S., Hopkinson, D.A., and Harris, H., The detection after electrophoresis of enzymes involved in ammonia metabolism using L-glutamate dehydrogenase as a linking enzyme, *Biochem. Genet.*, 15, 1023, 1977.

OTHER NAMES	Adenylate deaminase, adenylic acid deaminase, AMP aminase
REACTION	AMP + H_2O = IMP + NH_3
ENZYME SOURCE	Bacteria, fungi, vertebrates
SUBUNIT STRUCTURE	Tetramer[#] (mammals)

METHOD 1

Visualization Scheme

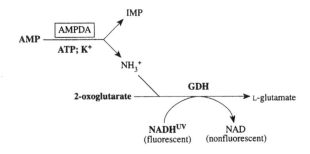

Staining Solution[1]

0.1 *M* Tris–HCl buffer, pH 7.6	5 ml
AMP (disodium salt; 6 H_2O)	100 mg
ATP (disodium salt; 3 H_2O)	10 mg
KCl	40 mg
2-Oxoglutaric acid (neutralized)	25 mg
NADH	10 mg
500 U/ml Glutamate dehydrogenase (GDH)	50 μl

Procedure

Apply the staining solution to the gel surface on a filter paper overlay. Incubate the covered gel in a humid chamber at 37°C and monitor under long-wave UV light. Dark (nonfluorescent) bands visible on a light (fluorescent) background indicate areas of AMPDA activity. Photograph the developed gel using a yellow filter.

Notes: To make AMPDA bands visible in daylight, cover the processed gel with a second filter paper overlay saturated with the PMS–MTT solution. Achromatic bands visible on a blue background appear almost immediately. Fix the negatively stained gel in 25% ethanol or 3% acetic acid.

To increase the apparent affinity of the enzyme for AMP, ATP and K[+] are included in the stain.

METHOD 2

Visualization Scheme

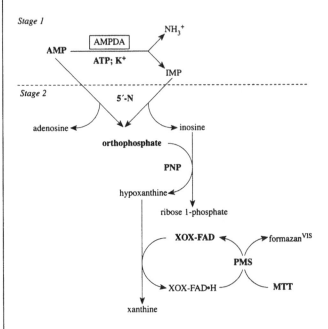

Staining Solution[2]

A. 0.1 *M* Tris–HCl buffer, pH 7.8	10 ml
AMP (disodium salt; 6 H_2O)	200 mg
ATP (disodium salt; 3 H_2O)	10 mg
KCl	40 mg
B. 2% Agar solution (60°C)	20 ml
C. 0.1 *M* Phosphate buffer, pH 7.5	10 ml
100 U/mg 5′-Nucleotidase (5′-N)	1 mg
4 U/ml Xanthine oxidase (XOX)	10 μl
25 U/ml Purine-nucleoside phosphorylase (PNP)	10 μl
MTT	2.5 mg
PMS	2.5 mg

Procedure

Mix A with 10 ml of B and pour the mixture over the gel surface. Incubate the gel at 37°C for 3 h. Remove the first agar overlay. Mix C with remaining the 10 ml of B and pour over the gel surface. Incubate the gel with a second agar overlay at 37°C in the dark until dark blue bands appear. Fix the stained gel in 25% ethanol.

Notes: Additional bands caused by adenosine deaminase activity (see 3.5.4.4 — ADA, Method 1) can also be developed by this method. This is due to production of adenosine by the linking enzyme 5′-nucleotidase. Adenosine is then converted by endogenous ADA into inosine, which is detected as a result of the action of two other linking enzymes (PNP and XOX) present in the staining solution. Thus, when this method is used, a control staining for ADA should also be made in parallel with AMPDA staining.

To increase the apparent affinity of the enzyme for AMP, ATP and K[+] are included in the stain.

METHOD 3

Visualization Scheme

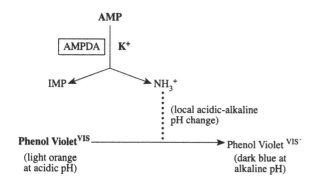

Staining Solution[3]

A. AMP   250 mg
    KCl   550 mg
    Phenol Violet (saturated water solution)   5 ml
    H₂O   20 ml
B. 2% Agar solution (60°C)   25 ml

Procedure

Mix A and B components of the staining solution and pour the mixture over the gel surface. Incubate the overlayed gel at room temperature or at 37°C until dark blue bands appear on a light orange background. Record or photograph the zymogram.

Notes: This method requires that the electrophoretic gel be only minimally buffered. All solutions should be prepared using bidistilled water lacking ammonia.

OTHER METHODS

A. The local pH change due to the production of ammonia can also be detected using the NBT (or MTT)–dithiothreitol system (e.g., see 3.5.4.5 — CDA, Method 1) or neutral AgNO₃ solution containing photographic developers (e.g., see 3.5.1.5 — UR, Method 2).

B. The areas of ammonia production can be detected using sodium tetraphenylborate, which precipitates in the presence of ammonia ions, resulting in the formation of white opaque bands visible in PAG (e.g., see 3.5.1.1 — ASP).

REFERENCES

1. Nelson, R.L., Povey, S., Hopkinson, D.A., and Harris, H., The detection after electrophoresis of enzymes involved in ammonia metabolism using L-glutamate dehydrogenase as a linking enzyme, *Biochem. Genet.*, 15, 1023, 1977.

2. Anderson, J.E., Teng, Y.-S., and Giblett, E.R., Stains for six enzymes potentially applicable to chromosomal assignment by cell hybridization, *Cytogenet. Cell Genet.*, 14, 465, 1975.

3. Daly, M.P. and Tully, E.R., Detection of ammonia-producing enzymes after electrophoresis: a screening procedure for adenosine deaminase in blood, *Biochem. Soc. Trans.*, 5, 1756, 1977.

OTHER NAMES	Pyrophosphatase
REACTION	Pyrophosphate + H_2O = 2 orthophosphate
ENZYME SOURCE	Bacteria, fungi, plants, invertebrates, vertebrates
SUBUNIT STRUCTURE	Dimer (invertebrates, vertebrates)

Method 1

Visualization Scheme

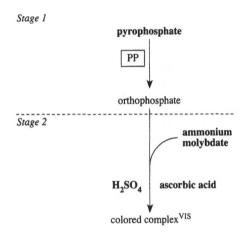

Staining Solution[1]

A.	0.05 M Tris–HCl buffer, pH 7.8	12.5 ml
	0.02 M Pyrophosphate	12.5 ml
	0.2 M MgCl$_2$	1.25 ml
B.	2% Agar solution (60°C)	50 ml
C.	Ascorbic acid	2.5 g
	2.5% Ammonium molybdate in 4 N H$_2$SO$_4$	25 ml

Procedure

Mix solution A with 25 ml of solution B and pour the mixture over the gel surface. Incubate the gel with an agar overlay for 30 to 60 min. Remove the agar overlay. Mix solution C with the remaining 25 ml of solution B and pour over the gel surface. Dark blue bands visible on a light blue background appear after 1 to 5 min. Record the zymogram or photograph because after about 1 h the PP bands disappear.

Notes: To obtain permanent PP zymograms with colored bands stable for several months, the Malachite Green–phosphomolybdate method should be used.[2] In this method the first stage of PP activity detection is the same as described above. The second stage is treatment of the gel with orthophosphate detection solution, which is prepared as follows. Initially stock solutions of 0.045% Malachite Green (oxalate salt) and 4.2% ammonium molybdate in 4 N HCl prepared using deionized water.

Then 90 ml of the Malachite Green solution and 30 ml of the ammonium molybdate solution are mixed for 20 min at room temperature, after which the mixture is passed through a Whatman No. 5 filter and 2.4 ml of 2% sterox (a detergent diluent used in flame photometry) is added. This stock solution may be prepared in advance and stored at 4°C for a week. The orthophosphate detection solution is prepared by mixing 60 ml of stock solution with 40 ml of deionized water immediately prior to use. After treatment with this solution, color development is usually complete in 10 to 20 min. The stained gel is washed with several changes of water and stored in water or 5% acetic acid containing 20% ethanol. The color is stable for several months. The only restriction related to the use of the Malachite Green is prevention of color formation when a citrate-containing staining buffer is used.

Method 2

Visualization Scheme

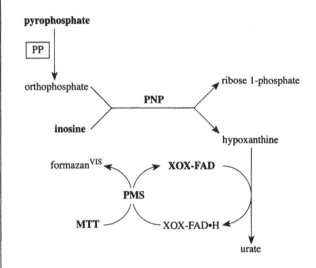

Staining Solution[3] (adapted)

A.	0.01 M Tris–HCl buffer, pH 7.5	23 ml
	Na$_4$P$_2$O$_7$	100 mg
	0.1 M Inosine	2 ml
	0.75 U/ml Xanthine oxidase (XOX)	10 μl
	1.67 U/ml Purine-nucleoside phosphorylase (PNP)	250 μl
	MTT	10 mg
	PMS	3 mg
B.	2% Agarose solution (60°C)	25 ml

Procedure

Mix A and B components of the staining solution and pour the mixture over the gel surface. Incubate the gel in the dark at 37°C until dark blue bands appear. Fix the stained gel in 25% ethanol.

3.6.1.1 — Inorganic Pyrophosphatase; PP (continued)

OTHER METHODS

A. A sensitive and simple method based on the use of firefly luciferase may be adapted to detect PP. Inorganic pyrophosphate inhibits the luciferase activity, but in the presence of PP the luciferase activity is restored. The method is based on PP-induced activation of the firefly luciferase activity in the presence of inorganic pyrophosphate.[4]

B. Some other methods of detection of the product orthophosphate are available (e.g., see 3.1.3.1 — ALP, Methods 6 and 7).

GENERAL NOTES

Orthophosphate should not be used as a component of any buffer used for electrophoresis or staining of PP.

Two major classes of PP are known, soluble and membrane bound.[5] Soluble PP is a ubiquitous enzyme that catalyzes specifically the hydrolysis of inorganic pyrophosphate to orthophosphate. Bacterial membrane-bound PPs catalyze the synthesis or hydrolysis of pyrophosphate and are coupled to proton translocation over a membrane. These PPs do not show any sequence similarity to soluble PPs.[6]

Specificity of the enzyme varies with the source and with the activating metal ion. The enzyme from some sources may be identical with alkaline phosphatase (see 3.1.3.1 — ALP) or glucose-6-phosphatase (see 3.1.3.9 — G-6-PH).[7]

REFERENCES

1. Fisher, R.A., Turner, B.M., Dorkin, H.L., and Harris, H., Studies of human erythrocyte inorganic pyrophosphatase, *Ann. Hum. Genet.*, 37, 341, 1974.
2. Zlotnick, G.W. and Gottlieb, M., A sensitive staining technique for the detection of phosphohydrolase activities after polyacrylamide gel electrophoresis, *Anal. Biochem.*, 153, 121, 1986.
3. Klebe, R.J., Schloss, S., Mock, L., and Link, C.R., Visualization of isozymes which generate inorganic phosphate, *Biochem. Genet.*, 19, 921, 1981.
4. Eriksson, J., Karamohamed, S., and Nyrén, P., Method for real-time detection of inorganic pyrophosphatase activity, *Anal. Biochem.*, 293, 67, 2001.
5. Sivula, T., Salminen, A., Parfenyev, A.N., Pohjanjoki, P., Goldman, A., Cooperman, B.S., Baykov, A.A., and Lahti, R., Evolutionary aspects of inorganic pyrophosphatase, *FEBS Lett.*, 454, 75, 1999.
6. Baltscheffsky, M., Schultz, A., and Baltscheffsky, H., H⁺-PPases: tightly membrane-bound family, *FEBS Lett.*, 457, 527, 1999.
7. NC-IUBMB, *Enzyme Nomenclature*, Academic Press, San Diego, 1992, p. 440 (EC 3.6.1.1, Comments).

3.6.1.3 — Adenosine Triphosphatase; ATPASE

OTHER NAMES	Adenylpyrophosphatase, ATP monophosphatase, triphosphatase, ATPase
REACTION	ATP + H_2O = ADP + orthophosphate
ENZYME SOURCE	Bacteria, fungi, plants, protozoa, invertebrates, vertebrates
SUBUNIT STRUCTURE	Unknown or no data available

METHOD 1

Visualization Scheme

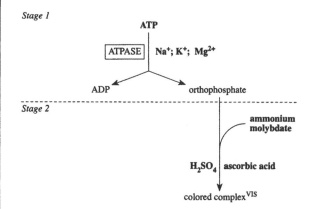

Staining Solution[1] (modified)

A. 0.1 M Tris–HCl buffer, pH 7.6
   0.1 M NaCl
   20 mM KCl
   6 mM $MgCl_2$
   1.5 mM ATP
B. 1.5% Agar solution (60°C)
C. 100 mg/ml Ascorbic acid
   2.5% Ammonium molybdate in 4 N H_2SO_4

Procedure

Mix equal volumes of A and B solutions and pour the mixture over the gel surface. Incubate the gel 1 to 2 h at 37°C. Remove the first agar overlay. Mix equal volumes of solutions B and C and pour over the gel surface. Dark blue bands visible on a light blue background appear after 1 to 5 min. Record or photograph the zymogram because after about 1 h the ATPASE bands disappear.

Notes: To obtain permanent ATPASE zymograms with colored bands stable for several months, the Malachite Green–phosphomolybdate method should be used (e.g., see 3.6.1.1 — PP, Method 1, *Notes*).

Orthophosphate should not be used as a component of any buffer used for electrophoresis or staining of ATPASE by methods based on orthophosphate detection.

METHOD 2

Visualization Scheme

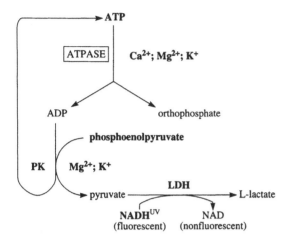

Staining Solution*

0.1 M Tris–HCl buffer, pH 7.6	5 ml
ATP	30 mg
Phosphoenolpyruvate (trisodium salt)	6 mg
NADH	6 mg
Pyruvate kinase (PK)	30 U
Lactate dehydrogenase (LDH)	50 U
CaCl$_2$	10 mg
MgCl$_2$	10 mg
KCl	20 mg

Procedure

Apply the staining solution to the gel surface on filter paper or dropwise and incubate at 37°C until dark (nonfluorescent) bands visible in long-wave UV light appear on a light (fluorescent) background. Record or photograph the zymogram using a yellow filter.

Notes: Areas of phosphoenolpyruvate phosphatase activity of alkaline phosphatase (see 3.1.3.1 — ALP, General Notes) and phosphoglycolate phosphatase (see 3.1.3.18 — PGP, Method 2) can also be detected by this method. These areas can be identified by incubating the control gel with staining solution lacking ATP and PK.

A similar staining procedure was used for detection of Mg^{2+}-dependent ATPASE in brain or eye extracts of poeciliid fish.[2]

OTHER METHODS

A. An immunoblotting procedure (for details, see Part II) based on the utility of monoclonal antibodies specific to the barley enzyme[3] can also be used for immuno-histochemical visualization of the enzyme protein on electrophoretic gels. Monoclonal antibodies specific to Na$^+$/K$^+$ (H$^+$/K$^+$) ATPASE (EC 3.6.1.37) from lamb, sheep, and rat kidneys, as well as from canine cardiac and pig gastric microsomes, are now available from Sigma.

B. The product orthophosphate can be detected enzymatically using two linking enzymes, purine-nucleoside phosphorylase and xanthine oxidase, coupled with the PMS–MTT system (e.g., see 3.6.1.1 — PP, Method 2).

C. Some other nonenzymatic methods of orthophosphate detection are also available (see 3.1.3.1 — ALP, Method 6; 3.1.3.2 — ACP, Method 5). The lead conversion technique was successfully used to detect alkaline muscle ATPASE in PAG, while the cobalt conversion technique proved inappropriate.[4]

GENERAL NOTES

Many enzymes previously listed under this number are now listed separately as EC 3.6.1.32–39. The remaining enzymes, not separately listed on the basis of some function coupled with hydrolysis of ATP, include ATPASEs dependent on Ca^{2+}, Mg^{2+}, anions, H$^+$, or DNA.[5] Some of these enzymes can also be detected by the methods described above.

Mitochondrial ATPASE can be prepared for electrophoresis by treating mitochondria with nonionic detergents such as Triton X-100, dodecyl maltoside, Nonidet P-40, Lubrol, octyl glucoside, or Hecameg. It is shown, however, that different ATPASE forms related to different complexes with contaminating proteins and lipids are generated as a result. All these forms are active in ATP hydrolysis as revealed by ATPASE staining directly on electrophoretic gels.[6]

REFERENCES

1. Brewer, G.J., *An Introduction to Isozyme Techniques*, Academic Press, New York, 1970, p. 129.

2. Morizot, D.C. and Schmidt, M.E., Starch gel electrophoresis and histochemical visualization of proteins, in *Electrophoretic and Isoelectric Focusing Techniques in Fisheries Management*, Whitmore, D.H., Ed., CRC Press, Boca Raton, FL, 1990, p. 23.

3. Chin, J.J., Monoclonal antibodies that immunoreact with a cation stimulated plant membrane ATPase, *Biochem. J.*, 203, 51, 1982.

4. Kirkeby, S. and Moe, D., Demonstration of activity for alkaline muscle ATPase in polyacrylamide gels using metal conversion methods, *Electrophoresis*, 4, 236, 1983.

5. NC-IUBMB, *Enzyme Nomenclature*, Academic Press, San Diego, 1992, p. 440 (EC 3.6.1.3, Comments).

6. GrandierVaseille, X. and Guerin, M., Separation by blue native and colorless native polyacrylamide gel electrophoresis of the oxidative phosphorylation complexes of yeast mitochondria solubilized by different detergents: specific staining of the different complexes, *Anal. Biochem.*, 242, 248, 1996.

* New; recommended for use.

3.6.1.7 — Acylphosphatase; AP

REACTION   Acylphosphate + H_2O = carboxylate + orthophosphate

ENZYME SOURCE   Vertebrates

SUBUNIT STRUCTURE   Unknown or no data available

METHOD

Visualization Scheme

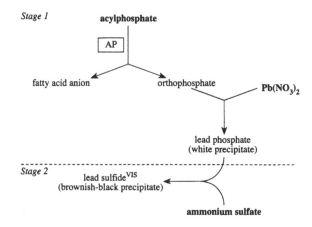

Staining Solution[1]

A. 0.1 *M* Sodium acetate buffer, pH 5.3   8 ml
    0.05 *M* Acetyl phosphate (dissolved in 0.1 *M*
       acetate buffer, pH 5.3)   5 ml
    H_2O (deionized)   8 ml
B. 2% $Pb(NO_3)_2$ 4 ml
C. 1% Ammonium sulfide

Procedure

Add solution B to solution A with stirring. After standing for about 5 min, centrifuge the mixture for 5 min at 1500 *g* to remove the white precipitate formed and transfer the supernatant to a transparent plastic tray. Rinse the electrophorized PAG with deionized water and place in the tray. Incubate the gel at room temperature with gentle rocking until white bands of lead phosphate are seen when the gel is viewed against a black background. Then wash the PAG for 30 to 40 min in repeatedly changed deionized water and immerse in solution C. Brownish black bands appear after about 10 min, which indicate areas of AP activity. Rinse the stained gel with deionized water and fix in 5% methanol–7.5% acetic acid.

Notes: Benzoyl phosphate and carbamoyl phosphate can also be used as substrates.

OTHER METHODS

A. Ammonium molybdate (e.g., see 3.1.3.1 — ALP, Method 5) or Malachite Green–ammonium molybdate (e.g., see 3.1.3.2 — ACP, Method 4) methods can also be used to detect the product orthophosphate.

B. Orthophosphate can also be detected using linking enzymes purine-nucleoside phosphorylase and xanthine oxidase coupled with the PMS–NBT system (e.g., see 3.1.3.1 — ALP, Method 8).

REFERENCES

1. Mizuno, Y., Ohba, Y., Fujita, H., Kanesaka, Y., Tamura, T., and Shiokawa, H., Activity staining of acylphosphatase after gel electrophoresis, *Anal. Biochem.*, 183, 46, 1989.

OTHER NAMES	Inosine triphosphatase, ITPase
REACTION	Nucleoside triphosphate + H_2O = nucleotide + pyrophosphate
ENZYME SOURCE	Vertebrates
SUBUNIT STRUCTURE	Dimer (vertebrates)

Method

Visualization Scheme

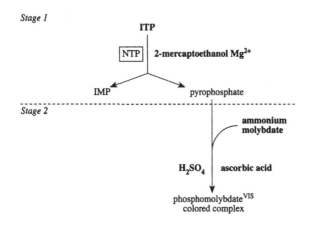

Staining Solution[1]

A.	0.2 M Tris–HCl buffer, pH 7.6	10 ml
	Inosine triphosphate (trisodium salt)	20 mg
	0.1 M MgCl$_2$	10 ml
	2-Mercaptoethanol	0.2 ml
B.	Ascorbic acid	1 g
	2.5% Ammonium molybdate in 4 N H$_2$SO$_4$	20 ml
C.	2% Agar solution (60°C)	20 ml

Procedure

Apply solution A to the gel surface on a filter paper overlay and incubate at 37°C for 1 to 2 h. Remove the filter paper overlay. Mix solutions B and C and pour the mixture over the gel surface. Blue bands of NTP activity appear after 1 to 5 min. Record or photograph the zymogram because the bands are ephemeral.

Notes: Buffer systems used for electrophoresis and staining NTP should be orthophosphate-free.

Other Methods

A. The calcium pyrophosphate method can also be used to detect NTP activity in transparent gels.[2] This method is based on the formation of white calcium pyrophosphate precipitate in gel areas where pyrophosphate-releasing enzymes are localized (e.g., see 2.7.7.9 — UGPP, Method 3).

B. Pyrophosphate can be detected by fluorogenic (or chromogenic) enzymatic methods that use at least three linked enzymatic reactions sequentially catalyzed by bacterial pyrophosphate-fructose-6-phosphate 1-phosphotransferase, aldolase, and glycerol-3-phosphate dehydrogenase (or glyceraldehyde-3-phosphate dehydrogenase). For details, see 2.7.1.90 — PFPPT.

C. The product IMP can be detected using an auxiliary enzyme, NAD-dependent IMP dehydrogenase, coupled with the PMS–MTT system (see 1.1.1.205 — IMPDH).

References

1. Harris, H. and Hopkinson, D.A., *Handbook of Enzyme Electrophoresis in Human Genetics*, North-Holland, Amsterdam, 1976 (loose-leaf, with supplements in 1977 and 1978).
2. Nimmo, H.G. and Nimmo, G.A., A general method for the localization of enzymes that produce phosphate, pyrophosphate, or CO$_2$ after polyacrylamide gel electrophoresis, *Anal. Biochem.*, 121, 17, 1982.

OTHER NAMES   Alkaline phosphodiesterase I
REACTION   NAD + H$_2$O = AMP + nicotinamide
                     mononucleotide
ENZYME SOURCE   Plants, invertebrates, vertebrates
SUBUNIT STRUCTURE   Unknown or no data available

METHOD 1

Visualization Scheme

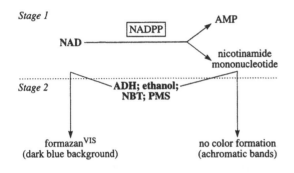

Staining Solution[1]

A. 50 mM Phosphate buffer, pH 6.5
    0.65 mM NAD
B. 0.1 M Pyrophosphate buffer, pH 8.8   50 ml
    Ethanol   0.3 ml
    NBT   10 mg
    PMS   1 mg
    Alcohol dehydrogenase (ADH)   25 U

Procedure

Incubate an electrophorized gel in solution A at 37°C for 1 to
2 h. Rinse the gel with water and incubate in solution B in the
dark at 37°C to detect NAD. Achromatic bands visible on a blue
background of the gel indicate areas occupied by NAD-degrading enzymes, including NADPP. Wash the negatively stained gel
with water and fix in 25% ethanol.

Notes: This method also detects activity bands of two other NAD-
catabolizing enzymes, the NAD nucleotidase (see 3.2.2.5 — NADN)
and the NAD phosphodiesterase (see 3.1.4.X′ — NADPDE). Methods
specific for NADPP are outlined below (see Other Methods).

METHOD 2

Visualization Scheme

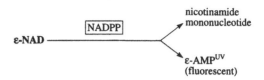

Staining Solution[2]

50 mM Tris–HCl buffer, pH 7.5
0.03% Lauryl dimethylamine N-oxide (LDAO)
150 μM ε-NAD (see *Notes*)

Procedure

The method is developed for an SDS-PAG. So, after separation
of proteins by SDS-PAG electrophoresis, wash the gel for 30
min in a solution containing 50 mM Tris–HCl buffer (pH 7.5)
and 0.5% LDAO.

To visualize NAD-converting activities, incubate the gel for
15 min in the staining solution and place on a UV transilluminator to observe fluorescent bands. Photograph the developed
gel using a 550-nm interference filter.

Notes: ε-NAD is a NAD derivative (1,N^6 etheno-NAD) containing a
modified adenine ring whose fluorescence is greatly enhanced after its
separation from the quenching nicotinamide moiety.

This method also detects activity bands of other NAD-catabolizing
enzymes, the NAD nucleotidase (see 3.2.2.5 — NADN) and the NAD
phosphodiesterase (see 3.1.4.X′ — NADPDE). After SDS-PAG electrophoresis of rat kidney and heart preparations, the fluorescent band of
NADPP (105,000 Da) is situated between fluorescent bands of NADN
(41,000 Da) and NADPDE (210,000 Da).[2] Because of a very large
difference between molecular masses of the three NAD-degrading
enzymes, the same order of their activity bands in the same tissues may
be expected for all mammals.

OTHER METHODS

A. The product 5′-AMP can be detected using auxiliary
    enzymes alkaline phosphatase, purine-nucleoside phos-
    phorylase, and xanthine oxidase (see 3.1.3.1 — ALP,
    Method 8). This method will result in positively stained
    NADPP bands visible in daylight.
B. The product 5′-AMP can also be detected using auxil-
    iary enzymes adenylate kinase, pyruvate kinase, and lac-
    tate dehydrogenase (see 3.1.4.47 — CNPE, Method 1).
    This method will result in negatively stained NADPP
    bands visible in UV light.

3.6.1.22 — NAD Pyrophosphatase; NADPP (continued)

GENERAL NOTES

Only methods outlined in the section Other Methods may be considered as specific for NADPP. These methods are recommended for a trial.

The enzyme also acts on NADP, 3-acetylpyridine, and the thionicotinamide analogues of NAD and NADP.[3]

REFERENCES

1. Ravazzolo, R., Bruzzone, G., Garrè, C., and Ajmar, F., Electrophoretic demonstration and initial characterization of human red cell NAD(P)+ase, *Biochem. Genet.*, 14, 877, 1976.
2. Hagen, T.H. and Ziegler, M., Detection and identification of NAD-catabolizing activities in rat tissue homogenates, *Biochim. Biophys. Acta*, 1340, 7, 1997.
3. NC-IUBMB, *Enzyme Nomenclature*, Academic Press, San Diego, 1992, p. 443 (EC 3.6.1.22, Comments).

3.6.1.X — Nucleoside-Diphosphosugar Pyrophosphatase; NDP

REACTION	Nucleoside diphosphosugar + H_2O = nucleotide + sugar 1-phosphate
ENZYME SOURCE	Vertebrates
SUBUNIT STRUCTURE	Unknown or no data available

METHOD

Visualization Scheme

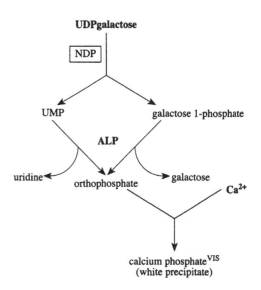

Staining Solution[1]

A. 0.15 M Tris–HCl buffer, pH 9.0
   5 mM CaCl$_2$
   0.1% Triton X-100
   3 U/ml Alkaline phosphatase (ALP; Boehringer, type I)
B. 60 mM UDPgalactose

Procedure

Apply solution A to the surface of an electrophorized PAG dropwise and allow the solution to enter the gel for 15 min. Then apply solution B and incubate the gel in a humid chamber at 37°C until opaque bands visible against a dark background appear. When an activity stain of sufficient intensity is obtained, store the gel in 50 mM glycine–KOH buffer (pH 10), containing 5 mM Ca^{2+}, at 5°C or at room temperature in the presence of an antibacterial agent. The stained gel can be photographed by reflected light against a dark background.

Notes: The precipitated calcium phosphate can be subsequently stained with Alizarin Red S. This does not increase the sensitivity of the staining method, although it is of advantage for more opaque gel systems, such as starch.[2]

UDP–N-acetylglucosamine and UDPglucuronate can also be used as substrates. Several other methods may be used to detect the product orthophosphate (e.g., see 3.1.3.1 — ALP).

3.6.1.X — Nucleoside-Diphosphosugar Pyrophosphatase; NDP (continued)

REFERENCES

1. Van Dijk, W., Lasthuis, A.-M., Koppen, P.L., and Muilerman, H.G., A universal and rapid spectrophotometric assay of CMP–sialic acid hydrolase and nucleoside-diphosphosugar pyrophosphatase activities and detection in polyacrylamide gels, *Anal. Biochem.*, 117, 346, 1981.
2. Nimmo, H.G. and Nimmo, G.A., A general method for the localization of enzymes that produce phosphate, pyrophosphate, or CO₂ after polyacrylamide gel electrophoresis, *Anal. Biochem.*, 121, 17, 1982.

3.11.1.X — Phosphonoacetate Hydrolase; PAH

REACTION	Phosphonoacetate + H₂O = acetate + orthophosphate
ENZYME SOURCE	Bacteria
SUBUNIT STRUCTURE	Dimer# (bacteria)

METHOD

Visualization Scheme

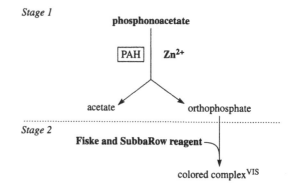

Staining Solution[1]

A. 50 mM Tris–HCl, pH 7.5   19.2 ml
    500 mM Phosphonoacetate   0.4 ml
    50 mM ZnSO₄·7H₂O   0.4 ml
B. Fiske and SubbaRow color reagent
    (formulation of 1-amino-2-naphthol-4-sulfonic acid, sodium sulfite, and bisulfite; prepared according to Sigma Technical Bulletin 670)

Procedure

Incubate a PAG in solution A for 60 min at 30°C. Following incubation, transfer the gel into 20 ml of premixed solution B. Dark blue bands against a clear background of the gel appear at the sites of PAH activity. The bands are ephemeral; thus the zymogram should be recorded, scanned, or photographed immediately.

OTHER METHODS

Several other methods may be used to detect the product orthophosphate (see 3.1.3.1 — ALP, Methods 6 to 8; 3.1.3.2 — ACP, Methods 4 and 5). All these methods may be used only when electrophoresis is carried out in a phosphate-free buffer system.

GENERAL NOTES

The mechanism of C–P bond cleavage by PAH is different from that of phosphonoacetaldehyde hydrolase (EC 3.11.1.1). The enzyme from *Pseudomonas fluorescens* is a dimer specific to phosphonoacetate; it displays twofold activation by 1 mM Zn²⁺ and complete inhibition by 5 mM EDTA.[2]

3.11.1.X — Phosphonoacetate Hydrolase; PAH (continued)

REFERENCES

1. McGrath, J.W. and Quinn, J.P., A plate assay for the detection of organophosphonate mineralization by environmental bacteria, and its modification as an activity stain for identification of the carbon–phosphorus bond cleavage enzyme phosphonoacetate hydrolase, *Biotechnol. Tech.*, 9, 497, 1995.
2. McGrath, J.W., Wisdom, G.B., McMullan, G., Larkin, M.J., and Quinn, J.P., The purification and properties of phosphonoacetate hydrolase, a novel carbon–phosphorus bond cleavage enzyme from *Pseudomonas fluorescens* 23F, *Eur. J. Biochem.*, 234, 225, 1995.

4.1.1.1 — Pyruvate Decarboxylase; PDC

OTHER NAMES	α-Carboxylase, pyruvic decarboxylase, α-ketoacid carboxylase
REACTION	2-Oxo acid = aldehyde + CO_2
ENZYME SOURCE	Bacteria, fungi, plants, vertebrates
SUBUNIT STRUCTURE	Heterodimer[#] (fungi – brewer's yeast, plants – *Pisum sativum*); see General Notes

METHOD 1

Visualization Scheme

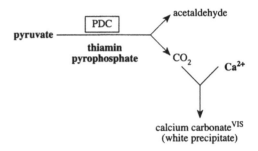

Staining Solution[1]

100 mM Tris–HCl buffer, pH 8.5
5 mM Sodium pyruvate
67 mM Thiamin pyrophosphate
2 mM Ca^{2+}

Procedure

Soak an electrophorized PAG in 100 mM Tris–HCl buffer (pH 8.5) at 37°C for 20 to 30 min and transfer to staining solution. Incubate the gel at room temperature until opaque bands visible against a dark background appear. Store the stained gel in 50 mM glycine–KOH (pH 10.0) at 5°C or at room temperature in the presence of an antibacterial agent.

Notes: It was supposed that this method is not adequate for PDC detection because the enzyme has almost no activity at pH values above 7.0, loses its cofactor (thiamin pyrophosphate) rapidly at pH values above 8.0, and displays a tendency to denature at room temperature under alkaline conditions.[2] It was also pointed out that at an acid pH (6.0 to 6.2) optimal for PDC activity, the formation of calcium carbonate is questionable.

METHOD 2

Visualization Scheme

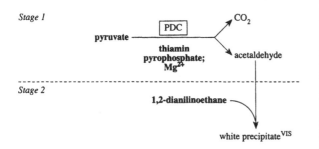

Stage 1

pyruvate ——— [PDC] ———→ CO_2
thiamin pyrophosphate; Mg^{2+} ——→ acetaldehyde

Stage 2

1,2-dianilinoethane ——→ white precipitate[VIS]

Staining Solution[2]

A. 0.3 *M* Sodium citrate buffer, pH 6.0
   30 m*M* Sodium pyruvate
   5 m*M* $MgSO_4$
   5 m*M* Thiamin pyrophosphate

B. 100 mg of 1,2-Dianilinoethane dissolved in 10 ml of glacial acetic acid and total volume adjusted to 35 ml with water

Procedure

Incubate an electrophorized PAG in 200 ml of solution A at room temperature for 50 min and then add 3 ml of solution B. Opaque bands clearly visible against a dark background appear after 5 to 15 min. Wash the stained gel with water and photograph.

Notes: The method is very sensitive and allows detection of about 10 mU of PDC activity per band.

A high concentration of staining buffer is used to overcome the alkaline pH of the electrophoretic gel.

OTHER METHODS

NAD-dependent aldehyde dehydrogenase coupled with the PMS–NBT system can also be used to detect the product acetaldehyde.[2] This method is of advantage when PDC electrophoresis is carried out in opaque gels, e.g., acetate cellulose or starch. However, the difference of pH optima for PDC (6.0) and aldehyde dehydrogenase (8.0) does not allow staining of PDC activity bands using a one-step procedure. So, zymograms obtained by this method display more diffuse PDC bands than do zymograms obtained by Method 2.

GENERAL NOTES

The yeast enzyme is a heterotetramer consisting of 2α and 2β subunits. The active sites of yeast PDC are comprised of amino acid residues from the both subunits.[3] A similar subunit structure is characteristic of the enzyme from pea seeds.[4] However, in the haploid yeast strain YSH 4.127–1A expressing only one of the three structural genes for PDC, only homotetrameric PDC molecules are formed.[5]

REFERENCES

1. Nimmo, H.G. and Nimmo, G.A., A general method for the localization of enzymes that produce phosphate, pyrophosphate, or CO_2 after polyacrylamide gel electrophoresis, *Anal. Biochem.*, 121, 17, 1982.
2. Zehender, H., Trescher, D., and Ullrich, J., Activity stain for pyruvate decarboxylase in polyacrylamide gels, *Anal. Biochem.*, 135, 16, 1983.
3. Sergienko, E.A. and Jordan, F., Yeast pyruvate decarboxylase tetramers can dissociate into dimers along two interfaces: hybrids of low-activity D28A (or D28N) and E477Q variants, with substitution of adjacent active center acidic groups from different subunits, display restored activity, *Biochemistry*, 41, 6164, 2002.
4. Mucke, U., Wohlfarth, T., Fiedler, U., Baumlein, H., Rucknagel, K.P., and Konig, S., Pyruvate decarboxylase from *Pisum sativum* properties, nucleotide and amino acid sequences, *Eur. J. Biochem.*, 237, 373, 1996.
5. Killenberg-Jabs, M., Konig, S., Hohmann, S., and Hubner, G., Purification and characterisation of the pyruvate decarboxylase from a haploid strain of *Saccharomyces cerevisiae*, *Biol. Chem. Hoppe-Seyler*, 377, 313, 1996.

OTHER NAMES	L-Glutamic acid decarboxylase, L-glutamic decarboxylase
REACTION	L-Glutamate = 4-aminobutanoate + CO_2
ENZYME SOURCE	Bacteria, plants, invertebrates, vertebrates
SUBUNIT STRUCTURE	Dimer# (bacteria – *Lactobacillus brevis*), hexamer# (bacteria – *E. coli*)

METHOD

Visualization Scheme

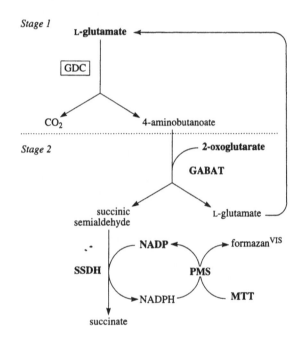

Stage 1

L-glutamate

GDC

CO_2 4-aminobutanoate

Stage 2

2-oxoglutarate

GABAT

succinic semialdehyde   L-glutamate

NADP → formazan^VIS

SSDH   **PMS**

NADPH   **MTT**

succinate

Staining Solution[1]

A. 0.2 *M* Citrate buffer, pH 4.7   40 ml
    20 m*M* L-Glutamic acid in 0.2 *M* citrate buffer, pH 4.7   5 ml

B. 1.0 *M* Tris–HCl buffer, pH 8.0   40 ml
    NADP   10 mg
    PMS   1.5 mg
    NBT   25 mg
    0.7% 2-Mercaptoethanol in 1.0 *M* Tris–HCl buffer, pH 8.0   0.5 ml
    0.02 *M* 2-Oxoglutaric acid in 1.0 *M* Tris–HCl buffer, pH 8.0   1 ml
    GABASE: a mixture of 4-aminobutyrate aminotransferase (GABAT; EC 2.6.1.19) and succinate-semialdehyde dehydrogenase (SSDH; EC 1.2.1.16)   1 ml

Procedure

Incubate the gel in solution A at 37°C for 30 min and then place in solution B. Incubate the gel in solution B in the dark at 37°C until dark blue bands appear. Wash the stained gel in water and fix in 25% ethanol.

Notes: Both enzymes, GDC and auxiliary GABAT, require pyridoxal 5′-phosphate as a cosubstrate. However, usually there is no need to add the cosubstrate in the staining solution because both enzymes contain sufficient quantities of pyridoxal 5′-phosphate bound to the enzyme molecules.

Buffers containing Cl^- should not be used for electrophoresis because Cl^- anions competitively inhibit GDC activity.

The brain enzyme also acts on L-cysteate, 3-sulfino-L-alanine, and L-aspartate.[2]

Several methods suitable for detection of carbonate ions also may be applied to visualize GDC activity bands (for details, see Part II). However, all these methods work well only at alkaline pH values, which are not compatible with the acidic pH values optimal for GDC activity. Thus, the use of these methods requires that GDC visualization be carried out in a two-step procedure.

OTHER METHODS

An immunoblotting procedure (for details, see Part II) may also be used to detect GDC. Monoclonal and polyclonal antibodies specific to the human enzyme are available from Sigma.

REFERENCES

1. Akers, E. and Aronson, J.N., Detection on polyacrylamide gels of L-glutamic acid decarboxylase activities from *Bacillus thuringiensis, Anal. Biochem.*, 39, 535, 1971.
2. NC-IUBMB, *Enzyme Nomenclature*, Academic Press, San Diego, 1992, p. 453 (EC 4.1.1.15, Comments).

4.1.1.20 — Diaminopimelate Decarboxylase; DAPD

REACTION   *meso*-2,6-Diaminoheptanedioate = L-lysine + CO_2

ENZYME SOURCE   Bacteria, plants

SUBUNIT STRUCTURE   Dimer[#] (bacteria)

METHOD

Visualization Scheme

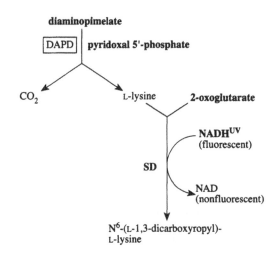

Staining Solution[1] (adapted)

100 μM Potassium phosphate buffer, pH 6.8

1 μM EDTA

1 μM 2,3-Dimercaptopropanol

0.1 μM Pyridoxal 5'-phosphate

0.3 μM NADH

4 μM *Meso*-2,6-Diaminopimelate

2.5 μM 2-Oxoglutarate

0.1 U/ml Saccharopine dehydrogenase (NAD, L-lysine forming) (SD)

Procedure

Apply the staining solution to the gel surface on a filter paper overlay and incubate the gel at 37°C in a moist chamber. After 20 to 30 min of incubation, monitor the gel under long-wave UV light. Dark (nonfluorescent) bands of DAPD are visible on a light (fluorescent) background. Record the zymogram or photograph using a yellow filter.

Notes: When a zymogram that is visible in daylight is required, counterstain the processed gel with MTT–PMS solution to develop white bands of DAPD on a blue background. Fix the negatively stained gel in 25% ethanol.

GENERAL NOTES

The enzyme is a pyridoxal-phosphate protein.[2]

REFERENCES

1. Laber, B. and Amrhein, N., A spectrophotometric assay for *meso*-diaminopimelate decarboxylase and L-α-amino-ε-caprolactam hydrolase, *Anal. Biochem.*, 181, 297, 1989.
2. NC-IUBMB, *Enzyme Nomenclature*, Academic Press, San Diego, 1992, p. 453 (EC 4.1.1.20, Comments).

OTHER NAMES	Dopa decarboxylase, tryptophan decarboxylase, hydroxytryptophan decarboxylase
REACTION	L-Tryptophan = tryptamine + CO_2 (also see General Notes)
ENZYME SOURCE	Bacteria, fungi, invertebrates, vertebrates
SUBUNIT STRUCTURE	Monomer[a] (fungi)

METHOD

Visualization Scheme

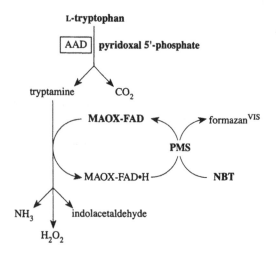

Staining Solution[1]

0.05 M Phosphate buffer, pH 7.6

0.1 mg/ml Pyridoxal 5'-phosphate

0.4 mg/ml NBT

0.04 mg/ml PMS

0.4 mg/ml L-Tryptophan (or L-tyrosin, or dihydroxyphenylalanine)

Monoamine oxidase (MAOX; EC 1.4.3.4; the enzyme concentration is determined experimentally)

Procedure

Incubate the gel in staining solution in the dark at 37°C until dark blue bands appear. Wash the stained gel in water and fix in 25% ethanol.

OTHER METHODS

A. The calcium carbonate precipitation method[2] may be used to detect the product carbon dioxide (for example, see 4.1.1.1 — PDC, Method 1). Phosphate-containing buffers must not be used in this case because calcium ions also interact with phosphate, resulting in the formation of calcium phosphate precipitate all over the gel.

B. The product carbon dioxide can also be detected using a linked reaction catalyzed by phosphoenolpyruvate carboxylase coupled with a diazonium salt (see 4.1.1.31 — PEPC, Method), or using two linked reactions sequentially catalyzed by phosphoenolpyruvate carboxylase and malate dehydrogenase (see 4.1.1.31 — PEPC, Other Methods, B).

C. An immunoblotting procedure (for details, see Part II) may also be used to detect AAD. Antibodies specific to the human enzyme are available from Sigma.

GENERAL NOTES

The enzyme is a pyridoxal-phosphate protein that also acts on 5-hydroxy-L-tryptophan and dihydroxy-L-phenylalanine.[3] The fungal enzyme has a broad substrate specificity utilizing L-tryptophan, L-tyrosine, and L-phenylalanine, as well as o-fluorophenylalanine and p-fluorophenylalanine.[4]

REFERENCES

1. Antonas, K.N., Coulson, W.F., and Jepson, J.B., A monoamine oxidase–tetrazolium reaction for locating aromatic amino acid decarboxylases in electrophoretic media, *Biochem. J.*, 121, 38, 1971.

2. Nimmo, H.G. and Nimmo, G.A., A general method for the localization of enzymes that produce phosphate, pyrophosphate, or CO_2 after polyacrylamide gel electrophoresis, *Anal. Biochem.*, 121, 17, 1982.

3. NC-IUBMB, *Enzyme Nomenclature*, Academic Press, San Diego, 1992, p. 454 (EC 4.1.1.28, Comments).

4. Niedens, B.R., Parker, S.R., Stierle, D.B., and Stierle, A.A., First fungal aromatic L-amino acid decarboxylase from a paclitaxel-producing *Penicillium raistrickii*, *Micologia*, 91, 619, 1999.

4.1.1.31 — Phosphoenolpyruvate Carboxylase; PEPC

REACTION	Orthophosphate + oxaloacetate = H_2O + phosphoenolpyruvate + CO_2
ENZYME SOURCE	Bacteria, fungi, plants, vertebrates
SUBUNIT STRUCTURE	Tetramer[#] (bacteria), dimer (plants)

METHOD

Visualization Scheme

Stage 1

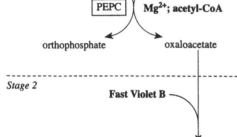

Stage 2

Fast Violet B

colored dye[VIS]

Staining Solution[1]

A. 100 mM Tris–HCl buffer, pH 8.5
   3 mM Phosphoenolpyruvate
   5 mM $MgCl_2$
   20 mM $KHCO_3$
   0.5 mM Acetyl-CoA (or propionyl-CoA)
B. 1 mg/ml Fast Violet B salt

Procedure

Incubate the gel in solution A at room temperature for 30 min, wash in distilled water, and place in solution B. Incubate the gel in the dark until colored bands appear. Wash the stained gel in water and fix in 50% glycerol or 5% acetic acid.

Notes: The plant enzyme is not activated by acetyl-CoA (or propionyl-CoA). Thus, when the enzyme from a plant source is detected, this expensive ingredient may be excluded from the staining solution.[2]

Fast Blue B or BB salts can also be used in place of Fast Violet B salt.

The two-step postcoupling procedure (as shown on the scheme) is recommended when a diazonium salt has an inhibitory effect on PEPC activity. If there is no obvious inhibition, a one-step detection procedure is preferable because it reduces diffusion of the product oxaloacetate and results in the development of sharper and more distinct bands.

OTHER METHODS

A. A general method of detection of orthophosphate[3] may also be used to detect PEPC activity. This method involves two linking enzymes, purine-nucleoside phosphorylase and xanthine oxidase, coupled with the PMS–MTT system (e.g., see 3.6.1.1 — PP, Method 2). It should be kept in mind that gel areas occupied by phosphoenolpyruvate phosphatase (see 3.1.3.18 — PGP, Method 2, *Notes*) can also be developed by this method.

B. The product oxaloacetate can be detected using the auxiliary enzyme malate dehydrogenase. This method results in the appearance of dark (nonfluorescent) bands of PEPC visible on a light (fluorescent) background in long-wave UV light (e.g., see 2.6.1.1 — GOT, Method 2).

REFERENCES

1. Scrutton, M.C. and Fatebene, F., An assay system for localization of pyruvate and phosphoenolpyruvate carboxylase activity on polyacrylamide gels and its application to detection of these enzymes in tissue and cell extracts, *Anal. Biochem.*, 69, 247, 1975.
2. Brown, A.H.D., Nevo, E., Zohary, D., and Dagan, O., Genetic variation in natural populations of wild barley (*Hordeum spontaneum*), *Genetica*, 49, 97, 1978.
3. Klebe, R.J., Schloss, S., Mock, L., and Link, C.R., Visualization of isozymes which generate inorganic phosphate, *Biochem. Genet.*, 19, 921, 1981.

OTHER NAMES — Fructose-bisphosphate aldolase (recommended name), fructose-1,6-bisphosphate triose-phosphate-lyase

REACTION — D-Fructose-1,6-bisphosphate = glycerone phosphate + D-glyceraldehyde-3-phosphate

ENZYME SOURCE — Bacteria, fungi, plants, protozoa, invertebrates, vertebrates

SUBUNIT STRUCTURE — Tetramer (plants, vertebrates)

METHOD 1

Visualization Scheme

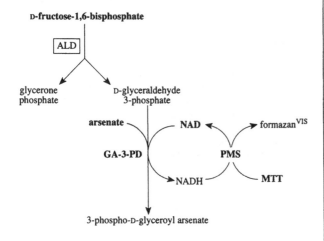

Staining Solution[1]

A. 0.1 M Tris–HCl buffer, pH 8.0 — 25 ml
Fructose-1,6-bisphosphate — 100 mg
NAD — 20 mg
Sodium arsenate — 60 mg
Glyceraldehyde-3-phosphate dehydrogenase (GA-3-PD; 800 U/ml) — 50 μl
MTT — 7.5 mg
PMS — 2.5 mg
B. 2% Agar solution (60°C) — 25 ml

Procedure

Mix A and B components of the staining solution and pour the mixture over the gel surface. Incubate the gel in the dark at 37°C until dark blue bands appear. Fix the stained gel in 25% ethanol.

Notes: The linking enzyme triose-phosphate isomerase may be added to the staining solution to convert glycerone phosphate into D-glyceraldehyde-3-phosphate, and thus to enhance the staining of low-activity ALD isozymes.[2]

PMS and MTT may be omitted from the staining solution and fluorescent bands of ALD activity observed in long-wave UV light.[3]

METHOD 2

Visualization Scheme

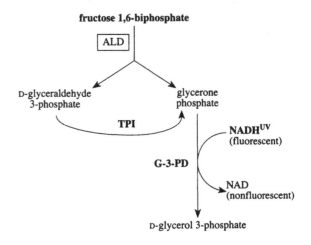

Staining Solution[4]

0.1 M Tris–HCl buffer, pH 7.8
1.1 mM NADH
2.5 mM Fructose-1,6-bisphosphate
1.5 U/ml Glycerol-3-phosphate dehydrogenase (G-3-PD)
111 U/ml Triose-phosphate isomerase (TPI)

Procedure

Cover the gel surface with filter paper and flood with the staining solution. Incubate the covered gel at 37°C and monitor under long-wave UV light. Dark (nonfluorescent) bands on a light (fluorescent) background indicate areas of ALD activity localization. Record the zymogram or photograph using a yellow filter.

Notes: This method is the most sensitive among the ALD activity detection methods so far described.

The linking enzyme triose-phosphate isomerase can be omitted where ALD activity is high.

To develop ALD bands visible in daylight, treat the processed gel with the PMS–MTT mixture. White bands visible on a blue background appear almost immediately.

METHOD 3

Visualization Scheme

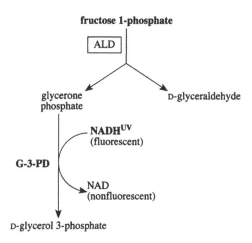

fructose 1-phosphate

ALD

glycerone phosphate   D-glyceraldehyde

NADH^UV (fluorescent)

G-3-PD

NAD (nonfluorescent)

D-glycerol 3-phosphate

Staining Solution[1]

0.1 *M* Tris–HCl buffer, pH 8.0	25 ml
Fructose 1-phosphate (disodium salt)	25 mg
NADH (disodium salt)	10 mg
Glycerol-3-phosphate dehydrogenase (G-3-PD; 80 U/ml)	50 μl

Procedure

Apply the staining solution to the gel surface on filter paper. Incubate the gel overlay in a humid chamber at 37°C and view under long-wave UV light. Dark (nonfluorescent) bands visible on a light (fluorescent) background indicate areas of ALD activity. Record the zymogram or photograph using a yellow filter.

Notes: Some ALD isozymes (e.g., human liver ALD-B) catalyze the breakdown of fructose 1-phosphate to glyceraldehyde and glycerone phosphate. The method described above was developed to detect just such ALD isozymes.

OTHER METHODS

The silver method may be used to detect orthophosphate liberated from the products glycerone phosphate and glyceraldehyde-3-phosphate after treatment with iodacetamide.[5] This method is not widely used because the intermediary orthophosphate is readily diffusable.

GENERAL NOTES

The bacterial and yeast enzymes are zinc-containing proteins.[6]

REFERENCES

1. Harris, H. and Hopkinson, D.A., *Handbook of Enzyme Electrophoresis in Human Genetics*, North-Holland, Amsterdam, 1976 (loose-leaf, with supplements in 1977 and 1978).
2. Richardson, B.J., Baverstock, P.R., and Adams, M., *Allozyme Electrophoresis: A Handbook for Animal Systematics and Population Studies*, Academic Press, Sydney, 1986, p. 167.
3. Susor, W.A. and Rutter, W.J., Method for the detection of pyruvate kinase, aldolase, and other pyridine nucleotide linked enzyme activities after electrophoresis, *Anal. Biochem.*, 43, 147, 1971.
4. Anderson, J.E. and Giblett, E.R., Intraspecific red cell enzyme variation in the pigtailed macaque (*Macaca nemestrina*), *Biochem. Genet.*, 13, 189, 1975.
5. Pietruszko, R. and Baron, D.N., A staining procedure for demonstration of multiple forms of aldolase, *Biochim. Biophys. Acta*, 132, 203, 1967.
6. NC-IUBMB, *Enzyme Nomenclature*, Academic Press, San Diego, 1992, p. 460 (EC 4.1.2.13, Comments).

OTHER NAMES   2-Dehydro-3-deoxyphosphohepto-
                   nate aldolase (recommended name),
                   phospho-2-keto-3-deoxyheptonate
                   aldolase, DHAP synthase, KDPH
                   synthetase

REACTION   2-Dehydro-3-deoxy-D-*arabino*-hep-
                   tonate 7-phosphate + orthophos-
                   phate = phosphoenolpyruvate +
                   D-erythrose 4-phosphate + H_2O

ENZYME SOURCE   Bacteria, fungi

SUBUNIT STRUCTURE Unknown or no data available

METHOD

Visualization Scheme

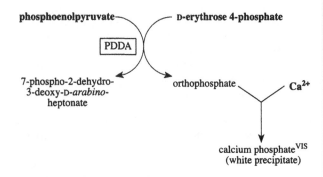

Staining Solution[1]

50 mM Glycine–KOH buffer, pH 10.0
0.1 mM Phosphoenolpyruvate
0.1 mM Erythrose 4-phosphate
10 mM Ca^{2+}

Procedure

Incubate an electrophorized PAG in the staining solution at 37°C until opaque bands visible against a dark background appear. Store the stained gel in 50 mM glycine–KOH (pH 10.0), containing 5 mM Ca^{2+}, at 5°C or at room temperature in the presence of an antibacterial agent (e.g., sodium azide).

Notes: The areas of calcium phosphate precipitation may be counterstained with Alizarin Red S. This does not increase the sensitivity of the method, but would be of advantage for opaque gels such as acetate cellulose or starch.

OTHER METHODS

The enzymatic method of detection of the product orthophosphate can also be adapted to visualize PDDA activity in electrophoretic gels. This method[2] is based on reactions catalyzed by two linking enzymes, purine-nucleoside phosphorylase and xanthine oxidase, coupled with the PMS–MTT system (e.g., see 3.6.1.1 — PP, Method 2).

REFERENCES

1. Nimmo, H.G. and Nimmo, G.A., A general method for the localization of enzymes that produce phosphate, pyrophosphate, or CO_2 after polyacrylamide gel electrophoresis, *Anal. Biochem.*, 121, 17, 1982.

2. Klebe, R.J., Schloss, S., Mock, L., and Link, C.R., Visualization of isozymes which generate inorganic phosphate, *Biochem. Genet.*, 19, 921, 1981.

4.1.2.32 — Trimethylamine-Oxide Aldolase; TMAOA

OTHER NAMES	Trimethylamine oxide demethylase
REACTION	$(CH_3)_3NO = (CH_3)_2NH$ + formaldehyde
ENZYME SOURCE	Vertebrates
SUBUNIT STRUCTURE	Unknown or no data available

METHOD

Visualization Scheme

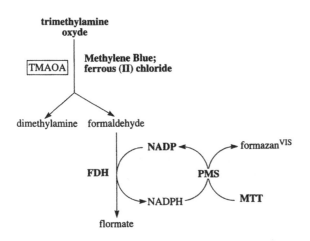

Staining Solution[1]

50 mM Sodium phosphate buffer, pH 6.0
100 mM Trimethylamine oxide
10 mM NAD
0.1 mM Methylene Blue
0.1 mM Ferrous (II) chloride tetrahydrate
4 U/ml Formaldehyde dehydrogenase (see 1.2.1.46 — FDH)
2.5 mg/ml MTT or NBT (see *Notes*)
1.5 mg/ml PMS

Procedure

After electrophoresis, wash the gel in 50 mM sodium phosphate buffer (pH 6.0) for 5 min and in water for 1 min. Incubate the gel in staining solution in the dark at 37°C for 15 to 60 min (depending on the enzyme activity) until colored bands appear. Wash the stained gel in 50% ethanol for 5 to 7 min for fixing and destaining the gel background.

Notes: The results obtained using NBT are not as good as those obtained with MTT. NBT is a ditetrazolium salt, which means it needs four electrons for complete reduction and conversion into colored formazan, and dehydrogenases (like FDH) transfer only two electrons. Thus, two substrate molecules should be dehydrogenated to reduce one molecule of NBT. Therefore, MTT is more sensitive and its use is preferable, especially when the activity of the enzyme is low.

The pH (6.0) of the staining solution is a compromise of the pH optima of TMAOA (pH 4.5 to 5.0) and of the auxiliary FDH (pH 7.8).

REFERENCES

1. Havemeister, W., Rehbein, H., Steinhart, H., Gonzales-Sotelo, C., Krogsgaard-Nielsen, M., and Jørgensen, B., Visualization of the enzyme trimethylamine oxide demethylase in isoelectric focusing gels by an enzyme-specific staining method, *Electrophoresis*, 20, 1934, 1999.

4.1.3.1 — Isocitrate Lyase; IL

OTHER NAMES	Isocitrase, isocitritase, isocitratase
REACTION	D-Isocitrate = succinate + glyoxylate
ENZYME SOURCE	Bacteria, fungi, plants, protozoa
SUBUNIT STRUCTURE	Tetramer# (bacteria, fungi, plants), trimer# (protozoa – *Euglena*)

METHOD 1

Visualization Scheme

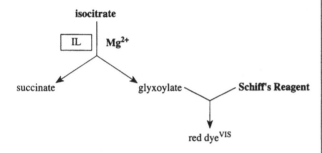

Staining Solution[1]

A. 50 mM Potassium phosphate buffer, pH 7.5
   1 mM EDTA
   3 mM MgCl$_2$
   30 mM Dithiothreitol
   10 mM D,L-Isocitrate

B. Schiff's reagent (see *Notes*)

Procedure

Incubate an electrophorized PAG in a mixture of 50 ml of A and 3 ml of B at 37°C. Dark red bands appear after 20 to 30 min of incubation. Record or photograph the zymogram.

Notes: Prepare modified Schiff's reagent as follows: Add 500 ml of warm distilled water to 0.25 g of fuchsin (basic); mix thoroughly and filtrate. Add to the filtered solution 5 ml of 1 N HCl and then 0.125 ml of concentrated HCl. Finally, add to the mixture 0.5 g of sodium metabisulfite. Store solution in a dark glass bottle at 4°C.

This method requires control staining of an additional gel in a mixture lacking isocitrate and MgCl$_2$ to be sure that developed bands are really caused by IL activity.

METHOD 2

Visualization Scheme

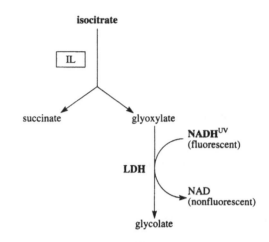

Staining Solution[2] (modified)

100 mM Phosphate buffer, pH 7.5
4 mg/ml D,L-Isocitric acid
80 U/ml Lactate dehydrogenase (LDH)
2 mg/ml NADH

Procedure

Apply staining solution to the gel surface as a filter paper or 1% agar overlay. Incubate the gel at 37°C and monitor under long-wave UV light. Dark (nonfluorescent) bands on a light (fluorescent) background indicate the sites of localization of IL activity in the gel. Record the zymogram or photograph using a yellow filter.

Notes: The described method is based on the ability of auxiliary LDH to metabolize the glyoxylate to glycolate in the presence of NADH. Originally, this method was used coupled with the MTT–dichlorophenol indophenol (DCIP) system to develop positively stained IL activity bands due to the reduction of MTT and the formation of dark blue formazan.[2] The MTT–DCIP system was used for the first time in 1967 by Kaplan and Beutler to obtain positively stained activity bands of NADH-dependent diaphorase.[3] Thus, the reduction of MTT via DCIP can only occur in the presence of diaphorase, and it is unclear how this system can work in the original method.[2] Because of this, the use of the modified method (lacking the MTT–DCIP system) is recommended.

4.1.3.1 — Isocitrate Lyase; IL (continued)

OTHER METHODS

A. The product glyoxylate can be detected by a negative fluorescent method using NADH-dependent glyoxylate reductase (EC 1.1.1.26) as a linking enzyme. When using this method, it should be taken into account that commercially available glyoxylate reductase from spinach leaves displays optimal activity at pH 6.4, while the pH optimum of IL is above 7.0. Thus, the detection of IL by this method should be carried out in a two-step procedure.

B. Succinate and glyoxylate can be used as substrates for the backward reaction of IL to produce isocitrate, which can be further detected using auxiliary NADP-dependent isocitrate dehydrogenase coupled with the PMS–NBT system.[4] The use of this method, however, is limited because IL from some sources (e.g., plants) works only in the forward direction.

GENERAL NOTES

The fungal enzyme requires Mg^{2+} and sulfhydryl compounds for optimal activity.

REFERENCES

1. Reeves, H.C. and Volk, M.J., Determination of isocitrate lyase activity in polyacrylamide gels, *Anal. Biochem.*, 48, 437, 1972.
2. Hou, W.-C., Chen, H.-J., Lin, Y.-H., Chen, Y.-C., Yang, L.-L., and Lee, M.-H., Activity staining of isocitrate lyase after electrophoresis on either native or sodium dodecyl sulfate polyacrylamide gels, *Electrophoresis*, 22, 2653, 2001.
3. Kaplan, J.-C. and Beutler, E., Electrophoresis of red cell NADH- and NADPH-diaphorases in normal subjects and patients with congenital methemoglobinemia, *Biochem. Biophys. Res. Commun.*, 29, 605, 1967.
4. Beeckmans, S., Khan, A.S., Van Driessche, E., and Kanarek, L., A specific association between the glyoxylic-acid-cycle enzymes isocitrate lyase and malate synthase, *Eur. J. Biochem.*, 224, 197, 1994.

4.1.3.2 — Malate Synthase; MS

OTHER NAMES	Malate condensing enzyme, glyoxylate transacetylase, malate synthetase
REACTION	L-Malate + CoA = acetyl-CoA + H_2O + glyoxylate
ENZYME SOURCE	Bacteria, fungi, plants
SUBUNIT STRUCTURE	Dimer[#] (bacteria), trimer[#] (fungi – yeast *Pichia methanolica*), octamer[#] (fungi – basidiomycete *Fomitopsis palustris*)

METHOD

Visualization Scheme

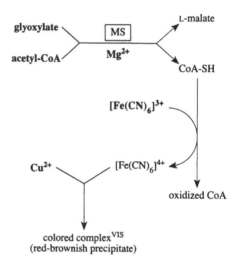

Staining Solution[1]

30 mM Phosphate buffer, pH 7.6	7.5 ml
100 mM Sodium phosphotartrate	2.5 ml
50 mM CuSO$_4$	2.5 ml
15 mM Potassium ferricyanide	12.5 ml
25 mM MgCl$_2$	25 ml
10 mM Acetyl-CoA	2.5 ml
50 mM Sodium glyoxylate	5 ml

Procedure

Incubate an electrophorized gel in a staining solution at 37°C for 30 to 60 min. Reddish brown bands indicate areas of MS activity.

Notes: This method and some others (see below) based on the detection of the free sulfhydryl group of the product CoA require the control staining of an additional gel in staining solution lacking glyoxylate to identify bands of acetyl-CoA hydrolase (see 3.1.2.1 — ACoAH).

4.1.3.2 — Malate Synthase; MS (continued)

OTHER METHODS

A. PAG prepared using *N*-[5-(hydroxyethyl)dithio-2-nitrobenzoylaminoethyl]acrylamide (iodide) can be used to visualize MS activity bands. The method is based on the reduction of disulfide bonds of the chromogenic group (dithio-2-nitrobenzene) by thiol reagents, including CoA-SH produced by MS (e.g., see 3.1.1.7 — ACHE, Method 2). Free dithiobis(2-nitrobenzoic acid) can also be used (e.g., see 1.6.4.2 — GSR, Method 3).

B. The product CoA-SH can be detected using redox indicator 2,6-dichlorophenolindophenol coupled with the tetrazolium system (e.g., see 4.1.3.7 — CS).

C. The reaction of a linking enzyme, NAD-dependent malate dehydrogenase, coupled with the PMS–MTT system, can be used to detect the product L-malate (see 1.1.1.37 — MDH, Method 1).

D. The NADH-dependent reaction of a linking enzyme, lactate dehydrogenase, can be used to detect glyoxylate produced by the forward reaction of MS (e.g., see 4.1.3.1 — IL, Method 2).

REFERENCE

1. Volk, M.J., Trelease, R.N., and Reeves, H.C., Determination of malate synthase activity in polyacrylamide gels, *Anal. Biochem.*, 58, 315, 1974.

4.1.3.6 — Citrate Lyase; CL

OTHER NAMES	[Citrase (*pro-3S*)-lyase] (recommended name), citrase, citratase, citritase, citridesmolase, citrate aldolase
REACTION	Citrate = acetate + oxaloacetate
ENZYME SOURCE	Bacteria
SUBUNIT STRUCTURE	Heterotrimer[a] (bacteria); see General Notes

METHOD 1

Visualization Scheme

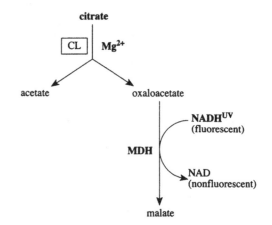

Staining Solution[1] (adapted)

0.1 M Tris–HCl buffer, pH 8.0	16 ml
Citric acid (trisodium salt)	50 mg
0.2 M MgCl$_2$	50 μl
Malate dehydrogenase (MDH)	55 U
NADH	10 mg

Procedure

Apply the staining solution to the gel surface on a filter paper overlay. Incubate the gel at 37°C and monitor under long-wave UV light. Dark (nonfluorescent) bands visible on a light (fluorescent) background indicate areas of CL activity. Record the zymogram or photograph using a yellow filter.

Notes: The processed gel can be counterstained with the PMS–MTT solution and achromatic bands observed on a blue background of the gel. Fix the negatively stained gel in 25% ethanol.

4.1.3.6 — Citrate Lyase; CL (continued)

METHOD 2

Visualization Scheme

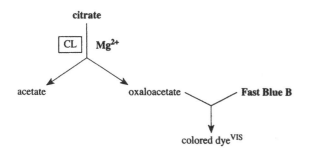

Staining Solution*

0.1 M Tris–citrate buffer, pH 7.6	100 ml
MgCl$_2$·6H$_2$O	50 mg
Fast Blue B salt	100 mg

Procedure

Incubate the gel in staining solution in the dark at 37°C until brown bands appear. Wash the stained gel with water and fix in 7% acetic acid.

Notes: This method is less expensive than the former one. It may be preferable when a double stain of the same gel is made.

GENERAL NOTES

The enzyme can be dissociated into components, two of which are identical with EC 2.8.3.10 and EC 4.1.3.34. The third component is an acyl carrier protein. The enzyme, citrate lyase deacetylase (EC 3.1.2.16), deacetylates and inactivates CL.[2]

REFERENCES

1. Slaughter, C.A., Hopkinson, D.A., and Harris, H., The distribution and properties of aconitase isozymes in man, *Ann. Hum. Genet.*, 40, 385, 1977.
2. NC-IUBMB, *Enzyme Nomenclature*, Academic Press, San Diego, 1992, p. 464 (EC 4.1.3.6, Comments).

---

* New; recommended for use.

4.1.3.7 — Citrate Synthase; CS

OTHER NAMES	Citrate (*si*)-synthase (recommended name), condensing enzyme, citrate condensing enzyme, citrogenase, oxaloacetate transacetase
REACTION	Citrate + CoA = acetyl-CoA + H$_2$O + oxaloacetate
ENZYME SOURCE	Bacteria, fungi, plants, invertebrates, vertebrates
SUBUNIT STRUCTURE	Dimer (vertebrates)

METHOD

Visualization Scheme

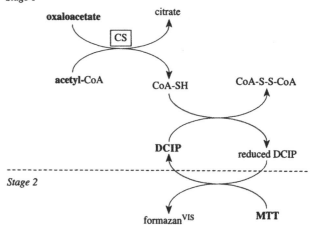

Staining Solution[1]

A.	100 mM Tris–HCl buffer, pH 7.6	125 μl
	10 mM Oxaloacetate in 100 mM Tris–HCl buffer, pH 7.6	50 μl
	10 mM Acetyl-CoA in 100 mM Tris–HCl buffer, pH 7.6	25 μl
	2,6-Dichlorophenolindophenol (DCIP)	0.3 mg
B.	0.5 mg/ml MTT in 100 mM Tris–HCl buffer, pH 7.6	

Procedure

Apply solution A dropwise to electrophorized acetate cellulose gel and incubate at room temperature in a moist chamber for 10 to 15 min or until white bands of CS activity appear on a blue gel background. Then counterstain the gel with solution B to obtain purple CS bands on a white background. Remove excess DCIP and MTT by extensive washing of the gel with water. Fix the stained gel in 25% ethanol.

4.1.3.7 — Citrate Synthase; CS (continued)

OTHER METHODS

A. PAG prepared using *N*-[5-(hydroxyethyl)dithio-2-nitro-benzoylaminoethyl]acrylamide (iodide) can be used to visualize CS activity bands. The method is based on the reduction of disulfide bonds of the chromogenic group (dithio-2-nitrobenzene) by thiol reagents, including CoA-SH produced by CS (e.g., see 3.1.1.7 — ACHE, Method 2). Free 2-nitrobenzoic acid can also be used (e.g., see 1.6.4.2 — GSR, Method 3).

B. The ferricyanide method developed for detection of malate synthase (see 4.1.3.2 — MS) can also be adapted for visualization of CS activity bands.

REFERENCES

1. Craig, I., A procedure for the analysis of citrate synthase (EC 4.1.3.7) in somatic cell hybrids, *Biochem. Genet.*, 9, 351, 1973.

4.1.3.27 — Anthranilate Synthase; ANS

REACTION — Chorismate + NH₃ (or L-glutamine) = anthranilate + pyruvate + (L-glutamate)

ENZYME SOURCE — Bacteria

SUBUNIT STRUCTURE — See General Notes

METHOD

Visualization Scheme

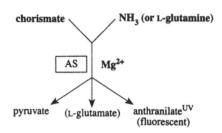

Staining Solution[1]

A. For detection of NH₃-dependent AS activity:
   50 m*M* Triethanolamine–HCl buffer, pH 8.3
   0.34 m*M* Chorismate
   50 m*M* Ammonium sulfate
   5 m*M* MgCl₂
   2 m*M* 2-Mercaptoethanol

B. For detection of glutamine-dependent AS activity:
   50 m*M* Potassium phosphate buffer, pH 7.4
   0.34 m*M* Chorismate
   5 m*M* L-Glutamine
   5 m*M* MgCl₂

Procedure

Depending on what ANS activity should be detected, incubate an electrophorized PAG in staining solution A or B at 37°C for 10 to 30 min and view fluorescent bands in long-wave UV light. Record the developed gel or photograph using Corning No. 18 filters and a Wratten 2B filter.

OTHER METHODS

A. The product of glutamine-dependent ANS activity, L-glutamate, may be detected using the linking enzyme NAD(P)-dependent glutamate dehydrogenase coupled with the PMS–MTT system (e.g., see 3.5.1.2 — GLUT).

B. The product, L-glutamate, can also be detected using three auxiliary enzymes, L-glutamate oxidase (LGO; EC 1.4.3.11), alanine transaminase (GPT; EC 2.6.1.2), and horseradish peroxidase (PER; EC 1.11.1.7).[2] This method involves two different enzymes coupled in the opposite direction, so that the substrate of one of them is the product of the other and vice versa. Under such conditions, the reaction turns out with no consumption of recycling substrate (L-glutamate), while other products of enzymatic reactions (in particular hydrogen peroxide) are accumulated with each turn of the cycle and are readily detected by colorometric PER reaction. This method is suitable for detecting low levels of ANS activity.

GENERAL NOTES

In some organisms, this enzyme is part of a multifunctional protein complex together with one or more other components of the system for biosynthesis of tryptophan (EC 2.4.2.18, EC 4.1.1.48, EC 4.2.1.20, EC 5.3.1.24), which can use either L-glutamine or NH_3. The native enzyme in the complex uses either NH_3 (less effectively) or glutamine. The enzyme separated from the complex uses NH_3 only.[3]

The enzyme from *Sulfolobus solfataricus* has been studied using x-ray crystallography and shown to be a heterotetramer of ANS and glutamine amidotransferase (GAT) subunits, in which two GAT–ANS protomers associate via the GAT subunits. This structure suggests a model in which chorismate binding triggers a relative movement of the two domain tips of the ANS subunit, activating the GAT subunit and creating a channel for passage of ammonia toward the active site of the ANS subunit. Tryptophan presumably blocks this rearrangement, thus stabilizing the inactive states of both subunits.[4]

REFERENCES

1. Grove, T.H. and Levy, H.R., Fluorescent assay of anthranilate synthetase–anthranilate 5-phosphoribozylpyrophosphate phosphoribozyltransferase enzyme-complex on polyacrylamide gels, *Anal. Biochem.*, 65, 458, 1975.
2. Valero, E. and Garcia-Carmona, F., A continuous spectrophotometric method based on enzymatic cycling for determining L-glutamate, *Anal. Biochem.*, 259, 265, 1998.
3. NC-IUBMB, *Enzyme Nomenclature*, Academic Press, San Diego, 1992, p. 467 (EC 4.1.3.27, Comments).
4. Knochel, T., Ivens, A., Hester, G., Gonzalez, A., Bauerle, R., Wilmanns, M., Kirschner, K., Jansonius, J.N., The crystal structure of anthranilate synthase from *Sulfolobus solfataricus*: functional implications, *Proc. Natl. Acad. Sci. U.S.A.*, 96, 9479, 1999.

4.2.1.1 — Carbonic Anhydrase; CA

OTHER NAMES	Carbonate dehydratase (recommended name)
REACTION	$H_2CO_3 = CO_2 + H_2O$
ENZYME SOURCE	Fungi, green algae, plants, invertebrates, vertebrates
SUBUNIT STRUCTURE	Monomer (plants, invertebrates, vertebrates)

METHOD 1

Visualization Scheme

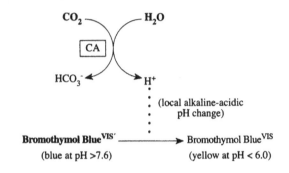

Staining Solution[1]

A. 0.1 M Veronal buffer, pH 9.0
0.1% Bromothymol Blue
B. CO_2

Procedure

Soak an electrophorized gel in solution A for about 10 min. Blot the gel with filter paper and place it in a tank containing dry ice, or pass a stream of CO_2 over the gel surface. Yellow bands on a blue background appear rapidly and then fade. Record the zymogram as quickly as possible.

Notes: Drawing on the gel (using a felt-tip pen) in the position of the major isozymes will allow permanent positions to be noted after the bands have faded.

It is possible to restain the gel by repeating the procedure described above.

Phenol Red pH indicator dye (yellow at pH < 6.8, red at pH > 8.2) may be used in place of Bromothymol Blue.[2] Bromocresol Blue, a pH indicator, becomes fluorescent at low pH values and also is used to detect CA activity.[3]

METHOD 2

Visualization Scheme

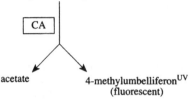

Staining Solution[4]

0.1 M Phosphate buffer, pH 6.5	25 ml
4-Methylumbelliferyl acetate (dissolved in a few drops of acetone)	2.5 mg

Procedure

Apply the staining solution to the gel surface on a filter paper overlay and inspect the gel under long-wave UV light after 5 to 10 min of incubation at 37°C. Fluorescent bands indicate the areas of CA activity. Record the zymogram or photograph using a yellow filter.

Notes: The fluorogenic substrate fluoresceine diacetate can also be used in place of 4-methylumbelliferyl acetate. Different isozymes can display different activities toward these synthetic substrates, e.g., human red cell CA_I isozyme preferentially cleaves 4-methylumbelliferyl acetate, whereas CA_{II} is more specific to fluoresceine diacetate.[4]

Some esterase isozymes (see 3.1.1 ... — EST) can also be visualized by this method. To identify CA activity bands, CA-specific Method 1 (see above) should be used initially. Specific CA inhibitors can also be used to distinguish CA from esterase. The most commonly used inhibitors (and their concentrations) are acetazolamide (7.8×10^{-9} M), sulfanilamide (1.3×10^{-5} M), sodium azide (3.9×10^{-5} M), and sodium chloride (5.6×10^{-2} M).[5]

METHOD 3

Visualization Scheme

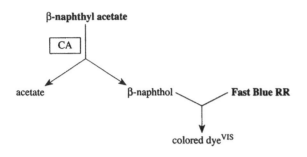

β-naphthyl acetate

CA

acetate   β-naphthol   Fast Blue RR

colored dye^{VIS}

Staining Solution[1]

0.1 M Phosphate buffer, pH 7.2	100 ml
β-Naphthyl acetate (dissolved in 20 mg of a few drops of acetone)	
Fast Blue RR salt	50 mg

Procedure

Incubate the gel in staining solution in the dark at 37°C until red bands appear. Wash the gel with water and fix in 7% acetic acid.

Notes: Esterase (3.1.1 ... — EST) activity bands also are visualized by this method. Thus, the control staining of an additional gel should be made initially using CA-specific Method 1 (see above) to identify true CA activity bands. The use of CA-specific inhibitors is also recommended in order to distinguish CA from esterase activity (see Method 2, *Notes*).

OTHER METHODS

A. CA activity bands may be detected on electrophorized PAGs by using conductivity measurements. In areas of CA activity the conductivity increases due to CO_2 hydratation.[6]

B. The positive fluorescent stain based on specific binding of the fluorochrome 5-(dimethylamino)naphthalene-1-sulfonamide (DNSA) with CA molecules is also available.[7] In this method DNSA (2.5 mg) is initially dissolved in 0.1 N NaOH (10 ml) and then mixed with 10 mM Tris–sulfate, pH 8.9 (190 ml). This mixture (from 0.1 to 0.5 ml) is mixed with a crude enzyme preparation (2 ml) and incubated for 15 min. Pretreated preparations are electrophorized as usual. After completion of electrophoresis the gel is viewed under longwave UV light for fluorescent CA bands.

C. The immunoblotting procedure (for details, see Part II) based on the utility of monoclonal antibodies specific to the human and cow enzyme[8] can also be used for immunohistochemical visualization of the enzyme protein on electrophoretic gels. This procedure is not quite appropriate for large-scale population studies, but it may be of great value in specific analyses of CA.

REFERENCES

1. Tashian, R.E., The esterases and carbonic anhydrases in human erythrocytes, in *Biochemical Methods in Red Cell Genetics*, Yunis, J.J., Ed., Academic Press, New York, 1969, p. 307.
2. Richardson, B.J., Baverstock, P.R., and Adams, M., *Allozyme Electrophoresis: A Handbook for Animal Systematics and Population Studies*, Academic Press, Sydney, 1986, p. 171.
3. Patterson, B.D., Atkins, C.A., Graham, D., and Wills, R.B.H., Carbonic anhydrase: a new method of detection on polyacrylamide gels using low-temperature fluorescence, *Anal. Biochem.*, 44, 388, 1971.
4. Hopkinson, D.A., Coppock, J.S., Mühlemann, M.F., and Edwards, Y.H., The detection and differentiation of the products of the human carbonic anhydrase loci, CA_I and CA_{II}, using fluorogenic substrates, *Ann. Hum. Genet.*, 38, 155, 1974.
5. Bundy, H.F. and Coté, S., Purification and properties of carbonic anhydrase from *Chlamidomonas reinhardii*, *Phytochemistry*, 19, 253, 1980.
6. Wiedner, G., Simon, B., and Thomas, L., Carbonic anhydrase (carbonate dehydratase): new method of detection on polyacrylamide gels by using conductivity measurements, *Anal. Biochem.*, 55, 93, 1973.
7. Drescher, D.G., Purification of blood carbonic anhydrase and specific detection of carbonic anhydrase isozymes on polyacrylamide gels with 5-dimethylaminonaphthalene-1-sulfonamide (DNSA), *Anal. Biochem.*, 90, 349, 1978.
8. Erickson, R.P., Kay, G., Hewett-Emmett, D., Tashian, R.E., and Claflin, J.L., Cross-reactions among carbonic anhydrase-I, anhydrase-II, and anhydrase-III studied by binding tests and with monoclonal antobodies, *Biochem. Genet.*, 20, 809, 1982.

4.2.1.2 — Fumarate Hydratase; FH

OTHER NAMES	Fumarase
REACTION	(S)-Malate = fumarate + H_2O
ENZYME SOURCE	Bacteria, fungi, plants, protozoa, invertebrates, vertebrates
SUBUNIT STRUCTURE	Tetramer (fungi, invertebrates, vertebrates)

METHOD 1

Visualization Scheme

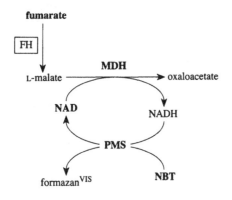

Staining Solution[1]

25 mM Phosphate buffer, pH 7.1	100 ml
Potassium fumarate	770 mg
NAD	80 mg
Malate dehydrogenase (MDH)	200 U
NBT	30 mg
PMS	1 mg

Procedure

Incubate the gel in staining solution in the dark at 37°C until dark blue bands appear. Wash the stained gel in water and fix in 25% ethanol.

Notes: Staining solution can also be applied as a 1% agar overlay.

METHOD 2

Visualization Scheme

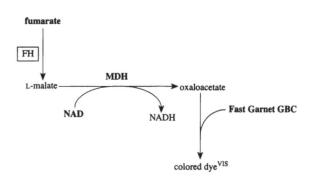

Staining Solution[2]

A. 0.2 M Tris–HCl buffer, pH 8.0	20 ml
100 mg/ml Fumaric acid, pH 8.0	2 ml
50 mg/ml NAD	1 ml
Fast Garnet GBC	120 mg
Malate dehydrogenase (MDH)	80 U
B. 2% Agar solution (60°C)	20 ml

Procedure

Mix A and B components of the staining solution and pour the mixture over the gel surface. Incubate the gel in the dark at 37°C until colored bands appear. Wash the stained gel in water and fix in 3% acetic acid.

Notes: Where enzyme activity is weak, greater sensitivity may be obtained using a postcoupling technique: apply liquid staining solution (A) lacking Fast Garnet GBC to the gel, incubate the gel for 15 to 30 min, and then add Fast Garnet GBC.

GENERAL NOTES

Method 1 is more frequently used and is more sensitive than Method 2. The latter method may be preferable when a double stain of the same gel is made. This method does not allow the staining of artifact bands caused by "nothing dehydrogenase" (see 1.X.X.X — NDH).

Maleic and citric acids concurrently inhibit FH. Thus, electrophoretic and staining buffers containing these substances should not be used.

REFERENCES

1. Shaw, C.R. and Prasad, R., Starch gel electrophoresis of enzymes: a compilation of recipes, *Biochem. Genet.*, 4, 297, 1970.
2. Richardson, B.J., Baverstock, P.R., and Adams, M., *Allozyme Electrophoresis: A Handbook for Animal Systematics and Population Studies*, Academic Press, Sydney, 1986, p. 178.

4.2.1.3 — Aconitase; ACON

OTHER NAMES	Aconitate hydratase (recommended name)
REACTION	Citrate = *cis*-aconitate + H$_2$O; also converts isocitrate into *cis*-aconitate
ENZYME SOURCE	Bacteria, fungi, plants, invertebrates, vertebrates
SUBUNIT STRUCTURE	Monomer (fungi, plants, invertebrates, vertebrates)

METHOD 1

Visualization Scheme

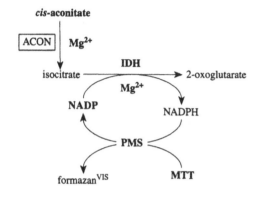

Staining Solution[1]

A. 0.2 *M* Tris–HCl buffer, pH 8.0	20 ml
20 mg/ml *cis*-Aconitic acid, pH 8.0	5 ml
MgCl$_2$·6H$_2$O	150 mg
NADP	10 mg
PMS	1.5 mg
MTT	7 mg
Isocitrate dehydrogenase (IDH)	10 U
B. 2% Agar solution (60°C)	25 ml

Procedure

Mix A and B components of the staining solution and pour the mixture over the gel surface. Incubate the gel in the dark at 37°C until dark blue bands appear. Fix the stained gel or agar overlay in 25% ethanol.

METHOD 2

Visualization Scheme

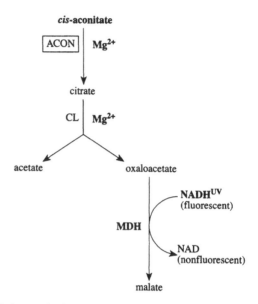

Staining Solution[2]

0.1 *M* Tris–HCl buffer, pH 8.0	16 ml
cis-Aconitic acid (neutralized)	20 mg
0.2 *M* MgCl$_2$	50 μl
Citrate lyase (CL)	38 U
Malate dehydrogenase (MDH)	55 U
NADH	10 mg

Procedure

Apply the staining solution to the gel surface on a filter paper overlay and incubate at 37°C. Monitor the gel under long-wave UV light. Dark (nonfluorescent) bands on a light (fluorescent) background indicate areas of ACON activity. Record the zymogram or photograph using a yellow filter.

Notes: To make ACON bands visible in daylight, counterstain the processed gel with the PMS–MTT solution. Achromatic ACON bands will appear on a blue background almost immediately. Fix the stained gel in 25% ethanol.

GENERAL NOTES

Magnesium ions are included in the staining solutions given in both methods to shift the reaction equilibrium toward citrate and isocitrate. *cis*-Aconitic acid may form the *trans* isomer upon prolonged storage. Thus, the use of fresh preparations of this substrate is recommended.

REFERENCES

1. Shaw, C.R. and Prasad, R., Starch gel electrophoresis of enzymes: a compilation of recipes, *Biochem. Genet.*, 4, 297, 1970.
2. Slaughter, C.A., Hopkinson, D.A., and Harris, H., The distribution and properties of aconitase isozymes in man, *Ann. Hum. Genet.*, 40, 385, 1977.

4.2.1.9 — Dihydroxy-Acid Dehydratase; DHAD

REACTION   2,3-Dihydroxy-3-methylbutanoate =
                                  3-methyl-2-oxobutanoate + H_2O

ENZYME SOURCE   Bacteria, fungi

SUBUNIT STRUCTURE   Dimer[a] (bacteria)

Method

Visualization Scheme

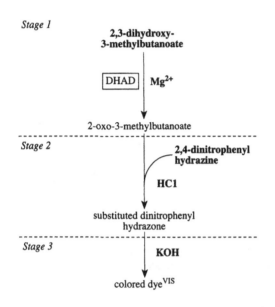

Stage 1

2,3-dihydroxy-3-methylbutanoate

DHAD Mg^{2+}

2-oxo-3-methylbutanoate

- -

Stage 2

2,4-dinitrophenyl hydrazine

HCl

substituted dinitrophenyl hydrazone

- -

Stage 3

KOH

colored dye[VIS]

Staining Solution[1]

A.   50 mM Tris–HCl buffer, pH 7.8
     5 mM MgSO$_4$

B.   0.5 M 2,3-Dihydroxy-3-methylbutanoate (sodium salt)

C.   0.025% 2,4-Dinitrophenyl hydrazine
     0.5 N HCl

D.   10% KOH

Procedure

Incubate an electrophorized PAG in solution A at 37°C for 5 min, and then add solution B to a 5 mM final concentration of 2,3-dihydroxy-3-methylbutanoate and incubate the gel further at 37°C with gentle shaking for 30 min. Transfer the gel into solution C and incubate at room temperature for 20 min with occasional agitation. Finally, transfer the gel into solution D. Brown bands, at the positions of the DHAD, begin to develop rapidly and reach maximum intensity in 30 min. Record the zymogram or photograph quickly, because the brown bands gradually become blurred and fade upon storage in the KOH.

Notes: The substrate, 2,3-dihydroxy-3-methylbutanoate (or DL-α,β-dihydroxyisovalerate), is not yet commercially available and should be synthesized in the laboratory.

3-Hydroxy-2-naphthoic acid hydrazide can also be used to detect the carbonyl ketonic group of the product 2-oxo-3-methylbutanoate (or α-ketoisovalerate). The hydrazones formed can then be cross-linked with each other by a tetrazonium salt (see 2.7.1.40 — PK, Method 3).

References

1. Kuo, C.F., Mashino, T. and Fridovich, I., An activity stain for dihydroxy-acid dehydratase, *Anal. Biochem.*, 164, 526, 1987.

4.2.1.11 — Enolase; ENO

OTHER NAMES	Phosphopyruvate hydratase (recommended name), 2-phosphoglycerate dehydratase
REACTION	2-Phospho-D-glycerate = phosphoenolpyruvate + H_2O
ENZYME SOURCE	Bacteria, fungi, plants, invertebrates, vertebrates
SUBUNIT STRUCTURE	Dimer (invertebrates, vertebrates)

Method 1

Visualization Scheme

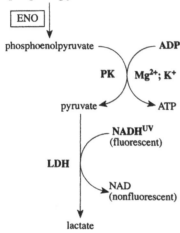

Staining Solution[1]

38 mM Triethanolamine buffer, pH 7.5
7.7 mM MgSO$_4$
3.2 mM EDTA
1.8 mM ADP
1.0 mM NADH
67 mM KCl
1 mM 2-Phospho-D-glycerate
1.5 U/ml Pyruvate kinase (PK)
3.6 U/ml Lactate dehydrogenase (LDH)

Procedure

Apply the staining solution to the gel surface on a filter paper overlay and incubate at 37°C for 1 h. Monitor the gel under long-wave UV light. Dark (nonfluorescent) bands on a light (fluorescent) background indicate areas of ENO activity. Record the zymogram or photograph using a yellow filter.

Notes: To make ENO bands visible in daylight, counterstain the processed gel with PMS–MTT solution. Achromatic ACON bands will appear on a blue background almost immediately. Fix the stained gel in 25% ethanol.

3-Phospho-D-glycerate and phosphoglycerate mutase (see 5.4.2.1 — PGLM) may be included in the staining mixture instead of 2-phospho-D-glycerate.[2]

Method 2

Visualization Scheme

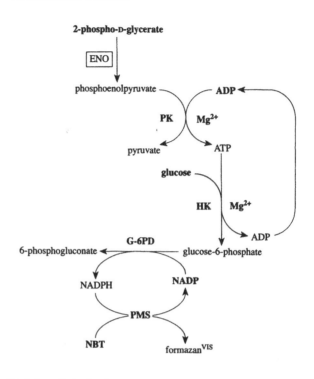

Staining Solution[3]

50 mM Tris–HCl buffer, pH 7.5
2 mM MgCl$_2$
1 mM Glucose
0.1 mM ADP
20 mM AMP
0.5 mM NADP
4 U/ml Pyruvate kinase (PK)
3 U/ml Hexokinase (HK)
1.4 U/ml Glucose-6-phosphate dehydrogenase (G-6-PD)
0.4 mg/ml NBT
1.4 mM 2-Phospho-D-glycerate
0.024 mg/ml PMS

Procedure

Apply the staining solution to the gel surface dropwise and incubate at 37°C in the dark until dark blue bands appear. Fix the stained gel in 25% ethanol.

Notes: The appearance of adenylate kinase (see 2.7.4.3 — AK) activity bands is prevented by inclusion of excess AMP and by use of catalytic levels of ADP, the latter being regenerated by hexokinase. Omission of AMP and use of a higher concentration of ADP lead to the appearance of extra bands, which are not dependent on the presence of 2-phospho-D-glycerate and are due to adenylate kinase.

4.2.1.11 — Enolase; ENO (continued)

OTHER METHODS

A. A spectrophotometric procedure for the qualitative and quantitative estimation of ENO on polyacrylamide gels is described.[4] The procedure is based on the acid hydrolysis (at 100°C) of the product phosphoenolpyruvate to pyruvate, which is then treated with phenylhydrazine and detected as phenylhydrazone by spectrophotometric scanning of PAG at 325 nm.

B. An immunoblotting procedure (for details, see Part II) may also be used to detect ENO. Antibodies specific to the enzyme from the human brain are available from Sigma.

REFERENCES

1. Rider, C.C. and Taylor, C.B., Enolase isoenzymes in rat tissues: electrophoretic, chromatographic, immunological and kinetic properties, *Biochim. Biophys. Acta*, 365, 285, 1974.

2. Harris, H. and Hopkinson, D.A., *Handbook of Enzyme Electrophoresis in Human Genetics*, North-Holland, Amsterdam, 1976 (loose-leaf, with supplements in 1977 and 1978).

3. Hullin, D.A. and Thompson, R.J., An improved nonfluorescent stain for enolase activity on cellulose acetate strips, *Anal. Biochem.*, 82, 240, 1977.

4. Sharma, H.K. and Rothstein, M., A new method for the detection of enolase activity on polyacrylamide gels, *Anal. Biochem.*, 98, 226, 1979.

4.2.1.13 — L-Serine Dehydratase; L-SD

OTHER NAMES	Serine deaminase, L-hydroxyaminoacid dehydratase
REACTION	L-Serine + H_2O = pyruvate + NH_3 + H_2O (also see General Notes)
ENZYME SOURCE	Bacteria, fungi, invertebrates, vertebrates
SUBUNIT STRUCTURE	Heterodimer($\alpha\beta$)[#] (bacteria – *Peptostreptococcus asaccharolyticus*), tetramer[#] (bacteria – *Lactobacillus fermentum*)

METHOD

Visualization Scheme

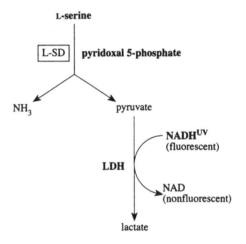

Staining Solution[1]

A. 0.1 *M* Potassium phosphate buffer, pH 8.0
   0.05 m*M* Pyridoxal 5-phosphate
   1 m*M* NADH
   3.5 U/ml Lactate dehydrogenase (LDH)
   0.15 *M* L-Serine

B. 3% Agar solution (60°C)

Procedure

Mix 19 ml of solution A with 5 ml of solution B and pour the mixture over the gel surface. Incubate the gel at 30°C for 10 to 15 min and observe nonfluorescent bands on a fluorescent background under long-wave UV light. Record the zymogram or photograph using a yellow filter.

Notes: The bands of threonine dehydratase can also be developed by this method (see General Notes).

4.2.1.13 — L-Serine Dehydratase; L-SD (continued)

OTHER METHODS

Several methods are also available to detect the product ammonia (e.g., see 4.2.1.16 — TD, Other Methods). However, all these methods are usually less sensitive and not as practical as the method described above.

GENERAL NOTES

The enzyme is a pyridoxal-phosphate protein. This reaction also may be catalyzed by threonine dehydratase (see 4.2.1.16 — TD) from a number of sources.[2]

REFERENCES

1. Yanagi, S., Tsutsumi, T., Saheki, S., Saheki, K., and Yamamoto, N., Novel and sensitive activity stains on polyacrylamide gel of serine and threonine dehydratase and ornithine aminotransferase, *Enzyme*, 28, 400, 1982.
2. NC-IUBMB, *Enzyme Nomenclature*, Academic Press, San Diego, 1992, p. 471 (EC 4.2.1.13, Comments).

4.2.1.16 — Threonine Dehydratase; TD

OTHER NAMES	Threonine deaminase, L-serine dehydratase, serine deaminase
REACTION	L-Threonine + H_2O = 2-oxobutanoate + NH_3 + H_2O
ENZYME SOURCE	Bacteria, fungi, invertebrates, vertebrates
SUBUNIT STRUCTURE	Tetramer[a] (bacteria)

METHOD 1

Visualization Scheme

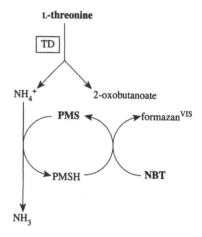

Staining Solution[1]

    0.1 *M* Phosphate buffer, pH 8.0
    20 m*M* L-Threonine
    0.2 mg/ml PMS
    1 mg/ml NBT

Procedure

Incubate the gel in staining solution in the dark at 37°C until dark blue bands appear. Fix the gel in 25% ethanol.

Notes: Pyridoxal 5-phosphate may be added to the staining solution and electrophoretic buffer to enhance TD activity.

METHOD 2

Visualization Scheme

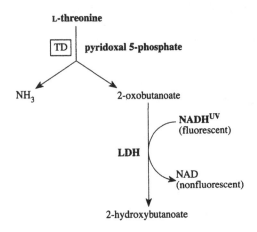

L-threonine

TD   pyridoxal 5-phosphate

NH₃   2-oxobutanoate

NADH^{UV}
(fluorescent)

LDH

NAD
(nonfluorescent)

2-hydroxybutanoate

Staining Solution[2]

A. 0.1 *M* Potassium phosphate buffer, pH 8.0
0.05 m*M* Pyridoxal 5-phosphate
1 m*M* NADH
3.5 U/ml Lactate dehydrogenase (LDH)
0.15 *M* L-Threonine
B. 3% Agar solution (60°C)

Procedure

Mix 19 ml of solution A with 5 ml of solution B and pour the mixture over the gel surface. Incubate the gel at 30°C for 10 to 15 min and observe dark (nonfluorescent) bands under long-wave UV light. Record the zymogram or photograph using a yellow filter.

Notes: The method is based on the ability of LDH to use 2-oxo-butanoate as a substrate in the backward reaction. It is about two orders more sensitive than Method 1.

OTHER METHODS

A. The areas of ammonia production may be detected in UV light using the backward reaction of linking enzyme glutamate dehydrogenase (e.g., see 3.5.3.1 — ARG, Method 2).
B. The areas of ammonia production may also be visualized using sodium tetraphenylborate, which precipitates in the presence of ammonia ions, resulting in the formation of white opaque bands visible in PAG (e.g., see 3.5.1.1 — ASP).
C. The local pH change due to production of ammonia may be detected using the pH indicator dye Phenol Violet (e.g., see 3.5.4.4 — ADA, Method 3), the NBT (or MTT)–dithiothreitol system (e.g., see 3.5.3.1 — ARG, Method 1), or neutral AgNO₃ solution containing photographic developers (e.g., see 3.5.1.5 — UR, Method 2).

GENERAL NOTES

The enzyme from some sources can use L-serine as a substrate and thus be functionally identical to L-serine dehydratase (see 4.2.1.13 — L-SD).[3]

REFERENCES

1. Feldberg, R.S. and Datta, P., Threonine deaminase: a novel activity stain on polyacrylamide gels, *Science*, 170, 1414, 1970.
2. Yanagi, S., Tsutsumi, T., Saheki, S., Saheki, K., and Yamamoto, N., Novel and sensitive activity stains on polyacrylamide gel of serine and threonine dehydratase and ornithine aminotransferase, *Enzyme*, 28, 400, 1982.
3. NC-IUBMB, *Enzyme Nomenclature*, Academic Press, San Diego, 1992, p. 472 (EC 4.2.1.16, Comments).

4.2.1.20 — Tryptophan Synthase; TS

OTHER NAMES	Tryptophan desmolase
REACTION	L-Serine + 1-(indol-3-yl)glycerol-3-phosphate = L-tryptophan + glyceraldehyde-3-phosphate
ENZYME SOURCE	Bacteria, fungi, plants
SUBUNIT STRUCTURE	Heterotetramer($2\alpha 2\beta$)[#] (bacteria)

METHOD

Visualization Scheme

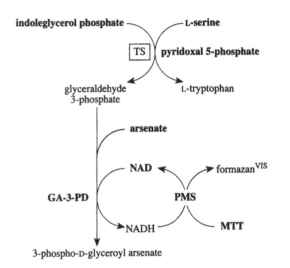

indoleglycerol phosphate — L-serine

TS — pyridoxal 5-phosphate

glyceraldehyde 3-phosphate — L-tryptophan

arsenate

NAD — formazan[VIS]

GA-3-PD — PMS

NADH — MTT

3-phospho-D-glyceroyl arsenate

Staining Solution[1] (modified)

A.	0.05 M Tris–HCl buffer, pH 7.8	25 ml
	Indoleglycerol-phosphate	30 mg
	L-Serine	30 mg
	Pyridoxal 5-phosphate	30 mg
	Sodium arsenate	100 mg
	Glyceraldehyde-3-phosphate dehydrogenase (GA-3-PD)	90 U
	NAD	30 mg
	MTT	8 mg
	PMS	1 mg
B.	2% Agar solution (60°C)	25 ml

Procedure

Mix A and B components of the staining solution and pour the mixture over the gel surface. Incubate the gel in the dark at 37°C until dark blue bands appear. Fix the stained gel in 25% ethanol.

GENERAL NOTES

The enzyme is a pyridoxal-phosphate protein. It catalyzes the conversion of L-serine and indole into L-tryptophan and H_2O, and of indoleglycerol-phosphate into indole and glyceraldehyde-3-phosphate. In some bacteria, TS is part of a multifunctional protein complex together with one or more other components of the system for biosynthesis of tryptophan (EC 2.4.2.18, EC 4.1.1.48, EC 4.1.3.27, EC 5.3.1.24).[2]

REFERENCES

1. Crawford, I.P., Ito, J., and Hatanaka, M., Genetic and biochemical studies of enzymatically active subunits of *E. coli* tryptophan synthetase, *Ann. N.Y. Acad. Sci.*, 151, 171, 1968.
2. NC-IUBMB, *Enzyme Nomenclature*, Academic Press, San Diego, 1992, p. 472 (EC 4.2.1.20, Comments).

OTHER NAMES — Cystathionine β-synthase (recommended name), β-thionase, methylcysteine synthase

REACTIONS —
1. L-Serine + L-homocysteine = cystathionine + H_2O
2. L-Cysteine + H_2O = L-serine + H_2S
3. See General Notes

ENZYME SOURCE — Fungi, vertebrates

SUBUNIT STRUCTURE — Unknown or no data available

METHOD

Visualization Scheme

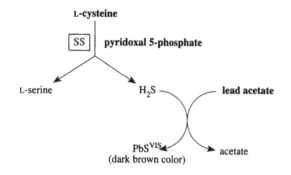

Staining Solution[1]

0.1 *M* Sodium phosphate buffer, pH 7.8
20 m*M* L-Cysteine
50 m*M* 2-Mercaptoethanol
0.1 m*M* Pyridoxal 5-phosphate
0.2 m*M* Lead acetate

Procedure

Incubate the gel in the staining solution at 37°C until dark brown bands appear. Record or photograph the zymogram.

GENERAL NOTES

The enzyme is a multifunctional pyridoxal-phosphate protein that catalyzes β-replacement reactions between L-serine, L-cysteine, cysteine thioethers or some other β-substituted α-L-amino acids, and a variety of mercaptans.[2]

REFERENCES

1. Willhardt, I. and Wiederanders, B., Activity staining of cystathionine-β-synthetase and related enzymes, *Anal. Biochem.*, 63, 263, 1975.
2. NC-IUBMB, *Enzyme Nomenclature*, Academic Press, San Diego, 1992, p. 472 (EC 4.2.1.22, Comments).

4.2.1.24 — Porphobilinogen Synthase; PBGS

OTHER NAMES   Aminolevulinate dehydratase

REACTION   2 5-Aminolevulinate = porphobilinogen + 2 H_2O

ENZYME SOURCE   Bacteria, fungi, plants, vertebrates

SUBUNIT STRUCTURE   Unknown or no data available

METHOD

Visualization Scheme

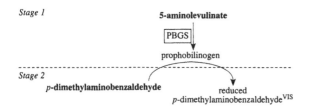

Stage 1

5-aminolevulinate

PBGS

prophobilinogen

Stage 2

p-**dimethylaminobenzaldehyde**

reduced
p-dimethylaminobenzaldehyde[VIS]

Staining Solution[1]

A.   0.1 *M* Potassium phosphate, pH 7.2
     13 m*M* 5-Aminolevulinic acid
     15 m*M* Reduced glutathione (or 2 m*M* dithiothreitol)

B.   Erlich's reagent:

     1.   *p*-Dimethylaminobenzaldehyde   0.1 g
          Acetic acid   3.4 ml

     2.   0.25 *M* $HgCl_2$ 1 ml
          70% Perchloric acid   1.6 ml

Procedure

Apply solution A to the gel surface on a filter paper overlay and incubate at 37°C for 2 to 3 h. Remove the filter paper and rinse the gel with water. Place the gel in solution B and incubate until pink bands appear.

Notes: $HgCl_2$ is added to Erlich's reagent to oxidize reducing agents before reaction between porphobilinogen and *p*-dimethyl-aminobenzaldehyde. Dithioerythritol may be included in solution A in place of reduced glutathione or dithiothreitol.[2]

OTHER METHODS

The immunoblotting procedure (for details, see Part II) based on the utility of monoclonal antibodies specific to the spinach enzyme[3] can also be used for immunohistochemical visualization of the enzyme protein on electrophoretic gels. This procedure is expensive, time consuming, and thus unsuitable for routine laboratory use. Monoclonal antibodies, however, may be of great value in special analyses of PBGS.

REFERENCES

1. Battistuzzi, G., Petrucci, R., Silvagni, L., Urbani, F.R., and Caiola, S., δ-Aminolevulinate dehydratase: a new genetic polymorphism in man, *Ann. Hum. Genet.*, 45, 223, 1981.

2. Meera Khan, P., Rijken, H., Wijnen, J.Th., Wijnen, L.M.M., and De Boer, L.E.M., Red cell enzyme variation in the orang utan: electrophoretic characterization of 45 enzyme systems in cellogel, in *The Orang Utan: Its Biology and Conservation*, De Boer, L.E.M., Ed., Dr. W. Junk Publishers, The Hague, 1982, p. 61.

3. Liedgens, W., Grutzmann, R., and Schneider, H.A., Highly efficient purification of the labile plant enzyme 5-aminolevulinate dehydratase (E.C.4.2.1.24) by means of monoclonal antibodies, *Z. Naturforsch.*, 35, 958, 1980.

4.2.1.46 — dTDPglucose 4,6-Dehydratase; TDPGD

REACTION	dTDPglucose = dTDP-4-dehydro-6-deoxy-D-glucose + H$_2$O
ENZYME SOURCE	Bacteria
SUBUNIT STRUCTURE	Unknown or no data available

METHOD

Visualization Scheme

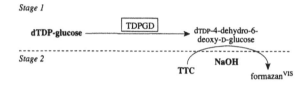

Staining Solution[1]

A. 20 m*M* Tris–HCl buffer, pH 8.0
   3 m*M* dTDPglucose
B. 0.1% 2,3,5-Triphenyltetrazolium chloride (TTC)
   1 *N* NaOH

Procedure

Rinse an electrophorized gel in distilled water and incubate in solution A at 37°C for 20 min. Rinse the gel again thoroughly and put in freshly prepared solution B. Incubate the gel in this solution in the dark until violet bands appear on a pink background. Fix the stained gel in 7% acetic acid.

Notes: The method is based on the ability of the product dTDP-4-dehydro-6-deoxy-D-glucose to reduce TTC in the presence of alkali.

GENERAL NOTES

The enzyme requires bound NAD.[2]

REFERENCES

1. Gabriel, O. and Wang, S.-F., Determination of enzymatic activity in polyacrylamide gels: I. Enzymes catalyzing the conversion of nonreducing substrates to reducing products, *Anal. Biochem.*, 27, 545, 1969.
2. NC-IUBMB, *Enzyme Nomenclature*, Academic Press, San Diego, 1992, p. 475 (EC 4.2.1.46, Comments).

4.2.1.49 — Urocanate Hydratase; UH

OTHER NAMES	Urocanase
REACTION	4,5-Dihydro-4-oxo-5-imidazolepropanoate = urocanate + H$_2$O
ENZYME SOURCE	Bacteria, vertebrates
SUBUNIT STRUCTURE	Unknown or no data available

METHOD

Visualization Scheme

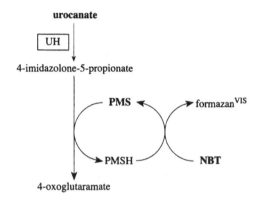

Staining Solution[1]

A. 0.1 *M* Phosphate buffer, pH 7.0
   0.01 mg/ml PMS
   0.04 mg/ml NBT
B. 0.1 *M* Phosphate buffer, pH 7.0
   0.01 mg/ml PMS
   0.04 mg/ml NBT
   10 m*M* Urocanate

Procedure

Incubate the gel in solution A in the dark at room temperature for 45 min and then place in solution B. Dark blue bands indicating UH activity appear after 20 to 60 min. Fix the stained gel in 25% ethanol.

REFERENCES

1. Bell, M.V. and Hassall, H., A method for the detection of urocanase on polyacrylamide gels and its use with crude cell extracts of *Pseudomonas*, *Anal. Biochem.*, 75, 436, 1976.

4.2.2.2 — Polygalacturonate Lyase; PGL

OTHER NAMES	Pectate lyase (recommended name), pectate transeliminase
REACTION	Eliminative cleavage of pectate to give oligosaccharides with 4-deoxy-α-D-gluc-4-enuronosyl groups at their nonreducing ends
ENZYME SOURCE	Bacteria
SUBUNIT STRUCTURE	Unknown or no data available

METHOD

Visualization Scheme

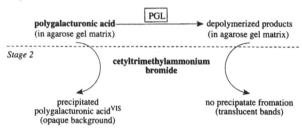

Stage 1

polygalacturonic acid ——[PGL]—→ depolymerized products
(in agarose gel matrix)   (in agarose gel matrix)

Stage 2

cetyltrimethylammonium bromide

precipitated polygalacturonic acid^{VIS} (opaque background)

no precipatate fromation (translucent bands)

Staining Solution[1]

A. 0.1 *M* Tris–HCl buffer, pH 8.6
   1% Polygalacturonic acid
B. 3% Agar solution (60°C)
C. 1% Cetyltrimethylammonium bromide

Procedure

Mix equal volumes of A and B solutions and form a reactive agarose plate of an appropriate size. Rinse an electrophorized PAG in 10 volumes of 0.05 *M* Tris–HCl buffer (pH 8.6). Place a reactive agarose plate on top of the PAG and incubate at 37°C for 1 to 4 h. Put the PAG and reactive agarose plate in solution C. The areas of PGL appear on the PAG and agarose plate as translucent bands visible on an opaque background. Photograph the developed gel against a black background.

Notes: The areas of high PGL activity are more clearly displayed on the reactive agarose plate, while the areas of low PGL activity are more clearly manifested on the PAG.

In principle, areas of galacturan 1,4-α-galacturonidase (see 3.2.1.67 — GG) activity can also be developed by the method described above; however, this enzyme displays maximal activity at pH 5.0.

Another enzyme, polygalacturonase (see 3.2.1.15 — PG), can also display its activity on PGL zymograms. However, in contrast to PGL, polygalacturonase also hydrolyzes pectin. This difference may be used to identify bands caused by PGL and PG.

GENERAL NOTES

The enzyme also acts on other polygalacturonides. It does not act on pectin.[2]

REFERENCES

1. Bertheau, Y., Madgidi-Hervan, E., Kotoujansky, A., Nguyen-The, C., Andro, T., and Coleno, A., Detection of depolymerase isoenzymes after electrophoresis or electrofocusing, or in titration curves, *Anal. Biochem.*, 139, 383, 1984.
2. NC-IUBMB, *Enzyme Nomenclature*, Academic Press, San Diego, 1992, p. 481 (EC 4.2.2.2, Comments).

OTHER NAMES	Alginate lyase, alginate lyase I
REACTION	Eliminative cleavage of polysaccharides containing β-ᴅ-mannuronate residues to give oligosaccharides with 4-deoxy-α-ʟ-*erythro*-hex-4-enopyranuronosyl groups at their ends (see also General Notes)
ENZYME SOURCE	Bacteria, fungi, algae, invertebrates
SUBUNIT STRUCTURE	Unknown or no data available

Method

Visualization Scheme

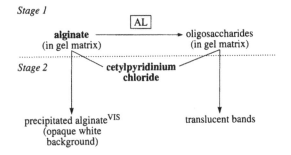

Staining Procedure[1]

The method is developed for an SDS-PAG[2] containing 2 mg/ml alginic acid. Alginic acid is insoluble, but swells in water. It is included in the gel solution just before polymerization and remains a uniform suspension in the gel.

After electrophoresis, wash the gel by three successive 30-min washes in cold 50 m*M* Tris–HCl buffer (pH 8.2) containing 1% (w/v) casein, 2 m*M* EDTA, and 0.01% sodium azide. Then submerse the gel in 10 m*M* phosphate buffer (pH 7.5) containing 0.2 *M* NaCl, 0.2 *M* KCl, and 0.01% sodium azide for 1 h at 30°C to promote the enzyme reaction. Finally, treat the gel with a 10% solution of cetylpyridinium chloride for 20 min at room temperature to precipitate alginic acid. Activity bands of AL are visible as clean hydrolytic halos.

Notes: The method results in development of sharp activity bands. Application of 8×10^{-7} U of AL gives a clear activity stain. Detection of AL activity bands can be obtained in 60 min after the completion of electrophoresis.

An alternative staining procedure is based on the use of 0.05% (w/v) aqueous Ruthenium Red instead of a 10% solution of cetylpyridinium chloride. This procedure results in the appearance of light pink clearing zones against a dark red background. It should be stressed that the contrast between the clearing zones and background is better with cetylpyridinium chloride.[3]

When an insoluble AL substrate (e.g., polyguluronate) is applied to the gel surface as an agar overlay, only agarose should be used as a medium for solidification. Microbiological agar is not suitable because it has a significant sulfate content and is stained with cetylpyridinium chloride and Ruthenium Red, thus masking the clearing zones produced by AL activity.[3]

General Notes

Alginates are (1,4)-linked glucuronans composed of β-ᴅ-mannuronate (M) and α-ʟ-guluronate (G) residues arranged in homopolymeric (polyM or polyG) or heteropolymeric (random sequence of polyMG) block structures.[4] Alginate lyase from some bacteria displays activity toward polyG only, while the enzyme from other bacterial species can use both polyM and polyG as substrates.[3]

References

1. Pecina, A. and Paneque, A., Detection of alginate lyase by activity staining after sodium dodecyl sulfate–polyacrylamide gel electrophoresis and subsequent renaturation, *Anal. Biochem.*, 217, 124, 1994.
2. Laemmli, U.K., Cleavage of structural proteins during the assembly of the head of bacteriophage T4, *Nature*, 227, 680, 1970.
3. Gacesa, P. and Wusteman, F.S., Plate assay for simultaneous detection of alginate lyases and determination of substrate specificity, *Appl. Environ. Microbiol.*, 56, 2265, 1990.
4. Haug, A., Larsen, B., and Smidsrod, O., Studies on the sequence of uronic acid residues in alginic acid, *Acta Chem. Scand.*, 21, 691, 1967.

4.2.2.8 — Heparitin-Sulfate Lyase; HSL

OTHER NAMES   Heparitinase, heparinase III, heparin-sulfate eliminase, heparin-lyase III

REACTION   Elimination of sulfate; appears to act on linkages between *N*-acetyl-D-glucosamine and uronate; product is an unsaturated sugar

ENZYME SOURCE   Bacteria, vertebrates

SUBUNIT STRUCTURE   Unknown or no data available

METHOD

Visualization Scheme

Stage 1

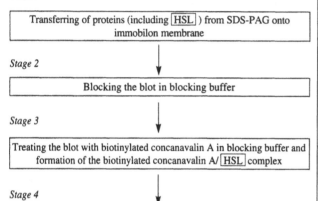

Stage 1. Transfer proteins from an electrophoretic SDS-PAG onto an Immobilon PVDF-P membrane using routine capillary blotting procedure (see Part II, Section 6).

Stage 2. Block the membrane containing blotted proteins for 1 h with phosphate-buffered saline with 1 mM CaCl$_2$, 1% gelatin, and 0.05% Nonidet P-40 (pH 7.4) (blocking buffer).

Stage 3. Incubate the blot for 2 h at room temperature with 0.2 µg/ml biotinylated concanavalin A in blocking buffer.

Stage 4. Wash the blot in blocking buffer and incubate for 1 h with avidin–alkaline phosphatase (ALP) (1 µg/ml) in blocking buffer. Wash the membrane five times for 5 min with blocking buffer and stain with 0.33 mg/ml NBT and 0.165 mg/ml 5-bromo-4-chloro-3-indolyl phosphate in 100 mM Tris–HCl, 100 mM NaCl, and 5 mM MgCl$_2$ (pH 9.5).

The dark blue bands on the developed membrane correspond to localization of HSL on the gel.

Notes: The method is based on the specific and strong binding of biotinylated concanavalin A with HSL enzyme protein and the subsequent histochemical detection of the HSL–biotinylated concanavalin A complex with ALP-conjugated avidin.

The method was used to detect purified HSL from human platelets after SDS-PAG electrophoresis. The convenience of the method for the in-gel detection of HSL after electrophoresis of crude enzyme preparations was not tested.

OTHER METHODS

The fluorochrome, tris(2,2′-bipyridine)-ruthenium(II) (Rubipy), was recently used to study catalytic activity of HSL by the in-gel monitoring of the decrease of fluorescence of the enzyme reaction products.[2] This method is based on the ability of Rubipy (a fluorescent cationic molecule) to bind to the negatively charged heparan sulfate (HS) groups. The Rubipy method can be adapted to detect HSL activity bands on electrophoretic gels using an HS-containing (10 µg/ml) agarose gel as an indicator. After incubation of combined electrophoretic and indicator gels and subsequent staining of the substrate-containing indicator gel with Rubipy solution (50 µg/ml), dark (nonfluorescent) bands of HSL can be observed in UV light on a light (fluorescent) gel background. Careful washing of the indicator gel stained with Rubipy is recommended to eliminate unbound fluorochrome. Two 10-min washes with distilled water may prove sufficient to develop HSL bands. However, differences in the water pH can modify the time needed to obtain maximum contrast. It is recommended, therefore, to monitor the gel in UV light during the washes. This method is recommended on trial.

GENERAL NOTES

The enzyme catalyzes elimination of sulfate in heparan sulfate, appears to act at the hexosamine–glucuronic acid linkage, and preferentially cleaves the antithrombin-III binding domain of heparan sulfate.

It does not act on *N,O*-desulfated glucosamine or *N*-acetyl-*O*-sulfated glucosamine linkages.[3]

Human HSL exhibits a pH optimum of 6.0, but is rapidly and reversibly converted to an inactive form at neutral pH. The enzyme from human platelets forms dimers and tetramers upon storage at 4°C.[1]

The Rubipy method[2] can be adapted for in-gel detection of activity bands of another enzyme with heparinase activity, heparin lyase (EC 4.2.2.7). In this case, however, heparin should be used as a substrate instead of heparan sulfate. Heparin is a linear, highly sulfated polysaccharide consisting of either L-iduronic acid or D-glucuronic acid and *N*-acetyl-D-glucosamine residues. HS has a lower degree of sulfatation and more glucuronic than iduronic acid residues. Heparin lyase preferentially acts at the hexosamine–iduronic acid linkages, and therefore HS is a poor substrate for heparin lyase. This difference may be used for discrimination between HSL and heparin lyase activity bands on electrophoretic gels.

4.2.2.8 — Heparitin-Sulfate Lyase; HSL (continued)

REFERENCES

1. Gonzalez-Stawinski, G.V., Parker, W., Holzknecht, Z.E., Huber, N.S., and Platt, J.L., Partial sequence of human platelet heparitinase and evidence of its ability to polymerize, *Biochim. Biophys. Acta*, 1429, 431, 1999.
2. Rosenberg, G.I., Espada, J., de Cidre, L.L., Eiján, A.M., Calvo, J.C., and Bertolesi, G.E., Heparan sulfate, heparin, and heparinase activity detection on polyacrylamide gel electrophoresis using the fluorochrome tris(2,2′-bipyridine) ruthenium (II), *Electrophoresis*, 22, 3, 2001.
3. NC-IUBMB, *Enzyme Nomenclature*, Academic Press, San Diego, 1992, p. 482 (EC 4.2.2.8, Comments).

4.2.2.10 — Pectin Lyase; PECL

REACTION	Eliminative cleavage of pectin to give oligosaccharides with terminal 4-deoxy-6-methyl-α-D-galact-4-enuronosyl groups
ENZYME SOURCE	Fungi
SUBUNIT STRUCTURE	Monomer[#] (fungi)

METHOD

Visualization Scheme

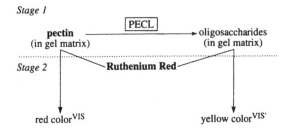

Staining Solution[1]

A. 100 mM Malic acid
B. 0.2% Ruthenium Red

Procedure

Incubate an electrophorized PAG containing 0.1% citrus pectin in 100 ml of solution A at room temperature for 90 min to cause a gradual pH change in the gel from the pH value of the gel buffer to pH 3.0, allowing PECL isozymes and different pectic enzymes (see *Notes*) to act on the pectin while passing through a suitable pH range. Following a brief rinse in distilled water, stain the gel in solution B for 30 to 120 min. Where PECLs active at high pH values are likely to occur, treat a control gel in the same way, but use solution B containing 2 mM CaCl$_2$. Wash the stained gel with several changes of distilled water for 1 h to overnight. Yellow bands visible on a pale red background of the gel correspond to localization of PECL.

Notes: The change in color of the stain from red to yellow at sites of PECL action at low pH is probably due to intermediate reaction products oxidizing Ruthenium Red. The continuation of enzyme activity during staining is necessary to produce the effect of color change. The color change is not produced when enzyme action is inhibited by chilling to 0°C during staining. To detect activity bands of PECL at high pH values, the Ca^{2+} requirement of PECL should be used.

This method also detects activity bands caused by pectinesterase (see 3.1.1.11 — PE, Method 3) and polygalacturonase (see 3.2.1.15 — PG, Method 2). In pectin-containing PAGs processed as described above, PE produces zones of dark red color and PG produces colorless or pale zones visible on a pale red background.

4.2.2.10 — Pectin Lyase; PECL (continued)

A colorimetric test was recently developed that allows differentiation between PG and PECL activities using N-(pyridin-2-yl)-thiobarbituric acid, which specifically interacts with products of PECL reaction and forms a pink fluorescent dye ($\lambda = 550$ nm). No interaction was observed with products of PE or PG reactions.[2] This method may be adapted to electrophoretic gels using the two-dimensional gel spectroscopy technique (see Part II, Section 5).

GENERAL NOTES

The enzyme does not act on deesterified pectin.[3]

REFERENCES

1. Cruickshank, R.H. and Wade, G.C., Detection of pectic enzymes in pectin-acrylamide gels, *Anal. Biochem.*, 107, 177, 1980.
2. Nedjma, M., Hoffmann, N., and Belarbi, A., Selective and sensitive detection of pectin lyase activity using a colorimetric test: application to the screening of microorganisms possessing pectin lyase activity, *Anal. Biochem.*, 291, 290, 2001.
3. NC-IUBMB, *Enzyme Nomenclature*, Academic Press, San Diego, 1992, p. 483 (EC 4.2.2.10, Comments).

4.3.1.8 — Uroporphyrinogen-I Synthase; UPS

OTHER NAMES	Hydroxymethylbilane synthase (recommended name), porphobilinogen deaminase, pre-uroporphyrinogen synthase
REACTION	4 Porphobilinogen + H_2O = uroporphyrinogen-I + 4 NH_3
ENZYME SOURCE	Vertebrates
SUBUNIT STRUCTURE	Monomer (vertebrates)

METHOD

Visualization Scheme

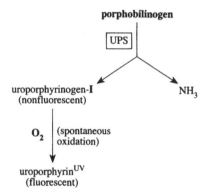

Staining Solution[1]

0.1 M Tris–HCl buffer, pH 8.2
90 mM Porphobilinogen

Procedure

Incubate the gel in staining solution in the dark at 45°C until red fluorescent bands become visible under long-wave UV light. Expose the gel to light and air for several minutes to complete the conversion of uroporphyrinogen to uroporphyrin. Record the zymogram or photograph using a yellow filter.

Notes: Staining solution may be applied to the gel surface on a cellulose acetate membrane and the final concentration of porphobilinogen reduced to 0.5 mM.[2]

4.3.1.8 — Uroporphyrinogen-I Synthase; UPS (continued)

OTHER METHODS

A. Areas of ammonia production may be detected in UV light using the backward reaction of the linking enzyme glutamate dehydrogenase (e.g., see 3.5.4.4 — ADA, Method 2).

B. The product ammonia may be visualized using sodium tetraphenylborate, which is precipitated in the presence of ammonia ions, resulting in the formation of white opaque bands visible in PAG (see 3.5.1.1 — ASP).

C. The local pH change due to production of ammonia may be detected using the pH indicator dye Phenol Violet (e.g., see 3.5.4.4 — ADA, Method 3), the NBT (or MTT)–dithiothreitol system (e.g., see 3.5.3.1 — ARG, Method 1), or neutral $AgNO_3$ solution containing photographic developers (e.g., see 3.5.1.5 — UR, Method 2).

GENERAL NOTES

In the presence of a second enzyme, uroporphyrinogen-III synthase (EC 4.2.1.75), the product uroporphyrinogen-I (hydroxymethylbilane) is cyclized to form uroporphyrinogen-III.[3]

REFERENCES

1. Meisler, M.H. and Carter, L.C., Rare structural variants of human and murine uroporphyrinogen I synthase, *Proc. Natl. Acad. Sci. U.S.A.*, 77, 2848, 1980.

2. Veser, J., Preparative free solution isoelectric focusing of human erythrocyte uroporphyrinogen I synthase in an ampholyte pH gradient, *Anal. Biochem.*, 182, 217, 1989.

3. NC-IUBMB, *Enzyme Nomenclature*, Academic Press, San Diego, 1992, p. 487 (EC 4.3.1.8, Comments).

4.3.2.1 — Argininosuccinate Lyase; ASL

OTHER NAMES	Argininosuccinase
REACTION	N-(L-Arginino)succinate = fumarate + L-arginine
ENZYME SOURCE	Bacteria, fungi, plants, invertebrates, vertebrates
SUBUNIT STRUCTURE	Tetramer[a] (plants, vertebrates)

METHOD 1

Visualization Scheme

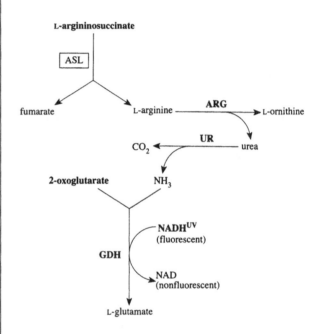

Staining Solution[1]

0.1 M Tris–HCl buffer, pH 7.6	5 ml
L-Argininosuccinic acid (Ba salt)	50 mg
Arginase (ARG)	40 U
Urease (UR)	20 U
2-Oxoglutarate	25 mg
NADH	10 mg
Glutamate dehydrogenase (GDH)	25 U

Procedure

Apply staining solution to the gel on a filter paper overlay. Incubate the gel at 37°C and monitor for defluorescent zones under long-wave UV light. Record the zymogram or photograph using a yellow filter.

METHOD 2

Visualization Scheme

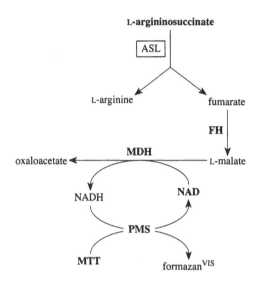

Staining Solution*

A. 0.1 *M* Phosphate buffer, pH 7.5	25 ml
L-Argininosuccinic acid (Ba salt)	50 mg
NAD	25 mg
Malate dehydrogenase (MDH)	30 U
Fumarate hydratase (FH)	60 U
PMS	1 mg
MTT	8 mg
B. 2% Agarose solution (60°C)	25 ml

Procedure

Mix A and B components of the staining solution and pour the mixture over the gel surface. Incubate the gel at 37°C in the dark until dark blue bands appear. Fix the stained gel in 25% ethanol.

METHOD 3

Visualization Scheme

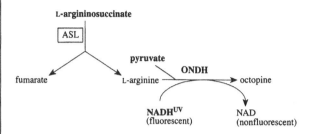

Staining Solution*

0.1 *M* Phosphate buffer, pH 7.5	10 ml
L-Argininosuccinic acid (Ba salt)	50 mg
NADH	10 mg
Octopine dehydrogenase (ONDH)	10 U
Sodium pyruvate	30 mg

Procedure

Apply staining solution to the gel on a filter paper overlay. Incubate the gel at 37°C and monitor for defluorescent zones under long-wave UV light. Record the zymogram or photograph using a yellow filter.

Notes: Lactate dehydrogenase activity bands can also be developed by this method. Therefore, an additional gel should be stained for LDH (see 1.1.1.27 — LDH, Other Methods, A) as a control.

OTHER METHODS

A bioautographic method (for details, see Part II) for ASL detection on an electrophoretic gel is also available.[2] However, this method is not as practical as the methods described above.

REFERENCES

1. Nelson, R.L., Povey, S., Hopkinson, D.A., and Harris, H., Detection after electrophoresis of enzymes involved in ammonia metabolism using L-glutamate dehydrogenase as a linking enzyme, *Biochem. Genet.*, 15, 1023, 1977.
2. Naylor, S.L., Klebe, R.J., and Shows, T.B., Argininosuccinic aciduria: assignment of the argininosuccinate lyase gene to the pter→q22 region of human chromosome 7 by bioautography, *Proc. Natl. Acad. Sci. U.S.A.*, 75, 6159, 1978.

* New; recommended for use.

OTHER NAMES	Lactoylglutathione lyase (recommended name), methylglyoxalase, aldoketomutase, ketone-aldehyde mutase
REACTION	*(R)-S*-Lactoylglutathione = glutathione + methylglyoxal
ENZYME SOURCE	Fungi, invertebrates, vertebrates
SUBUNIT STRUCTURE	Dimer (fungi, invertebrates, vertebrates)

METHOD 1

Visualization Scheme

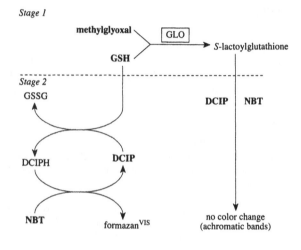

Staining Solution[1]

A. 0.2 *M* Phosphate buffer, pH 6.8
257 m*M* Methylglyoxal
16.3 m*M* Glutathione (reduced; GSH)
2.4 m*M* NBT

B. 0.1 *M* Tris–HCl buffer, pH 7.0 to 8.0
0.06 m*M* 2,6-Dichlorophenol indophenol (DCIP)

Procedure

Incubate the gel in solution A at 37°C for 15 to 45 min, and then place in solution B. Achromatic bands indicating areas of GLO activity develop on a blue background of the gel. Fix the stained gel in 25% ethanol.

Notes: This method was developed for starch gel. When it was applied to acetate cellulose gel with solution B prepared using Tris–HCl buffer (pH 7.8), dark blue bands were observed at the sites of GLO activity when the negatively stained gel was left in a moist chamber at room temperature.[2] This reversal is pH dependent because at pH levels higher than 8.0, white bands appear against a blue background also on an acetate cellulose gel.

A modified method that uses PMS instead of DCIP is also available.[3]

METHOD 2

Visualization Scheme

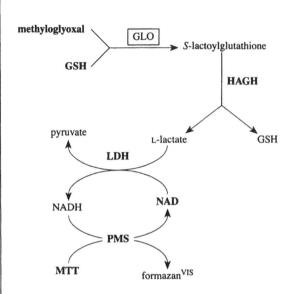

Staining Solution[3]

A. 0.2 *M* Phosphate buffer, pH 6.7	25 ml
Methylglyoxal (40% water solution)	0.5 ml
Glutathione (reduced; GSH)	30 mg
Hydroxyacylglutathione hydrolase (HAGH)	5 U
Lactate dehydrogenase (LDH)	20 U
NAD	25 mg
PMS	1 mg
MTT	8 mg
B. 2% Agarose solution (60°C)	25 ml

Procedure

Mix A and B components of the staining solution and pour the mixture over the gel surface. Incubate the gel at 37°C in the dark until dark blue bands appear. Fix the stained gel in 25% ethanol.

Notes: The method usually gives substantial staining of the gel background. This problem can be avoided by omission of PMS and MTT from the staining solution and viewing the gel for fluorescent bands under long-wave UV lamp.

4.4.1.5 — Glyoxalase I; GLO (continued)

METHOD 3

Visualization Scheme

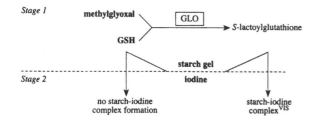

Stage 1

methylglyoxal + GSH → GLO → S-lactoylglutathione

starch gel / iodine

Stage 2

no starch-iodine complex formation | starch-iodine complex[VIS]

Staining Solution[4]

A.	0.2 M Phosphate buffer, pH 6.7	12 ml
	Glutathione (reduced; GSH)	40 mg
	Methylglyoxal (40% water solution)	0.5 ml
B.	0.7% Agar solution (45°C)	30 ml
	1% Iodine in 1% KI solution	1.3 ml

Procedure

Apply solution A to the surface of the starch gel on a filter paper overlay and incubate the gel at 37°C for 40 min. Remove the filter paper and blot the gel carefully to remove excess solution A; then pour solution B over the gel surface. Intense blue bands appear almost immediately in areas of GLO location due to starch–iodine complex formation. Record the zymogram or photograph because staining is not stable.

Notes: A starch gel preincubated with solution A may be further developed by placing it in a tank containing some crystals of iodine.[5]

This method is applicable to starch gels or PAGs containing soluble starch.

REFERENCES

1. Kömpf, J., Bissbort, S., Gussmann, S., and Ritter, H., Polymorphism of red cell glyoxalase I (EC 4.4.1.5): a new genetic marker in man, *Humangenetik*, 27, 141, 1975.
2. Meera Khan, P. and Doppert, B.A., Rapid detection of glyoxalase I (GLO) on cellulose acetate gel and the distribution of GLO variants in a Dutch population, *Hum. Genet.*, 34, 53, 1976.
3. Bagster, I.A. and Parr, C.W., Human erythrocyte glyoxalase I polymorphism, *J. Physiol.*, 256, 56P, 1976.
4. Parr, C.W., Bagster, I.A., and Welch, S.G., Human red cell glyoxalase I polymorphism, *Biochem. Genet.*, 15, 109, 1977.
5. Pflugshaupt, R., Scherz, R., and Bütler, R., Human red cell glyoxalase I polymorphism in the Swiss population: phenotype frequencies and simplified technique, *Hum. Hered.*, 28, 235, 1978.

4.4.1.13 — Cysteine-Conjugate β-Lyase; CCL

OTHER NAMES	Cysteine-*S*-conjugate β-lyase (recommended name)
REACTION	RS–CH₂CH(NH₃⁺)COO⁻ = RSH + pyruvate + ammonia; RH may represent aromatic compounds such as 4-bromobenzene and 2,4-dinitrobenzene
ENZYME SOURCE	Vertebrates
SUBUNIT STRUCTURE	Dimer[#] (vertebrates)

METHOD 1

Visualization Scheme

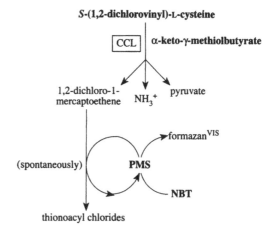

S-(1,2-dichlorovinyl)-L-cysteine

CCL → α-keto-γ-methiolbutyrate

1,2-dichloro-1-mercaptoethene + NH₃⁺ + pyruvate

(spontaneously) → thionoacyl chlorides

PMS → formazan[VIS] / NBT

Staining Solution[1]

0.1 M Potassium phosphate buffer, pH 7.2
2 mM *S*-(1,2-Dichlorovinyl)-L-cysteine
0.5 mM α-Keto-γ-methiolbutyrate
0.1 mM PMS
1 mM NBT

Procedure

Wash an electrophorized PAG twice with distilled water and incubate in the staining solution in the dark at 37°C until dark blue bands appear. Fix the stained gel in 25% ethanol.

Notes: Each of two identical subunits of CCL contains one equivalent of tightly bound pyridoxal 5′-phosphate. α-Keto-γ-methiolbutyrate is added to ensure maintenance of the enzyme in the pyridoxal 5′-phosphate form. In the absence of α-keto-γ-methiolbutyrate, the activity of CCL is greatly reduced. This is because with *S*-(1,2-dichlorovinyl)-L-cysteine as a substrate the enzyme catalyzes competing β-elimination (the products are pyruvate, ammonia, and an unstable thiol) and transamination (the product is *S*-(1,2-dichlorovinyl)-3-mercapto-2-oxopropionate) reactions. During the transamination reaction, the enzyme is converted to the pyridoxamine phosphate form, which cannot catalyze the β-elimination reaction. The added α-keto-γ-methiolbutyrate converts the pyridoxamine phosphate form of CCL to the pyridoxal phosphate form, which is competent to catalyze the β-elimination reaction.[2]

The mechanism of reduction of NBT is not known in detail. It may be more complex than a simple electron transfer from the sulfhydryl of the 1,2-dichloro-1-mercaptoethene via PMS molecules.

Faint bands caused by L-2-hydroxy acid oxidase (see 1.1.3.15 — GOX, General Notes) can also be developed by this method. Thus, an additional gel should be stained for GOX as a control.

METHOD 2

Visualization Scheme

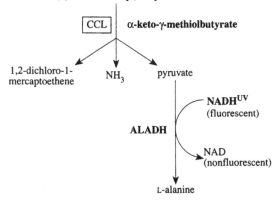

S-(1,2-dichlorovinyl)-L-cysteine

CCL   α-keto-γ-methiolbutyrate

1,2-dichloro-1-mercaptoethene   NH₃   pyruvate

NADH^UV (fluorescent)

ALADH

NAD (nonfluorescent)

L-alanine

Staining Solution[1] (adapted)

0.1 M Tris–HCl buffer, pH 8.0
1 mM α-Keto-γ-methiolbutyrate
2 mM S-(1,2-Dichlorovinyl)-L-cysteine
1 mM NADH
7 U/ml Alanine dehydrogenase (ALADH)

Procedure

Apply the staining solution to the gel surface with a filter paper or 1% agar overlay and incubate at 37°C. Monitor the gel under long-wave UV light for dark (nonfluorescent) bands visible on a light (fluorescent) background. Photograph the developed gel using a yellow filter.

Notes: This method is an adaptation of the spectrophotometric method used for detection of CCL activity in sections of PAG.

Lactate dehydrogenase (EC 1.1.1.27) is unsuitable as a linking enzyme for detection of the product pyruvate because it displays good activity toward α-keto-γ-methiolbutyrate.

GENERAL NOTES

The cystosolic form of CCL from the rat kidney was shown to be identical to a soluble form of glutamine transaminase K (see 2.6.1.64 — GTK).[2]

The enzyme is a pyridoxal-phosphate protein.[3]

REFERENCES

1. Abraham, D.G. and Cooper, A.J.L., Glutamine transaminase K and cysteine S-conjugate β-lyase activity stains, *Anal. Biochem.*, 197, 421, 1991.
2. Cooper, A.J.L. and Anders, M.W., Glutamine transaminase K and cysteine conjugate β-lyase, *Ann. N.Y. Acad. Sci.*, 585, 118, 1990.
3. NC-IUBMB, *Enzyme Nomenclature*, Academic Press, San Diego, 1992, p. 490 (EC 4.4.1.13, Comments).

4.6.1.1 — Adenylate Cyclase; AC

OTHER NAMES   Adenylyl cyclase, adenyl cyclase, 3′,5′-cyclic AMP synthetase

REACTION   ATP = 3′,5′-cyclic AMP + pyrophosphate

ENZYME SOURCE   Fungi, plants, invertebrates, vertebrates

SUBUNIT STRUCTURE   Oligomer^a (mammals); see General Notes

METHOD

Visualization Scheme

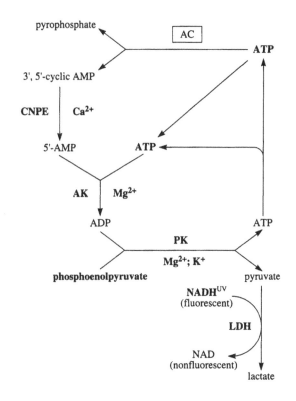

Staining Solution[1,2] (adapted)

A. 50 mM Tris–HCl buffer, pH 8.0
   20 mM ATP
   0.01 U/ml 3′,5′-Cyclic-nucleotide phosphodiesterase (CNPE; see *Notes*)
   2 U/ml Adenylate kinase (AK)
   2 U/ml Pyruvate kinase (PK)
   3 U/ml Lactate dehydrogenase (LDH)
   0.6 mg/ml Phosphoenolpyruvate
   1 mM NADH
   2 mM MgCl$_2$
   200 mM KCl
   0.06 mM CaCl$_2$
B. 1% Agarose solution (55°C)

Procedure

Mix equal volumes of A and B components of the staining solution and pour the mixture over the gel surface. Incubate the gel at 37°C and monitor under long-wave UV light for dark (nonfluorescent) bands visible on a light (fluorescent) background. Photograph the gel using a yellow filter.

Notes: Concentrations of auxiliary enzymes in solution A can require some adjustments.

Sigma preparation of CNPE (crude complex from bovine heart; P 0134) is recommended for use because it contains CNPE protein activator near saturation level and is the less expensive one.

OTHER METHODS

A fluorometric assay of AC activity can be adapted for detection of AC activity bands on electrophoretic gels. This method depends on the breakdown of cAMP generated by AC to AMP by auxiliary 3′,5′-cyclic-nucleotide phosphodiesterase (see 3.1.4.17 — CNPE) and the subsequent stimulation by AMP of auxiliary glycogen phosphorylase (see 2.4.1.1 — PHOS, General Notes). Activated glycogen phosphorylase produces glucose 1-phosphate, which can further be detected using auxiliary phosphoglucomutase and NAD- or NADP-dependent glucose-6-phosphate dehydrogenase.[3] Gel areas of NAD(P)H production may be observed in UV light or visualized using the PMS–MTT system. Theophylline (0.2 mM) can be used to inhibit endogenous 5′-nucleotidase that can destroy AMP (see 3.1.3.5 — 5′-N, Method 1) and thus interfere with development of AC activity bands by this method.

GENERAL NOTES

The enzyme is activated by NAD(P)–arginine ADP-ribosyltransferase (EC 2.4.2.31) in the presence of NAD.[4]

AC is a membrane-bound enzyme that plays a critical role in the signal transduction cascade of a number of fundamental neurotransmitters and hormones. Mammalian adenylate cyclases possess complex topologies, comprising two cassettes of six transmembrane-spanning motifs followed by a cytosolic, catalytic ATP-binding domain. It is established that the two cytosolic domains dimerize to form a catalytic core. The same is true for the two transmembrane cassettes. They also associate to facilitate the functional assembly and trafficking of the enzyme. Results are obtained that strongly suggest that adenylate cyclase molecules oligomerize via their hydrophobic domains. It is speculated that this property may allow AC to participate in multimeric signaling assemblies.[5]

REFERENCES

1. Moon, E. and Christiansen, R.O., Adenosine 3′,5′-monophosphate phosphodiesterase: multiple molecular forms, *Science*, 173, 540, 1971.

4.6.1.1 — Adenylate Cyclase; AC (continued)

2. Sugiyama, A. and Lurie, K.G., An enzymatic fluorometric assay for adenosine 3′,5′-monophosphate, *Anal. Biochem.*, 218, 20, 1994.
3. Wiegn, P., Dutton, J., and Lurie, K.G., An enzymatic fluorometric assay for adenylate cyclase activity, *Anal. Biochem.*, 208, 217, 1993.
4. NC-IUBMB, *Enzyme Nomenclature*, Academic Press, San Diego, 1992, p. 492 (EC 4.6.1.1, Comments).
5. Gu, C., Cali, J.J., and Cooper, D.M.F., Dimerization of mammalian adenylate cyclases: functional, biochemical and fluorescence resonance energy transfer (FRET) studies, *Eur. J. Biochem.*, 269, 413, 2002.

4.6.1.3 — 3-Dehydroquinate Synthase; DQS

REACTION	3-Deoxy-*arabino*-heptulosonate 7-phosphate = 3-dehydroquinate + orthophosphate
ENZYME SOURCE	Bacteria, fungi
SUBUNIT STRUCTURE	See General Notes

METHOD

Visualization Scheme

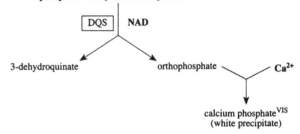

7-phospho-3-deoxy-*arabino*-heptulosonate

DQS | NAD

3-dehydroquinate   orthophosphate   Ca^{2+}

calcium phosphate^{VIS}
(white precipitate)

Staining Solution[1]

50 mM Glycine–KOH buffer, pH 10.0
1 mM 7-Phospho-3-deoxy-*arabino*-heptulosonate
50 μM NAD
10 mM Ca^{2+}

Procedure

Incubate an electrophorized PAG in the staining solution at 37°C until opaque bands visible against a dark background appear. Store the stained gel in 50 mM glycine–KOH (pH 10.0), containing 5 mM Ca^{2+}, at 5°C or at room temperature in the presence of an antibacterial agent (e.g., sodium azide).

Notes: The areas of calcium phosphate precipitation can be counterstained with Alizarin Red S. This does not increase the sensitivity of the method, but would be of advantage for opaque gels such as acetate cellulose or starch.

OTHER METHODS

The enzymatic method of detection of the product orthophosphate can also be used.[2] This method is based on reactions catalyzed by two linking enzymes, purine-nucleoside phosphorylase and xanthine oxidase, coupled with PMS–MTT (e.g., see 3.6.1.1 — PP, Method 2).

GENERAL NOTES

The enzyme from *Neurospora* is a domain in the multienzyme polypeptide "*arom*."[3]

4.6.1.3 — 3-Dehydroquinate Synthase; DQS (continued)

REFERENCES

1. Nimmo, H.G. and Nimmo, G.A., A general method for the localization of enzymes that produce phosphate, pyrophosphate, or CO_2 after polyacrylamide gel electrophoresis, *Anal. Biochem.*, 121, 17, 1982.
2. Klebe, R.J., Schloss, S., Mock, L., and Link, C.R., Visualization of isozymes which generate inorganic phosphate, *Biochem. Genet.*, 19, 921, 1981.
3. NC-IUBMB, *Enzyme Nomenclature*, Academic Press, San Diego, 1992, p. 563 (Nomenclature of Multienzymes, Symbolism).

4.6.1.4 — Chorismate Synthase; CHOS

REACTION	5-*O*-(1-Carboxyvinyl)-3-phosphoshikimate = chorismate + orthophosphate
ENZYME SOURCE	Bacteria, fungi, plants
SUBUNIT STRUCTURE	Tetramer[#] (bacteria, fungi), dimer[#] (plant plastids); see General Notes

METHOD

Visualization Scheme

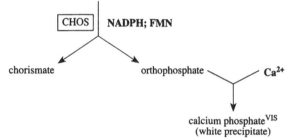

O^5-(1-carboxyvinyl)-3-phosphoshikimate

CHOS   NADPH; FMN

chorismate   orthophosphate Ca^{2+}

calcium phosphate[VIS]
(white precipitate)

Staining Solution[1]

100 mM Tris–HCl buffer, pH 8.0
50 μM 3-Enolpyruvylshikimate-5-phosphate
0.5 mM NADPH
10 μM FMN
10 mM Ca^{2+}

Procedure

Incubate an electrophorized PAG in a staining solution at 37°C until opaque bands visible against a dark background appear. Store the stained gel in 50 mM glycine–KOH (pH 10.0), containing 5 mM Ca^{2+}, at 5°C or at room temperature in the presence of an antibacterial agent (e.g., sodium azide).

Notes: The areas of calcium phosphate precipitation can be counterstained with Alizarin Red S. This would be of advantage for opaque gels such as acetate cellulose or starch.

GENERAL NOTES

The enzyme from *Neurospora* is a part of the multienzyme complex "*arom*" and is active only in the complex with NADPH-dependent flavine reductase. The purified enzyme requires Mg^{2+}, FAD, or FMN and diaphorase for its activity.

REFERENCES

1. Nimmo, H.G. and Nimmo, G.A., A general method for the localization of enzymes that produce phosphate, pyrophosphate, or CO_2 after polyacrylamide gel electrophoresis, *Anal. Biochem.*, 121, 17, 1982.

OTHER NAMES — Phosphoribulose epimerase, erythrose 4-phosphate isomerase

REACTION — D-Ribulose 5-phosphate = D-xylulose 5-phosphate (also see General Notes)

ENZYME SOURCE — Bacteria, fungi, vertebrates

SUBUNIT STRUCTURE — Dimer (vertebrates)

METHOD

Visualization Scheme

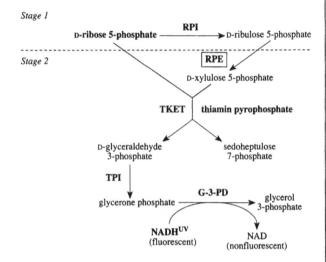

Stage 1

D-ribose 5-phosphate — RPI → D-ribulose 5-phosphate

Stage 2

RPE

D-xylulose 5-phosphate

TKET   thiamin pyrophosphate

D-glyceraldehyde 3-phosphate   sedoheptulose 7-phosphate

TPI

glycerone phosphate — G-3-PD → glycerol 3-phosphate

NADH^{UV} (fluorescent)   NAD (nonfluorescent)

Staining Solution[1]

A. 0.1 M Tris–HCl buffer, pH 8.0 — 10 ml
    0.1 M MgCl$_2$ — 40 ml
    0.25% Thiamin pyrophosphate, pH 7.0 — 40 ml
    Transketolase (TKET) — 1 U
    Triose-phosphate isomerase (TPI) — 50 U
    Glycerol-3-phosphate dehydrogenase (G-3-PD) — 17 U
    NADH — 20 mg
B. D-Ribose 5-phosphate — 10 mg
    Ribose 5-phosphate isomerase (RPI) — 50 U
    H$_2$O — 0.5 ml
C. 2% Agar solution (55°C) — 100 ml

Procedure

Prepare solution B and incubate at room temperature for 15 min before addition to solution A. Mix A and B solutions and add solution C. Pour the resulting mixture over the gel surface and incubate at 37°C until dark (nonfluorescent) bands visible in long-wave UV light appear on a light (fluorescent) background. Record the zymogram or photograph using a yellow filter.

OTHER METHODS

To obtain zymograms with RPE activity bands visible in daylight, glyceraldehyde-3-phosphate dehydrogenase (1.2.1.12 — GA-3-PD), sodium arsenate, NAD, PMS, and MTT should be included in solution A in place of TPI, G-3-PD, and NADH (e.g., see 5.4.2.7 — PPM, Method 1).

GENERAL NOTES

The enzyme also converts D-erythrose 4-phosphate into D-erythrulose 4-phosphate and D-threose 4-phosphate.[2]

REFERENCES

1. Spencer, N. and Hopkinson, D.A., Biochemical genetics of the pentose phosphate cycle: human ribose 5-phosphate isomerase (RPI) and ribulose 5-phosphate 3-epimerase (RPE), *Ann. Hum. Genet.*, 43, 335, 1980.
2. NC-IUBMB, *Enzyme Nomenclature*, Academic Press, San Diego, 1992, p. 496 (EC 5.1.3.1, Comments).

5.1.3.3 — Aldose 1-Epimerase; AE

OTHER NAMES | Mutarotase, aldose mutarotase
REACTION | α-D-Glucose = β-D-glucose
ENZYME SOURCE | Fungi, vertebrates (kidney)
SUBUNIT STRUCTURE | Unknown or no data available

METHOD

Visualization Scheme

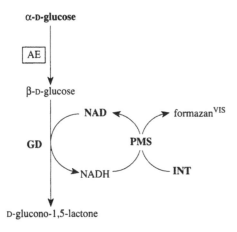

Staining Solution[1] (adapted)

A. 0.2 M Sodium phosphate buffer, pH 7.6 ... 20 ml
    α-D-Glucose ... 100 mg
    β-D-Glucose dehydrogenase (GD; EC 1.1.1.47) ... 150 U
    NAD ... 25 mg
    MTT (or INT) ... 8 mg
    PMS ... 1 mg
B. 1.5% Agarose solution (55°C) ... 20 ml

Procedure

Mix A and B components of the staining solution and pour the mixture over the gel surface. Incubate the gel in the dark at room temperature until dark blue bands appear. Fix the stained gel in 25% ethanol.

Notes: It is important to use α-D-glucose preparations that are free of the β-anomer.

OTHER METHODS

Two coupled reactions catalyzed by the auxiliary enzymes glucose oxidase and peroxidase can be used to detect the product β-D-glucose. Since glucose oxidase is an FAD-containing enzyme, the product β-D-glucose can also be detected using auxiliary glucose oxidase coupled with the MTT–PMS system (e.g., see 1.1.3.4 — GO).

GENERAL NOTES

The enzyme also acts on L-arabinose, D-xylose, D-galactose, maltose, and lactose.[2]

REFERENCES

1. Babczinsky, P., Fractionation of yeast invertase isozymes and determination of enzymatic activity in sodium dodecyl sulfate–polyacrylamide gels, *Anal. Biochem.*, 105, 328, 1980.
2. NC-IUBMB, *Enzyme Nomenclature*, Academic Press, San Diego, 1992, p. 496 (EC 5.1.3.3, Comments).

5.3.1.1 — Triose-Phosphate Isomerase; TPI

OTHER NAMES	Phosphotriose isomerase, triose-phosphate mutase
REACTION	D-Glyceraldehyde-3-phosphate = glycerone phosphate
ENZYME SOURCE	Bacteria, fungi, plants, invertebrates, vertebrates
SUBUNIT STRUCTURE	Dimer (fungi, plants, invertebrates, vertebrates)

METHOD 1

Visualization Scheme

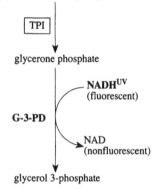

Staining Solution[1]

A. 0.1 *M* Triethanolamine–HCl buffer, pH 8.0
     (containing 5 m*M* EDTA)   20 ml
     30 m*M* D,L-Glyceraldehyde-3-phosphate
     (see *Notes*)   2 ml
     NADH   20 mg
     Glycerol-3-phosphate dehydrogenase (G-3-PD)   16 U
B. 2% Agar solution (60°C)   20 ml

Procedure

Mix A and B solutions and pour over the gel surface. Incubate the gel at 37°C until dark (nonfluorescent) bands visible in long-wave UV light appear on a light (fluorescent) background. Record the zymogram or photograph using a yellow filter.

Notes: D,L-Glyceraldehyde-3-phosphate (free acid) is commercially available; however, it is not sufficiently stable (stability studies indicate decomposition of as much as 10% in 3 days when exposed to room temperature). D-Glyceraldehyde-3-phosphate diethyl acetal (dicyclohexylammonium salt) available from Sigma is the more practical preparation, but it is very expensive. The optimal way is to generate D,L-glyceraldehyde-3-phosphate from the relatively cheap diethyl acetal barium salt, or from fructose-1,6-diphosphate using aldolase (see 1.2.1.12 — GA-3-PD, Method 1).

METHOD 2

Visualization Scheme

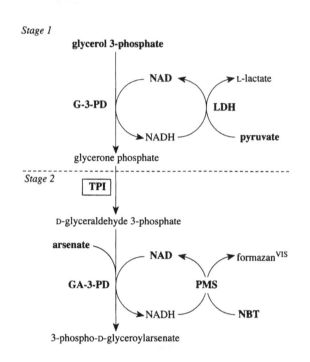

Staining Solution[2] (modified)

A. 0.1 *M* Tris–HCl buffer, pH 8.5   20 ml
     α-Glycerophosphate   600 mg
     NAD   20 mg
     Glycerol-3-phosphate dehydrogenase (G-3-PD)   20 U
     Sodium pyruvate   200 mg
     Lactate dehydrogenase (LDH)   60 U
B. Solution A (prepared as described in Procedure) 20 ml
     NAD   20 mg
     PMS   1 mg
     MTT   8 mg
     Sodium arsenate   70 mg
     Glyceraldehyde-3-phosphate dehydrogenase
     (GA-3-PD)   40 U
C. 2% Agar solution (55°C)   20 ml

Procedure

Incubate solution A at 37°C for 2 h and then stop reaction by dropwise addition of concentrated HCl until the pH is 2.0. Read-just solution A to pH 8.0 with NaOH. Prepare solution B and mix with solution C. Pour the resulting mixture over the gel surface and incubate the gel at 37°C in the dark until dark blue bands appear. Fix the stained gel in 25% ethanol.

5.3.1.1 — Triose-Phosphate Isomerase; TPI (continued)

Notes: The bands of G-3-PD and LDH also develop on TPI zymograms obtained by this method. Therefore, control stainings of two additional gels are necessary to identify activity bands caused by these two dehydrogenases.

A stable lithium salt of dihydroxyacetone phosphate (glycerone phosphate) is now available from Sigma Chemical Company. The use of this preparation in place of solution A makes control staining for G-3-PD and LDH activities unnecessary.[3]

REFERENCES

1. Harris, H. and Hopkinson, D.A., *Handbook of Enzyme Electrophoresis in Human Genetics*, North-Holland, Amsterdam, 1976 (loose-leaf, with supplements in 1977 and 1978).
2. Shaw, C.R. and Prasad, R., Starch gel electrophoresis of enzymes: a compilation of recipes, *Biochem. Genet.*, 4, 297, 1970.
3. Aebersold, P.B., Winans, G.A., Teel, D.J., Milner, G.B., and Utter, F.M., *Manual for Starch Gel Electrophoresis: A Method for the Detection of Genetic Variation*, NOAA Technical Report NMFS 61, U.S. Department of Commerce, National Marine Fisheries Service, Seattle, WA, 1987.

5.3.1.5 — Xylose Isomerase; XI

REACTION	D-Xylose = D-xylulose (see General Notes)
ENZYME SOURCE	Bacteria
SUBUNIT STRUCTURE	Tetramer[a] (bacteria); see General Notes

METHOD 1

Visualization Scheme

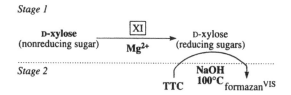

Staining Solution[1]

A. 100 mM Tricine–NaOH buffer, pH 8.0
   100 mM D-Xylose
   30 mM MgCl$_2$
B. 1 mg/ml 2,3,5-Triphenyltetrazolium chloride (TTC)
   1 M NaOH

Procedure

Incubate an electrophorized PAG in solution A at 37°C for 30 min. Wash the gel once with distilled water to remove excess substrate and incubate in the dark in solution B. Red bands appear in gel areas containing D-xylulose produced by XI. Terminate the staining reaction by soaking the gel in 2 M HCl.

Notes: The staining reaction may be accelerated by placing the gel in solution B heated in a boiling water bath.

METHOD 2

Visualization Scheme

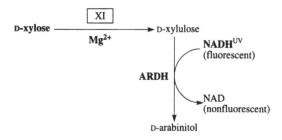

Staining Solution[1] (adapted)

100 mM Tris–HCl buffer, pH 8.0
100 mM D-Xylose
0.5 U/ml D-Arabinitol 4-dehydrogenase (ARDH; from
   Klebsiella aerogenes)
0.33 mM HADH
30 mM MgCl$_2$

511

Procedure

Apply the staining solution to the gel surface on a filter paper overlay and incubate at 37°C in a moist chamber. Monitor the gel under long-wave UV light. Dark (nonfluorescent) bands of XI are visible on a light (fluorescent) background. Record the zymogram or photograph using a yellow filter.

Notes: The auxiliary enzyme ARDH is not commercially available and should be obtained at laboratory conditions from an *E. coli* strain that harbors a bacteriophage causing super production of ARDH, as described.[2]

GENERAL NOTES

The enzyme from some sources demonstrates relatively high side specificity for glucose and converts D-glucose to D-fructose.[3]

It requires Mg^{2+}, Co^{2+}, or Mn^{2+} for activity; Mg^{2+} is best.

A trimeric subunit structure was suggested for the enzyme from *Bacillus*.[4]

REFERENCES

1. Smith, C.A., Rangarajan, M., and Hartley, B.S., D-Xylose (D-glucose) isomerase from *Arthrobacter* strain N.R.R.L. B3728, *Biochem. J.*, 277, 255, 1991.
2. Neuberger, M.S., Patterson, R.A., and Hartley, B.S., Purification and properties of *Klebsiella aerogenes* D-arabitol dehydrogenase, *Biochem. J.*, 183, 31, 1979.
3. NC-IUBMB, *Enzyme Nomenclature*, Academic Press, San Diego, 1992, p. 501 (EC 5.3.1.5, Comments).
4. Chauthaiwale, J. and Rao, M., Production and purification of extracellular D-xylose isomerase from an alkaliphilic, thermophilic *Bacillus* sp., *Appl. Environ. Microbiol.*, 60, 4495, 1994.

OTHER NAMES	Phosphopentoisomerase, phosphoriboisomerase
REACTION	D-Ribose 5-phosphate = D-ribulose 5-phosphate
ENZYME SOURCE	Bacteria, plants, invertebrates, vertebrates
SUBUNIT STRUCTURE	Tetramer[#] (fungi – *Saccharomyces cerevisiae*), dimer[#] (plants — spinach chloroplasts)

METHOD

Visualization Scheme

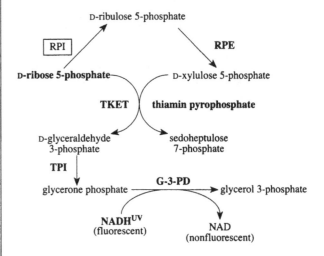

Staining Solution[1]

A.	0.1 *M* Tris–HCl buffer, pH 8.0	5 ml
	0.1 *M* MgCl$_2$	20 ml
	0.25% Thiamin pyrophosphate, pH 7.0	20 ml
	Transketolase (TKET)	0.5 U
	D-Ribose 5-phosphate	5 mg
	Ribulose-phosphate 3-epimerase (RPE)	0.5 U
	Triose-phosphate isomerase (TPI)	25 U
	Glycerol-3-phosphate dehydrogenase (G-3-PD)	8.5 U
	NADH	10 mg
B.	2% Agar solution (55°C)	45 ml

Procedure

Mix A and B components of the staining solution and pour the mixture over the gel surface. Incubate the gel at 37°C until dark (nonfluorescent) bands visible in long-wave UV light appear on a light (fluorescent) background. Record the zymogram or photograph using a yellow filter.

5.3.1.6 — Ribose 5-Phosphate Isomerase; RPI (continued)

Notes: To obtain a zymogram with RPI activity bands visible in daylight, glyceraldehyde-3-phosphate dehydrogenase (1.2.1.12 — GA-3-PD), sodium arsenate, NAD, PMS, and MTT should be included in solution A in place of TPI, G-3-PD, and NADH.

GENERAL NOTES

The enzyme also acts on D-ribose 5-diphosphate and D-ribose 5-triphosphate.[2]

REFERENCES

1. Spencer, N. and Hopkinson, D.A., Biochemical genetics of the pentose phosphate cycle: human ribose 5-phosphate isomerase (RPI) and ribulose 5-phosphate 3-epimerase (RPE), *Ann. Hum. Genet.*, 43, 335, 1980.
2. NC-IUBMB, *Enzyme Nomenclature*, Academic Press, San Diego, 1992, p. 501 (EC 5.3.1.6, Comments).

5.3.1.8 — Mannose-6-Phosphate Isomerase; MPI

OTHER NAMES	Phosphomannose isomerase, phosphohexomutase, phosphohexoisomerase
REACTION	D-Mannose 6-phosphate = D-fructose-6-phosphate
ENZYME SOURCE	Bacteria, fungi, plants, invertebrates, vertebrates
SUBUNIT STRUCTURE	Monomer (fungi, plants, invertebrates, vertebrates)

METHOD

Visualization Scheme

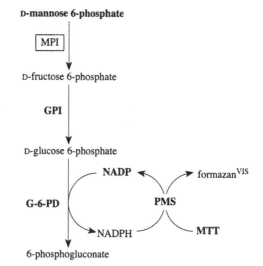

Staining Solution[1]

A. 0.35 *M* Tris–HCl buffer, pH 8.0   10 ml
    Mannose-6-phosphate (Ba salt; 3 H$_2$O);
      see *Notes*   10 mg
    PMS   0.8 mg
    MTT   3 mg
    NADP   5 mg
    Glucose-6-phosphate isomerase (GPI)   3.5 U
    Glucose-6-phosphate dehydrogenase (G-6-PD;
      see *Notes*)   1.5 U
B. 1.5% Agar solution (55°C)   20 ml

Procedure

Mix A and B components of the staining solution and pour the mixture over the gel surface. Incubate the gel in the dark at 37°C until dark blue bands appear. Fix the stained gel in 25% ethanol.

5.3.1.8 — Mannose-6-Phosphate Isomerase; MPI (continued)

Notes: The use of the NAD-dependent form of the linking enzyme G-6-PD (available from Sigma) is preferable because substitution of NAD for NADP in the staining solution results in cost savings.

A method for detecting MPI activity is described that utilizes mannose-6-phosphate produced *in vitro* by auxiliary hexokinase from mannose and ATP.[2] This method saves money because mannose-6-phosphate preparations (especially sodium salts) are expensive. The next (modified) reaction mixture is recommended to produce mannose-6-phosphate: 2 mg/ml D-mannose, 2 mg/ml ATP, 0.2 mg/ml MgCl$_2$, 0.1 mg/ml yeast hexokinase (lyophilized powder, Fluka), and 0.2 M Tris–HCl (pH 8.0). Incubate 10 ml of the reaction mixture for 0.5 h at 37°C and mix with 10 ml of solution A (lacking mannose-6-phosphate and prepared at double concentration). Finally, mix the resulting staining solution with an equal volume of 2.0% agar solution (60°C) and pour over the gel surface.

REFERENCES

1. Nichols, E.A. and Ruddle, F.H., A review of enzyme polymorphisms, linkage and electrophoretic conditions for mouse and somatic cell hybrids in starch gels, *J. Histochem. Cytochem.*, 21, 1066, 1973.
2. Zasypkin, M.Yu., Lapinskii, A.G., and Primak, A.A., A modified method for detection of post-phoretic activities in mannose- (MPI, EC 5.3.1.8) and glucose- (GPI, EC 5.3.1.9) 6-phosphate isomerases, *Genetika* (Moscow), 37, 708, 2001 (in Russian with English summary).

5.3.1.9 — Glucose-6-Phosphate Isomerase; GPI

OTHER NAMES	Phosphohexose isomerase, phosphohexomutase, oxoisomerase, hexosephosphate isomerase, phosphosaccharomutase, phosphoglucoisomerase, phosphohexoisomerase
REACTION	D-Glucose-6-phosphate = D-fructose-6-phosphate
ENZYME SOURCE	Bacteria, fungi, algae, plants, protozoa, invertebrates, vertebrates
SUBUNIT STRUCTURE	Dimer (fungi, algae, plants, protozoa, invertebrates, vertebrates)

METHOD

Visualization Scheme

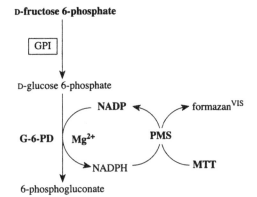

Staining Solution[1] (modified)

A.	0.1 M Tris–HCl buffer, pH 8.5	25 ml
	NADP	10 mg
	MTT	7 mg
	PMS	1 mg
	MgCl$_2$ (6H$_2$O)	40 mg
	D-Fructose-6-phosphate (Ba salt); see *Notes*	20 mg
	Glucose-6-phosphate dehydrogenase (G-6-PD; see *Notes*)	10 U
B.	2% Agar solution (60°C)	25 ml

Procedure

Mix A and B components of the staining solution and pour the mixture over the gel surface. Incubate the gel in the dark at 37°C until dark blue bands appear. Fix the stained gel in 25% ethanol.

5.3.1.9 — Glucose-6-Phosphate Isomerase; GPI (continued)

Notes: A method for detecting GPI activity is described that utilizes fructose-6-phosphate produced *in vitro* by auxiliary hexokinase from fruktose and ATP.[2] The next (modified) reaction mixture is recommended to produce fructose-6-phosphate: 2-mg/ml D-fructose, 2 mg/ml ATP, 0.2 mg/ml MgCl$_2$, 0.1 mg/ml yeast hexokinase (lyophilized powder, Fluka), and 0.2 M Tris–HCl (pH 8.5). Incubate 12.5 ml the reaction mixture for 0.5 h at 37°C and mix with an equal volume of solution A (lacking fructose-6-phosphate and prepared at double concentration). Finally, mix the resulting staining solution with an equal volume of 2% agar solution (60°C) and pour over the gel surface.

The use of the NAD-dependent form of the linking enzyme G-6-PD (available from Sigma) is preferable because substitution of NAD for NADP in the staining solution results in cost savings. The use of G-6-PD immobilized into the gel matrix is recommended when PAG is used for GPI electrophoresis.[3]

REFERENCES

1. De Lorenzo, R.J. and Ruddle, F.H., Genetic control of two electrophoretic variants of glucosephosphate isomerase, *Biochem. Genet.*, 3, 151, 1969.
2. Zasypkin, M.Yu., Lapinskii, A.G., and Primak, A.A., A modified method for detection of post-phoretic activities in mannose- (MPI, EC 5.3.1.8) and glucose- (GPI, EC 5.3.1.9) 6-phosphate isomerases, *Genetika* (Moscow), 37, 708, 2001 (in Russian with English summary).
3. Harrison, A.P., The detection of hexokinase, glucosephosphate isomerase and phosphoglucomutase activities in polyacrylamide gels after electrophoresis: a novel method using immobilized glucose-6-phosphate dehydrogenase, *Anal. Biochem.*, 61, 500, 1974.

5.3.3.12 — Dopachrome Δ-Isomerase; DI

OTHER NAMES	Dopachrome tautomerase, dopachrome conversion factor
REACTION	Dopachrome = 5,6-dihydroxyindole-2-carboxylate
ENZYME SOURCE	Invertebrates, vertebrates
SUBUNIT STRUCTURE	Unknown or no data available

METHOD 1

Visualization Scheme

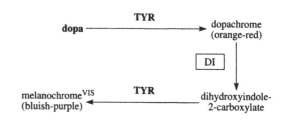

Staining Solution[1]

50 m*M* Sodium phosphate buffer, pH 6.0
10 m*M* L-Dopa

Procedure

Carry out electrophoresis in a PAG containing mushroom tyrosinase (Sigma; 3900 U/mg). Tyrosinase (1560 U/ml) is included in the gel mixture before polymerization.

After completion of electrophoresis, wash the tyrosinase-embedded PAG several times with distilled water and incubate in the staining solution at room temperature with constant shaking. DI activity bands appear within 3 to 10 min as bluish purple bands against an orange-red background, which is due to accumulation of dopachrome produced by tyrosinase. Frequent washing with water after 15 min incubation (to remove excess dopachrome) ensures clear background of the gel. Otherwise, the gel turns black due to continued formation of melanin.

Notes: The SDS-PAG containing tyrosinase can also be used to separate and detect DI; however, it must be washed extensively for 1 h after electrophoresis to remove the SDS. Otherwise, DI activity bands can not be clearly visualized. Even after removing the SDS, these gels require prolonged incubation (1 to 3 h) for development of DI bands. The background is much darker than that in native PAG.

The enzyme from mouse melanoma cells is a membrane-bound protein and requires solubilization by a detergent (e.g., Chaps). The buffer extracts do not show any activity bands of DI.

METHOD 2

Visualization Scheme

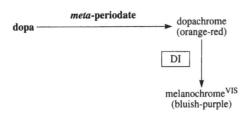

Staining Solution[2]

50 mM Sodium phosphate buffer, pH 6.0
10 mM L-Dopa
1% Sodium *meta*-periodate

Procedure

Prepare the staining solution immediately before use. Stain the gel with staining solution at room temperature for 3 min with constant shaking. Then wash the gel with 1% periodate in 50 mM sodium phosphate buffer (pH 6.0) for 3 min to continue visualization of bluish purple melanochrome bands against the orange-red background. Finally, wash the gel continuously in water (five to seven times, 100 ml each time). DI appears as bluish purple bands on a light background.

GENERAL NOTES

The tyrosinase Method 1 is two to three times more sensitive, but usually gives diffused DI bands in comparison with the periodate Method 2, probably because the dihydroxyindole produced by DI diffuses out before it is oxidized by tyrosinase to melanochrome. In the periodate Method 2, the dihydroxyindole suffers instantaneous oxidation with excess periodate. The melanochrome thus formed remains close to the enzyme molecules, thereby resulting in sharp bands. The tyrosinase Method 1 has some limitations, such as batch-to-batch variation in the specific activity of tyrosinase, loss of activity, and nonavailability of tyrosinase to certain investigators.[2]

REFERENCES

1. Nellaiappan, K., Nicklas, G., and Sugumaran, M., Detection of dopachrome isomerase activity on gels, *Anal. Biochem.*, 220, 122, 1994.
2. Nicklas, G. and Sugumaran, M., Detection of dopachrome isomerase on gels and membranes, *Anal. Biochem.*, 230, 248, 1995.

5.4.2.1 — Phosphoglycerate Mutase; PGLM

OTHER NAMES | Phosphoglyceric acid mutase, phosphoglycerate phosphomutase, phosphoglyceromutase

REACTION | 2-Phospho-D-glycerate = 3-phospho-D-glycerate (also see General Notes)

ENZYME SOURCE | Bacteria, fungi, plants, invertebrates, vertebrates

SUBUNIT STRUCTURE | Dimer (vertebrates)

METHOD 1

Visualization Scheme

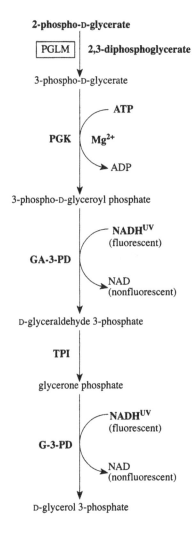

Staining Solution[1]

A. 0.2 M Tris–HCl buffer, pH 8.0   20 ml
   2-Phospho-D-glycerate (Na$_3$ salt; 6H$_2$O)   30 mg
   2,3-Diphosphoglycerate (pentacyclohexyl
   ammonium salt; 4H$_2$O)   15 mg
   Phosphoglycerate kinase (PGK)   60 U
   Glyceraldehyde-3-phosphate dehydrogenase
   (GA-3-PD)   100 U
   NADH   10 mg
   MgCl$_2$ (6H$_2$O)   12 mg
   EDTA (Na$_2$ salt; 2H$_2$O)   30 mg
   Triose-phosphate isomerase (TPI)   100 U
   Glycerol-3-phosphate dehydrogenase (G-3-PD)   40 U
B. 2% Agar solution (55°C)   20 ml

Procedure

Mix A and B solutions and pour the mixture over the gel surface. Incubate the gel at 37°C until dark (nonfluorescent) bands visible in long-wave UV light appear on a light (fluorescent) background. Record the zymogram or photograph using a yellow filter.

Notes: Auxiliary enzymes TPI and G-3-PD are used to intensify the progress of NADH-into-NAD conversion. These enzymes may be omitted from the staining solution when preparations with high PGLM activity are analyzed.

METHOD 2

Visualization Scheme

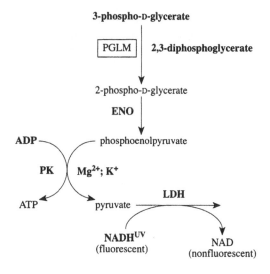

5.4.2.1 — Phosphoglycerate Mutase; PGLM (continued)

Staining Solution[2]

0.17 M Tris–HCl buffer, pH 8.0	10 ml
3-Phospho-D-glycerate (Na$_3$ salt)	10 mg
2,3-Diphosphoglycerate (pentacyclohexyl ammonium salt; 4H$_2$O)	8 mg
ADP (Na$_2$ salt)	15 mg
NADH	10 mg
MgCl$_2$ (6H$_2$O)	40 mg
KCl	30 mg
Enolase (ENO)	10 U
Pyruvate kinase (PK)	10 U
Lactate dehydrogenase (LDH)	15 U

Procedure

Apply the staining solution to the gel surface on a filter paper overlay and incubate the gel at 37°C until dark (nonfluorescent) bands visible in long-wave UV light appear on a light (fluorescent) background. Record the zymogram or photograph using a yellow filter.

GENERAL NOTES

The 2,3-diphosphoglycerate (DPG) may be firmly attached to the enzyme molecule during the catalytic cycle, or in other cases may be released so that free DPG is required as an activator. The enzyme from mammals and yeast requires DPG, while the enzyme from wheat, rice, insects, and some fungi has maximum activity in the absence of DPG.

Some mammalian PGLM isozymes also catalyze the reactions of bisphosphoglycerate phosphatase (see 3.1.3.13 — BPGP) and bisphosphoglycerate mutase (see 5.4.2.4 — BPGM). Comparative electrophoretic studies showed that in mammalian erythrocytes, PGLM, BPGP, and BPGM activities are determined by one and the same protein.[3]

REFERENCES

1. Rosa, R., Gaillardon, J., and Rosa, J., Characterization of 2,3-diphosphoglycerate phosphatase activity: electrophoretic study, *Biochim. Biophys. Acta*, 293, 285, 1973.
2. Chen, S.-H., Anderson, J., Giblett, E.R., and Lewis, M., Phosphoglyceric acid mutase: rare genetic variants and tissue distribution, *Am. J. Hum. Genet.*, 26, 73, 1974.
3. Rosa, R., Audit, I., and Rosa, J., Evidence for three enzymatic activities in one electrophoretic band of 3-phosphoglycerate mutase from red cells, *Biochimie*, 57, 1059, 1975.

5.4.2.2 — Phosphoglucomutase; PGM

OTHER NAMES	Glucose phosphomutase
REACTION	α-D-Glucose 1-phosphate = α-D-glucose 6-phosphate (also see General Notes)
ENZYME SOURCE	Bacteria, green algae, fungi, plants, protozoa, invertebrates, vertebrates
SUBUNIT STRUCTURE	Monomer (fungi, algae, plants, protozoa, invertebrates, vertebrates)

METHOD 1

Visualization Scheme

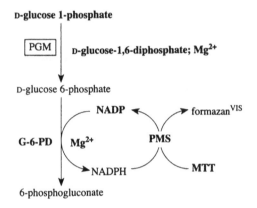

Staining Solution[1] (modified)

A.	0.05 M Tris–HCl buffer, pH 8.2	25 ml
	Glucose 1-phosphate (Na$_2$ salt, 4H$_2$O; containing about 1% of glucose-1,6-diphosphate)	100 mg
	MgCl$_2$ (6H$_2$O)	40 mg
	NADP	8 mg
	Glucose-6-phosphate dehydrogenase (G-6-PD; see *Notes*)	6 U
	PMS	1 mg
	MTT	7 mg
B.	2% Agar solution (60°C)	25 ml

Procedure

Mix A and B components of the staining solution and pour the mixture over the gel surface. Incubate the gel in the dark at 37°C until dark blue bands appear. Fix the stained gel in 25% ethanol.

Notes: The use of the NAD-dependent form of the linking enzyme G-6-PD (available from Sigma) is preferable because substitution of NAD for NADP in the staining solution results in cost savings.

METHOD 2

Visualization Scheme

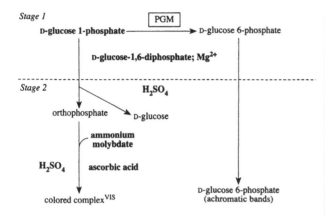

Stage 1

PGM

D-glucose 1-phosphate ⟶ D-glucose 6-phosphate

D-glucose-1,6-diphosphate; Mg^{2+}

Stage 2

H_2SO_4

orthophosphate ⟶ D-glucose

ammonium molybdate

H_2SO_4 | ascorbic acid

colored complex^{VIS}

D-glucose 6-phosphate (achromatic bands)

Staining Solution[2]

A. 25 mM Tris–2 mM cysteine hydrochloride buffer, pH 8.0
3 mM D-Glucose 1-phosphate
0.0147 mM D-Glucose-1,6-diphosphate
3 mM $MgCl_2$

B. 2% Agar solution (60°C)

C. 1.25% Ammonium molybdate in 2 N H_2SO_4 containing 50 mg/ml ascorbic acid

Procedure

Mix equal volumes of A and B solutions and pour the mixture over the gel surface. Incubate the gel at 37°C for 2 h. Remove the agar overlay, mix equal volumes of B and C solutions, and pour the mixture over the gel surface. Achromatic bands of PGM activity appear on a blue gel background almost immediately or 5 to 10 min after incubation at room temperature. Record or photograph the zymogram because staining is not stable.

GENERAL NOTES

The tetrazolium Method 1 gives better results. In the phosphate detection Method 2 the acid-labile phosphate tends to diffuse from the background into the clear areas of PGM activity.

The enzyme also, more slowly, catalyzes the interconversion of 1-phosphate and 6-phosphate isomers of many other α-D-hexoses, and the interconversion of α-D-ribose 1-phosphate and 5-phosphate. The last reaction is usually catalyzed by a separate enzyme, phosphopentomutase (see 5.4.2.7 — PPM). It was found, however, that some PGM isozymes exhibit strong phosphopentomutase activity.[2-4]

Maximum PGM activity is obtained only in the presence of α-D-glucose-1,6-diphosphate. This diphosphate is an intermediate in the reaction, being formed by transfer of a phosphate residue from the enzyme to the substrate, but the dissociation of diphosphate from the enzyme complex is much slower than the overall isomerization.[4]

REFERENCES

1. Spencer, N., Hopkinson, D.A., and Harris, H., Phosphoglucomutase polymorphism in man, *Nature*, 204, 742, 1964.
2. Quick, C.B., Fisher, R.A., and Harris, H., Differentiation of the PGM$_2$ locus isozymes from those of PGM$_1$ and PGM$_3$ in terms of phosphopentomutase activity, *Ann. Hum. Genet.*, 35, 445, 1972.
3. Quick, C.B., Fisher, R.A., and Harris, H., A kinetic study of the isozymes determined by the three human phosphoglucomutase loci PGM$_1$, PGM$_2$ and PGM$_3$, *Eur. J. Biochem.*, 42, 511, 1974.
4. NC-IUBMB, *Enzyme Nomenclature*, Academic Press, San Diego, 1992, p. 508 (EC 5.4.2.2, Comments).

OTHER NAMES	Diphosphoglycerate mutase, glycerate phosphomutase, bisphosphoglycerate synthase
REACTION	3-Phospho-D-glyceroyl phosphate = 2,3-bisphospho-D-glycerate (also see General Notes)
ENZYME SOURCE	Vertebrates
SUBUNIT STRUCTURE	Dimer (vertebrates)

METHOD

Visualization Scheme

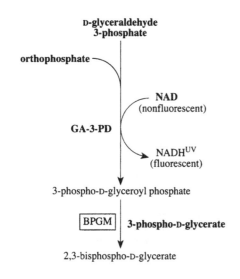

Staining Solution[1]

0.1 *M* Tris–HCl buffer, pH 7.8
3.5 m*M* D-Glyceraldehyde-3-phosphate
4 m*M* 3-Phospho-D-glycerate
10 m*M* K₂HPO₄
1.5 m*M* NAD
1 U/ml Glyceraldehyde-3-phosphate dehydrogenase (GA-3-PD)

Procedure

Apply the staining solution to the gel surface on a filter paper overlay. After 20 to 30 min of incubation at 37°C, monitor the gel for fluorescent bands under long-wave UV light. Record the zymogram or photograph using a yellow filter.

Notes: The method is based on removing the inhibitory effect of 3-phospho-D-glyceroyl phosphate on GA-3-PD activity via BPGM. As a result, NAD-to-NADH conversion is more pronounced in gel areas where BPGM activity is localized.

Arsenate may not be used as a substrate for GA-3-PD in place of K₂HPO₄ because of spontaneous arsenolysis of the product 3-phospho-D-glyceroyl arsenate and formation of free arsenate and 3-phospho-D-glycerate.

D-Glyceraldehyde-3-phosphate preparation may be obtained by 1 h of incubation at 37°C of the following mixture: 10 U of aldolase, 50 mg of fructose-1,6-diphosphate, and 1 ml of 0.1 *M* Tris–HCl buffer (pH 7.0).

OTHER METHODS

An alternative acid phosphomolybdate method for demonstration of BPGM that is based on bisphosphoglycerate phosphatase activity of the enzyme is also available (see 3.1.3.13 — BPGP, Method 2).

GENERAL NOTES

The enzyme is phosphorylated by 3-phospho-D-glyceroyl phosphate to give phosphoenzyme and 3-phospho-D-glycerate. The latter is rephosphorylated by the enzyme to yield 2,3-bisphospho-D-glycerate. This reaction, however, is slowed by dissociation of 3-phospho-D-glycerate from the enzyme, which is therefore more active in the presence of added 3-phospho-D-glycerate.[2]

The enzyme also catalyzes, slowly, the reaction of bisphosphoglycerate phosphatase (see 3.1.3.13 — BPGP) and phosphoglycerate mutase (see 5.4.2.1 — PGLM). Comparative electrophoretic studies showed that BPGM, BPGP, and PGLM activities in mammalian red cells are determined by one and the same protein.[3]

REFERENCES

1. Chen, S.-H., Anderson, J.E., and Giblett, E.R., 2,3-Diphosphoglycerate mutase: its demonstration by electrophoresis and the detection of a genetic variant, *Biochem. Genet.*, 5, 481, 1971.
2. NC-IUBMB, *Enzyme Nomenclature*, Academic Press, San Diego, 1992, p. 508 (EC 5.4.2.4, Comments).
3. Rosa, R., Audit, I., and Rosa, J., Evidence for three enzymatic activities in one electrophoretic band of 3-phosphoglycerate mutase from red cells, *Biochimie*, 57, 1059, 1975.

5.4.2.7 — Phosphopentomutase; PPM

OTHER NAMES   Phosphodeoxyribomutase
REACTION   D-Ribose 1-phosphate = D-ribose 5-phosphate (see also General Notes)
ENZYME SOURCE   Bacteria, invertebrates, vertebrates
SUBUNIT STRUCTURE Monomer (vertebrates)

METHOD 1

Visualization Scheme

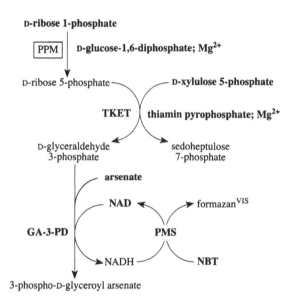

Staining Solution[1]

A. 30 mM Tris–HCl buffer, pH 8.0
   3 mM D-Ribose 1-phosphate
   0.0147 mM D-Glucose-1,6-diphosphate
   12 mM MgCl$_2$
   2 mM D-Xylulose 5-phosphate (however, see *Notes*)
   0.01 U/ml Transketolase (TKET)
   0.2 U/ml Glyceraldehyde-3-phosphate dehydrogenase (GA-3-PD)
   2 mM Sodium arsenate
   0.2 mM Thiamin pyrophosphate
   1.5 mM NAD
   0.02 mg/ml PMS
   0.2 mg/ml NBT
B. 2% Agar solution (55°C)

Procedure

Mix equal volumes of A and B components of the staining solution and pour the mixture over the gel surface. Incubate the gel in the dark at 37°C until dark blue bands appear. Fix the stained gel in 25% ethanol.

Notes: In the original description of the method,[1] D-ribulose 5-phosphate is included in the staining solution as the second substrate for linking enzyme transketolase. However, this enzyme catalyzes the reaction D-ribose 5-phosphate + D-xylulose 5-phosphate = D-glyceraldehyde-3-phosphate + sedoheptulose 7-phosphate. Therefore, D-xylulose 5-phosphate is given in this version of the staining solution in place of D-ribulose 5-phosphate. D-Ribulose 5-phosphate can be used in place of D-xylulose 5-phosphate only when coupled with ribulose-phosphate 3-epimerase (5.1.3.1 — RPE).

METHOD 2

Visualization Scheme

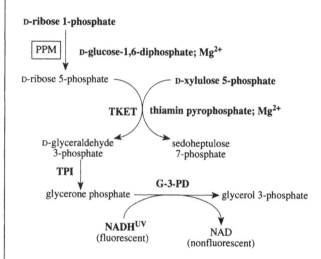

Staining Solution[1]

A. 25 mM Tris–20 mM cysteine hydrochloride buffer, pH 8.0
   3 mM D-Ribose 1-phosphate
   0.0147 mM D-Glucose-1,6-diphosphate
   2 mM MgCl$_2$
   2 mM D-Xylulose 5-phosphate (however, see Method 1, *Notes*)
   0.01 U/ml Transketolase (TKET)
   0.2 U/ml Glycerol-3-phosphate dehydrogenase (G-3-PD)
   1 U/ml Triose-phosphate isomerase (TPI)
   0.2 mM Thiamin pyrophosphate
   1.5 mM NADH
B. 2% Agar solution (55°C)

Procedure

Mix equal volumes of A and B solutions and pour the mixture over the gel surface. Incubate the gel at 37°C until dark (nonfluorescent) bands visible in long-wave UV light appear on a light (fluorescent) background. Record the zymogram or photograph using a yellow filter.

METHOD 3

Visualization Scheme

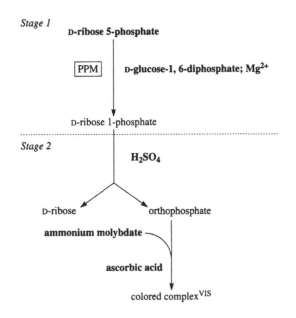

Staining Solution[1]

A. 2 mM Cysteine hydrochloride buffer, pH 8.0
    3 mM D-Ribose 5-phosphate
    2 mM MgCl$_2$
    0.0147 mM D-Glucose-1,6-diphosphate
B. 2% Agar solution (55°C)
C. 1.25% Ammonium molybdate in 2 N H$_2$SO$_4$ containing
    50 mg/ml ascorbic acid

Procedure

Mix equal volumes of A and B solutions and pour the mixture over the gel surface. Incubate the gel at 37°C for 1 to 2 h. Remove the first agar overlay, mix equal volumes of B and C solutions, and pour the mixture over the gel surface. The labile phosphate from the product D-ribose 1-phosphate is hydrolyzed from the sugar moiety under the acid conditions. The free phosphate then forms a phosphomolybdate complex with the molybdate, and this in turn is reduced by the ascorbic acid to form a blue compound in gel areas where PPM is localized. Record or photograph the zymogram because staining is not stable.

METHOD 4

Visualization Scheme

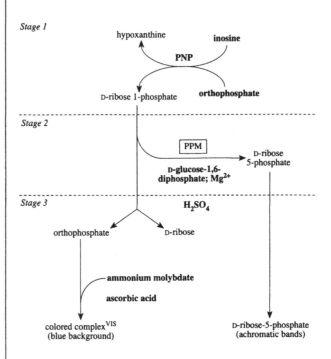

Staining Solution[1]

A. 20 mM Tris–2 mM cysteine hydrochloride buffer, pH 8.0
    2.5 mM K$_2$HPO$_4$ (see *Notes*)
    0.024 U/ml Purine-nucleoside phosphorylase (PNP)
    3 mM Inosine
    2 mM MgCl$_2$
B. D-Glucose-1,6-diphosphate (see Procedure)
C. 2% Agar solution (55°C)
D. 1.25% Ammonium molybdate in 2 N H$_2$SO$_4$ containing
    50 mg/ml ascorbic acid

Procedure

Prepare solution A and incubate at 37°C overnight. Then add component B (D-glucose-1,6-diphosphate) to solution A to give a final concentration of 0.0147 mM. Finally, add an equal volume of C and pour the resulting mixture over the gel surface. Incubate the gel at 37°C for 1 to 2 h. Remove the first agar overlay. Mix equal volumes of C and D solutions and pour the mixture over the gel surface again. White bands indicating PPM activity appear on a blue gel background. Record or photograph the zymogram because staining is labile.

Notes: If too much orthophosphate is added to solution A, the residual orthophosphate molecules can interact with ammonium molybdate in gel areas occupied by PPM and thus mask the development of achromatic PPM bands on a blue background.

5.4.2.7 — Phosphopentomutase; PPM (continued)

General Notes

The enzyme also converts 2-deoxy-D-ribose 1-phosphate into 2-deoxy-D-ribose 5-phosphate. The enzyme requires D-ribose-1,5-diphosphate or 2-deoxy-D-ribose-1,5-diphosphate as cofactors. D-Glucose-1,6-diphosphate also acts as a cofactor.[2]

It was found that some isozymes of mammalian red cell phosphoglucomutase (see 5.4.2.2 — PGM) exhibit strong PPM activity.[1,3]

References

1. Quick, C.B., Fisher, R.A., and Harris, H., Differentiation of the PGM$_2$ locus isozymes from those of PGM$_1$ and PGM$_3$ in terms of phosphopentomutase activity, *Ann. Hum. Genet.*, 35, 445, 1972.
2. NC-IUBMB, *Enzyme Nomenclature*, Academic Press, San Diego, 1992, p. 509 (EC 5.4.2.7, Comments).
3. Quick, C.B., Fisher, R.A., and Harris, H., A kinetic study of the isozymes determined by the three human phosphoglucomutase loci PGM$_1$, PGM$_2$ and PGM$_3$, *Eur. J. Biochem.*, 42, 511, 1974.

5.5.1.6 — Chalcone Isomerase; CI

OTHER NAMES	Chalcone-flavanone isomerase
REACTION	Chalcone = flavanone
ENZYME SOURCE	Plants
SUBUNIT STRUCTURE	Unknown or no data available

Method

Visualization Scheme

trihydroxychalcone[VIS] (yellow) $\xrightarrow{\boxed{\text{CI}}}$ dihydroxyflavanone (achromatic)

Staining Solution[1]

0.25% 2′,4,4′-Trihydroxychalcone in 0.002 N NaOH

Procedure

Spray an electrophorized PAG with the staining solution and incubate at room temperature. Within 10 to 60 min the yellow color of the chalcone disappears from the areas where CI activity is located. Record the zymogram.

Notes: 2′,4,4′,6′-Tetrahydroxychalcone can also be used as a colored CI substrate. The corresponding product of the CI reaction is a natural fluorochrome and can be observed under long-wave UV light.[2]

The enzyme from some sources (e.g., parsley) is inactive with chalcones lacking a 6′-hydroxyl group.

References

1. Hahlbrock, K., Wong, E., Schill, L., and Grisebach, H., Comparison of chalcone-flavanone isomerase heteroenzymes and isoenzymes, *Phytochemistry*, 9, 949, 1970.
2. Eigen, E., Blitz, M., and Gunsberg, E., The detection of some naturally occurring flavanone compounds on paper chromatography, *Arch. Biochem. Biophys.*, 68, 501, 1957.

OTHER NAMES	Tryptophanyl-tRNA synthetase
REACTION	ATP + L-tryptophan + tRNA^Trp = AMP + pyrophosphate + L-tryptophanyl-tRNA^Trp (see also *Notes*)
ENZYME SOURCE	Bacteria, fungi, plants, protozoa, invertebrates, vertebrates
SUBUNIT STRUCTURE	Dimer# (bacteria), dimer (vertebrates)

METHOD

Visualization Scheme

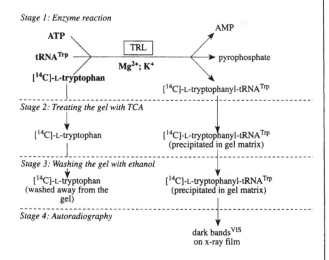

Stage 1: Enzyme reaction

ATP

tRNA^Trp   TRL   AMP

Mg²⁺; K⁺   pyrophosphate

[¹⁴C]-L-tryptophan

[¹⁴C]-L-tryptophanyl-tRNA^Trp

Stage 2: Treating the gel with TCA

[¹⁴C]-L-tryptophan

[¹⁴C]-L-tryptophanyl-tRNA^Trp
(precipitated in gel matrix)

Stage 3: Washing the gel with ethanol

[¹⁴C]-L-tryptophan
(washed away from the gel)

[¹⁴C]-L-tryptophanyl-tRNA^Trp
(precipitated in gel matrix)

Stage 4: Autoradiography

dark bands^VIS
on x-ray film

Staining Solution[1]

A. 50 mM Tris–HCL buffer, pH 7.4
   10 mM ATP
   40 mM Magnesium acetate
   10 mM Potassium chloride
   20 mM 2-Mercaptoethanol
   1 mg/ml Bovine serum albumin
   10⁻⁵ M [¹⁴C]-L-Tryptophan
   1 mg/ml crude yeast tRNA
B. 5% Trichloroacetic acid (TCA)
C. 95% Ethanol:1 M acetate buffer, pH 5.0   9:1

Procedure

Stage 1. Apply solution A (0.5 ml) by holding an electrophorized acetate cellulose gel (7.8 × 15 cm) at one end with forceps and placing the gel, porous side down, into solution A placed in a plastic tray. Invert the gel (porous side up), avoiding the formation of air bubbles between the gel and the tray. Blot excess liquid with filter paper. Moisten two strips of Whatman 3MM (4 × 8 cm) with 50 mM Tris–HCl buffer (pH 8.5) containing 20 mM 2-mercaptoethanol and 15% glycerol, and place lengthwise across both ends of the gel to prevent dessication. Place a tight-fitting cover on the tray and place the tray in a water bath at 37°C for 30 min.

Stage 2. After 30 min of incubation, flood the tray with 250 ml of cold solution B and expose for 10 min at 4°C with occasional agitation to stop the TRL reaction and precipitate tRNA in the gel matrix. Repeat this procedure twice.

Stage 3. Discard solution B and wash the gel twice for 10 min with 250 ml of cold solution C.

Stage 4. Dry the gel and mark the origin of each sample with red ink containing ³⁵S (2 × 10⁶ dpm/ml). Place the dry gel (porous side up) in contact with Kodak Blue Brand x-ray film and expose for a week at room temperature. Develop the exposed film using Kodak D 19 developer.

The dark bands on the developed x-ray film indicate areas occupied by TRL on the electrophoretic gel.

Notes: There is strong evidence that an aminoacyl-tRNA ligase reaction proceeds in two steps: (1) an enzyme + an amino acid + ATP = an enzyme·aminoacyl-AMP + pyrophosphate; and (2) an enzyme·aminoacyl-AMP + a tRNA = an enzyme + AMP + an aminoacyl-tRNA. It is obvious that the first step reaction can be used to detect aminoacyl-tRNA ligase activity in the absence of tRNA by the use of the product pyrophosphate.

OTHER METHODS

A. A sensitive method for the localization in PAG of enzymes that produce pyrophosphate is available.[2] This method is based on the formation of white calcium pyrophosphate precipitate. It was successfully used to detect leucine-tRNA ligase (6.1.1.4 — LRL), valine-tRNA ligase (6.1.1.9 — VRL), and cysteine-tRNA ligase (6.1.1.16 — CRL) activities after PAG electrophoresis.[3] The calcium pyrophosphate method can also be applied to TRL. This method, however, requires control staining of an additional gel to identify ATPase (see 3.6.1.3 — ATPASE) activity bands, which can also develop on TRL zymograms.

B. Fluorogenic reaction of L-7-azatryptophan (L-7-AT) with TRL, ATP, and Mg^{2+} in the presence of auxiliary inorganic pyrophosphatase results in the formation of a highly fluorescent L-7-AT-adenylate complex.[4] Detection of this complex is based on its enhanced fluorescence at 315-nm excitation and 360-nm emission. This method can be adapted for the in-gel detection of TRL.

GENERAL NOTES

It is well known that the formation of the TRL·aminoacyl-AMP complex results in the formation of inorganic pyrophosphate. The addition of excess inorganic pyrophosphate causes disruption of the complex and formation of ATP and the free amino acid. Thus, it is necessary to remove inorganic pyrophosphate in order to allow the formation of the aminoacyl-AMP to proceed to completion.[4]

REFERENCES

1. Denney, R.M. and Craig, I.W., Assignment of a gene for tryptophanyl–transfer ribonucleic acid synthetase (E.C.6.1.1.2) to human chromosome 14, *Biochem. Genet.*, 14, 99, 1976.
2. Nimmo, H.G. and Nimmo, G.A., A general method for the localization of enzymes that produce phosphate, pyrophosphate, or CO_2 after polyacrylamide gel electrophoresis, *Anal. Biochem.*, 121, 17, 1982.
3. Chang, G.-G., Deng, R.-Y., and Pan, F., Direct localization and quantitation of aminoacyl-tRNA synthetase activity in polyacrylamide gel, *Anal. Biochem.*, 149, 474, 1985.
4. Brennan, J.D., Hogue, C.W.V., Rajendran, B., Willis, K.J., and Szabo, A.G., Preparation of enantiomerically pure L-7-azatryptophan by an enzymatic method and its application to the development of a fluorimetric activity assay for tryptophanyl-tRNA synthetase, *Anal. Biochem.*, 252, 260, 1997.

OTHER NAMES	Leucyl-tRNA synthetase
REACTION	ATP + L-leucine + tRNA^{Leu} = AMP + pyrophosphate + L-leucyl-tRNA^{Leu} (see also *Notes*)
ENZYME SOURCE	Bacteria, fungi, plants, protozoa, invertebrates, vertebrates
SUBUNIT STRUCTURE	Dimer$^{\#}$ (fungi)

METHOD

Visualization Scheme

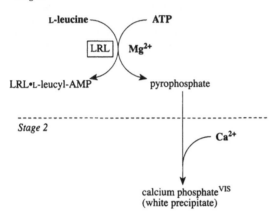

Staining Solution[1]

A. 50 mM Tris–HCL buffer, pH 7.7
   5 mM Magnesium acetate
   5 mM ATP
   5 mM L-Leucine
B. 100 mM CaCl$_2$, pH 8.9

Procedure

Incubate an electrophorized PAG in solution A at 37°C for 30 min and then in solution B at room temperature until white bands visible against a dark background appear. Store the stained gel in 50 mM glycine–KOH buffer (pH 10.0), containing 5 mM CaCl$_2$, either at 5°C or at room temperature in the presence of an antibacterial agent (e.g., sodium azide).

Notes: The reaction catalyzed by the aminoacyl-tRNA ligase proceeds in two steps: (1) enzyme + amino acid + ATP = enzyme·aminoacyl-AMP + pyrophosphate; and (2) enzyme·aminoacyl-AMP + tRNA = enzyme + AMP + aminoacyl-tRNA.

The first reaction, which does not require tRNA^{Leu}, is used in the method described above.

The calcium pyrophosphate method is simpler than the autoradiographic method, which uses labeled ATP.[1] This method, however, requires control staining of an additional gel to identify ATPase (see 3.6.1.3 — ATPASE) activity bands, which can also develop on LRL zymograms. When using this method, a two-step procedure of gel staining is needed because of differences in pH optimum of the enzyme and pH optimum of the calcium pyrophosphate formation.

REFERENCES

1. Chang, G.-G., Deng, R.-Y., and Pan, F., Direct localization and quantitation of aminoacyl-tRNA synthetase activity in polyacrylamide gel, *Anal. Biochem.*, 149, 474, 1985.

6.1.1.9 — Valine-tRNA Ligase; VRL

OTHER NAMES   Valyl-tRNA synthetase

REACTION   ATP + L-valine + tRNA^{Val} = AMP + pyrophosphate + L-valyl-tRNA^{Val} (see also *Notes*)

ENZYME SOURCE   Bacteria, fungi, plants, protozoa, invertebrates, vertebrates

SUBUNIT STRUCTURE Unknown or no data available

Method

Visualization Scheme

Stage 1

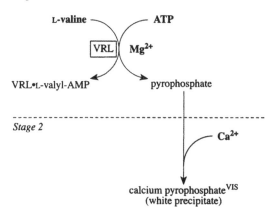

Stage 2

calcium pyrophosphate^{VIS}
(white precipitate)

Staining Solution[1]

A. 50 mM Tris–HCL buffer, pH 7.7
     5 mM Magnesium acetate
     5 mM ATP
     5 mM L-Valine
B. 100 mM CaCl$_2$, pH 8.9

Procedure

Incubate an electrophorized PAG in solution A at 37°C for 30 min and then in solution B at room temperature until white bands visible against a dark background appear. Store the stained gel in 50 mM glycine–KOH buffer (pH 10.0), containing 5 mM CaCl$_2$, either at 5°C or at room temperature in the presence of an antibacterial agent (e.g., sodium azide).

Notes: The reaction catalyzed by the aminoacyl-tRNA ligase proceeds in two steps: (1) enzyme + amino acid + ATP = enzyme·aminoacyl-AMP + pyrophosphate; and (2) enzyme·aminoacyl-AMP + tRNA = enzyme + AMP + aminoacyl-tRNA.

The first reaction, which does not require tRNA^{Val}, is used in the method described above.

The calcium pyrophosphate method is simpler than the autoradiographic method, which uses labeled ATP.[1] This method, however, requires control staining of an additional gel to identify ATPase (see 3.6.1.3 — ATPASE) activity bands, which can also develop on VRL zymograms. When using this method, a two-step procedure of gel staining is needed because of differences in pH optimum of the enzyme and pH optimum of the calcium pyrophosphate formation.

References

1. Chang, G.-G., Deng, R.-Y., and Pan, F., Direct localization and quantitation of aminoacyl-tRNA synthetase activity in polyacrylamide gel, *Anal. Biochem.*, 149, 474, 1985.

OTHER NAMES   Cystenyl-tRNA synthetase

REACTION   ATP + L-cysteine + tRNA^{Cys} = AMP + pyrophosphate + L-cystenyl-tRNA-^{Cys} (see also *Notes*)

ENZYME SOURCE   Bacteria, fungi, plants, protozoa, invertebrates, vertebrates

SUBUNIT STRUCTURE Unknown or no data available

METHOD

Visualization Scheme

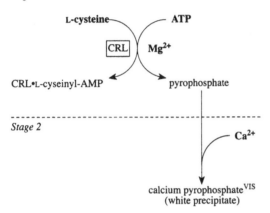

Stage 1

L-cysteine — ATP

CRL   Mg^{2+}

CRL•L-cyseinyl-AMP   pyrophosphate

Stage 2

Ca^{2+}

calcium pyrophosphate^{VIS}
(white precipitate)

Staining Solution[1]

A. 50 m*M* Tris–HCL buffer, pH 7.7
   5 m*M* Magnesium acetate
   5 m*M* ATP
   5 m*M* L-Cysteine
B. 100 m*M* CaCl$_2$, pH 8.9

Procedure

Incubate an electrophorized PAG in solution A at 37°C for 30 min and then in solution B at room temperature until white bands visible against a dark background appear. Store the stained gel in 50 m*M* glycine–KOH buffer (pH 10.0), containing 5 m*M* CaCl$_2$, either at 5°C or at room temperature in the presence of an antibacterial agent (e.g., sodium azide).

Notes: The reaction catalyzed by the aminoacyl-tRNA ligase proceeds in two steps: (1) enzyme + amino acid + ATP = enzyme·aminoacyl-AMP + pyrophosphate; and (2) enzyme·aminoacyl-AMP + tRNA = enzyme + AMP + aminoacyl-tRNA.

The first reaction, which does not require tRNA^{Cys}, is used to detect CRL by the method described above.

The calcium pyrophosphate precipitation method is simpler than the autoradiographic method, which uses labeled ATP.[1] This method, however, requires control staining of an additional gel to identify ATPase (see 3.6.1.3 — ATPASE) activity bands, which can also develop on CRL zymograms.

The two-step procedure of gel staining is needed because of differences in the pH optimum of the enzyme and the pH optimum of the calcium pyrophosphate formation.

REFERENCES

1. Chang, G.-G., Deng, R.-Y., and Pan, F., Direct localization and quantitation of aminoacyl-tRNA synthetase activity in polyacrylamide gel, *Anal. Biochem.*, 149, 474, 1985.

6.2.1.3 — Long-Chain Fatty Acid–CoA Ligase; ACAS

OTHER NAMES   Acyl-CoA synthetase, fatty acid thiokinase (long chain), acyl-activating enzyme, long-chain acyl-CoA synthetase, palmitoyl-CoA synthetase, lignoceroyl-CoA synthase, arachidonyl-CoA synthetase

REACTION   ATP + long-chain carboxylic acid + CoA = AMP + pyrophosphate + acyl-CoA (also see General Notes)

ENZYME SOURCE   Bacteria, vertebrates

SUBUNIT STRUCTURE Hexamer$^{\#}$ (vertebrates)

METHOD

Visualization Scheme

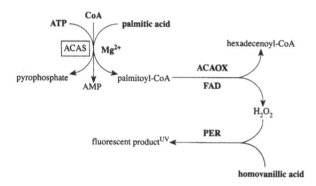

Staining Solution[1] (adapted)

0.01 M Tris–HCl buffer, pH 8.15
20 mM ATP
100 μM CoA
50 μM Palmitic acid
20 mM MgCl$_2$
2.5 U/ml Acyl-CoA oxidase (ACAOX; Sigma)
20 U/ml Horseradish peroxidase (PER)
50 μM FAD
1 mM Homovanillic acid

Procedure

Apply the staining solution to the gel surface with a filter paper or 1% agar overlay. Incubate the gel at 37°C and monitor under long-wave UV light for fluorescent bands. Record the zymogram or photograph using a yellow filter.

Notes: In a coupled peroxidase reaction, homovanillic acid reacts with hydrogen peroxide and a highly fluorescent dimer is formed that is excited at 313 nm and emits at 421 nm. Many other chromogenic substrates may be used in a coupled peroxidase reaction in place of homovanillic acid in order to generate products that are visible in daylight (see 1.11.1.7 — PER, Method 1).

It is quite probable that the PMS–MTT system also works well with auxiliary FAD-dependent acyl-CoA oxidase, as it does with other FAD-containing oxidases (e.g., see 1.1.3.4 — GO, Method 1; 1.1.3.22 — XOX, Method 1; 1.1.3.23 — TDH). If this is true, PMS and MTT (or NBT) may be included in the staining solution in place of peroxidase and homovanillic acid in order to generate blue formazan in gel areas where ACAS activity is localized.

OTHER METHODS

A. Exogenous adenylate kinase may be used as a linking enzyme to produce ADP from ATP present in the staining solution (see above) and AMP produced by ACAS. Using two additional linked reactions catalyzed by exogenous pyruvate kinase and lactate dehydrogenase, areas of ADP generation may be detected in long-wave UV light as dark bands on the fluorescent background of the gel (for example, see 2.7.1.11 — PFK, Method 3). When using this method, it should be taken into account that ADP may be produced from ATP by ATPase, for which activity bands can also be detected by this method (see 3.6.1.3 — ATPASE, Method 2).

B. The calcium pyrophosphate method may be used to detect pyrophosphate produced by ACAS in transparent electrophoretic gels. This method is based on the formation of white calcium pyrophosphate precipitate in gel areas where a pyrophosphate-releasing enzyme is localized (for example, see 2.7.7.9 — UGPP, Method 3).

C. Auxiliary inorganic pyrophosphatase may be used to convert pyrophosphate produced by ACAS into ortho-phosphate, which can subsequently be detected by the acid molybdate method (see 3.6.1.1 — PP, Method 1); the enzymatic tetrazolium method, which involves two linked reactions catalyzed by auxiliary enzymes purine-nucleoside phosphorylase and xanthine oxidase (see 3.6.1.1 — PP, Method 2); the modified calcium phosphate method, suitable for detection of orthophosphate in opaque gels (see 3.1.3.1 — ALP, Method 7); or the acid molybdate–Malachite Green method (see 3.1.3.2 — ACP, Method 4), which results in a stabler colored product than the routine acid molybdate method and is ideal for use in the assay of detergent-solubilized membrane-associated enzymatic activity.

GENERAL NOTES

The enzyme acts on a wide range of long-chain saturated and unsaturated fatty acids. Enzyme preparations from different tissues show some variation in specificity. For example, the liver enzyme acts on acids from C$_6$ to C$_{20}$, while that from the brain shows high activity up to C$_{24}$.[2]

REFERENCES

1. Lageweg, W., Steen, I., Tager, J.M., and Wanders, R.J.A., A fluorimetric assay for acyl-CoA synthetase activities, *Anal. Biochem.*, 197, 384, 1991.
2. NC-IUBMB, *Enzyme Nomenclature*, Academic Press, San Diego, 1992, p. 518 (EC 6.2.1.3, Comments).

6.2.1.X — Malonyl-CoA Synthase; MCAS

REACTION   ATP + malonate + CoA = malonyl-CoA + ADP + orthophosphate

ENZYME SOURCE   Bacteria, plants, vertebrates

SUBUNIT STRUCTURE Unknown or no data available

METHOD

Visualization Scheme

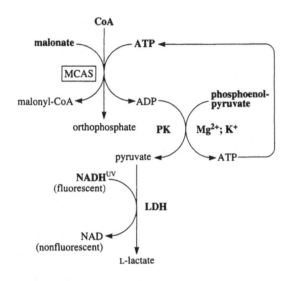

Staining Solution[1] (adapted)

A. 50 mM Potassium phosphate buffer, pH 7.2
     10 mM Sodium malonate
     5 mM MgCl$_2$
     1 mM Phosphoenolpyruvate
     0.2 mM ATP
     0.2 mM CoA
     0.1 mM NADH
     3.5 U/ml Pyruvate kinase (PK)
     5 U/ml Lactate dehydrogenase (LDH)

B. 2% Agarose solution (60°C)

Procedure

Mix equal volumes of solutions A and B and pour the mixture over the gel surface. Incubate the gel at 37°C until dark (nonfluorescent) bands visible in long-wave UV light appear on a light (fluorescent) background. Record the zymogram or photograph using a yellow filter.

REFERENCES

1. Kim, Y.S. and Bang, S.K., Assays for malonyl-coenzyme A synthase, *Anal. Biochem.*, 170, 45, 1988.

6.3.1.2 — Glutamine Synthetase; GS

OTHER NAMES   Glutamate–ammonia ligase (recommended name)

REACTION   ATP + L-glutamate + NH$_3$ = ADP + orthophosphate + L-glutamine

ENZYME SOURCE   Bacteria, fungi, plants, protozoa, vertebrates

SUBUNIT STRUCTURE Octamer[#] (bacteria, plant chloroplasts)

METHOD 1

Visualization Scheme

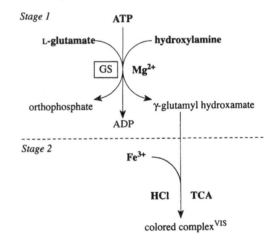

Staining Solution[1]

A. 0.1 M Tricine buffer, pH 7.8	100 ml
L-Glutamic acid (Na salt)	1.4 g
MgSO$_4$	250 mg
ATP (Na$_3$; 2H$_2$O)	500 mg
Hydroxylamine hydrochloride	50 mg
EDTA (Na$_4$; 3H$_2$O)	50 mg
B. 2.5 N HCl	100 ml
Trichloroacetic acid (TCA)	5 g
FeCl$_3$	10 g

Procedure

Incubate the gel in freshly prepared solution A at 37°C for 15 min to 3 h (see *Notes*). Then wash the gel in water and place in solution B. Greenish brown bands appear in a few minutes. Wash the stained gel in water and record or photograph the zymogram immediately because the stain is ephemeral.

Notes: Several slices from the same starch gel block should be used when assaying the enzyme for the first time. This is necessary to bracket the right incubation time in solution A, since the solution must be removed before one can see the stained bands.

METHOD 2

Visualization Scheme

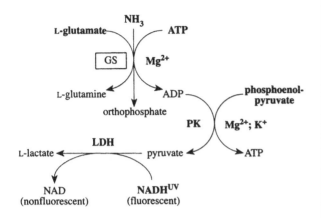

Staining Solution[2] (adapted)

0.1 M Tris–HCl buffer, pH 7.5
0.9 M KCl
50 mM MgCl$_2$
30 mM L-Glutamate
5 mM ATP
50 mM NH$_4$Cl
0.3 mM NADH
1 mM Phosphoenolpyruvate
1 U/ml Pyruvate kinase (PK)
5 U/ml Lactate dehydrogenase (LDH)

Procedure

Apply the staining solution to the gel surface on a filter paper overlay and incubate the gel in a humid chamber at 37°C until dark (nonfluorescent) bands visible in long-wave UV light appear on a light (fluorescent) background. Record the zymogram or photograph using a yellow filter.

Notes: When dark bands are clearly visible in UV light, treat the processed gel with PMS–MTT solution to obtain a permanent negative zymogram (white bands on a blue background).

In some organisms (e.g., mammals) additional bands caused by phosphoenolpyruvate phosphatase activity can also be developed by this method. Thus, an additional gel should be stained for activity of this phosphatase (see 3.1.3.18 — PGP, Method 2) as a control.

METHOD 3

Visualization Scheme

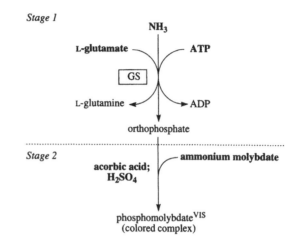

Staining Solution[3]

A.	0.2 M Tris–HCl buffer, pH 8.0	3 ml
	L-Glutamic acid	200 mg
	ATP	80 mg
	MgCl$_2$	2 mg
	NH$_4$OH	0.2 ml
B.	H$_2$O	89 ml
	Sulfuric acid	11 ml
	Ammonium molybdate	2.5 g
C.	Solution B	9 ml
	L-Ascorbic acid	1.2 g

Procedure

Prepare solution A and adjust the pH to 9.3. Apply the solution to the gel surface and incubate at 37°C for 1 h. Cover the gel with 50% acetone for 15 min at room temperature, and then rinse with deionized water. Prepare solution C and apply to the gel surface. Blue bands of GS activity appear after 1 to 5 min. Record or photograph the zymogram because the bands are ephemeral.

Notes: Do not add H$_2$O to acid when preparing solution B.

6.3.1.2 — Glutamine Synthetase; GS (continued)

OTHER METHODS

A. Using the reverse reaction, GS activity bands may be visualized by detecting the product ATP using two linking enzymes, hexokinase and glucose-6-phosphate dehydrogenase, coupled with the PMS–MTT system (see 2.7.1.1 — HK). This method, however, requires control staining of an additional gel for adenylate kinase (see 2.7.4.3 — AK), whose activity bands can also become apparent on GS zymograms.

B. The product of the reverse GS reaction, L-glutamate, may be detected using auxiliary glutamate dehydrogenase coupled with the PMS–MTT system (e.g., see 2.6.1.2 — GPT, Method 2).

C. An immunoblotting procedure (for details, see Part II) may also be used to detect GS. Antibodies specific to the mouse enzyme are now available from Sigma.

GENERAL NOTES

The enzyme from a variety of sources varies greatly in its ability to catalyze the reverse reaction. Thus, that from bacteria catalyzes the reverse reaction very slowly or not at all. With the mammalian enzyme, however, the forward rate relative to the reverse rate is about 1.

The enzyme acts, more slowly, on 4-methylene-L-glutamate.[4]

REFERENCES

1. Barratt, D.H.P., Method for the detection of glutamine synthetase activity on starch gels, *Plant Sci. Lett.*, 18, 249, 1980.
2. Soliman, A. and Nordlund, S., Purification and partial characterization of glutamine synthetase from the photosynthetic bacterium *Rhodospirillum rubrum*, *Biochim. Biophys. Acta*, 994, 138, 1989.
3. Morizot, D.C., Greenspan, J.A., and Siciliano, M.J., Linkage group VI of fish of the genus *Xiphophorus* (Poeciliidae): assignment of genes coding for glutamine synthetase, uridine monophosphate kinase, and transferrin, *Biochem. Genet.*, 21, 1041, 1983.
4. NC-IUBMB, *Enzyme Nomenclature*, Academic Press, San Diego, 1992, p. 522 (EC 6.3.1.2, Comments).

6.3.2.2 — Glutamate–Cysteine Ligase; GCL

OTHER NAMES	γ-Glutamylcysteine synthetase
REACTION	ATP + L-glutamate + L-cysteine = ADP + orthophosphate + γ-L-glutamyl-L-cysteine
ENZYME SOURCE	Bacteria, vertebrates
SUBUNIT STRUCTURE	Heterodimer[a] (mammals); see General Notes

METHOD

Visualization Scheme

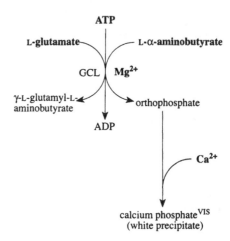

Staining Solution[1]

0.14 *M* Tris–HCl buffer, pH 8.0
0.14 *M* MgCl$_2$
14 m*M* L-Glutamate
14 m*M* L-α-Aminobutyrate
14 m*M* ATP
38 m*M* CaCl$_2$

Procedure

Incubate an electrophorized PAG in the staining solution at 37°C until white bands of calcium phosphate precipitation appear. Store the stained gel in 50 m*M* glycine–KOH buffer (pH 10.0), containing 5 m*M* Ca^{2+}, either at 5°C or at room temperature in the presence of an antibacterial agent.

Notes: Calcium phosphate precipitate may be counterstained with Alizarin Red S. This does not increase the sensitivity of the staining method, although it would be of advantage when more opaque gels are used (e.g., starch or acetate cellulose).

Some compounds, including L-α-aminobutyrate, can replace L-cysteine.

6.3.2.2 — Glutamate–Cysteine Ligase; GCL (continued)

OTHER METHODS

The product ADP can be detected by a negative fluorescent method using two linking enzymes, pyruvate kinase and lactate dehydrogenase (e.g., see 6.3.1.2 — GS, Method 2).

GENERAL NOTES

GCL can use L-aminohexanoate in place of L-glutamate.[2]

The enzyme from the developing mouse embryo is a heterodimer composed of a large catalytic subunit and a smaller modifying subunit, originating from different genes. The expression of mRNAs for different subunits is not always parallel in the embryo. Some tissues express one of the subunits preferentially, suggesting that corresponding genes are differentially expressed during mouse development.[3]

REFERENCES

1. Seelig, G.F., Simondsen, R.P., and Meister, A., Reversible dissociation of γ-glutamylcysteine synthetase into two subunits, *J. Biol. Chem.*, 259, 9345, 1984.
2. NC-IUBMB, *Enzyme Nomenclature*, Academic Press, San Diego, 1992, p. 523 (EC 6.3.2.2, Comments).
3. Diaz, D., Krejsa, C.M., and Kavanagh, T.J., Expression of glutamate-cysteine ligase during mouse development, *Mol. Reprod. Dev.*, 62, 83, 2002.

6.3.4.5 — Argininosuccinate Synthase; ARGS

OTHER NAMES	Citrulline–aspartate ligase
REACTION	ATP + L-citrulline + L-aspartate = AMP + pyrophosphate + *N*-(L-argino)succinate
ENZYME SOURCE	Bacteria, plants, vertebrates
SUBUNIT STRUCTURE	Unknown or no data available

METHOD

Visualization Scheme

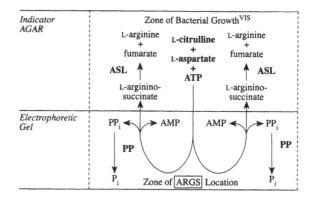

Indicator Agar[1]

1.5% indicator agar containing 100 μg/ml L-citrulline, 100 μg/ml L-aspartate, 100 μg/ml ATP, 1 U/ml argininosuccinate lyase (ASL), 0.5 U/ml inorganic pyrophosphatase (PP), and 10^8 bacteria/ml of *E. coli* 1115 (Arg G⁻) in minimal medium (see Appendix A-1) lacking L-arginine.

Procedure

Prepare indicator agar and pour it onto a sterile plate. After the bacteria-seeded indicator agar solidifies, a slice of electrophoretic starch gel (cut surface down), cellulose acetate strip, or PAG is laid over the agar, avoiding the formation of air bubbles between the indicator agar and electrophoretic gel. The bands of bacterial growth become visible in transmitted light after 6 to 12 h of incubation at 37°C.

Notes: As a control for bacterial growth, L-arginine should be spotted at one corner of the indicator agar. The origin and slot locations should be marked on the indicator agar before it is removed.

P. cerevisiae (ATCC 8081) with an arginine biosynthesis defect can also be used in place of *E. coli*. In this case, however, citrate medium (see Appendix A-2) lacking L-arginine should be substituted for minimal medium in the indicator agar.

Argininosuccinate lyase is used as a linking enzyme to produce L-arginine from L-argininosuccinate. Inorganic pyrophosphatase is added to destroy pyrophosphate (PPi), which inhibits ARGS activity.

6.3.4.5 — Argininosuccinate Synthase; ARGS (continued)

OTHER METHODS

A. A general method for the localization of enzymes that produce pyrophosphate is available.[2] The method is based on the insolubility of white calcium pyrophosphate precipitate forming as a result of interaction between Ca^{2+} ions included in the reaction mixture and pyrophosphate molecules produced by an enzyme reaction. This method can also be applied to ARGS electrophorized in clean gels (e.g., PAG or agarose gel). The zones of calcium pyrophosphate precipitation can be subsequently counterstained with Alizarin Red S. This procedure would be of advantage for starch or acetate cellulose gels.

B. The product AMP can be detected histochemically using four linked reactions catalyzed by auxiliary enzymes 5'-nucleotidase, adenosine deaminase, purine-nucleoside phosphorylase, and xanthine oxidase coupled with the PMS–MTT system (see 2.7.4.3 — AK, Method 3). Production of AMP can also be detected in UV light using three linked reactions catalyzed by three auxiliary enzymes, adenylate kinase, pyruvate kinase, and lactate dehydrogenase (see 3.1.4.17 — CNPE, Method 1).

C. The product L-argininosuccinate can be detected using two linking enzymes, argininosuccinate lyase (to produce L-arginine and fumarate from L-argininosuccinate) and octopine dehydrogenase (to convert NADH into NAD in areas where L-arginine is accumulated) (see 4.3.2.1 — ASL, Method 3). Two other, more complex methods of detection of L-argininosuccinate are also available (see 4.3.2.1 — ASL, Methods 1 and 2).

GENERAL NOTES

When methods B and C (see Other Methods above) are used, the addition of inorganic pyrophosphatase may prove beneficial to remove accumulating pyrophosphate, which inhibits AS activity.

REFERENCES

1. Naylor, S.L., Bioautographic visualization of enzymes, in *Isozymes: Current Topics in Biological and Medical Research*, Vol. IV, Rattazzi, M.C., Scandalios, J.G., and Whitt, G.S., Eds., Alan R. Liss, New York, 1980, p. 69.
2. Nimmo, H.G. and Nimmo, G.A., A general method for the localization of enzymes that produce phosphate, pyrophosphate, or CO_2 after polyacrylamide gel electrophoresis, *Anal. Biochem.*, 121, 17, 1982.

6.4.1.1 — Pyruvate Carboxylase; PC

OTHER NAMES	Pyruvic carboxylase
REACTION	ATP + pyruvate + HCO_3^- = ADP + orthophosphate + oxaloacetate
ENZYME SOURCE	Bacteria, fungi, plants, invertebrates, vertebrates
SUBUNIT STRUCTURE	Tetramer[a] (bacteria, vertebrates)

METHOD

Visualization Scheme

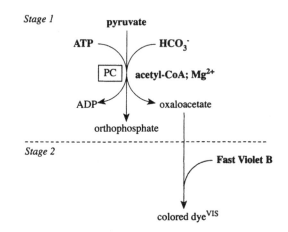

Staining Solution[1]

A. 100 mM Tris–HCl buffer, pH 7.8
5 mM Sodium pyruvate
2 mM ATP
5 mM $MgCl_2$
50 mM $KHCO_3$
0.3 mM Acetyl-CoA
B. 0.1% Fast Violet B

Procedure

Incubate the gel in solution A for 20 to 30 min at room temperature, wash with distilled water, and place in solution B. Incubate the gel in the dark until red bands appear. Wash the stained gel with water and fix in 5% acetic acid.

Notes: Propionyl-CoA also may be used in some cases in place of acetyl-CoA. Fast Blue B and Fast Garnet GBC can be used instead of Fast Violet B.

OTHER METHODS

A. The product pyruvate formed as a result of the reverse PC reaction can be detected by a negative fluorescent method using lactate dehydrogenase as a linking enzyme (e.g., see 2.7.1.40 — PK, Method 1).

6.4.1.1 — Pyruvate Carboxylase; PC (continued)

B. Another product of the reverse PC reaction, ATP, can be detected using two linking enzymes, hexokinase and NAD- or NADP-dependent glucose-6-phosphate dehydrogenase, coupled with the PMS–MTT system (e.g., see 2.7.1.40 — PK, Method 2).

C. There are several methods suitable for detection of orthophosphate produced by the forward PC reaction: the lead sulfide method (e.g., see 2.1.3.2 — ACT), the calcium phosphate method (e.g., see 3.1.3.2 — ACP, Method 5), the acid phosphomolybdate method (e.g., see 3.6.1.1 — PP, Method 1), and the tetrazolium enzymatic method (e.g., see 3.1.3.1 — ALP, Method 8). It should be taken into account that all these methods can also visualize the activity bands of acetyl-CoA carboxylase (see 6.4.1.2 — ACC) and propionyl-CoA carboxylase (when propionyl-CoA is used as the PC activator). In contrast to PC, these bands can develop in staining solution lacking pyruvate. Control staining for ATPase (see 3.6.1.3 — ATPASE) is also needed.

D. Because animal PC is a biotinyl-protein, the biotin–streptavidin (or avidin) conjugation method can also be used to detect the enzyme protein molecules (e.g., see 6.4.1.2 — ACC). This method is not specific for PC and detects some other biotin-containing proteins.

General Notes

The enzyme is a biotinyl-protein containing manganese (animal tissues) or zink (yeast). The animal enzyme is activated by acetyl-CoA.[2]

The enzyme from *Sinorhizobium meliloti* also is activated by acetyl-CoA.[3]

Most well-characterized forms of active PC consist of four identical subunits arranged in a tetrahedron-like structure. Each subunit includes three domains: the biotin carboxylation domain, the transcarboxylation domain, and the biotin carboxyl carrier domain.[4]

References

1. Scrutton, M.C. and Fatebene, F., An assay system for localization of pyruvate and phosphoenolpyruvate carboxylase activity on polyacrylamide gels and its application to detection of these enzymes in tissue and cell extracts, *Anal. Biochem.*, 69, 247, 1975.
2. NC-IUBMB, *Enzyme Nomenclature*, Academic Press, San Diego, 1992, p. 532 (EC 6.4.1.1, Comments).
3. Dunn, M.F., Araiza, G., and Finan, T.M., Cloning and characterization of the pyruvate carboxylase from *Sinorhizobium meliloti* Rm1021, *Arch. Microbiol.*, 176, 355, 2001.
4. Jitrapakdee, S. and Wallace, J.C., Structure, function and regulation of pyruvate carboxylase, *Biochem. J.*, 340, 1, 1999.

6.4.1.2 — Acetyl-CoA Carboxylase; ACC

REACTION	ATP + acetyl-CoA + HCO$_3$– = ADP + orthophosphate + malonyl-CoA
ENZYME SOURCE	Bacteria, plants, vertebrates
SUBUNIT STRUCTURE	Unknown or no data available

Method

Visualization Scheme (see *Notes*)

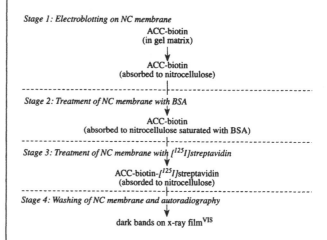

Stage 1: Electroblotting on NC membrane

ACC-biotin
(in gel matrix)
↓
ACC-biotin
(absorbed to nitrocellulose)

- -

Stage 2: Treatment of NC membrane with BSA

ACC-biotin
(absorbed to nitrocellulose saturated with BSA)

- -

Stage 3: Treatment of NC membrane with [^{125}I]streptavidin

ACC-biotin-[^{125}I]streptavidin
(absorbed to nitrocellulose)

- -

Stage 4: Washing of NC membrane and autoradiography

↓

dark bands on x-ray film[VIS]

Development by Autoradiography[1]

Stage 1. Transfer ACC molecules from the electrophoretic gel to a nitrocellulose membrane via electroblotting in 20 m*M* Tris–HCl (pH 8.3) buffer.

Stage 2. Incubate the nitrocellulose membrane overnight in 10 m*M* Tris–HCl (pH 7.4), 0.9% NaCl, and 3% BSA to saturate free protein-binding sites of nitrocellulose.

Stage 3. Place the nitrocellulose membrane in 3% BSA solution containing 8 × 10^5 cpm [^{125}I]streptavidin for 2 h.

Stage 4. Wash the nitrocellulose membrane in three changes of 10 m*M* Tris–HCl (pH 7.4) containing 0.9% NaCl to remove unbound [^{125}I]streptavidin. Dry the nitrocellulose membrane and expose to x-ray film at –90°C.

The dark bands on the developed x-ray film indicate areas of ACC localization in the electrophoretic gel.

Notes: The method is based on specific conjugation of labeled streptavidin molecules with biotinyl-protein molecules of ACC. Labeled avidin may be used in place of streptavidin.

Streptavidin and avidin can also be labeled by peroxidase, β-galactosidase, or alkaline phosphatase. Such conjugates are now commercially available (e.g., from Sigma). After specific conjugation with biotin-containing ACC molecules, these enzymes may be detected by routine histochemical procedures (see 1.11.1.7 — PER; 3.2.1.23 — β-GAL; 3.1.3.1 — ALP).

In plants, ACC is the only known biotin-containing enzyme. In animals some other enzymes are known to be biotinyl-proteins: pyruvate carboxylase (see 6.4.1.1 — PC), propionyl-CoA carboxylase (EC 6.4.1.3), methylcrotonoyl-CoA carboxylase (EC 6.4.1.4), and geranoyl-CoA carboxylase (EC 6.4.1.5). Thus, the biotin–streptavidin (or avidin) conjugation method is thought to be specific only for ACC from plant sources.

6.4.1.2 — Acetyl-CoA Carboxylase; ACC (continued)

OTHER METHODS

A. The product ADP can be detected by a negative fluorescent method using pyruvate kinase and lactate dehydrogenase as linking enzymes (e.g., see 2.7.1.35 — PNK).

B. The product orthophosphate can be detected by the lead sulfide method (e.g., see 2.1.3.2 — ACT), the calcium phosphate method (e.g., see 3.1.3.2 — ACP, Method 5), the acid phosphomolybdate method (e.g., see 3.6.1.1 — PP, Method 1), and the tetrazolium enzymatic method (e.g., see 3.1.3.1 — ALP, Method 8).

Both methods (A and B) can also detect ATPase activity bands. Thus, when these methods are used, a control staining of an additional gel for ATPase activity is needed (see 3.6.1.3 — ATPASE).

GENERAL NOTES

The enzyme is a biotinyl-protein. The plant enzyme also carboxylates propanoyl-CoA and butanoyl-CoA.[2]

REFERENCES

1. Nikolau, B.J., Wurtele, E.S., and Stumpf, P.K., Tissue distribution of acetyl-coenzyme A carboxylase in leaves, *Plant Physiol.*, 75, 895, 1984.
2. NC-IUBMB, *Enzyme Nomenclature*, Academic Press, San Diego, 1992, p. 532 (EC 6.4.1.2, Comments).

6.5.1.1 — DNA Ligase (ATP); DL(ATP)

OTHER NAMES	Polydeoxyribonucleotide synthase (ATP), polynucleotide ligase, sealase, DNA repair enzyme, DNA joinase
REACTION	(Deoxyribonucleotide)$_n$ + (deoxyribonucleotide)$_m$ + ATP = (deoxyribonucleotide)$_{n+m}$ + pyrophosphate + AMP
ENZYME SOURCE	Bacteriophages, bacteria, fungi, plants, protozoa, invertebrates, vertebrates
SUBUNIT STRUCTURE	Monomer[a] (bacteriophages, bacteria); see General Notes

METHOD 1

Visualization Scheme

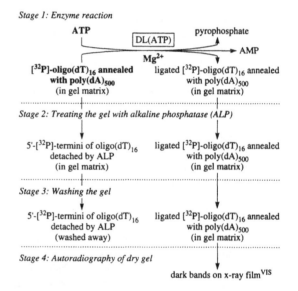

Substrate-Containing PAG[1]

Electrophoresis of DL(ATP) is carried out in SDS-PAG containing 1 to 3 nmol/ml of [^{32}P]oligo(dT)$_{16}$:poly(dA)$_{500}$ substrate (3×10^7 cpm/nmol). The substrate is prepared by labeling the oligo(dT)$_{16}$ with polynucleotide kinase (EC 2.7.1.78) and [γ-^{32}P]ATP (3000 Ci/mmol) and annealing the [^{32}P]oligo(dT) with poly(dA) (1:1).

Visualization Procedure[1]

After electrophoresis, renature proteins within the gel by washing it with five changes of 1 l of buffer containing 50 mM Tris–HCl (pH 7.8), 1 mM EDTA, 3 mM 2-mercaptoethanol, and 5 mM MgCl$_2$ at room temperature for 3 to 4 h. Then wash the gel twice for 15 min with 10 volumes of 20 mM Tris–HCl (pH 7.8), 5 mM MgCl$_2$, and 2 mM dithiothreitol.

Stage 1. Initiate the *in situ* ligase reaction by adding 1 mM ATP to the 20 mM Tris–HCl (pH 7.8), 5 mM MgCl$_2$, and 2 mM dithiothreitol. Incubate the gel in the reaction mixture at room temperature for 18 h with gentle agitation.

Stages 2 and 3. Wash the gel twice for 15 min with 100 ml of 80 mM Tris–HCl (pH 9.0), containing 5 mM MgCl$_2$, and treat twice for 2 h each with 10 U of calf intestinal alkaline phosphatase in 50 ml of phosphatase buffer (pH 9.0) in order to remove the unligated 5′-[^{32}P]-termini of oligo(dT). Finally, wash the gel twice with 100 ml of phosphatase buffer at room temperature for 15 min.

Stage 4. Autoradiograph the gel using Fuji x-ray film.

The dark bands on the developed x-ray film correspond to localization of DL(ATP) activity in the gel.

Notes: The electrophoresis of ligated [^{32}P]oligo(dT) using DNA sequencing gels is recommended to analyze the conversion of [^{32}P]-(dT) oligomers into labeled polymers of higher molecular mass, and thus to identify more precisely the product(s) of enzymatic reaction (see Method 2 below).

METHOD 2

Visualization Scheme

Stage 1: Electrophoretic separation of proteins in substrate-containing PAG (first dimension)

Stage 2: Vertical incision of the sample line of the first-dimension PAG and receipt of narrow (2 mm) substrate-containing gel strip

Stage 3: Enzyme reaction in the strip of the first-dimension PAG

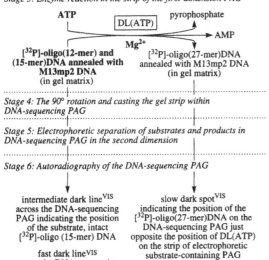

Stage 4: The 90° rotation and casting the gel strip within DNA-sequencing PAG

Stage 5: Electrophoretic separation of substrates and products in DNA-sequencing PAG in the second dimension

Stage 6: Autoradiography of the DNA-sequencing PAG

intermediate dark line^{VIS} across the DNA-sequencing PAG indicating the position of the substrate, intact [^{32}P]-oligo (15-mer) DNA

fast dark line^{VIS} across the DNA-sequencing PAG indicating the position of the substrate, intact [^{32}P]-oligo (12-mer)DNA

slow dark spot^{VIS} indicating the position of the [^{32}P]-oligo(27-mer)DNA on the DNA-sequencing PAG just opposite the position of DL(ATP) on the strip of electrophoretic substrate-containing PAG

Substrate-Containing PAG^{2,3}

Prepare a native PAG (2.5% stacking gel and 10% resolving gel) or SDS-PAG (4% stacking gel and 10% resolving gel) containing

a 1:1 mixture of [^{32}P]-oligo(12-mer)DNA and [^{32}P]-oligo(15-mer)DNA annealed with single-stranded M13mp2 DNA (see *Notes*).

Reaction Mixture^{2,3}

40 mM Tris–HCl buffer, pH 8.0
20 mM Dithiothreitol
400 µg/ml Bovine serum albumin
5% (w/v) Glycerol
10 mM MgCl$_2$
10 mM Ammonium sulfate
1 mM ATP

Procedure

Stage 1. After electrophoresis, immerse the resolving substrate-containing SDS-PAG in 33 gel volumes of SDS extraction buffer: 10 mM Tris–HCl (pH 7.5), 5 mM 2-mercaptoethanol, 25% (v/v) 2-propanol. Gently agitate the gel on a gyratory shaker (60 rpm) at 25°C. Discard the buffer after 30 min and replace with the same volume of fresh SDS extraction buffer. Extract SDS for an additional 30 min. Place the gel in 3 volumes of enzyme renaturation buffer (the reaction mixture lacking MgCl$_2$, ammonium sulfate, and ATP, but containing 50 mM KCl). After brief (30 sec) agitation, place the gel in 27 volumes of the enzyme renaturation buffer and incubate for 18 to 25 h with gentle agitation at 4°C.

Stage 2. Following renaturation of the enzyme within SDS-PAG or after native PAG electrophoresis, cut each line of the substrate-containing PAG into 2-mm-wide longitudinal strips.

Stage 3. Place an individual strip into a 16 × 100 mm test tube containing 5 ml of the reaction mixture. Perform the enzyme reaction at 25°C for 30 min in a stoppered (Parafilm) tube that is placed on its side and gently shaken (60 rpm). Terminate the reaction by substituting the reaction mixture with an equal volume of ice-cold renaturation buffer containing 10 mM EDTA and gently shaking the tube for 30 min at 4°C.

Stages 4, 5, and 6. Following the *in situ* enzyme reaction, cast the processed strip of substrate-containing PAG within a 20% DNA sequencing PAG containing 8.3 M urea, avoiding formation of air bubbles. After polymerization, carry out electrophoresis in the second dimension, transfer the electrophorized 20% PAG to filter paper, dry, and autoradiograph.

Three dark areas are observed on the developed x-ray film. These are two dark lines across the 20% PAG, indicating positions of nonligated 12-mer (fast mobility) and 15-mer (intermediate mobility) DNA oligomers, and a dark spot (slow mobility), indicating the position of the ligase reaction product, the 27-mer DNA oligomer. The position of DL(ATP) is detected via projection of the dark spot onto the strip of substrate-containing PAG.

Notes: The oligo(12-mer)DNA and oligo(15-mer)DNA are 5′-TAACGCCAGGGT-3′ and 5′-GGCGATTAAGTTGGG-3′ oligonucleotides, respectively, complementary to M13mp2 DNA (see below). Oligonucleotides are labeled at the 5′ end by T4 polynucleotide kinase with [γ-^{32}P]-ATP as described.[2,3] Labeled oligonucleotides (specific activities of 3.0×10^6 to 3.5×10^6 cpm/pmol of 5′ ends) are annealed with single-stranded M13mp2 DNA molecules (0.27 pmol of 5′ ends/μg of DNA) by heating to 70°C, with slow cooling to 25°C. M13mp2 DNA is a DNA isolated from bacteriophage M13mp2 grown in *E. coli* strain JM107.[4] The substrate, [^{32}P]-labeled 12-mer (final concentration of about 4 ng/ml), 15-mer (final concentration of about 4 ng/ml)/M13mp2 DNA (final concentration of about 2 μg/ml), is included in the resolving PAG before polymerization.[2,3]

GENERAL NOTES

The enzyme catalyzes the formation of a phosphodiester bond at the site of a single-strand break in a double-stranded DNA molecule. Its activity is essential in joining DNA chains during recombination, repair, and replication processes. RNA can also act as substrate to some extent.[5]

A single enzyme form is present in bacteria and yeast, while two molecular forms of the DL(ATP) are known in mammals, DL(ATP) I and DL(ATP) II, which seem to be involved in DNA replication and DNA reparation, respectively. The molecular heterogeneity of mammalian DL(ATP) detected by Method 1 is suggested to result from physiological or artifactual proteolysis; however, the real nature of this heterogeneity remains obscure.[1,6,7]

At least some DNA ligase isoforms of vertebrates are known to be localized in mitochondria.[8]

The enzyme from bacteriophage T7 is a two-domain protein. The larger domain is able to band shift both single- and double-stranded DNA, while the smaller domain is only able to bind to double-stranded DNA. It is suggested that the specificity of DL(ATP) for nick sites in DNA is produced by a combination of these different DNA-binding activities in the intact enzyme.[9]

The human DNA ligase I undergoes self-association to form a homotrimer. At temperatures over 18°C, three monomers of DNA polymerase , (see 2.7.7.7 — DDDP) are attached to the DNA ligase I trimer, thus forming a stable heterohexamer. This interaction appears to be responsible for the gap filling and nick ligation steps in short patch or simple base excision repair.[10]

REFERENCES

1. Mezzina, M., Sarasin, A., Politi, N., and Bertazzoni, U., Heterogeneity of mammalian DNA ligase detected on activity and DNA sequencing gels, *Nucleic Acids Res.*, 12, 5109, 1984.
2. Longley, M.J. and Mosbaugh, D.W., Characterization of DNA metabolizing enzymes *in situ* following polyacrylamide gel electrophoresis, *Biochemistry*, 30, 2655, 1991.
3. Longley, M.J. and Mosbaugh, D.W., *In situ* detection of DNA-metabolizing enzymes following polyacrylamide gel electrophoresis, in *Methods in Enzymology*, Vol. 218, *Recombinant DNA*, Part I, Wu, R., Ed., Academic Press, New York, 1993, p. 587.
4. Kunkel, T.A., Roberts, J.D., and Zakour, R.A., Rapid and efficient site-specific mutagenesis without phenotypic selection, *Methods Enzymol.*, 154, 367, 1987.
5. NC-IUBMB, *Enzyme Nomenclature*, Academic Press, San Diego, 1992, p. 533 (EC 6.5.1.1, Comments).
6. Mezzina, M., Franchi, E., Izzo, R., Bertazzoni, U., Rossignol, J.M., and Sarasin, A., Variation in DNA ligase structure during repair and replication processes in monkey kidney cells, *Biochem. Biophys. Res. Commun.*, 132, 857, 1985.
7. Mezzina, M., Rossignol, J.M., Philippe, M., Izzo, R., Bertazzoni, U., and Sarasin, A., Mammalian DNA ligase: structure and function in rat-liver tissue, *Eur. J. Biochem.*, 162, 325, 1987.
8. Perez-Jannotti, R.M., Klein, S.M., and Bogenhagen, D.F., Two forms of mitochondrial DNA ligase III are produced in *Xenopus laevis* oocytes, *J. Biol. Chem.*, 276, 48978, 2001.
9. Doherty, A.J. and Wigley, D.B., Functional domains of an ATP-dependent DNA ligase, *J. Mol. Biol.*, 285, 63, 1999.
10. Dimitriadis, E.K., Prasad, R., Vaske, M.K., Chen, L., Tomkinson, A.E., Lewis, M.S., and Wilson, S.H., Thermodynamics of human DNA ligase I trimerization and association with DNA polymerase β, *J. Biol. Chem.*, 273, 20540, 1998.

6.5.1.2 — DNA Ligase (NAD); DL(NAD)

OTHER NAMES	Polydeoxyribonucleotide synthase (NAD), polynucleotide ligase (NAD), DNA repair enzyme, DNA joinase
REACTION	(Deoxyribonucleotide)$_n$ + (deoxyribonucleotide)$_m$ + NAD = (deoxyribonucleotide)$_{n+m}$ + nicotinamide nucleotide + AMP
ENZYME SOURCE	Bacteria, fungi, plants, protozoa, invertebrates, vertebrates
SUBUNIT STRUCTURE	Unknown or no data available

Method 1

Visualization Scheme

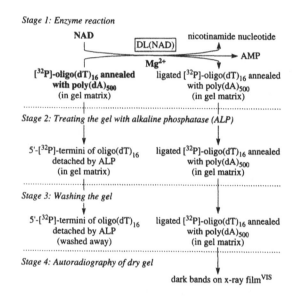

Stage 1: Enzyme reaction

Stage 2: Treating the gel with alkaline phosphatase (ALP)

Stage 3: Washing the gel

Stage 4: Autoradiography of dry gel

dark bands on x-ray film^{VIS}

Substrate-Containing PAG[1] (adapted)

Native or SDS-PAG electrophoresis of DL(NAD) is carried out in gels containing 1 to 3 nmol/ml of [^{32}P]oligo(dT)$_{16}$:poly(dA)$_{500}$ substrate (3×10^7 cpm/nmol). The substrate is prepared by labeling the oligo(dT)$_{16}$ with polynucleotide kinase (EC 2.7.1.78) and [γ-^{32}P]ATP (3000 Ci/mmol) and annealing the [^{32}P]oligo(dT) with poly(dA) (1:1).

Visualization Procedure[1]

After SDS-PAG electrophoresis, renaturate the proteins by washing the gel with five changes of 1 l of buffer containing 50 mM Tris–HCl (pH 7.8), 1 mM EDTA, 3 mM 2-mercaptoethanol, and 5 mM MgCl$_2$ at room temperature for 3 to 4 h. Then wash the gel twice for 15 min with 10 volumes of 20 mM Tris–HCl (pH 7.8), 5 mM MgCl$_2$, and 2 mM dithiothreitol.

Stage 1. Initiate the *in situ* ligase reaction by adding 1 mM NAD to the 20 mM Tris–HCl (pH 7.8), 5 mM MgCl$_2$, 10 mM (NH$_4$)$_2$SO$_4$, and 2 mM dithiothreitol. Incubate the gel in the reaction mixture at room temperature for 18 h with gentle agitation.

Stages 2 and 3. Wash the gel twice for 15 min with 100 ml of 80 mM Tris–HCl (pH 9.0), containing 5 mM MgCl$_2$, and treat twice for 2 h each with 10 U of calf intestinal alkaline phosphatase in 50 ml of phosphatase buffer (pH 9.0) in order to remove the unligated 5′-[^{32}P]-termini of oligo(dT). Finally, wash the gel twice with 100 ml of phosphatase buffer at room temperature for 15 min.

Stage 4. Autoradiograph the gel using Fuji x-ray film.

The dark bands on the developed x-ray film correspond to localization of DL(NAD) activity in the gel.

Notes: The electrophoresis of ligated [^{32}P]oligo(dT) using DNA sequencing gels is recommended to analyze the conversion of [^{32}P]-(dT) oligomers into labeled polymers of higher molecular mass and so to identify more precisely the product(s) of enzymatic reaction (see Method 2 below).

Method 2

Visualization Scheme

Stage 1: Electrophoretic separation of proteins in substrate-containing PAG (first dimension)

Stage 2: Vertical incision of the sample line of the first-dimension PAG and receipt of narrow (2 mm) substrate-containing gel strip

Stage 3: Enzyme reaction in the strip of the first-dimension PAG

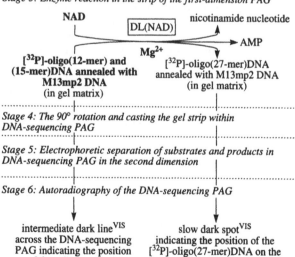

Stage 4: The 90° rotation and casting the gel strip within DNA-sequencing PAG

Stage 5: Electrophoretic separation of substrates and products in DNA-sequencing PAG in the second dimension

Stage 6: Autoradiography of the DNA-sequencing PAG

intermediate dark line^{VIS} across the DNA-sequencing PAG indicating the position of the substrate, intact [^{32}P]-oligo(15-mer) DNA

fast dark line^{VIS} across the DNA-sequencing PAG indicating the position of the substrate, intact [^{32}P]-oligo (12-mer)DNA

slow dark spot^{VIS} indicating the position of the [^{32}P]-oligo(27-mer)DNA on the DNA-sequencing PAG just opposite the position of DL(NAD) on the strip of electrophoretic substrate-containing PAG

Substrate-Containing PAG[2,3]

Prepare a native (2.5% stacking and 10% resolving) or denaturing (4% stacking and 10% resolving) SDS-PAG containing a 1:1 mixture of [^{32}P]-oligo(12-mer)DNA and [^{32}P]-oligo(15-mer)DNA annealed with single-stranded M13mp2 DNA (see *Notes*).

Reaction Mixture[2,3]

40 mM Tris–HCl buffer, pH 8.0
20 mM Dithiothreitol
50 mM KCl
400 μg/ml Bovine serum albumin
5% (w/v) Glycerol
40 mM MgCl$_2$
10 mM Ammonium sulfate
150 μM NAD

Procedure[2,3]

Stage 1. After electrophoresis, immerse the resolving substrate-containing SDS-PAG in 33 gel volumes of SDS extraction buffer: 10 mM Tris–HCl (pH 7.5), 5 mM 2-mercaptoethanol, and 25% (v/v) 2-propanol. Gently agitate the gel on a gyratory shaker (60 rpm) at 25°C. Discard the buffer after 30 min and replace with the same volume of fresh SDS extraction buffer. Extract the SDS for an additional 30 min. Place the gel in 3 volumes of enzyme renaturation buffer (the reaction mixture lacking MgCl$_2$, ammonium sulfate, and NAD). After brief (30 sec) agitation, place the gel in 27 volumes of the enzyme renaturation buffer and incubate for 18 to 25 h with gentle agitation at 4°C.

Stage 2. Following renaturation of the enzyme within SDS-PAG or after native PAG electrophoresis, cut each line of the substrate-containing PAG into 2-mm-wide longitudinal strips.

Stage 3. Place an individual strip into a 16 × 100 mm test tube containing 5 ml of the reaction mixture. Perform the enzyme reaction at 25°C for 30 min in a stoppered (Parafilm) tube that is placed on its side and gently shaken (60 rpm). Terminate the reaction by substituting the reaction mixture with an equal volume of ice-cold renaturation buffer containing 10 mM EDTA and gently shaking the tube for 30 min at 4°C.

Stages 4, 5, and 6. Following the *in situ* enzyme reaction, cast the processed strip of substrate-containing PAG within a 20% DNA sequencing PAG containing 8.3 M urea, avoiding formation of air bubbles. After polymerization, carry out electrophoresis in the second dimension, transfer the electrophorized 20% PAG to filter paper, dry, and autoradiograph.

Three dark areas are observed on the developed x-ray film. These are two dark lines across the 20% PAG, indicating positions of nonligated 12-mer (fast mobility) and 15-mer (intermediate mobility) DNA oligomers, and a dark spot (slow mobility), indicating the position of the ligase reaction product, the 27-mer DNA oligomer. The position of DL(NAD) is detected via projection of the dark spot onto the strip of substrate-containing PAG.

Notes: The oligo(12-mer)DNA and oligo(15-mer)DNA are 5′-TAACGCCAGGGT-3′ and 5′-GGCGATTAAGTTGGG-3′ oligonucleotides, respectively, complementary to M13mp2 DNA (see below). Oligonucleotides are labeled at the 5′ end by T4 polynucleotide kinase with [γ-^{32}P]-ATP as described.[2,3] Labeled oligonucleotides (specific activities of 3.0 × 10^6 to 3.5 × 10^6 cpm/pmol of 5′ ends) are annealed with single-stranded M13mp2 DNA molecules (0.27 pmol of 5′ ends/μg of DNA) by heating to 70°C, with slow cooling to 25°C. M13mp2 DNA is a DNA isolated from bacteriophage M13mp2 grown in *E. coli* strain JM107.[4] The substrate, [^{32}P]-labeled 12-mer (final concentration of about 4 ng/ml), 15-mer (final concentration of about 4 ng/ml)/M13mp2 DNA (final concentration of about 2 μg/ml), is included in the resolving PAG before polymerization.[2,3]

GENERAL NOTES

The enzyme catalyzes the formation of a phosphodiester bond at the site of a single-strand break in a double-stranded DNA molecule. Its activity is essential in joining DNA chains during recombination, repair, and replication processes. RNA can also act as a substrate to some extent.[5]

REFERENCES

1. Mezzina, M., Sarasin, A., Politi, N., and Bertazzoni, U., Heterogeneity of mammalian DNA ligase detected on activity and DNA sequencing gels, *Nucleic Acids Res.*, 12, 5109, 1984.
2. Longley, M.J. and Mosbaugh, D.W., Characterization of DNA metabolizing enzymes *in situ* following polyacrylamide gel electrophoresis, *Biochemistry*, 30, 2655, 1991.
3. Longley, M.J. and Mosbaugh, D.W., *In situ* detection of DNA-metabolizing enzymes following polyacrylamide gel electrophoresis, *Methods Enzymol.*, 218, 587, 1993.
4. Kunkel, T.A., Roberts, J.D., and Zakour, R.A., Rapid and efficient site-specific mutagenesis without phenotypic selection, *Methods Enzymol.*, 154, 367, 1987.
5. NC-IUBMB, *Enzyme Nomenclature*, Academic Press, San Diego, 1992, p. 533 (EC 6.5.1.2, Comments).

Appendix A-1
Minimal Medium
for *Escherichia coli*[1,2]

I. STOCK SOLUTIONS FOR MINIMAL MEDIUM

A. MINIMAL SALTS

NH_4Cl	20 g
NH_4NO_3	4 g
Na_2SO_4 (anhydrous)	8 g
K_2HPO_4 ($3H_2O$)	15.7 g
KH_2PO_4	4 g
$MgSO_4$ ($7H_2O$)	0.4 g

Dissolve salts in distilled water in the indicated order. Adjust pH to 7.2 with 1 *N* NaOH and bring total volume to 1000 ml with distilled water. Filtrate the solution through Whatman No. 1 filter paper and autoclave appropriate aliquots at 15 psi for 15 min.

B. NUTRIENTS

1. 0.2 g/ml Glucose
2. 2 mg/ml Amino acids
   0.1 mg/ml Vitamins
   2 mg/ml Purines and pyrimidines

Use amino acids, vitamins, purines, and pyrimidines required by particular *E. coli* mutant stocks. Prepare nutrient solutions using distilled water and sterilize by filtration through a 0.22-μm Millipore filter.

II. 2× STOCK MINIMAL MEDIUM (2× MM)

Mix 50 ml of A, 2 ml of B1, and 2 ml of B2 stock solutions under aseptic conditions. Bring the total volume to 100 ml with sterile water.

III. MINIMAL MEDIUM (MM)

Add an equal volume of sterile water to 2× stock minimal medium.

Notes: A minimal medium (MM) containing all nutrients required by a particular *E. coli* mutant is used for the maintenance of stock cultures of *E. coli* mutant strains.

To prepare an *E. coli* microbial reagent for bioautographic assay, transfer mutant *E. coli* from a stock culture into an MM containing all required nutrients and aerate at 37°C for 4 to 6 h. Centrifuge the bacteria at 5000 *g* for 10 min. Wash the bacteria by centrifugation and resuspend in an MM lacking the nutrient to be detected.

To prepare 1.5% indicator agar for bioautography, sterilize 3% Bacto agar (Difco) solution by autoclaving at 15 psi for 15 min. Cool the melted 3% Bacto agar solution to 47°C and mix with an equal volume of 2× MM (47°C). Store the agar-containing mixture at 47°C to prevent solidification. The substrates for the enzyme to be detected, which have been prepared as 100× stocks and sterilized by filtration through a 0.22-μm membrane, are added to the agar-containing mixture at this point, as well as the washed bacteria (final concentration of 10^8 bacteria/ml). After this the mixture is poured into sterile trays of appropriate size.

REFERENCES

1. Naylor, S.L. and Klebe, R.J., Bioautography: a general method for the visualization of isozymes, *Biochem. Genet.*, 15, 1193, 1977.
2. Naylor, S.L., Bioautographic visualization of enzymes, in *Isozymes: Current Topics in Biological and Medical Research*, Vol. IV, Rattazzi, M.C., Scandalios, J.G., and Whitt, G.S., Eds., Alan R. Liss, New York, 1980, p. 69.

Appendix A-2
Citrate Medium for
Pediococcus
cerevisiae[1,2]

I. STOCK SOLUTIONS FOR CITRATE MEDIUM

Glucose Buffer

Glucose	200 g
Trisodium citrate (2 H_2O)	200 g
Sodium acetate (3 H_2O)	30 g
KH_2PO_4	30 g
Ammonium sulfate	30 g
NaCl	10 g

Dissolve in distilled water and bring the total volume to 2 l with distilled water.

Amino Acid Mixture I

L-Isoleucine	1.5 g
L-Leucine	1.5 g
L-Valine	1.5 g
L-Phenylalanine	1.0 g
L-Proline	1.0 g

Dissolve in 250 ml of distilled water.

Amino Acid Mixture II

L-Alanine	1.5 g
L-Arginine hydrochloride	5.0 g
L-Glutamate	5.0 g
Glycine	3.0 g
L-Histidine hydrochloride (H_2O)	1.0 g
L-Serine	1.0 g
L-Threonine	1.0 g

Dissolve in 400 ml of distilled water. Adjust the pH to 6.0 with 1 N NaOH and bring the total volume to 500 ml with distilled water.

Amino Acid Mixture III

L-Asparagine	2.0 g
L-Aspartate	1.0 g
L-Cystine	2.0 g
L-Tyrosine	1.0 g
L-Tryptophan	1.0 g

Dissolve in 400 ml of distilled water. Adjust the pH to 11.0 with 1 N NaOH and bring the total volume to 500 ml with distilled water.

Amino Acid Mixture IV

L-Lysine hydrochloride	2.0 g
L-Methionine	1.0 g
L-Cysteine hydrochloride	2.0 g

Dissolve in 150 ml of distilled water. Adjust the pH to 2.0 with concentrated HCl and bring the total volume to 200 ml with distilled water.

Purines and Pyrimidine Solution

Adenine	0.1 g
Guanine hydrochloride	0.1 g
Uracil	0.1 g
Xanthine	0.1 g

Suspend in 90 ml of distilled water. Add 8 ml of 1 N NaOH and dissolve by stirring. Bring the total volume to 100 ml with distilled water.

Vitamin Mixture I

Thiamin hydrochloride	10 mg
Niacin	10 mg
Calcium pantothenane	10 mg
Pyridoxine hydrochloride	10 mg
Pyridoxamine	10 mg
Pyridoxal dihydrochloride	10 mg
Biotin (100 μg/ml in 50% EtOH)	1.0 ml

Dissolve the dry components of the mixture in 80 ml of distilled water. Add the biotin solution. Bring the total volume to 100 ml with distilled water.

Vitamin mixture II

Folic acid (0.1 mg/ml in 0.01 N NaOH)	1.0 ml
Folinic acid (0.1 mg/ml solution)	1.0 ml

Bring total volume to 100 ml with distilled water.

p-Aminobenzoic Acid Solution

p-Aminobenzoic acid	5 mg

Dissolve in 10 ml of 10% acetic acid.

Tween 80 and $CaCl_2$ Solution

Tween 80	2.0 ml
$CaCl_2$	1.0 g

Add Tween 80 to 90 ml of warm distilled water. Add $CaCl_2$ and dissolve by stirring. Bring the total volume to 100 ml with distilled water.

Riboflavin Solution

Riboflavin	10 mg

Add riboflavin to mixture of 10 ml of distilled water, 0.5 ml of glacial acetic acid, and 3 ml of EtOH. Heat the mixture gently to dissolve the riboflavin. Bring the total volume to 100 ml with distilled water.

Metal Salts Solution

$MgSO_4$ ($7H_2O$)	5.0 g
$MnSO_4$ ($4H_2O$)	1.5 g
$FeSO_4$	0.5 g

Dissolve metal salts in 100 ml of 0.03 N HCl.

Note: All stock solutions are stable for several months when stored at –20°C. Precipitation can occur in some stock solutions (e.g., in amino acid mixtures). To dissolve precipitate, heat the stock solution to 60°C.

II. 2× STOCK CITRATE MEDIUM

Glucose buffer	2000 ml
Amino acid mixture I	250 ml
Amino acid mixture II	500 ml
Amino acid mixture III	500 ml
Amino acid mixture IV	200 ml
Purines and pyrimidine solution	100 ml
Vitamin mixture I	100 ml
Vitamin mixture II	100 ml
p-Aminobenzoic acid solution	10 ml
Tween 80 and $CaCl_2$ solution	100 ml
Ascorbate	10 g

Add the stock solutions to 600 ml of distilled water in the indicated order. Adjust the pH to 6.8 with 12 N HCl and then add:

Riboflavin solution	100 ml
Metal salts solution	100 ml

Bring the total volume to 5000 ml with distilled water. Sterilize 2× stock citrate medium by filtration through a 0.22-μm Millipore filter and store at –20°C. It is stable for several months.

III. CITRATE MEDIUM

Add an equal volume of sterile water to 2× stock citrate medium.

Notes: Citrate medium is a complete medium suitable for maintenance of *P. cerevisiae* stock culture.

To maintain an exponentially growing *P. cerevisiae* culture, which may be used as a microbial reagent for bioautographic assay, transfer the *P. cerevisiae* into complete citrate medium at a 1:20 dilution three times per day. Wash the bacteria, 6 to 7 h after transfer, twice with citrate medium lacking the nutrient to be detected, by centrifugation at 5000 g for 10 min.

Preparation of 1.5% indicator agar for bioautographic assay using *P. cerevisiae* as a microbial reagent is essentially the same as described in Appendix A-1, except citrate medium is used in place of *E. coli* minimal medium.

REFERENCES

1. Naylor, S.L. and Klebe, R.J., Bioautography: a general method for the visualization of isozymes, *Biochem. Genet.*, 15, 1193, 1977.
2. Naylor, S.L., Bioautographic visualization of enzymes, in *Isozymes: Current Topics in Biological and Medical Research*, Vol. IV, Rattazzi, M.C., Scandalios, J.G., and Whitt, G.S., Eds., Alan R. Liss, New York, 1980, p. 69.

Appendix B
Alphabetical List
of Enzymes[1]

This list is provided to enable the location of certain enzymes arranged in the enzyme sheets in Part III according to their numbers and not names. Therefore, the list includes the enzyme names and corresponding nomenclature numbers used in the enzyme sheets, but arranged alphabetically. As a rule, these are the enzyme names recommended by the Nomenclature Committee of the International Union of Biochemistry and Molecular Biology. In some cases, however, names other than the recommended ones were used in the enzyme sheets. Usually these are widely used routine enzyme names. All the enzyme names (recommended and routine) used as titles in the enzyme sheets are given in the list in bold letters, for example, **"Alcohol dehydrogenase 1.1.1.1"** (recommended name), **"Glutamic–oxaloacetic transaminase 2.6.1.1"** (widely used routine name). Recommended enzyme names are also given in regular type for those enzymes for which routine names were used as titles in the enzyme sheets. For convenience and cross-reference purposes, these recommended names are given with special references to the corresponding routine names, for example, "Aspartate transaminase, see **Glutamic-oxaloacetic transaminase.**"

Acetylcholinesterase	3.1.1.7
Acetyl-CoA carboxylase	6.4.1.2
Acetyl-CoA hydrolase	3.1.2.1
β-N-Acetylgalactosaminidase	3.2.1.53
N-Acetyl-β-glucosaminidase, see **Hexosaminidase**	
N-Acetyllactosamine synthase	2.4.1.90
Acid phosphatase	3.1.3.2
Aconitase	4.2.1.3
Aconitate hydratase, see **Aconitase**	
Acrosin	3.4.21.10
Acyl-CoA dehydrogenase	1.3.99.3
Acyl-CoA oxidase, see **Palmitoyl-CoA oxidase**	
Acylphosphatase	3.6.1.7
Adenine phosphoribosyltransferase	2.4.2.7
Adenosine deaminase	3.5.4.4
Adenosine kinase	2.7.1.20
Adenosinetriphosphatase	3.6.1.3
Adenosylhomocysteinase	3.3.1.1
Adenylate cyclase	4.6.1.1
Adenylate kinase	2.7.4.3
Agarase	3.2.1.81
Alanine aminopeptidase	3.4.11.2
Alanine dehydrogenase	1.4.1.1
Alanine–glyoxylate transaminase	2.6.1.44
β-Alanine–pyruvate transaminase	2.6.1.18
Alanine transaminase, see **Glutamic–pyruvic transaminase**	

Alanopine dehydrogenase	1.5.1.17
β-Alanopine dehydrogenase	1.5.1.26
Alcohol dehydrogenase	1.1.1.1
Alcohol dehydrogenase (acceptor)	1.1.99.8
Alcohol dehydrogenase (NADP)	1.1.1.2
Alcohol oxidase	1.1.3.13
Aldehyde dehydrogenase (NAD)	1.2.1.3
Aldehyde oxidase	1.2.3.1
Aldolase	4.1.2.13
D-threo-Aldose dehydrogenase, see L-**Fucose dehydrogenase**	
Aldose 1-epimerase	5.1.3.3
Alkaline phosphatase	3.1.3.1
Allantoinase	3.5.2.5
Amine oxidase (copper containing)	1.4.3.6
Amine oxidase (flavin containing), see **Monoamine oxidase**	
D-**Amino acid oxidase**	1.4.3.3
L-**Amino acid oxidase**	1.4.3.2
Aminoacylase	3.5.1.14
Aminoaldehyde dehydrogenase	1.2.1.X'
4-Aminobutyrate transaminase	2.6.1.19
AMP deaminase	3.5.4.6
α-Amylase	3.2.1.1
Anthranilate synthase	4.1.3.27
α-N-Arabinofuranosidase	3.2.1.55
Arginase	3.5.3.1
Arginine aminopeptidase	3.4.11.6
Arginine kinase	2.7.3.3
Argininosuccinate lyase	4.3.2.1
Argininosuccinate synthase	6.3.4.5
Arginyl aminopeptidase, see **Arginine aminopeptidase**	
Aromatic-L-amino acid decarboxylase	4.1.1.28
Aromatic α-keto acid reductase	1.1.1.96
Aryl-alcohol dehydrogenase	1.1.1.90
Aryl-alcohol dehydrogenase (NADP)	1.1.1.91
Arylsulfatase	3.1.6.1
L-**Ascorbate oxidase**	1.10.3.3
L-**Ascorbate peroxidase**	1.11.1.11
Asparaginase	3.5.1.1
Aspartate carbamoyltransferase	2.1.3.2
Aspartate dehydrogenase	1.4.1.X
D-**Aspartate oxidase**	1.4.3.1
Aspartate transaminase, see **Glutamic–oxaloacetic transaminase**	
Autolysin	3.4.24.38
Betaine-aldehyde dehydrogenase	1.2.1.8
Biliverdin reductase	1.3.1.24
Bisphosphoglycerate mutase	5.4.2.4
Bisphosphoglycerate phosphatase	3.1.3.13
Branched-chain amino acid transaminase	2.6.1.42
Butyryl-CoA dehydrogenase	1.3.99.2
Ca²⁺/calmodulin-dependent protein kinase	2.7.1.123
Calf thymus ribonuclease H	3.1.26.4
cAMP-dependent protein kinase	2.7.1.X

Glucan endo-1,6-β-glucosidase	3.2.1.75
Glucan 1,4-α-glucosidase	3.2.1.3
Gluconate 2-dehydrogenase	1.1.1.215
Gluconate 5-dehydrogenase	1.1.1.69
Glucose 1-dehydrogenase	1.1.1.47
Glucose oxidase	1.1.3.4
Glucose-6-phosphatase	3.1.3.9
Glucose 1-phosphate adenylyltransferase	2.7.7.27
Glucose-6-phosphate 1-dehydrogenase	1.1.1.49
Glucose-6-phosphate isomerase	5.3.1.9
α-Glucosidase	3.2.1.20
β-Glucosidase	3.2.1.21
Glucuronate reductase	1.1.1.19
β-Glucuronidase	3.2.1.31
Glutamate–ammonia ligase,	
see Glutamine synthetase	
Glutamate–cysteine ligase	6.3.2.2
Glutamate decarboxylase	4.1.1.15
Glutamate dehydrogenase (NAD(P))	1.4.1.2–4
L-Glutamate oxidase	1.4.3.11
Glutamic–oxaloacetic transaminase	2.6.1.1
Glutamic–pyruvic transaminase	2.6.1.2
Glutaminase	3.5.1.2
Glutamine–phenylpyruvate transaminase,	
see Glutamine transaminase K	
Glutamine–pyruvate transaminase	2.6.1.15
Glutamine synthetase	6.3.1.2
Glutamine transaminase K	2.6.1.64
γ-Glutamylcyclotransferase	2.3.2.4
γ-Glutamyltransferase,	
see Glutamyl transpeptidase	
Glutamyl transpeptidase	2.3.2.2
Glutathione peroxidase	1.11.1.9
Glutathione reductase	1.6.4.2
Glutathione reductase (NADPH),	
see Glutathione reductase	
Glutathione transferase	2.5.1.18
Glyceraldehyde-3-phosphate dehydrogenase	1.2.1.12
Glyceraldehyde-3-phosphate dehydrogenase	
(NADP)	1.2.1.13
Glycerate dehydrogenase	1.1.1.29
Glycerol dehydrogenase	1.1.1.6
Glycerol dehydrogenase (NADP)	1.1.1.72
Glycerol kinase	2.7.1.30
Glycerol-3-phosphate dehydrogenase,	
see Glycerol-3-phosphate dehydrogenase (FAD)	
Glycerol-3-phosphate dehydrogenase (NAD)	1.1.1.8
Glycerol-3-phosphate dehydrogenase (FAD)	1.1.99.5
Glycine N-methyltransferase	2.1.1.20
Glycogen (starch) synthase	2.4.1.11
Glycolate oxidase	1.1.3.15
Glyoxalase I	4.4.1.5
Guanine deaminase	3.5.4.3
Guanylate kinase	2.7.4.8
L-Gulonate 3-dehydrogenase	1.1.1.45
Heparitin-sulfate lyase	4.2.2.8

Hexokinase	2.7.1.1
Hexosaminidase	3.2.1.52
Histidinol dehydrogenase	1.1.1.23
Homoserine dehydrogenase	1.1.1.3
Hyaluronoglucosaminidase	3.2.1.35
L-2-Hydroxy acid oxidase, see Glycolate oxidase	
3-Hydroxyacyl-CoA dehydrogenase	1.1.1.35
Hydroxyacylglutathione hydrolase	3.1.2.6
3-Hydroxybutyrate dehydrogenase	1.1.1.30
Hydroxymethylbilane synthase,	
see Uroporphyrinogen-I synthase	
15-Hydroxyprostaglandin dehydrogenase (NAD)	1.1.1.141
3(or 17)β-Hydroxysteroid dehydrogenase	1.1.1.51
3α (or 20β)-Hydroxysteroid dehydrogenase,	
see (R)-20-Hydroxysteroid dehydrogenase	
(R)-20-Hydroxysteroid dehydrogenase	1.1.1.53
3β-Hydroxy-Δ^5-steroid dehydrogenase	1.1.1.145
Hygromycin-B kinase	2.7.1.119
Hypoxanthine phosphoribosyltransferase	2.4.2.8
L-Iditol 2-dehydrogenase,	
see Sorbitol dehydrogenase	
IMP dehydrogenase	1.1.1.205
Indoleacetaldehyde dehydrogenase	1.2.1.X
Inorganic pyrophosphatase	3.6.1.1
Isoamylase	3.2.1.68
Isocitrate dehydrogenase (NAD)	1.1.1.41
Isocitrate dehydrogenase (NADP)	1.1.1.42
Isocitrate lyase	4.1.3.1
Isopropanol dehydrogenase (NADP)	1.1.1.80
Kanamycin kinase	2.7.1.95
Laccase	1.10.3.2
D-Lactaldehyde dehydrogenase	1.1.1.78
β-Lactamase, see Penicillinase	
L-Lactate dehydrogenase	1.1.1.27
Lactoylglutathione lyase, see Glyoxalase I	
Legumain	3.4.22.34
Leucine aminopeptidase	3.4.11.1
Leucine dehydrogenase	1.4.1.9
Leucine-tRNA ligase	6.1.1.4
Leucyl aminopeptidase,	
see Leucine aminopeptidase	
Lipoxygenase	1.13.11.12
Long-chain fatty acid–CoA ligase	6.2.1.3
Lysine dehydrogenase	1.4.1.15
L-Lysine-lactamase	3.5.2.11
D-Lysopine dehydrogenase	1.5.1.16
Lysozyme	3.2.1.17
Malate dehydrogenase	1.1.1.37
Malate dehydrogenase (decarboxylating)	1.1.1.39
Malate dehydrogenase (NADP)	1.1.1.40
Malate dehydrogenase (oxaloacetate decarboxylating)	
(NADP), see Malate dehydrogenase (NADP)	
Malate synthase	4.1.3.2
Malonyl-CoA synthase	6.2.1.X
Maltose phosphorylase	2.4.1.8
Mannitol 2-dehydrogenase	1.1.1.67

5-*enol*-Pyruvylshikimate-3-phosphate synthase	2.5.1.19
Retinol dehydrogenase	1.1.1.105
Ribonuclease I	3.1.27.5
Ribonuclease III	3.1.26.3
Ribose 5-phosphate isomerase	5.3.1.6
Ribose-phosphate pyrophosphokinase	2.7.6.1
Ribulose-phosphate 3-epimerase	5.1.3.1
RNA-directed DNA polymerase	2.7.7.49
RNA-directed RNA polymerase	2.7.7.48
Saccharopine dehydrogenase (NAD, L-lysine forming)	1.5.1.7
Sarcosine oxidase	1.5.3.1
Sedoheptulose-bisphosphatase	3.1.3.37
Serine carboxypeptidase	3.4.16.1
L-Serine dehydratase	4.2.1.13
Serine sulfhydrase	4.2.1.22
Serine-type carboxypeptidase, see **Serine carboxypeptidase**	
Shikimate 5-dehydrogenase	1.1.1.25
Sialidase	3.2.1.18
Sorbitol dehydrogenase	1.1.1.14
Sphingomyelin phosphodiesterase	3.1.4.12
Spleen exonuclease	3.1.16.1
Starch synthase	2.4.1.21
Strombine dehydrogenase	1.5.1.22
Submandibular proteinase A, see **Esteroprotease**	
Succinate dehydrogenase	1.3.99.1
Succinate-semialdehyde dehydrogenase	1.2.1.16
Succinate-semialdehyde dehydrogenase (NAD(P)), see **Succinate-semialdehyde dehydrogenase**	
Sucrose α-glucosidase	3.2.1.48
Sucrose phosphorylase	2.4.1.7
Sulfate adenylyltransferase	2.7.7.4
Sulfite oxidase	1.8.3.1
Superoxide dismutase	1.15.1.1
Tau-protein kinase	2.7.1.135
Tauropine dehydrogenase	1.5.1.23
Tau-tubulin kinase	2.7.1.X'
Thiamin dehydrogenase	1.1.3.23
Thiamin oxidase, see **Thiamin dehydrogenase**	
Thioglucosidase	3.2.3.1
Thiol oxidase	1.8.3.2
Thiosulfate sulfurtransferase	2.8.1.1
Thiosulfate–thiol sulfurtransferase	2.8.1.3
Threonine dehydratase	4.2.1.16
L-Threonine 3-dehydrogenase	1.1.1.103
Thrombin	3.4.21.5
Thymidine kinase	2.7.1.21
Transaldolase	2.2.1.2
Transketolase	2.2.1.1
α,α-Trehalase	3.2.1.28
Triacylglycerol lipase	3.1.1.3
Trimethylamine-oxide aldolase	4.1.2.32
Trimethylamine-*N*-oxide reductase	1.6.6.9
Triokinase	2.7.1.28
Triose-phosphate isomerase	5.3.1.1

Trypsin	3.4.21.4
Tryptophan synthase	4.2.1.20
Tryptophan-tRNA ligase	6.1.1.2
Type II site-specific deoxyribonuclease	3.1.21.4
Tyrosine transaminase	2.6.1.5
UDPglucose 6-dehydrogenase	1.1.1.22
UDPglucose–hexose-1-phosphate uridylyltransferase, see **Galactose-1-phosphate uridylyltransferase**	
Uracil-DNA glycosylase	3.2.2.X
Urate oxidase	1.7.3.3
Urease	3.5.1.5
Uridine phosphorylase	2.4.2.3
Urocanate hydratase	4.2.1.49
Urokinase	3.4.21.73
Uroporphyrinogen-I synthase	4.3.1.8
UTP–glucose-1-phosphate uridylyltransferase	2.7.7.9
Valine-tRNA ligase	6.1.1.9
Xanthine dehydrogenase	1.1.1.204
Xanthine oxidase	1.1.3.22
X-Pro dipeptidase, see **Proline dipeptidase**	
Xylan endo-1,3-β-xylosidase	3.2.1.32
Xylan 1,4-β-xylosidase	3.2.1.37
D-Xylose 1-dehydrogenase (NADP)	1.1.1.179
Xylose isomerase	5.3.1.5
L-Xylulose reductase	1.1.1.10

REFERENCES

1. NC-IUBMB, *Enzyme Nomenclature*, Academic Press, San Diego, 1992.

Appendix C
Buffer Systems Used for Enzyme Electrophoresis

The composition and pH values of buffer systems used for gel electrophoresis are crucial for the resolution of multiple molecular forms of enzymes. There are two types of buffer systems developed for enzyme electrophoresis. The first one is represented by continuous buffers, i.e., buffers having the same composition and pH values in both electrode tanks and gels. The second type includes buffer systems with different composition or pH values in electrode tanks and in gels. Such buffer systems are known as discontinuous. A great variety of continuous and discontinuous buffer systems are used for enzyme electrophoresis. Although most buffer systems are developed for the starch gel electrophoresis, every buffer system can be easily adapted to any supporting medium. Many buffer systems therefore are shared by different supporting media. A great variety of modifications of electrophoretic buffer systems have been accumulated during the more than four and a half decades since the advent of gel electrophoresis[1] and its combination with histochemical visualization of activity bands of specific enzymes on electrophoretic gels.[2,3] These modifications are so numerous that it is impossible even to list them.

None of an array of buffers introduced in gel electrophoresis is ideal for all enzymes detected in a single organism or for the same enzyme detected in different organisms. It should be stressed that variation among organisms is often less pronounced than that among enzymes. It is well established that even different allelic variants of the same enzyme can be differently resolved in different buffer systems. Different strategies can be used with respect to the selection of adequate buffer systems for enzyme electrophoresis, depending on the task to be solved and the amount of enzyme-containing preparation available. Some investigators prefer the strategy of testing many buffer systems with the purpose of selecting the most optimal one for each enzyme. However, when the amount of an available enzyme-containing preparation is a limiting factor (e.g., fruit fly individuals or somatic cell hybrids), the selection of a few buffer systems suitable for the resolution of a variety of different enzymes is a strategy of preference. There are obvious advantages and disadvantages of each of these two strategies, and selection of buffer systems remains largely a matter of individual preference, depending on the situation. The selection of a buffer system may also be a compromise between resolution of specific individual variants or interspecific differences and overall sharpness of banding patterns. The identification of superior buffer–enzyme systems ideally is accomplished by screening every enzyme involved in electrophoretic analysis on as many available buffer systems as possible. Such an approach always results in discovery of at least one superior buffer system for each enzyme involved in a survey. However, it is time consuming and costly, and therefore impractical in most cases. Nevertheless, the use of at least two buffer systems of differing pH values and buffer constituents is recommended to obtain reliable data on the intra- or interspecific variation of the electrophoretic banding pattern of an enzyme involved in electrophoretic analysis.[4] Some authors prefer to use a few buffer systems, ensuring satisfactory resolution of many enzymes.[5] A rational compromise between the two extremes is a common practice.[6] It is important to remember that there is no such thing as a single best buffer system for any situation and that the selection of buffers is a continuously developing process based on accumulating empiric experience.[7,8]

The use of fresh buffer solution is recommended for every electrophoretic run. It is a common practice, however, to use the same buffer several times after mixing the anodal and cathodal buffer solutions before every subsequent run.[7,8] Only continuous buffers, however, are suitable for the recycling.

It should be taken into account that some buffer components can serve as inhibitors of the enzyme under analysis (e.g., see 1.1.1.90 — AAD, *Notes*; 3.4.17.1 — CPA, *Notes*; 3.5.3.1 — ARG, General Notes; 3.11.1.X — PAH, General Notes; 4.2.1.2 — FH, General Notes). On the other hand, some buffer components or their contaminants can serve as substrates for enzymes under study. For example, citrate- and malate-containing buffers may respectively contain isocitrate and malate as contaminants. This may allow endogenous isocitrate dehydrogenase (see 1.1.1.42 — IDH) and malate dehydrogenases (see 1.1.1.37 — MDH; 1.1.1.40 — ME) to stain as "artifacts" along with other NAD- or NADP-dependent dehydrogenases under study.[8] In some cases it may prove beneficial to incorporate activators (e.g., Mg^{2+}), cofactors (e.g., NAD and NADP), nonionic detergents (e.g., Triton X-100), or disulfide bond–reducing reagents (e.g., 2-mercaptoethanol) into electrode or gel buffer solutions in order to enhance the staining intensity of enzyme activity bands, to improve their resolution, or to prevent the formation of secondary isozymes caused by oxidation of sulfhydryl groups in enzyme molecules.

Below are listed only those buffer systems that are most commonly used for enzyme electrophoresis in starch, cellulose acetate, and polyacrylamide gels.

I. STARCH GEL

CONTINUOUS BUFFER SYSTEMS

1. Histidine–citrate (pH 5.7)[8]
   Electrode: 65 mM L-histidine
                   19 mM citric acid
   Gel: 1:6 dilution of electrode buffer
   Modifications: The pH may be varied from 5.0 to 6.5 by modification of the citric acid concentration, and the ionic strength of the gel buffer may be altered using smaller or higher dilutions of the electrode buffer.

2. Amine–citrate (pH 6.0 to 8.0)[6,9]

   Electrode: 40 mM citric acid, titrated with N-(3-aminopropyl)-morpholine to desired pH

   Gel: 1:9 dilution of electrode buffer

   Note: N-(3-Aminopropyl)-morpholine is hazardous; handle gels prepared using this buffer with protective gloves.

   Modifications: Electrode and gel buffer pH values may be changed (to between 6.0 and 8.0) by varying the amount of N-(3-aminopropyl)-morpholine. A smaller dilution of electrode buffer may be used to prepare a gel buffer of higher ionic strength.

3. Phosphate (pH 5.8 to 7.5)[7]

   Electrode: 200 mM phosphate (sodium or potassium monobasic), titrated to desired pH with 10 N NaOH or 200 mM phosphate (sodium or potassium dibasic)

   Gel: 1:19 dilution of electrode buffer

   Modifications: There are numerous modifications of this buffer concerning the pH value, the ionic strength of both electrode and gel buffers, and the use of titrants.

4. Phosphate–citrate (pH 5.7 to 7.5)[7]

   Electrode: 245 mM phosphate (sodium monobasic) 150 mM citric acid

   Adjust with 10 N NaOH to desired pH

   Gel: 1:39 dilution of electrode buffer

   Modifications: Variations of this buffer mainly concern the pH and the rate of dilution of the gel buffer. Molarities of the buffer components can also vary.

5. Tris–citrate (pH 7.0)[10]

   Electrode: 135 mM Tris 43 mM citric acid

   Gel: 1:14 dilution of electrode buffer

   Modifications: There are numerous modifications of this buffer concerning the ionic strength of both electrode and gel buffers. The range of use of this buffer varies between pH 5.0 and pH 8.6.

6. Tris–maleate (pH 7.6)[11]

   Electrode: 100 mM Tris 10 mM EDTA 100 mM maleic acid 10 mM MgCl$_2$

   Adjust to pH 7.6 with NaOH

   Note: Dissolve buffer components in the indicated order to prevent formation of insoluble products.

   Gel: 1:9 dilution of electrode buffer

   Modifications: Most modifications of this buffer involve changing the pH in the range from 7.0 to 7.6.

7. Tris–borate–EDTA (pH 8.6)[6]

   Stock buffer: 900 mM Tris 500 mM boric acid 20 mM EDTA

   Adjust to pH 8.6 with NaOH

   Electrode: Anode: 1:6 dilution of stock buffer Cathode: 1:4 dilution of stock buffer

   Gel: 1:19 dilution of stock buffer

   Modifications: Usual modifications of this buffer concern minor changes of pH values of electrode and gel buffers and the rate of dilution of anode and cathode buffers.

   Note: For the first time, a similar buffer system was introduced in 1963 by Boyer et al.[12]

DISCONTINUOUS BUFFER SYSTEMS

1. Citrate (pH 7.0)/histidine (pH 7.0)[13]

   Electrode: 410 mM citric acid (trisodium salt), adjusted to pH 7.0 with free citric acid

   Note: Free citric acid may be used in place of its sodium salt. NaOH should be used as a titrant in this case. If sodium salt is used, it may be titrated with HCl.

   Gel: 5 mM L-histidine (free base), adjusted with NaOH to pH 7.0

   Modifications: This buffer is usually titrated in the pH range 7.0 to 8.0.

   Note: Gels prepared using this buffer should be monitored while running because the electrical current may sharply increase after a few hours.

2. Borate–NaOH (pH 8.1)/Tris–borate–EDTA (pH 7.9)[14]

   Electrode: 300 mM boric acid, adjusted to pH 8.1 with NaOH

   Gel: 50 mM Tris 75 mM boric acid 2 mM EDTA

   Adjust to pH 7.9

   Modifications: Gel buffer pH can be changed (to between 7.5 and 9.0) by proper modification of boric acid and Tris concentrations.

3. Borate–NaOH (pH 8.1)/Tris–citrate (pH 8.6)[15]

   Electrode: 300 mM boric acid, adjusted to pH 8.1 with NaOH

   Gel: 76 mM Tris, adjusted to pH 8.6 with citric acid

   Modifications: Electrode buffer varies in the range between pH 7.8 and pH 8.6 by adjusting the amount of NaOH. Gel buffer varies considerably with respect to concentration of Tris (in the range of 15 to 100 mM) and pH (in the range of 8.6 to 9.5).

Note: The gel buffer with a high pH can result in good resolution of otherwise fuzzy isozyme systems.[16]

4. Borate–LiOH (pH 8.1)/Tris–citrate (pH 8.5)[17]

Stock A: 300 mM boric acid, adjusted to pH 8.1 with LiOH

Stock B: 30 mM Tris
5 mM citric acid
Adjust to pH 8.5

Electrode: Undiluted stock A

Gel: 1:99 mixture of stocks A:B

Modifications: There are many modifications of this buffer system involving minor changes of the pH of the electrode and gel buffers and their ionic strength.

Note: For the first time, a similar buffer system (192 mM boric acid–38 mM LiOH (pH 8.3)/47 mM Tris–7 mM citric acid–19 mM boric acid–4 mM LiOH (pH 8.3)) was introduced by Ashton and Braden.[18]

II. CELLULOSE ACETATE GEL

Only continuous buffer systems are suitable for electrophoresis in cellulose acetate gel. The gel and the electrode buffers are usually of the same ionic strength.

1. Phosphate–citrate (pH 6.4)[8]

10 mM phosphate (sodium dibasic)
2.5 mM citric acid

Modifications: There are numerous modifications of this buffer concerning the pH value (varies from pH 5.0 to 7.5) and final concentrations of phosphate (varies from 5 to 100 mM) and citrate (varies from 1.5 to 35 mM) constituents.

2. Phosphate (pH 7.0)[19]

20 mM phosphate (sodium dibasic), adjusted to pH 7.0 with 20 mM phosphate (sodium monobasic)

Modifications: There are numerous modifications of this buffer concerning the pH value (varies from pH 6.5 to 7.5), the final concentration of phosphate (varies from 10 to 40 mM), and the presence of some additional components (e.g., magnesium chloride, EDTA, 2-mercaptoethanol, Triton X-100) used to activate or preserve an enzyme under analysis or to improve resolution of its isozymes.

3. Tris–maleate (pH 7.4)[20]

10 mM Tris
1 mM EDTA
10 mM maleic acid
1 mM MgCl$_2$

Note: Dissolve buffer components in the indicated order to prevent formation of insoluble products. Adjust to pH 7.4 using sodium hydroxide.

Modifications: There are modifications of this buffer concerning the pH value (varies from pH 6.0 to 7.8), the final concentration of Tris (varies from 15 to 100 mM), and the absence of EDTA or magnesium chloride.

4. Tris–citrate (pH 7.5)[19]

20 mM Tris, adjusted to pH 7.5 with saturated solution of citric acid

Modifications: There are numerous modifications of this buffer concerning the pH value (varies from pH 5.0 to 8.6), the final concentration of Tris (varies from 10 to 100 mM), and the presence of EDTA (varies from 2 to 4 mM).

5. Tris–borate–EDTA (pH 7.8)[8]

15 mM Tris
5 mM EDTA (disodium salt)
10 mM MgCl$_2$
5.5 mM boric acid

Note: Use boric acid to adjust pH to 7.8.

Modifications: Most modifications are minor and concern adjustments of pH (range from 7.8 to 9.0), the final concentration of Tris (varies from 15 to 130 mM), and the presence of magnesium chloride (varies from 1 to 10 mM).

6. Tris–glycine (pH 8.5)[21]

25 mM Tris
192 mM glycine

Modifications: Most modifications are minor and concern adjustments of pH (range from 8.5 to 9.0).

General Notes: It should be taken into account that the use of highly concentrated buffers can draw a large current that may cause overheating of the gel. Thus, the progress of an electrophoretic run should be carefully monitored, even when electrophoresis is carried out in a refrigerator at 4°C or in a cold room at nearly 0°C.

III. POLYACRYLAMIDE GEL

Unlike cellulose acetate gel electrophoresis, which involves only continuous gel and buffer systems, and starch gel electrophoresis, which involves only continuous gel systems but is able to use continuous and discontinuous buffer systems, PAG electrophoresis may involve discontinuities in both gel and buffer systems. However, the use of discontinuous polyacrylamide gels (i.e., combination of stacking and running gels) is preferable only when diluted enzyme-containing samples are used for electrophoresis. When concentrated preparations are available and small sample volumes may be applied, there is no need to use stacking gels. Then it is possible to employ rapid and simple PAG preparation using commercially available acrylamide–bis-acrylamide mixtures (e.g., from Sigma). In principle, various buffer systems described above for starch and cellulose acetate gels may be adapted for PAG electrophoresis. Those who need more details concerning PAG preparation and electrophoresis procedures are referred to special guides.[22,23]

PAG is not as commonly used for enzyme electrophoresis as starch or cellulose acetate gels. However, it is the most suitable medium for separation of enzyme protein molecules in denaturating conditions, e.g., in the presence of SDS. Because of this, below are given only the compositions of specific buffers and gels of the most commonly used method of SDS-PAG electrophoresis, described by Laemmli.[24]

SDS-PAG[24]

Stacking gel buffer:	125 mM Tris–HCl, pH 6.8	
	0.1% SDS	
Running gel buffer:	375 mM Tris–HCl, pH 8.8	
	0.1% SDS	
Cathode buffer:	25 mM Tris	
	192 mM glycine	
	0.1% SDS	
	pH 8.3	
Stacking gel:	Acrylamide	3 g
	Bis-acrylamide	80 mg
	Stacking gel buffer	to 100 ml
	TEMED	25 µl
	Ammonium persulfate	25 mg
Running gel:	Acrylamide	10 g
	Bis-acrylamide	267 mg
	Running gel buffer	to 100 ml
	TEMED	25 µl
	Ammonium persulfate	25 mg

Note: It is recommended that an SDS-PAG prepared according to Laemmli[24] be allowed to polymerize overnight to reduce the potential that free radicals remaining from the polymerization process could generate hydrogen peroxide.[25]

REFERENCES

1. Smithies, O., Zone electrophoresis in starch gels: group variations in the serum proteins of normal human adults, *Biochem. J.*, 61, 629, 1955.
2. Hunter, R.L. and Markert, C.L., Histochemical demonstration of enzymes separated by zone electrophoresis in starch gels, *Science*, 125, 1294, 1957.
3. Markert, C.L. and Möller, F., Multiple forms of enzymes: tissue, ontogenetic, and species specific patterns, *Proc. Natl. Acad. Sci. U.S.A.*, 45, 753, 1959.
4. Morizot, D.C. and Schmidt, M.E., Starch gel electrophoresis and histochemical visualization of proteins, in *Electrophoretic and Isoelectric Focusing Techniques in Fisheries Management*, Whitmore, D.H., Ed., CRC Press, Boca Raton, FL, 1990, p. 23.
5. Siciliano, M.J. and Shaw, C.R., Separation and visualization of enzymes on gels, in *Chromatographic and Electrophoretic Techniques*, Vol. 2, *Zone Electrophoresis*, Smith, I., Ed., Heineman, London, 1976, p. 185.
6. Murphy, R.W., Sites, J.W., Jr., Buth, D.G., and Haufler, C.H., Proteins I: isozyme electrophoresis, in *Molecular Systematics*, 2nd ed., Hillis, D.M., Moritz, C., and Mable, B.K., Eds., Sinauer, Sunderland, U.K., 1996, p. 51.
7. Harris, H. and Hopkinson, D.A., *Handbook of Enzyme Electrophoresis in Human Genetics*, North-Holland, Amsterdam, 1976 (loose-leaf, with supplements in 1977 and 1978).
8. Richardson, B.J., Baverstock, P.R., and Adams, M., *Allozyme Electrophoresis: A Handbook for Animal Systematics and Population Studies*, Academic Press, Sydney, 1986.
9. Clayton, J.W. and Tretiak, D.N., Amine-citrate buffers for pH control in starch gel electrophoresis, *J. Fish. Res. Board Can.*, 29, 1169, 1972.
10. Meizel, S. and Markert, C.L., Malate dehydrogenase isozymes of the marine snail *Ilyanassa obsoleta*, *Arch. Biochem. Biophys.*, 122, 753, 1967.
11. Spencer, N., Hopkinson, D.A., and Harris, H., Phosphoglucomutase polymorphism in man, *Nature*, 204, 742, 1964.
12. Boyer, S.H., Fainer, D.C., and Watson-Williams, E.J., Lactate dehydrogenase variant from human blood: evidence for molecular subunits, *Science*, 141, 642, 1963.
13. Fields, R.A. and Harris, H., Genetically determined variation of adenylate kinase in man, *Nature*, 209, 262, 1966.
14. Serov, O.L., Korochkin, L.I., and Manchenko, G.P., Methods of electrophoretic analysis of isozymes, in *Genetics of Isozymes*, Beljaev, D.K., Ed., Nauka, Moscow, 1977, p. 18 (in Russian).
15. Poulik, M.D., Starch gel electrophoresis in a discontinuous system of buffers, *Nature*, 180, 1477, 1957.
16. Chippindale, P.T., A high-pH discontinuous buffer system for resolution of isozymes in starch-gel electrophoresis, *Stain. Technol.*, 64, 61, 1989.
17. Ridgway, G.J., Sherburne, S.W., and Lewis, R.D., Polymorphisms in the esterases of Atlantic herring, *Trans. Am. Fish. Soc.*, 99, 147, 1970.
18. Ashton, G.C. and Braden, A.W.H., Serum β-globulin polymorphism in mice, *Aust. J. Biol. Sci.*, 14, 248, 1961.
19. van Someren, H., van Henegouwen, H.B., Los, W., Wurzer-Figurelli, E., Doppert, B., Vervloet, M., and Meera Khan, P., Enzyme electrophoresis on cellulose acetate gel: II. Zymogram patterns in man–Chinese hamster somatic cell hybrids, *Humangenetik*, 25, 189, 1974.
20. Meera Khan, P., Rijken, H., Wijnen, J.Th., Wijnen, L.M.M., and de Boer, L.E.M., Red cell enzyme variation in the orang utan: electrophoretic characterization of 45 enzyme systems in Cellogel, in *The Orang Utan: Its Biology and Conservation*, De Boer, L.E.M., Ed., Dr. W. Junk Publishers, The Hague, 1982, p. 61.
21. Holmes, R.S., Moxon, L.M., and Parsons, P.A., Genetic variability of alcohol dehydrogenase among Australian *Drosophila* species: correlation of ADH biochemical phenotype with ethanol resource utilization, *J. Exp. Zool.*, 214, 199, 1980.

22. Andrews, A.T., *Electrophoresis: Theory, Techniques, and Biochemical and Clinical Applications*, Clarendon Press, Oxford, 1988.

23. Gordon, A.H., *Electrophoresis of Proteins in Polyacrylamide and Starch Gels*, North-Holland, Amsterdam, 1980.

24. Laemmli, U.K., Cleavage of structural proteins during the assembly of the head of bacteriophage T4, *Nature*, 227, 680, 1970.

25. Liu, L., Dean, J.F.D., Friedman, W.E., and Eriksson, K.-E.L., A laccase-like phenoloxidase is correlated with lignin biosynthesis in *Zinnia elegans* stem tissues, *Plant J.*, 6, 213, 1994.

Printed and bound by CPI Group (UK) Ltd, Croydon, CR0 4YY

23/10/2024

01778248-0018